P9-CSB-290

A. P. DEACON ASSOCIATES LTD
MAR 1972

RELIABILITY HANDBOOK

OTHER McGRAW-HILL HANDBOOKS OF INTEREST

AMERICAN SOCIETY OF MECHANICAL ENGINEERS · ASME Handbooks:
- Engineering Tables
- Metals Engineering—Design
- Metals Engineering—Processes
- Metals Properties

AMERICAN SOCIETY OF TOOL AND MANUFACTURING ENGINEERS:
- Die Design Handbook
- Handbook of Fixture Design
- Manufacturing Planning and Estimating Handbook
- Tool Engineers Handbook

ARCHITECTURAL RECORD · Time-Saver Standards
BAUMEISTER · Marks' Mechanical Engineers' Handbook
BEEMAN · Industrial Power Systems Handbook
BRADY · Materials Handbook
CARROLL · Industrial Instrument Servicing Handbook
CONSIDINE · Process Instruments and Controls Handbook
CROCKER · Piping Handbook
DUDLEY · Gear Handbook
EMERICK · Heating Handbook
FACTORY MUTUAL ENGINEERING DIVISION · Handbook of Industrial Loss Prevention
FLÜGGE · Handbook of Engineering Mechanics
HARRIS · Handbook of Noise Control
HARRIS AND CREDE · Shock and Vibration Handbook
HEYEL · The Foreman's Handbook
KALLEN · Handbook of Instrumentation and Controls
KING AND BRATER · Handbook of Hydraulics
KNOWLTON · Standard Handbook for Electrical Engineers
KOELLE · Handbook of Astronautical Engineering
KORN AND KORN · Mathematical Handbook for Scientists and Engineers
LE GRAND · The New American Machinists' Handbook
MAGILL, HOLDEN, AND ACKLEY · Air Pollution Handbook
MANAS · National Plumbing Code Handbook
MANTELL · Engineering Materials Handbook
MAYNARD · Industrial Engineering Handbook
MORROW · Maintenance Engineering Handbook
PERRY, CHILTON, AND KIRKPATRICK · Chemical Engineers' Handbook
PERRY · Engineering Manual
ROSSNAGEL · Handbook of Rigging
ROTHBART · Mechanical Design and Systems Handbook
SHAND · Glass Engineering Handbook
STANIAR · Plant Engineering Handbook
STREETER · Handbook of Fluid Dynamics
TOULOUKIAN · Retrieval Guide to Thermophysical Properties Research Literature
TRUXAL · Control Engineers' Handbook

RELIABILITY HANDBOOK

W. GRANT IRESON, Editor-in-Chief

Executive Head
Department of Industrial Engineering
Stanford University

McGRAW-HILL BOOK COMPANY

New York San Francisco Toronto London Sydney

RELIABILITY HANDBOOK

 Library of Congress Catalog Card Number 65-18747

32040

1234567890(MP)721069876

CONTRIBUTORS

MARTIN BARBE, *IDEP Office, Reliability Department, Aerospace Corporation, El Segundo, California.* (SECTION 6, ACCESSION AND ORGANIZATION OF RELIABILITY DATA)

DAVID C. DELLINGER, *Major, U.S. Air Force, Air Force Institute of Technology, Wright-Patterson Air Force Base, Ohio.* (SECTION 3, SELECTING RELIABILITY TEST PLANS)

AARON FOGEL, *Retired, Reliability Control Department, General Dynamics/Astronautics, San Diego, California.* (SECTION 7, THE COFEC RELIABILITY DATA SYSTEM)

LAWRENCE J. FOGEL, *President, Decision Science, Inc., San Diego, California.* (SECTION 7, THE COFEC RELIABILITY DATA SYSTEM)

WALTER L. HURD, JR., *Reliability and Quality Engineering Division Manager, Missile System, Lockheed Missile and Space Company, Sunnyvale, California.* (SECTION 10, ENGINEERING DESIGN AND DEVELOPMENT FOR RELIABLE SYSTEMS)

JOHN H. K. KAO, *Associate Professor of Industrial Engineering, New York University, New York, New York.* (SECTION 2, CHARACTERISTIC LIFE PATTERNS AND THEIR USES)

E. W. KIMBALL, *Reliability Group Engineer, Martin Company Aerospace Division of Martin Marietta Corporation, Orlando, Florida.* (SECTION 9, MALFUNCTION AND FAILURE ANALYSIS)

L. G. KNIGHT, *Chief, Quality Control Group, Northrop Space Laboratories, Northrop Corporation, Hawthorne, California.* (SECTION 17, COST ASPECTS OF A RELIABILITY AND QUALITY ASSURANCE PROGRAM)

E. JACK LANCASTER, *Director of Quality Assurance, Kearfott Systems Division, General Precision, Inc., Wayne, New Jersey.* (SECTION 14, RELIABILITY SPECIFICATION AND PROCUREMENT)

JAMES A. MARSHIK, *Senior Quality Control Engineer, Aeronautical Division, Minneapolis-Honeywell, Minneapolis, Minnesota.* (SECTION 13, RELIABILITY CONSIDERATIONS FOR PRODUCTION)

J. Y. McCLURE, *Director, Reliability, Quality Control, Value Control, General Dynamics Corporation, New York.* (SECTION 16, ORGANIZING FOR RELIABILITY AND QUALITY CONTROL)

DAVID MEISTER, *Head, System Research Section, Systems Effectiveness Department, Bunker-Ramo Corporation, Canoga Park, California.* (SECTION 12, HUMAN FACTORS IN RELIABILITY)

O. B. MOAN, *Professor of Engineering, Arizona State University, Temple, Arizona.* (SECTION 4, APPLICATION OF MATHEMATICS AND STATISTICS TO RELIABILITY AND LIFE STUDIES)

EMIL M. OLSEN, *General Manager, Air Cruiser Division, The Garrett Corporation, Belmar, New Jersey.* (SECTION 14, RELIABILITY SPECIFICATION AND PROCUREMENT)

B. L. RETTERER, *Manager, Technique Research Program, ARINC Research Corporation, Annapolis, Maryland.* (SECTION 11, CONSIDERATION OF MAINTAINABILITY IN RELIABILITY PROGRAMS)

CLIFFORD M. RYERSON, *Senior Staff Engineer, Head, Components Assurance, Hughes Aircraft Company, Culver City, California.* (SECTION 15, ACCEPTANCE TESTING)

ROBERT W. SMILEY, *Commander, USN, Officer-in-Charge, Polaris Missile Facility Pacific, Bremerton, Washington.* (SECTION 8, TESTING PROGRAMS)

EVERETT L. WELKER, *Manager, System Effectiveness Analysis Program, The General Electric Company, TEMPO, Center for Advanced Studies, Santa Barbara, California.* (SECTION 1, SYSTEM EFFECTIVENESS)

PAUL H. ZORGER, *Vitro Laboratories, Inc., Silver Springs, Maryland.* (SECTION 5, RELIABILITY ESTIMATION)

PREFACE

Reliability, as a separate function and a formal discipline in design and production, is a relatively new development, yet consideration of the consequences of a failure of a system is as old as the factory system. The word "reliability" was not used then, but the designers of the first steamships were concerned about the ability of the boilers and engines to stand up to the long hours of demand placed on them by a transocean crossing. Redundancy, in the form of sails, was provided in case the steam engines should fail. Cities hesitated to install gas mains and gas street lights because it was feared that a failure of the system or cutting of the main might plunge the city into darkness. It was even suggested that criminals might be able to shut off the supply in order to carry out crimes under the protection of darkness. The Dodge Brothers were thinking about reliability when the slogan "Dodge Dependability" was coined many years ago. Long after electric starters were standard equipment on American automobiles, hand cranks were still provided. (The 1960 Triumph TR-3 came equipped with a hand crank.) The *Titanic* was described as an unsinkable ship, and overconfidence in its reliability probably played some part in that great disaster.

The questions raised by the cited examples are the same questions that are raised today in connection with reliability: Will the device work when needed? Will it work long enough to perform its intended function? Can the operator or user compensate for failures or malfunctions and accomplish the functional objective in spite of the malfunction? What are the costs or penalties associated with a failure, and what are the costs of reducing the probability of or preventing a failure? How long can the device be expected to operate satisfactorily without repair or maintenance? Part of the purpose of a reliability program today is to find the answers to such questions, and another part is to provide techniques, methods, information, procedures, and know-how which will tend to assure that the most economical balance of cost and benefits is accomplished in the initial design, development, and production of the devices.

In the unhurried days of the previous century and even in the first quarter of this century, the problems were serious but the consequences of failures were not as dramatic or as catastrophic as now. The advent of large, high-speed aircraft and extremely complex war machines and the necessity to reduce the development program time meant that there was not enough time to follow former, slow procedures of designing, testing, redesigning, and retesting until a completely satisfactory product could be attained. Within the span of five years, from 1946 to 1951, it became obvious that the then current practices in design, development, and manufacturing required a new orientation in order that amazingly complex systems, many requiring breakthroughs in science and technology, could be designed and produced in relatively short periods of time and still have a high probability of performing the assigned functions satisfactorily. "Reliability" was the result of an urgent need.

As in the development of any new or different concept, the objectives of reliability were reasonably clear but there was no agreement as to how it was to be attained, and yet industrial organizations and government agencies suddenly found themselves faced with the necessity to implement reliability programs. The programs required not only that superior systems be produced in record time but also that the progress and success be measurable. Quantification of the "reliability" of a system was absolutely essential, and a whole body of knowledge in the form of mathematical and statistical models was formulated to serve this need. The measurement of the success of the program did not, however, help very much in determining what elements should be included in the reliability program. Engineers, scientists, and manufacturing specialists received almost no guidance in determining how to go about the attainment of reliability. For that matter, the manager, responsible for the programs, received very little help in deciding how to allocate the funds he had available to improve the reliability of the product. Many different opinions were expressed and followed throughout the 1950's. Even today, arguments still rage among reliability and quality control engineers, design engineers, scientists, and managers as to how to accomplish the desired results.

This handbook attempts to alleviate these conditions by providing specific recommendations for complete reliability programs. While the handbook has been prepared as a reference book for reliability, quality, design and production engineers, managers, and scientists, sufficient expository material has been included so that it is suitable for senior or first-year graduate university students. The practitioner will find it a ready source of information on specific problems, and the student will find that it develops the fundamentals of reliability programs in an orderly fashion so that the relatively inexperienced person can understand the problems and the techniques of solution.

The handbook has been prepared with two primary purposes in mind. First, it was to be a comprehensive collection of reliability experience, arranged in an orderly fashion so that engineers, scientists, and managers could find out, with a minimum amount of reading, what others have done and what success they have had in obtaining high reliability in complex systems. It was intended that no single philosophy would be espoused but that conflicting viewpoints would be presented factually and objectively without editorial comment. Each reader could then evaluate the recorded experience in the light of his own problem and choose those methods which offer the best solution. Secondly, the handbook was to provide reference data of interest to all persons involved in reliability. Thus, the handbook consists of a large collection of principles, examples, tables, sources of reliability data, statistical and mathematical models and tables, data systems, forms, charts, and analysis techniques.

The first five sections present mathematical and statistical tools. These sections include System Effectiveness, Distributions and Life Patterns, Fundamental Statistical and Mathematical Techniques, Reliability Prediction, and Selection of Statistical Tests for Reliability Studies. The next eight sections present the principal elements of a reliability program: Reliability Data Systems, Test Programs, Malfunction and Failure Analysis, Engineering Design and Development, Maintainability, Human Factors, and Production. These sections develop the concepts and principles employed in a reliability program and illustrate these principles through recommended procedures and examples. Specific guidelines for the design of the different elements of the reliability program are provided. The remaining four sections are devoted to the managerial aspects of reliability: Specification, Procurement, and Acceptance of Supplies, Costs Aspects of Reliability and Quality, and the Organization of Reliability and Quality Assurance.

None of the elements of a reliability program are independent of other elements, so it is necessary to discuss the same topics in two or more sections in order to give complete coverage of the interrelated subject matter. Cross references have been provided in all sections to call the reader's attention to the fact that the same or related topics are discussed elsewhere in the handbook. The reader is reminded that a very detailed topical index has been provided at the end of the handbook. The index provides him with a rapid reference to all the entries under each subject and will save him much time in locating specific information he desires.

A handbook of this magnitude would be impossible without the cooperation of a large number of persons and organizations. The authors of the various sections have been recognized at the beginning of each section, but I would like to thank each of these men for the great care and time that each devoted to the preparation of outstanding sections. The authors were selected because they had previously made great contributions to the literature of reliability and because each is a specialist who has been faced with the kinds of problems discussed in his section. They know from first-hand experience as well as from theoretical study the different philosophies related to their areas of specialization and the results of experimenting with the different philosophies. Each has brought years of intensive experience to the reader through his contributions to this handbook.

The contributors have been encouraged and given valuable assistance in this project by their employers, and I would like to acknowledge with sincere thanks the organizations which have helped make this handbook possible. Special thanks are due the following organizations for assistance to the contributors, permissions to reproduce material, examples, etc.:

Aerospace Corporation
American Institute for Research
American Society for Quality Control
ARINC Research Corporation
Arizona State University
Bunker-Ramo Corporation
Cambridge University Press
The Garrett Corporation (Air Cruiser Company)
General Dynamics Corporation/Astronautics/Electric Boat
The General Electric Company (TEMPO)
General Precision, Inc.
Hughes Aircraft Company
Lockheed Missile and Space Company
The Martin Company
McGraw-Hill Book Company
Minneapolis-Honeywell Aero Division
New York University
Northrop Corporation
Prentice-Hall, Inc.
RCA Service Company
Sandia Corporation
Stanford University
Trident Laboratories
U.S. Air Force Aeronautical Systems Division
U.S. Air Force Institute of Technology
U.S. Naval Ordnance Test Station, China Lake
U.S. Navy, Special Projects
Virginia Polytechnic Institute

I am indebted to the Literary Executor of the late Sir Ronald A. Fisher, Cambridge, to Dr. Frank Yates, F. R. S. Rothamsted, and to

Oliver & Boyd Ltd., Edinburgh, for permission to reprint my table number A-5 from their book, "Statistical Tables for Biological, Agricultural, and Medical Research."

Finally, I would like to thank Eugene L. Grant, Emeritus Professor of Stanford University, for his valuable assistance and encouragement throughout the preparation of this handbook.

W. Grant Ireson

CONTENTS

Section 1

SYSTEM EFFECTIVENESS

EVERETT L. WELKER

MANAGER, SYSTEM EFFECTIVENESS ANALYSIS PROGRAM,
THE GENERAL ELECTRIC COMPANY,
TEMPO, CENTER FOR ADVANCED STUDIES
SANTA BARBARA, CALIFORNIA

Dr. Welker received the Ph.D. and M.A. degrees in mathematical statistics and the A.B. degree in mathematics from the University of Illinois. He then served that University as Associate Professor of Mathematics until 1947, when he joined the American Medical Association as a statistician in medical economic research and as a consultant on medical research studies.

In 1952, Dr. Welker moved to the Institute for Defense Analysis as an operations research analyst in the Weapons Systems Evaluation Group, which conducted effectiveness studies of various military units in both limited and total war for the Joint Chiefs of Staff and the Secretary of Defense. From 1957 to 1963 he was with the ARINC Research Corporation, where he served as manager of several programs, including course development and presentations in theory and practice of reliability and of corporate programs in the evaluation of satellites, Saturn space booster, and other programs for ARPA and NASA.

Since 1963, Dr. Welker has been in his present position working on the development of models and methods for analyzing, assessing, and predicting elements associated with system worth, including reliability, maintainability, operational readiness, and system effectiveness for complex systems. His work has included leadership in these areas and mathematical and statistical support to various TEMPO projects in operations research and special system studies.

SYSTEM EFFECTIVENESS

Everett L. Welker

CONTENTS

1.1 SYSTEM EFFECTIVENESS AS AN ELEMENT OF SYSTEM WORTH

Management decision in favor of one of a number of competing alternative products (equipments or systems) results from an evaluation of the product worth of each of the possible alternatives. This criterion, product worth, is a complex which includes many elements: engineering characteristics, human factors, and economic burdens. These elements are not independent; rather, they are interrelated intrinsically through the management controls over purchasing procedures, logistics systems, and methods of product operation in the user environment.

Those aspects of product or system worth most closely associated with engineering design and hardware characteristics are included under the title Systems Effectiveness. It is the purpose of this section to discuss system effectiveness, first in general as an element or portion of overall system worth and second in a detailed way to describe the basic elements of system effectiveness itself—the properties which ultimately determine it. We shall first define a number of concepts to avoid the many, well-known semantic problems in this general area. Then we shall consider interrelationships in terms of trade-offs between the various factors associated with system effectiveness in particular and system worth in general.

While it is not our intention to give complete coverage to the concept of system worth, it is useful to identify its basic elements and to place system effectiveness in its proper context. The factors which combine to determine system worth can be classified under four headings: dollar cost, manpower requirements, delivery schedule, and system effectiveness. Dollar cost includes original acquisition cost, maintenance cost, and amortization for replacement to cover depreciation. Manpower requirements include the numbers and skill levels of operators, maintenance, and management personnel. Delivery schedule relates the industry capacity to produce the system to the operational needs with respect to time. Finally, system effectiveness is concerned with the ability of the system to perform its intended function, including the frequency with which it fails, the difficulties involved in servicing and repair, and the suitability or adequacy of the system to perform its intended function when it is operating in accordance with design concept.

1.1a System Effectiveness—a Function of Use Requirements, Equipment Condition, and Performance Characteristics. We have previously indicated that system effectiveness is closely associated with hardware characteristics, specifically, the ability of the system to perform its intended function. It frequently happens that the designer, the salesman, and the user have three widely different ideas of the intended function. In final analysis the user's viewpoint must prevail; but the designer's ideas still have a significant influence, since they are paramount in fashioning the hardware which is delivered. We might say that the user supplies the environment, the designer supplies the heredity, and system effectiveness involves both, including the interactions between them.

A more precise description of effectiveness requires a consideration of time in addition to or as a common denominator of the user's environment and the designer's heredity. Thus we must analyze time on the basis of two different system viewpoints. We must analyze the use requirements which are placed upon the system in terms of time, and we must also study the equipment condition in terms of time as it is influenced by engineering characteristics, environment stresses, and performance demands. This dual time classification can be listed briefly as follows:

Use Requirements:

1. System operation required.
2. System operation not required.
   - 2.1 No operation scheduled.
   - 2.2 The system is in storage (as a spare).

Equipment Condition:

1. System is in an operable condition.
2. System is in an inoperable condition.
   - 2.1 Repair action is being programmed.
   - 2.2 Repair action is under way.
     - 2.2.1 Maintenance preparation.
     - 2.2.2 Fault location.
     - 2.2.3 Repair is being accomplished.
     - 2.2.4 Completed repair is being checked.
   - 2.3 Repair action is suspended.
     - 2.3.1 Awaiting parts.
     - 2.3.2 Maintenance shop closed by normal work schedule.
     - 2.3.3 Repair delay for higher-priority work.

It is recognized that the above listing of categories is based on the preference of the writer. For some purposes less detail is required; for others one might wish to make even finer subdivisions.

The third element in system effectiveness is the equipment performance capability. This element is merely the adequacy of the design to perform its intended function when it is operating properly. The criterion for determining proper operation is itself rather complex and not easily defined in general terms. Rather, one must specify a criterion for each system. It is immediately recognized that we are talking about failure, but it is important to realize that we also refer to performance level in a broader sense. We must consider degraded performance as well, and it is implicit that such degraded performance may mean reduced system effectiveness.

The primary objective of this section is, then, the development of a logical structure of the three elements—use requirements, equipment condition, and performance characteristics—related to time as the common denominator in order to obtain a more precisely identified concept which we call system effectiveness. The accomplishment of this objective is dependent upon a further investigation of the various states of the elements which determine system effectiveness, using the categories as listed above for the first two elements and a performance degradation index for the third element. This investigation of states will serve to introduce the concepts of operational readiness, availability, maintainability, reliability, etc. These terms have been responsible for many semantic problems in the past, since each has regularly been used in a variety of senses. Hence, this chapter will include a rather complete set of definitions of terms to cover the critical factors of importance in the overall concept system effectiveness.

1.1b System Effectiveness as a Combination of Operational Readiness, Mission Reliability, and Design Adequacy. As a general statement of the problem we face in analyzing system effectiveness, we wish to answer three basic questions:

1. Is the system ready to operate when demanded to do so?
2. Will the system continue to operate for the required duration, assuming an affirmative answer to the first question?
3. Will the system accomplish the desired mission objective, assuming affirmative answers to the first two questions?

The first question deals with operational readiness, the second with mission reliability, and the third with design adequacy. Each of these three concepts is dependent on use requirements, equipment condition, and performance characteristics. Time is a critical element in both operational readiness and mission reliability; it is of minor importance in design adequacy. Thus, it is necessary to see how time enters, and this can be accomplished by means of an analysis of use requirements and equipment condition jointly and separately.

1.2 TIME CATEGORIES BASED ON USE REQUIREMENTS AND EQUIPMENT CONDITION

A variety of time categories will be named and identified, and we shall later discuss the nature and effect of the overlaps and interactions which exist between them.

Use-requirement states lead to the following divisions of calendar time:

1. *Operational time* is that interval during which the equipment is in actual operation. This time is basic to reliability. Operational time is really a category related to equipment condition. It is included here because of its relation to item 2.
2. *Scheduled operating time* is that interval during which equipment operation is desired. In other words, the equipment will be in use during this time if it is in an operable condition.
3. *Free time* is that interval during which the equipment is scheduled off duty—the "office is closed"—as, for example, an office telephone at night or television-station cameras when the station is off the air.
4. *Storage time* is that interval during which the equipment is a spare—it is extra equipment intended for service only when required as a replacement.

Time categories related to equipment condition are listed below. It will turn out that the times during which equipment is "good" and during which it is "bad" are of little interest by themselves. The more critical categories are subdivisions which relate to the activities with which the equipment is involved. That is to say, if the equipment is bad, what are we doing to restore it to an operable state?

1. *Good* or *up time* is that interval during which the equipment is being or could be operated.
2. *Operating time,* as defined previously, is the interval of good time during which the equipment is being operated.
3. *Bad* or *down time* is that interval during which the equipment is in a failed state—it could not be operated without some repair or maintenance activity.
4. *Administrative time* is that interval during which maintenance orders are being prepared. It includes all administrative delays prior to and subsequent to actual initiation of maintenance action.
5. *Active repair time* is that interval during which activities associated with repair

are being carried out. It includes a variety of such activities: preparation, fault location, actual repair, equipment test, and checkout.

6. *Logistic time* is that interval during which active repair has been suspended while awaiting the arrival of spare parts.

1.3 CONCEPTS IN TERMS OF WHICH SYSTEM EFFECTIVENESS IS STUDIED AND EVALUATED

A complete study of the previously listed time categories is obviously essential in evaluating the effectiveness of any system—adding, of course, a treatment of the

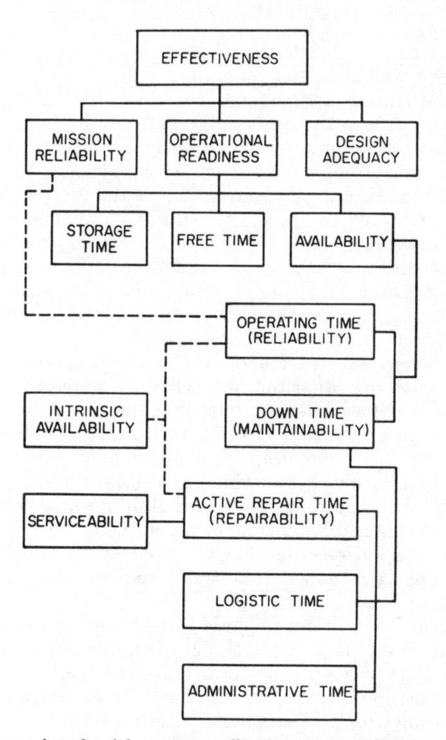

FIG. 1.1 Concepts associated with system effectiveness. (*Courtesy of ARINC Research Corporation.*)

design adequacy to include the capability of the system to accomplish mission objectives when it is operating properly. This study of the time categories will be facilitated by the introduction of terms which identify category characteristics like means, ratios, and distributions, including interrelationships between categories. Reference has already been made, for example, to the term "reliability," and we shall now introduce and define many other terms. While these terms will be defined in quantitative language, it is important to remember that a complete quantitative understanding of system effectiveness is not accomplished by evaluating all such quantities. Many trade-off relationships and special considerations are critical for each particular system. At the present time, we are merely identifying the concepts which must be analyzed; we are not doing the analysis itself.

1.3a A Graphic Presentation of the Concepts. An indication of the concept interrelationships is shown in Fig. 1.1, and reference to this figure can be of assistance in understanding the concept definitions which follow.

1.3b Concepts Based on Single Time Categories. Because the definition and understanding of reliability have been developed and standardized over many years of usage, we shall consider it first. The reliability of a system or item of equipment is the probability that the equipment will operate satisfactorily for at least a given interval of time[1] when used under specified conditions. The reliability function $R(t)$ is the expression of this probability as a function of the interval of time from 0 to t. Thus, reliability is defined in terms of a cumulative distribution function. The associated density is called the time-to-failure density function, denoted by $u(t)$. The two functions are related by the equation

$$u(t) = -dR(t)/dt$$

The functional notations $u(t)$ and $R(t)$ are becoming standard; the definition of reliability has already been well accepted generally.

We can now define similar terms for other time intervals of interest in system effectiveness. *Repairability* is the probability that a failed item of equipment will be restored to operability in not more than a specified interval of active repair time when maintenance is performed under specified conditions. There is an associated time-to-repair density function. *Maintainability* is the probability that a failed item of equipment will be restored to operability in not more than a specified interval of down time when maintenance and administrative conditions are stated. There is, of course, an associated down-time density function. While special names have not been proposed, it is obvious that similar probabilities and density functions can be considered for administrative time, logistic time, and other time subdivisions such as those mentioned previously for active repair time.

1.3c Concepts Based on Ratios of Time Categories. We have previously indicated that time-category distributions tell only a partial story. There are a variety of trade-offs to be considered, but, in addition, there are other important characteristics which are expressible as ratios of times. We shall emphasize three of them: *intrinsic availability*, *availability*, and *operational readiness*. Usage of these terms has not been standardized, and this has caused serious communication problems. For example, some persons use "availability" and "operational readiness" interchangeably, while others differentiate between them. To some, the term "intrinsic availability" may be entirely new. Concerted effort is essential to solve these semantic problems, and it is hoped that this discussion will contribute toward the accomplishment of that end.

Before we write formal definitions of these three terms, it will be helpful to describe the concepts in nonrigorous language. Each of the concepts is concerned with the relationship of some type of good time to a total time, perhaps the sum of this good time and a type of bad time. Thus, each concept deals with a proportion (or probability) of the time during which a "satisfactory" state exists. Obviously, in this sense good time could refer either to time during which the equipment is operating or to time during which it is operable but not operating, or to the sum of the two. Bad time could be active repair time, or down time, or some combination of other time categories. Or we might wish to think of total time—for example, all time except storage time, all time except storage and free time, or all calendar time. Hence, we shall define each concept by identifying the good time and the total time for each.

Note. Certain peculiar problems arise when there are periods of free time and/or storage time. Special treatment is necessary in these cases, and this will be discussed later. It will be recognized that this later discussion will actually reveal a certain amount of confusion in the definitions as presented here. This merely constitutes a delay in clearing up such confusion, an oversimplification made intentionally in the interests of ease of presentation.

In view of the above general treatment, we can write the concept definitions in a symbolic fractional notation as follows:

[1] For convenience, we shall discuss time-dependent cases only. The extension to cycle-dependent systems and one-shot devices is obvious.

$$\text{Intrinsic availability} = \frac{\text{operating time}}{\text{operating time plus active repair time}}$$

$$\text{Availability} = \frac{\text{operating time}}{\text{operating time plus down time}}$$

$$\text{Operational readiness} = \frac{\text{all good time}}{\text{total calendar time}}$$

where "all good time" is time during which the equipment is operating plus all time during which it is not being operated but is operable and could be used if required.

Two of the blocks of Fig. 1.1 reference concepts closely tied to the specific mission requirements of the user of the system; they are mission reliability and design adequacy. We define *mission reliability* as the probability that the system will operate in an appropriate mode for the duration of a mission, given the conditions of use and the fact of operability in this mode at the start of the mission. More simply, this says that mission reliability is the probability that no failure of the system occurs during a single mission, given that the system was working at the beginning. *Design adequacy* is the probability that the system successfully accomplishes the mission if the system does in fact operate within design specifications. Thus, design adequacy is concerned with inherent capability of the system to do its job when it is working. As an analog we can think in terms of the CEP as a property of design adequacy. The smaller the CEP, the more adequate the design and the greater the probability of being "on target."

1.3d Serviceability. We have previously described the time element relationships and the design performance characteristic, design adequacy, which together are determinants of system effectiveness. One important design attribute remains, and we choose to call it *serviceability*. It is defined in qualitative language as the ease or difficulty with which a system can be repaired. Obviously, conditions in the maintenance shop enter as a major influence in the determination of serviceability. Crudely, we can say that the relationship between serviceability and maintenance is about the same as that between design adequacy and mission performance requirements. This is a reasonable parallel, but there are a few important differences. Past experience and simplicity of concepts involved permit us to be quite quantitative in discussions of design adequacy. Later we may be able to do this for serviceability, but at the present time we must be satisfied with qualitative serviceability evaluation. Of course, the effect of serviceability is reflected in the repair-time distribution even though serviceability itself is not quantified as an element of system effectiveness.

We can now summarize these ideas by the following list of more formal definitions of concepts:

1.3e Definitions of Concepts

System effectiveness is the probability that the system can successfully meet an operational demand within a given time when operated under specified conditions.

System effectiveness (for a one-shot device such as a missile) is the probability that the system (missile) will operate successfully (kill the target) when called upon to do so under specified conditions.

Reliability is the probability that the system will perform satisfactorily for at least a given period of time when used under stated conditions.

Mission reliability is the probability that, under stated conditions, the system will operate in the mode for which it was designed (i.e., with no malfunctions) for the duration of a mission, given that it was operating in this mode at the beginning of the mission.

Operational readiness is the probability that, at any point in time, the system is either operating satisfactorily or ready to be placed in operation on demand when used under stated conditions, including stated allowable warning time. Thus, total calendar time is the basis for computation of operational readiness.

Availability is the probability that the system is operating satisfactorily at any point in time when used under stated conditions, where the total time considered includes operating time, active repair time, administrative time, and logistic time.

Intrinsic availability is the probability that the system is operating satisfactorily at any point in time when used under stated conditions, where the time considered is operating time and active repair time.

Design adequacy is the probability that the system will successfully accomplish its mission, given that the system is operating within design specifications.

Maintainability is the probability that, when maintenance action is performed under stated conditions, a failed system will be restored to operable condition within a specified total down time.

Repairability is the probability that a failed system will be restored to operable condition within a specified active repair time.

Serviceability is the degree of ease or difficulty with which a system can be repaired.

1.4 THE DUAL-CRITERIA TIME BREAKDOWN

We have previously noted that time categories were determined from two factors: use requirements and equipment conditions. In order to simplify the treatment and

Table 1.1 A Two-dimensional Time-category Classification

Equipment condition	Use requirements		
	Operation required	Operation not required	
		Free time	Storage time
System operable (up time)	No problems exist		
System inoperable (down time): Administrative time (repair planning) Active repair time Repair action suspended: Logistic time Administrative time (Shop closed—higher-priority work)	Problems exist and system effectiveness is degraded	Problems exist *but* no degradation of system effectiveness results	

permit us to introduce required terminology, we delayed consideration of a time breakdown based on the two factors simultaneously. We shall now describe this two-way time classification and then relate it to the concepts which we have previously defined. The dual-criteria time classes are shown in Table 1.1.

The use-requirements time classes are shown in the columns and those based on equipment condition are shown in the rows of Table 1.1. The body of the table shows the criticality of the dual time categories with respect to degradation of system effectiveness. If we were to collect data from field experience and compute observed percentages of total time and mean times for each of the dual time classes, we would certainly have many indices useful in studying system effectiveness.

Perhaps a few qualitative and obvious aspects of Table 1.1 will be worth noting before we look into the details of the quantitative side. Degradation of system effectiveness never occurs when system operation is not required, and, of course, it does not occur if the system is operable. Trouble arises only when the system is inoperable and when its use is required, i.e., when down time overlaps required operation. Perhaps the most significant aspect of this qualitative evaluation is that high operational readiness can and often does result primarily from excessive free time and storage time. Equipment which is not used can usually be kept in a state of operability! Lack of use can result from excessive free time and also from equipment redundancy which provides storage time, and these can be used as "crutches" to permit us to "live with" poor hardware. Thus, a high level of operational readiness is not *necessarily* an indication that no hardware problems exist.

On the positive side, Table 1.1 points out the importance of planning repair activities. Major problems are solved if we can keep down time from overlapping time when operation is required. This can be accomplished by scheduling maintenance during free time and by providing adequate spares to permit transfer of operable equipment from supply to use and failed equipment from use to storage.

In view of the above observation, it is apparent that *all* of the dual time categories must be considered in order to give a fair evaluation of the effectiveness of a system. Too frequent failure and too long repair time are bad in any case. The ideal answer is direct improvement—fewer failures and more efficient repair methods. The substitute answer is more free time or more spares to provide storage time. Another poor substitute is reduction in performance requirements and/or improved operating procedures.

1.5 QUANTIFICATION OF CONCEPTS

There are a number of concept indices which we would wish to compute for the dual time classification. We would certainly want to know the probability that system use requirements and equipment condition would place the system in a specified time class. Given that it was in the time class, we would wish to know the probability that it would remain there for at least a specified duration—a cumulative distribution function. From the cumulative functions we would compute densities, means, variances, etc. We would also want to measure ratios of the times in various dual-criteria time categories—measures related to such concepts as intrinsic availability and operational readiness. We shall discuss some of these indices and measurements in more detail, noting the mathematical problems and the use of the quantitative elements as criteria for making management decisions.

1.5a Quantification of Time for a Single Time Category. As an example, consider active repair time regardless of the use requirements. This time would be determined by the equipment itself, the quantity and quality of maintenance personnel, and the maintenance facilities. Its description would be complete if we gave an expression for the active repair time density function, the percentage of total time represented by active repair time, and any correlation which may exist between this and each of the other times, separately and collectively. It is noteworthy that not all of these descriptions are usually presented in system analyses. Some depend only on mean active repair time, others depend on proportion of total time devoted to active repair, and still other studies give both. A few even show the associated density function.

The most frequently observed density for active repair time has characteristics like the log-normal distribution: starting at the origin, rising to a peak at the modal value, and extending with a long tail to the right. It is not uncommon to find these densities with an additional complication or modification of the standard log-normal, namely, a starting point not at time zero, but at some minimum repair time, say, 10 minutes.

In maintenance language, then, there would be no repairs completed in less than some minimum time, a gradual increase in the probability of repair completion until some modal value, then a gradual decrease and tail-off to indicate that some repairs are expected to take an extremely long time. There are many distribution types which exhibit these general characteristics, the gamma distribution, for example.

Naturally, one would expect to encounter other active repair time densities. Some have more than one modal value, such as failures concentrated around a low value, or quick repair time, and others concentrated around a much longer value, or long repair time. In this case repair is relatively quick and easy or else it is much more difficult and time consuming.

1.5b Quantification of Ratios of Times in More Than One Category. There is no simple index for describing a time ratio, e.g., the ratio of operating time to the sum of operating time and active repair time, a ratio called intrinsic availability. Indeed, it turns out that more than one index is absolutely essential to describe all of the system engineering, operational, and administrative characteristics involved in this one ratio. We shall try to describe these characteristics and illustrate some indices as examples of the types of thing one must do in evaluating system-effectiveness attributes in terms of ratios of time categories. This description is intended to show that the effectiveness of a system is a complex multidimensional vector and is not capable of direct simple analysis in terms of a small number of clearly defined measures. Rather, system-effectiveness analysis must be approached in each case as a new and challenging statistical problem requiring uniqueness of analytical method and ingenuity on the part of the analyst. No simple, cut-and-dried format exists.

Now, one has the general impression or hope that intrinsic availability is determined primarily by engineering or hardware properties. This is true, of course, if proper operational demands and use conditions are imposed upon the system and if repair facilities and maintenance personnel are adequately provided. Hence intrinsic availability is supposed to be an attribute which identifies the built-in capability of the system to provide continuity of service under proper use and maintenance conditions. As such, it is a critical measure of the success of the designer in meeting the objectives which have been claimed—an index of the "best" the customer has a right to expect, exclusive of the adequacy of the design to accomplish its assigned mission when it is working properly. Intrinsic availability is also the best the customer can expect in the sense that all administrative and logistic times are excluded from consideration—a view directly descriptive of military equipment in a hot-war situation.

There is one obvious index of intrinsic availability: the ratio of mean operating time to the sum of mean operating time and mean active time. This ratio is equivalent to the probability that the system is operable, where the only time considered is the total of operating and active repair time categories. However, the existence of free and storage times complicates even this simple index. In reality, active repair time is not a source of degradation if it occurs during free time or if availability of a spare system can provide either uninterrupted system operation or operation with only a minor delay involved in procuring a spare from supply. Thus, we are perhaps more interested in a modification of the ratio: Replace active repair time in the formula by only that portion of active repair time during which system use is required.

On the other hand, it must be recognized that the true engineering problem is more completely identified by the first form of the ratio. This can be visualized as follows. If continuous operation is required and if no spares are available, then the two ratios are equivalent. In this case, intrinsic availability is truly descriptive of a critical aspect of system quality determined predominantly by engineering characteristics and stresses imposed by operational conditions (see Secs. 5 and 11).

Now, hardware characteristics can be included in quite a different but closely related manner; the engineering can influence the interaction between operating and active repair times in a way not described by either of the ratios discussed above. Furthermore, this interaction can be important in estimating the probability of system failure during a specific mission by considering prior active repair time experience. This can arise as follows.

Equipments can, in some cases, exhibit a correlation between the time required for a repair action and the subsequent failure pattern. For example, one would hope that the family lawn mower, refrigerator, automobile, or other equipment would work with relatively little trouble for some time after major overhaul.

In mathematical language, we can describe these ideas as follows. Let t_i denote the operating time between the $(i - 1)$st and the ith repair and let x_i denote the active repair time involved in the maintenance occasioned by the ith failure. We can postulate a number of correlations between t's and x's. The most obvious ones are:

1. The correlation between t_i and x_{i-1}
2. The correlation between t_i and x_i

The first correlation relates the length of an operating time to the immediately preceding active repair time, while the second uses the immediately succeeding repair time. A complete analysis of intrinsic availability, especially from the viewpoint of predicting failure-free mission probability, would involve consideration of all such correlations. Unfortunately, the number of studies along these lines is far too small to furnish any significant information.

Now, the usual approach has been merely to consider the only index of intrinsic availability to be the ratio

$$\bar{t}/(\bar{t} + \bar{x})$$

where $\bar{t}$ and $\bar{x}$ are mean values of operating time between failures t_i and active repair time x_i. This is indeed a very useful index, but we could obtain additional information from knowledge of the density function of the variable y defined by

$$y = t/(t + x)$$

where t and x refer to specifically related operating and active repair times as illustrated above. That is, t might be t_i and x might be either x_{i-1} or x_i. It is obvious that the mean value of y, say, $\bar{y}$, is not usually equal to the index based on means $\bar{t}$ and $\bar{x}$,

$$\bar{t}/(\bar{t} + \bar{x})$$

To illustrate this density-function approach to the analysis of intrinsic availability, let us assume two simple and no doubt unrealistic density properties:

1. Operating time and active repair time are uncorrelated.
2. Each of these times is exponentially distributed. Denoting operating time by t and active repair time by x, the problem to be considered is the derivation of the density function of y, which is defined as

$$y = t/(t + x)$$

For this purpose, let the means be

$$\bar{t} = a \qquad \bar{x} = b$$

giving the joint density function of t and x in differential form

$$(1/ab)e^{-t/a}e^{-x/b}\,dx\,dt$$

The density function of y can be obtained by making the transformation of variables

$$y = t/(t + x) \qquad z = x$$

obtaining the joint distribution of y and z, and integrating on z. The steps of this process are as follows:

The transformation can be written as

$$t = yz/(1 - y) \qquad x = z$$

and the Jacobian is

$$z/(1 - y)^2$$

Hence the density of y, $g(y)$, is given by

$$\begin{aligned}
g(y) &= \int_0^\infty \frac{1}{ab} \exp\left\{\left[-\frac{1}{a}\left(\frac{yz}{1-y}\right) - \frac{z}{b}\right]\right\} \frac{z}{(1-y)^2}\,dz \\
&= \frac{1}{ab(1-y)^2}\int_0^\infty z \exp\left\{-z\left[\frac{y}{a(1-y)} + \frac{1}{b}\right]\right\} dz \\
&= \frac{1}{ab(1-y)^2}\,\frac{1}{[1/a(1-y) + 1/b]^2} \\
&= \frac{ab}{[by + a(1-y)]^2} \\
&= \frac{ab}{[a + (b-a)y]^2}
\end{aligned}$$

the range being $0 \leq y \leq 1$.

The mean value of y, say, $\bar{y}$, is

$$\begin{aligned}\bar{y} &= \int_0^1 \frac{aby}{[a+(b-a)y]^2}\,dy \\ &= \frac{ab}{b-a}\int_0^1 \left[\frac{-a}{[a+(b-a)y]^2} + \frac{1}{a+(b-a)y}\right] dy \\ &= \frac{ab}{(b-a)^2}\left[\frac{a}{a+(b-a)y} + \log\,[a+(b-a)y]\right]\int_0^1 \\ &= \frac{ab}{(b-a)^2}\left[\frac{a}{b} - 1 + \log b - \log a\right] \\ &= \frac{a}{a-b} + \frac{ab}{(a-b)^2}\log\frac{b}{a}\end{aligned}$$

Actually, $g(y)$ is a function of the ratio of b to a. Thus, if $ac = b$,

$$g(y) = \frac{c}{[1+(c-1)y]^2} \qquad \text{and} \qquad \bar{y} = \frac{1}{1-c} + \frac{c}{(1-c)^2}\log c$$

Now, the operations involved in the above derivation required that $a \neq b$ and that $c \neq 1$, since division by $b - a$ was involved. Hence, to find $g(y)$ if $a = b$, it is necessary to go back to the original expression. In this case

$$\begin{aligned}g(y) &= \int_0^\infty \frac{1}{a^2}\left\{\exp\left[-\frac{z}{a}\left(\frac{y}{1-y}+1\right)\right]\right\}\frac{z}{(1-y)^2}\,dz \\ &= \frac{1}{a^2}\int_0^\infty \left\{\exp\left[-\frac{z}{a(1-y)}\right]\right\}\frac{z}{(1-y)^2}\,dz \\ &= 1\end{aligned}$$

and

$$\bar{y} = 0.5$$

When the mean of y, $\bar{y}$, is compared with the index based on the ratios of means, that is,

$$\bar{y} = \frac{1}{1-c} + \frac{c}{(1-c)^2}\log c \qquad c \neq 1$$

$\bar{y} = 0.5$, $c = 1$, and the index based on means $1/(1+c)$, it is apparent that the two indices are equal only when $c = 1$. In all other cases, $1/(1+c)$ is a biased estimate of $\bar{y}$.

Graphs of $g(y)$ for various values of the parameter c are shown in Fig. 1.2. The shapes of these curves can be described by examining the following function characteristics. The derivative of $g(y)$ with respect to y is

$$\frac{dg(y)}{dy} = \frac{2c(1-c)}{[1+(c-1)y]^3}$$

The slope of the curve changes from $2c(1-c)$ at $y = 0$ to $2c^{-2}(1-c)$ at $y = 1$. The slope is negative for $c > 1$ and positive for $c < 1$ throughout the range $0 \leq y \leq 1$. The curve for $c = k$ is merely a reflection of the curve for $c = 1/k$ about the line $y = \frac{1}{2}$. Hence, the curves in Fig. 1.2 can be used to read density-function values for $c = \frac{1}{2}, \frac{1}{3}, \frac{1}{4}$, and $\frac{1}{5}$ as well as for $c = 2, 3, 4$, and 5 merely by reversing the scale of the abscissa.

1.5c Defining the Random Variable. The preceding discussion of intrinsic availability highlighted one of the most important steps in system-effectiveness analysis: that of defining the random variable which measures the system characteristic of interest. When this is accomplished, it is easy to plan a useful data-collection program and to develop indices, density functions, means, and so on to describe the characteristic.

For single category times the variable is simple. We would want to define the random variable as the interval of time starting with entry of the system into the category and terminating on its departure therefrom. In some cases we might want to delete all time spent by the system in certain other categories. For example, in studying reliability we are concerned with operating time: the system is needed and it is in an operable state. We might well wish to disregard free time: interruption of operation not due to system failure, but due merely to a cessation in the requirement for use of the system. This is usually defined in terms of "operating time between failures." Similarly, active repair time must be described in terms which exclude interruption of repair for reasons not associated with the system when we are interested in repairability as a system engineering characteristic. Thus, even when the random variable is easily spotted, it is still essential to give it precise definition in order to relate it in a meaningful way to the hardware and/or the operational environment or use condition.

As indicated above, the random variable is not easily identified when looking at ratios of times. Many choices can be made when we study even a simple ratio like

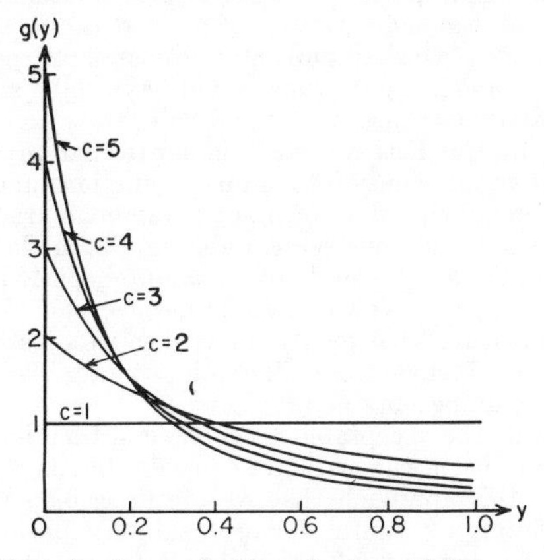

Fig. 1.2 The distribution of y for a selection of values of the parameter c:

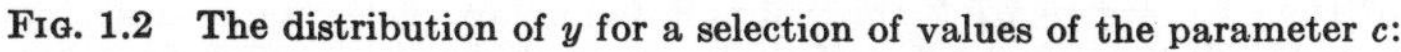

$$g(y) = c[1 + (c - 1)y]^{-2}$$

intrinsic availability. Different random variables are associated with different system characteristics, and a complete system-effectiveness study would treat them all—at least all that have any significant descriptive value.

1.6 PREDICTION PROBLEMS

We are usually more interested in predicting than in measuring system effectiveness and the factors which combine to produce it. This follows naturally from our desire to select the best product or system from the group of competing systems before we buy; system effectiveness and the associated factors are selection criteria. When applied to research and development stages—before production—prediction is a necessity and measurement is impossible. Indeed, such predictions form the basis for approval or disapproval of the design concept before development funds are committed. The prediction technique can be most easily described in terms of the research and development case. Assume no completed system exists and consider the problem of forecasting each of the factors involved in system effectiveness. (See Sec. 5.)

At the outset it is obvious that we must find some way to use past experience on a new system, a system which has not yet been built and therefore one for which no system data exist. We must apply the knowledge which has been developed on similar systems and/or which has been derived by theoretical studies on the new system. Thus we must extrapolate and interpolate available data to fit new and often unknown stresses which are imposed by the unbuilt hardware and the untried operating conditions. We usually do so by breaking the system up into functional subunits, analyzing the characteristics of these subunits, and synthesizing the system prediction from the results of the subunit analyses. The logic behind this approach is that many new systems are made up largely of novel combinations of familiar subunits. Often only a few of the subunits are of radically novel design, and they can be given special treatment, perhaps even special testing to collect the necessary data. To make this approach more clearly understood, let us first look at the reliability prediction problem, the prediction with which we have had the most experience. (See Secs. 6 to 8.)

1.6a Reliability Prediction. Implicit in the general discussion of prediction is the existence of two basic steps: development of a formula for the system characteristic in terms of subunit characteristics and collection of past experience or data gathering on subunits for substitution in the formula. We shall emphasize the first step here; details of the entire process will be covered in later sections. (See Sec. 5.) For reliability prediction, the characteristic of interest is operating time as shown in Fig. 1.1. For the moment, we shall omit any discussion of special conditions which apply to operating time under study; they would have to be stated very precisely in any application to a particular case.

The theory of reliability prediction is very similar to that of performance prediction in which we substitute input values into transfer functions and compute the output parameters. Usually, nominal values of input parameters are used in such performance predictions. If we would consider variable inputs in the form of performance probability density functions, we would obtain simultaneously a reliability prediction and a performance prediction. Now in actual practice, we do not commonly use this method because the mathematics is too complex when we substitute input densities in place of nominal values. Instead, we split the input variable into two classes:

1. The input is inside an acceptable range.
2. The input is outside the acceptable range, a condition defined as failure.

With this one simplification, the arithmetic usually becomes quite manageable.

It is apparent from this approach that we are describing a system failure-effect analysis. We are defining a "failure" for each subunit considered (maybe even for each part), and then we are relating this subunit failure to the system—a system response as "failure" or "satisfactory performance," including degradation of performance level. It is not our intent to minimize the problems involved in accomplishing such a failure-effect analysis. There are difficulties in determining what constitutes part or component failure, in analyzing the design to determine part and subunit interactions, in allocating system failures to specific sources, in handling the voluminous detail involved in performing such an analysis for a complex system, etc. However, it can be argued that a design is not understood and perhaps not even completed until a failure-effect analysis has been prepared. Certainly, we must accept the fact that it is an engineering task which constitutes an essential step in reliability prediction.

The derivation of the system reliability-prediction formula is facilitated by preparing a pictorial representation of the failure-effect analysis—the so-called reliability block diagram. This diagram identifies those subunits whose failure assures system failure, the series elements, and those subunits whose failure merely enhances chance of system failure, the parallel elements. For the parallel case, system failure results only from combinations of subunit failures. In other words, the reliability block diagram is a probability problem in pictorial form rather than in words. The solution of this probability problem is the expression of the system failure probability in terms of the failure probabilities of the subunits represented—the ones which were listed in the failure-effect analysis.

As indicated earlier, we are not particularly concerned in this section with data problems per se. We have emphasized the fact that the data must be collected in a form compatible with the objectives; they must reflect the characteristics of interest. In reliability prediction this usually means merely that the appropriate internal and external stress levels be documented, in addition, of course, to the use of statistically sound experiment design methods. Documentation of stress levels suggests the necessity for a certain amount of extrapolation and interpolation to fit new design conditions. The necessity for such adjustment of input data is one of the major elements which differentiates prediction from measurement or assessment.

1.6b The Allocation of System Failures to Parts. In discussing reliability prediction it was implied that every system failure could be traced to or attributed to some specific part or subunit. Obviously this is an oversimplification; the situation is much more complex. Some system failures result from performance drifts or failures of a number of parts, with repair being accomplished by adjustments and/or replacements of the parts involved. In essence, this is the problem of defining part, subunit, and system failure, and we need to consider the way or ways in which we wish to state these definitions to facilitate adequate system-effectiveness evaluation.

There are essentially two objectives to be considered: (1) estimation of the frequency with which system service is interrupted by malfunction and (2) estimation of the logistic load or spare-part requirements. These two objectives are accomplished by two different failure definitions. System failures which are repaired only by adjustment involving no part replacement are counted for objective 1 but not for objective 2. On the other hand, multiple part replacements in a single maintenance action count as a single system malfunction for objective 1 but as a multiple failure when we are considering objective 2. These are only indications of the scope of the failure definition problem. Every system reliability analysis must include its own statement of failure criteria appropriate to the immediate purpose of the analysis, and the best we can do here is to state some of the general principles which govern the formulation of these failure definitions for parts, for subunits, and even for the system itself. (See Sec. 5.)

For predicting system failure frequency, input data must be made to reflect all system failures, including those repairable by adjustment and without part replacement. This means in some cases a rather arbitrary allocation of system failures to parts or some larger subunit so that a synthesized failure estimate for some new system will be unbiased. If the number of adjustment-type repairs is rather large, the allocation to parts can tend to misrepresent sources of trouble. Some parts may be unjustly charged with causing trouble, while others may not be red-flagged when they should be. In spite of this fact, the procedure of allocating all system failures to parts has proved to be useful and resulting biases not too serious for most purposes. If special requirements demand it, predictions can be based on data adjusted according to other criteria.

1.6c Predictions for Other Single Time Categories. Predictions for single time categories other than reliability can be discussed in terms of any one of them, say, maintainability as a property of the down-time classification. We defined maintainability as the probability that, when maintenance is performed under stated conditions, a failed system will be restored to operable condition within a specified total down time. In order to predict system maintainability, we must synthesize a cumulative distribution function or a density function or some other probability representation from past experience. If we use a method analogous to the reliability-prediction technique, we shall use subunit divisions of the system and apply down-time experience from identical or similar subunits used in other systems.

Now in reliability prediction we focused attention on one thing: the time-dependent failure probability of each subunit in terms of its influence on the system failure probability. In maintainability we must consider two things: the probability that the failure of the system was caused by the failure of a particular subunit and the probability density function of the down time given the identity of the subunit which has failed.

We can immediately write a system down-time prediction formula for the simple case in which failure of the system is always due to only one of the N subunits which make up the entire system. For this purpose a set of parallel subunits can be defined as a single subunit. Given that the system has failed, let p_i represent the probability that the system failure was due to the failure of the ith subunit. Let the down-time density for this subunit be denoted by $f_i(t)$. Then the system down-time density function is

$$\sum_{i=1}^{i=N} p_i f_i(t)$$

This can be described as a weighted average of the subunit down-time densities, the weights being the subunit failure probability, conditional on the existence of system failure.

The assumption that each system failure involves the failure of only one of the subunits is quite easily removed. The only information really pertinent is the relative frequency of subunit failures, the p_i in the above formula. If collected data can give estimates of such relative frequencies of subunit failures per system failure, then the same formula applies. We merely have

$$\sum_{i=1}^{i=N} p_i = 1$$

if single subunit failure is assumed, while

$$\sum_{i=1}^{i=N} p_i > 1$$

if multiple subunit failures are possible. In this latter case the down-time densities must reflect the times for both single and multiple subunit failures.

1.6d Ratio-variable Prediction Problems. It might be assumed that prediction of the density for a variable which describes time ratios would be accomplished by the same method as was used in predicting the density for a single time category. In one sense this is indeed true, but in a more general sense the ratio variable is much more complex. The primary difficulty is, as indicated previously, the definition of the appropriate random variable. Indeed, it is usually necessary to consider a number of random variables determined by the ratio characteristics of interest.

For example, consider a system which is required to operate continuously, and assume that repair action is initiated immediately upon system failure. In this case, there is no free time, and, assuming no spares, there is no storage time. Then operational readiness is the same as availability, and an appropriate index might be A, defined by the formula

$$A = t_o/(t_o + t_r)$$

where t_o is total operating time and t_r is total repair time in a specified time interval.

The system availability is indeed indicated by the value of A, but there are certain critical characteristics or assumptions peculiar to A; it is an index of availability in a restricted but very meaningful sense. If we assume that repair actions can and do restore the system to "like-new" condition, A does describe the probability that the system is operable at a randomly selected point in time in the interval specified. This implies that A is time independent (except for the slight variation introduced if we do not consider a well-chosen time interval, say, one starting and stopping immediately after failure). If, on the other hand, we assume that repair does not restore to like-new condition, but rather that failure probability increases with time, i.e., with each repair, then A is time dependent and its value for the time interval from t to $t + h$ decreases with t for h fixed and increases with h for t fixed. Thus, in general, A is really a function of t and h, where t is the start of the time interval and h is the length of the interval over which system availability is being considered.

It is customary to use the index A under the assumptions of time independence and restoration of failed equipment to like-new condition. Under these conditions, prediction is indeed simple; for we merely need to observe that A can be computed as

$$A = \bar{t}_o/(\bar{t}_o + \bar{t}_r)$$

where $\bar{t}_o$ and $\bar{t}_r$ are means of operating time between failures and repair time per failure, respectively. Then the prediction is accomplished by establishing values of $\bar{t}_o$ and $\bar{t}_r$ from past experience, as discussed above, and substituting these means in the formula.

This availability index is an estimate of the probability that the system is operable at any randomly selected point in time under the assumptions noted above.

However, we have observed that the time to repair a malfunction may well be correlated with the operating time in a number of ways. Let $t_{o,i}$ be the operating time between the $(i - 1)$st and the ith failure and let $t_{r,i}$ be the time to repair the system immediately after the ith failure. We can consider many correlations such as those for the following pairs of variables:

1. $t_{o,i}$ and $t_{r,i-1}$
2. $t_{o,i}$ and $t_{r,i}$
3. $t_{o,i+1}$ and $\bar{t}_{r,i}$
   etc.

where $\bar{t}_{r,i}$ is the mean repair time observed for the first i repairs.

For each such correlated pair of variables we can define a random variable y given by

$$y = t_o/(t_o + t_r)$$

where the t_o and t_r refer to the combination of interest: (1), (2), or (3) above. We would then look at the density function of the variable y and interpret this density in terms of a forecast for the system under study. Prediction of the form of the density would be based on data collection indicated by the definition of the variable y being used. The variable for the first case is of interest in describing availability, given information on the length of time required for the last system repair. In the second case, variable y would be useful in describing estimated repair time, knowing preceding operating time—not actually an availability characteristic. The third variable y will assist in estimating operating time, given information on all previous repair times.

We wish to reemphasize that the mean value of y for any case like one of the above is not in general the same as the availability index A given by $A = \bar{t}_o/(\bar{t}_o + \bar{t}_r)$. Hence, it is suggested that A constitutes only a partial description of system-availability characteristics. However, little use has been made of y-type variables, and we are unable at this time to present techniques and examples in an acceptable form. This can perhaps be a fruitful research area. Hence, availability prediction now rests solely on our ability to synthesize system reliability and repair-time density functions and compute means for substitution in the formula for A.

1.7 THE INCLUSION OF DESIGN ADEQUACY

It is common to visualize a number of operational modes, each with its own performance level. This implies variation in the level of success with which the system mission is accomplished. Selection of the operational mode may depend on the free choice of the operator or on the existence of certain system failures or both. The inclusion of these various levels of adequacy depends on measurement of both the level and the frequency of occurrence of the mode which produces the level. Certainly, then, design adequacy would be treated like any other gain-and-loss function or payoff function. We would be interested in computing the sum of a series of products each involving two factors: a numerical measure of the payoff or degree of mission success and the probability of the accomplishment of the specified level. Thus, we are concerned with an expected value in the usual mathematical sense.

In addition to the expected value, the systems engineer would also be very much interested in the individual terms which were added together. Redesign effort to increase design adequacy would be planned on the basis of the potential gain due to the reduction or increase in the frequency of occurrence of a given operational mode. Thus, a low degree of mission success in a mode is not critical if the mode is extremely unlikely, and redesign to reduce the probability of this mode would be assigned low priority. On the other hand, high priority would be given to redesign if the low (high) degree of mission success were coupled with high (low) probability of occurrence of the associated mode. Therefore, it is apparent that a complete analysis of design adequacy requires considerations of these operational modes in terms of mission success level and relative frequency of occurrence both separately and together.

1.8 THE INCLUSION OF SERVICEABILITY

Serviceability is defined as the degree of ease or difficulty with which a system can be repaired. Thus, serviceability is not itself a time category, though it does exert a major influence on active repair time: the better the serviceability, the shorter the time needed to make repairs. Therefore, we can use active repair time functions as indicative of the serviceability of an equipment; but when we describe the serviceability itself, we are more directly concerned with those hardware design features which enhance or degrade the ease with which the repair is accomplished. For example, these design features might include built-in test points, failure indicators, accessibility, modular design, plug-in connectors, etc. (See Sec. 8, Test Planning.)

Attempts to quantify serviceability have been only partially successful, and we have chosen to omit them from consideration here. It is suggested, however, that competing equipments be compared by means of checklists of critical design features and that such checklists also be used in evaluating preproduction models. Some features which enhance serviceability can be provided at minimal cost in money, weight, etc.; some involve no cost at all. Others may involve costs, and trade-off functions must be developed to assist management in deciding whether or not such features should be incorporated in the design.

Efforts to quantify serviceability should definitely be continued. It is reasonable to assume that adequate techniques can indeed be developed. In the mean time, it is worthwhile to try to incorporate in every design as many of the serviceability features as possible within limits of time, budget, etc. Too many equipments have been marketed with unnecessarily difficult maintenance routines, and this is costly to the customer in time, money, and patience.

1.9 INTERPRETATION OF TIME-CATEGORY RELATIONSHIPS

We have indicated that the effectiveness of a system is determined by many characteristics, and we have defined the associated concepts and briefly discussed their quantification. We now need to look at the interrelationships between them and the way in which management can use them in formulating decision criteria. Obviously, the effectiveness of a system will be determined by the composite of all of these characteristics, but management must consider much more than the composite; it must look at the characteristics separately, the trade-offs between them, the redesign improvement potential, and the other factors which determine overall system worth. It is also essential to consider the enhancement of effectiveness through improved conditions of use, maintenance, and administration. In other words, management must use these concepts to isolate and quantify areas of trouble and to assign responsibility for accomplishing improvement. We shall mention a few of the ways in which this is done.

At the outset, it is useful to rephrase the meaning of system effectiveness not as a definition, but as a customer objective of system operation. From the user's viewpoint, system effectiveness is determined by the answer to the following question: How well does the system do the intended job, disregarding costs, manpower, and troubles of getting delivery from the manufacturer? In these terms it is obvious that the user of the system will make trade-offs of many kinds. For example, he may be able to accept rather frequent failure if repair is simple and fast. On the other hand, if interruption of service is serious even for a short time or if system repair is difficult and time consuming, even infrequent failure is a cause for concern and a degradation in acceptability of the system to the customer. For example, frequent failure of the family lawn mower is not serious if repair is simple, but failure of the landing gear on a commercial aircraft is catastrophic regardless of the ease of repair.

Now these examples lead one to system-effectiveness characteristics such as operational readiness, availability, and design adequacy. High operational readiness is desirable, but it is important to know how it is achieved. If it results from high availability, this is good; for it means down time is low relative to operating time—repairs do not cause undue delay. If, however, we achieve high operational readiness primarily by infrequent equipment use (free time) or by maintaining available redundancy at the system level (storage time) and we have low availability, then we are using free and storage times as crutches to permit us to live with poor equipment.

With respect to design adequacy we must rate the quality of or degree of user satisfaction with system performance. Thus, failure of the lawn mower to cut properly is annoying, and it degrades design adequacy to a limited extent; but failure of the aircraft landing gear reduces adequacy to nothing. The latter is a totally unacceptable situation; there is really no alternative mode of operation. Variation in operational mode, either discrete or continuous, generates various levels of design adequacy. We may well accept degradation in performance quality as a compensation for more reliability or ease of repair or both.

If down time is excessive, we must examine the cause. Engineering design is an important influence on active repair time. Logistic time is largely dependent on inventory policy, which in turn is set by management, perhaps the purchasing agent. Administrative time is also a management responsibility in most cases. Thus, we shall want to make a detailed analysis of all types of time, recognizing that system effectiveness can be enhanced or degraded by system design, management policy, maintenance methods, and use conditions.

The statistical problems involved in interpreting measures of factors affecting system effectiveness are actually quite similar to those which we encounter in any statistical study. The numbers must be interpreted properly, which in this case means that they must be translated into the language of the system-effectiveness concepts. For example, a system might have an availability of 90 per cent and be totally unsatisfactory, while another system with an 80 per cent availability might be quite acceptable even for the same mission. The mission might be a 12-hour bombing mission, and the system under study might be the bombing-navigation system. The 90 per cent availability might be due to low reliability and short repair time. Hence, with the 80 per cent system, probability of mission success might actually be the higher, given operability at the start. We cannot interpret any number out of context.

One especially tricky statistical problem is associated with the inclusion and interpretation of free time and storage time. These two times often "synthetically" make a system look good. This is not as serious as the problem of assuring good use of these times to improve system effectiveness. If maintenance can be planned to overlap free and storage times, the degrading effect of repair actions can be minimized. Thus, we would wish to consider an availability measure based on down time when operation is required, in addition, perhaps, to a measure which includes all down time. This is best accomplished by a data-collection system which uses the dual-criteria time breakdown. Of course it can be approximated by some acceptable prorating of down time to the use-required category.

Mathematical techniques in Secs. 2 to 4 can, of course, clarify the statistical problems indicated here. This brief discussion is not intended to cover all of the methods which the analyst may wish to apply.

1.10 AN EXAMPLE

In order to give a better idea of the numerical aspects of system-effectiveness analysis, this section includes an extremely abbreviated example of computations based on actual field data for a military system: the AN/APS-20E radar. This radar system is a relatively high-powered airborne pulsed-power search radar which has been in use by the Air Force and the Navy for a number of years. The data cover the experience with 24 systems in field use. The assigned mission in this case was not especially appropriate to the system, and the surveillance was not conducted in a manner to support a valid evaluation of the system design. Hence, the example is illustrative of method only and not of actual effectiveness measurement.

Figure 1.3 shows the breakdown of total observed calendar time into various subdivisions of the type previously discussed. Table 1.2 presents this same basic time information, but with time expressed as percentages of the several major time categories. Table 1.3 gives certain supplementary information on the numbers of missions

Table 1.2 Time Elements Expressed as Percentages of Major Time Categories
(Percentages correspond to hours shown on Fig. 1.3)

Time element	Per cent of calendar time	Per cent of nonflight time	Per cent of nonflight down time
Nonflight time:			
Admin. time	5.02	5.47	35.98
Logistic time	8.65	9.43	61.96
Active work time	0.29	0.32	2.06
Total down time	13.96	15.22	100.00
Total up time	77.79	84.78	
Total nonflight time	91.75	100.00	
Time element	**Per cent of calendar time**	**Per cent of flight time**	**Per cent of in-flight down time**
Flight time:*			
Admin. time	0.60	7.21	50.06
Logistic time	0.59	7.15	49.62
Active work time	0.003	0.05	0.32
Total down time	1.19	14.41	100.00
Total up time	7.06	85.59	
Total flight time	8.25	100.00	
Time element	**Per cent of calendar time**	**Per cent of flight time**	
Flight time:*			
Radar not used	3.94	47.76	
Radar in standby only	2.06	24.98	
Radar in operate	2.25	27.26	
Total flight time	8.25	100.00	
Total calendar time	100.00		

* *Note.* Either of, *not* both of, these flight-time divisions is to be used in accounting for 100% of calendar time.

Table 1.3 Malfunction and Flight Data

Number of flights or missions	2,068
Number of missions on which radar was used or its use was attempted	1,067
Mean length of mission, hours	4.70
Number of malfunctions detected in flight by radar operator	96
Number of malfunctions detected on ground by maintenance technician	86
Total number of system malfunctions, all types	182
Total number of repair actions	135

and malfunctions observed in the surveillance interval. Figure 1.4 illustrates one of the associated curves—that of the active work or repair times. The figure shows both the observed data and the fitted log-normal curve.

Table 1.4 presents some additional numerical values of interest in the evaluation of the system. The mean time between malfunctions is shown for several types of time bases, and the mean values of the elements of down time are also included. The

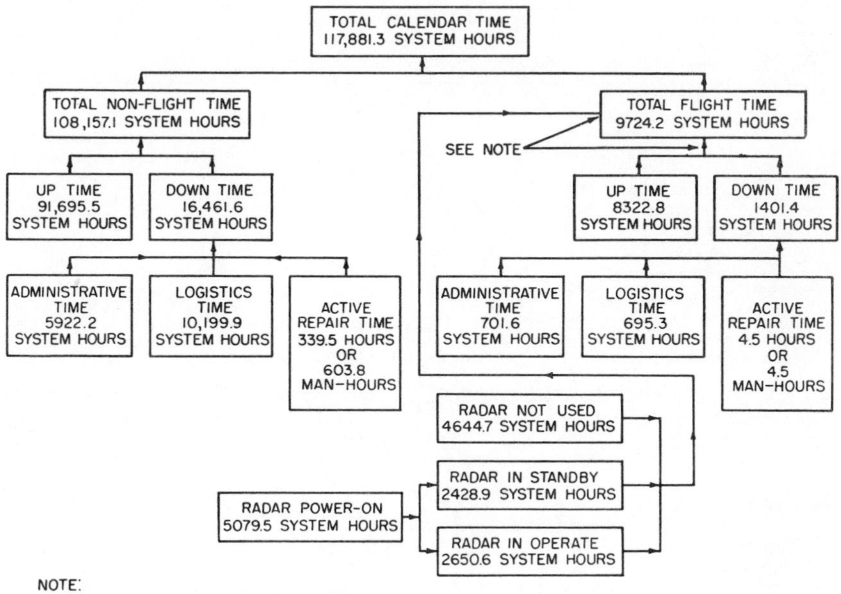

Fig. 1.3 Apportionment of total surveillance time for 24 AN/APS-20E radar systems used in two VP squadrons. Observation period August, 1958, to March, 1959.

Table 1.4 Reliability and Maintainability Indices
(Computations based on data in Fig. 1.3 and Table 1.3)

Mean time between malfunctions:	
MTBM, calendar hours	647.7
MTBM, airframe or flight hours	53.4
MTBM, heater power-on hours	27.9
MTBM, radiate hours	14.6
Nonflight down time per repair action:	
Mean administrative time, hours	43.9
Mean logistic time, hours	75.6
Mean active work time, hours	2.5
Mean active man time, man-hours	4.5
Median active work time, hours	1.0
Maintenance man-hour requirements:	
Man-hours per flight hour	0.062
Man-hours per radar-used mission	0.57
Maintenance support index:	
Man-hours per 1,000 power-on hours	118.9
Man-hours per 1,000 radiate hours	227.8

maintenance man-hour requirements are expressed in terms of man-hours per flight hours, man-hours per mission on which the radar was used, and the maintenance support index (MSI). The MSI is shown in terms of power-on hours and radiate hours. All of these values are descriptive of some phase of the operation or maintenance of the system, and the investigator must decide which of the several indices are applicable to his particular purpose in an evaluation of system effectiveness.[1]

[1] Information for this example is reproduced from "Maintainability and Reliability of the AN/APS-20E Radar System," Publication 101-33-180, Sept. 1, 1960, by permission of the publisher, ARINC Research Corporation.

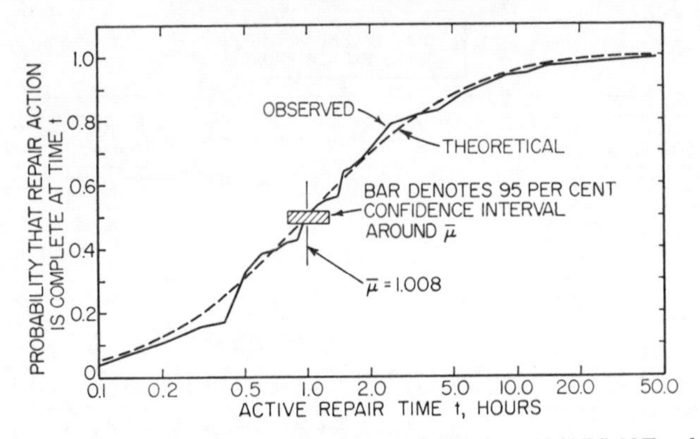

FIG. 1.4 Observed and theoretical repairability functions for AN/APS-20E radar systems. Theoretical repairability function

$$M_R(t) = 1/(\sqrt{2\pi}\,\sigma) \int_{-\infty}^{x} e^{-(x-\xi)^2/2\sigma^2}\,dx$$

where $x = \log_{10} t$
ξ = mean value of x
σ = standard deviation of x
$\bar{\mu}$ = antilog ξ

95% confidence interval around $\bar{\mu}$ = antilog $(\xi \pm 1.96\sigma/\sqrt{n})$

where n = number of observations
$\bar{\mu} = 1.008$
$\xi = 0.0037$
$\sigma = 0.5891$
n = 135 repair actions
$\bar{t} = (\Sigma ti)/n = 2.52$ hours

1.11 FUTURE RESEARCH REQUIREMENTS

Much remains to be done in the quantitative analysis of system effectiveness and the associated concepts. Even with the historical emphasis on reliability we are still in need of further methods research and data collection in order to understand part, component, and system failure patterns. Methods and data shortages are even more critical in other areas—repair times, free times, availability, design adequacy, etc. We have very little information about the types of mathematical functions involved for the single time categories, and even less for the ratios. This complicates the theoretical aspects of the sampling problems to an extreme degree.

It would appear that nonparametric methods offer the best promise for immediate use. They are particularly helpful in describing both basic concepts and underlying analytical methods. Particular emphasis should be placed on the development and study of trade-off relationships. Previous studies have depended largely on arithmetic means instead of analysis of density functions, and hence on some rather crude approximations. Computer simulation using discrete distributions is very promising, and the recent trend toward this approach should be continued. Finally, we must put system effectiveness into the context of system worth. We have chosen to include four dimensions under system worth; others may wish to select more. In any case, the overall concept of worth and its associated concepts must be organized by appropriate analysis into a composite decision criterion which can be of value to management.

Section 2

CHARACTERISTIC LIFE PATTERNS AND THEIR USES

JOHN H. K. KAO

ASSOCIATE PROFESSOR OF INDUSTRIAL ENGINEERING,

NEW YORK UNIVERSITY, NEW YORK, NEW YORK

Dr. Kao, now a naturalized United States citizen, served as the engineer in charge of purchasing and specifications at the official purchasing agency of the Republic of China. He has served as consultant to the U.S. Army Signal Corps, Bell Laboratories, Corning Glass, Electra, General Electric, and Westinghouse, as well as to numerous aerospace firms. He was Associate Professor of Mechanical Engineering at Cornell University before joining the New York University faculty.

Dr. Kao is a member of the editorial board of the Journal of the ASQC Electronics Division and is Chairman of a Special Task Group on the Application of the Weibull Distribution in Fatigue Testing under ASTM Committee E-9. He is author of about thirty papers and reports on mechanical engineering and reliability problems. For a number of years he has been conducting research, under a contract with the Office of Naval Research, on the statistical aspects of life testing and reliability. He is coauthor with H. P. Goode of the 1962–1963 ASQC Electronics Division Award Paper for significant contributions in the area of reliability and quality control.

He is a senior member of ASQC, a director of CIE, and a member of the Operations Research Society of America, Sigma Xi, ASA, AAAS, and AAUP.

CHARACTERISTIC LIFE PATTERNS AND THEIR USES

John H. K. Kao

CONTENTS

2.1 INTRODUCTION

There are two areas in which it is important to conduct failure studies to analyze the characteristic life patterns: (1) when a new product is in its development stage and (2) when an old product is finding new applications. In both situations engineers are interested in the product life patterns because of the inadequacy in the current product specifications in guaranteeing the future product performance. In fact, one of the many objectives of failure studies is to gain feedback information for the purpose of adjusting or modifying the product life pattern so that an economic balance in the production of the product may be achieved.

Good economic balance in production means that the product meets its functional requirements, has satisfactory life pattern, and is reasonable in cost. The product should be neither underdesigned, so that it gives poor immediate and future performances, nor overdesigned, so that cost or weight or size becomes excessive. In order to make full use of this feedback information, engineers must know how to interpret failure data through the product's characteristic life patterns.

Another important objective of failure studies is to obtain the so-called trade-off characteristics of the product when used under various environmental conditions. This problem occurs frequently in specifying products in military equipment. Although, in this application, components are manufactured under very stringent specifications and often are the best a manufacturer can offer in the present state of knowledge, the components sometimes are still not good enough Under such circumstances engineers in applying the components in the design of a piece of military equipment would usually try to relax the operating stress to the component in order to trade for an incremental increase in performance necessary for the job. The important question is this: How much stress should be relaxed on the component and how much of an increment in performance would be expected as a result? The complete answer to this question is in the valid trade-off curves (for example, *S-N* curves in metal fatigue), which are obtainable only from the characteristic life patterns of the component under various stresses.

2.2 CONCEPT OF LIFE-LENGTH DISTRIBUTION—BASIS FOR ESTIMATING RELIABILITY FROM SAMPLE DATA

If a random sample of items (the item may be a component or a system; see next paragraph) taken from a statistically stable population is put to use under a set of fixed environmental conditions, members of the sample will fail successively in time. The sample data so obtained will be a set of nonnegative numbers representing the life length of each item.

For mathematical convenience, items are taken as systems if they are repairable and as components if they are not. Under this definition for items, molded capacitors and resistors, transistors, receiving-type electron tubes, etc., are components, while washing machines, automobiles, airplanes, etc., are systems. However, it might be noted, in contrast to the ordinary usage, that a solenoid relay which may be revived by adjusting its springs should be classified as a system, since it is repairable, and a launched missile should be classified as a component, since it is nonrepairable once launched.

Distinguishing components from systems in this manner emphasizes the fact that the salient principles in design and the underlying mathematics in analysis for repairable items are entirely different from those for nonrepairable items. For example, a

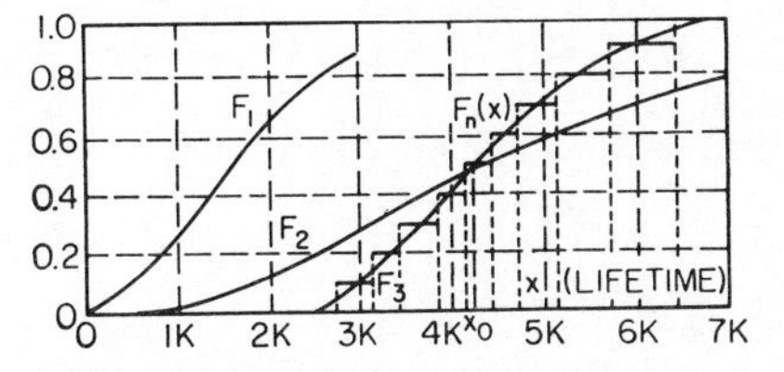

Fig. 2.1 Population and sample c.d.f.'s.

good design for a repairable item is characterized by the requirement of little maintenance (and easy maintenance when needed); as such, the life length is measured by the mean time between failures (MTBF). However, a good design of a component is characterized by the longevity under extremely severe environment, and therefore the life length is measured by the mean time to failure (MTTF). The terms MTBF and MTTF cited here, although obviously different in meaning, are often misused as synonyms. The concept of life-length distribution in Sec. 2 will hold only for the case of nonrepairable items (components, devices, etc.); and Sec. 11, dealing with maintainability using stochastic (renewal) process, will take up the characteristic life patterns and their uses for repairable items (systems, subassemblies, etc.).

If one waits until all members in the sample fail, then the sample data will constitute a random sample of observations, although they become available by order of their magnitudes. These data give an "image" of the cumulative distribution function (c.d.f.) (see Sec. 4) of the life length of the component. This image, called sample c.d.f., is defined by $\hat{F}(x)$ as follows:

$$\hat{F}(x) = i/n \qquad \text{if } x_i < x < x_{i+1} \text{ for } i = 0, 1, \ldots, n$$

where

$$x_0 = 0 \qquad x_{n+1} = \infty \qquad 0 \leqq x_1 \leqq x_2 \leqq \cdots \leqq x_n < \infty \tag{2.1}$$

For example, consider the following set of sample data (hypothetical) consisting of 10 observations: 2,750, 3,100, 3,400, 3,800, 4,100, 4,400, 4,700, 5,100, 5,700, and 6,400. The sample c.d.f. is shown in Fig. 2.1. Suppose one increases the sample size n (here $n = 10$) without limit; then $\hat{F}(x)$ would tend to the population c.d.f. $F(x)$, which is defined as (Sec. 4):

$$F(x) = P\{X \leqq x\} \tag{2.2}$$

The population c.d.f. (here F_3 in Fig. 2.1) is the characteristic life pattern and is the totality of the information available on the life length of the above items. For example, in Fig. 2.1, the c.d.f. F_2 is everywhere below F_1; hence, we infer that the items associated with F_2 are uniformly more reliable than those associated with F_1. However, this is not the case when we compare F_2 with F_3, which cross each other at x_0. For life length less than x_0, F_3 is more reliable than F_2, while for life length more than x_0, the reverse is true. Later we shall see how one would make an intelligent comparison in this case.

In practice, unfortunately, it is not physically possible or economically feasible always to use large sample sizes to get "good" approximation to the characteristic life pattern from the sample data. It turns out that if one is willing to make some assumptions on the function form of the c.d.f., a lot can be said about the life length even with a relatively small sample. To assume that the true c.d.f. belongs to a specified family of distributions indexed by a number of parameters requires some theoretical considerations. The following section will take up this matter.

2.3 SOME POSSIBLE FAMILIES OF LIFE-LENGTH DISTRIBUTIONS

Since life length is a continuous variable, there exists a function $f(x)$, which is the first derivative of $F(x)$ and is known as the probability density function p.d.f. (Sec. 4). Thus,

$$F(x) = \int_{-\infty}^{x} f(t)\,dt = 1 - R(x) \tag{2.3}$$

where $R(x)$ is called the reliability function. For $a < b$,

$$P\{a < X < b\} = \int_{a}^{b} f(t)\,dt = F(b) - F(a) = R(a) - R(b) \tag{2.4}$$

2.3a Hazard-rate Consideration. For an arbitrary life-length distribution defined over $\gamma \leqq x < \infty$ (γ is the threshold or location parameter) with c.d.f. $= F(x)$ and p.d.f. $= f(x)$, the failure rate $G(x,T)$ at time x in fraction per T time units is

$$G(x,T) = \frac{\int_{x}^{x+T} f(t)\,dt}{T[1 - F(x)]} = \frac{F(x+T) - F(x)}{T} \cdot \frac{1}{R(x)} \tag{2.5}$$

In Eq. (2.5) the integral represents the fraction of the original items expected to fail from time x to time $x + T$ and $1 - F(x) = R(x)$ represents the fraction of the original items expected to survive to time x. Hence, its ratio divided by T represents the

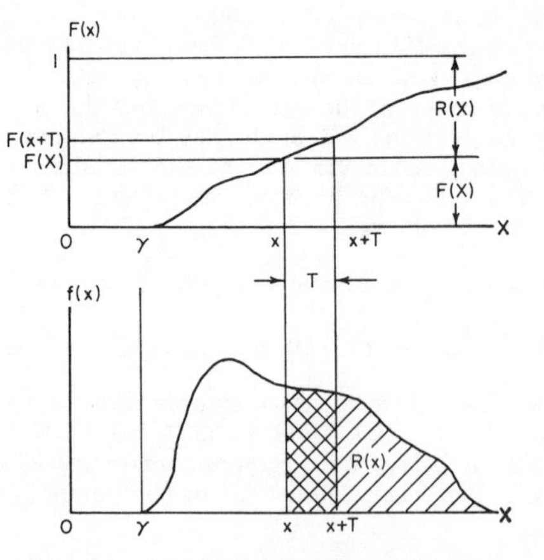

FIG. 2.2 Pictorial definition of $G(x,T)$.

expected failure rate of time x for the next T time units. The introduction of γ, some finite time prior to which no failure can occur, means when $x \leqq \gamma$, $F(x) = 0$. A negative value for γ means that the item can fail prior to use, i.e., the item has shelf life and can fail during storage, a common phenomenon for batteries, electrolytic capacitors, electron tubes, photographic films, canned foods, and the like. When $x = \gamma$, $G(\gamma,T)$ is called the initial failure rate. Here $F(\gamma) = 0$ and $R(\gamma) = 1$

$$G(\gamma,T) = F(\gamma + T)/T \tag{2.6}$$

For $\gamma = 0$, the initial failure rate is

$$G(0,T) = F(T)/T \tag{2.7}$$

which is a very special kind of failure rate. This expression, intuitively appealing, is erroneously used as the definition of failure rate in general by many writers. As will be seen later, Eq. (2.7) for failure rate is true for all values of x only if the life-length distribution is of the exponential type.

The instantaneous failure, or hazard, rate $Z(x)$ is the limiting value of $G(x,T)$ as $T \to 0$. Thus,

$$Z(x) = \lim_{T\to 0} \left[\frac{F(x + T) - F(x)}{T}\right] \cdot \frac{1}{R(x)} = \frac{f(x)}{R(x)} = \frac{-R'(x)}{R(x)} \tag{2.8}$$

Equation (2.8) may be used to express the life-length distribution in terms of $Z(x)$ as follows:

$$-Z(x) = \frac{R'(x)}{R(x)} = \frac{d}{dx} [\ln R(x)] \tag{2.9}$$

Transpose dx and integrate both sides over the interval $(-\infty,x)$:

$$-\int_{-\infty}^{\gamma} Z(t)\, dt - \int_{\gamma}^{x} Z(t)\, dt = \int_{-\infty}^{\gamma} d[\ln R(x)] + \int_{\gamma}^{x} d[\ln R(x)] \tag{2.10}$$

Since, for $-\infty < x \leqq \gamma$, $Z(x) = 0$ and $R(x) = 1$, one has

$$-\int_{\gamma}^{x} Z(t)\, dt = \ln R(x) \qquad \text{for } x \geqq \gamma \tag{2.11}$$

or,

$$R(x) = \exp\left[-\int_{\gamma}^{x} Z(t)\, dt)\right] \tag{2.12}$$

Hence, the cumulative distribution function of life length is

$$F(x) = 1 - \exp\left[-\int_{\gamma}^{x} Z(t)\, dt\right] \qquad \text{for } x \geqq \gamma \tag{2.13}$$

and its first derivative with respect to x, the probability density function of life length for $Z(\gamma) = 0$, is

$$f(x) = Z(x) \exp\left[-\int_{\gamma}^{x} Z(t)\, dt\right] \qquad \text{for } x \geqq \gamma \tag{2.14}$$

If the hazard rate $Z(x)$ is constant over time, say, equal to λ, then

$$\lambda \int_{\gamma}^{x} dt = \lambda(x - \gamma)$$

and the resulting c.d.f. and p.d.f. become the well-known two-parameter exponential distribution, for $\lambda = 1/\theta$,

$$F(x) = 1 - \exp[-(x - \gamma)/\theta] \qquad \text{for } x \geqq \gamma \tag{2.15}$$

and

$$f(x) = [1/\theta] \exp[-(x - \gamma)/\theta] \qquad \text{for } x \geqq \gamma \tag{2.16}$$

However, if $Z(x)$ is some power function of time such that $\int_\gamma^x Z(t)\,dt = [(x-\gamma)/\eta]^\beta$, the resulting c.d.f. and p.d.f. become the three-parameter Weibull distribution function, thus,

$$F(x) = 1 - \exp\left[-\left(\frac{x-\gamma}{\eta}\right)^\beta\right] \qquad \text{for } x \geqq \gamma \tag{2.17}$$

and

$$f(x) = \frac{\beta}{\eta}\left(\frac{x-\gamma}{\eta}\right)^{\beta-1} \exp\left[-\left(\frac{x-\gamma}{\eta}\right)^\beta\right] \qquad \text{for } x \geqq \gamma \tag{2.18}$$

The three parameters of the Weibull distribution are

η = scale parameter
β = shape parameter
γ = location parameter

The hazard rate $Z(x)$ for the Weibull case is $(\beta/\eta)[(x-\gamma)/\eta]^{\beta-1}$, which is a decreasing (increasing) function in $x-\gamma$ if $\beta < 1$ $(\beta > 1)$ and a constant equal to $1/\eta$ if $\beta = 1$, in which case the Weibull distribution specializes to the exponential distribution. The foregoing consideration for $Z(x)$ as simple monotone function of x makes the Weibull (including exponential) distribution analytically appealing.

2.3b Weakest-link Theory. Under this theory of failure, each component is treated as consisting of many subcomponents which form the component itself in much the same manner as links form a chain. Then the characteristic life pattern of the component (chain) is equivalent to the characteristic life pattern of the weakest subcomponent (link). Assuming the life lengths of all n subcomponents are independently and identically distributed with some p.d.f. $= f(x)$ and c.d.f. $= F(x)$, the life length of the aggregated component would be distributed according to the smallest-order statistic; thus,

$$F_1(x) = 1 - [1 - F(x)]^n \tag{2.19}$$

and

$$f_1(x) = n[1 - F(x)]^{n-1} f(x) \tag{2.20}$$

For example, if the subcomponents have life-length distribution of the Weibull form [Eqs. (2.17) and (2.18)], then the distribution for the life length of the component would be

$$F_1(x) = 1 - \exp\left[-n\left(\frac{x-\gamma}{\eta}\right)^\beta\right] \tag{2.21}$$

$$f_1(x) = \frac{n\beta}{\eta}\left(\frac{x-\gamma}{\eta}\right)^{\beta-1} \exp\left[-n\left(\frac{x-\gamma}{\eta}\right)^\beta\right] \tag{2.22}$$

Equations (2.21) and (2.22) indicate the self-reproducing property for the Weibull distribution (including the exponential distribution) when the smallest-order statistic is considered. It will be recalled that the normal distribution also has the self-reproducing property when the sample mean (as a statistic) is considered.

In applying the weakest-link theory of failure, the number n of subcomponents can be quite small, such as the number of layers of dielectric in a paper capacitor. In this case the capacitor (component) fails when any one of the layers (subcomponent) shorts. If the flaws which cause premature shorts can be assumed as randomly distributed over time, i.e., constant hazard rate, then the life length of the paper capacitor as a whole has an exponential distribution. On the other hand, the number n of subcomponents could be very large, as in the case of the fatigue life of a metal part. Here the subcomponent may be two layers of metal grains of minute thickness between which the fatigue crack propagates. For large values of n the distribution of the smallest-order statistic can be approximated in the following manner.

We make use of a result due to Cramer which states for $Y = nF(X)$, the p.d.f. of the smallest extreme for Y, $g_1(y) \rightarrow e^{-y}$ as n tends to infinity. The p.d.f. of the smallest extreme for X may then be found by changing the variable from Y to X through the relationship given above, which is order-preserving. For example, suppose the life length of the subcomponent follows the c.d.f. $F(x) = [(x - \gamma)/\eta]^{\beta-1}$ for $\beta > 0$ and $\gamma < x < \gamma + \eta$, which is a special beta distribution. As the number of subcomponents becomes large, $g_1(y) \to e^{-y}$, for $y > 0$ and $f_1(x) = g_1(y) \cdot |dy/dx|$. Since $y = nF(x)$, $|dy/dx| = nf(x) = (n\beta/\eta)[(x - \gamma)/\eta]^{\beta-1}$; hence,

$$f_1(x) = \frac{n\beta}{\eta}\left(\frac{x-\gamma}{\eta}\right)^{\beta-1} \exp\left[-n\left(\frac{x-\gamma}{\eta}\right)^{\beta}\right] \quad \text{for } x > \gamma \tag{2.23}$$

It will be noted that Eqs. (2.22) and (2.23) are the same, although they come from two different underlying distributions for the subcomponent. Equation (2.22) is true for any n if the life length of the subcomponent follows a Weibull distribution, while Eq. (2.23) is true only for large n and if the life length of the subcomponent follows a beta distribution. This raises the following speculation. Are there any other subcomponent populations which will also lead to the Weibull distribution for the component under the weakest-link theory? The answer is yes, and it was provided, as a by-product, in a paper by Fisher and Tippett [1]. They have shown that, under some regularity conditions, many distributions (including normal distribution) have their smallest extremes distributed as one of the following three possible types, given that n is sufficiently large. They are

Type I: $F_1(x) = 1 - \exp\{-\exp[(x - \gamma)/\eta]\}$ for $-\infty < x < \infty$ (2.24)
Type II: $F_1(x) = 1 - \exp\{[(x - \gamma)/\eta]^{-\beta}\}$ for $-\infty < x < \gamma$ (2.25)
Type III: $F_1(x) = 1 - \exp\{-[(x - \gamma)/\eta]^{\beta}\}$ for $\gamma < x < \infty$ (2.26)

Since for Type II $-\infty < x < \gamma$, Type II is not likely to be useful in failure studies. Type III can be recognized as the Weibull distribution. Fisher and Tippett called Type II and III the penultimate form and Type I the ultimate form, because under the normality assumption the distribution of the extremes, as n increases, first assumes the form of Type II and III and then passes to Type I. Gumbel has written an excellent text [2] on all three types and has applied Type I (for the largest extreme) extensively. Furthermore, if the underlying distribution of X is already in the forms given by Eqs. (2.24) and (2.26), it is easy to see that the distribution of the smallest-order statistic for X for any n will also possess the same form with only a change in location by the amount of $\ln n$ (Type I) or a change in scale by the factor of $n^{-1/\beta}$ (Type III). This self-reproductive property explains why these three distributional types are called extreme-value distributions.

2.3c Parallel-strand Theory. In contrast to the weakest-link theory of failure, the parallel-strand theory assumes that each component consists of many subcomponents in the manner of a multistrand rope. The rope is not broken until all its strands are broken. Then the characteristic life pattern of the component (rope) is the sum or convolution of the characteristic life pattern of all subcomponents (strands). Assuming the life lengths of all n subcomponents are independently and identically distributed with some p.d.f. $= f(x)$, then the life length p.d.f. $g(x)$ for the aggregated component would be the n-fold convolution of $f(x)$; thus

$$g(x) = [f(x)]^{n*}$$

where
$$[f(x)]^{n*} = [f(x)] * [f(x)]^{(n-1)*}$$

and
$$[f(x)]^{(n-1)*} = [f(x)] * [f(x)]^{(n-2)*} \quad \text{etc.} \tag{2.27}$$

Lastly,
$$[f(x)]^{2*} = [f(x)] * [f(x)]$$

$$= \int_{-\infty}^{+\infty} f(t)f(x - t)\,dt \tag{2.28}$$

For example, suppose the life length of the subcomponent follows the p.d.f.

$$f(x) = (1/\theta)\exp[-x/\theta] \quad \text{for } x > 0$$

which is a one-parameter exponential density. Since

$$[f(x)]^{2*} = (1/\theta^2)\int_0^x e^{-t/\theta}e^{-(x-t)/\theta}\,dt = xe^{-x/\theta}/\theta^2$$

hence,

$$[f(x)]^{3*} = (1/\theta^3)\int_0^x te^{-t/\theta}e^{-(x-t)/\theta}\,dt = x^2e^{-x/\theta}/2\theta^3$$

and

$$[f(x)]^{4*} = (1/2\theta^4)\int_0^x t^2e^{-t/\theta}e^{-(x-t)/\theta}\,dt = x^3e^{-x/\theta}/3!\theta^4$$

we may deduce

$$g(x) = [f(x)]^{n*} = x^{n-1}e^{-x/\theta}/\Gamma(n)\theta^n \qquad \text{for } x > 0 \tag{2.29}$$

which is a gamma density with shape parameter $= n$ and the scale parameter $= \theta$, the same as that for the original exponential density for the subcomponents. When the number of subcomponents becomes large, the life-length distribution $g(x)$ of the component tends to a normal density with mean $= n\theta$ and variance $= n\theta^2$. This is due to the fact that gamma distribution tends to normality as its shape parameter increases. As a matter of fact, by virtue of the central-limit theorem, given any (not only exponential) independently and identically distributed subcomponent life-length variable X with $EX = \mu_x$ and Var $X = \sigma_x^2$, the standardized variable $(X - \mu_x)/\sigma_x$ has a standard normal distribution as $n \to \infty$. Therefore, in comparing the parallel-strand theory with the weakest-link theory of failure, the gamma and the Weibull distributions play the central role of penultimate distributions and the normal and the Gumbel Type I distributions are the respective ultimate distributions.

The fact that both Weibull and gamma distributions (penultimate) are multishaped and both Gumbel Type I and normal distributions (ultimate) are fixed-shaped is an additional interesting feature in comparing these two theories of failure. Finally, all four distributions are shape-preserving, i.e., self-reproductive under their respective failure theories. As noted before, the smallest extreme from a Weibull (Gumbel Type I) distribution is still a Weibull (Gumbel Type I) variable. And so the convolution of a sequence of independent observations from a gamma (normal) distribution is still a gamma (normal) variable.

2.3d Proportional-effect Theory. Let $X_1 < X_2 < \cdots < X_n$ be a sequence of random variables to denote the sizes of, say, a fatigue crack at successive stages of its growth. When the size of the crack gets to the value of X_n, the component fails. The proportional-effect theory of failure states that the crack growth $X_i - X_{i-1}$ at stage i is proportional to the crack size X_{i-1} of the preceding stage for all stages. Thus,

$$X_i - X_{i-1} = \delta_i X_{i-1} \qquad i = 1, 2, \ldots, n \tag{2.30}$$

where X_0 = initial size of the crack and is to be interpreted as the size of minute flaws, inclusions, voids, etc. in the component and $\delta_1, \delta_2, \ldots$ are independently distributed positive random variables which may or may not have an identical distribution for all i.

From Eq. (2.30) one may write

$$X_i = (1 + \delta_i)X_{i-1} = (1 + \delta_i)(1 + \delta_{i-1}) \cdots (1 + \delta_1)X_0 \tag{2.31}$$

Since the component is assumed to fail when the size of the crack reaches the value of X_n, the characteristic life pattern of the component is the distribution of X_n. By replacing i by n in Eq. (2.31), it is seen that X_n is the product of many independently distributed positive random variables. Its logarithm ln X_n is then the sum of these variables. The central-limit theorem will state that ln X_n is approximately normally distributed; hence, X_n follows a log-normal distribution with the following density·

$$f(x) = \frac{1}{x\sigma\sqrt{2\pi}} \exp\left[-\frac{1}{2}\left(\frac{\ln x - \mu}{\sigma}\right)^2\right] \qquad x > 0 \tag{2.32}$$

Reference to Aitchison and Brown [3] on several plots of Eq. (2.32) reveals that μ, although a location parameter for the normal variable ln X, behaves as if it were a scale parameter for the log-normal variable X and that σ, originally a scale parameter for ln X, behaves as a shape parameter for X. These shifts in behavior of parameter from normal to log-normal variables also hold for changing variables from the Gumbel

Type I to the Weibull distribution. As a matter of fact, it can be shown that if ln X has a Gumbel Type I distribution, then X has a Weibull distribution. This is illustrated as follows.

Suppose X has the Weibull distribution given by Eq. (2.17). If $Y = \ln(X - \gamma)$, $a = \ln \eta$, and $b = 1/\beta$, then $[(x - \gamma)/\eta]\beta = \exp[(y - a)/b]$ and the distribution of Y becomes

$$F(y) = 1 - \exp\{-\exp[(y - a)/b]\} \qquad -\infty < y < +\infty \tag{2.33}$$

Equation (2.33) can be seen to be identical with Eq. (2.24), which is the Gumbel Type I. The parameter for location a (or for scale b) in Eq. (2.33) behaves respectively as the scale ln η (or shape $1/\beta$) parameters in Eq. (2.17). Here one will also note that the log-normal and Weibull (penultimate) distributions are multishaped and their ultimate counterparts, both the normal and the Gumbel Type I, are fixed-shaped.

The foregoing four failure theories serve to indicate the important roles to be played by the log-normal, gamma (including the exponential when $n = 1$), Weibull (including the exponential when $\beta = 1$ and the Rayleigh when $\beta = 2$), normal, and Gumbel Type I distributions as the reasonable choices for the characteristic life patterns of nonrepairable items which we defined as components. We shall discuss their uses in the following paragraphs.

2.4 THE CONSTRUCTION AND USES OF PROBABILITY PAPERS BASED ON THE LIFE-LENGTH MODELS

2.4a General Consideration. As seen in preceding paragraphs, one may characterize a life-length distribution by its location, scale, and shape parameters. Some

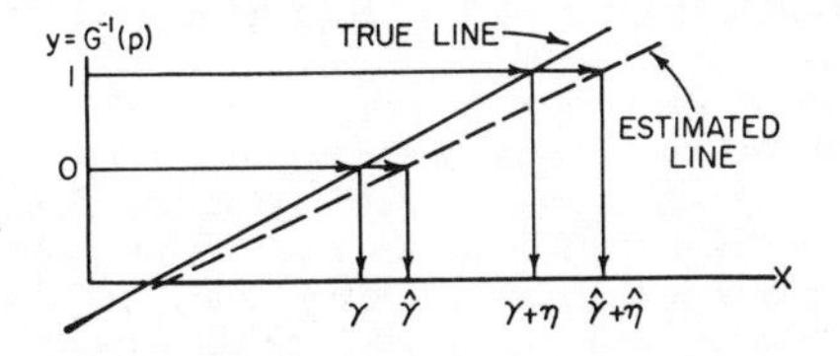

FIG. 2.3 Construction of probability paper.

distributions such as normal, Gumbel Type I, exponential, and Rayleigh are fixed-shaped and do not have an explicit shape parameter; others such as log-normal, Weibull, gamma, Student t, F, and beta possess one or more shape parameters and hence are more flexible for the purpose of fitting to the observed sample data. Whether a distribution has shape parameter or not, one may always introduce location and scale parameters in order to make the model (distribution) more flexible. This is done by the following linear transformation:

$$Y = (X - \gamma)/\eta \qquad \text{or} \qquad X = \gamma + \eta Y \tag{2.34}$$

By this process, if we put $p = F(x) = G(y)$ for $0 < p < 1$ and G is free of γ and η, then γ is the location parameter, η is the scale parameter, and G is the standard distribution. Therefore, one may write for G^{-1}, the inverse function of G,

$$y = G^{-1}(p) = G^{-1}[F(x)] \tag{2.35}$$

Instead of plotting x against $F(x)$, which will look like Fig. 2.1, one may choose to plot x against a function of $F(x)$, say, $G^{-1}[F(x)]$, which is equal to y But from Eq. (2.34) we observe that x and y are linearly related. On a graph paper with the y axis so calibrated, called probability paper (Fig. 2.3), every straight line with positive slope represents a c.d.f. $F(x)$ with some scale and location parameters. Therefore, one may plot the same data on the probability paper for a rough test of goodness of fit and at the same time obtain graphical estimates of scale and location parameters by passing a straight line through the sample points.

In plotting the sample points, $p = F(x)$, being unknown, must be estimated. One large sample estimate for p is the sample c.d.f. given by Eq. (2.1). However, this estimate has its shortcomings. If $F(\infty) = G(\infty) = 1$, which is true for all life-length models mentioned in this section, one has $y = G^{-1}(1) = \infty$. This means that the value for y is infinite at $p = 1$, which is the sample c.d.f. value $\hat{F}(x) = n/n = 1$ for $x \geqq x_n$, the largest-order statistic. Thus, the largest observation in the observed sample cannot be plotted on the probability paper. The following is a list of some of the more widely used estimates for p:

1. Sample c.d.f., i/n
2. Symmetrical sample c.d.f., $(i - 1/2)/n$
3. Mean of c.d.f., $i/(n + 1)$, also known as mean rank
4. Mode of c.d.f., $(i - 1)/(n - 1)$
5. Median of c.d.f., $H^{-1}(\frac{1}{2}|i, n - i + 1)$
6. C.d.f. of $\text{E}X_i$, $F(\text{E}X_i) = G(\text{E}Y_i)$
7. Blom's [6] estimate, $(i - \alpha_i)/(n - \alpha_i - \beta_i + 1)$

The estimates for p given by estimates 1 to 4 are very simple to use, since they do not require the use of any table. Estimate 5 needs an incomplete beta-function table by K. Pearson [4] or a Hartley and Fitch chart [5]. Estimate 6 gives unbiased estimates of both location and scale parameters. Unfortunately, tables for $\text{E}Y_i$ are available only for a few distributions (exponential, gamma, normal, and Gumbel Type I to a limited extent). Estimate 7, proposed by Blom [6], is a corrected version of estimate 3 and possesses many "good" statistical properties, among them near unbiasedness and minimum mean squared error. In a modified Blom's version the α_i and β_i corrections chosen are independent of n and i. Then estimate 7 specializes to estimate 1 for $\alpha_i = 0$, $\beta_i = 1$, to estimate 2 for $\alpha_i = \beta_i = \frac{1}{2}$, to estimate 3 for $\alpha_i = \beta_i = 0$, to estimate 4 for $\alpha_i = \beta_i = 1$.

For certain distributions Blom also gives the "optimum" values for α_i and β_i. For example, $\alpha_i = \beta_i = 3/8$ for the normal distribution; $\alpha_i = 0$, $\beta_i = 1/2$ for the exponential distribution; $\alpha_i = 1/4$, $\beta_i = 1/2$ for the Gumbel Type I distribution; and $\alpha_i = 0.52(1 - b)$, $\beta_i = 0.5 - 0.2(1 - b)$ for the Weibull distribution with shape parameter $= 1/b$. Note that these values for α_i and β_i for the Weibull distribution include the special case of $\alpha_i = 0$ and $\beta_i = 1/2$ for the exponential case whose shape parameter is unity.

Another interesting fact about the α_i and β_i values for the Weibull distribution is found when one equates $\alpha_i = \beta_i$. That is, $0.52(1 - b) = 0.5 - 0.2(1 - b)$; then one finds the Weibull shape parameter $1/b = 3.27$, a value for which the Weibull density function has the appearance of a normal density (see Fig. 1, page 1*a* of Ref. 12 for $\beta = 3\frac{1}{3}$). When the Weibull shape parameter is 3.27, $\alpha_i = \beta_i \cong 3/8.3$, which is very close to Blom's recommendation of $\alpha_i = \beta_i = 3/8$ for the normal case. Of course, estimate 7 has the advantage over estimates 5 and 6 in that it, like estimates 1 to 4, needs no special tables. Compared with 6, estimate 7 still partly retains the unbiasedness property of 6.

It should be pointed out that in order to write down G^{-1} in Eq. (2.35), the shape parameter (if any) of the distribution F [and hence G, since Eq. (2.34) is a shape-preserving transformation] must be known. That is to say, for each given value of shape parameter there is a different calibration of y scale. This is true for the gamma and the Weibull distribution. When the experimenter is unwilling to assume a specific value for the shape parameter, he may construct, per Eq. (2.35), several y scales each corresponding to a specific shape-parameter value. By connecting points of equal p value on these properly spaced y scales, a range of infinitely many y scales are available. In this way, the shape parameter may also be estimated by choosing the value corresponding to the probability plot of the best linearity. This procedure will be illustrated later.

2.4b Exponential Distribution. For the exponential distribution function $F(x)$ as given by Eq. (2.15), the application of the linear transformation specified by

Eq. (2.34) yields

$$G(y) = 1 - \exp(-y) \qquad \text{for } y > 0 \tag{2.36}$$

Setting $G(y) = p$ and solving for y, we have

$$y = G^{-1}(p) = -\ln(1 - p) \tag{2.37}$$

Figure 2.4, with y axis given by Eq. (2.37), shows a mean-rank plot of the data points given in Fig. 2.1. The fact that the plot shows a single curvature with small scatter among points leads us to conclude that there exist better distributions than the exponential distribution for the data at hand. A forced straight line among the points gives estimates for the location and scale parameters as $\hat{\gamma} = 2{,}800$ hours and $\hat{\eta} = 4{,}400 - 2{,}800 = 1{,}600$ hours.

Since the mean and variance of an exponential density with location parameter γ and scale parameter η are

$$\text{Exponential mean} = \gamma + \eta \tag{2.38}$$
$$\text{Exponential variance} = \eta^2 \tag{2.39}$$

their estimates for the data at hand are

$$\text{Exponential mean} = 4{,}400 \text{ hours} \tag{2.40}$$
$$\text{Exponential variance} = 256 \times 10^4 (\text{hours})^2 \tag{2.41}$$

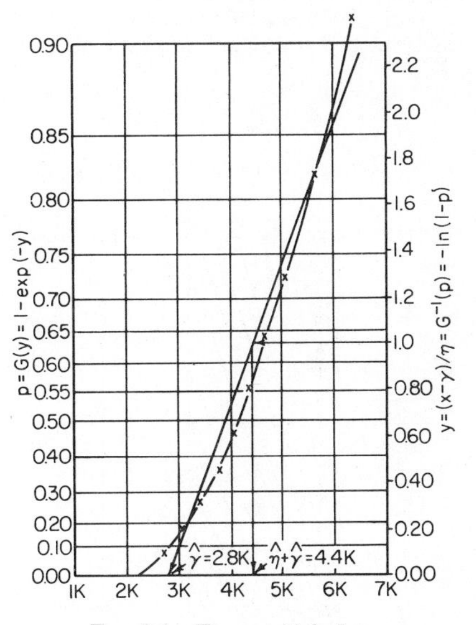

FIG. 2.4 Exponential plot.

2.4c Gamma Distribution. The gamma density function with location parameter γ, scale parameter η, and shape parameter β is

$$f(x) = \frac{(x - \gamma)^{\beta-1} \exp[-(x - \gamma)/\eta]}{\eta^\beta \Gamma(\beta)} \qquad \text{for } x > \gamma,\ \beta, \eta > 0 \cdots \tag{2.42}$$

One can easily see that Eq. (2.29) is a special case of Eq. (2.42) for $\gamma = 0$ and that Eq. (2.16), the exponential density, is a special case of Eq. (2.42) for $\beta = 1$. By applying the transformation specified by Eq. (2.34) to Eq. (2.42), one gets the standard gamma density

$$g(y) = \frac{y^{\beta-1} \exp(-y)}{\Gamma(\beta)} \qquad \text{for } y > 0,\ \beta > 0 \tag{2.43}$$

The standard gamma distribution function is then

$$G(y) = \int_0^y \frac{z^{\beta-1} \exp(-z)}{\Gamma(\beta)}\, dz \tag{2.44}$$

For β equal to positive integers, the above integral is

$$\begin{aligned} G(y) &= 1 - \sum_{k=0}^{\beta-1} \exp(-y)\, y^k/k! \\ &= 1 - e^{-y}[1 + y + y^2/2! + y^3/3! + \cdots + y^{\beta-1}/(\beta - 1)!] \end{aligned} \tag{2.45}$$

Thus it can be seen again that $G(y) = 1 - \exp(-y)$, the standard exponential distribution function when β is set to unity in Eq. (2.45) The inverse function G^{-1} cannot be put in closed form (with the exception of $\beta = 1$). However, the values for $y = G^{-1}(p)$ are tabulated for various values of shape parameter β against the percentage $100 \times p$ in percentage point tables of gamma and chi-square distributions, notably by Harter [7]. With the aid of these tables one may calibrate the y axis and plot the data. Figure 2.5 shows such a plot (again using mean ranks) of the data points given in Fig. 2.1. Here one would have to choose a value for the shape parameter β for the plot of "best" linearity through trial and error. For data at hand, β values of 1 and 5 show a reversal of curvature and $\beta = 3$ renders the plot approximately linear. Hence, the estimates for the gamma parameters are $\hat{\gamma} = 2{,}000$ hours, $\hat{\eta} = 2{,}800 - 2{,}000 = 800$ hours, and $\hat{\beta} = 3$. The estimated mean and variance are

$$\text{Gamma mean} = \hat{\gamma} + \hat{\beta}\hat{\eta} = 2{,}000 + 3(800) = 4{,}400 \text{ hours} \tag{2.46}$$

$$\text{Gamma variance} = \hat{\beta}\hat{\eta}^2 = 3(800)^2 = 192 \times 10^4 \text{ (hours)}^2 \tag{2.47}$$

2.4d Normal Distribution. The normal density function with location parameter μ and scale parameter σ is well known.

$$f(x) = \frac{1}{\sigma\sqrt{2\pi}} \exp\left[-\frac{1}{2}\left(\frac{x-\mu}{\sigma}\right)^2\right] \qquad \text{for } \sigma > 0,\ -\infty < x, \mu < +\infty \tag{2.48}$$

By applying the transformation $Y = (x - \mu)/\sigma$ specified by Eq. (2.34) to Eq. (2.48),

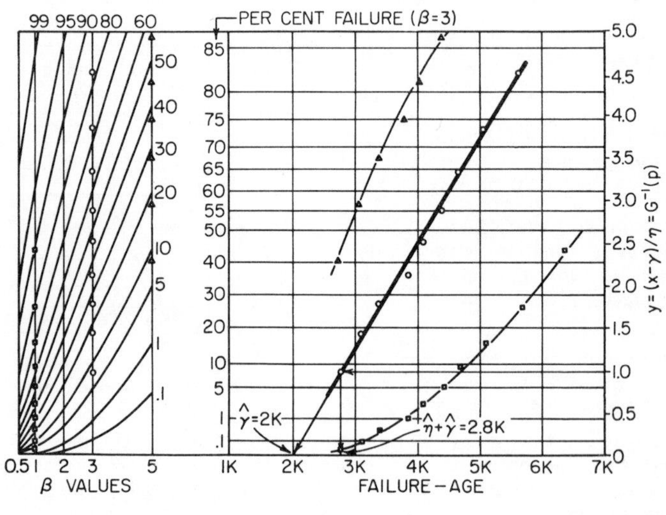

Fig. 2.5 Gamma plot.

one gets the standard normal density

$$g(y) = (1/\sqrt{2\pi}) \exp(-y^2/2) \qquad \text{for } -\infty < y < +\infty \tag{2.49}$$

Similarly to the case of gamma distribution, the inverse function G^{-1} cannot be put in closed form and tables for $y = G^{-1}(p)$ will have to be used in constructing the y axis for the plot. Fisher and Yates [8] (see their Table IX) tabulated the values $Y + 5$, which are known as probits. An abridged version of this probit table can be found in Hald (9) (see his Table III). However, normal-probability papers are commercially available (e.g., Keuffel & Esser No. 359–23). Unless these commercial papers are not of the desired size, there is no need to construct one's own. Figure 2.6 shows a mean-rank plot of the data points given in Fig. 2.1 on the normal-probability paper. A forced straight line gives the estimates of normal parameters, which happen to be the mean and standard deviation:

$$\text{Normal mean} = \hat{\mu} = 4{,}330 \text{ hours} \tag{2.50}$$

$$\begin{aligned}\text{Normal variance} = \hat{\sigma}^2 &= (5{,}700 - 4{,}330)^2 = 1{,}370^2 \\ &= 188 \times 10^4 \text{ (hours)}^2\end{aligned} \tag{2.51}$$

Another estimate of standard deviation from the plot is (4,330 − 2,960) = 1,370 hours, which happens to check with Eq. (2.51).

2.4e Log-normal Distribution. If $Z = \ln X$ and Z is normally distributed with location parameter μ and scale parameter σ, then the distribution of X is known as log-normal, which can be found from $f_X(\cdot) = f_Z(\cdot)|J|$. The Jacobian J of this transformation is $dz/dx = 1/x$; hence, the log-normal density is

$$f(x) = \frac{1}{x\sigma\sqrt{2\pi}} \exp\left[-\frac{1}{2}\left(\frac{\ln x - \mu}{\sigma}\right)^2\right] \qquad x > 0, \sigma > 0 \tag{2.52}$$

as given by Eq. (2.32). Since $(\ln x - \mu)/\sigma$ can be rewritten as $\ln [(x/e^{\mu})^{-\sigma}]$, μ would behave as a "scale" parameter and σ would behave as a shape parameter. (See Plots on page 10 of Aitchison and Brown [3].) In fact, a new location parameter τ may be

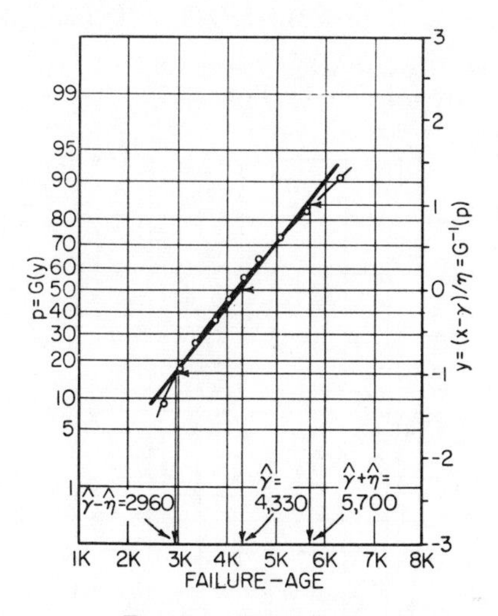

FIG. 2.6 Normal plot.

introduced to get a three-parameter log-normal distribution whose density would be

$$f(x) = \frac{1}{(x-\tau)\sigma\sqrt{2\pi}} \exp\left\{-\frac{1}{2}\ln^2\left[\left(\frac{x-\tau}{e^{\mu}}\right)^{-\sigma}\right]\right\} \qquad \text{for } x > \tau, \sigma > 0 \tag{2.53}$$

and Eq. (2.52) is then a special case of Eq. (2.53) with $\tau = 0$. A transformation of $Y = (x - \tau)/e^{\mu}$ suggested by Eqs. (2.34) to (2.53) would yield a standard log-normal density (with shape parameter σ),

$$g(y) = (1/y\sigma\sqrt{2\pi}) \exp\{-\tfrac{1}{2}[(\ln y)/\sigma]^2\} \qquad \text{for } y > 0, \sigma > 0 \tag{2.54}$$

The inverse function G^{-1} cannot be put in closed form. Worse yet, no table of $G^{-1}(p) = y$ is even available. Hence it is not easy to construct probability paper for the log-normal case which provides the estimates of the location parameter τ and of the "scale" parameter μ through $\exp(\mu)$ for each selected value of the shape parameter σ. However, if one is willing to assume $\tau = 0$ (or is known), then $\ln X = Z$, by definition, is normally distributed and the probability grid for the normal distribution given by Fig. 2.6 may be used provided that the axis for the random variable is calibrated in the logarithmic scale. In fact, such graph paper is commercially available (e.g., Keuffel and Esser No. 359-24).

Figure 2.7 shows a mean-rank plot of the data points given in Fig. 2.1 on the log-normal probability paper. Here the location parameter τ may be estimated by trial and error. If the "best" linearity seems to occur when $\hat{\tau}$ is subtracted from each observation and the adjusted data are plotted, then $\hat{\tau}$ is the estimate for the location parameter. Here $\hat{\tau} = 0$ and the estimates for the remaining parameters are

$$\hat{\mu} = \ln 42.3 = 3.74 \text{ and } \hat{\sigma} = \ln 58 - \ln 42.3 = 4.06 - 3.74 = 0.32$$
$$(\text{or } \hat{\sigma} = \ln 42.3 - \ln 32 = 3.74 - 3.43 = 0.31)$$

which yield the estimates for

$$\begin{aligned} \text{Log-normal mean} &= \exp\,[\hat{\mu} + \hat{\sigma}^2/2] \\ &= \exp\,(3.79) = 44.3 \text{ (or 4,430 hours)} \qquad (2.55) \\ \text{Log-normal variance} &= \exp\,[2(\hat{\mu} + \hat{\sigma}^2)] - (\exp\,[\hat{\mu} + \hat{\sigma}^2/2])^2 \\ &= \exp\,(7.68) - (44.3)^2 \\ &= 210 \text{ (or } 210 \times 10^4 \text{ (hours)}^2) \qquad (2.56) \end{aligned}$$

2.4f Gumbel Type I Distribution. The Gumbel Type I smallest-extreme distribution function was given before by Eq. (2.24). Its density function, by differ-

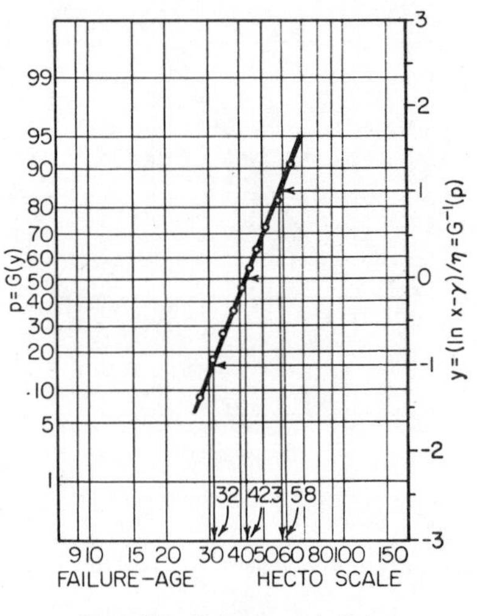

FIG. 2.7 Log-normal plot.

entiation, is

$$f(x) = \frac{1}{\eta} \exp\left[\left(\frac{x-\gamma}{\eta}\right) - \exp\left(\frac{x-\gamma}{\eta}\right)\right] \qquad \text{for } x > \gamma,\ \eta > 0 \qquad (2.57)$$

The transformation specified by Eq. (2.34) applied to Eq. (2.57) would yield

$$g(y) = \exp\,[y - \exp\,(y)] \qquad \text{for } -\infty < y < +\infty \qquad (2.58)$$

Hence the standard Gumbel Type I distribution function is

$$G(y) = 1 - \exp\,[-\exp\,(y)] \qquad \text{for } -\infty < y < +\infty \qquad (2.59)$$

On setting $G(y) = p$ and solving for y, we have

$$y = \ln\,[-\ln\,(1-p)] \qquad (2.60)$$

It is possible to get values of y given by Eq. (2.60) from a National Bureau of Standards table [10]. The inverse of the cumulative probability function of extremes tabulates the function

$$Y = -\ln(-\ln \Phi\hat{y}) \tag{2.61}$$

By substituting $1 - p$ for $\Phi\hat{y}$, the negative value of Eq. (2.60) is obtained. Figure 2.8, with y axis given by Eq. (2.60), shows a mean-rank plot of the data points given in Fig. 2.1. A forced straight line among the points gives estimates for the location and scale parameters as $\hat{\gamma} = 5{,}000$ hours and $\hat{\tau} = 6{,}100 - 5{,}000 = 1{,}100$ hours. Hence, the estimated mean and variance are (C = Euler's constant $= 0.5771 \cdots$):

$$\text{Gumbel mean} = \hat{\gamma} - C\hat{\tau} = 5{,}000 - (0.577)1{,}100 = 4{,}370 \text{ hours} \tag{2.62}$$

$$\text{Gumbel variance} = (\pi^2/6)\hat{\tau}^2 = (9.9/6)1{,}100^2 = 199 \times 10^4 \text{ (hours)}^2 \tag{2.63}$$

2.4g Weibull Distribution. For the Weibull distribution function $F(x)$, as given by Eq. (2.17), the application of the linear transformation specified by Eq. (2.34) yields

$$G(y) = 1 - \exp[-y^\beta] \qquad \text{for } y > 0, \beta > 0 \tag{2.64}$$

On setting $G(y) = p$ and solving for y, we have

$$y = [-\ln(1 - p)]^b \qquad \text{where } b = 1/\beta \tag{2.65}$$

A table by Plait [11] is available for Eq. (2.65) for $\beta = 0.1(0.1)4.0$. Figure 2.9 is

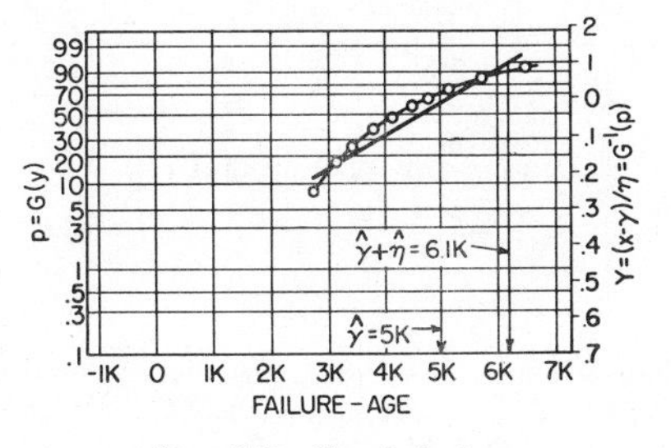

Fig. 2.8 Gumbel plot.

similar to Fig. 5, page 56 of the U.S. Defense Department QC & R TR-3 [12] for such a plot. Since the y values of Eq. (2.65) depend on β, it is necessary to give a different calibration for each β value. For β values of 1 and 3 (curves D and E), the reversal of curvature in the two plots is noted. By trial and error, curve F for $\hat{\beta} = 1.85$ is approximately linear. Extending curve F toward the bottom gives $\hat{\gamma} = 2{,}000$ hours and $\hat{\tau} = 4{,}750 - 2{,}000 = 2{,}750$ hours. Hence, the estimate for the mean and variance are

$$\text{Weibull mean} = \hat{\gamma} + \hat{\tau}\Gamma(\hat{b} + 1) = 2{,}000 + 2{,}750(0.888) = 4{,}442 \text{ hours} \tag{2.66}$$

$$\text{Weibull variance} = \hat{\tau}^2[\Gamma(2\hat{b} + 1) - \Gamma^2(\hat{b} + 1)] = 2{,}750^2(0.25) = 189 \times 10^4 \text{ (hours)}^2 \tag{2.67}$$

An alternative method is given below.

As pointed out in Sec. 2.3*d*, the Weibull distribution may be considered as the logarithmic Gumbel Type I distribution if the Weibull parameter γ is equal to zero or is known. That is to say, if $Z = \ln X$ and Z has a Gumbel Type I (smallest-extreme) distribution with location parameter $= a$ and scale parameter $= b$, then the distribution of X is Weibull with scale parameter $\eta = e^a$ and shape parameter $\beta = 1/b$. This is true, as seen from Eq. (2.33), because

$$\exp[(z - a)/b] = \exp[(\ln x - \ln \eta)\beta] = (x/\eta)^\beta \tag{2.68}$$

For this reason, the probability grid for the Gumbel Type I distribution given by Fig. 2.8 may be used provided that the axis for the random variable is calibrated in the logarithmic scale. A graph paper with such scales and additional scales for ease of

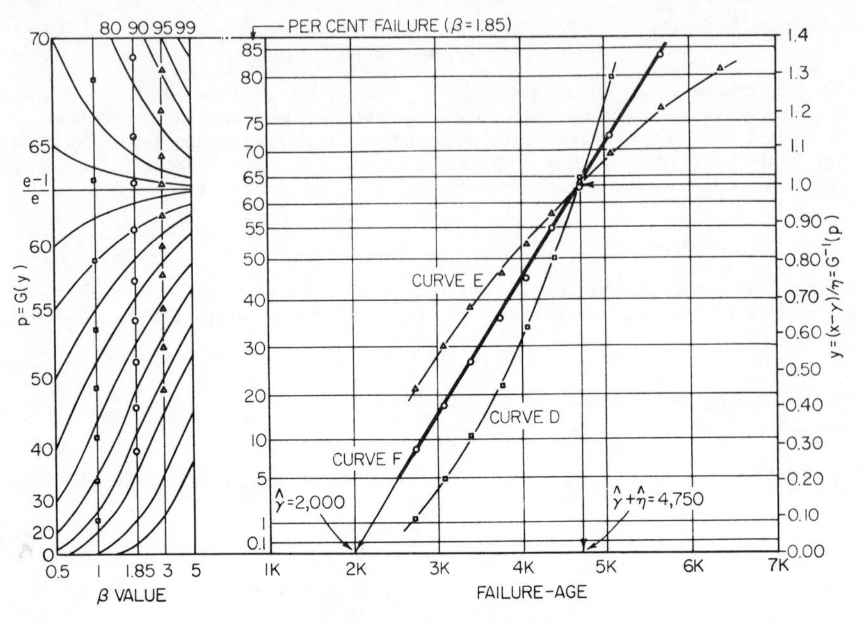

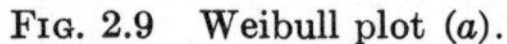
Fig. 2.9 Weibull plot (*a*).

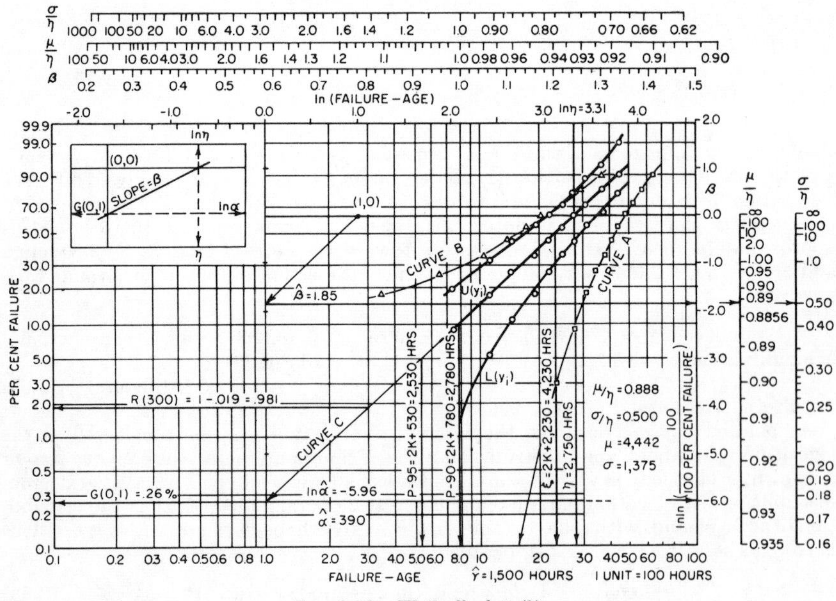

Fig. 2.10 Weibull plot (*b*).

computing Weibull mean and variance was designed by Kao [13]. Figure 2.10 shows this plot using mean ranks for the data points given in Fig. 2.1. In the figure, curve A represents the plot of raw data and curve B represents the plot of adjusted raw data—subtracting trial value for $\gamma = 2{,}750$ hours.

The fact that curves A and B have opposite curvatures indicates that the true location parameter γ lies somewhere between 2,750 and 1,500, which is the abscissa value of curve A when it is extended downward. Upon several trials, $\hat{\gamma} = 2{,}000$ hours gives a nearly straight line (curve C) with the estimates $\hat{a} = \ln \hat{\eta} = 3.31$ (with x in hecto-hours) and $\hat{\eta} = 2{,}750$ hours and slope $\hat{b} = 1/\beta = 1/1.85$ or $\beta = 1.85$. These results check exactly with those obtained by the first method discussed previously in this section. Kao [13] gave a simpler explanation for this alternative method which is based on the fact that Eq. (2.17) can be written as the following upon taking the natural logarithm twice:

$$\ln \{-\ln [1 - F(x)]\} = -\beta \ln \eta + \beta \ln (x - \gamma) = -\ln \alpha + \beta \ln (x - \gamma) \quad (2.69)$$

Hence, on graph paper with ln versus ln ln coordinates, the Weibull distribution function will appear as a straight line. Figure 2.11 depicts such a straight line. With

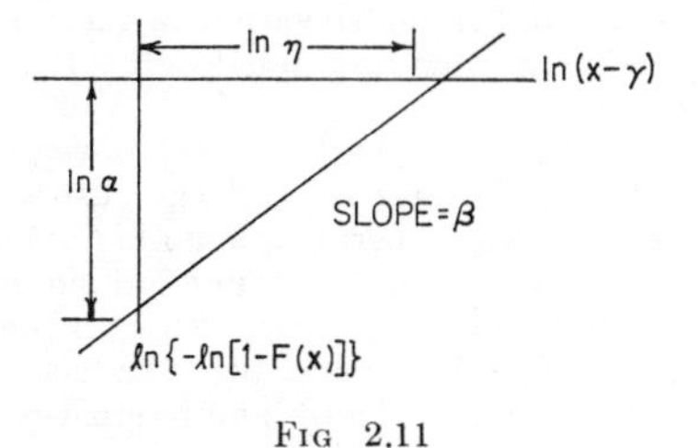

FIG 2.11

the estimates $\hat{\beta} = 1.85$ and $\hat{\eta} = 2{,}750$ hours, a few further calculations give the following (refer to Fig. 2.10):

Mean $= 2{,}000 + 0.888(2{,}750) = 4{,}442$ hours
Standard deviation $= 0.5(2{,}750) = 1{,}375$ hours
Reliability function at 2,300 $= R(2{,}300 - 2{,}000) = R(300) = 1 - 0.019 = 0.981$
Reliable life at 90 per cent $= 2{,}000 + 780 = 2{,}780$ hours
Reliable life at 95 per cent $= 2{,}000 + 530 = 2{,}530$ hours
Median life $= 2{,}000 + 2{,}230 = 4{,}230$ hours
Initial failure rate per 100 hours $= 0.26$ per cent

Since $x - \hat{\gamma}$ are plotted in Fig. 2.10, the initial failure rate of the last entry actually refers to failure rate at 2,000 hours. The curves labeled $L(y_i)$ and $U(y_i)$ are lower and upper confidence bands on $F(x)$ for a confidence coefficient = 80 per cent (see Kao [14].)

2.5 GENERATING THE TRADE-OFF CHARACTERISTICS FROM THE ESTIMATED LIFE-LENGTH DISTRIBUTION

To say that an item has a certain trade-off characteristic means that the life quality of the item depends on the severity of the stress applied to it. The relationship between *life quality* and *applied stress* is a complex one. In general (but not always) better life quality is expected of an item with reduced applied stress. In application, however, lack of concise definition for both life quality and applied stress has made the analysis a difficult one. This section has proposed defining the life quality of an item by its life pattern; now we shall deal with its relationship with the applied stress.

2.5a Stationary Life Test. Because of its simplicity, most analysis belongs to this category. Here items are life-tested under operating and environmental conditions (applied stress) which are identical with or at least similar to those in which the items are intended to be used. Since the applied stress is held at the so-called normal level and the life test is merely a simulation of the field use, *no* trade-off curve can be derived. Although the data here provide the needed reliability information directly, they have the disadvantage that the results are applicable to only one single stress level.

2.5b Single-factor Life Test. Some of the so-called accelerated or increased-severity life tests belong to this group. Here factors of applied stress are deliberately varied, one at a time, above and below the normal level. While one factor is varied, all other factors are kept stationary at the normal level. From this it is possible to answer questions of the following kind. For example, if an equipment designer uses electron tubes at a higher-than-normal bulb temperature, what is the quantitative effect on life quality of the tubes? The single factor involved in this question is bulb temperature. This type of life test is useful in any exploratory stage in conserving testing time. However, it is rather inefficient.

2.5c Multifactor Life Test. The single-factor life test cannot be made to answer questions involving two or more factors. For example, we might take the above question and extend it. If the equipment designer is forced to operate the tubes at an elevated bulb temperature and he is willing to sacrifice some output power from the tubes (i.e., trade off power for elevation in temperature), is it possible to maintain the life quality value at some desirable level? If so, how much sacrifice of output power is necessary?

Here two factors are involved: temperature and power. In practice, questions involving even more factors may be posed. But the answers can come only from a multifactor life test. In multifactor experiments, all factors affecting the life quality of the item are varied at all levels, not one at a time but simultaneously, thus making it possible to evaluate the effects one at a time, two at a time, three at a time, etc. The layout (or design) of such experiments is called the factorial design. The reader is directed to Sec. 4 of this handbook for details and references.

References

1. Fisher, R. A., and L. H. C. Tippett, "Limiting Forms of the Frequency Distribution of the Largest or Smallest Member of a Sample." *Proc. Cambridge Phil. Soc.*, vol. 24, part 2, p. 180, 1928. *Reprinted in Fisher's Contributions to Mathematical Statistics*, John Wiley & Sons, Inc., New York, 1950.
2. Gumbel, E. J., *Statistics of Extremes*, Columbia University Press, New York, 1958.
3. Aitchison, J., and J. A. C. Brown, *The Lognormal Distribution*, Cambridge University Press, London, 1957.
4. Pearson, K., *Tables of the Incomplete Beta Function*, Cambridge University Press, London, 1934.
5. Pearson, E. S., and H. O. Hartley, *Biometrika Tables for Statisticians*, 2d ed., table 17, Cambridge University Press, London, 1958. (Hartley and Fitch chart.)
6. Blom, G., *Statistical Estimates and Transformed Beta Variables*, John Wiley & Sons, Inc., New York, 1958.
7. Harter, H. L., *New Tables of the Incomplete Gamma Function Ratio and of Percentage Points of the Chi Square and Beta Distributions*, GPO, 1964.
8. Fisher, R. A., and F. Yates, *Statistical Tables*, Oliver & Boyd Ltd., Edinburgh and London, 1938.
9. Hald, A., *Statistical Tables and Formulas*, John Wiley & Sons, Inc., New York, 1952.
10. "Probability Tables for the Analysis of Extreme Value Data," *Natl. Bur. Standards Appl. Math. Ser.* 22, 1953.
11. Plait, A., "The Weibull Distribution with Tables," *Indus. Quality Control*, vol. 19, no. 5, November, 1962.
12. "Sampling Procedures and Tables for Life and Reliability Testing based on the Weibull Distribution (Mean Life Criterion)," *Quality Control and Reliability Tech. Rept. TR-3*, Office of the Assistant Secretary of Defense (Installation and Logistics), GPO, 1961.
13. Kao, J. H. K., "A Summary of Some New Techniques on Failure Analysis," *Proc. Sixth Natl. Symp. Reliability Quality Control*, 1960, p. 196.
14. Kao, J. H. K., "The Beta Distribution in R and Q.C.," *Seventh Natl. Symp. Reliability Quality Control*, 1961, p. 496.

Section 3

SELECTING RELIABILITY TEST PLANS

DAVID C. DELLINGER

U.S. AIR FORCE, AIR FORCE INSTITUTE OF TECHNOLOGY
WRIGHT-PATTERSON AIR FORCE BASE, OHIO

Major Dellinger is presently Associate Professor of Operations Research and Statistics in the Department of Systems Management, AFIT. He is responsible for the development of the quantitative portion of the Graduate Systems Management curriculum. He also shares responsibility for curriculum development for the Graduate Reliability Engineering Program. He is a member of the Air Force Weapon Systems Effectiveness Industry Advisory Committee.

Major Dellinger was recalled to active duty with the Air Force in 1951 after he had received the B.S. degree in mechanical engineering at Duke University. He served in various capacities in the operational forces (primarily fighter operations) until 1956, when he entered Stanford University. After receiving the M.S. degree in industrial engineering, he was assigned to the Office of the Chief, Quality Control, in the Air Materiel Command. In that capacity he was engaged in the development of the Air Force Quality Control Engineering Program as well as being a consultant to operating level QC organizations.

In 1961, Major Dellinger returned to Stanford, where he received the Ph.D. degree in industrial engineering with specialization in operations research. His doctoral dissertation was concerned with the application of operations research to reliability testing programs. He has written a technical report on the subject and coauthored a technical article on incentive contracting.

SELECTING RELIABILITY TEST PLANS

David C. Dellinger

CONTENTS

3.1 INTRODUCTION

In this section, the problem of selecting the optimal test plan for each particular application is discussed. It is a problem with which every person who prescribes reliability tests must come to grips, and, unfortunately, it is a very perplexing problem. It involves consideration of factors which are often ill defined and difficult to measure or estimate. Moreover, the interrelationships between these factors are not usually known with any degree of precision. Normally, one does have some information relative to these factors, and it is desirable to introduce this information, however skimpy it may be, into the test-plan-selection process. It is the purpose of this section to examine this problem and to suggest ways for utilizing whatever information is available in test-plan selection.

A test plan, as discussed in this section, refers to a procedure which prescribes the amount of testing to be performed and the decision rules which will be utilized to decide whether the reliability of a group (lot) of items meets some predetermined standard. The discussion in this section does not generally apply to tests which are conducted in order to estimate the reliability of a given lot. This section is primarily concerned with those cases where the testing is incorporated into the decision-making process.

The problem of test-plan selection is not a new one to industrial concerns. It has been present in the use of acceptance sampling procedures utilized in statistical quality control for years. In the application of statistical testing procedures to reliability, however, the problem takes on added significance. In general, reliability-testing costs are higher than quality-inspection costs, and the consequences of an erroneous decision are often much more severe. From an economic point of view, it is certainly more important to find the optimal, or near-optimal, test plan when the costs involved are so great.

In general, the likelihood of making an optimal test-plan selection is somewhat reduced by the fact that the necessity for making certain decisions in the selection process is not made clear in much of the literature relative to testing. While most sets of test plans published include a description of the assumptions upon which the plans are based, they rarely emphasize the fact that, if reasonable test results are to be obtained, one must determine if these assumptions are reasonably true for the particular application. Moreover, it is even more unusual for these sets of test plans to provide guidance for the user in his efforts to determine the reasonableness of the assumptions. Unless one faces these critical questions in the process of making the selection of a test plan, he is apt to obtain results which are either useless or, worse, misleading. (See Sec. 8.)

In this section the view is taken that the proper selection of a statistical test plan involves three key decisions: (1) defining the lot (or group) of items to which the decision will apply, (2) determining the proper model for the underlying distribution of time to failure, and (3) selecting the particular test plan from among those available which are based on the selected distribution. Each of these decisions is discussed at length in the remainder of this section, and, where possible, procedures are suggested to assist in making the required decisions.

As indicated earlier, these decisions are not easy to make. Consequently, the procedures suggested are not precise, and in some cases they may not even lead to a decision. They do, however, provide general guidance for test-plan selection and focus attention on the relevant factors in the process.

3.2 HOMOGENEOUS-LOT SELECTION

In lot-by-lot reliability testing, most standard tests are based on the assumption that the time to failure (or time between failures) characteristic for all items in the lot is identically distributed, i.e., failures are caused by the same chance-cause system. Such a lot is said to be statistically homogeneous. Quite often in practice, lots which are homogeneous in regard to time between failures, or reasonably close to being homogeneous, can be formed. It is usually reasonable to assume that items off the same production line made of parts obtained from the same sources would possess this desired property. In such cases statistical test plans based on this assumption can be readily applied. If, however, the lot is not statistically homogeneous, conclusions based on an assumption of homogeneity can be erroneous. If accept/reject decisions are made on this basis, good items can be rejected with bad items and bad items can be accepted with good ones.

The question of homogeneity in reliability testing is largely a question of the nature of the causes of failure and would be difficult, if not impossible, to answer in the engineering sense. It seems reasonable, however, to expect that items manufactured under essentially identical conditions and by using identical parts would suffer from the same failure causes. Where these identical conditions do not exist, one may suspect lack of statistical homogeneity. For example, when parts are basically identical from a design point of view but are manufactured on different production lines or by different producers, an assumption of homogeneity should be viewed with suspicion. Complex items, such as radar sets, which are subject to many failure and repair cycles can also develop individual failure characteristics.

Unfortunately, as in the case of sampling from any infinite population, no method is available for determining precisely the form of the underlying distribution or distributions. Statistically, the problem is one of determining whether the variation of the time between failures observed on different items can be attributed to chance variation expected from a homogeneous lot or must be attributed to different chance-cause systems (different failure characteristics possessed by the items in the lot). Under certain assumptions, statistical tests are available for the hypothesis of homogeneity. When the underlying distribution is normal, one can apply the standard tests described in most statistical handbooks. A procedure which can apply when the underlying distribution is, or distributions are, exponential is given below.

It is essential that the person applying reliability tests which are based on the assumption of homogeneous lots be aware of this assumption if he is to obtain reasonable results from the test. He must guard against the arbitrary mixing of items in order to obtain large lots (and the accompanying economy in testing). Only when there is an affirmative answer to the question, "Can these items reasonably be expected to have the same chance-failure patterns?" should items be mixed to form homogeneous lots.

3.2a Test for Homogeneity of Lots, Exponential Distribution. The following test can be used to test for the homogeneity of a lot which is made up of several subgroups which are in themselves homogeneous. If the test is negative, i.e., the lot is not homogeneous, a basis for combining the subgroups into smaller, homogeneous lots is provided. The procedure should be applied only when the underlying distribution of all of the subgroups is exponential. The procedure is as follows:

1. Identify the homogeneous subgroups. Items are segregated into n subgroups each of which is believed to be homogeneous. For example, all the items from a single production line or the same manufacturer are placed in the same subgroup. When complex items which can be repaired and returned to test after each failure are being tested, each item may be identified as a subgroup. If one has no rational basis for subgrouping items, there is no point in applying this test.

2. Test all subgroups for the same period of time t. The amount of testing is arbitrary; however, the more testing performed, the better the results of the test. A good rule of thumb in selecting the test time is to select t such that the product nt will equal the amount of testing which would be required to accept or reject the lot if the lot were homogeneous. If the lot is homogeneous, one can immediately utilize the test results to make a reject/accept decision (see the example below).

3. Observe the number of failures occurring in each subgroup during the test period. Designate this number by x_i, where the subscript i refers to the number of the subgroup, $i = 1, 2, \ldots, n$. Assign subscripts so that subgroups are placed in descending order of the magnitude of the number of failures, that is, $x_1 \geq x_2 \geq \cdots \geq x_n$.

Table 3.1 Critical Values of D for n Means at α Level of Confidence

n	2	3	4	5	6	7	8	9	10	11	12
$\alpha = 0.01$	3.64	3.19	3.00	2.71	2.57	2.47	2.39	2.33	2.27	2.23	2.19
$\alpha = 0.05$	2.77	2.48	2.28	2.13	2.02	1.93	1.86	1.81	1.76	1.72	1.69

Reprinted by permission from H. A. David, "The Ranking of Variance in Normal Populations," *J. Am. Statist. Assoc.*, vol. 51, 1956.

4. Compute $D_i = 2\sqrt{x_i + \frac{1}{2}} - 2\sqrt{x_{i+1} + \frac{1}{2}}$.

5. For the desired level of significance and the number of subgroups in the test find the critical value of D from Table 3.1.

6. Compare the values of D_i computed in step 4 with the critical values of D selected from Table 3.1. Any value of D_i which exceeds D indicates that subgroups 1, 2, . . . , i should not be mixed with subgroups $i + 1, i + 2, \ldots, n$ to form a single lot. If all values of D_i are less than the critical value of D, there is no evidence that the subgroups should not be mixed to form a lot which is homogeneous.

Example 3.1

Suppose one has 10 airborne radar sets which are to be tested to determine if they as a group meet a reliability requirement. It is suspected that, even though experience indicates that the time to failure of the items is exponentially distributed, the individual component distributions of time to failure may not necessarily have the same mean value. It is desired to apply the procedure described in Sec. 3.2*a* to test for the homogeneity of the lot of 10 items. If it can be reasonably well established that times between failures for sets are exponentially distributed (or nearly so), the test plan described below will be followed to determine if the lot should be accepted or rejected. The selected test plan requires that each of the 10 items be placed on test for a period of 65 hours, so that a total of 650 test hours will be accumulated. When failures occur, repairs will be made and the items will immediately again be placed on test. If more than 30 failures are observed during the test, the entire lot will be rejected. If 30 or fewer failures are observed, the lot will be accepted. If the 10 items cannot be grouped into a single lot, smaller lots will be formed and each lot will be subjected to a total test time of 650 hours in order to meet the required operating characteristics of the test.

In this case it will be desirable to test each item for a period of 65 hours in order to test for the homogeneity of the lot. If the test indicates homogeneity, an accept/reject decision can be made immediately. If not, additional testing will be required.

Suppose that, after 65 hours of testing on each set, the data in Table 3.2 are obtained.

Table 3.2 Data for Example 3.1

Order number i	Number of failures x_i	$2\sqrt{x_i + \frac{1}{2}}$	D_i
1	11	6.80	1.36
2	7	5.44	0.34
3	6	5.10	1.94*
4	2	3.16	0
5	2	3.16	0.71
6	1	2.45	0
7	1	2.45	0
8	1	2.45	0
9	1	2.45	0
10	1	2.45	0

Table 3.3 Data for Example 3.1

Order number i	Number of failures x_i	$2\sqrt{x_i + \frac{1}{2}}$	D_i
1	11	6.80	1.36
2	7	5.44	0.76
3	5	4.68	0.44
4	4	4.24	0.49
5	3	3.75	1.30
6	1	2.45	0
7	1	2.45	0
8	1	2.45	1.03
9	0	1.42	0
10	0	1.42	0

The data as shown have been ordered and the item numbers established in accordance with step 3 of the procedure, and the computations required by steps 4 and 5 have been completed and are shown in the last two columns of the table.

If it had been decided that a level of significance of 0.05 was desirable, the critical value of D for $n = 10$ from Table 3.1 is found to be 1.76. An examination of Table 3.2 shows that $D_3 = 1.94$, which is greater than the critical value of $D = 1.76$. The results of this test indicate that it would be undesirable to group items 1 to 3 with the remaining items. However, there is no evidence to indicate that items 1 to 3 cannot be placed in one group and the remaining items in another. Each of the two lots formed in this way can be tested separately, and individual reject/accept decisions can be made for the two lots.

Suppose, on the other hand, that the data in Table 3.3 had been obtained in the test. None of the values of D_i are greater than the critical value of D, and one would conclude from this test that the 10 items could be grouped to form a homogeneous lot.

3.3 DETERMINING THE UNDERLYING DISTRIBUTION

It is to be emphasized that considerable care is required in the selection of the appropriate underlying distribution for reliability testing. The validity of the test results depends to a large degree upon how well the selected probability distribution represents the actual distribution of the time to failure upon which observations are being made. This is in contrast to the use of routine statistical quality control procedures which are relatively insensitive to the actual form of the underlying distribution. This point is emphasized because the widespread (perhaps indiscriminate) use of the exponential distribution as a model of failure patterns may lead one to believe that failure times in general may be adequately represented by such a distribution.

It is unfortunate that the exponential distribution cannot be universally applied to reliability testing because of its many convenient mathematical properties. The use of other distributions introduces complications into the testing procedures; however, if one is to draw reasonable conclusions from tests, he must be willing to face these other problems. (See Sec. 2.)

One of the most likely candidates for use as a failure distribution is the Weibull distribution. As indicated earlier, no particular distribution can be justified on theoretical grounds alone; however, empirical evidence indicates that the Weibull can be fitted to many failure patterns by the proper selection of the shape parameter. The gamma and normal distributions have also been used as models for reliability testing. Whatever model is selected, the user must keep in mind that the validity of test results depends upon how well the selected distribution represents the actual distribution.

A specific procedure for determining the actual underlying distribution is not available; however, one can minimize the danger of making an erroneous selection if he utilizes all the information available in the selection process. The following steps should aid in making such a selection.

3.3a Utilize Available Engineering Knowledge. Often, the engineer most familiar with the item being tested can provide some general information relative to failure patterns. For example, he might be able to indicate whether an increasing or decreasing hazard function is reasonable in a particular case. While judgments of this nature are not too precise, they may prevent gross errors in selecting a distribution.

3.3b Utilize Graphical Estimation Procedures. When test data for the device are available (or can be attained), graphical curve-fitting procedures can be utilized to aid in the selection of the distribution. For example, if one is satisfied that a Weibull distribution is suitable as a model, he can use the graphical procedures described in Sec. 2 to estimate the location and shape parameters of the distribution. Testing can then be based on the resulting one-parameter Weibull distribution. In using such a procedure, one should keep in mind that the implication is that future failure patterns of the device will be identical with the failure patterns observed in the past except that the scale parameter may vary.

3.3c Apply Goodness-of-fit Tests. There exist statistical procedures which can be utilized to test the goodness of fit of a particular set of data to an hypothesized distribution. Once a distribution has been selected, either graphically or by some other method, such tests can be applied. Two simple tests which can be readily utilized are the chi-square goodness-of-fit test and the Kolmogorov-Smirnov test.[1] The Kolmogorov-Smirnov test is particularly well suited for testing when the hypothesized distribution is continuous, as it usually is in reliability testing.

3.3d Utilize Distribution-free Tests. If one is unable to determine satisfactorily the form of the underlying distribution, distribution-free (sometimes called nonparametric) tests can be utilized and may be the best alternative in many applications. Distribution-free tests can be conducted merely by establishing the test time so that it is equal to the time the device will be required to operate in use. This

[1] For complete description of this test, see F. J. Massey, Jr., "The Kolmogorov-Smirnov Test for Goodness of Fit," *J. Am. Statist. Assoc.*, vol. 51, pp. 68–78, March, 1956.

eliminates time as a factor in the test, so that one has only to observe the attributes of success or failure. Mil-Std-105D or any other acceptance sampling procedure for attributes can be applied. Such a procedure, of course, can be quite expensive if the operating time is large, and in some cases it may even be impossible. When such tests can be applied and there is a reasonable doubt as to the form of the underlying distribution, the additional expense of conducting the nonparametric test in order to achieve meaningful results may be justified. This is especially true when the results of the test are particularly sensitive to the form of the distribution. (See Sec. 3.6 for a discussion of sensitivity tests.)

3.4 THE USE OF OPERATING CHARACTERISTICS IN TEST-PLAN SELECTION

Having made the decision as to the appropriate underlying distribution (or to use nonparametric test procedures), one is then faced with the problem of selecting a particular test plan from among the many available. It is assumed that the test plan is to be used for the purpose of determining whether a particular lot is to be accepted or rejected for some particular purpose. The selection of the test plan involves determining the amount of testing to be performed (the number of items to be tested and the amount of time each item is to be tested) and the decision rules (the acceptance or reject number). A particular test plan can be specified by arbitrarily selecting a number of test parameters; in some cases standard test plans are indexed by arbitrarily selected test parameters. For example, many procedures are indexed according to the number of failures one is to allow on test, the number of items to be placed on test, or the maximum amount of testing to be performed on each item. When restrictions are imposed on these factors because of practical considerations, such an indexing system may have merit. In many cases, however, this form of indexing can cloud the basic issues involved in the selection of a test plan.

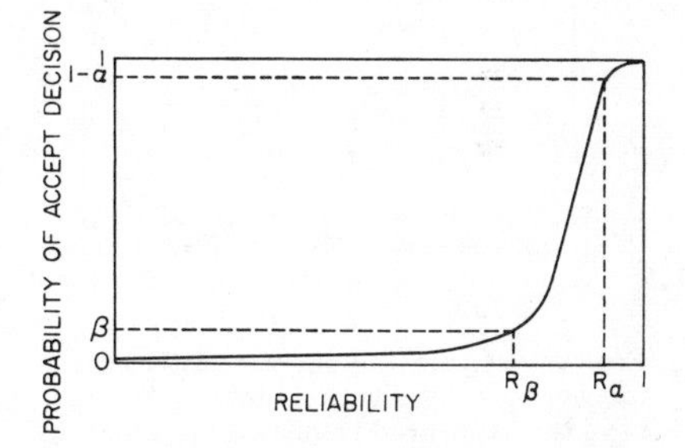

FIG. 3.1 Typical operating-characteristic curve for a reliability test plan.

A much more desirable method of test-plan selection is to make use of the operating characteristics of the test plans. The operating characteristics of a test plan indicate the probability of reaching an accept decision when utilizing the test plan for a lot having any given level of reliability. From an examination of these operating characteristics, one can determine the differentiating ability of the test procedures, i.e., the ability of the test procedure to distinguish between good and bad lots. In practically all cases, some form of the operating characteristics of test plans is provided along with the set of plans and is available for use in test-plan selection.

The operating-characteristic curve for a test plan is the probability of reaching an accept decision plotted as a function of the actual reliability of the lot being subjected to test. Figure 3.1 shows a typical operating-characteristic curve for a reliability test plan. For low values of reliability the probability of an accept decision is low. For higher levels of reliability the probability of an accept decision is also higher, approaching 1 as the reliability approaches 1.

For indexing purposes a system of identifying two particular points on the operating-characteristic curve has been developed. The two points are usually referred to as (1) the producer's risk [the point $(R_\alpha, 1 - \alpha)$ in Fig. 3.1] and (2) the consumer's risk [the point (R_β,β) in Fig. 3.1]. The producer's risk α is the probability of rejecting the lot when the actual reliability is R_α, a level of reliability which is viewed as acceptable for the particular application. In other words, it is the probability of making an error which would be undesirable from the producer's point of view, i.e., even though he produced a satisfactory product, it is rejected by the test. The consumer's risk β is the probability of accepting the lot when the actual reliability is R_β, a level of reliability which is considered as unsatisfactory for the particular application.

Figure 3.2 shows several operating-characteristic curves for a set of test plans all of which provide the same level of producer's risk but different levels of consumer's risk. The significant difference between these test plans is that a different amount of testing is required for each. This difference in the amount of testing is reflected in the different levels of consumer's risk. The more testing performed, the lower the consumer's risk for a fixed level of R_β (or the greater the value of R_β for a fixed level of β).

The test plans requiring a greater amount of testing are said to provide better differentiation between desirable and undesirable levels of reliability. This difference in differentiating ability is the significant practical difference between test plans requiring different amounts of testing and should be the dominant consideration in test-procedure selection. Viewed in this way, one can relate the value of a particular test plan in terms of its differentiation ability with its cost in terms of test time. Such a view provides the most rational basis for direct selection of a reliability test plan.

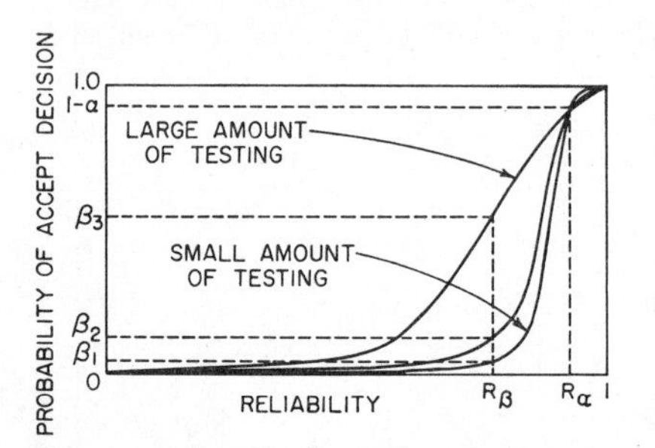

FIG. 3.2 Operating-characteristic curves for several reliability test plans. All plans have the same producers' risk but different consumers' risk and require different amounts of testing.

Since reliability is a function of some operating time as well as the parameters of the assumed underlying distribution (distribution-free plans excepted), it is usually impractical to provide operating-characteristic curves in terms of the actual reliability. In most publications the operating characteristics are provided in terms of the unknown parameter of the assumed distribution. For example, operating characteristics of test plans based on the exponential distribution are usually given in terms of the parameter which is the mean time between failures for the distribution. In such cases it is necessary to convert the desired reliability into the appropriate parameter values in order to make a rational selection. In any case it is desirable to think in terms of the actual reliability in making such a selection, since it is difficult to relate a parameter value to an application.

In practice, the problem of selecting a test procedure on the basis of operating characteristics can be formulated as that of selecting two points on the operating-characteristic curve and then finding a test plan which meets this specification. One would normally want to examine several test plans in order to make a comparison between their differentiating ability and cost before making the decision. In this way one can compare the incremental value of the better differentiating ability with the incremental cost (in additional testing) of the better test plans.

3.5 ECONOMIC TEST-PLAN SELECTION

In Sec. 3.4 it was pointed out that the most desirable method of selecting a test plan from a set is by a comparison of the operating characteristics of the available procedures. The use of operating characteristics does not, however, provide a complete solution to the plan-selection problem. One must be able to relate, in some rational manner, the value of the differentiating ability of a plan, as indicated by the operating-characteristic curve, to the cost of applying the plan to a particular case. In general, the better a testing procedure differentiates between lots having desirable and undesirable characteristics, the more testing is required and, consequently, the greater are the testing costs. The economic test-plan-selection problem can be viewed as one of achieving a balance between the value of the better differentiating ability and the cost of more testing.

When testing is to be performed so that a decision either to scrap or to utilize a lot of items in a particular application can be made, the decision-theory approach may provide some guidance in test-plan selection. Generally, this approach is not appropriate when testing is performed to determine if a specified value of reliability has been attained. The basic underlying assumption in this approach is that the objective is to minimize the sum of the test costs and the expected cost of the action resulting from the test. Intuitively, this means that if one were to select test plans in the same manner for many tests, the average cost would be minimized over the long run. Clearly, the cost of the action to be taken in the future involves a prediction, and the accuracy of the result is dependent upon the accuracy of the prediction.

In the decision-theory approach to economic test-plan selection one must predict future costs resulting from the decision and make some assumptions about the a priori distribution of the reliability of the device being tested. It is likely that in many cases one will be able to make gross estimates about these two factors, but he will rarely be able to make precise estimates about them. The procedure to be outlined here provides a means for incorporating general knowledge about these two factors into the test-plan-selection process. The resulting guidance is, of course, only general; however, this is preferable to the alternative, i.e., ignoring whatever information one has relative to these two factors.

3.5a General Description of the Decision-theory Approach. In general, decision theory is concerned with determining the optimal experiment and decision rule when one does not know the true circumstances to which the decision will apply. The problem at hand provides for a somewhat simplified application, since the selection of a standard testing procedure jointly determines the experiment and the decision rule for the case of only two alternatives. The true circumstance, usually referred to as the state of nature, is the true reliability of the lot to which the decision will apply. The consequences of any decision will be a function of the true state of nature and the action taken. If one can assign a cost to the consequences of any decision, given the true state of nature, the problem can be viewed as one of selecting the testing procedure which will minimize the expected cost of the experiment and the resulting action. Normally, costs are converted into losses to simplify the problem, and the testing procedure is selected on the basis of minimum expected loss. A more precise statement of the problem follows. Let

d_1 = decision to accept the lot
d_2 = decision to reject the lot
R = true reliability of the lot
$L(d_i,R)$ = loss for decision d_i for given R
$p[d_i|R]$ = probability of d_i given R
$s(t)$ = cost of conducting the test as a function of t
t = amount of testing required

Note that $p[d_i|R]$ is the operating-characteristic function of a testing procedure, since it is merely the probability of an accept decision expressed as a function of the actual reliability. Since this is a two-decision problem,

$$p[d_2|R] = 1 - p[d_1|R]$$

The expected loss for any given value of R and a particular testing procedure is

$$E\,[\text{Loss}|R] = \sum_{i=1}^{2} L(d_i,R)p[d_i|R] + s(t)$$

This is sometimes called the risk function. The total expected loss for a particular testing procedure requires a priori knowledge of the distribution of R which we indicate by $p(R)$, the probability function of R. The total expected loss for the discrete case is

$$E[\text{Loss}] = \sum_{\text{all } R} \sum_{i=1}^{2} L(d_i,R)p[d_i|R]p(R) + s(t)$$

A simplified example will serve to illustrate the application. Suppose R can take on only the two values R_1 and R_2 and it is known that $p(R_1) = 0.8$ and $p(R_2) = 0.2$. Also, suppose that the expected costs for given states and decisions are as given in Table 3.4. Since the best decision we could make for state R_1 is d_1 (even if we knew without testing the true reliability), the most we can reduce cost is \$60 − \$50 = \$10. This is normally called the loss, or the cost of making the wrong decision for state R_1. A loss table is obtained by subtracting the lowest cost for each state from the costs of the other decisions for that state. The loss table is given in Table 3.5. If the

Table 3.4 Cost Table

Decision	States	
	R_1	R_2
d_1	\$50	\$160
d_2	\$60	\$ 60

Table 3.5 Loss Table

Decision	States	
	R_1	R_2
d_1	0	\$100
d_2	\$10	0

cost of testing is given as a linear function of the number of items n on test, say $s(t) = (0.1)n$, the total expected loss is

$$E[\text{Loss}] = \sum_{j=1}^{2} \sum_{i=1}^{2} L(d_i, R_j) p[d_i | R_j] p(R_j) + 0.1n$$

which expands to

$$\begin{aligned} E[\text{Loss}] &= (0)p[d_1|R_1](0.8) + (100)p[d_1|R_2](0.2) \\ &\qquad + (10)p[d_2|R_1](0.8) + (0)p[d_2|R_2](0.2) + 0.1n \\ &= 20p[d_1|R_2] + 8p[d_2|R_1] + 0.1n \end{aligned}$$

Suppose that all of the testing procedures from which the selection is to be made have the property that $p[d_2|R_1] = 0.05$ and $p[d_1|R_2]$ is as given in the Table 3.6. The expected loss for each plan is computed and included in Table 3.6. Under the minimum expected loss criterion, the best testing procedure is easily seen to be 3, Table 3.6.

The application of the decision-theory model to this problem depends upon a knowledge of the loss function and the a priori distribution of the reliability. In practice, these functions are not always known. In fact, they are rarely known with the precision indicated by the above example. One may be inclined to discard the entire idea of a solution dependent upon such elusive quantities; however, much can be said for the decision-theory model as a general approach to the problem, particularly since no more satisfactory model is available. The best alternative, apparently, is the direct selection of an OC curve based on experience, intuition, or judgment. In general, the decision-theory model tends to focus attention on the relevant issues in the selection problem and provides a method for formally introducing economic and a priori knowledge into the decision-making process. It is true that it is difficult, if not impossible, to make precise estimates of the cost involved or the a priori distribution, but it would seem to be more desirable to introduce crude estimates of relevant quantities than to make a judgment based on precise estimates of irrelevant quantities.

3.5b Determining the Loss Function. While, in practice, it may be quite difficult to determine the exact economic values to assign the loss function, in principle, there should be no difficulty. In fact, one should be able to obtain reasonable estimates of these costs by applying the principles of engineering economy. The major problem is that of eliminating irrelevant costs and of concentrating on those differences in costs which will result from the selection of two mutually exclusive alternatives. One should consider only those costs which will arise in the future because of the decision to be made as a result of the procedure. This eliminates immediately the consideration of the cost of producing the lot in question as a relevant cost. Taking this view, it is clear that one must have a knowledge of the complete action which will result from either of the decisions in order to make appropriate estimates.

Table 3.6 E**[Loss] for Available Plans**

Procedure number	Sample size	$p[d_1\|R_2]$	E[loss]
1	10	0.20	$5.40
2	15	0.10	3.90
3	20	0.05	3.40*
4	25	0.03	3.50
5	30	0.02	3.80
6	40	0.01	4.60

* Best testing procedure.

Specifically, the costs to be determined are:

1. The cost of a failure of the item once installed in its operational environment. In the following discussion, the cost will be designated by M.
2. The cost of a reject decision.
3. The cost of testing expressed as a function of some index of the amount of testing required for the test plan.

From the definition of reliability, "the probability of successful operation," it can be seen that the consequences of an accept decision involve the possibility of failure and, consequently, the cost expected to arise as a result of such failure. An incremental increase in reliability can then be evaluated as a decrease in the expected cost due to failure. That is, if the cost of a failure is estimated to be M, then the expected cost function for an accept decision would be $(1 - R)M$. The real problem in estimating the cost of an accept decision arises in attempting to estimate the cost of a failure. To do so requires that one know, or be able to estimate, the consequences of a failure and the action which will be taken as a result.

For example, if it is known that the failure of a particular device will result in the destruction of a missile and that the missile will be replaced if it fails, a minimum cost of failure would be the cost of the replacement. This is a minimum cost since it ignores the possible cost associated with the delay in obtaining the replacement missile, the cost of procurement, the intangible costs of failure, etc. In other applications, it may be that failure will only result in unscheduled maintenance and loss of up time on the device. In such cases the increased cost of the unscheduled maintenance or the loss of revenue, if any, would be the appropriate cost to be considered.

In commercial applications, failure costs may be estimated from an examination of the cost of customer service. In all cases of failure, there probably will be undesirable consequences which cannot be reduced to economic terms. This is not unusual in economic studies, and such consequences should be recognized as irreducibles and treated subjectively in the final decision.

In some cases, the determination of the costs of a reject decision may not be so difficult. If one is aware of the precise actions to be taken as a result of a reject decision, one can determine the associated costs. For example, if the lot can be reprocessed or reworked in some way which will assure its satisfactory utilization, the cost of the reprocessing may be the appropriate cost. In the more likely case, however, when rejected lots are to be disposed of and replaced by additional production, the cost of a reject decision can be quite troublesome. The appropriate cost in this case is the cost which will be incurred in fulfilling the commitment which the rejected lot was to have filled. It is the cost of producing and using the replacement lot. Obviously, such a cost depends upon future accept/reject decisions and the reliability of the lot which is eventually accepted for this commitment.

Techniques which can be applied in such cases are available; however, these techniques depend heavily on knowing precisely the prior distribution of R and on the assumption of a stationary process. Since the procedure suggested here assumes only a general knowledge of these factors, the cost of a reject decision must be estimated directly. The average cost of accepted lots depends upon (1) the variable cost of producing the replacement lot, (2) the reliability of the accepted lot, and (3) the testing procedure adopted. The variable cost of production is always an element of the total cost and, consequently, can be used as a lower bound on the cost of rejection. The other two elements depend upon the reliability of the lot produced and tend to zero as the reliability approaches 1. When the reliability is high, the variable cost of production may be a suitable estimate of the cost of a reject decision.

The cost of conducting the test must be estimated. While this is probably the easiest cost to determine, care must be taken to assure that only the appropriate costs are considered. Since the question will normally be to determine how much testing to do, not whether or not to test, only the incremental cost of increasing the amount of testing should be considered. It would be inappropriate to charge the average cost per item tested or average cost per test hour when the average includes fixed costs which would arise regardless of the amount of testing.

3.5c Selection of the a Priori Distribution. The selection of the appropriate a priori distribution is a much more difficult problem and probably the major reason for the limited application of decision theory to problems of this nature. If one has no knowledge of either the cost involved or the a priori distribution of R, then he might just as well make his selection in the normal manner, i.e., application of intuition, judgment, etc. However, there are many cases when one will have some knowledge of the distribution of the parameter in question. In fact, many statistical sampling plans depend upon an estimate of the process average for their effective use. The point to be made is that if one has any knowledge of the underlying a priori distribution, then decision theory provides a means for introducing this knowledge into the selection process. It would seem preferable to make decisions based on whatever knowledge is available, even though it may be skimpy, than to make decisions ignoring this information.

Several sources of information relative to the a priori distribution are likely to be available to the decision maker. The most likely form would be the results of previous reliability tests conducted on the same item, if any. Another possible source of such information is the predicted reliability of a device made up of several components for which the reliability is known. In such cases, it may be reasonable to assume an a priori distribution which masses the reliability in a predicted confidence interval based on a system analysis. Reliability-growth models may in some cases provide information upon which to base an estimate of the a priori distribution. While none of these sources of information are near perfect, they are better than none at all; and the information they provide is relevant to the problem. The better the information, of course, the better the selection of a testing procedure will be.

3.5d The Model. A simplified model of the cost involved is utilized in this selection procedure. It is assumed (1) that the cost of a failure is known, (2) the cost of a reject decision is known and is a linear function of the number of items in the lot, (3) that the cost of testing is a known linear function of some test parameter such as the amount of testing performed, the number of items placed on test, or the number of failures observed, and (4) tested items are returned to the lot or replaced with items having the same reliability as the lot at no additional cost. Let:

M = cost incurred if an accepted item fails in use
c = cost per item of rejecting the lot
s = cost of testing per unit of the appropriate test parameter
N = number of items in the lot
n = significant parameter linearly related to the cost of testing, such as number of items tested, number of hours tested, or number of failures observed during test
R = actual reliability of the lot submitted to test (unknown)

The cost function for this model is shown in Fig. 3.3. From an examination of this function it can readily be seen that the cost would be minimized if one would reject all lots having an actual value of reliability less than R_0 and accept all lots having an

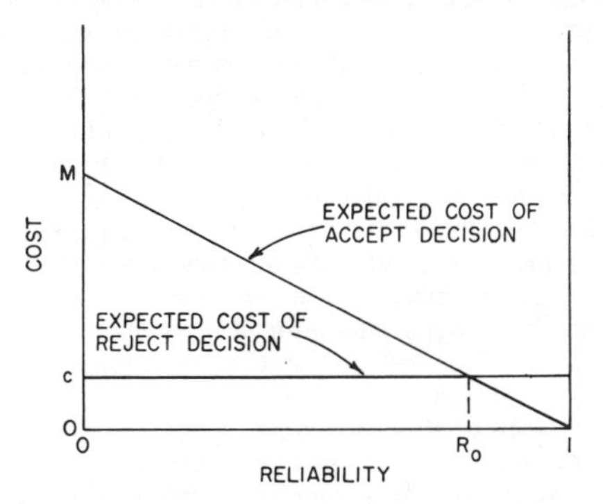

FIG. 3.3 Cost function for model described in Sec. 3.5*d*.

actual reliability greater than R_0. Since the disposition of lots having reliability equal to R_0 has no effect on costs, R_0 is called the break-even reliability. It can easily be shown that the value of R_0 for this model is $R_0 = 1 - c/M$.

The loss function, i.e., the loss expressed as a function of the actual reliability, for this model is

$$\text{Loss} = NM(R - R_0) + ns \qquad \text{for } R \text{ greater than } R_0$$

and

$$\text{Loss} = NM(R_0 - R) + ns \qquad \text{for } R \text{ less than } R_0$$

The loss function expresses the loss which would be suffered if one were to make the wrong decision for any given value of reliability. The loss function can be modulated by the probability of an erroneous decision for each value of R, and the resulting expected loss can be plotted as a function of R. Such a plot, called a risk curve, can be developed for each testing procedure for which an operating-characteristic curve is available. The procedure outlined later utilizes such curves as a guide to test-procedure selection.

In general, reliability test plans can be specified by two factors, i.e., the amount of testing to be performed and a point on the operating-characteristic curve. The two are not independent, but it is generally true that the amount of testing performed determines the differentiating ability (the general slope of the OC curve) of the test plan and the point on the operating-characteristic curve establishes the horizontal position of the curve. In this test-plan-selection procedure, a point on the operating-characteristic curve is arbitrarily established as a starting point. A comparison of the risk curves is made for several test plans having operating curves which pass through this point but which require different amounts of testing. From this comparison one selects a particular test plan which is consistent with his a priori knowledge of actual reliability. An optional feature of this procedure provides for reexamining the arbitrarily selected point on the operating-characteristic curve by plotting several risk curves for test plans which require approximately the same amount of testing.

3.5e Test-plan-selection Procedure. In this section a general procedure is given for selecting a test plan utilizing only a general knowledge of the a priori distribution of the reliability. It is assumed that one has determined the form of the underlying distribution of the time to failure, has estimates of the cost parameters M, c, and s, and has available a set of test plans from which one is to be selected. It is also assumed that the model discussed above is a suitable representation of the actual cost structure involved in the problem. It is emphasized that the procedure outlined here is intended to serve only as a general guide in the selection process. Because of the nature of the problem, a more detailed and specific procedure cannot be given without oversimplification.

The first step in the selection process is to reduce the number of test plans being considered. This is accomplished by selecting a subset of test plans which meet the basic practical limitations of the situation, such as limitations on the amount of time one can devote to testing and the number of items one can place on test, and which have the property that the probability of an accept decision when the actual reliability is R_0 (the breakeven reliability) is 0.50. These plans should be of such form that the plans can be indexed in terms of the test parameter which determines the cost of testing. These last two features warrant comment.

The selection of the 0.50 point on the operating-characteristic curve for R_0 is purely arbitrary. Experience has indicated that for this model one is apt to get a more "even spread" of the risk if this point is selected. This characteristic makes the selection of this point useful as a starting point for test-plan selection. If one's a priori knowledge of the reliability makes it desirable to minimize the risk over selected ranges of the reliability, the selection of other points may be appropriate. This point is considered in the last step of this procedure.

The second point, i.e., indexing the plans according to the parameter related to test cost, may be more of a problem unless one has a group of plans which are indexed in several different ways. Quite often, the operating characteristics of a test plan are determined from a known function of several parameters so that one parameter can be held fixed while the other is allowed to vary. For example, in test plans based on the exponential distribution, the sum of the test time on all items submitted to test is directly related to the operating characteristics of the plan. One can, therefore, fix either the number of items on test or the amount of testing on each item and let the other vary to obtain a series of test plans having different operating characteristics. Quite often, one will find plans indexed in terms of either of these variables with the other variable held fixed. In any case, one must be able to compare the costs of all tests in order to make a rational selection.

For each of the plans selected in step 1, risks must be calculated in order to make the desired comparison. The following steps can be followed to make these calculations.

1. Determine plotting points. First determine the value of $R_0 = 1 - c/M$ and select plotting points so that the density of points is greater near R_0. The risk curve is fast-changing in the vicinity of R_0 but stabilizes at points more distant from R_0. Therefore, it is desirable to have many plotting points in the vicinity of R_0.

2. Determine the loss for these selected plotting points, i.e., the loss as defined in Sec. 3.5*d*.

3. From the operating-characteristic curves, determine the probability of making an erroneous decision (i.e., probability of a reject decision if the reliability is greater than R_0 or the probability of making an accept decision when the reliability is less than R_0) for each of the plans.

4. Determine the total risk for each point. The variable portion of the risk (the product of the values found in steps 2 and 3) is added to the cost of testing for each of the plotting points.

5. Plot selection risk curves. Plot risk curves for selected sets of the plans in order to bracket the plans which are most economical. For example, it may be desirable to plot three curves which represent the greatest, the smallest, and the average test cost in order to "get a feel" for the test cost which is nearly optimal. One would then plot curves for test plans whose cost are in the area of optimality indicated by the first curves. In this way, one can bracket the "best" plan.

One must then make a selection based on joint consideration of the risk curves and one's a priori knowledge of the distribution of the reliability. In making this selection one can eliminate all test plans which are dominated, i.e., have a higher risk for all values of reliability than some other test plan. After all dominated test plans are eliminated, one can select from those remaining the procedure which offers a lower risk over the range of values of reliability which his subjective knowledge of the a priori distribution indicates are most likely to occur. When several test plans are considered approximately equivalent on the basis of this analysis, one would probably select the test plan requiring the lowest testing cost. No set rules can be given for this selection, but the examples to follow serve to illustrate the general approach recommended here.

Finally, if the general shape of the selected risk curve is not of the desired form, one can plot risk curves for several other test plans having the same test costs but different values of the α error. As illustrated in Example 3.3, this shifts the mass of the risk to different values of reliability. One may then select a plan which minimizes the risk over a desired range of reliability.

Example 3.2

The following example is intended to demonstrate the use of the procedure outlined in Sec. 3.5*e*. Suppose one is faced with the problem of selecting a test for determining the economic disposition of a lot of 100 items. It has been decided that an exponential distribution can be utilized as a model for the time-to-failure distribution and that the cost relationships are adequately represented by the cost model discussed in Sec. 3.5*e*. The cost parameter estimates are $M = \$1{,}000$, $c = \$100$, and $s = \$10$. When placed in use, the device is expected to operate for periods of 55 hours without failure. For practical reasons it is desired to reach a decision as to the disposition of the lot in a period not to exceed 120 hours. Facilities are available for placing all of the items on test if necessary, and for practical purposes the cost of testing may be considered a linear function of the number of items placed on test. The cost of restoring failed items to their original condition is insignificant.

The first step one must take is to select a set of test plans which meet the practical considerations in this example. There are in H-108† test plans which are terminated at a preselected time. Since a deadline for completing the tests has been established, it appears to be desirable to utilize this form of test procedure. Table 2C-2(*e*), page 2.49, of H-108 contains 180 test plans of this general form all of which have a producer's risk (α error) of 0.50. Furthermore, these plans can be indexed according to the number of items placed on test, which, in this case, is directly related to the cost of testing. The plans are also indexed according to the ratio of the maximum test time T and the value of the parameter θ_0, the mean time to failure corresponding to the 0.50 point on the operating-characteristic curve.

In order to find a subset of test plans having the property that the probability of acceptance when reliability is R_0 is 0.50, we must determine the value of θ_0 corresponding to R_0 and, subsequently, the value of the index T/θ_0 appropriate for this case. In this model

$$R_0 = 1 - c/M = 1 - 100/1{,}000 = 0.90$$

and the relationship for the exponential distribution

$$R_0 = \exp(-55/\theta_0)$$

† "Sampling Procedures and Tables for Life and Reliability Testing (Based on Exponential Distribution)," *Quality Control and Reliability Handbook* (*Interim*) *H*-108, Office of the Assistant Secretary of Defense (Supply and Logistics), GPO, 1960.

can be utilized to find that the appropriate value of θ_0 is 525 hours. The ratio T/θ_0 is 120/525, or 0.238. One can then enter Table 2C-2(e), H-108, and select those test plans for which the index ratio is approximately 0.238. Twelve such test plans which require sample sizes of 100 or fewer have been taken from this table of H-108 and are listed in Table 3.7. The first column of Table 3.7 gives the code which identifies the operating-characteristic curve associated with the plan in H-108. It is included here for the convenience of those who might like to refer to the source of the information relative to operating characteristics. The selection of this set of plans completes the first step in the selection procedure.

The next step in the procedure is to make risk calculations for the selected test plans and plot curves to aid in making the selection. The results of these calculations are given in Table 3.8. The following is an explanation of the calculations required to obtain the numbers in each column.

Column *a*. Plotting points are so selected as to cluster about $R_0 = 0.90$. The risk curves are fast-changing in the vicinity of R_0, but they become more stable at points more distant from R_0.

Table 3.7 Test Plans Selected for Consideration in Example 3.2

Code	Sample size	Reject number	Ratio T/θ_0
E-2	8	2	0.210
E-3	12	3	0.223
E-4	16	4	0.230
E-5	20	5	0.234
E-6	24	6	0.236
E-7	28	7	0.238
E-8	32	8	0.240
E-9	36	9	0.241
E-10	40	10	0.242
E-11	60	15	0.244
E-12	80	20	0.246
E-13	100	25	0.247

Column *b*. In H-108 the operating-characteristic curves are plotted against the ratio of θ/θ_0 in which $R_0 = \exp(-t/\theta_0)$. Column *b* gives this ratio for the values of reliability selected as plotting points. This ratio is useful for determining the probability of acceptance for the selected values of reliability given in column *d*.

Column *c*. This column gives the loss in dollars for each value of reliability selected as a plotting point. This is the loss which would occur if the wrong decision were to be made for each value of reliability. The loss function is discussed in Sec. 3.5*d*.

Column *d*. For each plan, the probability of an erroneous decision is taken from the operating-characteristic curves and entered in column *d* for the plan. This is the probability of an accept decision if the actual reliability is less than R_0 and the probability of a reject decision if the actual reliability is greater than R_0.

Column *e*. The product of the value in column *c* and the value in column *d* yields the expected loss, or risk, for the plan exclusive of the testing costs. This is entered in column *e* for each test plan being considered.

Column *f*. This column shows the total risk for each value of reliability selected as a plotting point. It is the sum of the testing cost and the value in column *e*.

No obvious advantage of one plan over the other is indicated by a comparison of the total-risk calculations; therefore, the curves in Fig. 3.4 were drawn to obtain an indication of the approximate optimal test plan. The exact analysis of these curves depends, of course, upon one's a priori knowledge of the distribution of the actual reliability. For the purposes of this example, we assume that all values of reliability

Table 3.8 Risk Calculations for Reliability Testing Plans, Example 3.2*

Selected plotting points, reliability	Ratio θ/θ_0	Loss, $	Plan E-3 sample size 12 reject No. 3 testing cost $120			Plan E-7 sample size 28 reject No. 7 testing cost $280			Plan E-11 sample size 60 reject No. 15 testing cost $600			Plan E-5 sample size 20 reject No. 5 testing cost $200			Plan E-9 sample size 36 reject No. 9 testing cost $360		
a	*b*	*c*	*d*	*e*	*f*	*d*	*e*	*f*	*d*	*e*	*f*	*d*	*e*	*f*	*d*	*e*	*f*
1.000	∞	10,000	0.0	0	120	0.0	0	280	0.0	0	600	0.0	0	200	0.0	0	360
0.975	4.20	7,500	0.03	225	345	0.01	75	355	0.0	0	600	0.01	75	275	0.0	0	360
0.950	2.10	5,000	0.14	600	720	0.05	250	530	0.005	25	625	0.08	400	600	0.03	150	510
0.925	1.40	2,500	0.30	750	870	0.20	500	780	0.12	300	925	0.24	600	800	0.17	425	785
0.910	1.10	1,000	0.44	440	560	0.40	400	680	0.36	360	960	0.42	420	620	0.39	390	750
0.900	1.00	0	0.50	0	120	0.50	0	280	0.50	0	600	0.50	0	200	0.50	0	360
0.890	0.91	1,000	0.47	470	590	0.40	400	680	0.34	340	940	0.42	420	620	0.39	390	750
0.875	0.80	2,500	0.35	875	995	0.26	650	930	0.18	450	1,050	0.30	750	950	0.23	577	937
0.850	0.65	5,000	0.28	1,400	1,520	0.12	600	880	0.04	200	800	0.14	700	900	0.08	400	760
0.800	0.50	10,000	0.10	1,000	1,120	0.02	200	480	0.005	50	650	0.04	400	600	0.01	100	460
0.750	0.36	15,000	0.01	150	270	0.0	0	280	0.0	0	600	0.0	0	200	0.0	0	360
0.700	0.30	20,000	0.005	100	220	0.0	0	280	0.0	0	600	0.0	0	200	0.0	0	300
0.600	0.21	30,000	0.0	0	120	0.0	0	280	0.0	0	600	0.0	0	200	0.0	0	360

* Columns *a*, *b*, and *c* are self-explanatory. Columns *d* give the probability of making an erroneous decision when the plan is utilized and reliability is that given in column *a*. Columns *e* give the variable portion of the risk obtained by multiplying columns *c* by columns *d*. Columns *f* give the total risk, i.e., the sum of columns *d* and *e*.

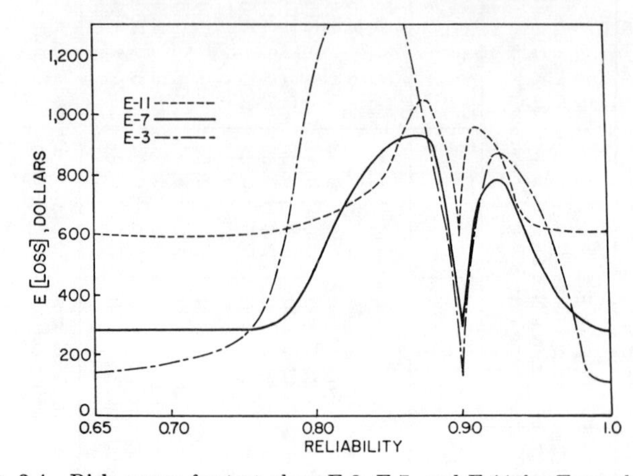

FIG. 3.4 Risk curves for test plans E-3, E-7, and E-11 for Example 3.2.

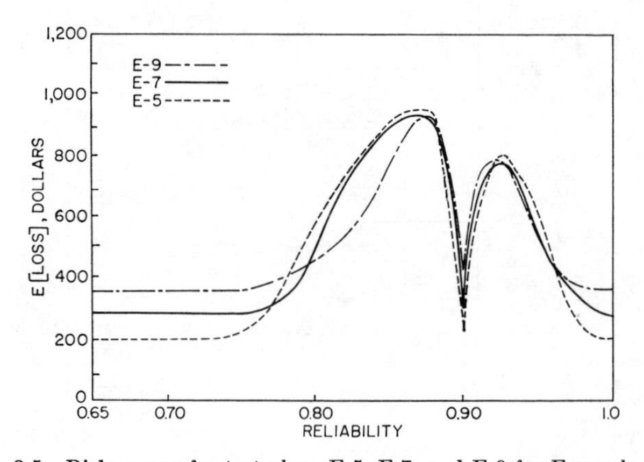

FIG. 3.5 Risk curves for test plans E-5, E-7, and E-9 for Example 3.2.

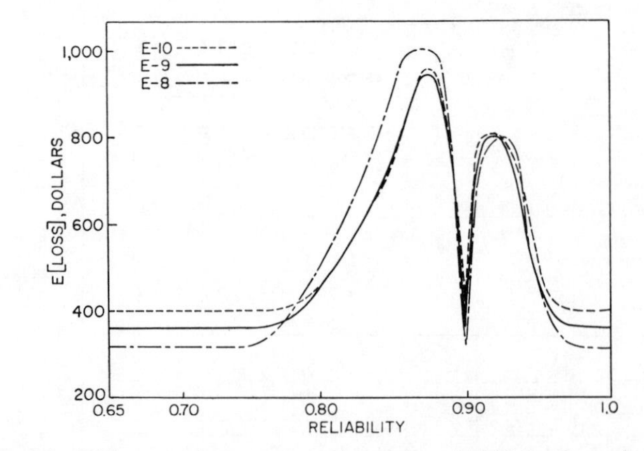

FIG. 3.6 Risk curves for test plans E-8, E-9, and E-10 for Example 3.2.

greater than 0.6 are equally likely and that the probability of an actual reliability less than 0.6 is negligible. On this basis, one would probably select a plan with sample size around that of E-7. Figure 3.5 shows risk curves for plans E-5, E-7, and E-9 which indicate that E-9 might be near the optimal test plan. To further refine the selection, Fig. 3.6 was drawn; it shows risk curves for test plans E-8, E-9, and E-10. An examination of these curves shows that E-8 is inferior to both E-9 and E-10. No clear advantage is indicated when E-9 and E-10 are compared. However, one would feel confident that the selection of test plan E-9 is reasonable and certainly "in the ballpark" of optimal test plans. If one had knowledge about the actual reliability, which, of course, differs from that assumed above, a different conclusion may have been drawn. For example, if one believed that the actual reliability was almost certain to lie in the interval between 0.9 and 1.0, he would probably desire to minimize the risk in that range. He could do so by changing the value of the α error. This point is illustrated in Example 3.3.

Example 3.3

In this example, it is assumed that the cost of a failure of the device in use is $1,000 and the cost of rejecting an item is $100. The lot under consideration contains 100 items, and testing facilities are available so that all can be placed on test if desired.

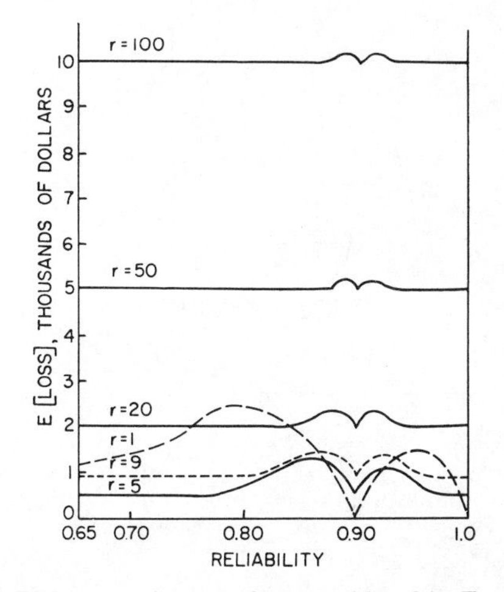

FIG. 3.7 Risk curves for test plans considered in Example 3.3.

The only significant cost associated with testing occurs when an item on test fails. This cost is $100 per failure; it includes the cost of restoring (or replacing) the failed item to its original state. Upon completion of the test, all items are returned to the lot and disposed of in the same manner as the rest of the lot. An exponential distribution of the time to failure is assumed. The principal difference between this and the preceding example is the fact that the cost of testing is related to a different test parameter. Also, in this example the last step of the selection procedure, i.e., checking for different values for α, is illustrated.

Table 2B-1 of H-108 contains 90 test plans which are terminated on the occurrence of a preselected number of failures. Since the cost of testing is related to the number of failures, this form of the test procedure will be utilized. The process of making up the risk curves is identical with that utilized in Example 3.2 and is not illustrated here.

A system analysis of the device indicates that the reliability is greater than 0.85 with a very high probability. It is assumed, therefore, a priori, that the distribution of the reliability is massed in the interval between 0.85 and 1.0.

Figure 3.7 shows the risk for several of the procedures available, all of which have the property that $\alpha = 0.50$. The problem is that of selecting the r value (and the corresponding differentiating ability and cost) which will minimize the risk in the interval of interest. An examination of the curves shows immediately that all procedures for which r is greater than or equal to 9 can be eliminated from consideration, since the risk is less for all values of R for the procedure having $r = 5$ than for any of these procedures. One might say that these procedures are dominated by the procedure having $r = 5$. The distinction between the procedure having $r = 1$ and that having $r = 5$ is not so clear. One may be inclined to select the procedure having $r = 5$, since it dominates the procedure having $r = 1$ over most of the range of interest. From these curves one could safely conclude that the best procedure for this application would have an r value of the order of 5. Figure 3.8 shows the risk for $r = 3$ and $r = 7$ along with that for $r = 5$ on a blown-up scale. No clear choice of the best procedure is indicated by these curves, and one would consider the irreducibles in the problem to make the final selection. It should be emphasized that this procedure

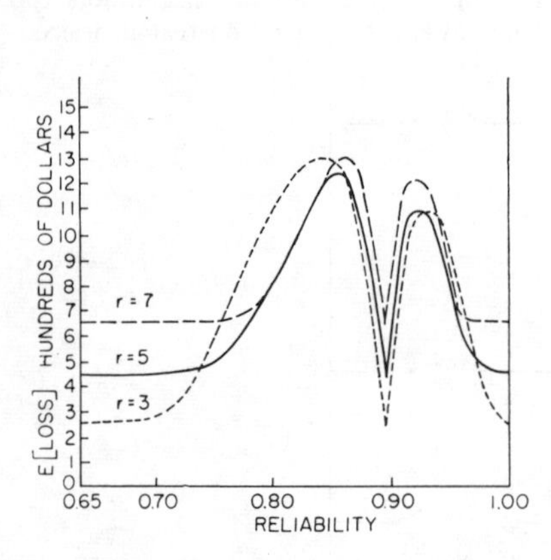

FIG. 3.8 Risk curves for test plans considered in Example 3.3 with blown-up scale.

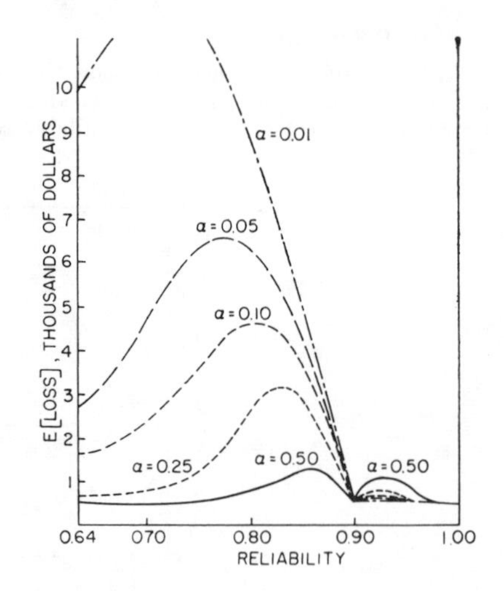

FIG. 3.9 Risk curves for test plans having the same testing cost but different producers' risk, Example 3.3.

has reduced the field of possible procedures to a rather small set and has greatly reduced the likelihood of making a gross error in the selection.

Upon completion of this portion of the analysis, it may be desirable to investigate whether some other value of the producer's risk (α error) might not be more appropriate. In order to make such a comparison, the curves in Fig. 3.9 were drawn for test plans having the same termination number, i.e., five failures, but different values for α. Since the a priori knowledge of the distribution of reliability indicates that the reliability is very likely to be in the interval from 0.80 to 1.00, one would probably be satisfied with the plan for which $\alpha = 0.50$. If, however, one were quite sure that the reliability were greater than 0.90, he would probably desire a test plan for which the $\alpha = 0.01$. (If one is quite sure that the reliability is greater than the break-even point, the necessity for testing is questionable.)

In general, the principal advantage of utilizing a selection procedure such as this is that the operating characteristics are presented in economic terms and should be more meaningful to the person making the selection than normal OC curves would be. If one has some general knowledge of the distribution of R, he is able to utilize this information in selecting a testing procedure. While the procedure will not always clearly indicate the one optimal procedure, it will indicate a subset of procedures containing the optimal procedure, which greatly reduces the selection problem and practically eliminates the likelihood of gross errors in making a selection. It is evident that a considerable amount of time and effort is required to utilize this procedure. However, when the costs are quite high, as they usually are in reliability testing, the effort may be more than justified.

3.6 SENSITIVITY ANALYSIS

Quite often one is unable to determine with an acceptable degree of precision the factors necessary for the analysis suggested above. For example, he may find that the cost involved can only be estimated as a large range of values, rather than a single number, or that the data for determining the form of the underlying distribution are not available. In such cases, the technique of sensitivity analysis can be utilized to pinpoint the factors to which the selection process is sensitive or to determine the range of values for the particular factor which will result in the selection of essentially the same test plan. In many instances, this form of analysis is more practical than expending a great effort to find precise estimates of the costs or parameter values for distributions. This section is devoted to a brief discussion of sensitivity analysis.

When one is not confident that the cost estimates utilized in his selection procedure are reasonably accurate, he can determine the sensitivity of the selection procedure to the particular cost in question. Suppose, for example, that one had made an estimate of $100 for the unit cost of a reject decision in Example 3.2. Actually, however, he would be just as confident of any estimate between $50 and $200. To apply sensitivity analysis to this case, he would repeat the entire selection process twice, one time using the $50 estimate and the second time using the $200 estimate. If the direct costs of the test plan (i.e., testing cost) selected for the three estimates were approximately the same, he could conclude that the selection procedure was relatively insensitive to this cost estimate. Furthermore, he could confidently utilize the test plan selected even though he had little confidence in the accuracy of the reject cost estimate upon which the selection was based.

If one were unsure of the cost estimate and were also unable to pin down a range of values which he could confidently use as an interval estimate, he could apply sensitivity analysis to determine the range of values for which testing plans having essentially the same cost would be selected. Suppose, for example, as in the preceding case, he was unsure of the validity of the $100 estimate of the unit reject cost utilized in the selection process and was unable to make an interval estimate of that cost. A reasonable approach to this problem is to go through the selection process and utilize a number of values for this estimate in order to determine the range of values for which the test-plan cost is essentially constant. If this range is large, he can confidently utilize the test plan even though he is unsure of one of the cost estimates utilized in the selection process.

Sensitivity analysis can also be applied when one is doubtful about the underlying distribution of the time to failure assumed. This analysis can be applied regardless of whether one is making a selection directly by the use of operating-characteristic curves or on an economic basis as suggested in Sec. 3.5. Generally, the approach is, as indicated in the preceding examples, a matter of determining the sensitivity of the selection process to different assumptions relative to the distribution of time to failure.

In order to obtain useful results, it is desirable to express the different distributions in terms of some continuous parameter. This can easily be done if one is willing to accept the Weibull distribution with zero position parameter as a general model. By varying the shape parameter, different distributions can be introduced. Since so many failure patterns can be fitted by such a Weibull distribution (including the exponential distribution), its use as a general model is reasonable. For a discussion of the Weibull distribution, see Sec. 2.

If one is utilizing operating-characteristic curves or the producer-consumer risk approach as a basis for selecting a test plan, the analysis is simply a matter of determining how the selection would differ if different values of the shape parameter were selected. The analysis should be made on the basis of the actual reliability, however, rather than the scale parameter alone. Test plans for which the scale parameter is relatively insensitive to changes in the assumed distribution may be extremely sensitive when comparisons of the effect on the actual reliability are made. Or the reverse might be true.

One might, for example, find that the scale parameter of the Weibull distribution is very sensitive to changes in the scale parameter around a value of 1. However, for certain values of the operating-time requirement, it may be that the computed values of the reliability based on these scale parameters are actually insensitive to such changes in the scale parameter. Since the variable of interest is always the reliability, not one of the parameters, it makes sense to make sensitivity tests based on the reliability. In order to make such a comparison, of course, it is necessary to know the time period (required failure-free operating time) for which the actual reliability is to be determined.

As one would expect, sensitivity analysis when one is selecting a test plan utilizing risk curves is merely a matter of repeating the selection process for several values of the shape parameter. From these repeated selections, one can determine how the cost of testing varies with the selected parameter values. If the testing cost is insensitive to changes in the value of the shape parameter, one can afford to utilize a crude estimate in the test-plan-selection process. If, however, testing costs are sensitive to the parameter value, he should expend more effort to obtain reasonable estimates of the shape parameter, or resort to the more expensive nonparametric testing.

Section 4

APPLICATION OF MATHEMATICS AND STATISTICS TO RELIABILITY AND LIFE STUDIES

O. B. MOAN

PROFESSOR OF ENGINEERING, ARIZONA STATE UNIVERSITY
TEMPE, ARIZONA

Dr. Moan has served on the faculties as lecturer or professor at Purdue University, University of California, University of Southern California, and now at Arizona State University. He has spent about twelve years in various positions as consulting scientist, reliability manager, quality control staff engineer, and technical staff of Lockheed Missile and Space Company, Hughes Aircraft, International Business Machines, and Julius Hyman and Company. He has served as consultant to Litton Systems, Sprague Electric Company, and Motorola Semiconductor Division.

Dr. Moan received his Ph.D. degree from Purdue University, M.S. degree from the University of Minnesota, and B.S. degree from Purdue. He is a member of Sigma Xi, Tau Beta Pi, American Society for Engineering Education, Institute of Mathematical Statistics, Society of Industrial and Applied Mathematics, American Association for the Advancement of Science, and Mathematical Association of America. He is a Fellow of ASQC.

He has written a large number of technical reports and several technical articles for professional journals.

APPLICATION OF MATHEMATICS AND STATISTICS TO RELIABILITY AND LIFE STUDIES

O. B. Moan

CONTENTS

4.1 BASIC SET THEORY

Definition. A *set* is a collection of objects each of which possesses a certain property which can be used to determine its membership in the set.

Definition. The individual objects of the set are called the *elements* of the set.

Notation. $a \in A$ means that a is an element of the set. $a \notin A$ means a is not an element of the set.

Example 4.1. $A = \{1,2,3,4,5,6\}$ means that the set A consists of the integers 1, 2, 3, 4, 5, and 6. Thus, $2 \in A$, but $8 \notin A$.

Example 4.2. $A = \{0 \leq x \leq 1\}$ means that the set A consists of the points for which $0 \leq x \leq 1$. Thus, $\frac{3}{4} \in A$, but $\frac{3}{2} \notin A$.

Definition. A set is called a *finite* set when it contains a finite number of elements (Example 4.1) and an *infinite* set otherwise (Example 4.2).

Definition. The *null set* ϕ is the set which contains no elements.

Definition. The *total* or *universal set* Ω is the set which contains all the elements under consideration.

Definition. When every element of the set A_1 is contained in the set A_2, the set A_1 is a *subset* of the set A_2 and written $A_1 \subset A_2$.

Note. The null set ϕ is a subset of every set. If $A_1 \subset A_2 \subset A_3$, then $A_1 \subset A_3$.

Example 4.3. Let $A_1 = \{1,2,3,4\}$ and $A_2 = \{1,2,3,4,5,6\}$. Then $A_1 \subset A_2$, since every element in A_1 is also in A_2.

Definition. Two sets are *equal* when both sets have exactly the same elements. If $A_1 \subset A_2$ and $A_2 \subset A_1$, then $A_1 = A_2$.

Definition. The *union* of the two sets A_1 and A_2, denoted by $A_1 \cup A_2$, is the set of all elements of either set, that is, $x \in (A_1 \cup A_2)$ means $x \in A_1$ or $x \in A_2$ or both.

Definition. The *intersection* of the two sets A_1 and A_2, denoted by $A_1 \cap A_2$, is the set of all elements common to both A_1 and A_2, that is, $x \in (A_1 \cap A_2)$ means $x \in A_1$ and $x \in A_2$.

Note. If $A_1 \subset A_2$ and $A_3 \subset A_2$, then $(A_1 \cup A_3) \subset A_2$ and $(A_1 \cap A_3) \subset A_2$.

Example 4.4. Let $A_1 = \{1,2,3,4\}$ and $A_2 = \{3,4,5,6\}$. Then

$$A_1 \cup A_2 = \{1,2,3,4,5,6\}$$

and $A_1 \cap A_2 = \{3,4\}$.

Definition. If $A_1 \subset A_2$, then the difference $A_2 - A_1$ is the set of all elements of A_2 not contained in A_1; that is, $x \in (A_2 - A_1)$ means $x \in A_2$ and $x \notin A_1$, is called the *complement* of A_1, and is denoted by $\bar{A}_1$.

Note. $A_1 - A_1 = \phi$ and $A_1 - \phi = A_1$.

The following are the operation rules associated with the union and intersection. Additional laws or relations of sets are illustrated in Example 4.5.

Associative law:

$$A_1 \cup (A_2 \cup A_3) = (A_1 \cup A_2) \cup A_3$$
$$A_1 \cap (A_2 \cap A_3) = (A_1 \cap A_2) \cap A_3$$

Commutative law:

$$A_1 \cup A_2 = A_2 \cup A_1$$
$$A_1 \cap A_2 = A_2 \cap A_1$$

Distributive law:

$$A_1 \cap (A_2 \cup A_3) = (A_1 \cap A_2) \cup (A_1 \cap A_3)$$
$$A_1 \cup (A_2 \cap A_3) = (A_1 \cup A_2) \cap (A_1 \cup A_3)$$

Example 4.5. Let $\Omega = \{a,b,c,d,e\}$, $A_1 = \{a,b\}$, $A_2 = \{a,c,d\}$.

Identity laws:

$A_1 \cup \Omega = \Omega$	$A_1 \cup \Omega = \{a,b,c,d,e\} = \Omega$
$A_1 \cup \phi = A_1$	$A_1 \cup \phi = \{a,b\} = A_1$
$A_1 \cap \Omega = A_1$	$A_1 \cap \Omega = \{a,b\} = A_1$
$A_1 \cap \phi = \phi$	$A_1 \cap \phi =$ no elements $= \phi$

Idempotent laws:

$A_1 \cup A_1 = A_1$	$A_1 \cup A_1 = \{a,b\} = A_1$
$A_1 \cap A_1 = A_1$	$A_1 \cap A_1 = \{a,b\} = A_1$

Complement laws:

$\bar{\bar{A}}_1 = A_1$	$\bar{A}_1 = \{c,d,e\}$, thus $\bar{\bar{A}}_1 = \{a,b\} = A_1$
$A_1 \cup \bar{A}_1 = \Omega$	$A_1 \cup \bar{A}_1 = \{a,b,c,d,e\} = \Omega$
$A_1 \cap \bar{A}_1 = \phi$	$A_1 \cap \bar{A}_1 =$ no elements $= \phi$

Dualization laws:

$\overline{A_1 \cap A_2} = \bar{A}_1 \cup \bar{A}_2$	$\bar{A}_1 = \{c,d,e\}$, $\bar{A}_2 = \{b,e\}$
	$A_1 \cap A_2 = \{a\}$, $\overline{A_1 \cap A_2} = \{b,c,d,e\}$
	$\bar{A}_1 \cup \bar{A}_2 = \{b,c,d,e\} = \overline{A_1 \cap A_2}$
$\overline{A_1 \cup A_2} = \bar{A}_1 \cap \bar{A}_2$	$A_1 \cup A_2 = \{a,b,c,d\}$, $\overline{A_1 \cup A_2} = \{e\}$
	$\bar{A}_1 \cap \bar{A}_2 = \{e\} = \overline{A_1 \cup A_2}$

The *Venn diagram* (Euler diagram) is a useful device to illustrate the relations in the algebra of sets, see Fig. 4.1.

Example 4.6. Let A_i represent the proper functioning of an electronic part and $\bar{A}_i$ the failure of the part; then representation of circuits using algebra of sets is

A_1 — A_2: $A_1 \cap A_2$, BOTH MUST FUNCTION

A_1 ∥ A_2: $A_1 \cup A_2$, AT LEAST ONE MUST FUNCTION

(A_1 — A_2) ∥ A_3, — A_4: $[(A_1 \cap A_2) \cup A_3] \cap A_4$

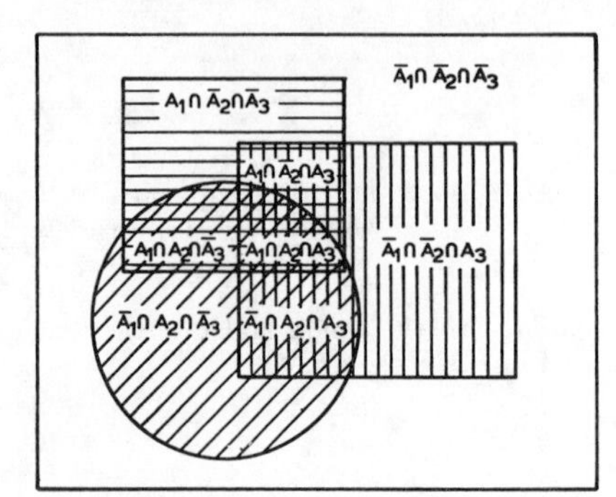

FIG. 4.1 Venn diagram. The universal set Ω is represented by the large rectangle, and the sets A_1, A_2, and A_3 are represented by the areas shaded with horizontal lines, slanting lines, and vertical lines, respectively.

4.2 PROBABILITY

Probability is the measure of the likelihood of occurrence of a chance event. The natural relationships between set theory and probability theory are as follows: The sample space is the universal set; elements are sample points; and the events are the subsets of the sample space.

Definition. A sample space is *discrete* if it contains only a finite or countably infinite number of sample points.

Example 4.7. Finite discrete sample space. Three missiles are fired, the possible outcomes are *SSS, SSF, SFS, FSS, SFF, FSF, FFS,* and *FFF*, where S denotes success and F a failure.

Example 4.8. Discrete sample space with countably infinite number of sample points. A resistor is selected from an infinitely large quantity of resistors until a defective resistor is found. The sample points are *D, GD, GGD, GGGD,* . . . , where D denotes a defective resistor and G a good resistor.

Definition. A sample space is called a *continuous* sample space when it contains a nondenumerable infinity of sample points.

Example 4.9. Continuous sample space. The set of all positive real numbers which are the possible outcomes of an experiment in life testing, that is,

$$S = \{\text{real numbers } x : 0 < x < \infty\}$$

Definition (Classical). If an event can occur in N mutually exclusive and equally likely ways n of which possess the characteristic A, then the probability of A occurring is n/N; symbolically, $P(A) = n/N$.

Example 4.10. Consider a lot of 100 transistors 5 of which are defective. There are 100 ways that a transistor may be selected, and of these 5 possess the characteristic, defective transistor. Thus, the probability of selecting a defective transistor is $5/100 = 0.05$.

Definition (Relative Frequency). If in a series of n trials, during which the event A occurs f_A times, the value f_A/n approaches a limit P as n increases, then P is the probability of the event A.

Definition (Mathematical). Given a sample space S consisting of the events E_i. *Probability* is a function of nonnegative real numbers assigned to each event E_i denoted by $P(E_i)$, the probability of realizing the event E_i, with the properties

1. $P(E_i) \geq 0$ for every event $E_i \subset S$.

2. $P\left(\sum_{\text{all } E_i \subset S} E_i\right) = 1$ for the certain event. *Note.* $\sum_{\text{all } E_i \subset S} E_i$ means the sum of all $E_i \subset S$.

3. $P[E_i \cup E_j] = P(E_i) + P(E_j)$, if $E_i \cap E_j = \phi$. (The probability of the union of two mutually exclusive events is the sum of their respective probabilities.)

Theorem. $P(\phi) = 0$, the probability of the impossible event is 0.

Theorem. $P(E) + P(\bar{E}) = 1$, the probability of realizing the event plus the probability of not realizing the event equals 1.

Theorem. For two arbitrary events E_1 and E_2

$$P(E_1 \cup E_2) = P(E_1) + P(E_2) - P(E_1 \cap E_2)$$

is the probability of realizing at least one of the two events E_1 and E_2.

Note. For n events

$$P(E_1 \cup E_2 \cup \cdots \cup E_n) = \sum_{i=1}^{n} P(E_i) - \sum_{i=1}^{n-1} \sum_{j=i+1}^{n} P(E_i \cap E_j) + \sum_{i=1}^{n-2} \sum_{j=i+1}^{n-1} \sum_{k=j+1}^{n} P(E_i \cap E_j \cap E_k) + \cdots + (-1)^{n+1} P(E_1 \cap E_2 \cap \cdots \cap E_n) \quad (4.1)$$

Example 4.11. Two dice are thrown. What is the probability of at least one 6? Let E_1 be the event of a 6 on die 1, and E_2 a 6 on die 2. Then from Eq. (4.1) we have

$$P(E_1 \cup E_2) = P(E_1) + P(E_2) - P(E_1 \cap E_2) = 1/6 + 1/6 - 1/36 = 11/36$$

Example 4.12. Suppose a failure of a component could be caused by the failure of a transistor or a resistor or both. Let $P(E_T)$ = probability of a transistor failure and $P(E_R)$ = probability of a resistor failure, then

$$P \text{ (failure due to transistor failure only)} = P(E_T) - P(E_T \cap E_R)$$
$$P \text{ (failure due to both transistor and resistor failure)} = P(E_T \cap E_R)$$

Theorem. For two mutually exclusive events E_1 and E_2

$$P(E_1 \cup E_2) = P(E_1) + P(E_2)$$

Note. For n mutually exclusive events the probability that at least one of the events will occur is

$$P(E_1 \cup E_2 \cup \cdots \cup E_n) = \sum_{i=1}^{n} P(E_i) \quad (4.2)$$

Example 4.13. A single die is thrown. What is the probability of throwing a 4 or 6? Let $P(E_1)$ = probability of throwing a 4 and $P(E_2)$ = probability of throwing a 6; then

$$P(E_1 \cup E_2) = P(E_1) + P(E_2) = 1/6 + 1/6 = 1/3$$

Definition. Let E_1 be an event in the sample space S such that $P(E_1) \neq 0$. Let E_2 be any event in the same sample space. The *conditional probability* that the event E_2 occurs, knowing that the event E_1 has occurred, is defined as

$$P(E_2|E_1) = P(E_1 \cap E_2)/P(E_1) \quad (4.3)$$

Note. By symmetry, $P(E_1|E_2) = P(E_1 \cap E_2)/P(E_2)$.

Example 4.14. Consider an assembly consisting of two subassemblies A_1 and A_2. Estimated probabilities based on test results are

$$\begin{aligned} P(A_1 \text{ failing}) &= 0.05 \\ P(A_2 \text{ failing}) &= 0.10 \\ P(A_1 \cap A_2 \text{ failing}) &= 0.02 \end{aligned}$$

Using Eq. (4.3), the conditional probabilities are

$$\begin{aligned} P(A_1 \text{ fails}|A_2 \text{ failed}) &= 0.02/0.10 = 0.20 \\ P(A_2 \text{ fails}|A_1 \text{ failed}) &= 0.02/0.05 = 0.40 \end{aligned}$$

Theorem. For two arbitrary events E_1 and E_2

$$P(E_1 \cap E_2) = P(E_2|E_1)P(E_1) = P(E_1|E_2)P(E_2)$$

Note. For n events $E_1, E_2, \ldots, E_n$ for which $P(E_1 \cap E_2 \cap \cdots \cap E_n) > 0$,

$$P(E_1 \cap E_2 \cap \cdots \cap E_n) = P(E_1)P(E_2|E_1)P(E_3|E_1 \cap E_2) \cdots P(E_n|E_1 \cap E_2 \cap \cdots \cap E_{n-1}) \tag{4.4}$$

Example 4.15. Consider a lot of 10 relays 2 of which are defective. The probability that a sample of 2 relays will not contain a defective relay is equal to the probability that first relay selected is good times the conditional probability that the second relay is good, given that the first relay was good:

$$P = (8/10)(7/9) = 28/45$$

Example 4.16. Relays are checked by selecting a relay at random and testing it. Using the data of Example 4.15, what is the probability that the second defective relay will be found at the third test? Let B denote a defective relay, G denote a good relay, and the subscripts 1, 2, and 3 denote tests 1, 2, and 3, respectively. Using Eqs. (4.2) and (4.4),

$$\begin{aligned} &P\text{ (second defective relay will be found on the third test)} \\ &\quad = P(G_1)P(B_2|G_1)P(B_3|B_2 \cap G_1) + P(B_1)P(G_2|B_1)P(B_3|B_1 \cap G_2) \\ &\quad = (8/10)(2/9)(1/8) + (2/10)(8/9)(1/8) = 2/45 \end{aligned}$$

Definition. Two events E_1 and E_2 are *statistically independent* if $P(E_1|E_2) = P(E_1)$ and $P(E_2|E_1) = P(E_2)$.

Theorem. Two events E_1 and E_2 are statistically independent if and only if $P(E_1 \cap E_2) = P(E_1)P(E_2)$.

Note. Three events E_1, E_2, and E_3 are mutually independent if the three events are pairwise independent, i.e.,

$$\begin{aligned} P(E_1 \cap E_2) &= P(E_1)P(E_2) \\ P(E_1 \cap E_3) &= P(E_1)P(E_3) \\ P(E_2 \cap E_3) &= P(E_2)P(E_3) \\ P(E_1 \cap E_2 \cap E_3) &= P(E_1)P(E_2)P(E_3) \end{aligned}$$

Theorem. The probability that n statistically independent events $E_1, E_2, \ldots, E_n$ will occur is equal to the product of the probabilities of the independent events, i.e.,

$$P(E_1 \cap E_2 \cap E_3 \cap \cdots \cap E_n) = P(E_1)P(E_2)P(E_3) \cdots P(E_n) \tag{4.5}$$

Note. To assure that n events are statistically independent requires

$$\begin{aligned} P(E_i \cap E_j) &= P(E_i)P(E_j) \\ P(E_i \cap E_j \cap E_k) &= P(E_i)P(E_j)P(E_k) \\ P(E_1 \cap E_2 \cap \cdots \cap E_n) &= P(E_1)P(E_2) \cdots P(E_n) \end{aligned}$$

for all combinations $1 \leq i < j < k \cdots \leq n$.

Example 4.17. The above theorem is commonly called the "product rule," and Eq. (4.5) is one of the widely used equations in reliability. Consider a system of three subsystems where probabilities of successfully completing a mission of length t are given by

$$P(\text{No. 1}) = e^{-\lambda_1 t_1} \qquad P(\text{No. 2}) = e^{-\lambda_2 t_1} \qquad P(\text{No. 3}) = e^{-\lambda_3 t_1}$$

Then the probability that the system successfully completes the mission is

$$P(\text{success}) = P(\text{No. 1})P(\text{No. 2})P(\text{No. 3}) = e^{-(\lambda_1+\lambda_2+\lambda_3)t}$$

assuming independence and that the three subsystems must function for the total mission.

Theorem (Bayes). Let $F_1, F_2, \ldots, F_n$ be mutually exclusive events whose union is the total sample space. Let E be an arbitrary event in the same sample space such that $P(E) \neq 0$. Then for every pair of events F_i and E

$$P(F_i|E) = \frac{P(F_i \cap E)}{P(E)} = \frac{P(F_i)P(E|F_i)}{\sum_{j=1}^{n} P(F_j)P(E|F_j)} \tag{4.6}$$

Example 4.18. Bertrand's paradox. Three boxes contain two coins each. Box 1 contains two gold coins; box 2, one gold and one silver coin; and box 3, two silver coins. A box is selected at random, and then a coin is selected at random from the box. The coin turns out to be gold. What is the probability that the other coin in the box is gold? Using Eq. (4.6),

$$P(\text{box 1}|\text{gold coin}) = \frac{P(\text{box 1})P(\text{gold}|\text{box 1})}{\sum_{i=1}^{3} P(\text{box } i)P(\text{gold}|\text{box } i)}$$

$$= \frac{(\tfrac{1}{3})1}{(\tfrac{1}{3})1 + (\tfrac{1}{3})(\tfrac{1}{2}) + (\tfrac{1}{3})(0)} = \tfrac{2}{3}$$

Definition. If an event E_1 can occur in n_1 ways and the event E_2 in n_2 ways which are independent of n_1, then both E_1 and E_2 can occur in $n_1 n_2$ ways.

Example 4.19. Consider the toss of two dice. Die 1 can fall in 6 different ways and die 2 can fall in 6 independent ways. The total number of ways the two dice may fall is $6 \times 6 = 36$.

Definition. If the event E_1 can occur in n_1 ways and the event E_2 can occur in n_2 ways which are mutually exclusive of n_1, then either E_1 or E_2 can occur in $n_1 + n_2$ ways.

Example 4.20. Consider the toss of one die. Either a 4 or 6 can occur in $1 + 1$ ways.

Definition. A k *permutation* of n things is an ordered selection of k of them.

Definition. The number of *permutations* of n things taken k at a time is

$$P(n,k) = n(n-1) \cdots (n-k+1) = n!/(n-k)! \tag{4.7}$$

Note. $n!$, read n factorial, means $1 \cdot 2 \cdot 3 \cdot 4 \cdots (n-1)n = \prod_{i=1}^{n} i$, where Π is the symbol for the product. $0! = 1$.

Stirlings Formula.[1] $n! \simeq \sqrt{2\pi}\, n^{n+1/2} e^{-n}$, where $\simeq$ indicates that the limit of the ratio of the two sides approaches 1 as n approaches infinity.

Definition. A k *combination* of n things is a selection of k of them without regard to order.

[1] For further discussion of factorials see W. Feller, *An Introduction to Probability Theory and Its Application*, vol. 1, 2d ed., p. 50, John Wiley & Sons, Inc., New York, 1951.

Definition. The number of *combinations* of n things taken k at a time is

$$C(n,k) = n!/(n-k)!k! \qquad \text{binomial coefficient} \tag{4.8}$$

Example 4.21. Consider the four letters a, b, c, d. The number of permutations of these four letters taken two at a time is $P(4,2) = 4 \times 3 = 12$. The permutations are *ab*, *ba*, *ac*, *ca*, *ad*, *da*, *bc*, *cb*, *bd*, *db*, *cd*, and *dc*. The number of combinations of these four letters taken two at a time is $C(4,2) = 4!/2!2! = 6$. The combinations are *ab*, *ac*, *ad*, *bc*, *bd*, and *cd*.

Note. *ab* and *ba* are two permutations but only one combination.

Table 4.1 Some Properties of the Binomial Coefficients

(a) If n and k are positive integers,

$$C(n,k) = P(n,k)/k!$$
$$C(n,k) = 0 \qquad \text{for } k < 0 \text{ or } k > n$$
$$C(n,k) = C(n, n-k)$$
$$C(n,n) = 1 \qquad C(n, n-1) = C(n,1) = n$$
$$\sum_{i=0}^{n} C(n,i) = 2^n \qquad \sum_{i=0}^{n} (-1)^i C(n,i) = 0$$
$$C(n, k-1) + C(n,k) = C(n+1, k)$$
$$\sum_{i=0}^{n} [C(n,i)]^2 = C(2n,n)$$

(b) If N, n, k are positive integers, such that $N \geq n$, $n \geq k$,

$$\sum_{i=0}^{k} C(n,i)C(k,i) = C(n+k, k)$$
$$\sum_{i=0}^{n-k} C(n, i+k)C(N,i) = C(N+n, n-k)$$

Definition. The permutations of n things taken n at a time when $n_1, n_2, \ldots, n_k$ are alike, where $\sum_{i=1}^{k} n_i = n$, is

$$P(n_1 \text{ alike}, n_2 \text{ alike}, \ldots, n_k \text{ alike}) = n! \Big/ \prod_{i=1}^{k} (n_i)! \tag{4.9}$$

4.3 PROBABILITY DISTRIBUTIONS

Definition. A *random variable* is a real-valued function defined over a sample space.

Note. The value of a random variable is a random phenomenon.

Definition. The *cumulative distribution function* (c.d.f.) of the random variable X, denoted by $F(x)$, is defined as

$$F(x) \equiv P(X \leq x)$$

Definition. X is a *discrete random variable* and is said to have a distribution of the discrete type when the probability density function (p.d.f.) of the random variable

X is $P(X = x)$, that is,

$$f(x) = P(X = x)$$

Note. Values of the random variable which are impossible have probability zero.

Example 4.22. Consider a lot of N fuses D of which are defective. If we select a random sample of n fuses, what is the probability that exactly x of the fuses will be defective?

$$f(x) = C(D,x)C(N - D, n - x)/C(N,n) \qquad x = 0, 1, 2, \ldots, n; \quad 0 \leq x \leq D; \quad 0 \leq n - x \leq N - D$$

$$= 0 \qquad \text{otherwise}$$

Definition. If X is a *continuous random variable*, the *continuous probability density function* (p.d.f.) is given by

$$f(x) = dF(x)/dx$$

Table 4.2 The Cumulative Distribution Function

For a discrete random variable X

$$F(x) = \sum_{\text{all } X \leq x} f(X) \qquad \text{a step function}$$

For a continuous random variable X

$$F(x) = \int_{-\infty}^{x} f(x)\, dx$$

Properties:

1. $F(-\infty) = \lim_{x \to -\infty} F(x) = 0$
2. $F(+\infty) = \lim_{x \to +\infty} F(x) = 1$
3. $F(x_2) \leq F(x_1) \qquad \text{if } x_2 \leq x_1$

Table 4.3 The Probability Density Function

For a discrete random variable X

$f(x) = P(X = x)$

For a continuous random variable X

$f(x) = dF(x)/dx$

Properties:

1. $f(x) \geq 0$
2. $\sum_{\text{all } x} f(x) = 1 \qquad$ discrete random variable X

   $\int_{-\infty}^{\infty} f(x)\, dx = 1 \qquad$ continuous random variable X

Note. For discrete random variable X

$$P(x_1 < X \leq x_2) = F(x_2) - F(x_1) = \sum_{x_1 < x \leq x_2} f(x)$$

For continuous random variable X

$$P(x_1 \leq X \leq x_2) = F(x_2) - F(x_1) = \int_{x_1}^{x_2} f(x)\, dx$$

Example 4.23. Let the random variable T represent the life, in hours, of an electronic component. The p.d.f. is given by

$$f(t) = 0 \qquad t < 0$$
$$= 0.002e^{-0.002t} \qquad t \geq 0$$

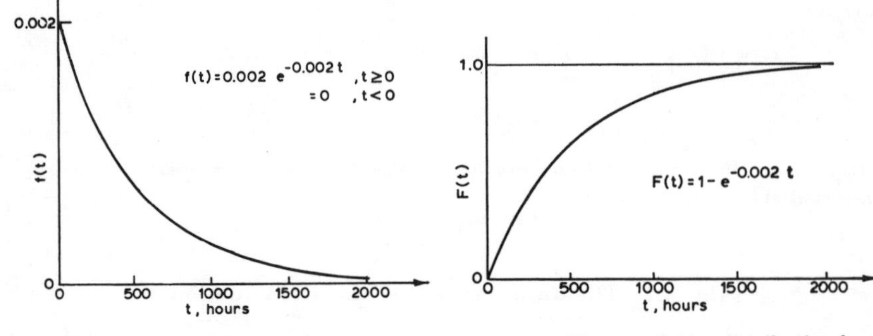

FIG. 4.2 The probability density function.

FIG. 4.3 The cumulative distribution function.

The c.d.f. is given by

$$F(t) = 0 \qquad t < 0$$
$$= \int_0^t 0.002e^{-0.002x}\,dx = 1 - e^{-0.002t} \qquad t \geq 0$$

The probability that the life of the electronic component will exceed 500 hours is

$$P(T \geq 500) = 1 - P(T < 500)$$
$$= 1 - (1 - e^{-1}) = e^{-1} \cong 0.37$$

Definition. The *expected value* of a function $g(x)$ of a discrete or continuous variable is

$$E[g(X)] = \sum_{\text{all } x} g(x)f(x) \qquad X \text{ a discrete random variable}$$

$$E[g(X)] = \int_{-\infty}^{+\infty} g(x)f(x)\,dx \qquad X \text{ a continuous random variable}$$

Table 4.4 Expected Values

1. $E(k) = k$ where k is a constant
2. $E[kg(X)] = kE[g(X)]$ where k is a constant
3. $E[g_1(X) + g_2(X)] = E[g_1(X)] + E[g_2(X)]$

Definition. The kth moment of the random variable X about the origin (*kth ordinary moment*) is defined as

$$\nu_k = E(X^k) = \sum_{\text{all } x} x^k f(x) \qquad X \text{ a discrete random variable}$$
$$\nu_k = E(X^k) = \int_{-\infty}^{\infty} x^k f(x)\,dx \qquad X \text{ a continuous random variable} \tag{4.10}$$

Definition. The kth moment of the random variable X about ν_1 (kth *central moment*) is defined as

$$\mu_k = E[(X - \nu_1)^k] = \sum_{\text{all } x} (x - \nu_1)^k f(x) \qquad X \text{ a discrete random variable}$$

$$\mu_k = E[(X - \nu_1)^k] = \int_{-\infty}^{\infty} (x - \nu_1)^k f(x) \qquad X \text{ a continuous random variable} \tag{4.11}$$

Table 4.5 Relationship between First Four Moments

$$\begin{aligned}
\mu_0 &= \nu_0 = 1 \\
\mu_1 &= 0 \\
\mu_2 &= \nu_2 - \nu_1^2 \\
\mu_3 &= \nu_3 - 3\nu_2\nu_1 + 2\nu_1^3 \\
\mu_4 &= \nu_4 - 4\nu_3\nu_1 + 6\nu_2\nu_1^2 - 3\nu_1^4
\end{aligned}$$

Note.

$$\mu_k = \sum_{i=0}^{k} C(k,i)(-1)^i \nu_{k-i}\nu_1^i \tag{4.12}$$

Example 4.24. Consider the hypergeometric distribution as given in Example 4.22,

$$f(x) = \frac{C(D,x)C(N - D, n - x)}{C(N,n)} \qquad x = 0, 1, 2, \ldots, n \quad 0 \le x \le D \quad 0 \le n - x \le N - D$$

$$= 0 \qquad \text{elsewhere}$$

$$\nu_1 = E(X) = \sum_{x=0}^{n} x \frac{C(D,x)C(N - D, n - x)}{C(N,n)} = \frac{nD}{N}$$

$$\nu_2 = E(X^2) = \sum_{x=0}^{n} x^2 \frac{C(D,x)C(N - D, n - x)}{C(N,n)} = \frac{nD}{N}\left[\left(\frac{N - n}{N - 1}\right)\left(1 - \frac{D}{N}\right) + \frac{nD}{N}\right]$$

$$\nu_3 = E(X^3) = \sum_{x=0}^{n} x^3 \frac{C(D,x)C(N - D, n - x)}{C(N,n)} = \frac{nD}{N}\left[\left(\frac{N - D}{N}\right)\left(\frac{N - 2D}{N}\right)\left(\frac{N - n}{N - 1}\right)\left(\frac{N - 2n}{N - 2}\right) + 3\frac{nD}{N}\left(\frac{N - n}{N - 1}\right)\left(\frac{N - D}{N}\right) + \left(\frac{nD}{N}\right)^2\right]$$

$$\mu = \frac{nD}{N}$$

$$\mu_2 = \nu_2 - \nu_1^2 = \frac{nD}{N}\left(1 - \frac{D}{N}\right)\left(\frac{N - n}{N - 1}\right)$$

$$\mu_3 = \nu_3 - 3\nu_2\nu_1 + 2\nu_1^3 = \frac{nD}{N}\left(1 - \frac{D}{N}\right)\left(1 - \frac{2D}{N}\right)\left(\frac{N - n}{N - 1}\right)\left(\frac{N - 2n}{N - 2}\right)$$

Example 4.25. For a continuous random variable consider the life of an electronic component as given in Example 4.23,

$$f(t) = 0 \qquad t < 0$$
$$= 0.002e^{-0.002t} \qquad t \geq 0$$

$$\nu_1 = E(T) = \int_0^\infty 0.002te^{-0.002t}\,dt = 500 \text{ hours}$$

$$\nu_2 = E(T^2) = \int_0^\infty 0.002t^2e^{-0.002t}\,dt = 5 \times 10^5 \text{ (hours)}^2$$

$$\nu_3 = E(T^3) = \int_0^\infty 0.002t^3e^{-0.002t}\,dt = 75 \times 10^7 \text{ (hours)}^3$$

$$\nu_4 = E(T^4) = \int_0^\infty 0.002t^4e^{-0.002t}\,dt = 15 \times 10^{11} \text{ (hours)}^4$$

$$\mu = \nu_1 = 500 \text{ hours}$$
$$\mu_2 = \nu_2 - \nu_1^2 = 25 \times 10^4 \text{ (hours)}^2$$
$$\mu_3 = \nu_3 - 3\nu_2\nu_1 + 2\nu_1^3 = 25 \times 10^7 \text{ (hours)}^3$$
$$\mu_4 = \nu_4 - 4\nu_3\nu_1 + 6\nu_2\nu_1^2 - 3\nu_1^4 = 5{,}625 \times 10^8 \text{ (hours)}^4$$

Note. Expected values are not functions of the random variable. They are numerical values—descriptive measures—of the distribution of the random variable X.

Table 4.6 Descriptive Measures (Parameters) of the Distribution of the Random Variable X

Fractiles:
The P fractile x_P is that value of the random variable which satisfies
$P(X \leq x_P) = F(x_P) = P \qquad 0 < P < 1$
Special fractiles:

$x_{0.25}$ and $x_{0.75}$	quartiles
$x_{0.50}$	median
$x_{0.10}, x_{0.20}, \ldots, x_{0.90}$	deciles
$x_{0.01}, x_{0.02}, \ldots, x_{0.99}$	percentiles

Central tendency:
1. Median: $x_{0.50}$ (see above)

$$\sum_{x \leq x_{0.50}} f(x) = 0.50 \qquad X \text{ a discrete random variable}$$

$$\int_{-\infty}^{x_{0.50}} f(x)\,dx = 0.50 \qquad X \text{ a continuous random variable}$$

2. Mean: $E(X) = \mu_X = \nu_1$
3. Mode: That value of x for which $f(x)$ is a relative maximum
Note. $f(x)$ may have more than one mode

Dispersion:
1. Variance: $\sigma^2 = E[(X - \mu_X)^2] = \mu_2$
2. Standard deviation: $\sigma = \sqrt{\text{variance}}$, the positive square root of the variance
3. Coefficient of variation: $\eta = \sigma/\mu_X$, the ratio of the standard deviation and the mean
4. Mean deviation (absolute deviation): $E[|X - \mu_X|]$
5. Range: The difference between the largest and smallest value of the random variable

Skewness:
Coefficient of skewness: $\alpha_3 = \mu_3/\mu_2^{3/2}$

Kurtosis:
Coefficient of kurtosis: $\alpha_4 = \mu_4/\mu_2^2$

Definition. The function

$$M_X(t) = E(e^{tX})$$

is called the *moment-generating function* (MGF) of the random variable X. The MGF does not exist for all distributions. However, when it does exist, it is unique and it completely determines the distribution.

Theorem. If the moment-generating function of a random variable approaches the moment-generating function of another variable, the two probability density functions approach each other.

Table 4.7 Properties of the Moment-generating Function

1. $\nu_k = \partial^k M_X(t)/\partial t^k|_{t=0}$, where $\partial^k M_X(t)/\partial t^k|_{t=0}$ is the kth partial derivative of $M_X(t)$ with respect to t evaluated at $t = 0$.
2. The moment-generating function of the sum of a finite number of independent random variables equals the product of their moment-generating functions, e.g.,

$$M_{X+Y}(t) = M_X(t)M_Y(t)$$

3. If $Y = a + bX$, then $M_Y(t) = e^{at}M_X(bt)$.

Example 4.26. Let X have the p.d.f.

$$\begin{aligned} f(x) &= C(n,x)p^x(1-p)^{n-x} \qquad x = 0, 1, 2, \ldots, n \\ &= 0 \qquad \text{elsewhere} \\ M_X(t) &= E(e^{tX}) = \sum_{x=0}^{n} e^{tx}C(n,x)p^x(1-p)^{n-x} \\ &= [pe^t + (1-p)]^n \qquad \text{for all real values of } t \end{aligned}$$

$$\begin{aligned} \nu_1 &= \partial M_X(t)/\partial t|_{t=0} = np \\ \nu_2 &= \partial^2 M_X(t)/\partial t^2|_{t=0} = np + n(n-1)p^2 \\ \nu_3 &= \partial^3 M_X(t)/\partial t^3|_{t=0} = np + 3n(n-1)p^2 + n(n-1)(n-2)p^3 \\ \nu_4 &= \partial^4 M_X(t)/\partial t^4|_{t=0} = np + 7n(n-1)p^2 + 6n(n-1)(n-2)p^3 \\ &\qquad + n(n-1)(n-2)(n-3)p^4 \end{aligned}$$

$$\begin{aligned} \mu &= np \\ \mu_2 &= \nu_2 - \nu_1^2 = np(1-p) \\ \mu_3 &= \nu_3 - 3\nu_2\nu_1 + 2\nu_1^2 = np(1-p)(1-2p) \\ \mu_4 &= \nu_4 - 4\nu_3\nu_1 + 6\nu_2\nu_1^2 - 3\nu_1^3 = np(1-p)[1 - 3p(1-p)(n-2)] \\ \alpha_3 &= \mu_3/\mu_2^{3/2} = (1-2p)/\sqrt{np(1-p)} \\ \alpha_4 &= \mu_4/\mu_2^2 = 3 - \frac{6}{n} + \frac{1}{np(1-p)} \end{aligned}$$

Example 4.27. Let X have the p.d.f.

$$\begin{aligned} f(x) &= (1/\sqrt{2\pi}\sigma)e^{-[(x-\mu)/\sigma]^2/2} \qquad -\infty < x < \infty \\ M_X(t) &= E(e^{tX}) = \int_{-\infty}^{+\infty} e^{tx}f(x)\,dx \\ &= e^{\mu t + (\sigma t)^2/2} \qquad \text{for real values of } t \\ \nu_1 &= \partial M_X(t)/\partial t|_{t=0} = \mu \\ \nu_2 &= \partial^2 M_X(t)/\partial t^2|_{t=0} = \mu^2 + \sigma^2 \\ \nu_3 &= \partial^3 M_X(t)/\partial t^3|_{t=0} = 3\mu\sigma^2 + \mu^3 \\ \nu_4 &= \partial^4 M_X(t)/\partial t^4|_{t=0} = 3\sigma^4 + 6\mu^2\sigma^2 + \mu^4 \end{aligned}$$

$$\mu = \nu_1 \qquad \mu_2 = \sigma^2 \qquad \mu_3 = 0 \qquad \mu_4 = 3\sigma^4$$

$$\alpha_3 = 0 \qquad \alpha_4 = 3$$

4.3a Bivariate Probability Distributions

Note. For discrete random variables X_1 and X_2

$$\begin{aligned} P(a < X_1 \le b, c < X_2 \le d) &= \sum_{a<x_1\le b,\ c<x_2\le d} \sum f(x_1,x_2) \\ &= F(b,d) - F(b,c) - F(a,d) + F(a,c) \end{aligned}$$

For continuous random variables X_1 and X_2

$$P(a < X_1 \leq b, c < X_2 \leq d) = \int_a^b \left[\int_c^d f(x_1,x_2)\, dx_2 \right] dx_1$$
$$= F(b,d) - F(b,c) - F(a,d) + F(a,c)$$

Theorem. Two random variables X_1 and X_2 having the joint probability density function $f(x_1,x_2)$ and marginal probability density functions $f_1(x_1)$ and $f_2(x_2)$, respectively, are statistically independent if $f(x_1,x_2) = f_1(x_1)f_2(x_2)$.

Table 4.8 The Joint Cumulative Distribution Function

For discrete random variables X_1 and X_2

$$F(x_1,x_2) = P(X_1 \leq x_1, X_2 \leq x_2) = \sum_{X_1 \leq x_1} \sum_{X_2 \leq x} f(X_1,X_2)$$

For continuous random variables X_1 and X_2

$$F(x_1,x_2) = \int_{-\infty}^{x_1} \int_{-\infty}^{x_2} f(x_1,x_2)\, dx_2\, dx_1$$

Properties:

1. $F(-\infty,x_2) = \lim_{x_1 \to -\infty} F(x_1,x_2) = 0$

   $F(x_1,-\infty) = \lim_{x_2 \to -\infty} F(x_1,x_2) = 0$

   $F(-\infty,-\infty) = \lim_{\substack{x_1 \to -\infty \\ x_2 \to -\infty}} F(x_1,x_2) = 0$

2. $F(\infty,\infty) = \lim_{\substack{x_1 \to \infty \\ x_2 \to \infty}} F(x_1,x_2) = 1$

3. $F(\infty,x_2) = \lim_{x_1 \to \infty} F(x_1,x_2) = F_2(x_2)$, the *marginal cumulative function* of the random variable X_2

4. $F(x_1,\infty) = \lim_{x_2 \to \infty} F(x_1,x_2) = F_1(x_1)$, the *marginal cumulative function* of the random variable X_1

Table 4.9 The Joint Probability Density Function

For discrete random variables X_1 and X_2

$$f(x_1,x_2) = P(X_1 = x_1, X_2 = x_2)$$

For continuous random variables X_1 and X_2

$$f(x_1,x_2) = \partial^2 F(x_1,x_2)/\partial x_1\, \partial x_2$$

Properties:

1. $f(x_1,x_2) \geq 0$

2. $\sum_{\text{all } x_1} \sum_{\text{all } x_2} f(x_1,x_2) = 1$ discrete random variables X_1 and X_2

   $\int_{-\infty}^{\infty} \left[\int_{-\infty}^{\infty} f(x_1,x_2)\, dx_2 \right] dx_1 = 1$ continuous random variables X_1 and X_2

Table 4.10 Joint Conditional Probability Density Functions

$$f(x_1|x_2) = f(x_1,x_2)/f_2(x_2)$$
$$f(x_2|x_1) = f(x_1,x_2)/f_1(x_1)$$
$$F(x_1|x_2) = F(x_1,x_2)/F_2(x_2)$$
$$F(x_2|x_1) = F(x_1,x_2)/F_1(x_1)$$

Example 4.28. Let the joint probability density function of T_1 and T_2 be

$$f(t_1,t_2) = e^{-(t_1+t_2)} \qquad t_1 > 0,\ t_2 > 0$$
$$= 0 \qquad \text{elsewhere}$$

then

$$f_1(t_1) = \int_0^\infty f(t_1,t_2)\,dt_2 = e^{-t_1} \qquad t_1 > 0$$
$$= 0 \qquad \text{elsewhere}$$
$$f_2(t_2) = \int_0^\infty f(t_1,t_2)\,dt_1 = e^{-t_2} \qquad t_2 > 0$$
$$= 0 \qquad \text{elsewhere}$$

Note. T_1 and T_2 are statistically independent, since $f(t_1,t_2) = f_1(t_1)f_2(t_2)$.

$$f(t_1|t_2) = f(t_1,t_2)/f_2(t_2) = e^{-t_1} \qquad t_1 > 0$$
$$= 0 \qquad \text{elsewhere}$$
$$f(t_2|t_1) = f(t_1,t_2)/f_1(t_1) = e^{-t_2} \qquad t_2 > 0$$
$$= 0 \qquad \text{elsewhere}$$

Definition. The product moments of the random variables X_1 and X_2 are defined by

$$\nu_{ij} = E(X_1{}^i X_2{}^j)$$

Definition. The central product moments of the random variables X_1 and X_2 are defined by

$$\mu_{ij} = E\{[X_1 - E(X_1)]^i[X_2 - E(X_2)]^j\} \tag{4.13}$$

Definition. The *covariance* of the two random variables X_1 and X_2 is defined by

$$\sigma_{X_1X_2} = E\{[X_1 - E(X_1)][X_2 - E(X_2)]\} = \mu_{11}$$

Definition. The *product-moment correlation coefficient* between the two random variables X_1 and X_2 is defined by

$$\rho_{X_1X_2} = \frac{E\{[X_1 - E(X_1)][X_2 - E(X_2)]\}}{\sqrt{E[X_1 - E(X_1)]^2}\,\sqrt{E[X_2 - E(X_2)]^2}} = \frac{\sigma_{X_1X_2}}{\sigma_{X_1}\sigma_{X_2}} \tag{4.14}$$

Note. $-1 \le \rho \le +1$.

Bivariate Normal Distribution. *Definition.* If the joint probability density function of the random variables X_1 and X_2 is given by

$$f(x_1,x_2) = \frac{1}{2\pi\sigma_{X_1}\sigma_{X_2}\sqrt{1-\rho^2}} \exp\left\{-\frac{1}{2(1-\rho^2)}\left[\left(\frac{x_1-\mu_{X_1}}{\sigma_{X_1}}\right)^2 - 2\rho\left(\frac{x_1-\mu_{X_1}}{\sigma_{X_1}}\right)\left(\frac{x_2-\mu_{X_2}}{\sigma_{X_2}}\right) + \left(\frac{x_2-\mu_{X_2}}{\sigma_{X_2}}\right)^2\right]\right\} \qquad \begin{array}{c} -\infty < x_1 < \infty \\ -\infty < x_2 < \infty \\ -1 < \rho < +1 \end{array} \tag{4.15}$$

the random variables X_1 and X_2 have a *bivariate normal distribution.*

Transformation of Variables; Discrete. *Theorem.* Let X be a discrete random variable with probability density function $f(x)$, which is greater than 0 for the set of discrete points A_1. Let $y = g(x)$ define a one-to-one transformation that maps A_1 into the set of discrete points A_2. If $x = h(y)$, then

$$f(y) = P(Y = y) = P[X = h(y)] = f[h(y)] \qquad y \in A_2$$
$$= 0 \qquad \text{elsewhere}$$

is the probability density function of the random variable Y.

Transformation of Variables; Continuous. *Theorem.* Let X be a continuous random variable with probability density function $f(x)$ which is greater than zero for $-\infty < x < +\infty$. Let $y = g(x)$ define a one-to-one transformation which maps the set on the x axis into the set on the y axis. If $x = h(y)$ and $dx/dy = h'(y)$, then

$$f(y) = f[h(y)][h'(y)] \qquad y \in A_2 \text{ (the set on the } y \text{ axis)}$$
$$= 0 \qquad \text{elsewhere}$$

is the probability density function of the random variable Y.

Example 4.29. Let the random variable T have the probability density function

$$f(t) = \lambda e^{-\lambda t} \qquad t > 0$$
$$= 0 \qquad \text{elsewhere}$$

and let T_1, T_2 be a random sample of size 2 from this distribution. Let $Y_1 = T_1 + T_2$ and $Y_2 = T_2/(T_1 + T_2)$; find the joint probability density function of the random variables Y_1 and Y_2.

$$y_1 = t_1 + t_2 \qquad t_1 = y_1(1 - y_2)$$
$$y_2 = t_2/(t_1 + t_2) \qquad t_2 = y_1 y_2$$

$$|J| = \begin{vmatrix} \partial t_1/\partial y_1 & \partial t_1/\partial y_2 \\ \partial t_2/\partial y_1 & \partial t_2/\partial y_2 \end{vmatrix} = \begin{vmatrix} 1 - y_2 & -y_1 \\ y_2 & y_1 \end{vmatrix} = y_1 \neq 0$$

The random variables T_1 and T_2 are independent; hence,

$$f(t_1,t_2) = f(t_1)f(t_2) = \lambda^2 e^{-(t_1+t_2)\lambda} \qquad t_1 > 0,\ t_2 > 0$$
$$= 0 \qquad \text{elsewhere}$$

the joint probability density function of Y_1 and Y_2 is

$$f(y_1,y_2) = \lambda^2 y_1 e^{-y_1\lambda} \qquad y_1 > 0,\ 0 < y_2 < 1$$
$$= 0 \qquad \text{elsewhere}$$

Note.

$$f_1(y_1) = \int_0^1 \lambda^2 y_1 e^{-y_1\lambda}\, dy_2 = \lambda^2 y_1 e^{-y_1\lambda} \qquad y_1 > 0$$
$$= 0 \qquad \text{elsewhere}$$
$$f_2(y_2) = \int_0^\infty \lambda^2 y_1 e^{-y_1\lambda}\, dy_1 = 1 \qquad 0 < y_2 < 1$$
$$= 0 \qquad \text{elsewhere}$$

hence the random variables Y_1 and Y_2 are statistically independent since

$$f(y_1,y_2) = f_1(y_1)f_2(y_2)$$

Reliability Function. *Definition.* For the random variable X, if $P(X \leq x) \equiv F(x{:}\theta_1,\theta_2, \ldots)$, the cumulative distribution function, where θ_i, $i = 1, 2, 3, \ldots$ are parameters of the distribution function, then if a and b are the limits defining the success event, the *reliability function* R is defined by

$$R = R(\theta_1,\theta_2, \ldots) = P(a \leq X \leq b)$$
$$= F(b{:}\theta_1,\theta_2, \ldots) - F(a{:}\theta_1,\theta_2, \ldots) \qquad (4.16)$$

Note. (Continuous case)

Reliability:
$$P(T > t) = \int_t^\infty f(x)\, dx = R(t)$$

Unreliability:
$$P(T < t) = \int_{-\infty}^t f(x)\, dx = F(t)$$
$$R(t) = 1 - F(t)$$

Example 4.30. Let the random variable X represent the performance parameter with L and U, the lower and upper specification limits, respectively. If the p.d.f. of the random variable is given by $f(x{:}\theta_1,\theta_2)$, we have

$$R = P(L \leq X \leq U) = \int_L^U f(x{:}\theta_1,\theta_2)\, dx$$
$$= F(U{:}\theta_1,\theta_2) - F(L{:}\theta_1,\theta_2)$$

with the event success defined as having the performance parameter between the specification limits.

Example 4.31. Let p.d.f. of time to failure of the random variable T be

$$f(t) = \lambda e^{-t\lambda} \qquad 0 < t < \infty$$
$$= 0 \qquad t \leq 0$$

If success is defined as operating beyond t_1, then

$$R = P(T \geq t_1) = \int_{t_1}^{\infty} \lambda e^{-t\lambda}\, dt = e^{-t_1\lambda} = R(t_1:\lambda)$$

Example 4.32. For a discrete random variable X let $1 - p$ be the probability of a component functioning successfully. If we define reliability as the probability that at least k components of n components function successfully, then

$$R = \sum_{x=k}^{n} C(n,x)p^{n-x}(1 - p)^k = R(k:n,p)$$

Definition. The *instantaneous failure-rate distribution* or the *hazard function* $h(t)$ of a p.d.f. of life length is defined as the conditional p.d.f. of life length, given that the unit has not failed prior to time t.

Note. Also known as the *force of mortality* [1]

$$h(t:\theta_1,\theta_2, \ldots) = h(t) = \frac{f(t:\theta_1,\theta_2, \ldots)}{1 - F(t:\theta_1,\theta_2, \ldots)} = \frac{f(t)}{1 - F(t)} \tag{4.17}$$

where

$$F(t:\theta_1,\theta_2, \ldots) = \int_0^t f(x:\theta_1,\theta_2, \ldots)\, dx$$

also,

$$f(t) = \{1 - F(0)\}h(t) \exp\left(-\int_0^t h(x)\, dx\right)$$

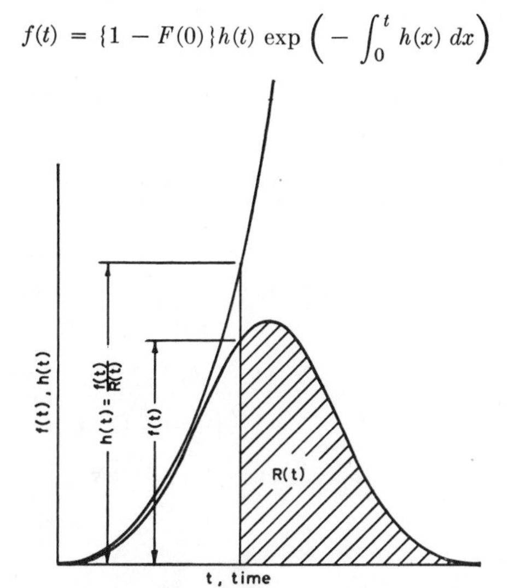

FIG. 4.4 Relationship between $f(t)$, $R(t)$, and $h(t)$.

Example 4.33. Let $f(t) = \lambda e^{-\lambda t} \qquad \lambda > 0,\ t > 0$

$$= 0 \qquad t \leq 0$$

then

$$R(t) = \int_t^{\infty} \lambda e^{-x}\, dx = e^{-\lambda t}$$

Thus,

$$h(t) = \lambda e^{-\lambda t}/e^{-\lambda t} = \lambda$$

interpreted as a constant *failure rate* and the reciprocal of the *mean time to failure.*

Example 4.34. Let $h(t) = (\beta/\alpha)t^{\beta-1}$, $\alpha > 0$, $\beta > 0$.

Then

$$\int_0^t (\beta/\alpha)x^{\beta-1}\,dx = t^\beta/\alpha$$

Therefore,

$$\begin{aligned} f(t) &= (\beta/\alpha)t^{\beta-1}e^{-t^\beta/\alpha} \qquad && t > 0,\ \alpha > 0,\ \beta > 0 \\ &= 0 && \text{elsewhere} \end{aligned}$$

This is the Weibull distribution which is discussed in Sec. 4.5*h*.

4.4 SPECIAL DISCRETE PROBABILITY DENSITY FUNCTIONS

4.4a Hypergeometric Distribution. Let the total set consist of N elements D of which possess a given property. Then the probability that a random sample of size n, without replacement, will contain exactly x elements which possess this property is

$$\begin{aligned} f(x{:}N,D,n) &= \frac{C(D,x)C(N-D,\,n-x)}{C(N,n)} \\ &= \frac{D!}{x!(D-x)!} \times \frac{(N-D)!}{(n-x)!(N-D-n+x)!} \times \frac{n!(N-n)!}{N!} \qquad (4.18) \\ &\qquad\qquad x = 0, 1, 2, \ldots, n \\ &\qquad\qquad N \geq n > 0 \\ &\qquad\qquad N - D \geq 0 \\ &= 0 \qquad \text{elsewhere} \end{aligned}$$

Table 4.11 Some Properties of the Hypergeometric Distribution

Property	Formula
Mean	$\mu = nD/N$
Variance	$\sigma^2 = \frac{nD}{N}\left(1 - \frac{D}{N}\right)\left(\frac{N-n}{N-1}\right)$
Third central moment	$\mu_3 = \frac{nD}{N}\left(1 - \frac{D}{N}\right)\left(1 - \frac{2D}{N}\right)\frac{(N-n)(N-2n)}{(N-1)(N-2)}$
Fourth central moment	$\mu_4 = \frac{nD}{N}\left(1 - \frac{D}{N}\right)\frac{(N-n)}{(N-1)(N-2)(N-3)} \left\{N(N+1) - 6n(N-n) + \frac{3D}{N}\left(1 - \frac{D}{N}\right)[n(N-n)(N+6) - 2N^2]\right\}$
Coefficient of variation	$\eta = \sqrt{\frac{N-D}{nD}\left(\frac{N-n}{N-1}\right)}$
Coefficient of skewness	$\alpha_3 = \frac{N-2D}{\sqrt{nD(N-D)}}\sqrt{\frac{N-1}{N-n}}\left(\frac{N-2n}{N-2}\right)$
Coefficient of kurtosis	$\alpha_4 = \frac{N^2(N-1)}{(N-2)(N-3)n(N-n)} \left[\frac{N(N+1) - 6N(N-n)}{D(N-D)} + \frac{3n(N-n)(N+6)}{N^2} - 6\right]$

Remarks

1. As $N \to \infty$ with n and D/N remaining constant, the hypergeometric distribution approaches the binomial distribution (Sec. 4.4*b*) with parameters n and $\theta = D/N$. The approximation is fairly satisfactory for $10n < N$.
2. Calculations of terms

$$f(x+1{:}N,D,n) = f(x{:}N,D,n)\left(\frac{n-x}{x+1}\right)\left(\frac{D-x}{N-n-D+x+1}\right)$$

Table 4.12 Some Properties of the Binomial Distribution
(See Fig. 4.5)

Moment-generating function	$M_X(t) = (\theta e^t + 1 - \theta)^n$
Mean	$\mu = n\theta$
Variance	$\sigma^2 = n\theta(1-\theta)$
Third central moment	$\mu_3 = n\theta(1-\theta)(1-2\theta)$
Fourth central moment	$\mu_4 = n\theta(1-\theta)[3n\theta(1-\theta) - 6\theta(1-\theta) + 1]$
Coefficient of variation	$\eta = \sqrt{(1-\theta)/n\theta}$
Coefficient of skewness	$\alpha_3 = (1-2\theta)/\sqrt{n\theta(1-\theta)}$
Coefficient of kurtosis	$\alpha_4 = 3 - 6/n + 1/n\theta(1-\theta)$

4.4b Binomial Distribution. Let θ denote the probability of the event occurring at each of n observations. Then the probability that the event will occur exactly

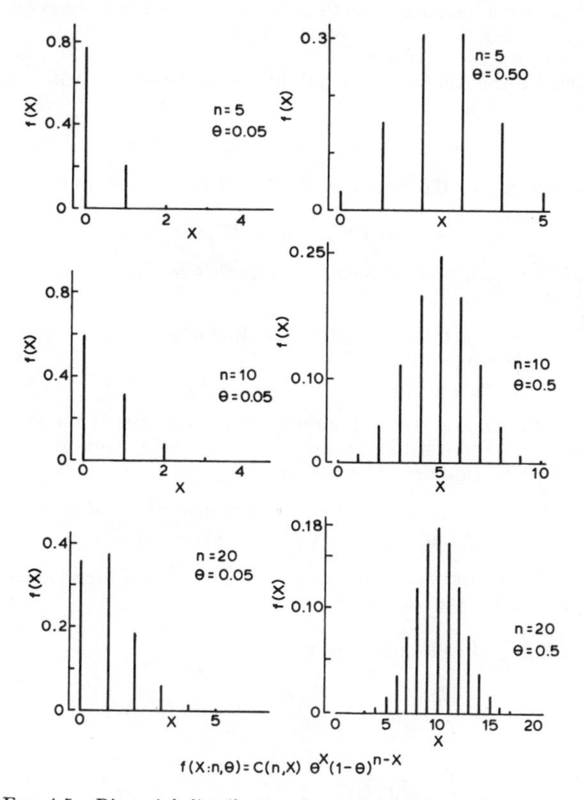

FIG. 4.5 Binomial distribution for various values of n and θ.

x times in the n trials is given by the binomial distribution

$$f(x{:}n,\theta) = C(n,x)\theta^x(1-\theta)^x \qquad x = 0, 1, 2, \ldots, n \quad 0 \le \theta \le 1$$
$$= 0 \qquad \text{elsewhere} \tag{4.19}$$

$$F(x) = \sum_{i=0}^{x} C(n,i)\theta^i(1-\theta)^{n-i} \tag{4.20}$$

Remarks

1. $f(x{:}n,\theta)$ has its maximum for values of x for which $\theta(n+1) - 1 \le x \le \theta(n+1)$.
2. As $n \to \infty$ and $\theta \to 0$ in such a way that $n\theta$ remains constant, the binomial distribution approaches the Poisson distribution (Sec. 4.4*e*) with parameter $\lambda = n\theta$. The approximation is fairly satisfactory for $n > 10$ and $\theta < 0.10$.
3. As $n \to \infty$, the binomial distribution approaches the normal distribution (Sec. 4.5*c*) with parameters $\mu = n\theta$ and $\sigma^2 = n\theta(1-\theta)$; good for $\theta = \frac{1}{2}$ and poor for $\theta < 1/(n+1)$, $\theta > n/(n+1)$ and outside 3σ. A discussion of the accuracy of the approximation is found in Refs. 2 and 3.
4. Calculation of terms

$$f(x+1{:}\,n,\theta) = f(x{:}n,\theta)\left(\frac{n-x}{x+1}\right)\left(\frac{\theta}{1-\theta}\right)$$

5. The binomial distribution is a discrete distribution; hence it is not a function of time. It may be applicable to situations where we classify items as defective and not defective. It may also be applicable to units where operating times are so short that time is not a governing factor, e.g., pyrotechnics.
6. For tables of the binomial see

a. "Tables of the Binomial Probability Distribution," *Natl. Bur. Standards, Appl. Math. Ser.* 6, 1950.

$$n = 2(1)49 \qquad p = 0.01(0.01)0.50$$

b. H. G. Romig, 50–100 *Binomial Tables*, John Wiley & Sons, Inc., New York, 1953.

$$n = 50(5)100 \qquad p = 0.01(0.01)0.99$$

c. Cumulative Binomial Probability Distribution, Harvard University Press, Cambridge, Mass., 1955.

$$n = 2(1)50(2)100(10)200(20)500(50)1{,}000$$
$$p = 0.01(0.01)0.50$$
$$p = \tfrac{1}{16}, \tfrac{1}{12}, \tfrac{1}{8}, \tfrac{1}{6}, \tfrac{3}{16}, \tfrac{5}{16}, \tfrac{1}{3}, \tfrac{3}{8}, \tfrac{5}{12}, \tfrac{7}{16}$$

Estimation of θ. The ratio of the number of occurrences of the event to the total number of trials is an unbiased and maximum-likelihood estimator for θ; for example, an estimate of the fraction defective is given by

$$\hat{\theta} = \frac{\text{number of observed defectives}}{\text{total number of units observed}} = \frac{d}{n} \tag{4.21}$$

Example 4.35. If 100 resistors are tested and 10 are found defective, then an estimate of θ is $\hat{\theta} = 10/100 = 0.10$.

Confidence Limits for θ. For a *two-sided*, 100γ per cent confidence interval the following equations must be solved for θ:

$$\sum_{x=d}^{n} C(n,x)\theta^x(1-\theta)^{n-x} = (1-\gamma)/2 \tag{4.22}$$

$$\sum_{x=0}^{d} C(n,x)\theta^x(1-\theta)^{n-x} = (1-\gamma)/2 \tag{4.23}$$

Example 4.36. Suppose, when 200 relays are tested, 20 are defective. Thus $\hat{\theta} = 20/200 = 0.10$. To obtain 90 per cent confidence in our limits, we would need to solve

$$\sum_{x=20}^{200} C(200,x)\theta^x(1-\theta)^{200-x} = 0.05$$

$$\sum_{x=0}^{20} C(200,x)\theta^x(1-\theta)^{200-x} = 0.05$$

From Chart B1[1] we find $\theta_L = 0.06$, $\theta_U = 0.15$. Thus, we are 90 per cent confident that the true value of θ is between 0.06 and 0.15. Using the normal approximation, we have

$$P\left[\frac{\theta_L - 1/2n - \theta}{\sqrt{\theta(1-\theta)/n}} \le Z \le \frac{\theta_U + 1/2n - \theta}{\sqrt{\theta(1-\theta)/n}}\right] = 1 - \gamma \tag{4.24}$$

From Table A4 of the areas for the normal curve we find for $1 - \gamma = 0.90$, $Z = 1.645$. Since our estimate of θ is $\hat{\theta} = 0.10$, we have

$$P\left[\frac{\theta_L - 0.0025 - 0.10}{0.30/\sqrt{200}} \le Z \le \frac{\theta_U + 0.0025 - 0.10}{0.30/\sqrt{200}}\right] \cong 0.90$$

from which we find $\theta_L = 0.063$ and $\theta_U = 0.137$.

Note. For $n \le 30$, Table A4 may be used in place of Chart B1. Usually, in the case of reliability estimates of a system, only lower confidence limits are computed. Since reliability is the probability of success, let θ be the probability of a single successful event. If at least k successes in n trials are required, then $\hat{\theta}_L$ is the value of θ which satisfies

$$\sum_{x=k}^{n} C(n,x)\theta^k(1-\theta)^{n-k} = 1 - \gamma$$

For zero failures, $k = n$, and we have $\theta^n = 1 - \gamma$ or $n = \log(1-\gamma)/\log\theta$.

Example 4.37. In 500 trials 50 failures were observed, thus our best reliability estimate is $\hat{\theta} = 450/500 = 0.95$. The lower confidence limit at a chosen confidence level of 90 per cent is $\theta_L = 0.88$ for $n = 500$, $F = 50$ failures, per Chart B2.

Table 4.13 Binomial Parameter θ Is Equal to a Specified Value θ_0 (Normal Approximation)

H_0: $\theta = \theta_0$	H_1: $\theta \ne \theta_2$	
Significance level	α	
Statistic	$Z = \dfrac{\bar{\theta} + 1/2n - \theta_0}{\sqrt{\theta_0(1-\theta_0)/n}}$	$\bar{\theta} < \theta_0$
	$Z = \dfrac{\bar{\theta} - 1/2n - \theta_0}{\sqrt{\theta_0(1-\theta_0)/n}}$	$\bar{\theta} > \theta_0$
Reject H_0 if	$Z \le z_{\alpha/2}$ $\quad Z \ge z_{1-\alpha/2}$	

Table 4.14 Two Binomial Parameters Are Equal (Normal Approximation)

H_0: $\theta_1 = \theta_2$	H_1: $\theta_1 \ne \theta_2$
Significance level	α
Statistic	$Z = \dfrac{\bar{\theta}_1 - \bar{\theta}_2}{\sqrt{\bar{\theta}_1(1-\bar{\theta}_1)/n_1 + \bar{\theta}_2(1-\bar{\theta}_2)/n_2}}$
Reject H_0 if	$Z \le z_{\alpha/2}$ $\quad Z \ge z_{1-\alpha/2}$

[1] Tables and charts identified by letters A and B are found in Appendix of the Handbook.

For a discussion on the confidence intervals for the product of two binomials the reader is referred to Lloyd and Lipow [4] and Buehler [5].

Test of Hypothesis

Example 4.38. In evaluating parts purchased from two suppliers, random samples are selected from lots submitted by each supplier and tested with the following results:

Supplier 1: $n_1 = 200$ number of defectives = 20
Supplier 2: $n_2 = 300$ number of defectives = 15

Using $\alpha = 0.05$, is there a significant difference in the quality of the parts? From Table A4, $z_{0.025} = -1.96$ and $z_{0.975} = 1.96$.

$$\bar{\theta}_1 = \frac{20}{200} = 0.10 \qquad \bar{\theta}_2 = \frac{15}{300} = 0.05$$

$$Z = \frac{0.10 - 0.05}{\sqrt{0.10(0.90)/200 + 0.05(0.95)/300}} = 2.03$$

Since $2.03 \geq 1.96$, we reject the hypothesis that quality of the parts is the same.

Table 4.15 A 2 × 2 Contingency Table for Binomial Distribution

	Defective	Not defective	Total
No. 1	d_1	$n_1 - d_1$	n_1
No. 2	d_2	$n_2 - d_2$	n_2
Totals	$d_1 + d_2$	$n_1 - d_1 + n_2 - d_2$	$n_1 + n_2$

Note. In evaluating data of the type in Example 4.37, an approximate method may be used. Construct a 2 × 2 contingency table as in Table 4.15. The computed value

$$\chi^2 = \frac{(n_1 + n_2)[|d_1(n_2 - d_2) - d_2(n_1 - d_1)| - (n_1 + n_2)/2]^2}{n_1 n_2 (d_1 + d_2)(n_1 - d_1 + n_2 - d_2)} \tag{4.25}$$

is distributed as χ^2 (Sec. 4.6*c*) with one degree of freedom. For the data of Example 4.37, $\chi^2 = 3.87$, which is larger than $\chi^2_{0.95,1} = 3.84$. Hence, hypothesis of equal quality is rejected.

For the use of sequential analysis with the binomial distribution see Sec. 4.10.

4.4c Negative Binomial Distribution. Let θ denote the probability of the event occurring at each trial. Consider repeating each observation until the event has occurred exactly k times. The probability that x trials must be made is given by

Pascal's distribution

$$\begin{aligned} f(x{:}k,\theta) &= C(x-1, k-1)\theta^k(1-\theta)^{x-k} && x = k, k+1, \ldots \\ & && 0 \leq \theta \leq 1 \\ &= 0 && \text{elsewhere} \end{aligned} \tag{4.26}$$

It sometimes is more convenient to consider the number of trials in addition to the k required trials. If we let $Z = X - k$, then we obtain the *negative binomial distribution*.

$$\begin{aligned} f(z{:}k,\theta) &= C(z+k-1, k-1)\theta^k(1-\theta)^Z && k = 0, 1, 2, \ldots \\ &= C(-k,z)\theta^k(1-\theta)^Z && z = 0, 1, 2, \ldots \\ &= 0 && \text{elsewhere} \end{aligned} \tag{4.27}$$

The probability of waiting at most x trials for the kth failure is given by

$$F(x) = \sum_{i=k}^{x} C(i-1, k-1)\theta^k(1-\theta)^{i-k}$$
$$= \sum_{i=k}^{x} C(n,i)\theta^i(1-\theta)^{x-i} \tag{4.28}$$

which can be evaluated from the cumulative sums of the binomial distribution.

Table 4.16 Some Properties of Negative Binomial Distribution
(See Fig. 4.6)

Moment-generating function	$M_X(t) = \theta^k/[1 - e^t(1-\theta)]^k$
Mean	$\mu = k(1-\theta)/\theta$
Variance	$\sigma^2 = k(1-\theta)/\theta^2$
Third central moment	$\mu_3 = k(1-\theta)(2-\theta)/\theta^3$
Fourth central moment	$\mu_4 = [k(1-\theta)/\theta^4][3k(1-\theta) + 6(1-\theta) + \theta^2]$
Coefficient of variation	$\eta = 1/\sqrt{k(1-\theta)}$
Coefficient of skewness	$\alpha_3 = (2-\theta)/\sqrt{k(1-\theta)}$
Coefficient of kurtosis	$\alpha_4 = 3 + 6/k + \theta^2/k(1-\theta)$

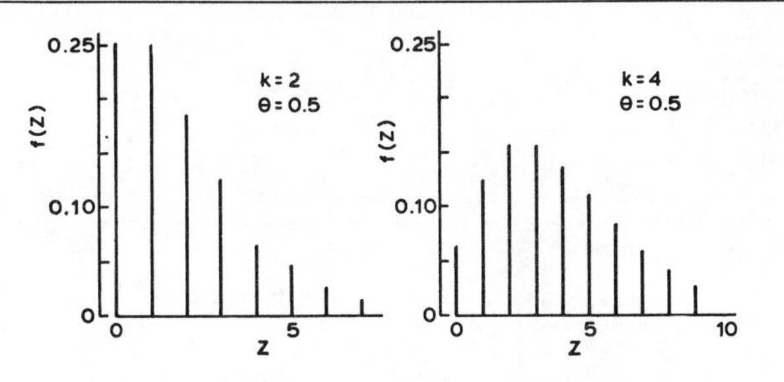

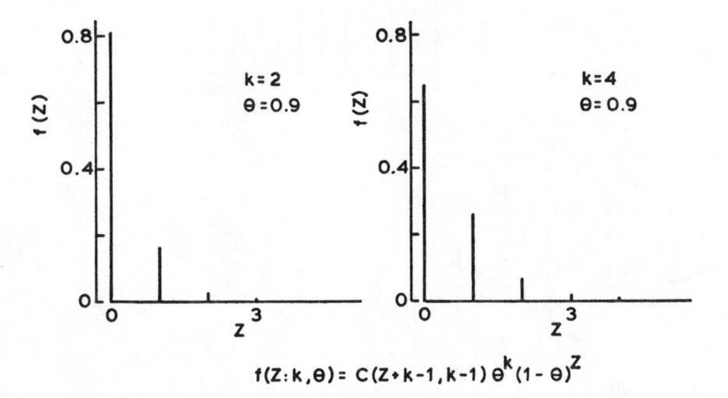

FIG. 4.6 Negative binomial distribution for various values of k and θ.

Example 4.39. If $\theta = 0.2$ and $k = 1$, we should expect $1(0.8)/0.2 = 4$ successes before a first failure or five trials to the first failure.

Example 4.40. If the probability of the failure of a device is 0.05, what is the probability that we must make at most 15 trials until the second failure occurs?

$$F(15) = \sum_{i=2}^{15} C(15,i)(0.05)^i(0.95)^{15-i} = 0.171$$

Estimation of θ. The unbiased estimator of θ is $\hat{\theta} = (k-1)/(n-1)$, $k > 1$, where n is the number of trials made. When $k = 1$, we have the geometric distribution, Sec. 4.4*d*. For additional discussion, see Lloyd and Lipow [4].

4.4d Geometric Distribution. Let θ denote the probability of the event occurring at each trial. Consider repeating each observation until the event occurs for the first time. The probability that x trials must be made is given by the *geometric distribution.*

$$f(x,\theta) = \theta(1-\theta)^{x-1} \qquad x = 1, 2, \ldots \quad 0 \le \theta \le 1$$
$$= 0 \qquad \text{elsewhere} \tag{4.29}$$

$$F(x) = \sum_{i=1}^{x} \theta(1-\theta)^{i-1} = 1 - (1-\theta)^x \tag{4.30}$$

Example 4.41. If $\theta = 0.2$, we should expect $1/0.2 = 5$ trials to the first failure. (See Example 4.39.)

Note. The unbiased estimator of θ is $\hat{\theta} = 1$ for $n = 1$ or $\hat{\theta} = 0$ for $n > 1$.

Table 4.17 Some Properties of the Geometric Distribution
(See Fig. 4.7)

Moment-generating function	$M_X(t) = \theta/[e^{-t} - (1-\theta)]$
Mean	$\mu = 1/\theta$
Variance	$\sigma^2 = (1-\theta)/\theta^2$
Third central moment	$\mu_3 = (1-\theta)(2-\theta)/\theta^3$
Fourth central moment	$\mu_4 = 9(1-\theta)^2/\theta^4 + (1-\theta)/\theta^2$
Coefficient of variation	$\eta = \sqrt{1-\theta}$
Coefficient of skewness	$\alpha_3 = (2-\theta)/\sqrt{1-\theta}$
Coefficient of kurtosis	$\alpha_4 = 9 + \theta^2/(1-\theta)$

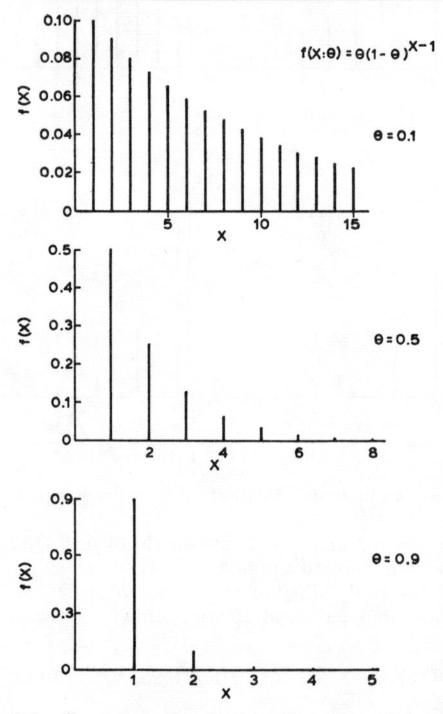

FIG. 4.7 Geometric distribution for various values of θ.

4.4e Poisson Distribution. The Poisson distribution is a useful approximation to the binomial and hypergeometric distributions and also one which arises when the

Table 4.18 Some Properties of the Poisson Distribution
(See Fig. 4.8)

Moment-generating function	$M_X(t) = e^{\lambda(e^t-1)}$
Mean	$\mu = \lambda$
Variance	$\sigma^2 = \lambda$
Third central moment	$\mu_3 = \lambda$
Fourth central moment	$\mu_4 = \lambda(3\lambda + 1)$
Coefficient of variation	$\eta = 1/\sqrt{\lambda}$
Coefficient of skewness	$\alpha_3 = 1/\sqrt{\lambda}$
Coefficient of kurtosis	$\alpha_4 = 3 + 1/\lambda$

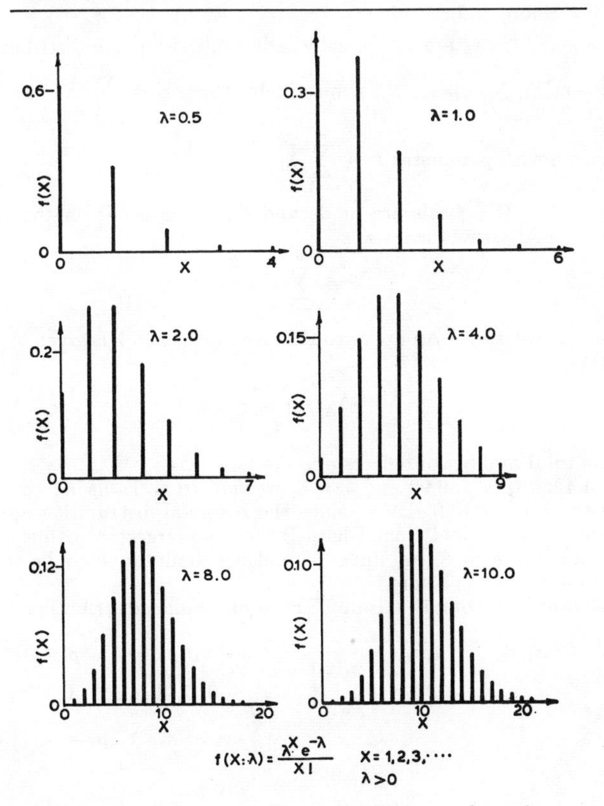

FIG. 4.8 Poisson distribution for various values of λ.

number of possible events is large but the probability of occurrence over a given area or interval is small, e.g., defects, waiting lines.

$$f(x,\lambda) = \lambda^x e^{-\lambda}/x! \qquad x = 0, 1, 2, \ldots \quad \lambda > 0$$
$$= 0 \qquad \text{elsewhere} \tag{4.31}$$

$$F(x) = \sum_{i=0}^{x} \lambda^i e^{-\lambda}/i! \tag{4.32}$$

Remarks.

1. $f(x,\lambda)$ has its maximum for $x \leq [\lambda]$ (the largest integer equal to or less than λ).
2. For small values of λ the distribution is J shaped. As λ increases, it changes to an unsymmetrical bell-shaped distribution. For large values of $\lambda(\lambda > 9)$, it may be approximated by the normal distribution with parameters $\mu = \lambda$ and $\sigma^2 = \lambda$.
3. Calculation of terms

$$f(x + 1, \lambda) = f(x,\lambda) \left(\frac{\lambda}{x + 1}\right)$$

4. For tables of the Poisson distribution, see:

a. Defense Systems Department, General Electric Company, *Tables of the Individual and Cumulative Terms of the Poisson Distribution,* D. Van Nostrand Company, Inc., Princeton, N.J., 1962.

b. E. C. Molina, *Tables of Poisson's Exponential Limit,* D. Van Nostrand Company, Inc., Princeton, N.J., 1945.

5. If $X_1, X_2, \ldots, X_n$ are statistically independent Poisson-distributed variables with parameters $\lambda_1, \lambda_2, \ldots, \lambda_n$, respectively, then $Y = \sum_{i=1}^{n} X_i$ is a Poisson-distributed variable with parameter $\lambda = \sum_{i=1}^{n} \lambda_i$.

Estimation of λ. If n trials are made and d_i events occur on the ith trial, the maximum-likelihood estimator for λ is

$$\hat{\lambda} = \sum_{i=1}^{n} d_i/n \tag{4.33}$$

Confidence Interval on λ. An upper confidence interval on λ is given by the solution of the equation

$$\sum_{d=0}^{k} e^{-\lambda}\lambda^d/d! = 1 - \gamma \tag{4.34}$$

where k is the total number of defects observed.

Example 4.42. If $\gamma = 0.90$ and $k = 8$, we find from Table A1 the upper confidence limits to be $\lambda_U = 13.0$. Also, since the binomial distribution approaches the Poisson distribution, we find from Chart B2 for the largest n, namely, 1,000, and, by interpolating for $F = 8$, an upper confidence limit on θ to be 0.013. Thus, $\lambda_U = 0.013(1{,}000) = 13$ as before.

4.4f Multinomial Distribution. The multinomial distribution is defined by

$$f(x_1,x_2, \ldots ,x_k) = \frac{n!}{x_1!x_2! \cdots x_k!} p_1^{x_1}p_2^{x_2} \cdots p_k^{x_k} \tag{4.35}$$

$$x_i = 0, 1, 2, \ldots, \quad 0 < p_i < 1$$

$$\sum_{i=1}^{k} x_i = n, \quad \sum_{i=1}^{k} p_i = 1$$

$$= 0 \quad \text{elsewhere}$$

Remarks.

1. Applicable to situations where items can be assigned to more than two classes, e.g., the four classes: critical defect, major defect, minor defect, good.
2. To test the hypothesis that p_i = given p_i, p_{0i}, $i = 1, 2, \ldots, k$, use

$$\chi^2 = \sum_{i=1}^{k} (0_i - np_{0i})^2/np_{0i}$$

where 0_i is the observed number of occurrences in the ith class

$$\sum_{i=1}^{k} p_i = \sum_{i=1}^{k} p_{0i} = 1 \qquad \sum_{i=1}^{k} 0_i = \sum_{i=1}^{k} np_{0i} = n$$

The quantity is distributed as chi-square with $k - 1$ degrees of freedom. Reject the hypothesis if $\chi^2 \geq \chi^2_{1-\alpha,k-1}$.

4.5 SPECIAL CONTINUOUS PROBABILITY DENSITY FUNCTIONS

4.5a Uniform Distribution. Uniform distribution is defined by the function

$$f(x) = 1/(b - a) \qquad a < x < b \tag{4.36}$$
$$= 0 \qquad \text{elsewhere}$$
$$F(x) = 0 \qquad x \leq a$$
$$= (x - a)/(b - a) \qquad a < x < b \tag{4.37}$$
$$= 1 \qquad x \geq b$$

Table 4.19 Properties of the Uniform Distribution
(See Fig. 4.9)

Moment-generating function	$M_X(t) = (e^{bt} - e^{at})/(b - a)t$
Mean	$\mu = (b + a)/2$
Variance	$\sigma^2 = (b - a)^2/12$
Third central moment	$\mu_3 = 0$
Fourth central moment	$\mu_4 = (b - a)^4/80$
Coefficient of variation	$\eta = (1/\sqrt{3})[(b - a)/(b + a)]$
Coefficient of skewness	$\alpha_3 = 0$
Coefficient of kurtosis	$\alpha_4 = 1.8$

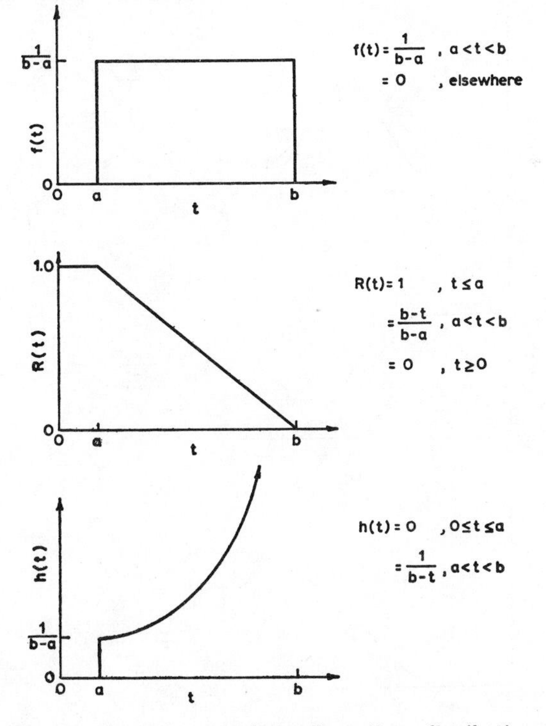

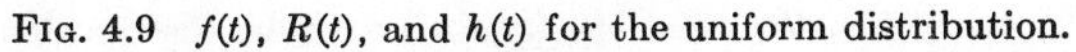

FIG. 4.9 $f(t)$, $R(t)$, and $h(t)$ for the uniform distribution.

Reliability Functions. Since $t \geq 0$, let $a \geq 0$.

$$R(t) = 1 \qquad 0 \leq t \leq a$$
$$= (b - t)/(b - a) \qquad a < t < b \tag{4.38}$$
$$= 0 \qquad t \geq b$$
$$h(t) = 0 \qquad 0 \leq t \leq a \tag{4.39}$$
$$= 1/(b - t) \qquad a < t < b$$

4.5b Triangular Distribution. Triangular distribution is defined by the function

$$f(x) = 0 \qquad x \leq a \tag{4.40}$$
$$= 4(x - a)/(b - a)^2 \qquad a < x \leq (a + b)/2$$
$$= 4(b - x)/(b - a)^2 \qquad (a + b)/2 < x < b$$
$$= 0 \qquad x \geq b$$
$$F(x) = 0 \qquad x \leq a \tag{4.41}$$
$$= 2(x - a)^2/(b - a)^2 \qquad a < x \leq (a + b)/2$$
$$= 1 - 2(b - x)^2/(b - a)^2 \qquad (a + b)/2 < x < b$$
$$= 0 \qquad x \geq b$$

Table 4.20 Properties of the Triangular Distribution
(See Fig. 4.10)

Moment-generating function	$M_X(t) = 4(e^{bt/2} - e^{at/2})^2/t^2(b - a)^2$
Mean	$\mu = (a + b)/2$
Variance	$\sigma^2 = (b - a)^2/24$
Third central moment	$\mu_3 = 0$
Fourth central moment	$\mu_4 = (b - a)^4/240$
Coefficient of variation	$\eta = (1/\sqrt{6})[(b - a)/(b + a)]$
Coefficient of skewness	$\alpha_3 = 0$
Coefficient of kurtosis	$\alpha_4 = 1.4$

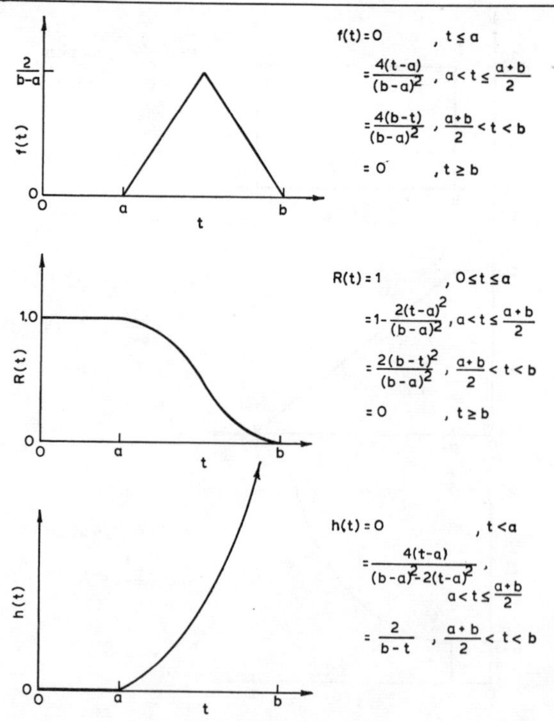

Fig. 4.10 $f(t)$, $R(t)$, and $h(t)$ for the triangular distribution.

Reliability Functions. Since $t \geq 0$, let $a \geq 0$.

$$\begin{aligned} R(t) &= 1 && 0 \leq t \leq a && (4.42)\\ &= 1 - 2(t-a)^2/(b-a)^2 && a < t \leq (a+b)/2 \\ &= 2(b-t)^2/(b-a)^2 && (a+b)/2 < t < b \\ &= 0 && t \geq b \\ h(t) &= 0 && 0 \leq t \leq a && (4.43)\\ &= 4(t-a)/[(b-a)^2 - 2(t-a)^2] && a < t \leq (a+b)/2 \\ &= 2/(b-t) && (a+b)/2 < t < b \end{aligned}$$

4.5c Normal Distribution. A two-parameter function defined by

$$f(x:\mu,\sigma) = (1/\sqrt{2\pi}\sigma)\, e^{-[(x-\mu)/\sigma]^2/2} \qquad -\infty < x < \infty,\ \sigma > 0 \qquad (4.44)$$

$$F(x) = \int_{-\infty}^{x} (1/\sqrt{2\pi}\sigma)\, e^{-[(x-\mu)/\sigma]^2/2}\, dx \qquad -\infty < x < \infty \qquad (4.45)$$

Table 4.21 Some Properties of the Normal Distribution
(See Fig. 4.11)

Moment-generating function	$M_X(t) = e^{\mu t + \sigma^2 t^2/2}$
Mean	$\mu = \mu$
Variance	$\sigma^2 = \sigma^2$
Third central moment	$\mu_3 = 0$
Fourth central moment	$\mu_4 = 3\sigma^4$
Coefficient of variation	$\eta = \sigma/\mu$
Coefficient of skewness	$\alpha_3 = 0$
Coefficient of kurtosis	$\alpha_4 = 3$

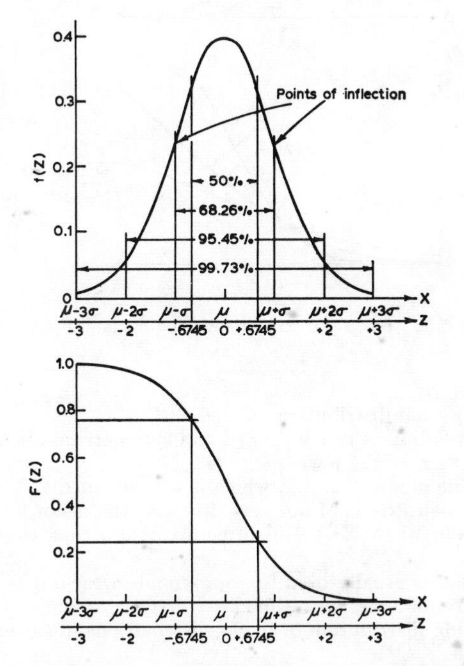

FIG. 4.11 The normal probability density function and cumulative distribution function.

For $Z = (X - \mu)/\sigma$, the standardized normal variable, we obtain

$$f(z) = (1/\sqrt{2\pi})\, e^{-z^2/2} \qquad -\infty < z < \infty \tag{4.46}$$

$$\mu_Z = 0 \qquad \sigma_Z{}^2 = 1 \qquad \alpha_{3Z} = 0 \qquad \alpha_{4Z} = 3$$

$$M_Z(t) = e^{t^2/2}$$

$$F(z) = (1/\sqrt{2\pi}) \int_{-\infty}^{z} e^{-z^2/2}\, dz \tag{4.47}$$

which is tabulated in Table A4.

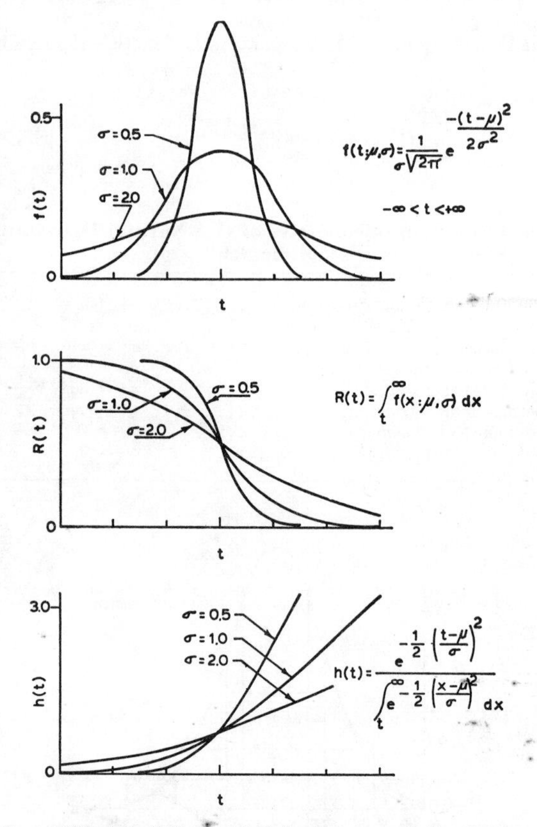

FIG. 4.12 $f(t)$, $R(t)$, and $h(t)$ for the normal distribution.

Remarks

1. Notation for normal distribution: $N(\mu,\sigma^2)$.
2. The normal distribution is bell-shaped and symmetrical about $x = \mu$, and it has points of inflection at $x = \mu \pm \sigma$.
3. There is a single mode at $x = \mu$, which is also the median.
4. Since the function extends from $-\infty$ to $+\infty$, the mean μ must be sufficiently above 0, $\mu - 3\sigma > 0$, when $X \geq 0$ in order to assure that the distribution is not truncated.
5. The distribution is usually used to approximate wear-out failures
6. The failure rate increases with time.
7. It is the limiting distribution of many distributions, for example, binomial and gamma distributions.

8. For comprehensive tabulation of values of $f(x)$ and $2F(x) - 1$ to 15 decimal places for $x = 0(0.0001)1(0.001)7.80$ see "Tables of Probability Functions," *Natl. Bur. Standards, Appl. Math. Ser.* 23, 1953.

9. If X_i is a random sample from a normal population with mean μ_i and variance σ_i^2 $(i = 1,2,3, \ldots ,n)$, then

$$Y = \sum_{i=1}^{n} a_i X_i$$

is normally distributed with mean

$$\mu_Y = \sum_{i=1}^{n} a_i \mu_i$$

and variance

$$\sigma_Y^2 = \sum_{i=1}^{n} a_i^2 \sigma_i^2$$

Note. If all the random samples are from the same normal population with mean μ and variance σ^2, then Y is normally distributed with mean

$$\mu \sum_{i=1}^{n} a_i$$

and variance

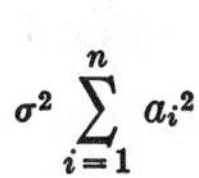

$$\sigma^2 \sum_{i=1}^{n} a_i^2$$

Therefore, $Y = \bar{x} = \sum_{i=1}^{n} x_i/n$, the sample mean, is normally distributed with mean $\mu_{\bar{x}} = \mu$ and variance $\sigma_{\bar{x}}^2 = \sigma_x^2/n$.

Example 4.43. Suppose that the life of an assembly is normally distributed with $\mu = 800$ hours and $\sigma = 50$ hours. What is the probability that an assembly will last at least 875 hours?

$$z = (t - \mu)/\sigma = (875 - 800)/50 = 1.5$$

From Table A4, we find $P(Z < 1.5) = 0.9332$. Thus,

$$P(t > 875) = 1 - 0.9332 = 0.0668 = R$$

Estimation of μ. The sample mean is an unbiased and maximum-likelihood estimator for μ, that is,

$$\hat{\mu} = \Sigma x_i/n = \bar{x} \tag{4.48}$$

Estimation of σ^2. The sample variance is an unbiased estimator for σ^2, that is,

$$\hat{\sigma}^2 = \left[\sum_{i=1}^{n} (x_i - \bar{x})^2\right]/(n - 1) = s^2 \tag{4.49}$$

The maximum-likelihood estimator $\sum_{i=1}^{n} (x_i - \bar{x})^2/n$ is a biased estimator.

Note. Neither the quantity s nor $\sqrt{\Sigma(x_i - \bar{x})^2/n}$ is an unbiased estimator for σ.

Confidence Limits for μ. The confidence limits for μ depend upon whether σ^2 is known or unknown. For the formulas to use see Table 4.22.

Table 4.22 Confidence Limits for the Mean of a Normal Population

Type	σ^2	Confidence Limits	
		μ_L	μ_U
Two-sided	Known......	$\bar{X} - z_{(1+\gamma)/2}\sigma/\sqrt{n}$	$\bar{X} + z_{(1+\gamma)/2}\sigma/\sqrt{n}$
	Unknown...	$\bar{X} - t_{(1+\gamma)/2,n-1}s/\sqrt{n}$	$\bar{X} + t_{(1+\gamma)/2,n-1}s/\sqrt{n}$
One-sided (upper only)	Known......	...	$\bar{X} + z_\gamma\sigma/\sqrt{n}$
	Unknown...	...	$\bar{X} + t_{\gamma,n-1}s/\sqrt{n}$
One-sided (lower only)	Known......	$\bar{X} - z_\gamma\sigma/\sqrt{n}$	
	Unknown...	$\bar{X} - t_{\gamma,n-1}s/\sqrt{n}$	

Note. 1. z is the standardized normal variate.
2. t is Student's t with $n - 1$ degrees of freedom.

Example 4.44. A sample of five capacitors, rated at 10 μf, was selected at random. The individual capacitances were measured with the following results: 9.9, 10.2, 10.3, 10.3, 9.8 μf. What are the 95 per cent confidence limits for μ?

$$\bar{x} = (9.9 + 10.2 + 10.3 + 10.3 + 9.8)/5 = 10.1\ \mu\text{f}$$

$$s^2 = [(9.9 - 10.1)^2 + (10.2 - 10.1)^2 + (10.3 - 10.1)^2 + (10.3 - 10.1)^2 + (9.8 - 10.1)^2]/4 = 0.055\ (\mu\text{f})^2$$

From Table A5, $t_{0.975,4} = 2.776$ Using Table 4.22,

$$\mu_L = 10.1 - 2.776(0.2345/\sqrt{5}) = 10.1 - 0.3 = 9.8\ \mu\text{f}$$
$$\mu_U = 10.1 + 2.776(0.2345/\sqrt{5}) = 10.1 + 0.3 = 10.4\ \mu\text{f}$$

Thus, we are 95 per cent confident that the true mean lies between 9.8 and 10.4 μf.

Confidence Limits for σ^2 The chi-square distribution is used to obtain the confidence limits for σ^2.

Two-sided

$$\sigma_L{}^2 = \frac{(n-1)s^2}{\chi^2{}_{(1+\gamma)/2,n-1}} = \frac{\sum_{i=1}^{n}(x_i - \bar{x})^2}{\chi^2{}_{(1+\gamma)/2,n-1}} \tag{4.50}$$

$$\sigma_U{}^2 = \frac{(n-1)s^2}{\chi^2{}_{(1\ \gamma)/2,n-1}} = \frac{\sum_{i=1}^{n}(x_i - \bar{x})^2}{\chi^2{}_{(1-\gamma)/2,n-1}} \tag{4.51}$$

One-sided

$$\sigma_U{}^2 = \frac{(n-1)s^2}{\chi^2{}_{1-\gamma,n-1}} = \frac{\sum_{i=1}^{n}(x_i - \bar{x})^2}{\chi^2{}_{1-\gamma,n-1}} \tag{4.52}$$

$$\sigma_L{}^2 = \frac{(n-1)s^2}{\chi^2{}_{\gamma,n-1}} = \frac{\sum_{i=1}^{n}(x_i - \bar{x})^2}{\chi^2{}_{\gamma,n-1}} \tag{4.53}$$

The one-sided lower limit is seldom of any interest, since one is usually interested in the largest value of the variance. Confidence limits for σ may be obtained by taking the square root of the confidence limits for σ^2. This is not an exact solution, but it is usually of sufficient accuracy.

Example 4.45. The 90 per cent two-sided confidence limits for σ^2 of Example 4.44 are

$$\sigma_L^2 = 0.22/\chi^2_{0.95,4} = 0.22/9.49 = 0.023\ (\mu f)^2$$
$$\sigma_U^2 = 0.22/\chi^2_{0.05,4} = 0.22/0.0711 = 3.09\ (\mu f)^2$$

The values of χ^2 are found in Table A6.

Test of Hypotheses

Table 4.23 The Normal Distribution Parameter μ Is Equal to a Specified Value μ_0 (Significance Level α)

H_0: $\mu = \mu_0$	H_1: $\mu \neq \mu_0$
	σ^2 Known
Statistic.	$z = (\bar{X} - \mu_0)/(\sigma/\sqrt{n})$
Reject H_0 if.	$\lvert z\rvert > z_{1-\alpha/2}$
	σ^2 Unknown
Statistic.	$t = (\bar{X} - \mu_0)/(s/\sqrt{n})$
Reject H_0 if.	$\lvert t\rvert > t_{1-\alpha/2,n-1}$

Table 4.24 The Normal Population Parameter μ with a One-sided Alternative (Significance Level α)

H_0: $\mu \leq \mu_0$	H_1: $\mu > \mu_0$
	σ^2 Known
Statistic.	$z = (\bar{X} - \mu_0)/(\sigma/\sqrt{n})$
Reject H_0 if.	$z > z_{1-\alpha}$
	σ^2 Unknown
Statistic.	$t = (\bar{X} - \mu_0)/(s/\sqrt{n})$
Reject H_0 if.	$t > t_{1-\alpha\ n-1}$
H_0: $\mu \geq \mu_0$	H_1: $\mu < \mu_0$
	σ^2 Known
Statistic.	$z = (\bar{X} - \mu_0)/(\sigma/\sqrt{n})$
Reject H_0 if.	$z < z_\alpha = -z_{1-\alpha}$
	σ^2 Unknown
Statistic.	$t = (\bar{X} - \mu_0)/(s/\sqrt{n})$
Reject H_0 if.	$t < t_{\alpha,n-1} = -t_{1-\alpha,n-1}$

Example 4.46. Suppose the life of a certain type of tube is normally distributed with $\mu = 200$ hours. A random sample of five new tubes is life-tested, and the following values are obtained: $\bar{x} = 240$ hours and $s = 40$ hours. Is there any evidence that these new tubes have average length of life which will exceed 200 hours?

$$H_0: \mu \leq 200 \qquad H_1: \mu > 200 \text{ hours}$$

Let $\alpha = 0.05$. In accordance with Table 4.24, the statistic to be used is

$$t = (\bar{x} - \mu)/(s/\sqrt{n}) = (240 - 200)/(40/2) = 2.0$$

From Table A5, $t_{0.95,4} = 2.132$. Since $t < t_{0.95,4}$, accept H_0; there is no evidence that the life of the new tubes exceeds 200 hours.

The method in Table 4.25 assumes equal but unknown variances. If there is some doubt about them being equal, we should test the hypothesis of equality in accord-

Table 4.25 Means of Two Normal Populations Are Equal (Equal Variances but Unknown)

H_0: $\mu_1 = \mu_2$	H_1: $\mu_1 \neq \mu_2$
Significance level	α
Statistic	$t = (\bar{X}_1 - \bar{X}_2)/(s_{\bar{X}_1 - \bar{X}_2})$ where $s^2_{\bar{X}_1-\bar{X}_2} = s^2/n_1 + s^2/n_2$ and $s^2 = \dfrac{(n_1 - 1)s_1^2 + (n_2 - 1)s_2^2}{n_1 + n_2 - 2}$
Reject H_0 if	$\lvert t \rvert > t_{1-\alpha/2, n_1+n_2-2}$

Table 4.26 Means of Two Normal Populations Are Equal (Unequal Variances and Unknown)

H_0: $\mu_1 = \mu_2$	H_1: $\mu_1 \neq \mu_2$
Significance level	α
Statistic	$t = \dfrac{\bar{X}_1 - \bar{X}_2}{\sqrt{s_1^2/n_1 + s_2^2/n_2}}$
Reject H_0 if	$\lvert t \rvert > \dfrac{(s_1^2/n_1)t_{1-\alpha/2,n_1-1} + (s_2^2/n_2)t_{1-\alpha/2,n_2-1}}{s_1^2/n_1 + s_2^2/n_2}$

Note. This an approximate but fairly satisfactory method.

ance with the procedure given in Table 4.28. If the hypothesis of equal variances is rejected, the approximate method of Table 4.26 should be used to test the equality of two means.

Example 4.47. Suppose the sampling results are $\bar{x}_1 = 42.3$, $s_1^2 = 2.25$, $n_1 = 15$, and $\bar{x}_2 = 41.1$, $s_2^2 = 12.25$, $n_2 = 20$. Assuming normal distributions, test the hypothesis that $\mu_1 = \mu_2$ at $\alpha = 0.05$. The test of equal variances results in rejection of H_0: $\sigma_1^2 = \sigma_2^2$. (The method is given in Table 4.28.) Using Table 4.26:

$$t_{\text{weighted}} \cong 2.10 \qquad t = (42.3 - 41.1)/0.88 = \mathbf{1.38} < 2.10$$

Hence, we accept H_0: $\mu_1 = \mu_2$.

Note. In case of paired observations, use the difference between the pairs as the variable and test H_0: $\mu_D = 0$ versus H_1: $\mu_D \neq 0$. Compute the average difference and the standard deviation of differences. Reject H_0: $\mu_1 = \mu_2$ if $t > t_{1-\alpha/2,n-1}$ or $t < -t_{1-\alpha/2,n-1}$

where
$$t = \frac{\bar{D}}{s_{\bar{D}}} = \frac{\bar{x}_1 - \bar{x}_2}{s_D/\sqrt{n}} = \frac{\bar{x}_1 - \bar{x}_2}{\sqrt{\dfrac{\sum_{i=1}^{n} [(x_{1_i} - x_{2_i}) - (\bar{x}_1 - \bar{x}_2)]^2}{n(n-1)}}} \tag{4.54}$$

Example 4.48. Use the data of Example 4.47, with $\alpha = 0.10$, to test H_0: $\sigma_1^2 = \sigma_2^2$ versus H_1: $\sigma_1^2 \neq \sigma_2^2$. $F = 2.25/12.25 = 0.18 < F_{0.05,14,19} = 0.417$, from Table A7. Therefore, H_0 is rejected.

4.5d Log-normal Distribution. The *log-normal distribution* is defined by the three-parameter function

$$f(x{:}\gamma,\mu,\sigma) = \frac{1}{(x-\gamma)\sqrt{2\pi}\,\sigma} e^{-\{\ln (x-\gamma)-\mu\}^2/2\sigma^2} \qquad x > \gamma > 0 \tag{4.55}$$
$$= 0 \qquad x \leq \gamma \qquad\qquad \sigma > 0$$

Table 4.27 Variance of a Normal Population Is Equal to a Specified Value σ_0^2

$H_0: \sigma^2 = \sigma_0^2$	$H_1: \sigma^2 \neq \sigma_0^2$
Level of significance	α
Statistic	$\chi^2 = \dfrac{\left[\sum_{i=1}^{n} (X_i - \bar{X})^2\right]}{\sigma_0^2}$
Reject H_0 if	$\chi^2 < \chi^2_{\alpha/2,n-1}$ or $\chi^2 > \chi^2_{1-\alpha/2,n-1}$

Table 4.28 Variances of Two Normal Populations Are Equal

$H_0: \sigma_1^2 = \sigma_2^2$	$H_1: \sigma_1^2 \neq \sigma_2^2$
Significance level	α
Statistic	$F = s_1^2/s_2^2$
Reject H_0 if	$F < F_{\alpha/2,n_1-1,n_2-1}$
	$F > F_{1-\alpha/2,n_1-1,n_2-1}$

Note. An alternative to the above is to always divide the larger variance by the smaller variance. In this case reject H_0 if $F \geq F_{1-\alpha/2,\nu_1,\nu_2}$ where ν_1 is degrees of freedom associated with the numerator and ν_2 with the denominator.

Table 4.29 Some Properties of the Log-normal Distribution ($\gamma = 0$) (See Fig. 4.13)

Mean	$E(X) = e^{\mu+\sigma^2/2}$
Variance	$\mu_2 = e^{2\mu+\sigma^2}(e^{\sigma^2} - 1)$
Third central moment	$\mu_3 = e^{3\mu+3\sigma^2/2}(e^{\sigma^2} - 1)^2(e^{\sigma^2} + 2)$
Fourth central moment	$\mu_4 = e^{4\mu+2\sigma^2}(e^{\sigma^2} - 1)^2 \times (e^{4\sigma^2} + 2e^{3\sigma^2} + 3e^{2\sigma^2} - 3)$
Coefficient of variation	$\eta = \sqrt{e^{\sigma^2} - 1}$
Coefficient of skewness	$\alpha_3 = \sqrt{e^{\sigma^2} - 1}\,(e^{\sigma^2} + 2)$
Coefficient of kurtosis	$\alpha_4 = 3 + (e^{\sigma^2} - 1)(e^{3\sigma^2} + 3e^{2\sigma^2} + 6e^{\sigma^2} + 6)$

where γ is the location parameter. If we let $\gamma = 0$, then the kth moment about the origin is given by

$$\nu_k = e^{k\mu+k^2\sigma^2/2}$$

$$F(x) = \frac{1}{\sqrt{2\pi}\,\sigma}\int_0^x \frac{1}{x} e^{-(\ln x-\mu)^2/2\sigma^2}\,dx \qquad x > 0 \tag{4.56}$$

$$= 0 \qquad x < 0$$

Reliability Functions

$$R(t) = 1 \qquad t \leq \gamma$$

$$= \frac{1}{\sqrt{2\pi}\,\sigma}\int_t^\infty \frac{1}{x-\gamma} e^{-\{\ln(x-\gamma)-\mu\}^2/2\sigma^2}\,dx \qquad t > \gamma \tag{4.57}$$

$$h(t) = \frac{[1/(t-\gamma)]e^{-\{\ln(t-\gamma)-\mu\}^2/2\sigma^2}}{\int_t^\infty [1/(x-\gamma)]e^{-\{\ln(x-\gamma)-\mu\}^2/2\sigma^2}\,dx} \qquad t > \gamma \tag{4.58}$$

$$= 0 \qquad t \leq \gamma$$

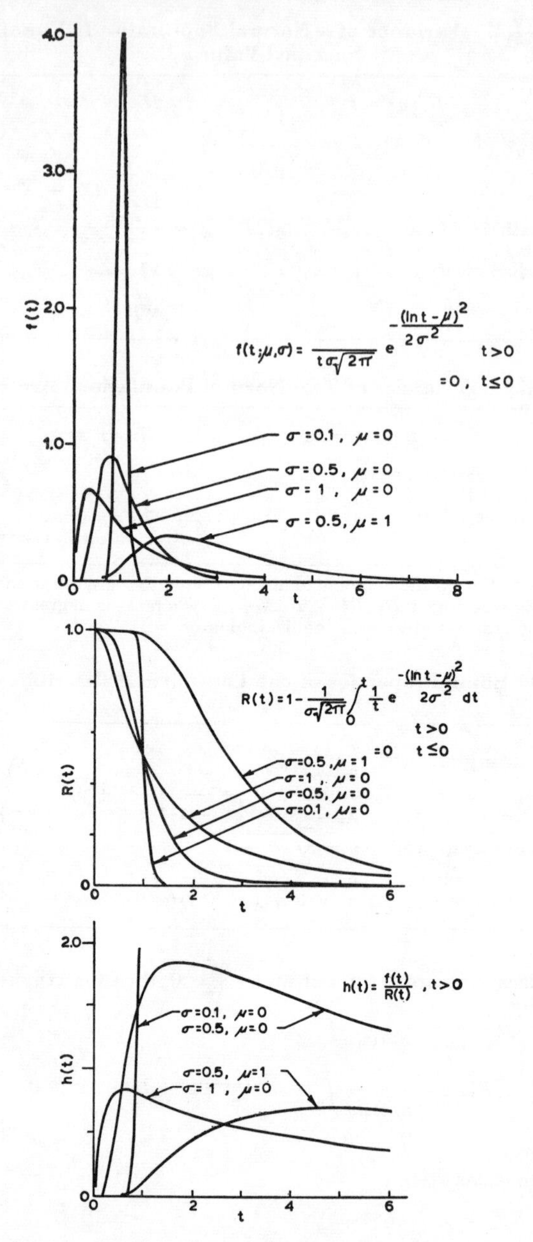

FIG. 4.13 $f(t)$, $R(t)$, and $h(t)$ for the log-normal distribution.

Remarks

1. As noted in Ref. 6, log-normal distributions can be mistaken for exponential distributions.

2. The log-normal distribution has a single mode at $x = \gamma + e^{\mu-\sigma^2}$ and the median at e^{μ}. The distribution is positively skewed.

3. If X_1 and X_2 are independent log-normal random variables, then the product of the random variables $Y = X_1X_2$ is also a log-normal variable.

4. $\hat{\mu} = \left(\sum_{i=1}^{n} \ln x\right)/n$ and $\hat{\sigma}^2 = \left[\sum_{i=1}^{n} (\ln x - \hat{\mu})^2\right]/(n-1)$ may be used for estimators for μ and σ, respectively.

5. Further details of the distribution function may be found in Aitchison and Brown [7].

6. The distribution has been used to approximate wear-out failures.

7. The failure rate increases with time.

Example 4.49. Suppose that the life of a component is log-normally distributed with median equal to 1,000 hours, $\gamma = 0$, and $\sigma = 1$. Using Table 4.29,

$$\begin{aligned}
\text{Median} &= e^{\mu} = 1{,}000 \text{ hours, or } \mu = 6.908 \text{ hours} \\
\text{Mode} &= e^{\mu}e^{-\sigma^2} = 367.88 \text{ hours} \\
\text{Variance} &= e^{2\mu+\sigma^2}(e^{\sigma^2} - 1) = 4.67 \times 10^6 \text{ hours}^2
\end{aligned}$$

What is the probability that the component will last at least 1,500 hours?

$$P(T > 1{,}500) = (1/\sqrt{2\pi}) \int_{1{,}500}^{\infty} (1/x)e^{-(\ln x - 6.908)^2/2}\, dx$$

Let $y = \ln x - 6.908$, then

$$P(T > 1{,}500) = (1/\sqrt{2\pi}) \int_{\ln 1{,}500 - 6.908}^{\infty} e^{-y^2/2}\, dy$$

Using Table A4,

$$P(T > 1{,}500) = 0.34$$

4.5e Gamma Distribution. The *gamma distribution* is defined by the two-parameter function

$$\begin{aligned}
f(x{:}\alpha,\beta) &= (1/\alpha!\beta^{\alpha+1})x^{\alpha}e^{-x/\beta} \qquad x > 0 \\
&= 0 \qquad x \leq 0
\end{aligned} \tag{4.59}$$

where the scale parameter $\beta > 0$ and the shape parameter $\alpha > -1$.

$$\begin{aligned}
F(x) &= \int_0^x (1/\alpha!\beta^{\alpha+1})x^{\alpha}e^{-x/\beta}\, dx \\
&= (1/\alpha!)\Gamma_{x/\beta}(\alpha + 1)
\end{aligned} \tag{4.60}$$

where $\Gamma_{x/\beta}(\alpha + 1)$ is the incomplete gamma function tabulated in Karl Pearson, *Tables of the Incomplete Gamma Function,* Cambridge University Press, London, 1922.

Table 4.30 Some Properties of the Gamma Distribution
(See Fig. 4.14)

Moment-generating function.	$M_X(t) = (1 - \beta t)^{-(\alpha+1)}$	$t < 1/\beta$
Mean.	$\mu = \beta(\alpha + 1)$	
Variance.	$\sigma^2 = \beta^2(\alpha + 1)$	
Third central moment.	$\mu_3 = 2\beta^3(\alpha + 1)$	
Fourth central moment.	$\mu_4 = 3\beta^4$	
Coefficient of variation.	$\eta = 1/\sqrt{\alpha + 1}$	
Coefficient of skewness.	$\alpha_3 = 2/\sqrt{\alpha + 1}$	
Coefficient of kurtosis.	$\alpha_4 = 3 + 6/(\alpha + 1)$	

Reliability Functions

$$R(t) = 1 \qquad t \leq 0$$

$$\begin{aligned}
R(t) &= \int_t^{\infty} (1/\alpha!\beta^{\alpha+1})x^{\alpha}e^{-x/\beta}\, dx \\
&= 1 - (1/\alpha!)\Gamma_{x/\beta}(\alpha + 1) \qquad t > 0
\end{aligned} \tag{4.61}$$

$$h(t) = t^{\alpha}e^{-t/\beta} \Big/ \int_t^{\infty} x^{\alpha}e^{-x/\beta}\, dx \tag{4.62}$$

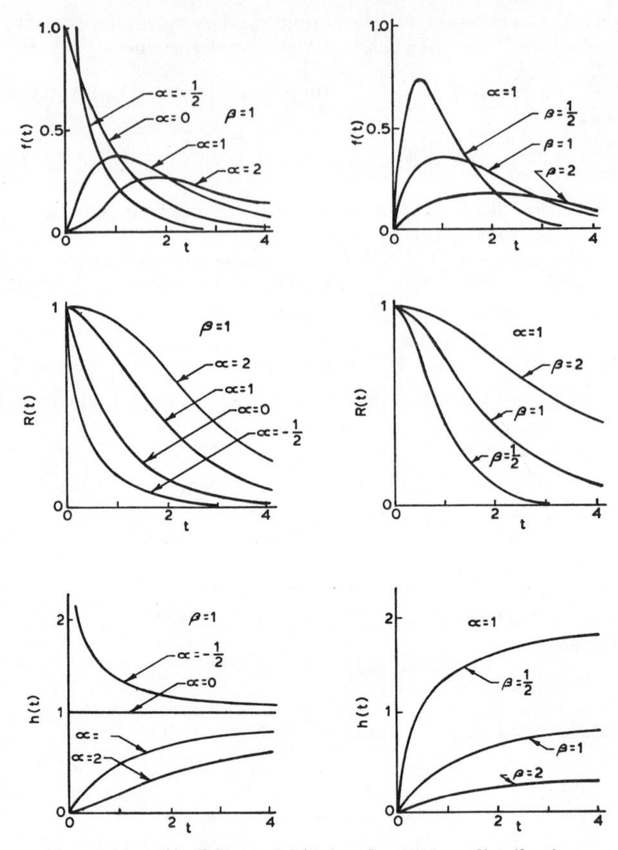

FIG. 4.14 $f(t)$, $R(t)$, and $h(t)$ for the gamma distribution.

Remarks

1. The distribution has a single mode at $x = \alpha\beta$, $\alpha \geq 0$.
2. The distribution becomes the exponential distribution (Sec. 4.5*g*) when $\alpha = 0$.
3. The failure rate is decreasing for $\alpha < 0$, constant for $\alpha = 0$, and increasing for $\alpha > 0$.
4. The incomplete gamma function is very difficult to evaluate; however, if α's are integers,

$$\int_t^\infty (1/\alpha!\beta^{\alpha+1})x^\alpha e^{-x/\beta}\,dx = \sum_{i=0}^{\alpha} e^{-t/\beta}(t/\beta)^i/i!$$

5. The sum of n independent gamma random variables with parameters β and α_i is a gamma random variable with parameter β and $\sum_{i=1}^{n} \alpha_i$.

Estimation of Parameters. Method of Moments. Using Table 4.30, we have

$$\beta(\alpha + 1) = \sum_{i=1}^{n} x_i/n = \bar{x}$$

$$\beta^2(\alpha + 1) = \sum_{i=1}^{n} (x_i - \bar{x})^2/(n - 1) = s^2$$

Solving for α and β, we have

Estimator for α: $$\hat{\alpha} = (\bar{x}/s)^2 - 1 \tag{4.63}$$
Estimator for β: $$\hat{\beta} = s^2/\bar{x} \tag{4.64}$$

Example 4.50. Suppose that the time to failure of an electrical component follows a gamma distribution. In testing five such components to failure without replacement, the results are 50, 75, 125, 250, and 300 hours. Using the method of moments find the estimators for α and β

$$\bar{x} = (50 + 75 + 125 + 250 + 300)/5 = 160 \text{ hours}$$

$$s^2 = \sum_{i=1}^{n} (x_i - \bar{x})^2/4 = 48{,}250/4 = 12{,}062.5 \text{ (hours)}^2$$

Thus,

$$\hat{\alpha} = 160^2/12062.5 - 1 = 1.12 \qquad \hat{\beta} = 12062.5/160 = 75.4$$

Method of Maximum Likelihood. The maximum-likelihood estimators of α and β are obtained by solving Eqs. (4.65) and (4.66).

$$-n \ln \beta - n \frac{\partial}{\partial \alpha} [\ln \Gamma(\alpha + 1)] + \sum_{i=1}^{n} \ln x_i = 0 \tag{4.65}$$

$$-n\beta(\alpha + 1) + \sum_{i=1}^{n} x_i = 0 \tag{4.66}$$

The solution of these two equations can be accomplished only by trial and error. A first approximation can be obtained by using the method of moments. Also, when α is not too small,

$$\frac{\partial}{\partial \alpha} [\ln \Gamma(\alpha + 1)] = \psi(\alpha + 1)$$
$$\cong \ln \left(\alpha + \frac{1}{2}\right) + 1 \Big/ 24 \left(\alpha + \frac{1}{2}\right)^2$$

If we assume that α is known, then from Eq. (4.66) we find

$$\hat{\beta} = \bar{x}/(\alpha + 1)$$

Example 4.51. Using the data of Example 4.50 and assuming that $\alpha = 1$, what is the maximum-likelihood estimator for β?

$$\hat{\beta} = \bar{x}/(\alpha + 1) = 160/2 = 80$$

Estimator and Confidence Limit for $R(t)$. When α is not known, approximate confidence limits for $R(t)$ can be obtained. For further discussion see Lloyd and Lipow [4]. However, when α is known, exact confidence limits for β, as well as for the reliability R, can be determined. By Eq. (4.66),

$$\hat{\beta} = \bar{t}/(\alpha + 1)$$

$2n\bar{t}/\beta$ has a χ^2 distribution with $2n(\alpha + 1)$ degrees of freedom. Thus,

$$P[1/\beta \leq \chi^2_{\gamma, 2n(\alpha+1)}/2n\bar{t}] = \gamma \tag{4.67}$$

With α known and $\hat{\beta} = \bar{t}/(\alpha + 1)$

$$\hat{R}(t) = \int_{t/\hat{\beta}}^{\infty} (u^\alpha e^{-u}/\alpha!)\, du$$

and, furthermore, if α is an integer,

$$\hat{R}(t) = \sum_{i=0}^{\alpha} e^{-t/\hat{\beta}}(t/\hat{\beta})^i/i!$$

The lower confidence limit on $R(t)$ is given by

$$\hat{R}_L(t) = \int_{t\,\chi^2_{\gamma,2n(\alpha+1)}/2n\bar{t}}^{\infty} (u^{\beta-1}e^{-u}/\alpha!)\,du \tag{4.68}$$

Example 4.52. Use the data of Example 4.51 to find a point estimate and 95 per cent lower confidence limit for $R(t)$ at $t = 400$ hours

$$\hat{R} = \sum_{i=0}^{1} e^{-t/\hat{\beta}}(t/\hat{\beta})^i/i! = \sum_{i=0}^{1} e^{-400/805}{}^i/i!$$
$$= 0.040$$

From Table A6

$$\chi^2_{\gamma,2n(\alpha+1)} = \chi^2_{0.95,20} = 31.410$$

therefore

$$\hat{R}_L = \sum_{i=0}^{1} e^{-400(31.410)/1600}[400(31.410)/1600]^i/i!$$
$$= 0.004$$

4.5f Beta Distribution. The *beta distribution* is defined by the two-parameter function

$$f(x{:}\alpha,\beta) = \frac{(\alpha+\beta+1)!}{\alpha!\beta!}x^{\alpha}(1-x)^{\beta} \qquad 0 < x < 1, \quad \alpha > -1, \quad \beta > -1$$
$$= 0 \qquad \text{elsewhere} \tag{4.69}$$

The generalized p.d.f. can be defined as

$$f(y{:}\alpha,\beta,\gamma,\eta) = \frac{(\alpha+\beta+1)!}{\alpha!\beta!\eta^{\alpha+\beta-1}}(y-\gamma)^{\alpha}(\gamma+\eta-y)^{\beta} \qquad \gamma < y < \gamma+\eta, \quad \gamma \text{ real}, \eta > 0$$
$$= 0 \qquad \text{elsewhere}$$

This generalized function covers a range of η; and since γ can be any real number, the distribution can be made to range over any arbitrary finite range. We need only study the properties of the random variable X, because $X = Y - \gamma/\eta$. The kth moment about the origin is

$$\nu_k = \frac{(\alpha+\beta+1)!}{(\alpha+\beta+k+1)!}\,\frac{(\alpha+k)!}{\alpha!}$$

$$F(x) = 0 \qquad x \le 0$$
$$= \int_0^x \frac{(\alpha+\beta+1)!}{\alpha!\beta!}x^{\alpha}(1-x)^{\beta}\,dx \qquad 0 < x < 1$$
$$= 1 \qquad x \ge 1 \tag{4.70}$$

Note

$$\int_0^x x^{\alpha}(1-x)^{\beta}\,dx = B_x(\alpha+1,\beta+1)$$

the incomplete beta function which is tabulated by Karl Pearson, *Tables of Incomplete Beta Function,* Cambridge University Press, London, 1932.

Table 4.31 Some Properties of Beta Distribution
(See Fig. 4.15)

Mean.	$\mu = \dfrac{\alpha + 1}{\alpha + \beta + 2}$
Variance.	$\sigma^2 = \dfrac{(\alpha + 1)(\beta + 1)}{(\alpha + \beta + 2)^2(\alpha + \beta + 3)}$
Third central moment. . . .	$\mu_3 = \dfrac{2(\alpha + 1)(\beta + 1)(\beta - \alpha)}{(\alpha + \beta + 4)(\alpha + \beta + 3)(\alpha + \beta + 2)^3}$
Fourth central moment. . .	$\mu_4 = \dfrac{3(\alpha + 1)(\beta + 1)}{(\alpha + \beta + 4)(\alpha + \beta + 3)(\alpha + \beta + 2)^3} \times \left[\dfrac{(\alpha + 2)(-\alpha + 2\beta + 1)}{\alpha + \beta + 5} + \dfrac{(\alpha + 1)(\alpha - \beta)}{\alpha + \beta + 2}\right]$
Coefficient of variation. . .	$\eta = \sqrt{\dfrac{\beta + 1}{(\alpha + 1)(\alpha + \beta + 3)}}$
Coefficient of skewness. . . .	$\alpha_3 = \dfrac{2(\beta - \alpha)}{\alpha + \beta + 4}\sqrt{\dfrac{\alpha + \beta + 3}{(\alpha + 1)(\beta + 1)}}$
Coefficient of kurtosis. . . .	$\alpha_4 = \dfrac{3(\alpha + \beta + 2)(\alpha + \beta + 3)}{(\alpha + \beta + 4)(\alpha + 1)(\beta + 1)} \times \left[\dfrac{(\alpha + 2)(-\alpha + 2\beta + 1)}{\alpha + \beta + 5} + \dfrac{(\alpha + 1)(\alpha - \beta)}{\alpha + \beta + 2}\right]$

Also,

$$\sum_{x=0}^{k} C(n,x)p^x q^{n-x} = \int_p^1 C(k + 1, n - k)y^k(1 - y)^{n-k-1}\,dy$$

for k an integer.

Reliability Function

$$\begin{aligned} R(t) &= 1 \qquad x \leq 0 &(4.71)\\ &= \int_t^1 \frac{(\alpha + \beta + 1)!}{\alpha!\beta!} x^\alpha(1 - x)^\beta\,dx = 1 - \frac{(\alpha + \beta + 1)!}{\alpha!\beta!} B_t(\alpha + 1, \beta + 1) \qquad 0 < t < 1\\ &= 0 \qquad x \geq 1 \end{aligned}$$

$$\begin{aligned} h(t) &= 0 \qquad x \leq 0 &(4.72)\\ &= \frac{t^\alpha(1 - t)^\beta}{\int_t^1 x^\alpha(1 - x)^\beta\,dx} \qquad \text{otherwise} \end{aligned}$$

Remarks

1. The mode of the distribution is at $\alpha/(\alpha + \beta)$.
2. Other tables of the incomplete beta function are:

a. C. M. Thompson, "Tables of Percentage Points of the Incomplete Beta Function," *Biometrika,* vol. 32, p. 151, 1941.

b. H. O. Hartley and E. R. Fitch, "A Chart for the Incomplete Beta Function and the Cumulative Binomial Distribution," *Biometrika,* vol. 38, p. 423, 1951.

c. The skewness of the distribution depends upon α and β as follows:

$$\begin{array}{ll} \alpha = \beta & \alpha_3 = 0\\ \alpha < \beta & \alpha_3 > 0\\ \alpha > \beta & \alpha_3 > 0 \end{array}$$

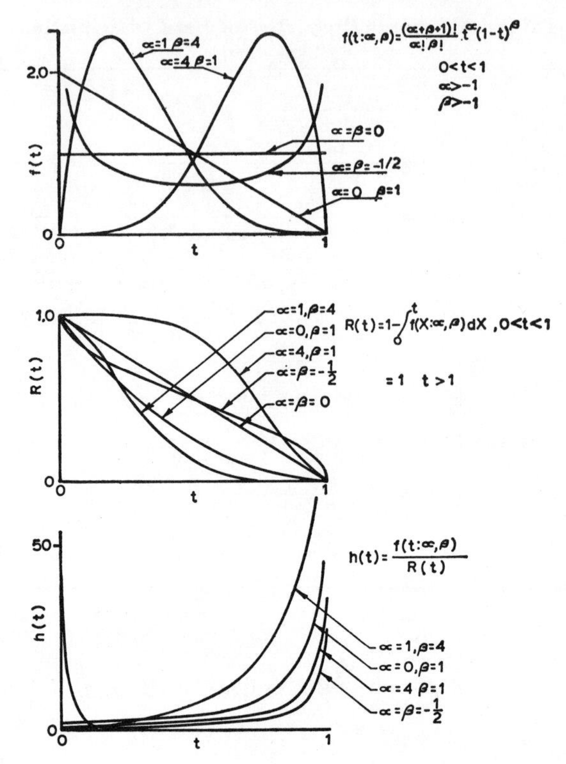

FIG. 4.15 $f(t)$, $R(t)$, and $h(t)$ for the beta distribution.

d. When $\alpha = \beta = 0$, the beta distribution becomes the uniform distribution (Sec. 4.5*a*).

e. With $\alpha = \beta = -\frac{1}{2}$, the p.d.f. is the arc sine.

f. Further details of the distribution may be found in Kao [8].

Estimation of Parameters α and β. Estimates of the beta distribution may be found by the methods of moments or maximum likelihood (Sec. 4.7*a*).

4.5g Exponential Distribution. One of the most widely used distributions in the field of reliability is the one-parameter exponential function defined by

$$f(x:\theta) = \frac{1}{\theta} e^{-x/\theta} \qquad x \geq 0 \tag{4.73}$$
$$= 0 \qquad \text{elsewhere}$$

$$F(x) = \frac{1}{\theta}\int_0^x e^{-x/\theta}\,dx = 1 - e^{-x/\theta} \qquad x > 0 \tag{4.74}$$
$$= 1 \qquad x \leq 0$$

Reliability Functions

$$R(t) = \int_t^\infty (1/\theta)e^{-x/\theta}\,dx = e^{-t/\theta} \qquad t \geq 0 \tag{4.75}$$
$$= 1 \qquad t < 0$$

$$h(t) = (1/\theta)e^{-t/\theta}/e^{-t/\theta} = 1/\theta \tag{4.76}$$

Table 4.32 Some Properties of the Exponential Distribution
(See Fig. 4.16)

Moment-generating function	$M_X(t) = (1 - \theta t)^{-1}$	$t < 1/\theta$
Mean	$\mu = \theta$	
Variance	$\sigma^2 = \theta^2$	
Third central moment	$\mu_3 = 2\theta^3$	
Fourth central moment	$\mu_4 = 9\theta^4$	
Coefficient of variation	$\eta = 1$	
Coefficient of skewness	$\alpha_3 = 2$	
Coefficient of kurtosis	$\alpha_4 = 9$	

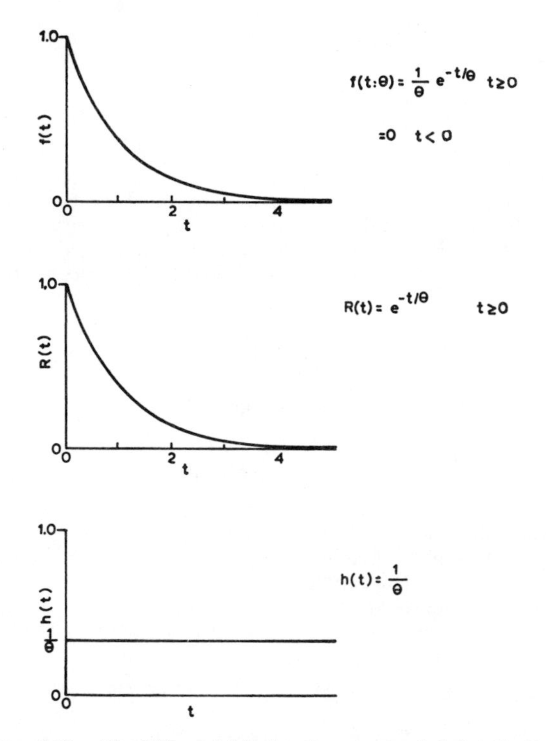

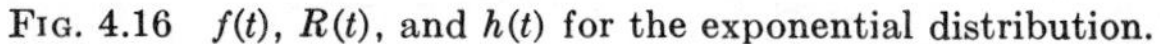

FIG. 4.16 $f(t)$, $R(t)$, and $h(t)$ for the exponential distribution.

Remarks

1. The distribution is a special case of both the gamma and Weibull distributions.
2. It is characterized by a constant hazard function $1/\theta$, which is also the parameter of the distribution; i.e., if a system has survived to time t, the probability of survival for the next increment of time is the same as if it had just been placed into service. This assumption neglects degradation failures.
3. The reciprocal of the hazard function is the mean time between failures (MTBF).
4. It is the distribution which is expected when the mechanisms are so complex that many deteriorations with different failure rates are operable.
5. When parts have an exponential failure distribution function, the equipment consisting of these parts also has an exponential distribution function and the hazard function is the sum of the part hazard functions. *Note.* Must assume independence of part failures.

Estimation of Parameter θ. The maximum-likelihood estimator for θ is

$$\hat{\theta} = \sum_{i=1}^{n} x_i/n = \bar{x} \tag{4.77}$$

When units which are selected at random are placed on test, the test may be terminated prior to all units having failed. Also, the units which fail may or may not be replaced. If n units are placed on test and time-to-failure data are arranged in increasing order, that is, $0 \leq t_1 \leq t_2 \cdots \leq t_r$ (r units failing), the maximum-likelihood estimators for θ are as follows.

No Replacement of Failed Units

$$\hat{\theta} = \sum_{i=1}^{n} t_i/n \qquad \text{for } r = n \tag{4.78}$$

$$= \left[\sum_{i=0}^{r} t_i + (n - r)t_r\right]/r \qquad \text{for } r < n \tag{4.79}$$

Note. The numerator in both cases is the total time accumulated by all items.

Replacement of Failed Units. Let X_i represent the time between the $(i - 1)$st and ith failure. The X_i's are independent identically exponentially distributed random variables. (If individual time to failure t_i is exponentially distributed, then time between failures X_i is also exponentially distributed.) The maximum-likelihood estimator for θ is

$$\hat{\theta} = \left(n \sum_{i=1}^{r} X_i\right)/r = T_{nt_r/r} \tag{4.80}$$

The *confidence interval*[1] *for the parameter θ* is given by

$$P\left[\frac{2r\hat{\theta}}{\chi^2_{(1+\gamma)/2,\,2r}} < \theta < \frac{2r\hat{\theta}}{\chi^2_{(1-\gamma)/2,\,2r}}\right] = \gamma \tag{4.81}$$

Note. If test is terminated at a predetermined time t, the degree of freedom associated with the lower limit is $2r + 2$ instead of $2r$.

Example 4.53. Ten units have been placed on life test and the failed units were not replaced. The first five failures were 50, 75, 125, 250, and 300 hours. Assuming the life of the product follows an exponential distribution, determine the point estimate and the 95 per cent lower confidence interval estimate for the reliability of the units at $t = 400$ hours. From Eq. (4.79)

$$\hat{\theta} = [50 + 75 + 125 + 250 + 300 + 5(300)]/5 = 460 \text{ hours}$$

Therefore,

$$\hat{R} = e^{-t/\hat{\theta}} = e^{-400/460} = 0.419$$

For lower confidence limit

$$P[\theta > 2r\hat{\theta}/\chi^2_{0.95,10} = (10 \times 460)/18.307 = 251.3] = 0.95$$

Therefore,

$$\hat{R}_L = e^{-t\frac{\chi^2_{0.95,10}}{2r\theta}} = e^{-400/251.3}$$
$$= e^{-1.59} = 0.204$$

Note. If $\hat{\theta}$ is a maximum-likelihood estimate of θ, then $\hat{R}(t) = e^{-t/\hat{\theta}}$ is a maximum-likelihood estimate provided $R(t)$ is monotonic. Substituting the lower confidence limit for θ in the reliability function results in a lower confidence limit for $R(t)$. When the reliability function contains two or more unknown parameters, we can obtain a point estimate for θ by using the maximum-likelihood estimates for the parameters in the reliability. In general, however, we cannot substitute the lower confidence limits for the parameters to obtain a lower confidence limit on reliability.

[1] B. Epstein, "Testing of Hypotheses," *Wayne State Univ. Tech. Rept.* 3, 1958.

Example 4.54. Suppose the five failed units of Example 4.53 were replaced. From Eq. (4.80)

$$\hat{\theta} = n \sum_{i=1}^{5} X_i/r = 10[50 + 25 + 50 + 125 + 50]/5 = 600 \text{ hours}$$

Therefore,

$$\hat{R} = e^{-400/600} = 0.512 \qquad \text{and} \qquad \hat{R}_L = e^{-(400\times 18.307/2\times 5\times 600)} = 0.293$$

Since χ^2 is approximately normally distributed for large values of degrees of freedom, it is possible to use the normal distribution for approximating the confidence limits. Let $\hat{\theta} = \bar{t}$; then $\bar{t}$ is asymptotically $N(\theta,\theta^2/n)$ and $(\bar{t} - \theta)\sqrt{n}/\theta$ is approximately $N(0,1)$. Thus,

$$P[(\bar{t} - \theta)\sqrt{n}/\theta < z_\gamma] = \gamma$$

and

$$P[\bar{t}/(1 + z_\gamma/\sqrt{n}) < \theta] = \gamma$$

$$\therefore\ \hat{R}_L(t) = e^{-(t/\bar{t})(1+z_\gamma/\sqrt{n})} \tag{4.82}$$

Test of Hypothesis. The χ^2 distribution is used to test the hypothesis that the parameter θ is equal to a value.

$$H_0{:}\theta = \theta_0 \qquad H_1{:}\theta \neq \theta_0$$

Let α = significance level of the test. Calculate

$$\chi^2 = \hat{\theta}2r/\theta$$

and if

$$\chi^2 < \chi^2_{\alpha/2,2r} \qquad \text{or} \qquad \chi^2 > \chi^2_{1-\alpha/2,2r}$$

reject H_0. In the field of reliability, one is interested in life of the item for a period T, that is, demonstration of $\hat{R}_L$ with coefficient γ.

$$\hat{R}_L = e^{-t\chi^2_{\gamma,2r}/2r\hat{\theta}} \qquad \text{or} \qquad \hat{\theta}/t = \chi^2_{\gamma,2r}/2r \ln(1/\hat{R}_L) \tag{4.83}$$

Example 4.55. Suppose $\hat{R}_L = 0.90$, $\gamma = 0.90$, $t = 100$ hours. Eight units are tested until the fourth failure is observed. Failures for a given test were observed at 175, 250, 500, 600 hours.

$$\chi^2_{\gamma,2r}/2r = 13.4/8 \qquad \ln(1/\hat{R}_L) = 0.10535$$

$$\hat{\theta}_c \geq 13.4/8 \times 100/0.10535 \cong 1{,}595 \text{ hours}$$

$$\hat{\theta} = [175 + 250 + 500 + 600 + 4(600)]/4 = 981.25 \text{ hours}$$

Since $\hat{\theta}$ (observed) $\leq \hat{\theta}_c$, we reject the hypothesis that the reliability of the units for 100 hours is greater than 0.90.

Note. The sample mean of the times to failure must exceed 1,595 hours to demonstrate reliability of 0.90 with 90 per cent confidence. In testing n units until r units have failed, what is the *expected waiting times* for the rth failure? Epstein has shown that for the *without-replacement* case[1]

$$E(t) = \theta \sum_{i=1}^{r} 1/(n - i + 1)$$

and for the *replacement case*[2]

$$E(t) = \theta r/n$$

The sequential probability ratio test for parameter θ is discussed in Sec. 4.10.

[1] B. Epstein and N. Sobel, "Life Testing," *J. Am. Statist. Assoc.*, vol. 48, p. 486, 1953.

[2] B. Epstein, "Truncated Life Tests in the Exponential Case," *Ann. Math. Statist.*, vol. 25, p. 555, 1954.

Reliability of Parallel Systems. The reliability of n components in parallel is

$$R(t) = 1 - \prod_{i=1}^{n} [1 - R_i(t)] \tag{4.84}$$

where $R_i(t)$ is the reliability of the ith component. If $R_i(t) = e^{-t/\theta_i}$, then

$$R_i(t) = 1 - \frac{t}{\theta_i} + \left(\frac{t}{\theta_i}\right)^2 \frac{1}{2!} - \left(\frac{t}{\theta_i}\right)^3 \frac{1}{3!} + \cdots$$

and

$$1 - R_i(t) = \frac{t}{\theta_i} - \left(\frac{t}{\theta_i}\right)^2 \frac{1}{2!} + \left(\frac{t}{\theta_i}\right)^3 \frac{1}{3!} - \cdots$$

If t/θ_i is much smaller than 1,

$$1 - R_i(t) \simeq t/\theta_i$$

Thus

$$R(t) \simeq 1 - \prod_{i=1}^{n} t/\theta_i \tag{4.85}$$

For three parallel components with parameters θ_1, θ_2, θ_3,

$$R(t) = \sum_{i=1}^{3} e^{-t/\theta_i} - \sum_{i=1}^{2} \sum_{j=i+1}^{3} e^{-t(1/\theta_i + 1/\theta_j)} + e^{-t(1/\theta_1 + 1/\theta_2 + 1/\theta_3)}$$

If the $\theta_i = \theta$, $i = 1, 2, 3$,

$$R(t) = 3e^{-t/\theta} - 3e^{-2t/\theta} + e^{-3t/\theta}$$

Reliability of Standby Systems. A standby system is an arrangement in which one or more systems is standing by to take over the operation when each succeeding system fails. Such an arrangement would normally require failure-sensing and switching mechanisms. The reliability of a system with one standby and

$$R_i(t) = e^{-\lambda_i t} \qquad i = 1, 2$$

is

$$R(t) = e^{-\lambda_1 t} + [\lambda_1/(\lambda_2 - \lambda_1)](e^{-\lambda_1 t} - e^{-\lambda_2 t}) \tag{4.86}$$

If the reliability of the sensing and switching mechanism is R_s instead of 1,

$$R(t) = e^{-\lambda_1 t} + R_s[\lambda_1/(\lambda_2 - \lambda_1)](e^{-\lambda_1 t} - e^{-\lambda_2 t}) \tag{4.87}$$

If λ_i's are equal, Eq. (4.86) becomes

$$R(t) = e^{-\lambda t}(1 + \lambda t) \tag{4.88}$$

and Eq. (4.87) becomes

$$R(t) = e^{-\lambda t}(1 + R_s \lambda t) \tag{4.89}$$

In general, when there are $n - 1$ standby units with $R_i(t) = e^{-\lambda t}$, $i = 1, 2, \ldots, n$

$$R(t) = e^{-\lambda t}[1 + \lambda t + (\lambda t)^2/2! + \cdots + (\lambda t)^n/n!] \tag{4.90}$$

Example 4.56. Suppose the failure rate of a component is $\lambda = 0.006$. For a 10-hour mission and one standby unit

$$R = e^{-\lambda t}(1 + \lambda t) = e^{-0.06}(1.06) = 0.998$$

For two components in parallel

$$R = 1 - (1 - e^{\lambda t})^2 = 0.997$$

4.5h Weibull Distribution. The Weibull distribution[1] is defined by the three-parameter function

$$f(x{:}\alpha,\beta,\gamma) = (\beta/\alpha)(x-\gamma)^{\beta-1}e^{-(x-\gamma)^{\beta}/\alpha} \qquad x \geq \gamma \qquad (4.91)$$

$$\gamma \geq 0,\ \beta > 0$$
$$\alpha > 0$$

$$= 0 \qquad \text{elsewhere}$$

with α = scale parameter (Goode and Kao [9] use $\eta = \alpha^{1/\beta}$)
β = the shape parameter
γ = the location parameter

With $\gamma = 0$, the kth moment about the origin is given by

$$\nu_k = \alpha^{k/\beta}\Gamma(1 + k/\beta)$$

$$F(x) = 0 \qquad x < \gamma \qquad (4.92)$$
$$= 1 - e^{-(x-\gamma)^{\beta}/\alpha} \qquad x \geq \gamma \geq 0,\ \alpha > 0,\ \beta > 0$$

Table 4.33 Some Properties of the Weibull Distribution ($\gamma = 0$)
(See Fig. 4.17)

Mean	$\mu = \alpha^{1/\beta}\Gamma(1 + 1/\beta)$
Variance	$\sigma^2 = \alpha^{2/\beta}[\Gamma(1 + 2/\beta) - \{\Gamma(1 + 1/\beta)\}^2]$
Third central moment	$\mu_3 = \alpha^{3/\beta}[\Gamma(1 + 3/\beta) - 3\Gamma(1 + 2/\beta) \times \Gamma(1 + 1/\beta) + 2\{\Gamma(1 + 1/\beta)\}^3]$
Fourth central moment	$\mu_4 = \alpha^{4/\beta}[\Gamma(1 + 4/\beta) - 4\Gamma(1 + 3/\beta) \times \Gamma(1 + 1/\beta) + 6\Gamma(1 + 2/\beta)\{\Gamma(1 + 1/\beta)\}^2 - 3\{\Gamma(1 + 1/\beta)\}^3]$
Coefficient of variation	$\eta = \sqrt{\Gamma(1 + 2/\beta)/\{\Gamma(1 + 1/\beta)\}^2 - 1}$
Coefficient of skewness	$\alpha_3 = \dfrac{\Gamma(1 + 3/\beta) - 3\Gamma(1 + 2/\beta)\Gamma(1 + 1/\beta) + 2\{\Gamma(1 + 1/\beta)\}^2}{[\Gamma(1 + 2/\beta) - \{\Gamma(1 + 1/\beta)\}^2]^{3/2}}$
Coefficient of kurtosis	$\alpha_4 = \dfrac{\Gamma(1 + 4/\beta) - 4\Gamma(1 + 3/\beta)\Gamma(1 + 1/\beta) + 6\Gamma(1 + 2/\beta)\{\Gamma(1 + 1/\beta)\}^2 - 3\{\Gamma(1 + 1/\beta)\}^3}{[\Gamma(1 + 2/\beta) - \{\Gamma(1 + 1/\beta)\}^2]^2}$

Reliability Functions

$$R(t) = 1 \qquad t < \gamma \qquad (4.93)$$
$$= e^{-(t-\gamma)^{\beta}/\alpha} \qquad t \geq \gamma$$
$$h(t) = (\beta/\alpha)(t-\gamma)^{\beta-1} \qquad t \geq \gamma \qquad (4.94)$$

Remarks

1. The distribution has a single mode at

$$x = \gamma + [\alpha(1 - 1/\beta)]^{1/\beta} \qquad \beta > 1$$

2. The distribution becomes the exponential distribution when $\beta = 1$.
3. γ is usually assumed to be zero. A value of γ less than zero could indicate failure in storage.
4. Since $E(X) = \alpha^{1/\beta}\Gamma(1 + 1/\beta)$, the mean time to failure, has two parameters, the reliability cannot be obtained directly as in the case of the exponential.
5. One of the main attractions of the Weibull distribution is the many shapes of the failure-rate function (Fig. 4.17).

[1] Originally proposed by a Swedish investigator of metallic fatigue phenomena, W. Weibull, "A Statistical Representation of Fatigue Failures in Solids," *Roy. Inst. Technology (Stockholm)*, November, 1954.

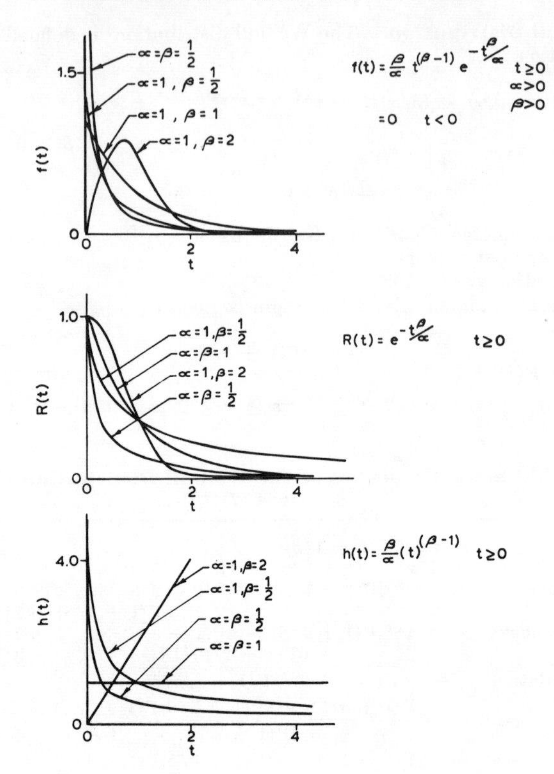

FIG. 4.17 $f(t)$, $R(t)$, and $h(t)$ for the Weibull distribution.

Estimation of Parameters. The maximum-likelihood estimates of α and β are obtained by solving

$$n\alpha - \sum_{i=1}^{n} t_i^{\beta} = 0 \tag{4.95}$$

$$n/\beta + \sum_{i=1}^{n} \ln t_i - (1/\alpha) \sum_{i=1}^{n} t_i^{\beta} \ln t_i = 0 \tag{4.96}$$

An initial value for β may be obtained by using Eq. (4.95) and $E(T) = \alpha^{1/\beta}\Gamma(1 + 1/\beta)$. Then find $\hat{\alpha}$ and $\hat{\beta}$ by an iteration procedure. Most investigations of parameter estimation assume β known. If β is known, then

$$\hat{\alpha} = \left(\sum_{i=1}^{r} t_i^{\beta} + (n - r)t_r^{\beta} \right) / r \tag{4.97}$$

and $2r\hat{\alpha}/\alpha$ is distributed as χ^2 with $2r$ degrees of freedom. Thus, the lower 100γ per cent confidence interval on α is given by

$$P\left[\frac{2r\hat{\alpha}}{\alpha} \leq \chi^2_{\gamma,2r}\right] = \gamma \tag{4.98}$$

and

$$\hat{R}_L(t) = e^{-t^{\beta}\chi^2_{\gamma,2r}/2r\hat{\alpha}} \tag{4.99}$$

Example 4.57. Assume that the life-length distribution of Example 4.50 is a Weibull distribution with $\beta = 1.2$. Determine a point estimate and the 95 per cent lower confidence interval estimate for the reliability of the units at $t = 400$ hours. From Eq. (4.97)

$$
\begin{aligned}
\hat{\alpha} &= \left[\sum_{i=1}^{5} t_i^{1.2} + (n - r)t_r^{1.2}\right]/5 \\
&= (2308.5 + 4{,}693.7)/5 = 1{,}400 \\
R &= e^{-t^{\beta}/\hat{\alpha}} = e^{-400^{1.2}/1{,}400} = 0.387 \\
\hat{R}_L &= e^{-t^{\beta}\chi^2_{\gamma,2r}/2r\hat{\alpha}} = e^{-400^{1.2}(18.307)/10\times 1{,}400} \\
&= 0.176
\end{aligned}
$$

A simpler method of estimating these parameters uses the Weibull probability paper, which is developed from the cumulative distribution function as follows

$$
\ln \ln 1/[1 - F(X)] = \beta \ln (X - \gamma) - \ln \alpha \qquad (4.100)
$$

on ln versus ln ln graph paper. Equation (4.100) is sketched in Fig. 4.18.

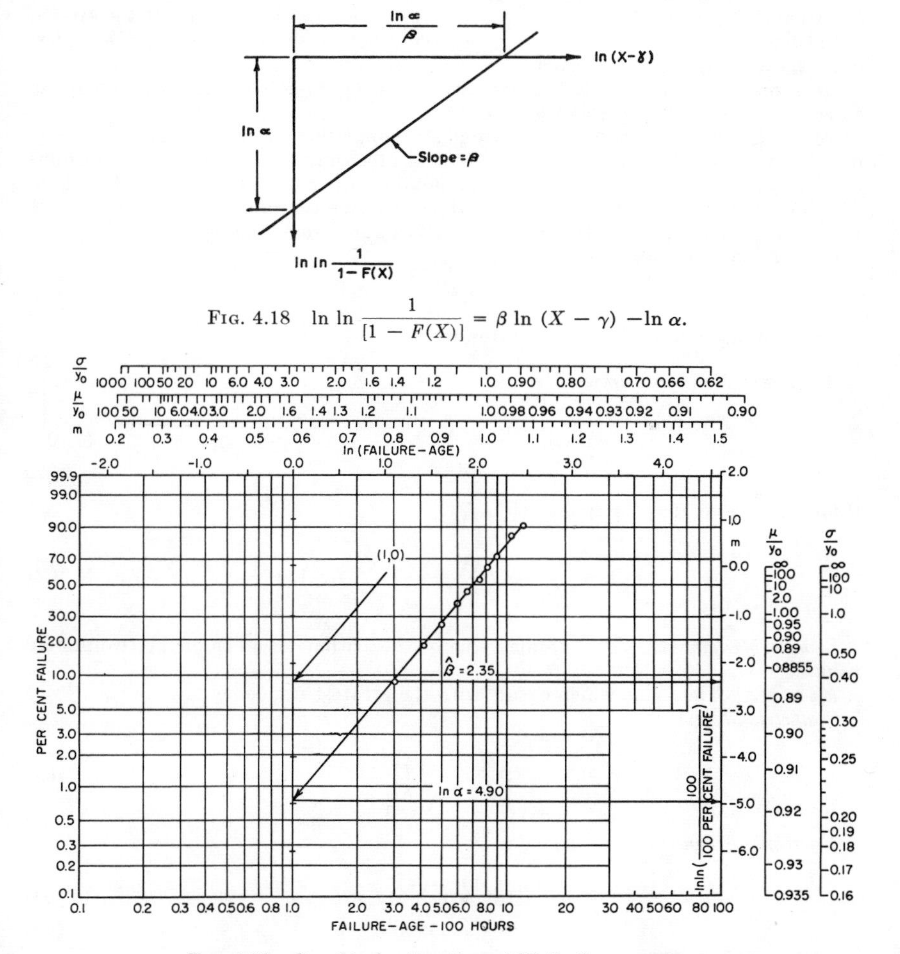

FIG. 4.18 $\ln \ln \dfrac{1}{[1 - F(X)]} = \beta \ln (X - \gamma) - \ln \alpha$.

FIG. 4.19 Graphical estimation of Weibull parameters.

Example 4.58. Suppose the 10 ordered observations in a life test are 300, 410, 500, 600, 660, 750, 825, 900, 1,050, and 1,200 hours. If the Weibull distribution fits the data, what are the estimates of the parameters of the distribution using Weibull probability paper? The corresponding plotting positions are given by $i/(10 + 1)$, that is, 0.09, 0.18, 0.27, 0.36, 0.46, 0.55, 0.64, 0.73, 0.82, and 0.91, respectively. Since these points plotted on Weibull probability paper (Fig. 4.19) result in a straight line, $\hat{\gamma} = 0$. The line gives $\hat{\beta} = 2.35$ and

$$\ln \hat{\alpha} = 4.90 \qquad \hat{\alpha} = 134.3 \text{ (X in hundreds of hours)}$$

For complete details on the use of Weibull Probability paper, see Ref. 9.

4.6 SAMPLING

When investigating the reliability of equipment, it becomes necessary to observe and record data. Since in most instances it is not possible to obtain the data for the total population, only a part of the population is selected and analyzed. The data are basically *descriptive* if we are dealing with the population. In case of *samples*, i.e., part of the population, there are not only the descriptive data about the sample but also informative data about the population. There are several types of samples, but *random* samples are preferred.

Definition. A *random sample* is one such that every member of the population has an equal chance of being in the sample.

Note. The set of *variates* picked from the population as observations are designated as x_i, $i = 1, 2, 3, \ldots, n$. As noted in preceding sections, the constant numbers, parameters, which describe the population are designated by Greek letters. The corresponding numbers, statistics, which describe the sample are designated by Roman letters (χ^2 is an exception). Statistics vary from sample to sample. Statistics are functions of x_i

$$S = f(x_1, x_2, \ldots, x_n)$$

4.6a Measures of Central Tendency

Arithmetic Mean or Sample Mean $\bar{x}$

$$\bar{x} = \sum_{i=1}^{n} x_i/n \tag{4.101}$$

If for each value of x_i the frequency is f_i

$$\bar{x} = \sum_{i=1}^{k} f_i x_i \Big/ \sum_{i=1}^{k} f_i = \Big(\sum_{i=1}^{k} f_i x_i\Big)\Big/n \tag{4.102}$$

Sample Median M_e. The middle value or the arithmetic mean of the two middle values.

Mode M_o. The value which occurs with the greatest frequency.

Geometric Mean

$$\text{GM} = \sqrt[n]{\prod_{i=1}^{n} x_i} \tag{4.103}$$

Harmonic Mean

$$\text{HM} = n \sum_{i=1}^{n} 1/x_i \tag{4.104}$$

Note. $\text{HM} \le \text{GM} \le \bar{x}$

Example 4.59. Given that five units failed at 75, 100, 110, 120, 130 hours, calculate the measures of central tendency.

$$\bar{x} = 535/5 = 107 \text{ hours} \qquad \text{GM} = \sqrt[5]{\Pi x_i} = 105.2 \text{ hours}$$

$$M_e = 110 \text{ hours} \qquad \text{HM} = 5 \Big/ \sum_{i=1}^{n} 1/x_i = 103.2 \text{ hours}$$

$$M_o = \text{no unique mode}$$

4.6b Measures of Dispersion. *Range.* Let x_{min} denote the smallest of the x_i and x_{max} the largest; then the *sample range* is

$$R = x_{max} - x_{min} \tag{4.105}$$

Mean or Average Deviation

$$\text{AD} = \left(\sum_{i=1}^{n} |x_i - \bar{x}| \right) / n \tag{4.106}$$

Sample Variance

$$s^2 = \left[\sum_{i=1}^{n} (x_i - \bar{x})^2 \right] / (n - 1) \tag{4.107}$$

If for each value of x_i the frequency is f_i,

$$s^2 = \sum_{i=1}^{k} f_i(x_i - \bar{x})^2 / \left(\sum_{i=1}^{k} f_i - 1 \right) = \left[\sum_{i=1}^{k} f_i(x_i - \bar{x})^2 \right] / (n - 1) \tag{4.108}$$

Sample Standard Deviation

$$s = \sqrt{s^2} \tag{4.109}$$

Example 4.60. Compute the measures of dispersion for the data of Example 4.59:

$$R = 130 - 75 = 55 \text{ hours} \qquad s^2 = 1{,}780/4 = 445 \text{ (hours)}^2$$
$$\text{AD} = 78/5 = 15.6 \text{ hours} \qquad s = 21.1 \text{ hours}$$

Theorem. If $x_i = u_i \pm k$, $i = 1, 2, \ldots, n$, then $\bar{x} = \bar{u} \pm k$ and $s_x = s_\mu$.
Theorem. If $x_i = ku_i$, $i = 1, 2, \ldots, n$, then $\bar{x} = k\bar{u}$ and $s_x = ks_u$.
Theorem. If

$$u_i = (x_i \pm k)/c \qquad i = 1, 2, \ldots, n$$

then $\bar{x} = c\bar{u} \mp k$ and $s_x = cs_u$.
Theorem. For x_i, $i = 1, 2, \ldots, n$

$$s_x^2 = \left[\sum_{i=1}^{n} (x_i - \bar{x})^2 \right] / (n - 1) = \left[\sum_{i=1}^{n} x_i^2 - (\Sigma x_i)^2/n \right] / (n - 1) \tag{4.110}$$

Theorem. If $\bar{x}_1$ and s_1^2 are the mean and variance of x_{1i}, $i = 1, 2, \ldots, n_1$, and $\bar{x}_2$ and s_2^2 are the mean and variance of x_{2i}, $i = 1, 2, \ldots, n_2$, then the mean and variance of the combined set are given by

$$\bar{x} = \frac{n_1\bar{x}_1 + n_2\bar{x}_2}{n_1 + n_2} \tag{4.111}$$

$$s^2 = \frac{(n_1 - 1)s_1^2 + (n_2 - 1)s_2^2 + n_1(\bar{x}_1 - \bar{x})^2 + n_2(\bar{x}_2 - \bar{x})^2}{n_1 + n_2 - 1} \tag{4.112}$$

where $\bar{x}$ is defined by Eq. (4.111).

Example 4.61. Compute the mean and variance for the following failure data

x_i, hours	f_i	$u_i = (x_i - 350)/50$	$f_i u_i$	$f_i u_i^2$
250	1	−2	−2	4
300	2	−1	−2	2
350	4	0	0	0
400	3	+1	+3	3
450	2	+2	+4	8
Totals	12		+3	17

$$\bar{u} = 3/12 = 0.25 \qquad s_u^2 = (17 - 9/12)/11 = 16.25/11$$
$$\bar{x} = 50\bar{u} + 350 \qquad s_x^2 = 50^2(16.25/11)$$
$$= 362.5 \text{ hours} \qquad = 3{,}693.18 \text{ hours}^2$$

4.6c Sampling Distributions. *Definition.* When all possible samples of size n are selected from a population (with or without replacement) and a statistic, e.g., the mean, is computed for each sample, the resulting distribution of the statistic is called its *sampling distribution.*

Theorem. If

$$\bar{x} = \sum_{i=1}^{n} x_i/n$$

then the sampling distribution of the means has for infinite population or sampling with replacement

$$\mu_{\bar{x}} = \mu_x \quad \text{and} \quad \sigma_{\bar{x}}^2 = \sigma_x^2/n \tag{4.113}$$

and for sampling from a finite population of size N

$$\mu_{\bar{x}} = \mu_x \quad \text{and} \quad \sigma_{\bar{x}}^2 = (\sigma^2/n)[(N - n)/(N - 1)] \tag{4.114}$$

Note. As long as the population has a finite variance, the larger the sample, the more certain we are that the sample mean will be close to the unknown population mean.

Theorem (Central Limit). If a population has a mean μ and a finite variance σ^2, then the sampling distribution of means approaches the normal distribution with mean μ and variance σ^2/n as n increases; i.e., the sampling distribution of means is asymptotically normal.

Note. The approximation is reasonably good for $n \geq 30$ and $N > 2n$.

Tchebycheff's inequality states that

$$P[|X - \mu| \geq k\sigma] \leq 1/k^2$$

with no assumption concerning the population except existence of μ and σ^2.

Note. If nothing is known about the population, then at most $1/4 = 0.25$ or 25 per cent of the population is more than 2σ from the mean.

Theorem. If a random sample of size n_1 is selected from a population with mean μ_1 and variance σ_1^2 and if a random sample of size n_2 is selected from a population with mean μ_2 and variance σ_2^2, then $y = \bar{x}_1 - \bar{x}_2$ has

$$\mu_y = \mu_{\bar{x}_1 - \bar{x}_2} = \mu_1 - \mu_2$$

and

$$\sigma_y^2 = \sigma^2_{\bar{x}_1 - \bar{x}_2} = \sigma_{\bar{x}_1}^2 + \sigma_{\bar{x}_2}^2 = \sigma_1^2/n_1 + \sigma_2^2/n_2 \tag{4.115}$$

where $\bar{x}_1$ is mean of the first sample and $\bar{x}_2$ is mean of the second sample.

Note. If sampled populations are normal, so is the distribution of y. If sampled populations are not normal but sample sizes are sufficiently large, the central-limit theorem may be used.

Chi-square Distribution. Let $x_1, x_2, \ldots, x_\nu$ be independent random variables normally distributed with zero mean and unit variance. Let

$$\chi^2 = x_1^2 + x_2^2 + \cdots + x_\nu^2 = \sum_{i=1}^{\nu} x_i^2$$

then χ^2 is a random variable whose distribution function is

$$f(\chi^2) = \frac{1}{2^{\nu/2}\Gamma(\nu/2)} (\chi^2)^{(\nu-2)/2} e^{-\chi^2/2} \qquad \chi^2 \geq 0 \tag{4.116}$$
$$= 0 \qquad \text{elsewhere}$$

Table 4.34 Some Properties of the χ^2 Distribution
(See Fig. 4.20)

Mean	$E[\chi^2] = \nu$
Variance	$\sigma^2_{\chi^2} = 2\nu$
Third central moment	$\mu_3 = 8\nu$
Fourth central moment	$\mu_4 = 12\nu(\nu + 4)$
Coefficient of variation	$\eta = \sqrt{2/\nu}$
Coefficient of skewness	$\alpha_3 = 2\sqrt{2/\nu}$
Coefficient of kurtosis	$\alpha_4 = 3 + 12/\nu$

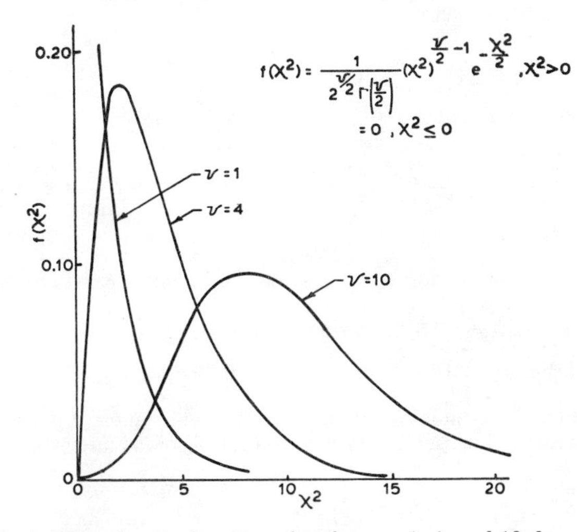

FIG. 4.20 Probability density function of χ^2 for $\nu = 1$, 4, and 10 degrees of freedom.

(ν = degrees of freedom: the number of independent random variables whose squares are being added.) The kth moment about the origin is

$$\nu_k = \nu(\nu + 2) \cdots (\nu + 2k - 2)$$

$$F(\chi^2_{p,\nu}) = P(\chi^2 \leq \chi^2_{p,\nu}) = \int_0^{\chi^2_{p,\nu}} f(\chi^2)\, d\chi^2 = p \tag{4.117}$$

which is tabulated in Table A6. For example, $\nu = 8$

$$P(\chi^2 \leq \chi^2_{0.95,8}) = P(\chi^2 \leq 15.5) = 0.95$$

Remarks

1. The distribution has a single mode at $\chi^2 = \nu - 2$ for $\nu \geq 2$.
2. If χ_1^2 and χ_2^2 are independent chi-square random variables with ν_1 and ν_2 degrees of freedom, respectively, then the sum $\chi_1^2 + \chi_2^2$ is a chi-square distribution with $\nu_1 + \nu_2$ degrees of freedom.
3. χ^2 is asymptotically normal as $\nu \to \infty$ with mean ν and variance 2ν.
4. $\sqrt{2\chi^2}$ is asymptotically normal as $\nu \to \infty$ with mean $\sqrt{2\nu - 1}$ and variance 1.

Thus, a useful approximation is

$$\chi^2_{p,\nu} \cong \tfrac{1}{2}[z_p + \sqrt{2\nu - 1}]^2 \qquad \nu > 30$$

For small or large p use

$$\chi^2_{p,\nu} \cong \nu[1 - 2/9\nu + z_p \sqrt{2/9\nu}]^3$$

5. If the random variable X is $N(\mu,\sigma^2)$, then for a random sample of size n the quantity

$$U = \sum_{i=1}^{n} (x_i - \mu)^2/\sigma^2$$

is distributed as χ^2 with n degrees of freedom.

Theorem. If x_i, $i = 1, 2, \ldots, n$, are random samples from a normal population with mean μ and variance σ^2, then $(n - 1)s^2/\sigma^2$ is distributed as chi-square with $n - 1$ degrees of freedom, and

$$f(s^2) = \frac{1}{\Gamma[(n-1)/2]} \left(\frac{n-1}{2\sigma^2}\right)^{(n-1)/2} (s^2)^{(n-3)/2} e^{-(n-1)s^2/2\sigma^2} \qquad s > 0$$
$$= 0 \qquad \text{elsewhere} \tag{4.118}$$

and

$$f(s) = \frac{2}{\Gamma[(n-1)/2]} \left(\frac{n-1}{2\sigma^2}\right)^{(n-1)/2} s^{n-2} e^{-(n-1)s^2/s\sigma^2} \qquad s > 0$$
$$= 0 \qquad \text{elsewhere} \tag{4.119}$$

Note. $\bar{X}$ and s^2 of a random sample from a normal population are statistically independent.

Example 4.62. Let $n = 15$; then from Table A6

$$P(14s^2/\sigma^2 \geq 21.1) = 0.10 \qquad \text{or} \qquad P(14s^2/21.1 \geq \sigma^2) = 0.10$$

a statement about the population variance. The chi-square statistic, Eq. (4.116), is also used to measure the discrepancy between the observed and expected frequencies

$$\chi^2 = \sum_{i=1}^{k} (f_{oi} - f_{ei})^2/f_{ei} \tag{4.120}$$

where k = number of classes
f_{oi} = observed frequency of ith class
f_{ei} = expected frequency of ith class; should be greater than 5

When $\chi^2 = 0$, the observed and expected frequencies agree exactly; when $\chi^2 > 0$, they do not agree. The larger the value of χ^2 the greater the discrepancy. The number of degrees of freedom is given by

$$\nu = k - 1$$

when no estimated population parameters are used in the computation of the expected frequencies, and by

$$\nu = k - 1 - m$$

when m estimated population parameters are used in the computation of the expected frequencies.

Student's t Distribution. If Z is a normally distributed random variable with zero mean and unit variance and the independent random variable χ^2 follows a chi-square distribution with ν degrees of freedom, then the random variable

$$t = Z/\sqrt{\chi^2/\nu}$$

has a t distribution with ν degrees of freedom. The p.d.f. of t is given by

$$f(t) = \frac{\Gamma[(\nu+1)/2]}{\sqrt{\pi\nu}\,\Gamma(\nu/2)}(1 + t^2/\nu)^{-(\nu+1)/2} \qquad -\infty < t < \infty \tag{4.121}$$

Table 4.35 Some Properties of the t Distribution
(See Fig. 4.21)

Mean	$E(t) = 0$	
Variance	$\sigma_t^2 = \nu/(\nu - 2)$	$\nu > 2$
Third central moment	$\mu_3 = 0$	
Fourth central moment	$\mu_4 = 3\nu^2/(\nu - 2)(\nu - 4)$	$\nu > 4$
Coefficient of skewness	$\alpha_3 = 0$	
Coefficient of kurtosis	$\alpha_4 = 3(\nu - 2)/(\nu - 4)$	$\nu > 4$

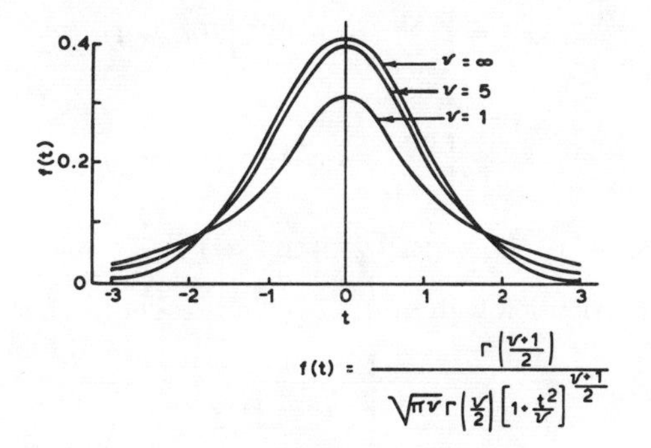

Fig. 4.21 Probability density function of Student's t for ν = 1, 5, and ∞ degrees of freedom.

The $2k$th moment about the origin is

$$\nu_{2k} = \frac{1 \cdot 3 \cdot 5 \cdots (2k-1)}{(\nu-2)(\nu-4)(\nu-2k)}\nu^k \qquad 2k < \nu$$

$$F(t_{p,\nu}) = P(t \le t_{p,\nu}) = \int_{-\infty}^{t_{p,\nu}} f(t)\,dt = p \tag{4.122}$$

which is tabulated in Table A5.

Remarks

1. The distribution has a single mode at $t = 0$.
2. The distribution is symmetric about $t = 0$.
3. As $\nu \to \infty$, t is asymptotically normal with mean 0 and variance 1. For $\nu \geq 30$, $t_{p,\nu} \simeq z_p$.
4. If x_i, $i = 1, 2, \cdots, n$, are independent normally distributed random variables with mean μ and variance σ^2, then $z = (\bar{x} - \mu)/(\sigma/\sqrt{n})$ is $N(0,1)$. Furthermore, $t = (\bar{x} - \mu)/(s/\sqrt{n})$ has a t distribution with $n - 1$ degrees of freedom.

Example 4.63. Let $n = 15$; then from Table A5

$$P[-1.761 \leq (\bar{x} - \mu)/(s/\sqrt{15}) \leq 1.761] = 0.90$$

or

$$P[-1.761s/\sqrt{15} \leq \bar{x} - \mu \leq 1.761s/\sqrt{15}] = 0.90$$

5. Consider the two random samples, $x_{11}, x_{12}, \ldots, x_{1n}$, from $N(\mu_1,\sigma^2)$ and $x_{21}, x_{22}, \ldots, x_{2n_2}$ from $N(\mu_2,\sigma^2)$. The distribution of the random variable

$$\bar{x}_1 = \sum_{i=1}^{n_1} x_{1i}/n_1$$

is $N(\mu_1,\sigma^2/n_1)$ and the distribution of the random variable

$$\bar{x}_2 = \sum_{i=1}^{n_2} x_{2i}/n_2$$

is $N(\mu_2,\sigma^2/n_2)$.

If σ is known, which is not the usual case, the distribution of

$$[\bar{x}_1 - \bar{x}_2 - (\mu_1 - \mu_2)]/\sigma^2\sqrt{1/n_1 + 1/n_2}$$

is $N(0,1)$.

If

$$s_1{}^2 = \left[\sum_{i=1}^{n_1} (x_{1i} - \bar{x}_1)^2\right]/(n_1 - 1)$$

and

$$s_2{}^2 = \left[\sum_{i=1}^{n_2} (x_{2i} - \bar{x}_2)^2\right]/(n_2 - 1)$$

then

$$(n_1 - 1)s_1{}^2/\sigma^2 + (n_2 - 1)s_2{}^2/\sigma^2$$

has a chi-square distribution with $n_1 + n_2 - 2$ degrees of freedom. Furthermore,

$$\frac{\bar{x}_1 - \bar{x}_2 - (\mu_1 - \mu_2)}{\sqrt{\dfrac{(n_1 - 1)s_1{}^2 + (n_2 - 1)s_2{}^2}{n_1 + n_2 - 2}}\sqrt{\dfrac{1}{n_1} + \dfrac{1}{n_2}}}$$

has a t distribution with $n_1 + n_2 - 2$ degrees of freedom.

F Distribution. Given two independently distributed chi-square variables $\chi_1{}^2$ with ν_1 degrees of freedom and $\chi_2{}^2$ with ν_2 degrees of freedom, the random variable

$$F = (\chi_1{}^2/\nu_1)/(\chi_2{}^2/\nu_2)$$

has an F distribution with ν_1 and ν_2 degrees of freedom. The p.d.f. of F is given by

$$f(F) = \frac{\Gamma[(\nu_1 + \nu_2)/2]}{\Gamma(\nu_1/2)\Gamma(\nu_2/2)}\left(\frac{\nu_1}{\nu_2}\right)^{\nu_1/2}\frac{F^{\nu_1/2-1}}{[1 + (\nu_1/\nu_2)F]^{[(\nu_1+\nu_2)/2]}} \qquad F > 0 \qquad (4.123)$$
$$= 0 \qquad \text{elsewhere}$$

The kth moment about the origin is

$$\nu_k = \frac{\Gamma(\nu_1/2 + k)\Gamma(\nu_2/2 - k)}{\Gamma(\nu_1/2)\Gamma(\nu_2/2)}\left(\frac{\nu_2}{\nu_1}\right)^k$$

$$F(F_{p,\nu_1,\nu_2}) = P(F \leq F_{p,\nu_1,\nu_2}) = \int_0^{F_{p,\nu_1,\nu_2}} f(F)\,dF = p \qquad (4.124)$$

which is tabulated in Table A7.

Table 4.36 Some Properties of the F Distribution
(See Fig. 4.22)

Mean	$E(F) = \nu_2/(\nu_2 - 2)$	$\nu_2 > 2$
Variance	$\sigma_F{}^2 = 2\nu_2{}^2(\nu_1 + \nu_2 - 2)/\nu_1(\nu_2 - 2)^2(\nu_2 - 4)$	$\nu_2 > 4$
Third central moment	$\mu_3 = \dfrac{8\nu_2{}^3(\nu_1 + \nu_2 - 2)(2\nu_1 + \nu_2 - 2)}{\nu_1{}^2(\nu_2 - 2)^3(\nu_2 - 4)(\nu_2 - 6)}$	$\nu_2 > 6$
Coefficient of variation	$\eta = \sqrt{2(\nu_1 + \nu_2 - 2)/\nu_1(\nu_2 - 4)}$	$\nu_2 > 4$
Coefficient of skewness	$\alpha_3 = \dfrac{2\nu_1 + \nu_2 - 2}{\nu_2 - 6}\sqrt{\dfrac{8(\nu_2 - 4)}{\nu_1 + \nu_2 - 2}}$	$\nu_2 > 6$

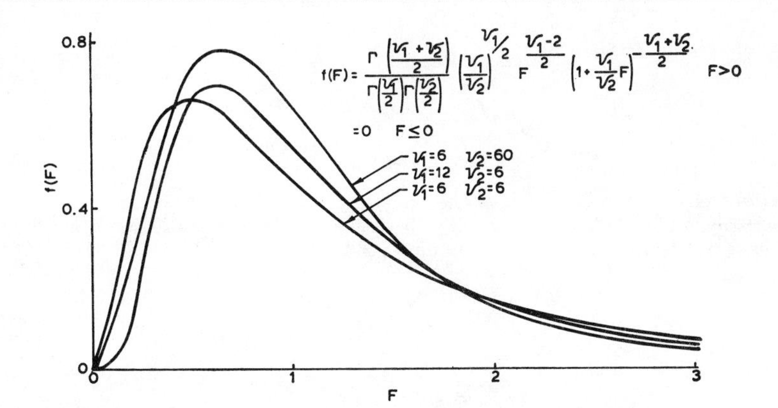

FIG. 4.22 Probability density function of F for various values of ν_1 and ν_2.

Remarks

1. The mode of the distribution is at

$$F = \nu_2(\nu_1 - 2)/\nu_1(\nu_2 + 2)$$

2. $F_{1-\alpha,\nu_2,\nu_1} = 1/F_{\alpha,\nu_1,\nu_2}$

Example 4.64. Let $\alpha = 0.05$, $\nu_1 = 4$, $\nu_2 = 8$, $F_{0.05,8,4} = 0.261$,

$$F_{0.95,4,8} = 3.84 = 1/0.261$$

3. Consider two random samples, one from each of two normal populations, then $(s_1{}^2/\sigma_1{}^2)/(s_2{}^2/\sigma_2{}^2)$ is distributed as F distribution with $n_1 - 1$ and $n_2 - 1$ degrees of freedom.

4.6d Order Statistics. *Definition.* When observations are ordered according to magnitude, the derived statistics are called *order statistics*. Consider a population $f(x)$, $a \leq x \leq b$. Let x_1 be the smallest observation and x_n be the largest observation in a random sample of size n. Then

$$\begin{aligned} f(x_1,x_n) &= n(n-1)f(x_1)f(x_n)[F(x_n) - F(x_1)]^{n-2} \qquad a \leq x_1 \leq x_n \leq b \\ &= 0 \qquad \text{elsewhere} \end{aligned} \tag{4.125}$$

The marginal p.d.f.'s for x_1 and x_n are

$$\begin{aligned} f_1(x_1) &= nf(x_1)[1 - F(x_1)]^{n-1} \qquad a \leq x_1 \leq b \\ &= 0 \qquad \text{elsewhere} \end{aligned} \tag{4.126}$$

and

$$\begin{aligned} f_n(x_n) &= nf(x_n)[F(x_n)]^{n-1} \qquad a \leq x_n \leq b \\ &= 0 \qquad \text{elsewhere} \end{aligned} \tag{4.127}$$

Equations (4.126) and (4.127) constitute the basis of the exact theory of *extreme values.* If the initial distribution is known, the distributions of the extreme values can readily be calculated and the associated moments obtained. In many cases the resulting integrals can only be approximated by numerical methods. The interested reader is referred to a monograph [10] which is based on four lectures given by Gumbel at the National Bureau of Standards.

An order statistic which is used extensively in the field of quality control is the *sample range.* Let $X_{max} - X_{min} = R$, the sample range. Let $x_n = x_1 + R$; then

$$\begin{aligned} f(R) &= \int_a^{b-R} f(x_1,R)\,dx_1 \qquad 0 \le R \le b-a \\ &= 0 \qquad \text{elsewhere} \end{aligned} \tag{4.128}$$

Or let $x_1 = x_n - R$; then

$$\begin{aligned} f(R) &= \int_{a+R}^{b} f(x_n,R)\,dx_n \qquad 0 \le R \le b-a \\ &= 0 \qquad \text{elsewhere} \end{aligned} \tag{4.129}$$

Equations (4.128) and (4.129) give the same result.

Example 4.65. Let Y_1, Y_2, Y_3, Y_4, Y_5 denote the order statistics of a random sample of five from a distribution having a p.d.f.

$$\begin{aligned} f(x) &= (1/\theta)e^{-x/\theta} \qquad x > 0,\ \theta > 0 \\ &= 0 \qquad \text{elsewhere} \end{aligned}$$

Determine the distribution of

$$R = Y_5 - Y_1$$

$$\begin{aligned} f(y_1,y_5) &= (5!/\theta^5)\int_{y_1}^{y_5}\int_{y_2}^{y_5}\int_{y_3}^{y_5} \exp\left(-\sum_{i=1}^{5} y_i/\theta\right) dy_4\,dy_3\,dy_2 \\ &= (20/\theta^2)e^{-(y_1+y_5)/\theta}\,[e^{-y_1/\theta} - e^{-y_5/\theta}]^3 \qquad 0 \le y_1 \le y_5 < \infty \\ &= 0 \qquad \text{elsewhere} \end{aligned}$$

Let $R = Y_5 - Y_1$ and $Y_5 = Y_5$, then

$$f(R,y_5) = (20/\theta^2)e^{-(y_5-R)/\theta - y_5/\theta}\,[e^{-(y_5-R)/\theta} - e^{-y_5/\theta}]^3 \qquad 0 \le R \le y_5 < \infty$$

and

$$\begin{aligned} f(R) &= (20/\theta^2)\int_R^{\infty} f(R,y_5)\,dy_5 \\ &= (4/\theta)e^{-R/\theta}[1 - e^{-R/\theta}]^3 \qquad R > 0 \\ &= 0 \qquad \text{elsewhere} \end{aligned}$$

4.7 ESTIMATION

One of the important problems of industrial experimentation is to estimate the parameters of a distribution. For example, what is the average life of an electronic component?

Definition. An estimator is some function of the sample values which provides an estimate of the population parameter.

Definition. A single-value estimate of the population parameter is called a *point estimate.*

Some Properties of Point Estimators. Let the statistic $\hat{\theta}$ calculated from a random sample of size n be an estimator for θ.

1. $\hat{\theta}$ is an *unbiased estimator* of θ if $E(\hat{\theta}) = \theta$.

Note. Since $E(\bar{x}) = \mu_x$, the sample mean $\bar{x}$ is an unbiased estimator of μ.

2. $\hat{\theta}$ is a *consistent estimator* of θ if $\hat{\theta}$ converges to θ in probability as the sample size increases.

Note. The sample mean $\bar{x}$ is a consistent estimator of μ_x, since $P(|\bar{x} - \mu| > \epsilon) \leq \sigma^2/n\epsilon^2$, which tends to 0 as $n \to \infty$.

3. If $\hat{\theta}_1$ and $\hat{\theta}_2$ are two different unbiased estimators of θ and if $E[(\hat{\theta}_1 - \theta)^2] < E[(\hat{\theta}_2 - \theta)^2]$, then $\hat{\theta}_1$ is an *efficient estimator* of θ and $\hat{\theta}_2$ is an *inefficient estimator* of θ.

Note. Both the mean and median are estimators of μ. However,

$$E[(\bar{x} - \mu)^2] = \sigma_x^2/n$$

is less than $E[(x_{0.50} - \mu)^2] \simeq 1.57\sigma_x^2/n$; hence, $\bar{x}$ is an efficient estimator of μ and the median is an inefficient estimator of μ.

4. $\hat{\theta}$ is a *sufficient estimator* if no other independent estimate based on the sample is able to yield any further information about the parameter which is being estimated.

Derivation of Point Estimates

1. *Method of moments.* The point estimates of the parameters may be obtained by equating the sample moments to the corresponding population moments.

Example 4.66. Consider the exponential distribution

$$\begin{aligned} f(t) &= (1/\theta)e^{-t/\theta} \qquad & x > 0 \\ &= 0 & \text{elsewhere} \end{aligned}$$

Since there is only one population parameter, we need only the first population moment $\mu = \theta$. Therefore,

$$\hat{\theta} = \sum_{i=1}^{n} t_i/n$$

the first sample moment.

2. *Method of maximum likelihood.* Let x_i, $i = 1, 2, \ldots, n$ be a random sample from the distribution whose p.d.f. is $f(x{:}\theta)$. What is a function of the sample values, $f(x_1,x_2, \ldots ,x_n)$, which will be a "good" estimator of θ? Define the likelihood function of the sample as

$$\begin{aligned} L(x_1,x_2, \ldots ,x_n{:}\theta) &= f(x_1{:}\theta)f(x_2{:}\theta) \cdots f(x_n{:}\theta) \\ &= \prod_{i=1}^{n} f(x_i{:}\theta) \end{aligned} \tag{4.130}$$

The method of maximum likelihood consists of the finding the value of θ which maximizes $L(x_1,x_2, \ldots ,x_n{:}\theta)$. Since $L(x_1,x_2, \ldots ,x_n{:}\theta)$ is a product, it is customary to take the logarithm of the likelihood function. Thus, by solving the *likelihood equation*

$$\frac{\partial \ln L(x_1,x_2, \ldots ,x_n{:}\theta)}{\partial \theta} = 0$$

for θ we obtain the maximum-likelihood estimator for θ, which is a function of x_1, $x_2, \ldots, x_n$. The maximum-likelihood estimators are consistent, sufficient, and approximately normally distributed, and they have maximum efficiency in large samples. They are not always unbiased. As noted in Lloyd and Lipow [4], the

$$\text{Var}\ (\hat{\theta}) = -\ (\partial^2 \ln L/\partial\theta^2)^{-1}$$

which may be used when sample values are replaced by their expected values. A discussion of variances and covariances of maximum-likelihood estimators may also be found in Mood [11].

Example 4.67. Let

$$f(x:\mu,\sigma^2) = (1/\sqrt{2\pi}\,\sigma)e^{-(x-\mu)^2/2\sigma^2} \qquad -\infty < x < \infty$$

then

$$L(x_1,x_2, \ldots ,x_n:\mu,\sigma^2) = 1/[(2\pi)^{n/2}\sigma^n] \exp\left[-\sum_{i=1}^{n} (x_i - \mu)^2/2\sigma^2\right]$$

$$\ln L = -(n/2)\ln(2\pi) - (n/2)\ln\sigma^2 - \sum_{i=1}^{n} (x_i - \mu)^2/2\sigma^2$$

$$\frac{\partial \ln L}{\partial \mu} = -\sum_{i=1}^{n} (x_i - \mu)/\sigma^2 = 0 \qquad (4.131)$$

$$\frac{\partial \ln L}{\partial \sigma^2} = -\frac{n}{2\sigma^2} + \frac{1}{2(\sigma^2)^2}\sum_{i=1}^{n} (x_i - \mu)^2 = 0 \qquad (4.132)$$

By solving Eqs. (4.131) and (4.132) simultaneously, we have

$$\hat{\mu} = \sum_{i=1}^{n} x_i/n \qquad \text{and} \qquad \hat{\theta}^2 = \sum_{i=1}^{n} (x_1 - \hat{\mu})^2/n$$

Notice that $\hat{\sigma}^2$ is a biased estimator of σ^2.

Confidence-interval Estimation. A point estimator of θ should have associated with it an interval-type estimate $[\theta_U,\theta_L]$, where θ_U is the upper bound and θ_L is the lower bound. The interval estimates will be functions of the sample, $x_1, x_2, \ldots, x_n$. Further, there should be some measure of assurance that the true parameter lies within the interval. The mean life for a component may be 400 ± 20 hours; i.e., the true mean life probably falls between 380 and 420 hours. The method for determining confidence intervals consists of:

1. Determining the random variable, say, $Y(\theta)$, which involves the desired parameter θ but whose distribution does not depend upon other parameters.

2. Finding, from the probability distribution of the estimator, two numbers C_1 and C_2 such that

$$P[C_1 < Y(\theta) < C_2] = \gamma$$

the *confidence coefficient.*

3. Constructing the probability statement

$$P[\theta_L < \theta < \theta_U] = \gamma$$

a two-sided confidence interval. In many instances, only one confidence limit is of interest. The probability statements for one-sided confidence intervals are for an upper limit

$$P[\theta < \theta_U] = \gamma$$

and for a lower limit

$$P[\theta_L < \theta] = \gamma$$

Example 4.68. Let x be $N(\mu,\sigma^2)$. Consider a random sample of size n. The 100γ confidence interval for μ is given by

$$P[-t_{(1+\gamma)/2,n-1} \leq (\bar{x} - \mu)/(s/\sqrt{n}) \leq +t_{(1+\gamma)/2,n-1}] = \gamma$$

or

$$P[\bar{x} - t_{(1+\gamma)/2,n-1}s/\sqrt{n} \leq \mu \leq \bar{x} + t_{(1+\gamma)/2,n-1}s/\sqrt{n}] = \gamma$$

Thus

$$\theta_L = \bar{x} - t_{(1+\gamma)/2,n-1}s/\sqrt{n} \qquad \theta_U = \bar{x} + t_{(1+\gamma)/2,n-1}s\sqrt{n}$$

where $t_{(1+\gamma)/2,n-1}$ is Student's t with $n - 1$ degrees of freedom. If 95 per cent intervals are obtained, then it is expected that approximately 95 per cent of these intervals will include μ. That is, you are 95 per cent confident that μ lies between θ_L and θ_U. The statement should *not* be made that the probability is 0.95 that μ lies between θ_L and θ_U, because μ either falls within the interval or does not; thus the probability is either 0 or 1 (Fig. 4.23).

Note. Point and interval estimates for the parameters of a particular distribution are discussed in Secs. 4.4 and 4.5, which pertain to that distribution.

Tolerance Limits for Normal Distribution. If μ and σ^2 are known, then tolerance limits are formed by $\mu \pm z\sigma$, where z is obtained from Table A4. It is true that 95 per cent of the population lies between $\mu \pm 1.96\sigma$; this same statement cannot be made regarding the interval $\bar{x} \pm 1.96s$. The latter limits depend upon the random variables $\bar{X}$ and s and how well they estimate μ and σ. It is possible to determine K such that in a long series of samples from a normal population 100γ per cent of the intervals, $\bar{x} \pm Ks$, will include $100(1 - \alpha)$ per cent or more of the population.

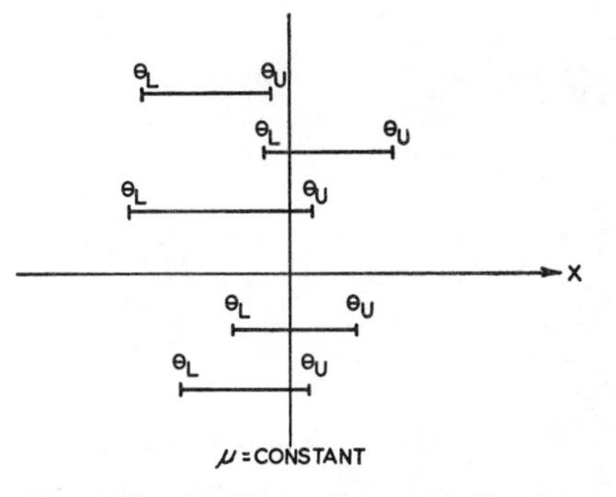

FIG. 4.23 Confidence-interval estimates.

Example 4.69. An engineer wishes to estimate the tolerance limits within which he can be reasonably certain ($\gamma = 0.90$) that at least 95 per cent of the distribution of the lifetimes of a certain electron tube will lie. Ten samples were measured to failure with $\bar{X} = 140$ hours and $s = 15$ hours. From Table A9 the value of K for $n = 10$, $\gamma = 0.90$, $\alpha = 0.05$ is 3.018. Thus, the tolerance limits are

$$140 \pm 3.018(15) = 94.7, 185.3] \text{ hours}$$

(assuming a normal distribution). In many instances one-sided tolerance limits are more appropriate. The single limit $\bar{x} - Ks$ (or $\bar{x} + Ks$) is such that a given percentage will be greater (or smaller) than this limit. Some values of K for one-sided tolerance limits for the normal distribution are given in Table A8.

4.8 TESTS OF HYPOTHESES

Statistical hypotheses are usually statements about the probability distributions of the population, e.g., the mean life of a component is 400 hours or the variate follows a particular distribution.

Definition. The hypothesis which states there is no difference between procedures except sampling fluctuations is called the *null hypothesis*, designated H_0. Any hypothesis which differs from this is called the *alternative hypothesis*, designated H_1.

Definition. *Type I error*, designated α, is the error made in rejecting a hypothesis which is true. 100α is the significance level of the test.

Definition. *Type II error*, designated β, is the error made in accepting a hypothesis which is false. $1 - \beta$ is called the power of the test. It is expressed as a function of the parameter and called the *power function*. It is obvious the selections of α and β must depend upon the consequences of making Type I and II errors, respectively. The only way to reduce both types of errors is to increase the sample size. The sample space of all possible values of the test statistic is divided into acceptance and critical (rejection) regions (Fig. 4.24).

Procedure for Testing Hypothesis

1. Statement of H_0 and H_1.
2. Selection of α and β. (In some cases n is selected instead of β.)
3. Selection of sample (test) statistic.

4. Determination of critical region.
5. Computation of statistic from the sample.
6. Acceptance or rejection of H_0.

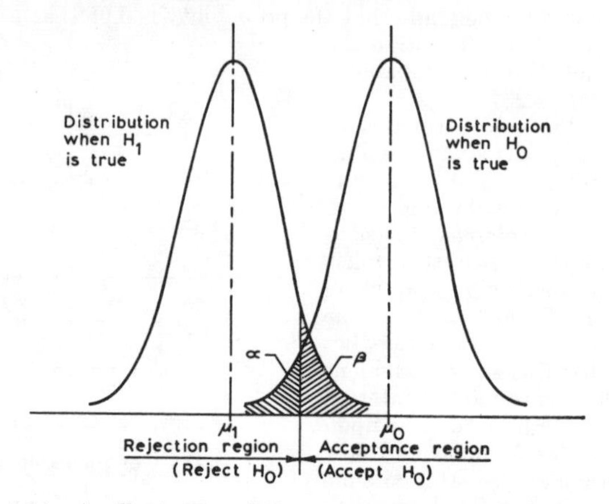

FIG. 4.24 An illustration of the acceptance and rejection regions.

Example 4.70. The mean lifetime of a sample of 100 tubes produced by a certain company is computed to be 2,100 hours with a known standard deviation of 300 hours. If μ is the mean lifetime of all tubes produced by the company, test the hypothesis that $\mu = 2{,}200$ hours against the alternative that $\mu \neq 2{,}200$ hours, using $\alpha = 0.05$ and assuming a normal distribution.

$$H_0\colon \mu = 2{,}200 \qquad H_1\colon \mu \neq 2{,}200 \qquad \alpha = 0.05$$

Statistic to be used: $z = (\bar{x} - \mu)/(\sigma/\sqrt{n})$

Critical region: $|z| > 1.96$

$$z = (2{,}100 - 2{,}200)/(300/\sqrt{100}) = -3.33$$

Therefore, we reject the hypothesis that $\mu = 2{,}200$. Note the similarity between the two-sided hypothesis test and the confidence-interval estimation. The two-sided test consists of finding limits between which $1 - \alpha$ of the sample statistics would fall under repeated random sampling. Obtaining confidence limits consists of finding an interval which would contain the true parameter 100γ per cent of the time under repeated sampling. For the same sample size and $1 - \alpha = \gamma$ the arithmetic is the same, but the interpretations are different. The tests of hypothesis for the parameters of a particular distribution are discussed in Secs. 4.4 and 4.5, which pertain to that distribution.

4.9 REGRESSION AND CORRELATION ANALYSIS

In many instances a relationship exists between several variables, e.g., hardness and tensile strength of an alloy. It may be desirable to express this relationship in functional form and also to determine the strength of the relationship. The method of *regression analysis* is used to determine the "best" functional relation among the variables. The technique used to measure the degree of association between variables is called *correlation analysis*. To decide which curve to fit to the data, it is beneficial to plot a scatter diagram of the data. If the scatter diagram indicates a linear relationship on (1) rectangular coordinate paper, the equation has the form $y = \alpha + \beta x$; (2) semilog graph paper, the equation has the form $y = \alpha\beta^x$; and (3) log-log graph paper, the equation has the form $y = \alpha x^\beta$.

Once we have decided on the function which is to be fitted to the data, it is necessary to estimate the parameters of the function. The determination of the estimates is called *curve fitting.*

Definition. Considering all the curves approximating the data, the curve that minimizes the squares of deviations from the curve is called the *best-fitting curve;* i.e., the estimates of the parameters are found by minimizing

$$\sum_{i=1}^{n} (y_i - \hat{y}_i)^2$$

the *method of least squares* (Fig. 4.25).

Simple Linear Regression. Let

$$S = \sum_{i=1}^{n} (y_i - \alpha - \beta x_i)^2$$

and a and b be the values of α and β which make $\partial S/\partial \alpha = 0$ and $\partial S/\partial \beta = 0$. The estimators for α and β are determined by solving the normal equations:

$$(n)a + \left(\sum_{i=1}^{n} x_i\right) b = \sum_{i=1}^{n} y_i \qquad (4.133)$$

$$\left(\sum_{i=1}^{n} x_i\right) a + \left(\sum_{i=1}^{n} x_i^2\right) b = \sum_{i=1}^{n} x_i y_i \qquad (4.134)$$

Therefore,

$$a = \bar{y} - b\bar{x} \qquad (4.135)$$

$$b = \sum_{i=1}^{n} (x_i - \bar{x})(y_i - \bar{y}) \Big/ \sum_{i=1}^{n} (x_i - \bar{x})^2 \qquad (4.136)$$

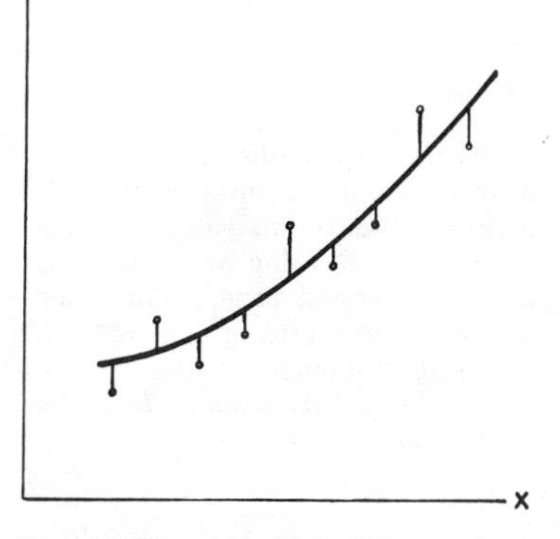

FIG. 4.25 An illustration of a scatter diagram showing the vertical deviations from a curve whose sum of squares is to be minimized.

These estimates give the regression equation, $\hat{y} = a + bx$.

Note. If the x's are assumed to be constant (measured without error), both a and b are linear functions of the y's.

The method of least squares requires *no* assumption regarding the y's. However, to construct confidence intervals, the additional assumptions required are:

1. $E(y) = \alpha + \beta x$.
2. Variance for each y is the same for all x; denoted by σ_E^2.
3. The distribution of y is normal for each given x.
4. Random sample.

If the above conditions are satisfied, then the estimates a and b are normally distributed, with means α and β, respectively. The following are the associated estimates of error.

Residual mean square:

$$s_E^2 = \sum_{i=1}^{n} (y - \hat{y})^2/(n - 2) \qquad (4.137)$$

Regression coefficient b:

$$s_b^2 = s_E^2 \Big/ \sum_{i=1}^{n} (x_i - \bar{x})^2 \qquad (4.138)$$

Regression coefficient a:

$$s_a^2 = s_E^2\left[1/n + \bar{x}^2/\sum_{i=1}^{n}(x_i - \bar{x})^2\right] \tag{4.139}$$

Estimated mean y for given x:

$$s_{\hat{y}}^2 = s_E^2[1/n + (x - \bar{x})^2/\Sigma(x_i - \bar{x})^2] \tag{4.140}$$

Estimated individual y for given x:

$$s_{\hat{y}}^2 = s_E^2\left[1 + 1/n + (x - \bar{x})^2/\sum_{i=1}^{n}(x_i - \bar{x})^2\right] \tag{4.141}$$

Thus, the following are the 100γ confidence intervals.

For β: $\quad b \pm t_{(1+\gamma)/2,n-2}s_b \quad$ (4.142)

For α: $\quad a \pm t_{(1+\gamma)/2,n-2}s_a \quad$ (4.143)

Simple Exponential Regression. In fitting a curve of the type $y = \alpha\beta^x$ the method of least squares results in normal equations which are difficult to solve. Therefore, an approximate procedure, namely, $\log y = \log \alpha + x \log \beta$, is used. Let $w = \log y$, $A = \log \alpha$, $B = \log \beta$; the equation then becomes $w = A + Bx$, which is the simple linear regression case. This least-squares solution is not identical with the least-squares solution using $y = \alpha\beta^x$. However, the approximation is of sufficient accuracy for most problems.

Correlation Analysis. Definition. The *sample linear correlation coefficient* r is defined as

$$r = \left[\sum_{i=1}^{n}(x_i - \bar{x})(y_i - \bar{y})\right]\Big/\sqrt{\sum_{i=1}^{n}(x_i - \bar{x})^2 \sum_{i=1}^{n}(y_i - \bar{y})^2} \tag{4.144}$$

Note. $-1 \leq r \leq +1$.

Definition. r^2 = coefficient of determination, $1 - r^2$ = coefficient of nondetermination, and $\sqrt{1 - r^2}$ = coefficient of alienation. If the random sample is from a bivariate normal population (Sec. 4.3*a*), then r is an estimate of $\rho = \sigma_{xy}/\sigma_x\sigma_y$. Thus, to test the hypothesis, H_0: $\rho = 0$ versus H_1: $\rho \neq 0$, calculate

$$t = r/s_r = r\sqrt{n - 2}/\sqrt{1 - r^2}$$

and reject H_0 if

$$|t| > t_{1-\alpha/2,n-2}$$

which is tabulated in Table A5. Since

$$t = r/s_r = b/s_b$$

this test is equivalent to the test of H_0: $\beta = 0$ versus H_1: $\beta \neq 0$. To test the hypothesis

$$H_0: \rho = \rho_0 \neq 0 \qquad \text{versus} \qquad H_1: \rho \neq \rho_0$$

use Fisher's Z transformation [12]

$$Z = \tfrac{1}{2} \ln [(1 + r)/(1 - r)]$$

which is approximately normally distributed with mean and variance

$$\mu_Z = \tfrac{1}{2} \ln [(1 + \rho_0)/(1 - \rho_0)] + \rho_0/2(n - 1) \qquad \sigma_Z^2 = 1/(n - 3)$$

4.10 SEQUENTIAL ANALYSIS

In a sequential test, the decision is to (1) accept the test hypothesis, (2) reject the test hypothesis, or (3) delay the decision by making another observation. Sequential tests are usually more economical than the nonsequential tests. Designating the hypothesis under test, the null hypothesis, by H_0, the alternative hypothesis by H_1, and α and β the risks associated with Type I and Type II errors, respectively, the sequential procedure is to

1. Accept H_0 if $L \leq \beta/(1 - \alpha) = 1/B$.
2. Accept H_1 if $L \geq (1 - \beta)/\alpha = A$.
3. Select another observation if $1/B \leq L \leq A$.

Here L is defined as the *likelihood ratio* P_1/P_0, where P_1 is probability of obtaining the selected observations on the basis that H_1 is true and P_0 is the probability when H_0 is true. The probability ratio test was developed by Wald in 1943.[1]

Basis for sequential plans:

1. Distribution function is known.
2. H_0 and H_1 specified.
3. Advance selection of α and β.

Recommended values for α and β are 0.10.[2]

Sequential Test for the Binomial Distribution

$$H_0: \theta = \theta_0 \qquad H_1: \theta = \theta_1$$

From a population with proportion defective θ, the probability of obtaining d_n defectives in a sample of size n is given by

$$P = C(n,d_n)\theta^{d_n}(1 - \theta)^{n-d_n}$$

Thus,

$$\frac{1}{B} \leq L = \frac{P_1}{P_0} = \frac{\theta_1^{d_n}(1 - \theta_1)^{n-d_n}}{\theta_0^{d_n}(1 - \theta_0)^{n-d_n}} \leq A$$

Taking logarithms and rearranging,

$$\frac{\ln [\beta/(1 - \alpha)] + n \ln [(1 - \theta_0)/(1 - \theta_1)]}{\ln (\theta_1/\theta_0) + \ln [(1 - \theta_0)/(1 - \theta_1)]} \leq d_n \leq \frac{\ln (1 - \beta)/\alpha + n \ln [(1 - \theta_0)/(1 - \theta_1)]}{\ln (\theta_1/\theta_0) + \ln [(1 - \theta_0)/(1 - \theta_1)]}$$

Let

$$b = \ln [(1 - \alpha)/\beta] \qquad g_1 = \ln (\theta_1/\theta_0)$$
$$a = \ln [(1 - \beta)/\alpha] \qquad g_2 = \ln [(1 - \theta_0)/(1 - \theta_1)]$$
$$h_1 = b/(g_1 + g_2) \qquad h_2 = a/(g_1 + g_2) \qquad s = g_2/(g_1 + g_2)$$

Then,

Acceptance line: $$a_n = -h_1 + sn \tag{4.145}$$
Rejection line: $$r_n = h_2 + sn \tag{4.146}$$

Formulas for the probability of acceptance $L(\theta)$ of H_0 and the average sample number (ASN) for five values of θ are given by:

$$\begin{array}{lll}
\theta = 0 & L(\theta) = 1 & \bar{n}_0 = h_1/s \\
\theta = \theta_0 & L(\theta) = 1 - \alpha & \bar{n}_{\theta_0} = [(1 - \alpha)h_1 - \alpha h_2]/(s - \theta_0) \\
\theta = s & L(\theta) = h_2/(h_1 + h_2) & \bar{n}_s = h_1 h_2/s(1 - s) \\
\theta = \theta_1 & L(\theta) = \beta & \bar{n}_{\theta_1} = [(1 - \beta)h_2 - \beta h_1]/(\theta_1 - s) \\
\theta = 1 & L(\theta) = 0 & \bar{n}_1 = h_2/(1 - s)
\end{array}$$

[1] A. Wald, *Sequential Analysis*, John Wiley & Sons, Inc., New York, 1947.

[2] *Reliability of Military Electronic Equipment*, Report of Advisory Group on Reliability of Electronic Equipment, Office of the Assistant Secretary of Defense (R&E), GPO, 1957.

Sequential Test for the Poisson Distribution

$$H_0: \theta = \theta_0 \qquad H_1: \theta = \theta_1 \qquad \theta_0 > \theta_1$$

For parameter θ probability of r failures in time[1] t:

$$P = (t/\theta)^r(e^{-t/\theta}/r!)$$

Hence,

$$1/B \leq (\theta_0/\theta_1)^r e^{-(1/\theta_1 - 1/\theta_0)T} \leq A$$

where r = number of failures
T = total test time
Taking logarithms and rearranging

$$\frac{\ln [\beta/(1-\alpha)] + r \ln (\theta_1/\theta_0)}{1/\theta_0 - 1/\theta_1} > T > \frac{\ln [(1-\beta)/\alpha] + r \ln (\theta_1/\theta_0)}{1/\theta_0 - 1/\theta_1}$$

or

$$\frac{\ln [\beta/(1-\alpha)] + (1/\theta_1 - 1/\theta_0)T}{\ln (\theta_0/\theta_1)} < r < \frac{\ln [(1-\beta)/\alpha] + (1/\theta_1 - 1/\theta_0)T}{\ln (\theta_0/\theta_1)} \tag{4.147}$$

Example 4.71. Let θ_0 = 1,500 hours, θ_1 = 500 hours, $\alpha = \beta = 0.10$.

$$\ln [\beta/(1-\alpha)] = \ln \tfrac{1}{9} = -2.19722$$
$$\ln [(1-\beta)/\alpha] = \ln 9 = 2.19722$$
$$\ln (\theta_1/\theta_0) = \ln \tfrac{1}{3} = -1.0986$$
$$1/\theta_0 - 1/\theta_1 = \tfrac{1}{1500} - \tfrac{1}{500} = -\tfrac{1}{750}$$

Thus, $-1{,}642.9 + 824.0r < T < 1{,}642.9 + 824.0r$.

The total average test times to reach a decision for four values of θ are given by

$$\theta = 0 \qquad \bar{T} = 0$$

$$\theta = \theta_0 \qquad \bar{T} = \frac{\beta \ln [\beta/(1-\alpha)] + (1-\beta) \ln [(1-\beta)/\alpha]}{\ln (\theta_0/\theta_1) - (1 - \theta_0/\theta_1)}\,\theta_0$$

$$\theta_s = \frac{\ln (\theta_0/\theta_1)}{1/\theta_1 - 1/\theta_0} \qquad \bar{T} = \frac{\ln [(1-\beta)/\alpha] \ln [(1-\alpha)/\beta]}{\ln (\theta_0/\theta_1)^2}\,\theta_s$$

$$\theta = \theta_1 \qquad \bar{T} = \frac{(1-\alpha) \ln [\beta/(1-\alpha)] + \alpha \ln [(1-\beta)/\alpha]}{\ln (\theta_0/\theta_1) - (\theta_0/\theta_1 - 1)}\,\theta_1$$

If truncation is desired, a parallel line through the origin represents the truncation line. If the test is to be truncated at $\theta = T_T$ when no prior decision has been made, the truncation rule says accept if

$$r \leq \frac{1/\theta_1 - 1/\theta_0}{\ln (\theta_0/\theta_1)}\,T_T$$

otherwise, reject. If T_T is large compared to minimum time required for acceptance, the effect on α and β is very small. A graphical presentation of the procedure and the truncation line is given in Fig. 4.26. The plan is represented by two parallel lines (Fig. 4.26) which divide the sample space into the three regions acceptance, rejection, and indecision.

4.11 ANALYSIS OF VARIANCE

The t test was used in the statistical decision procedure to determine whether the means of two distributions were considered equal. The more general problem which

[1] A method for sequential reliability tests for the replacement case is given by B. Epstein and M. Sobel, "Sequential Life Tests in the Exponential Case," *Ann. Math. Statist.*, vol. 26, p. 82, 1955.

faces the experimenter is the comparison of several means. A possible solution is to test every possible pair of means by using the t test. However, when there are several sample means, even though the distribution means are equal, we would expect some of the sample means to be significantly large or significantly small. Thus, the experimenter may be faced with the question: which sample means are indicative of a true difference? The appropriate analysis for a test of several means is the *analysis of variance* (ANOVA). Since the mathematical model which is frequently assumed in ANOVA is a linear model, the assumptions are basically the same as those associated

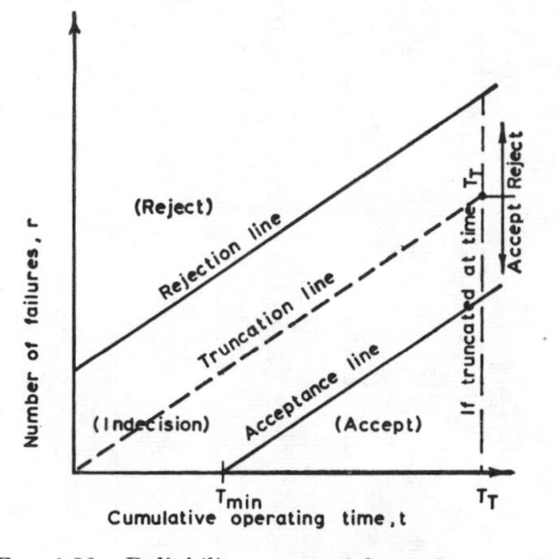

FIG. 4.26 Reliability sequential test for $\alpha = \beta$.

with linear regression. The consequences when the assumptions are not satisfied are discussed by Cochran [13].

4.11a One-way Classification. *Fixed-effects Model.* In this model the experimenter is concerned only with the treatments which are present in his experiment. Suppose he is interested in studying the effect of k different storage conditions (treatments) on the activated life of thermal batteries. If n_i batteries were subjected to the ith storage condition, then the data would appear as in Table 4.37. The

Table 4.37 Activated Life of Thermal Batteries for Different Storage Conditions

	Storage condition				Totals
	1	2	...	k	
	X_{11} X_{12} ... X_{1n_1}	X_{21} X_{22} ... X_{2n_2}		X_{k1} X_{k2} ... X_{kn_2}	
Totals	T_1	T_2	...	T_k	$T = \sum_{i=1}^{k} T_i$
Number of observations	n_1	n_2	...	n_k	$n = \sum_{i=1}^{k} n_i$

Table 4.38 ANOVA Associated with Table 4.37

Source of variation	Degrees of freedom	Sum of squares	Mean square	Expected mean square	F ratio
Between storage conditions...	$k - 1$	$\sum_{i=1}^{k} T_i^2/n_i - T^2/n = \mathrm{SS}_c$	$\mathrm{SS}_c/(k-1)$	$\sigma^2 + \sum_{i=1}^{k} n_i\alpha_i^2/(k-1)$	$\frac{\mathrm{SS}_c/(k-1)}{\mathrm{SS}_E/(n-k)}$
Within storage conditions....	$\sum_{i=1}^{k} (n_i - 1) = n - k$	$\sum_{i=1}^{k} \sum_{j=1}^{n_i} X_{ij}^2 - \sum_{i=1}^{k} T_i^2/n_i = \mathrm{SS}_E$	$\mathrm{SS}_E/(n-k)$	σ^2	
Total....................	$\sum_{i=1}^{n} n_i - 1 = n - 1$	$\sum_{i=1}^{k} \sum_{j=1}^{n_i} X_{ij}^2 - T^2/n = \mathrm{SS}$	....		

mathematical model for the data of Table 4.37 is

$$X_{ij} = \mu + \alpha_i + \epsilon_{ij} \qquad i = 1, 2, \ldots, k \quad j = 1, 2, \ldots, n_i \tag{4.148}$$

where μ = true mean

α_i = true effect of the ith storage condition and $\sum_{i=1}^{k} n_i\alpha_i = 0$

ϵ_{ij} = random effect which is assumed to be independent and $N(0,\sigma^2)$

The ANOVA for Table 4.37 are summarized in Table 4.38. In the fixed-effects model the derived conclusions extend only to the treatments under consideration.

Random-effects Model. The mathematical model is the same as the fixed-effects model. However, the interpretation of α_i is different. In the random-effects model the α_i are random independent variables and $N(0,\sigma_\alpha^2)$.

Note. We have selected k treatments from the distribution of treatments. The conclusions which are reached in the experiment are extended to all of the treatments. The ANOVA is the same as Table 4.38, except the expected mean square for between storage conditions (treatments) is $\sigma^2 + n_0\sigma_\alpha^2$, where

$$n_0 = \left[\sum_{i=1}^{k} n_i - \sum_{i=1}^{k} n_i^2 \Big/ \sum_{i=1}^{k} n_i \right] \Big/ (k - 1) \tag{4.149}$$

For the fixed-effects model, H_0: $\alpha_i = 0$, $i = 1, 2, \ldots, k$. If H_0 is true, then

$$\frac{\text{SS}_c/(k-1)}{\text{SS}_E/(n-k)}$$

is distributed as the F distribution with $k - 1$ and $n - k$ degrees of freedom. If this value is greater than $F_{1-\alpha,k-1,n-k}$, where α is the significance level, H_0 will be rejected and the conclusion will be that there are significant differences between the storage conditions (treatments). For the random-effects model, H_0: $\sigma_\alpha^2 = 0$. The same test procedure would be used. However, in more complex analyses, different test procedures may be indicated for the two models. Therefore, expected mean squares should be shown in an ANOVA table.

Example 4.72. An experiment was conducted to evaluate the effect of a constant humidity on the effective resistance of 10-ohm resistors from four different companies. The data are tabulated in Table 4.39. Is there a significant difference between the

Table 4.39 Resistance Values

Company			
A	*B*	*C*	*D*
8	12	9	10
11	11	13	10
6	7	12	9
7	6	14	7

resistors of the four companies at $\alpha = 0.05$? Since the four companies were not selected from a population of companies, the appropriate model is the fixed-effects model. The sums of squares are:

$$
\begin{aligned}
SS_T &= \sum_{i=1}^{4}\sum_{j=1}^{4} X_{ij}^2 - \Big(\sum_{i=1}^{4}\sum_{j=1}^{4} X_{ij}\Big)^2/16 \\
&= 1{,}540 - 152^2/16 = 96 \\
SS_c &= \tfrac{1}{4}(32^2 + 36^2 + 48^2 + 36^2) - 152^2/16 = 36 \\
SS_E &= 1{,}540 - 1{,}480 = 60
\end{aligned}
$$

The ANOVA is shown in Table 4.40. Since $F_{0.95,3,12} = 3.49 > 2.4$, there is no reason to reject the hypothesis, H_0: $\alpha_i = 0$; that is, there is no difference between the effective resistance of these 10-ohm resistors when subjected to this constant humidity.

Table 4.40 ANOVA for Table 4.40

Source of variation	Degrees of freedom	Sum of squares	Mean square	F
Between companies............	3	36	12	2.4
Within companies.............	12	60	5	
Total.......................	15	96		

4.11b Two-way Classification. Experiments may be conducted to study the effect of several variables at several levels with one or more observations at each level.

Definition. When observations are made for all possible combinations, the experiment is called a *factorial experiment.* Suppose we are interested in studying the effective resistance of a particular resistor for c different levels of temperature and r levels of humidity. If n observations were made for each combination, then the data would appear as in Table 4.41. The ANOVA for Table 4.41 is given in Table 4.42. The

Table 4.41 Effective Resistance of a Particular Resistor for Different Levels of Temperature and Humidity
(n Observations per Cell)

Humidity	Temperature				Totals
	T_1	T_2	...	T_c	
H_1	X_{111} X_{112} ... X_{11n}	X_{121} X_{122} ... X_{12n}	... ...	X_{1c1} X_{1c2} ... X_{1cn}	R_1
H_2	X_{211} X_{212} ... X_{21n}	X_{221} X_{222} ... X_{22n}	... ...	X_{2c1} X_{2c2} ... X_{2cn}	R_2
...	...	...	...	...	...
H_r	X_{r11} X_{r12} ... X_{r1n}	X_{r21} X_{r22} ... X_{r2n}	... ...	X_{rc1} X_{rc2} ... X_{rcn}	R_r
Totals	C_1	C_2	...	C_c	T

Table 4.42 ANOVA Associated with Table 4.41

Source of variation	Degrees of freedom	Sum of squares	Mean square
Temperature..........................	$c - 1$	$SS_C = (1/nr) \sum_{i=1}^{c} C_i^2 - T^2/rcn$	$SS_C/(c - 1)$
Humidity..........................	$r - 1$	$SS_R = (1/nc) \sum_{i=1}^{r} R_i^2 - T^2/rcn$	$SS_R/(r - 1)$
Temperature humidity (interaction)......	$(c - 1)(r - 1)$	$SS_I = (1/n) \sum_{i=1}^{r} \sum_{j=1}^{c} \left(\sum_{k=1}^{n} X_{ijk} \right)^2 - T^2/rcn - SS_C - SS_R$	$SS_I/[(c - 1)(r - 1)]$
Error..........................	$rc(n - 1)$	$SS_E = \sum_{i=1}^{r} \sum_{j=1}^{c} \sum_{k=1}^{n} X_{ijk}^2 - (1/n) \sum_{i=1}^{r} \sum_{j=1}^{c} \left(\sum_{k=1}^{n} X_{ijk} \right)^2$	$SS_E/[rc(n - 1)]$
Total..........................	$rcn - 1$	$SS_T = \sum_{i=1}^{r} \sum_{j=1}^{c} \sum_{k=1}^{n} X_{ijk}^2 - T^2/rcn$	

mathematical model for this type of experiment is

$$X_{ijk} = \mu + \alpha_i + \beta_j + (\alpha\beta)_{ij} + \in_{ijk} \qquad \begin{aligned} i &= 1, 2, \ldots, r \\ j &= 1, 2, \ldots, c \\ k &= 1, 2, \ldots, n \end{aligned} \qquad (4.150)$$

where μ = true mean effect

α_i = true effect of the ith level of humidity

β_j = true effect of the jth level of temperature

$(\alpha\beta)_{ij}$ = true effect of the interaction of the ith level of humidity and jth level of temperature, i.e., the true effect for the (ij)th cell not explained by α_i and β_j and $\in_{i,k}$ is the true effect of the kth observation in the (ij)th cell; independent $N(0,\sigma^2)$.

The following are the assumptions regarding the model.

Fixed-effects Model. (Concerned only with the levels of temperature and humidity present in the experiment.)

$$\sum_{i=1}^{r} \alpha_i = \sum_{j=1}^{c} \beta_j = \sum_{i=1}^{r} (\alpha\beta)_{ij} = \sum_{j=1}^{c} (\alpha\beta)_{ij} = 0$$

Random-effects Model. (Concerned with a population of levels of temperature and humidity from population of these two factors.)

α_i are independent $N(0,\sigma_\alpha^2)$.

β_j are independent $N(0,\sigma_\beta^2)$.

$(\alpha\beta)_{ij}$ are independent $N(0,\sigma_{\alpha\beta}^2)$.

Mixed Model. (Concerned with only the levels present for one factor and with the population of levels for the other.) Suppose humidity levels are fixed and temperature levels are random; then

$$\sum_{i=1}^{r} \alpha_i = \sum_{i=1}^{r} (\alpha\beta)_{ij} = 0$$

β_j is independent $N(0,\sigma_\beta^2)$.

The expected mean squares for these three models are shown in Table 4.43. The proper F tests can be determined from the expected mean squares given in Table 4.43. In the case of interaction, this effect exists if the joint effect of the two factors taken together is different from the sum of their individual effects.

Note. In the fixed-effects model, if we have only one observation per cell, it is necessary to assume that the interaction is zero in order to get a reasonable test on humidity and temperature effects.

4.11c Latin Square. This design is frequently used in industrial experimentation. It is best illustrated by the design which is given in Table 4.44.

Note. There are three factors at four levels each. The machines and operation are set up as columns and rows, respectively. The materials are assigned at random to the machines and the operators with the restriction that each machine and each material be used only once by each operator. For one observation per cell the mathematical model for Table 4.44 is

$$X_{ijk} = \mu + \alpha_i + \beta_j + \gamma_k + \in_{ijk} \qquad \begin{aligned} i &= 1, 2, 3, 4 \\ j &= 1, 2, 3, 4 \\ k &= A, B, C, D \text{ (k is not independent of i and j)} \end{aligned} \qquad (4.151)$$

where μ = true mean

α_i = true effect of the ith level of operator

β_j = true effect of the jth level of machine

γ_k = true effect of the kth level of material

and $\in_{ijk}$ is the true random effect; independent $N(0,\sigma^2)$.

The ANOVA for Table 4.44 is given in Table 4.45. The appropriate F tests are the ratios of the mean squares of the factors to error mean square. The Latin square

Table 4.43 Expected Mean Squares for ANOVA of Table 4.42

Source of variation	Fixed-effects model	Random-effects model	Mixed effects (humidity levels fixed; temperature levels random)
Temperature	$\sigma^2 + nr \sum_{j=1}^{c} \beta_j^2/(c-1)$	$\sigma^2 + n\sigma_{\alpha\beta}^2 + nr\sigma_\beta^2$	$\sigma^2 + nr\sigma_\beta^2$
Humidity	$\sigma^2 + nc \sum_{i=1}^{r} \alpha_i^2/(r-1)$	$\sigma^2 + n\sigma_{\alpha\beta}^2 + nc\sigma_\alpha^2$	$\sigma^2 + n\sigma_{\alpha\beta}^2 + nc \sum_{i=1}^{r} \alpha_i^2/(r-1)$
Temperature humidity	$\sigma^2 + n \sum_{i=1}^{r} \sum_{j=1}^{c} (\alpha\beta)_{ij}^2/(r-1)(c-1)$	$\sigma^2 + n\sigma_{\alpha\beta}^2$	$\sigma^2 + n\sigma_{\alpha\beta}^2$
Error	σ^2	σ^2	σ^2
Total	...	...	...

Table 4.44 A 4 = 4 Latin Square

Operator	Machine				Totals
	1	2	3	4	
1	D	A	B	C	$T_{1.}$
2	A	C	D	B	$T_{2.}$
3	B	D	C	A	$T_{3.}$
4	C	B	A	D	$T_{4.}$
Totals	$T_{.1}$	$T_{.2}$	$T_{.3}$	$T_{.4}$	T

Note. A, B, C, D represent four levels of materials with T_A, T_B, T_C, and T_D their respective totals.

Table 4.45 ANOVA for Table 4.44

Source of variation	Degrees of freedom	Sum of squares	Mean square	Expected mean square
Machine.........	3	$\sum_{j=1}^{4} T^2_{.j}/4 - T^2/16 = SS_C$	$SS_c/3$	$\sigma^2 + 4/3 \sum_{j=1}^{4} \beta_j^2$
Operator........	3	$\sum_{i=1}^{4} T^2_{i.}/4 - T^2/16 = SS_R$	$SS_R/3$	$\sigma^2 + 4/3 \sum_{i=1}^{4} \alpha_i^2$
Material.........	3	$\sum_{k=A}^{D} T^2_k - T^2/16 = SS_M$	$SS_M/3$	$\sigma^2 + 4/3 \sum_{k=A}^{D} \gamma_k^2$
Error............	6	$SS_E = SS_T - SS_C - SS_R - SS_M$	$SS_E/6$	σ^2
Total..........	15	$\sum_{i=1}^{4} \sum_{j=1}^{4} X_{ij}^2 - T^2/16 = SS_T$		

is an efficient design if the assumptions are met. It requires reduced sample sizes. This design can be used only when the interactions are zero. When information regarding interactions is lacking, a factorial design should be used.

Among the recent contributions to the field of statistical methods are the *response surface techniques*. They provide an economical procedure for locating a set of experimental conditions which will yield a maximum or minimum response by using a sequential feature. Both the theory and the application of these techniques are discussed by Davies [14], Hunter [15], and Box and Hunter [16]. The details and precautions in applying and analyzing the vast number of available designs are discussed in many excellent books and articles. Some of the plans may be found in the appendix. Be careful in selecting a specific plan. A proper experimental design depends upon many factors. Consultation with a competent statistician is advised.

References

1. Pieruschka, E., *Principles of Reliability*, Prentice-Hall, Inc., Englewood Cliffs, N.J., 1963.

2. Bowker, A., and G. Lieberman, *Engineering Statistics*, Prentice-Hall, Inc., Englewood Cliffs, N.J., 1959.
3. Hald, A., *Statistical Theory with Engineering Applications*, John Wiley & Sons, Inc., New York, 1952.
4. Lloyd, D., and M. Lipow, *Reliability: Management, Methods, and Mathematics*, Prentice-Hall, Inc., Englewood Cliffs, N.J., 1962.
5. Buehler, R. J., "Confidence Intervals for the Product of Two Binomial Parameters," *J. Am. Statist. Assoc.*, vol. 52, p. 482, 1953.
6. Goldthwaite, L. R., "Failure Rate Study for the Lognormal Lifetime Model," *Proc. Seventh Natl. Symp. Reliability Quality Control*, Philadelphia, 1961, p. 208
7. Aitchison, J., and J. Brown, *The Lognormal Distribution*, Cambridge University Press, London, 1951.
8. Kao, J. H. K., "The Beta Distribution in Reliability and Quality Control," *Tech. Rept.* 2, Department of Industrial and Engineering Administration, Cornell University, Ithaca, N.Y.
9. Goode, H. P., and J. H. K. Kao, "Sampling Procedures and Tables for Life and Reliability Testing Based on the Weibull Distribution (Hazard Rate Criterion)," *Tech. Rept. TR*4, Office of the Assistant Secretary of Defense (Installations and Logistcis), GPO, 1962.
10. Gumbel, E. J., "Statistical Theory of Extreme Values and Some Practical Applications," *Appl. Math. Series* 33, *Natl. Bur. Standards*, 1954.
11. Mood, A. M., and F. A. Graybill, *Introduction to the Theory of Statistics*, 2d ed., McGraw-Hill Book Company, New York, 1963.
12. Fisher, R. A., "On the Probable Error of a Coefficient of Correlation Deduced from Small Samples," *Metron*, vol. 1, no. 4, p. 3, 1921.
13. Cochran, W. G., "Some Consequences When the Assumptions for the Analysis of Variance Are Not Satisfied," *Biometrics*, vol, 3, p. 22, 1947.
14. Davies, O. L., *Design and Analysis of Industrial Experiments*, Hafner Publishing Company, Inc., New York, 1954.
15. Hunter, J. S., "Determination of Optimum Operating Conditions by Experimental Methods," *Indus. Quality Control*, vol. 15, 1958; vol. 16, 1959.
16. Box, G. E. P., and J. S. Hunter, "Experimental Designs for Exploring Response Surfaces," in V. Chew (ed.), *Experimental Designs in Industry*, John Wiley & Sons, Inc., New York, 1958.

Section 5

RELIABILITY ESTIMATION

PAUL H. ZORGER

VITRO LABORATORIES, INC., SILVER SPRINGS, MARYLAND

Dr. Zorger received a B.S. degree in electrical engineering, a B.S. degree in industrial education, an M.S. degree in industrial management, and a Ph.D. in economics from University of Minnesota, Pennsylvania State College, and University of Pennsylvania, respectively.

He has been associated with the electronics industry for the past twenty years, particularly in the areas of electronic system design and system reliability. He has worked with CEIR, Inc., Martin-Marietta, Thompson-Ramo-Wooldridge, the Bendix Corporation, RCA, U.S. Air Force, and Hamilton Watch Company. He taught in Pennsylvania for three years, and he is a member of the Army Reserve. He is presently a senior technical staff member of Vitro Laboratories, Inc., in Silver Springs, Maryland.

Dr. Zorger is well known in the field of reliability, having served as representative to the Battelle Memorial Institute ECRC program, the Electronic Industries Association, and the Aerospace Industries Association. He is a senior member of the Institute of Electrical and Electronic Engineers, the American Society for Quality Control, and the American Ordnance Association. He has contributed many articles, pamphlets, and books to the field of electronic system design and system reliability.

RELIABILITY ESTIMATION

Paul H. Zorger

CONTENTS

5.1 INTRODUCTION

The discussion in this section covers engineering reliability methods that help attain the maximum product reliability inherent in a system. It is primarily to be used by the design engineer. However, it should be of use to all segments of an organization which are involved in producing products of known reliability. All segments of management, including manufacturing, quality, contracts, procurement, and logistics personnel, should understand the concepts used by the design engineer in order to fulfill their responsibilities in producing the designed system.

Most of the technology developed thus far in reliability has been applied to the electronics area. Data of the type collected in the electronics area have not been applied to mechanical devices or parts. This deficit is indicated by the methods and data presented here. Basic principles of reliability given here are applicable to electronics areas, but not necessarily to the area of structures, mechanical or hydraulic. There is a need for collection of reliability data on mechanical parts similar to the data collected on electronic parts.

Technology of reliability is increasing as knowledge of parts and materials increases. Significant results have been achieved in a few cases. More application of existing knowledge in reliability must be achieved in such a way that results are widely significant.

This section on Reliability Estimation is presented in such a way that the design engineer may achieve reliability in design and contribute to producing reliable products.

5.2 DEFINITION OF RELIABILITY

To a design engineer reliability is defined as "the probability of performing a function, under specified conditions for which designed, for a specific period of time." It should be noted that reliability is measured in terms of probability, expressed in meaningful quantitative terms and evaluated through applicable statistical methods.

Reliability, being a probability, is often stated as 0.999 or 0.95 or some other positive number less than 1. This quantitative expression of reliability is not complete. It is not required that all the functions of a device perform satisfactorily at the same time, but the functions must operate for the length of time required individually to assure the device accomplishes its intended purpose. Time is an essential part of the definition of reliability, and it must be stated in specifying reliability in many cases.

Another part of the definition of reliability is "specified conditions" under which the functions are to be performed. A system which could perform satisfactorily in all extremes of heat, humidity, vibration, altitude, and all other conditions would be ideal. However, it would not be practical, so systems are designed to operate within specified limits of environment. These conditions are explicitly stated, generally in a specification. A computer that was designed to operate in an air-conditioned environment and has a high rate of function failures only because it is not operating in that environment, i.e., without air conditioning, is not considered to be unreliable.

5.3 INHERENT PRODUCT RELIABILITY

Product reliability encompasses two main program phases: design and manufacture. Only the design phase of a program will be considered in this section. Inherent design reliability consists of three main areas: the reliability of the design, parts, and processes. The product of these three reliability determinations provides the overall or inherent reliability of the product. It represents the highest quantitative reliability figure that the product can ever attain. Thus, every operation henceforth performed tends to modify this inherent quantitative figure by some amount less than 1 (implementation of the design by manufacturing, initiation of degradation caused by human errors, test-equipment errors, variance of material parameters, etc.).

An approach to inherent product reliability is to determine the overall design (inherent) reliability P_I by evaluating the individual design reliability P_d, part reliability P_c, and process reliability P_p and then using the relationship $P_I = P_d P_c P_p$.

The inherent design reliability P_d is defined as the probability that the required circuit output will remain within specified tolerances if no catastrophic failures occur. The probability of catastrophic failure is included in P_c. Degradation of part parameters with time, circuit interactions, and transient fluctuations all contribute to output variation. When these effects produce beyond-tolerance outputs, overall reliability is degraded and evaluation of these factors produces a value for P_d. Actual evaluation of P_d involves determining and isolating the controlling factors of the circuit in terms of part-parameter variations.

The inherent part reliability P_c is defined as the probability that the parts used in the circuit will function for a specified time without catastrophic failure under the expected environmental and electrical stresses encountered in test or service. Actual evaluation of P_c is determined from the relationship

$$P_c = GF_r K_A K_{op}$$

where GF_r = generic failure rate
K_A = application factor (stress-environment)
K_{op} = use factor (ground, airborne, etc.)

The process reliability P_p is defined as the probability that all the elementary processes involved in the fabrication of an assembly have been completed without producing a crucial defect. Actual evaluation of P_p constitutes a measurement of the workmanship level maintained in carrying out the process steps, each of which has an inherent defect rate that is a function of the adequacy and variance of the process controls.

5.4 APPROACH TO THE RELIABILITY DETERMINATION

The increasingly stringent and demanding reliability requirements placed on today's products for military use require new application of analytical techniques to design. Many factors must be integrated into such analyses by the design engineers if quantitative reliability descriptions of the design are to be meaningful and valid. With the high reliability requirements imposed and with the requirements of time, speed, etc., an impasse is eventually reached. Such analysis of complex systems may demand the use of high-speed digital computers to ascertain, in integrated form, the quantitative reliability while considering the high functional requirements, i.e., light weight, minimum volume, and low cost. The output of these computers provides an objective, detailed analysis of the product which allows for design improvement and includes information on the part parameters (drift, stability per unit time, etc.) required for a product to perform within specified limits. Visual graphs depicting the interdependence of part parameters as related to circuit performance, data on the voltage, and power stresses imposed on the parts in the circuit as voltage and part parameter values change may be used. A check of the characteristics of the required circuit design parameters against other parameters, i.e., size and weight vs. voltage and wattage of various piece parts, may be made to ascertain any deficiencies per allocated requirements. Further refinement of these techniques allows the determination of the design reliability P_d and provides optimization of nominal part values for specific circuit configurations.

5.5 STATISTICAL APPROACH TO CIRCUIT ANALYSIS

The design problems associated with mass-produced equipment are different from those associated with a single model because of parts tolerances. The engineer may achieve functional reliability by selecting the optimum value for each part to provide for equipment performance, but mass-produced equipment must achieve this performance with parts that have values that vary throughout tolerance ranges (distributions) under conditions of use.

For maximum producibility and reliability the designer tries to design the equipment to perform properly with all parts operating simultaneously near, but not exceeding, their tolerance limits and, hopefully, in such a direction as to produce the greatest variance of the nominal function. It has been found that such attempts fail because the best and highest-precision parts do not have sufficiently small tolerances. The problem may be solved, as far as the above tolerance considerations are concerned, by making the circuitry complex to the extent that its total reliability is lower (because of the increased probability of failure of a part in a series) than that for equipment of fewer parts, which will not satisfy the extreme tolerance buildup. The probability that all the parts will exist at their maximum tolerances simultaneously is very low. The manner in which the individual parts tolerances affect the overall tolerance forms a basis for application of statistical approaches to circuit design.

Most production parts have a normal (or Gaussian) frequency distribution. For example, a histogram as shown in Fig. 5.1 will usually result from plotting the capacitance of a number of capacitors from one manufacturing lot.

As the quantity of capacitors measured increases and the capacitance interval decreases, the envelope of the histogram will form a normal distribution curve symmetrical about the average and asymptotic to the base. The total area under the curve represents all the capacitors, while the area bounded by 1σ covers 68.3 per cent of the total area (that is, 68.3 per cent of the capacitors are included by $\pm 1\sigma$, 99.5 per cent are included by $\pm 2\sigma$, and 99.7 per cent are included by $\pm 3\sigma$). The manufacturing tolerance specified will usually correspond to $\pm 3\sigma$ or less, depending on the degree of assurance desired. If $\pm 3\sigma$ limits are exactly attained, 0.3 per cent of the parts or fewer will usually be out of tolerance, with additional variations resulting from operating and environmental conditions. Although additional parts values may not be normally distributed, their associated circuit output variations may be considered nearly normally distributed because of combination effects.

A very high percentage of the circuits built must be designed to meet tolerance specifications, both from production and reliability standpoints. To illustrate, consider a circuit which must meet a production tolerance of 97- to 103-volt output. If the design permits production to meet this specification to the 3σ limits of a typical normal distribution, as shown in Fig. 5.2*a* with $\sigma = 1$ volt, only 3 out of 1,000 circuits will require parts changes. However, in the many instances when design allows production to try to meet the specification with $\sigma = 3$ volts, as in Fig. 5.2*b*, over 100 times as many circuits (31.7 per cent) will require costly parts changes involving

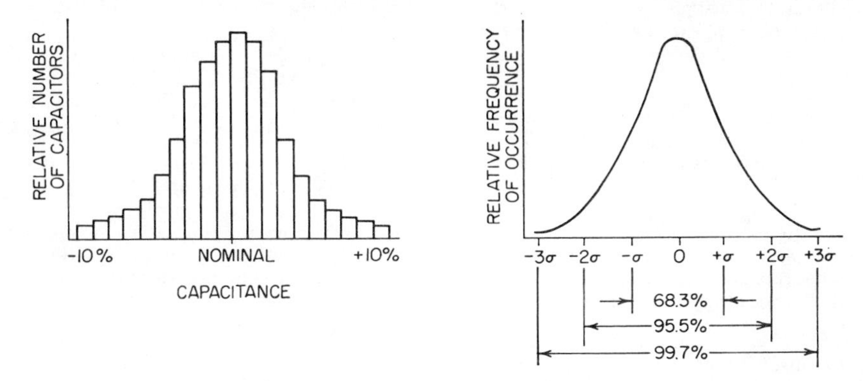

FIG. 5.1 Distribution of capacitor values.

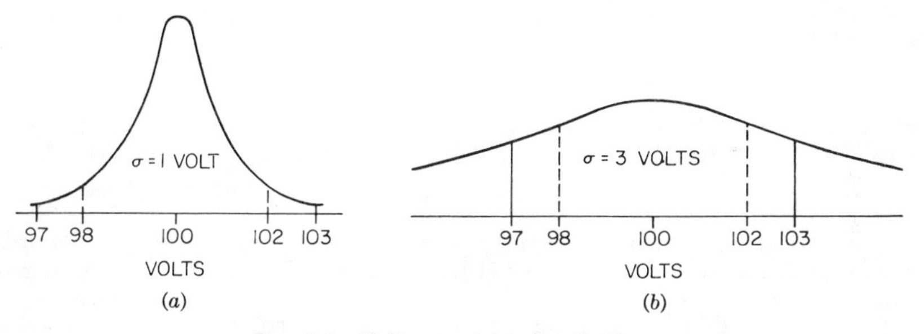

FIG. 5.2 Voltage-output distributions.

special selection to meet the specification. Thus, many more circuits are near the specification limits of Fig. 5.2*b* and much higher costs are incurred to meet the specification. This illustrates the necessity for designing to meet at least the $\pm 3\sigma$ limits, both from production and reliability standpoints.

Most circuits can be represented analytically to relate the functional variable to various part values. The functional variable and the part values in such an expression are continuous variables. When the part values are normal in frequency distribution and randomly combined in production to form functional circuits, the functional or performance variable will have a normal frequency distribution. Applying statistical methods of tolerances to the analytical expression yields a realistic tolerance for the performance or functional variable that will be obtained in actual use. (See Sec. 4 for statistical background.)

If the analytical expression is of the form

$$y = f(x_1, x_2, \ldots, x_n)$$

the standard deviation of the performance or functional variable is obtained from

$$\sigma_y^2 = (\partial y/\partial x_1)^2\sigma_{x_1}^2 + (\partial y/\partial x_2)^2\sigma_{x_2}^2 + \cdots + (\partial y/\partial x_n)^2\sigma x_n^2$$

and the nominal value of the functional variable is obtained from using the nominal values of the part in the analytical expression.

This technique is based upon assumptions that are usually obtained. Even when the assumptions are not fully realized, results by use of this technique are more realistic than simultaneous addition of part limits if part tolerances are small compared to the nominal value of a part and parts are independent of interactions and are randomly selected. Statistical techniques may be used to determine analytically any violation of the assumptions. These techniques become involved; and for application to most electronic engineering problems, sufficient accuracy is obtained by following applications as outlined herein.

Sum (or difference) tolerances, a general technique for statistical treatment of tolerances, readily yields more widely known and often useful basic equations. Consider the analytical expression for the simple sum

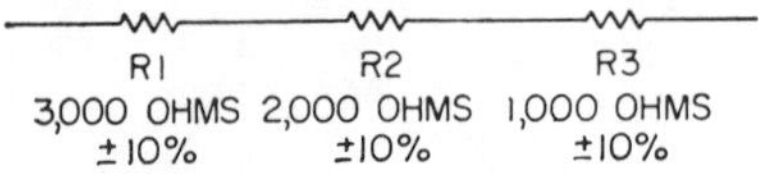

$$y = x_1 + x_2 + x_3 + \cdots + x_n$$

FIG. 5.3 Series circuit.

The well-known basic equation for the standard deviation of sums is obtained when the general technique for the computation of the performance-variable standard deviation is applied. Applying the general technique yields

$$\sigma_y^2 = \sigma_{x_1}^2 + \sigma_{x_2}^2 + \cdots + \sigma_{x_n}^2$$

because

$$\partial y/\partial x_1 = \partial y/\partial x_2 = \cdots = \partial y/\partial x_n = 1$$

Tolerances of sums can be combined if the tolerances contain the same number of sigmas. That is, if

$$t_y = A\sigma_{y_1} \qquad t_{x_1} = A\sigma_{x_1} \qquad \ldots \qquad t_{x_n} = A\sigma_{x_n}$$

where A denotes the same number of sigmas, such as $\pm 3\sigma$, and t denotes tolerances containing the same number of sigmas, then

$$t_y^2 = t_{x_1}^2 + t_{x_2}^2 + \cdots + t_{x_n}^2$$

When items are combined in series, the expression "square root of the sum of the squares" refers to actual values of the distribution and not percentages of the nominal value.

Tolerances of differences or combinations of sums and differences are similarly combined. An example of a typical application of sum or difference tolerances is resistors, $R1$, $R2$, and $R3$, combined in series (Fig. 5.3). The resistor values have normal distributions, and the tolerances are the limits of the $\pm 3\sigma$ distribution. Therefore, the nominal resistance sum is

$$\begin{aligned} R_t &= R_1 + R_2 + R_3 \\ &= 3{,}000 + 2{,}000 + 1{,}000 = 6{,}000 \text{ ohms} \end{aligned}$$

and the distribution of the sum is

$$\begin{aligned} t_{R_t}^2 &= t_{R_1}^2 + t_{R_2}^2 + t_{R_3}^2 \\ &= 300^2 + 200^2 + 100^2 \\ t_{R_t} &= 374 \text{ ohms} \end{aligned}$$

The nominal value and 3σ distribution limits of the sum are 6,000 ± 374 ohms, or 6,000 ohms ± 6.2 per cent. This illustrates that when items having a normal distribution are combined in series, the resultant sum distribution will exhibit a tolerance advantage, e.g., resistors with a ±10 per cent tolerance combined with a sum of a ±6.2 per cent tolerance.

Tolerances are combined to meet specific objectives. For example, the three resistors (3,000, 2,000, and 1,000 ohms) are connected in series and the total combined resistance (assuming that the tolerance of each resistor is ±10 per cent in 99.9 per cent of the cases) will be

$$\begin{aligned} R_S &= 6{,}000 \pm \sqrt{300^2 + 200^2 + 100^2} \\ &= 6{,}000 \pm 374 \text{ ohms} \\ &= 6{,}000 \pm 6.2 \text{ per cent} \end{aligned}$$

Although the sum will be 6,000 ± 374 ohms in 99.9 per cent of the cases presented in the preceding equation, overall tolerance improvement will be greater if more resistors are combined.

$$\text{Combined tolerance} = t_S = \sqrt{t_1^2 + t_2^2 + t_3^2 + \cdots + t_n^2}$$

where t_1, t_2, etc., are the individual tolerances, each of which must contain the same number of sigmas. As a result, the combined tolerance will also contain this same number of sigmas, i.e., if

$$t_1 = A\sigma_1, \qquad t_2 = A\sigma_2 \qquad t_n = A\sigma_n$$
$$\therefore t_{sum} = A\sigma_{sum}$$

Thus, it can be seen that the equation for combined tolerance is derived from the basic equation

$$\sigma_S^2 = \sigma_1^2 + \sigma_2^2 + \sigma_3^2 + \cdots + \sigma_n^2$$

Wherever possible, it is preferable to express a result as a sum of values by using the logarithms of values which are to be multiplied or divided and by using the reciprocal of values which combine like parallel resistances, thereby more easily utilizing the mathematics of probability.

Product (or quotient) tolerances, an analytical expression for simple products, yields a basic equation when a general technique for computation of the standard deviation of the performance of functional variable is applied. When the analytical expression is

$$y = x_1 x_2 \cdots x_n$$

application of a general technique yields

$$\sigma_y^2 = (x_2 x_3 \cdots x_n)^2 \sigma_{x_1}^2 + (x_1 x_3 \cdots x_n)^2 \sigma_{x_2}^2 + \cdots + (x_1 x_2 \cdots x_{n-1})^2 \sigma_{x_n}^2$$

Dividing the above expression by y^2 and canceling yields

$$(\sigma_y/y)^2 = (\sigma_{x_1}/x_1)^2 + (\sigma_{x_2}/x_2)^2 + \cdots + (\sigma_{x_n}/x_n)^2$$

Tolerances of products can be combined directly if the tolerances contain the same number of standard deviations and are expressed as percentages. That is, if

$$T_y = A\frac{\sigma_y}{y}100 \qquad T_{x_1} = A\frac{\sigma_{x1}}{x_1}100 \qquad \cdots \qquad T_{x_n} = A\frac{\sigma_{x_n}}{x_n}100$$

where A denotes the same number of sigmas, $y, x_1, \ldots, x_n$ are nominal values, and T denotes tolerances that contain the same number of sigmas expressed as percentages of the nominal value, the basic equation for products is

$$T_y^2 = T_{x_1}^2 + T_{x_2}^2 + \cdots + T_{x_n}^2$$

For products the familiar expression "square root of the sum of the squares" refers to tolerances expressed as "percentages" and should not be compared with the case applied to sums, where the tolerances are the "actual values."

Tolerances of quotients or combinations of products and quotients are similarly combined. The tolerance of the output of an idealized transformer (Fig. 5.4) is used as an example of this technique. The tolerances shown are the $\pm 3\sigma$ normal distribution limits. The nominal output voltage is

$$E_o = NE_i = 4 \times 110 = 440 \text{ volts}$$

and the distribution of the output voltage is

$$\begin{aligned} T_{E_o}{}^2 &= T_N{}^2 + T_{E_i}{}^2 \\ &= 5^2 + 15^2 \\ T_{E_o} &= 15.8 \text{ per cent} \end{aligned}$$

The nominal value and 3σ distribution limits of the output voltage are 440 $\pm$ 69.5 volts or 440 volts $\pm$ 15.8 per cent.

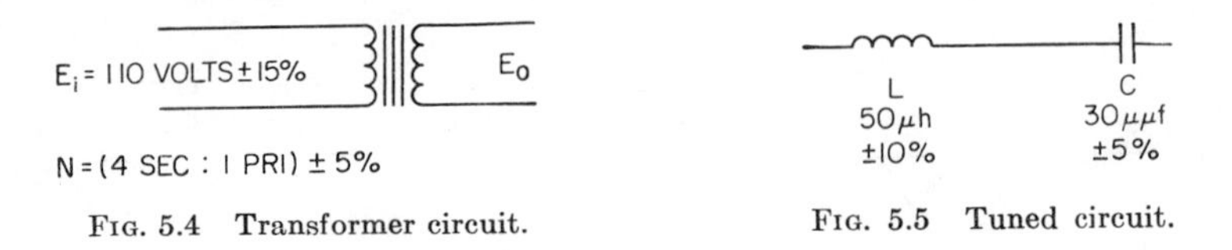

FIG. 5.4 Transformer circuit.

FIG. 5.5 Tuned circuit.

When the analytical expression relating the performance variable to the various part values is of a different form from the sum or difference and product or quotient, a general expression may be written as

$$\sigma_y{}^2 = (\partial y/\partial x_1)^2 \sigma_{x_1}{}^2 + (\partial y/\partial x_2)^2 \sigma_{x_2}{}^2 + \cdots + (\partial y/\partial x_n)^2 \sigma_{x_n}{}^2$$

and must be used to obtain the distribution limits of the performance variable. For example, a tuned circuit has a resonant frequency of

$$f = 1/2\pi \sqrt{LC}$$

Values and tolerances, with tolerances at the 3σ normal distribution limits, may be shown as in Fig. 5.5. The nominal resonant frequency is

$$\begin{aligned} f &= 1/6.28 \sqrt{50 \times 10^{-6} \times 30 \times 10^{-12}} \\ &= 4.12 \times 10^6 \text{ cps} \end{aligned}$$

To determine the tolerance of the frequencies, the general expression is used

$$\begin{aligned} \sigma_f{}^2 &= (\partial f/\partial C)^2 \sigma_C{}^2 + (\partial f/\partial L)^2 \sigma_L{}^2 \\ (3\sigma_f)^2 &= (-1/4\pi \sqrt{LCL})^2 (3\sigma_C)^2 + (-1/4\pi \sqrt{LCL})^2 (3\sigma_L)^2 \\ 3\sigma_f &= 23 \times 10^4 \end{aligned}$$

and the nominal value and 3σ limits of the resonant frequency are $4.12 \times 10^6 \pm 0.23 \times 10^6$ cps or 4.12×10^6 cps $\pm$ 5.6 per cent.

Thus far, only initial variations due to tolerance distributions have been considered. Under typical and more realistic operating conditions, part variations will result from both negative- and positive-coefficient effects, and distribution variations, over a range of operating conditions. The low and high conditions resulting from coefficient effects in the range of operating conditions are separately treated, and distributions are computed for each.

As another illustration, consider the gain function of a seven-stage *IF* amplifier, where each stage uses the same tube type with an optimum transconductance of 5,000- to 1,000-ohm composition load resistor for very light loading. The tube and resistor variations used are those given in Table 5.1, the $\pm$ values being the 3σ limits for normal distributions centered about the values preceding them.

Table 5.1 Tube and Resistor Variations

	Resistor Allen-Bradley composition, %		Tube Transconductance change, %	
	Low	High	Low	High
Manufacturing tolerance	± 5	± 5	−8 ± 10	+8 ± 10
Change at +85°C	± 4			
Change at −54°C	....	+12		
Aging change at 100 hours	−5 ± 2	....	−6 ± 5	
Soldering change	−2*	−2*		
Filament voltage change	....	....	−5 ± 3	+5 ± 3
Random change when first energized	....	....	± 10	± 10
Totals	−7 ± 6.7	+10 ± 5	−19 ± 15	+13 ± 15

* A fixed change that will affect both the low and high conditions.

The effects of the operating conditions must be considered for the proper low- and high-condition combination of coefficients and distribution tolerances. Consider a seven-stage amplifier, where the

$$\text{Nominal gain per stage} = G_m R_1$$
$$\text{Nominal } G_m = 5{,}000\ \mu\text{mhos}$$
$$\text{Nominal } R_1 = 1{,}000 \text{ ohms}$$

Lowest Expected Gain

Tube contribution: G_m low by 19% ± 15%,

$$G_m = 5{,}000(0.81 \pm 0.15) = 4{,}050 \pm 750\ \mu\text{mhos}$$

Resistor contribution: R_1 low by 7% ± 6.7%,

$$R_1 = 1{,}000(0.93 \pm 0.067) = 930 \pm 67 \text{ ohms}$$

Nominal gain per stage:

$$A = G_m R_1 = (4{,}050 \times 10^{-6})930 = 3.77$$

Seven-stage nominal gain:

$$A^7 = 3.77^7 = 10{,}800$$

Gain variability per stage:

$$T = \sqrt{T_{G_m}{}^2 + T_{R_1}{}^2}$$
$$= \sqrt{(750/4{,}050 \times 100)^2 + (67/930 \times 100)^2} = 19.9\%$$

Seven-stage variability:

$$T = \sqrt{7(19.9)^2} = 52.6\%$$

The nominal value and 3σ distribution limits of the gain considering low conditions in the operating range are 10,800 ± 5,680 or 10,800 ± 52.6 per cent. Expressed in another manner, the $\pm 3\sigma$ limits of the i-f amplifier gain distribution are 74.2 and 84.3 db.

While the distribution limits of the gain of the amplifier can be computed, the standard deviation of the gain cannot be computed from the following expressions.

Seven-stage nominal gain:

$$A^7 + (G_m R_1)^7 \qquad \sigma_A{}^2 = (\partial A/\partial G_m)^2 \sigma_{G_M}{}^2 + (\partial A/\partial R_1)^2 \sigma_{R_1}{}^2$$

because

$$G_{m_1} \neq G_{m_2} \neq \cdots \neq G_{m_7} \qquad \text{and} \qquad R_{1_1} \neq R_{1_2} \neq \cdots \neq R_{1_7}$$

The distribution of the total gain is computed as follows.

Seven-stage gain:

$$A = G_{m_1} R_{1_1} \times G_{m_2} R_{1_2} \times \cdots \times G_{m_7} R_{1_7}$$

$$\sigma_A{}^2 = (\partial A/\partial G_{m_1})^2 \sigma_{G_{m_1}}{}^2 + (\partial A/\partial R_{1_1})^2 \sigma_{R_{1_1}}{}^2 + \cdots + (\partial A/\partial G_{m_7})^2 \sigma_{G_{m_7}}{}^2 + (\partial A/\partial R_{1_7})^2 \sigma_{R_{1_7}}{}^2$$

and each tube and load resistor is an independent variable. Substituting in the previous equations the seven-stage gain is found to be 5,650 ± 750 μmhos.

Resistor contribution: R_1 high by 10% ± 5%,

$$R_1 = 1{,}000(1.10 \pm 0.05) = 1{,}100 \pm 50 \text{ ohms}$$

Nominal gain per stage:

$$\begin{aligned} A &= G_m R_1 \\ &= (5{,}650 \times 10^{-6})6{,}100 = 6.22 \end{aligned}$$

Seven-stage nominal gain:

$$A^7 = 6.22^7 = 360{,}000$$

Variability per stage:

$$\begin{aligned} T &= \sqrt{T_{G_m}{}^2 + T_{R_1}{}^2} \\ &= \sqrt{(750/5{,}650 \times 100)^2 + (50/1{,}100 \times 100)^2} = 14\% \end{aligned}$$

Seven-stage variability:

$$T = \sqrt{7(14)^2} = 37\%$$

The nominal value and 3σ distribution limits of the gain at the high conditions in the operating range are 360,000 ± 133,000 or 360,000 ± 37 per cent. The 3σ limits of the i-f amplifier gain distribution are 107 and 114 db at the high conditions. The gain of this amplifier under typical operating conditions is expected to lie between 74.2 and 114 db.

5.6 TYPES OF FAILURE

Excluding design, there are two types of circuit malfunction: drift and catastrophic. Drift failure, a time-related function, is caused by component variations (stresses and molecular changes) which produce changes in the performance characteristics of a circuit beyond those limits for which the circuit was designed. For example, a resistor in an amplifier may drift from the nominal value, owing to stresses invoked per unit time, to a value which renders the function marginal; and the amplifier will then exhibit spurious oscillations. In this type of failure the component performs a useful primary function but the circuit fails because it does not operate within specified design limits. Catastrophic failures, on the other hand, are caused by changes in part values of such magnitude that the component can be considered inoperable. This type of failure occurs, for example, when a broken resistor in a circuit causes an infinite resistance across two terminals. Then the part no longer performs the primary function for which it was intended.

Three basic principles generally can be applied to circuit design when considering drift and catastrophic failures: (1) drift failures can always be eliminated by considering the drift characteristic of a part for a specified time; (2) catastrophic failures can be neither predicted nor eliminated; (3) parameter drift and catastrophic failures are not necessarily unrelated. Considering these basic principles, trade-offs may be made and the reliability of a circuit may then be determined within very close values, with the reliability being improved by the use of redundancy if warranted.

It is assumed, therefore, that a circuit failure will inevitably result from a part failure of the catastrophic type. Furthermore, if redundancy is excluded, a single catastrophic part failure may result in a system failure.

Two principles apply to catastrophic part failures: Catastrophic failures cannot be prevented, and the rates at which catastrophic failures occur are related to the electrical and environmental stresses to which the parts are subjected. Although a finite probability of catastrophic failure will always remain, a decrease in the uncertainty in the manufacturing process by control of both materials and manufacturing techniques will reduce the failure rate. Moreover, stress-related failures, a confirmed relationship, can be decreased by the use and application of derating procedures.

As a system becomes more complex, the finite component catastrophic failure rates increase the probability of failure during a specified period of time. Thus, when

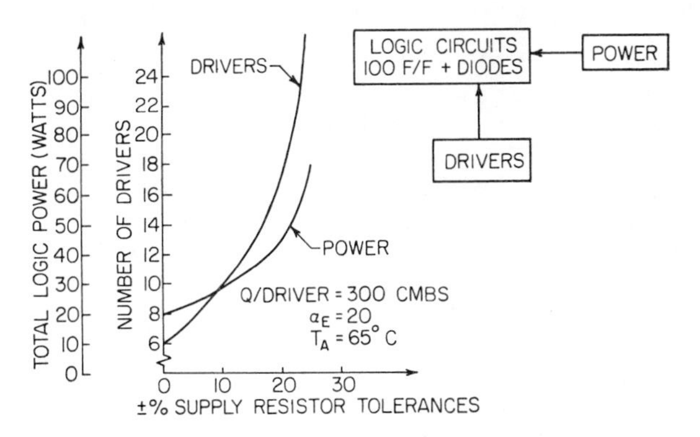

FIG. 5.6 Driver circuits and power vs. logic-circuit tolerances.

P_C is the probability that a single part will fail in a certain period of time, the failure probability of a system comprised of n such components for the same time period is

$$P_T = 1 - (1 - P_C)^n$$

Therefore, as n increases, the probability of a system failure also increases.

Although designing for the worst case (Sec. 5.7) reduces their probability, drift failures may increase the probability of a system failure owing to an increase in the number of components in a given system and, consequently, the probability of a single catastrophic failure. Moreover, an increase in the power level of a circuit increases the thermal stress on certain parts and the part catastrophic probability P_C, since some parts (e.g., transistors) cannot be replaced with higher-power units to preserve derating margins. An example of the dependence of catastrophic-failure rates upon drift-failure probabilities may be derived from the system-part tolerance relationships shown in Fig. 5.6.

Four transistors per driver are required to trigger the logic flip-flops (i.e., two transistors comprising a one-shot multivibrator followed by two transistors in a complementary emitter-follower cascade). Assuming a hypothetical transistor catastrophic-failure rate of 2 per 1,000 for a given time period, a failure-rate increase of 0, and 1 transistor per 1,000 for each per cent increase in allowable resistor and supply tolerances (in order to account for the increased thermal stress of the transistors as the tolerance level of the circuit is increased), the failure probability associated with each transistor may be shown to be

$$P_1(T) = 0.002 + 0.01d^5$$

where d = changes in resistor and supply tolerances

In addition, the drift-failure probability (logic flip-flop) decreases exponentially with increased design tolerances. If zero tolerances are applied in the circuit design, the drift-failure probability is 50 per cent; however, if 10 per cent tolerances are assumed, the drift-failure probability is reduced an order of magnitude. The probability of a logic-circuit failure owing to part drift is then

$$P_1(C) = 0.5 \exp(-25d)$$

Considering transistor catastrophic and circuit-drift failures, it is possible to calculate the probability of a system failure by using the curves of Fig. 5.7, where $P_N(T)$ is the probability of a single transistor catastrophic failure in a system composed of logic circuits and drivers and $P_M(C)$ is the probability of a single circuit-drift failure among the logic circuits only. Thus, with increasing parameter tolerances, $P_N(T)$ increases owing to additional transistors and the increasing thermal stresses applied

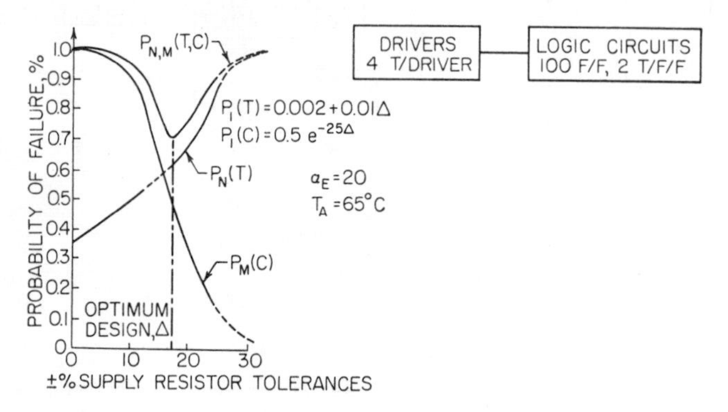

FIG. 5.7 Failure probabilities vs. logic-circuit tolerances.

as the circuit power level goes up, while $P_M(C)$ decreases owing to the increasing part-tolerance margins of the logic circuits.

The probability of a system malfunction by transistor catastrophic or circuit-drift failure is

$$P_{N,M}(T,C) = 1 - [1 - P_N(T)][1 - P_M(C)]$$

This illustrates the existence of an optimum-tolerance design point ($d = \pm 18$ per cent), from a reliability viewpoint, between catastrophic and drift failures. Worst-case circuit design carried beyond this point will reduce the system reliability despite the increase obtained in individual circuit tolerance margins.

5.7 WORST-CASE TECHNIQUE APPLICATION

Worst-case criteria generally state that a circuit must perform its function within a specified tolerance and according to specified functional parameters, all part parameters (including signal, power, and environmental) being at their extreme values. Worst-case values, defined as those values which are not mutually exclusive as related to tolerance limits but which affect a functional parameter of the part in an adverse manner, may be adjusted under certain conditions, considering the specified apportioned failure rate from tolerance buildup and the effects which critical parameters will have on the specified output function. This criterion specifies that no part shall be subjected to stresses greater than those specified for the part under the fulfilled reliability requirement, considering the factors of derating for electrical and environmental stresses.

Although the severity of worst-case analysis in terms of realistic stresses is occasionally questioned, numerous practical illustrations approached under different conditions show a high degree of correlation between critical parameters of a given component and a successful circuit, particularly in the extremes of a specified operating region. For example, assuming that a linear amplifier composed of a number of stages with a power source performing at the lower extreme of tolerance (the gain of each stage being voltage sensitive, hence all stages simultaneously exhibiting low gain) is operated at low temperatures, all the betas of the transistor in each stage will approach the lower limit and a worst-case situation will result.

Worst-case analysis (1) results in such beneficial gains as a rapid evaluation of the quality of performance of a specified circuit configuration, (2) is an exceptionally valuable tool in circuit development and part-value optimization, (3) simplifies designs and rarely imposes unrealistic stresses, (4) considers tolerance and end-of-life values, (5) is a necessary technique in determining minimum part reliability, utilizing statistically valid and available part-reliability data, and (6) facilitates the calculation of part performance limits. Overall development-test time and cost are reduced and expensive redesign is eliminated because fewer but statistically integrated planned part tests are required to verify results, part-parameter variations having been considered in the worst-case design. (Part information may be obtained from Battelle ECRC reports, from tests conducted and reported through the IDEP program, and from MIL-HDBK 217, "Reliability Stress Failure Rate Data," Aug. 8, 1962. Refer to Sec. 6.) Thus, the use and application of worst-case analysis increase the efficiency of the engineering and manufacturing departments and enhance product reliability.

A circuit may be worst-case-designed by using several techniques. For example, because all components exhibit a certain spread in values around their nominal ratings, the designer never works with ideal parts, and consideration of part variations in the initial design of a circuit (not always a simple procedure) necessitates that many circuits be designed with nominal part values and tested experimentally to determine operational limits. Experimentally derived profiles (schmoo diagrams), obtained by determining operational limits as a function of part variations, are commonly used to determine the tolerance of a circuit to part and supply-voltage changes. By iterating the circuit design so that the circuit performance falls in the center of the profile curves when all the circuit units are at nominal value, a worst-case design is achieved empirically. As an alternative, the effect of part variations on the performance of a circuit may be determined analytically by assuming part deviations which at maximum values degrade the circuit performance and by synthesizing the circuit so that it still meets its specifications. Thus, although a circuit may be worst-case-designed by empirical or analytical methods, basic drift-characteristic data determine the maximum expected deviations of the parts and may be stated as per cent of maximums on some deviation from nominal, for example, x sigmas.

Part deviations result from the manufacturing process as well as from normal attrition. The manufacturing spreads can be determined by measuring large samples (preselected either at the point of manufacture or at the point of use) to satisfy initial tolerance requirements (refer to Table 5.1). Determination of the deviations due to attrition is more difficult, but it may be accomplished by accelerated life tests utilizing linear discriminant analysis and by collection of historical data. By interpretation of the best part information available (manufacturer's tolerance guarantees and statistical life-test curves showing part-value degradation as a function of environmental stress and normal attrition), the designer can obtain worst-case part deviations, assuming that tolerances are assigned to parts. Problems occur in determining the overall system effects when tolerances are incorporated into the circuit design. Moreover, in designing to worst-case criteria, reasonable safety factors which are all too rarely applied in electronic design are automatically incorporated without overdesign, a normal and often valid criticism of worst-case design.

If worst-case criteria are met, then determination of part reliability must be considered. The following three items (one or more of which are usually not available) may be combined by means of multiple integrals to yield an accurate value for the probability of part failure. (The number of integrals equals the number of specified system parameters, and the order of each integral equals the number of independent parameters making up the system parameter under consideration.)

1. Functions describing the component parameters as functions of all part and input parameters.
2. The required limits on all part parameters.
3. The statistical distributions of all part and input parameters.

As a simplifying alternative, the probability of intrinsic failure for each part, if known, over the system operating life can be converted to an intrinsic failure rate if the system meets worst-case criteria. At this point the failure rates of all parts can then be added to yield a failure rate which is greater than or equal to the actual system failure rate. (The "greater than" is included because frequently a part or input parameter can be slightly outside its specified tolerance without causing a system failure.) Note that tolerance narrowing tends to minimize the differences between the failure rate determined by using the sum of intrinsic failure rates and the actual failure rates which can be expected.

Designing to worst-case criteria does not necessarily result in overcomplicated circuits. Circuits which are capable of performing their functions under nominal conditions and which are chosen at random and designed by means of nominal analysis, breadboarding, and subsequent testing can readily be modified to pass worst-case criteria by circuit or part changes. It is possible to remove superfluous parts, although circuits which exhibit failure tendencies because of overstressed parts sometimes require additional parts to limit excess voltage, current, or power rating.

Application of worst-case criteria benefits the parts, design, and system engineers, the administration of the engineering organization, and the manufacturer. Part (application) and design engineers profit from the following desirable features of this technique: (1) Worst-case criteria establish a minimum acceptable standard for part design which is clearly defined in the procurement documents and is possible to adhere to. (2) Use of poor circuits, untried or unspecified devices, and misunderstood designs are eliminated because worst-case criteria cannot be established under these conditions. (3) Worst-case-analysis calculations are relatively simple, because end-point calculations do not involve statistical or nonlinear theory. (4) Programs for digital computers which may be used in calculation of the various worst-case techniques by successive insertion of the appropriate parameter tolerance limits can readily be developed for worst-case solutions. (This point will be described in detail later.) The system design engineer benefits by receiving more accurate data on individual part performance, thereby forcing system design decisions involving the necessary trade-offs and compromises and facilitating more accurate determination of overall system accuracy and performance.

The administration of an engineering organization which applies worst-case analysis (in which the control of engineering methods and procedures is an important objective) benefits in terms of the following: (1) more standard and reliable designs, (2) reliability of a design less dependent upon an individual engineer's capability, (3) more effectively trained engineers, (4) utilization of engineers with relatively little experience to analyze portions of the design, (5) more accurate evaluation by supervision of the abilities of engineers performing analyses, (6) more accurate records of detailed design analyses, and (7) cost estimates for development of new parts, maintenance, and manpower loading.

Worst-case analysis benefits the manufacturer by (1) more careful application of specifications by the development engineer (who also will request and utilize more part-parameter data from the vendor) and the reduction in cost of performance and degradation tests by the user to obtain such data, (2) greater assurance that the vendor's part will be applied properly by the customer, thereby resulting in compatibility of the parts application and reliability figures, (3) establishment of more meaningful and optimized specifications and manufacturing checks by considering the functions of participating departments (i.e., quality, procurement), which results in receipt of worst-case part-parameter limits and therefore simplified checkout and analysis of system behavior.

Allowing greater part tolerances in a circuit design is analogous to allowing greater randomness in the selection of circuit parts, which, in turn, corresponds to an entropy increase in the circuit from a thermodynamic point of view. Thus, some measure of increased energy expenditure is necessary if the second law of thermodynamics is to apply, since the designer cannot permit the greater degree of part randomness to be reflected in the performance of the circuit. This principle is illustrated in Fig. 5.8, which shows relay coil Y in series with battery E and current-limiting resistor R. For relay operation, current I must exceed some threshold value I_T. The battery may vary by $\pm d_E$ per cent; and assuming a perfect resistor which may vary by $\pm d_R$ per cent, R must be selected as follows to satisfy worst-case conditions:

$$R \leq (E/I_T)(1 - d_E)/(1 + d_R)$$

For maximum efficiency, the nominal value of R is equal to the right-hand side of the preceding equation; and assuming all other circuit parts at nominal values, the relay will operate for the minimum expected battery voltage and for the maximum expected resistance, the steady-state power dissipation of the relay circuit being

$$P_N = EI_T(1 + d_R)/(1 - d_E)$$

The battery voltage may increase while the resistor voltage simultaneously decreases.

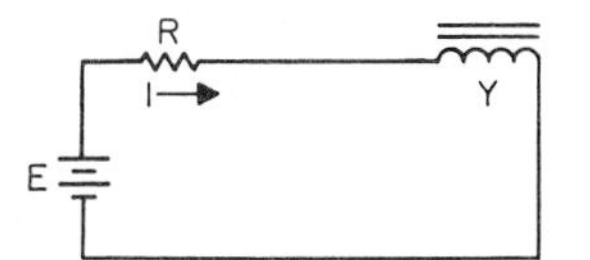

FIG. 5.8 Thermodynamic-principle application.

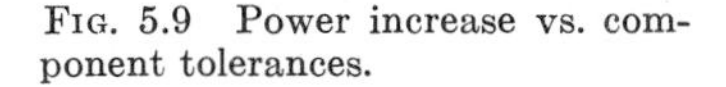

FIG. 5.9 Power increase vs. component tolerances.

The circuit must be designed to operate at a maximum power expenditure as shown in the following equation and illustrated in Fig. 5.9, where average power P/EI_T is plotted as a function of the design tolerances assuming that $d_E = d_R = d$.

$$P_{max} = EI_T(1 + d_E)^2(1 + d_R)/(1 - d_E)(1 - d_R)$$

Thus, the circuit of Fig. 5.8 will dissipate 50 per cent more power at nominal part values designed for ± 20 per cent part variations rather than ideal parts (zero tolerance). The rapidly increasing load on the power supply as increased part tolerances are demanded in circuit design is illustrated by the fact that circuits must be designed to function at power dissipation extremes because maximum worst-case dissipation could be as high as 270 per cent above part values when ± 20 per cent part variations are allowed.

It is apparent from this example that increased power expenditure is required to ensure a circuit against drift failure, thereby necessitating a trade-off. In addition, if unlimited power is available from sources of unlimited reliability, any part tolerance may be considered in the design procedure and circuit-drift failures may be eliminated entirely, indicating that the burden of reliability shifts from the circuit under design to its power supply when worst-case design practice is followed. In more complex circuits, an increase in power level due to an increase in design tolerances also increases the number of parts required to perform a given function. The part complexity of the circuit did not increase, because resistor R was assumed to be in any power rating. (Several resistors in parallel would have been necessary had the resistor been limited to some comparatively moderate finite power dissipation.)

A flip-flop (which must be capable of driving OR and AND loads and simultaneously must be center-point-triggered through a pulse gate) is illustrated in Fig. 5.10. This circuit is typical of those used in shift registers or counters where the stages are loaded by diode matrices or by arbitrary diode logic networks, the standby power dissipation (power dissipation of the circuit, not including power into the load) of these logic-driving flip-flops being a function of the resistor tolerances for various values of transistor current-amplification factors. Employing worst-case procedures in designing the flip-flop, the load requirements first are converted into terminal voltage and current specifications and then the component deviations which reduce the available

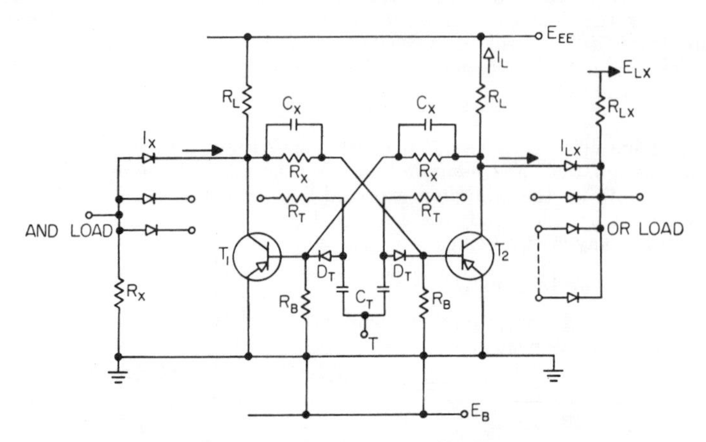

FIG. 5.10 Logic-driving flip-flop.

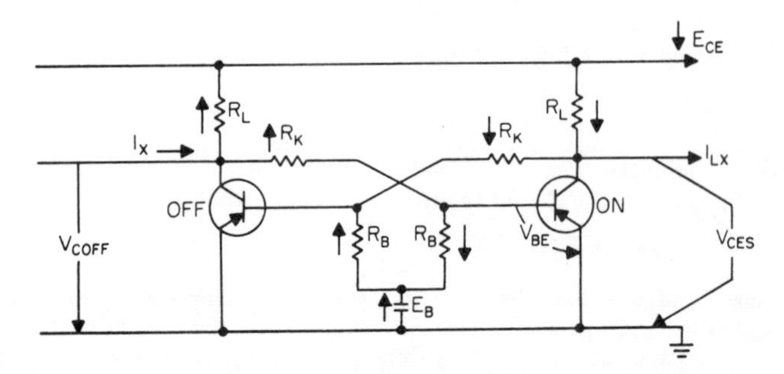

FIG. 5.11 Logic-driving flip-flop.

terminal currents and voltages in a manner most detrimental to meeting the load requirements are determined.

Figure 5.11 represents a steady-state d-c circuit where the worst-case deviation of each circuit part is indicated by the direction of the arrows next to the resistors and voltage supplies. The load requirements are given in the form of a minimum current I_{LX} for the OR load and a minimum voltage V_{COFF} for the AND load. In addition, stability requirements relate to the resistor tolerances R_i, the supply-voltage tolerances E_i, and the ambient-temperature range T_A. Transistor parameters and the variations which must be considered include the current-amplification factor I_E, the collector-saturation voltage V_{CES}, the base-to-emitter voltage of the saturated transistor V_{BE}, the junction temperature (the maximum allowable temperature $T_{j_{max}}$, in particular), the thermal-dissipation factor K, and the collector-to-base saturation (leakage) current of the OFF transistor I_{CBO}. The static design problem is to determine the nominal values for resistors R_L, R_B, and R_K which permit the load requirement to be met under the worst-case deviations of the circuit and transistor parameters.

Although the design procedure for the circuit of Fig. 5.11 is not derived in this discussion, Fig. 5.12 illustrates the variation of circuit power dissipation and number of components as a function of the component tolerances for a specific circuit design. The specific design values of Fig. 5.12 include a maximum ambient temperature of

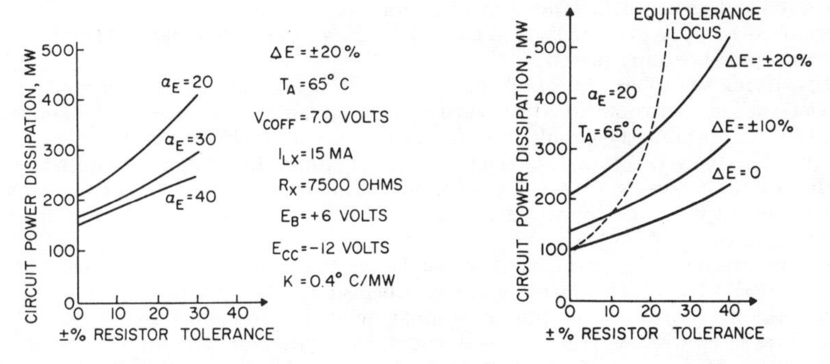

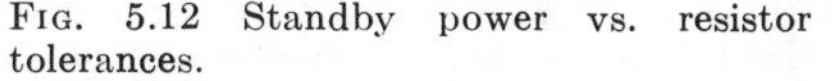
FIG. 5.12 Standby power vs. resistor tolerances.

FIG. 5.13 Standby power vs. resistor tolerances.

65°C, an OR load of 15 ma, and an AND load of 7 volts into 7,500 ohms. Figure 5.13 illustrates the relationship between circuit power dissipation and resistor tolerances for different power-supply tolerances, the dotted line "equitolerance locus" referring to the circuit power dissipation as a function of equal resistor and supply voltages.

Figure 5.14, where trigger charge, in coulombs, is plotted as a function of resistor tolerances for the circuit previously discussed, illustrates the additional energy required to trigger the flip-flop as the power level of the flip-flop increases with increasing parameter tolerances. Referring to the equitolerance locus, it can be seen that the charge required to trigger the flip-flop at a resistor and supply-voltage tolerance level of ±15 per cent is twice what would be required if the tolerances could be held to zero. Thus, an increased energy is used both statically and dynamically for increasing the tolerance values in the circuit.

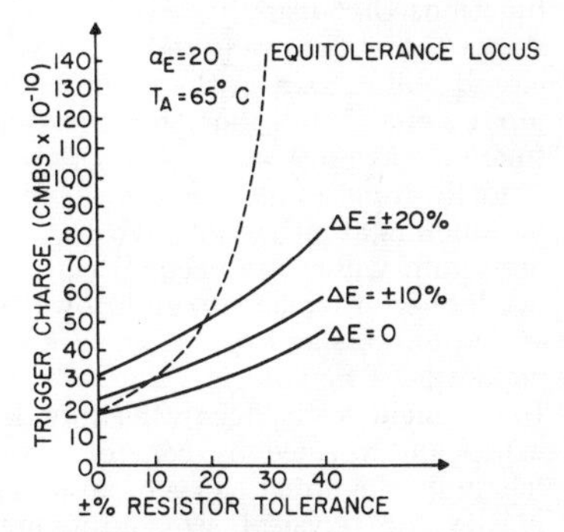

FIG. 5.14 Maximum trigger charge vs. resistor tolerances.

Because the driver circuits used to trigger flip-flops (e.g., monostable multivibrators or blocking oscillators) are energy-limited by the power-dissipation ratings of the transistors, it is apparent that additional drivers might be required in a large system as part tolerances are increased. For example, consider a system of 100 flip-flops of the same kind as those shown in Fig. 5.10 that require a number of regenerative pulse drivers to trigger them in parallel and are connected (as a function of resistor and supply-voltage tolerances) in a shift register or parallel-counter configuration. It can be seen that the number of parts in the system may increase considerably as a result of worst-case designing for greater part tolerances when the drivers and logic circuits are considered as a single system. Consequently, because system complexity (in terms of number of parts and allowable part tolerances) is not an independent consideration but may be strongly affected by the degree to which circuits are designed to reduce drift failures, consideration must be given to trade-offs during system design.

Worst-case criteria include the following:

1. A part must perform every requirement within tolerance when all basic parameters necessary and sufficient for proper system operation (part parameters), as well as signal, power, and environmental input parameters, are at worst-case values. (Worst-case values are not mutually exclusive and do not lie within the tolerance limits but tend to affect the operating parameter of a part in the most adverse manner.)
2. No part shall be subjected to stresses greater than those specified for the part under any ideal or nonideal conditions of operation.

Worst-case analysis is an analytical tool which aids in achieving designs that satisfy worst-case criteria and requires:

1. Specifying the tolerances of all the basic functions required of the circuit.
2. Determining all conditions (primarily tolerances, stresses, noise, and handling) which tend to prohibit the circuit from performing each function as required.
3. Quantitatively verifying that the circuit will perform a function as required, even though all failure-causing tendencies occur simultaneously at their maximum values. (If some should be mutually exclusive, each set of failure-causing tendencies is considered separately.)

The effectiveness of this analysis depends on the accuracy, completeness, and reasonableness with which functions are specified and failure modes are determined. Determination of basic circuit functions must include all possible circuit states. A circuit state is one set of voltage and current conditions for which only one set of matrix equations, loop equations, node equations, simultaneous equations, or signal-flow graphs can be written. (Signal-flow theory is a separate technique which is used in reliability analysis.) Because this set of equations and/or graphs applies to one state only, all reverse-biased diodes must remain reverse-biased, all relays and mechanical switches must remain in their required ON or OFF states, all saturated transistors must remain in saturation, etc., for a typical circuit to remain in one particular state. If one or more circuit elements change so that a different set of equations or graphs applies, the circuit changes state at what is known as a transition point.

A circuit can exhibit various states which, in turn, display specific input-output functions that usually cannot be met if the circuit is in the wrong state. Thus, all states (functions) required of a circuit must be listed, and it must be shown that the circuit will remain in the required state in spite of those values of circuit parameters, input-signal conditions, power-supply voltages, etc. which must be selected within tolerance as most adverse to the required state.

Each worst-case calculation must be made with all part, circuit, and environmental parameters at either the high (in the case of a maximum required value) or low (when minimum values are needed) extreme of their tolerance limits, except when considering higher-order effects and power dissipation in an active element. In these cases all parameters and conditions are selected at those values within tolerance which will maximize a maximum-value calculation or minimize a minimum-value calculation. In all calculations, accurate equivalent circuits that yield useful results must be established and verified by control tests. As a result, when use of worst-case values is felt to be the direct cause of an undesirable increase in power dissipation, circuit complexity, or physical size, adjustment (narrowing) of worst-case values should be considered.

Analysis can often be simplified by examining each of the system's several subsystems separately and determining the particular input-output requirements and equivalent circuits of each. (For example, a superhet receiver can be divided into r-f section, converter, i-f section, second detector, audio, and power supply.) This analysis consists of detailed consideration of the following:

1. Examine part specifications for reasonableness and completeness.
2. Examine functional test specifications for accurate and realistic operating conditions and completeness in those measurements required of all pertinent parameters.
3. Determine tolerances of all applicable part and environmental input parameters after aging and use, and determine stress ratings for all parts.
4. Define all circuit states, and determine the circuit state of each part.

5. Verify that each part will be in the required circuit state, and determine all active-element d-c operating points for each state under worst-case conditions.

6. Determine the values and tolerances of all such functional parameters as gain; phase shift; phase margin; stability with feedback; loop gain during a transition state; frequency; load impedance; input impedance; output impedance; voltage; current; power; rise time, waveform; d-c offset; balance; noise generated within one or more parts; regulation; stability of all adjustments depending upon tolerance, temperature, environment, aging, etc.; detection level for a threshold detector; timing; special logic and protective circuitry.

7. Demonstrate that each functional parameter will be within required tolerance under worst-case conditions.

8. Define the effects caused by system noises external to the circuit and coupled into the various input lines.

9. Compute the maximum stress possible on each part for all normal operating conditions or for such abnormal operating conditions as open or shorted load, reactive load feedback, turn-on and turn-off transients, abnormal signal condition, plug-in, and power-supply overvoltage.

10. Examine the design for susceptibility to pickup noise and intercoupling.

11. Investigate poorly mounted parts to discover whether they dissipate heat, are isolated from heat-generating parts, and are capable of withstanding vibration, handling, and other environmental conditions (a consideration of the P_p part of reliability determination of a design).

If worst-case criteria are not completely fulfilled, the sum of the intrinsic failure rates of all parts will be less than, equal to, or greater than the actual system failure rate and thus unbounded and useless as a value for a determined failure rate. (Because the excessive drift rates which result when all worst-case criteria are not met tend to cause premature part failures, inclusion of the "less than" appears justified.)

In applying worst-case criteria, borderline situations exist in which the probability of increasing or decreasing overall component failure must be determined from the standpoint of adding parts to stress-limit other parts. (The probability of failure increases with the addition of parts.) Judgments in such cases are based on the effects of stresses on the statistical distribution of the part's parameters as determined by a series of statistically valid tests for that part.

The values and tolerances of all part and input parameters ultimately will be reflected in the determination of the values and tolerances of all system parameters as either a continuous or discontinuous function. In addition, each part, signal, power, and environmental input parameter of a given part can be described, at any instant, by a statistical distribution around which limits may be fixed to define the tolerance of the parameter. Thus, the probability of an intrinsic failure (the condition under which one or more parameter values lie outside the tolerance limits) decreases as the tolerance range of each part and input parameter over which the component must function properly increases.

After the tolerance limits of each part and input parameter have been defined by using the required probability of system failure (apportionment), the dependence of system failure on the value of a particular part parameter, and the statistical distribution of any part parameter over the operating life of the system, the component must operate within specification and with all possible combinations of part and input parameters within tolerance or the probability of system failure will be greater than specified.

Worst-case tolerance narrowing is a method which retains the simplification of a worst-case type of analysis but introduces maximum narrowing of part tolerance ranges without causing the specified probability of part failure from tolerance buildup to be exceeded. It can be utilized to define the limits on part and input parameters, and it tends to optimize part tolerances.

For example, the tolerance of a parameter y which is dependent upon the values of n parameters $X_1, X_2, \ldots, X_n$ is presented in the form of matrix equations. Associated with the X's, the frequency distributions of which are not necessarily Gaussian, are standard deviations $(\sigma_1, \sigma_2, \ldots, \sigma_n)$ which occur because of a random parts selection process and not because of variances within a given part. If all X's are random and statistically independent, the standard deviation in y, σ_y, is given by

$$\sigma_y{}^2 = \sum_{i=1}^{n} \left(\frac{\partial y}{\partial X_i} \sigma_i \right)^2$$

This reveals that if the $\partial y/\partial X_i$ are not constant but depend on the values of the X's, substitution of average values of the $\partial y/\partial X_i$ for the $\partial y/\partial X_i$ will result in an exact solution. However, determination of these average values results in a long series of multiple cross-partial derivatives, each evaluated with all X's at nominal, the computation of which is prohibitive in length and complexity even with a computer. Therefore, if the magnitude of σ_i does not approach or exceed the magnitude of S_1 (true for any practical electronic part), it can be assumed with negligible error that $\partial y/\partial X_i$ (evaluated with all X's at nominal) is equal to the average value of $\partial y/\partial X_i$.

This method relies on the derivation of a parameter p which is so defined that

$$K\sigma_y = |\partial y/\partial X_i| pK\sigma_1 + |\partial y/\partial X_2| pK\sigma_2 + \cdots + |\partial y/\partial X_n| pK\sigma_n$$

holds where K, dependent on probability of tolerance failure, is a constant to be determined and represents the allowed probability of failure from tolerance buildup on one end of one operational parameter. Thus, if $\bar{X}_i \pm pK\sigma_i$ is substituted for X_i in the matrix equations (the $\pm$ sign being chosen according to the sign of $\partial y/\partial X_i$), the computed value of y will be equal to its upper or lower tolerance extreme, $\bar{y} \pm K\sigma_y$. If these tolerance extremes are within the limits required of the system, the circuit is acceptable from a tolerance standpoint. In addition, the probability that y lies outside the required limits then also satisfies the requirements.

Determined according to the preceding paragraph, p is given by

$$p = \sqrt{\sum_{i=1}^{n} \left(\frac{\partial y}{\partial X_i} \sigma_i \right)^2} \Bigg/ \sum_{i=1}^{n} |\partial y/\partial X_i| \sigma_i$$

and is always in the range $0 < p \leq 1$. The partials can be evaluated from the matrix equations with all X_i nominal; all σ_i can be determined either directly or indirectly from parts data; and a numerical value for p can be determined.

If the number of terms increases and the relative contribution of the largest term decreases, the frequency distribution of y will approach the Gaussian, regardless of the frequency distributions of the S_1. In this case the probability P_f that y will lie outside one of its tolerance extremes $y \pm K\sigma_y$ is obtained by integrating the Gaussian curve as follows:

$$P_f = \int_K^\infty \frac{1}{2\pi} e^{-X^2/2}\,dx$$

Thus, if the probability of failure of one parameter at one tolerance extreme P_f is given, a value for K can be determined.

In summary, the statistical distribution of each part parameter, conventionally determined by subjecting a quantity of parts to known stresses for known periods of time and measuring the part-parameter drift, is generally dependent on the part history of the stresses to which the part has been subjected, especially if the fatigue or breaking stress is approached. Although the statistical distribution increases in proportion to the stresses to which a part is subjected, submission of a part to stresses greater than those specified for a particular level will result in an increase in the component's probability of failure. That is, worst-case criteria must be met.

Although a single model may be designed to provide the necessary equipment performance by selecting the optimum value for each part, reliable mass-produced equipment must be designed to perform properly with parts which have values that vary throughout their tolerance ranges under conditions of use. In many cases the attempt to design the equipment to perform with all parts simultaneously at their tolerance limits and in such a direction as to produce the greatest deviation in nominal performance will fail, because even the best and highest-precision parts will not have sufficiently small tolerances for the circuit involved. In other cases tolerance considerations may be resolved by complicating the circuitry to the extent that its overall reliability is considerably lowered.

5.8 RELIABILITY ANALYSIS OF FUNCTIONS

In addition to considering part-failure rates in estimating the reliability of a design, fluctuations (parameter variations) of part values and input-signal levels are considered to ascertain the true reliability levels of functional circuits, subsystems, and systems. Circuits that may be "reliable" from a part-failure standpoint can produce undesirable outputs because their input signals vary widely or the values of parts drift or because the interactions of parts with one another vary when functionally designed. Such undesirable outputs are just as detrimental to overall system failure as circuit failure or part failure. Out-of-limits circuit outputs are considered circuit failures when referring to reliability.

One method of estimating out-of-limits outputs is to build a breadboard circuit and measure outputs as the values of its parts are individually varied above and below their nominal design values. When the circuit output exceeds its specified limits, the upper and lower part values are noted and specified as limits. With all parts set at nominal values, input-signal-level limits can be determined in the same way and used as output limits for the preceding stage. This method, in addition to being time consuming and expensive, does not include:

1. The effect(s) of simultaneously varying two or more part values.
2. The effect(s) of simultaneously varying input-signal levels and part values. [Both input-signal-level and part-value fluctuations are called input-parameter variations relative to the output fluctuation (output-parameter variation).]

5.9 EMPIRICAL METHODS OF ESTIMATING RELIABILITY OF CIRCUITS

The empirical method of circuit analysis, which uses statistical techniques to evolve a regression equation describing output variations as functions of input and part (circuit-parameter) variations, makes possible the determination of circuit performance if the circuit-part characteristics are known. Transistor rise and fall time, distributed capacitance, noise, and other measurable transient effects can also be included for consideration. The limiting condition to this method of analysis is that the circuit parameters must be linearly related to the output. However, this condition is not too restrictive.

The empirical-method equation is of the general form

$$Y = a_0 + a_1x_1 + a_2x_2 + \cdots + a_kx_k + e$$

where Y = output characteristic being investigated
k = number of independent variables
e = random variable representing such things as experimental errors
x = some function of the independent variables
a = coefficient relating its part parameter to Y

The a terms in the above equation must be empirically determined. The necessary data for computation of the coefficients may be obtained by approximating the analysis of the circuit, and the characteristic being investigated may be measured a number of times. Part measurements must be made in an essentially different circuit and by changing the circuit elements before each measurement. Parts used in the breadboard must be representative of parts that will be used in production (i.e., parts whose quality is controlled and whose parameter variations are known).

When the equation coefficients are determined, solutions of the equation with circuit input parameters set at various points within their ranges will show the output-parameter variations to be expected. This method of circuit analysis is quite tedious, time consuming, and prone to human errors. To eliminate these undesirable elements, five computer methods of circuit analysis have been developed.

The parameter-variation methods used in circuit analysis and described herein have been developed to eliminate the arduous calculations required in the empirical method by utilizing the capabilities of digital computers. A helpful method of circuit analysis is that which attempts to simulate the circuit mathematically on a digital computer and to show how the circuit will behave as its parts deteriorate during life. The analysis is accomplished by programming the circuit equations into the computer and methodically varying the values of the circuit part parameters. Extensive data are not required from large-sample life tests on the electronic parts used in the circuit; information obtained from small-sample accelerated tests is quite useful. This analytical method has been programmed in a general manner so that it may be applied to many circuits with only minor modifications to the program. This has been done by expressing the circuit equations in matrix form and using a general matrix-solution subroutine to solve them.

5.10 EQUIVALENT-CIRCUIT SOLUTION

Very briefly, the equivalent-circuit method of analysis entails the following steps (Fig. 5.15):

1. Draw an accurate equivalent circuit.
2. Write the circuit equations and circuit requirements in terms of the part parameters.
3. Incorporate the equations, the circuit requirements, and the desired part-parameter variation ranges into the general computer program.
4. Debug and run the computer program.
5. Plot and analyze the computer output.

The circuit designer is usually capable of performing the entire analysis with the aid of a computer programmer and part specialist as required. These procedures may vary to suit the organization of the design team. At present the analysis is suited primarily to a-c and d-c steady-state conditions (Fig. 5.16). It is not capable of evaluating behavior during a switching period, transients, noise, and the like

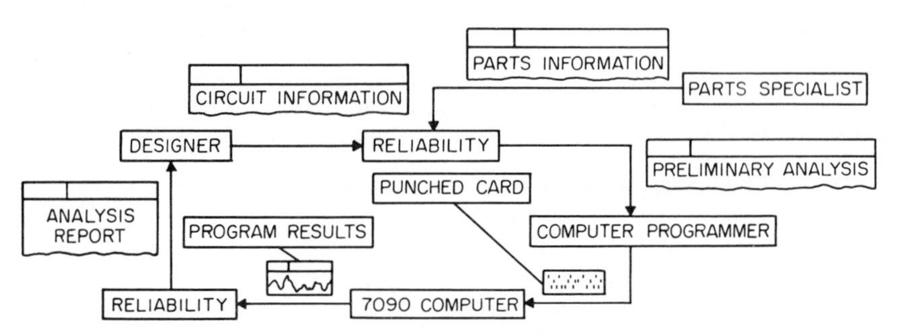

FIG. 5.15 Circuit-analysis cycle.

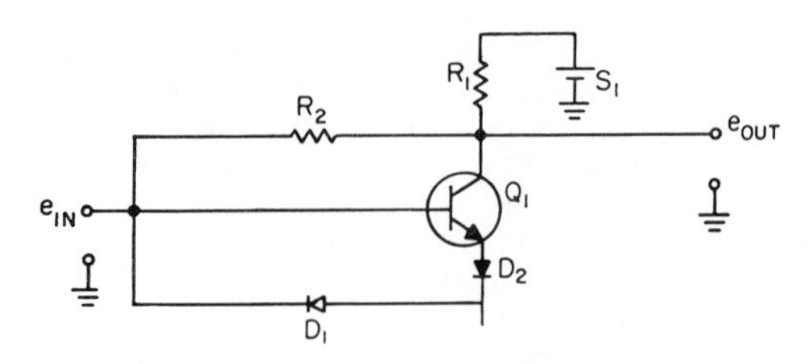

FIG. 5.16 Logic-circuit schematic.

because these data cannot be readily represented by linear circuit equations, although it is planned to incorporate these factors into an empirical analysis in the future.

The circuit is a hypothetical single-transistor switch, a typical element of pulse and logic networks. When the input is positive, the output of the circuit should be 1 volt; when the input is negative, the output should be S_1 volts. Depending on the state of the circuit, the transistor Q_1 operates in the saturation or cutoff region. The equivalent circuits shown in Figs. 5.17 and 5.18 represent the two possible states of the switch, the closed position (or positive input) and the open position (or negative input).

The equivalent circuits depicted are linear representations of the physical circuit. The equivalencies used for Q_1 reflect the operating region of the transistor for the particular state of the circuit (i.e., a positive input is reflected in saturated transistor, low collector-emitter resistance, etc.).

Notice that diode D_1 has two possible conditions, one for conducting and the other for nonconducting. The condition that exists at any particular time is a function of the base voltage V_1 of the transistor. Special computer logic must be included. Consider first the case when a diode is operating above the knee of its characteristic curve. The dynamic impedance at that point is relatively small as compared to the

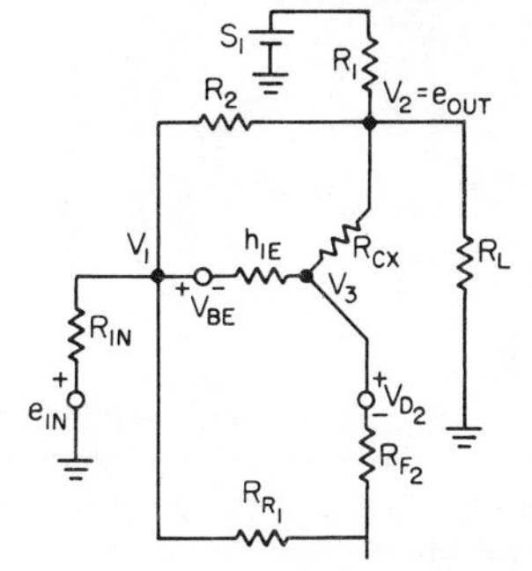

FIG. 5.17 Conducting state.

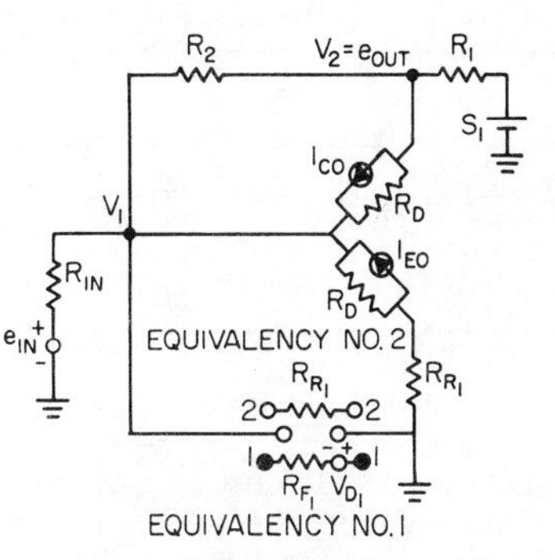

FIG. 5.18 Nonconducting state.

impedance below the knee. It is obvious then that the same linear equivalent circuit will not suffice for both cases. Therefore, a test of the voltage across the diode must be made to determine if the proper equivalency has been used. If the test indicates the voltage is correct, the program may continue; if the voltage is incorrect, a new equivalent circuit must be used, changing the matrix and necessitating a solution of the new matrix. This method may be applied to any part in the circuit by using any test procedure desired.

The step-by-step procedure for special diode logic is as follows:

1. Use equivalency 1 (Fig. 5.18) and solve the matrix equations.
2. Check the voltage across the diode to determine if the proper equivalency (No. 1) has been used.
3. If the proper equivalency has been used, continue the program. If equivalency 1 is incorrect, use equivalency 2 (Fig. 5.18) and resolve the matrix.
4. Check the voltage across the diode to determine if equivalency 2 is proper.
5. If equivalency 2 is correct, continue the program. If equivalency 2 is not correct, stop the program and reevaluate the equivalent circuits chosen.

The equivalent circuit should include the inputs and outputs of the circuit and should consider the load impedance, input signal, and input-generator impedance. When the equivalent circuit has been drawn, the circuit equations should be written and reduced to matrix form. The equations, in matrix form, for the equivalent circuits appearing in Figs. 5.17 and 5.18 are presented in Tables 5.2 and 5.3.

Although the nodal method of writing circuit equations was used in this example (the loop-current method could be used equally well), the method that describes the circuit in the fewest equations will achieve considerable savings in computer time later in the analysis. These circuit equations must be verified by experimental data as derived from the breadboard model of the circuit (assuming the proper measurements are made) before proceeding with the analysis. If comparison between the measurements and the solutions to the circuit equations indicates discrepancies rather than compatible results, it may be necessary to alter the equivalent circuit to represent the physical circuit more accurately.

Table 5.2 Conducting State

V_1	V_2	V_3	Vector
$1/R_{IN} + 1/R_2 + 1/h_{IE} + 1/R_{RI}$	$-1/R_2$	$-1/h_{IE}$	$e_{IN}/R_{IN} + V_{BE}/h_{IE}$
$-1/R_2$	$1/R_1 + 1/R_2 + 1/R_{CX} + 1/R_L$	$-1/R_{CX}$	S_1/R_1
$-1/h_{IE}$	$-1/R_{CX}$	$1/h_{IE} + 1/R_{CX} + 1/R_{F_2}$	V_{D_2}/R_{F_2}

Table 5.3 Nonconducting State

V_1	V_2	Vector
$1/R_{IN} + 1/R_2 + 1/R_D + 1/R_D + R_{R_2} + 1/R_{F_1}$ or $1/R_{R_1}$	$-1/R_2 - 1/R_D$	$I_{EO} + I_{CO} - e_{IN}/R_{IN} - V_{D_1}/R_{F_1}$ or 0
$-1/R_D\ 1/R_2$	$1/R_2 + 1/R_1 + 1/R_D$	$S_1/R_1 - I_{CO}$

To determine if the circuit is functioning properly, circuit-failure criteria [which describe conditions that the circuit must meet at a minimum or maximum output voltage (or both)] must be written in terms of the equivalent-circuit part parameters and the solutions to the circuit equations. These failure criteria must be defined with discretion because of their vital role in the parameter-variation logic selected for the computer program (i.e., criteria more restrictive than required establish tolerances on parts that are tighter than necessary, and vice versa).

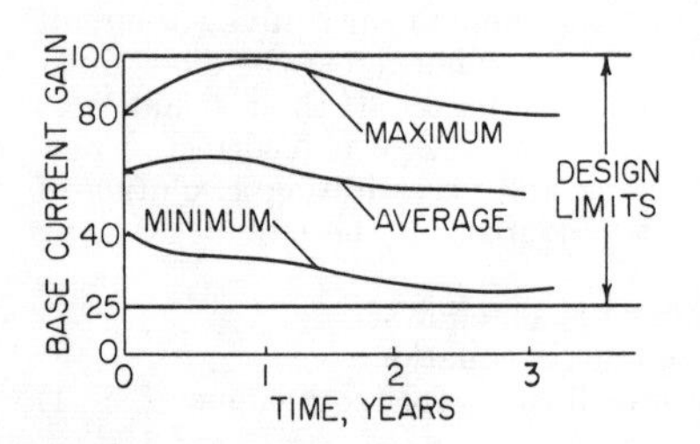

FIG. 5.19 Drift limitations. *Note.* Design limits must account for parameter variation.

Circuit-failure criteria can be developed for any part in the circuit which can successfully operate in only one of its two or more stable states. (That is, assuming that a circuit will function only when a particular transistor is operating in its saturation region, a failure criterion of this transistor's base current or base-emitter voltage can be developed.) However, when a circuit performs successfully in either state, special logic which changes the matrix to represent the part in its alternate state must be incorporated into the computer program and a recheck of the failure criteria must be conducted.

After development of circuit-failure criteria, parameter-variation routines require establishment of a variation range and the number of points to be checked within this range for each part parameter (Fig. 5.19). Although the limits selected should be wider than the allowable degradation limits so that enough information will be available for redesign, they should not be so wide as to waste computer time. Moreover, the number of points to be checked should adequately present time-failure-line areas. For example, if a carbon-composition resistor in a circuit (R_1 or R_2 in Fig. 5.16) has allowable drift limits of +5, −8 per cent and the variation limits chosen for this analysis are ±20 per cent, the selection of 40 points would give a 1 per cent step size and, therefore, sufficient resolution (1 per cent). However, in the case of a non-conducting diode, because the impedance (R_{R_1} in Fig. 5.17) should be high and thus nothing would be gained by further increasing this resistance during the analysis, the variation limits on this part parameter possibly could be +0, −99 per cent.

The heart of the analysis concerns the stress functions to which the circuit parts will be subjected during operation. These stress-function equations, fed to the computer in terms of the part parameters and the matrix solutions, are computed, and the maximum values of each stress (stored by the computer and printed out at the completion of the computer program as well as at every failure point during the analysis) are analyzed at the completion of the program to determine if parts will be overstressed during operation as the various parameters drift. For example, any overstress in the transistor can be determined by examination of its collector dissipation (Fig. 5.17)

$$P_C = (V_3 - V_2)^2/R_{CX}$$

and the emitter current would be

$$I_E = (V_3 - V_{D_2})/R_{F_2}$$

When the computer program is complete, analysis of the graphs (schmoo plots) constructed for each parameter from plots of the designated failure points reveals the following:

1. Which parameters are interdependent and the extent of their interdependence
2. The independently considered allowable drift limits for each parameter

A typical schmoo plot or composite failure-line graph, as shown in Fig. 5.20, is generated by the computer with each line representing circuit failure due to a combination of parameter drift and any other specified parameter. Referring to Fig. 5.20:

1. Parameters 2 to 4 are interdependent with parameter 1.
2. Parameter 5 is independent of parameter 1 and will cause circuit failure at a 30 per cent decrease of its nominal value.
3. Parameter 1 will cause circuit failure (assuming all other parameters at their nominal values) when it decreases by 20 per cent.
4. Parameter 5 will cause circuit failure (all other parameters at their nominal values) when it increases by 90 per cent.

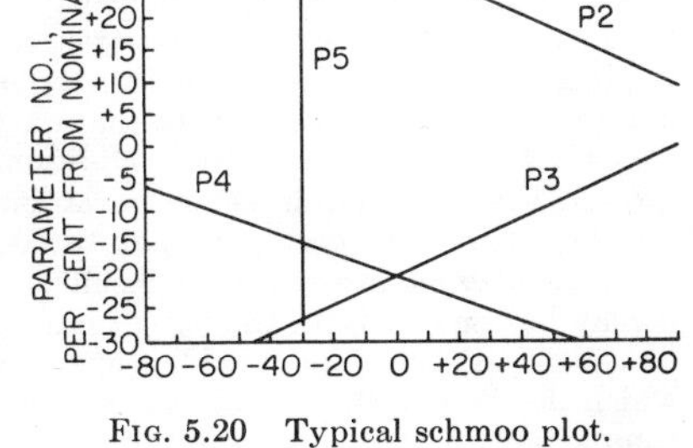

FIG. 5.20 Typical schmoo plot.

The allowable parameter drift must be an optimum point which depends upon the particular part parameters in question and how they will behave during life, considering the interdependency of the parameters. For example, if parameter 1 is allowed a −19 per cent variation from nominal, parameter 3 could vary +70 per cent while parameter 1 would be allowed only −5 per cent variation. Instead, choosing parameter 1 at −17 per cent and parameter 3 at +15 per cent results in a −18 per cent limit on parameter 4. An optimum point for parameters 1 and 2 might be +24 and +25 per cent, respectively.

The tables which are prepared for each schmoo plot and which list the parameter limits (an example is Table 5.4) are compared and the most restrictive limit for each parameter is determined. If adjustments in some limits are necessary because of extremely restrictive limits on interdependent parameters, reexamination of all of the plots is then required to ensure that other limits are not affected.

Examination of the stress values along the failure lines in terms of the selected drift limits supplements the schmoo plots and provides an effective evaluation of the circuit design; if the limits are tight, it may be necessary to redesign or to substitute better parts, whereas a decrease in the nominal value of the part may be advisable with a narrow upper tolerance limit and a wide lower limit. Derivation of the schmoo plots is described in the following paragraph.

Parameter 1 (the dependent variable) is set and maintained at its nominal value, and parameter 2 (the independent variable) is discretely incremented, the size of these increments depending upon the range of variation and the number of steps previously established for this parameter. The matrix is resolved, and the failure criteria are checked after each increment until the entire range of the independent parameter is completed. If a circuit failure is encountered, the success point prior to the failure point is printed out to be on the safe side of the actual failure line, the dependent variable is stepped one increment, and the routine proceeds as before until the entire range of this parameter is covered. At this point the independent variable is replaced with the next parameter and the entire routine is repeated. When every parameter has been varied against the dependent variable, the next highest parameter is established as the dependent variable and the entire variation routine continues until every parameter has been varied against every other one.

Table 5.4 Parameter Limits from Schmoo Plot

Parameter	Limits, %	
	Upper	Lower
$P1$	+24	−17
$P2$	+25	
$P3$	+15	
$P4$	....	−18
$P5$	....	−30

Although this means of circuit analysis uses a combination of well-known techniques, it is not the ultimate, in that only two parameters at a time are being varied. (This is a compromise between additional information and cost, since investigation showed that the simultaneous variation of more than two parameters resulted in excessive time and cost and that three-at-a-time and higher interrelations may be determined from two-at-a-time results.) The method does have advantages, because it may be used early in the design phase before statistical parts data (i.e., frequency distributions) become available, thus giving the designer the set of drift limits within which the circuit parts should remain as well as an objective analysis of the circuit by comparison of these limits to the part-parameter drift-limit estimates. However, methods are being developed to replace the parameter-variation logic of this technique with a statistically based means of combining the data and the circuit equations to provide predictions of the distributions of the circuit functional parameters.

5.11 CIRCUIT ANALYSIS PROGRAM DESCRIPTION

The circuit analysis program for the IBM-709 can be divided as follows:

1. One-at-a-time parameter-variation program (OAAT).
2. Two-at-a-time extreme-point parameter-variation program (TAAT).
3. Two-at-a-time variation of interdependent parameters program (TAATI).

The purpose of programs OAAT and TAAT is to eliminate all unnecessary parameter variations (thereby saving computer time) except those forming failure lines (straight lines other than vertical or horizontal), while program TAATI will generate failure lines for all the parameter combinations not eliminated by programs OAAT and TAAT. Each program can be divided into two sections:

1. The parameter-variation logic.
2. The particular circuit-equation subroutine.

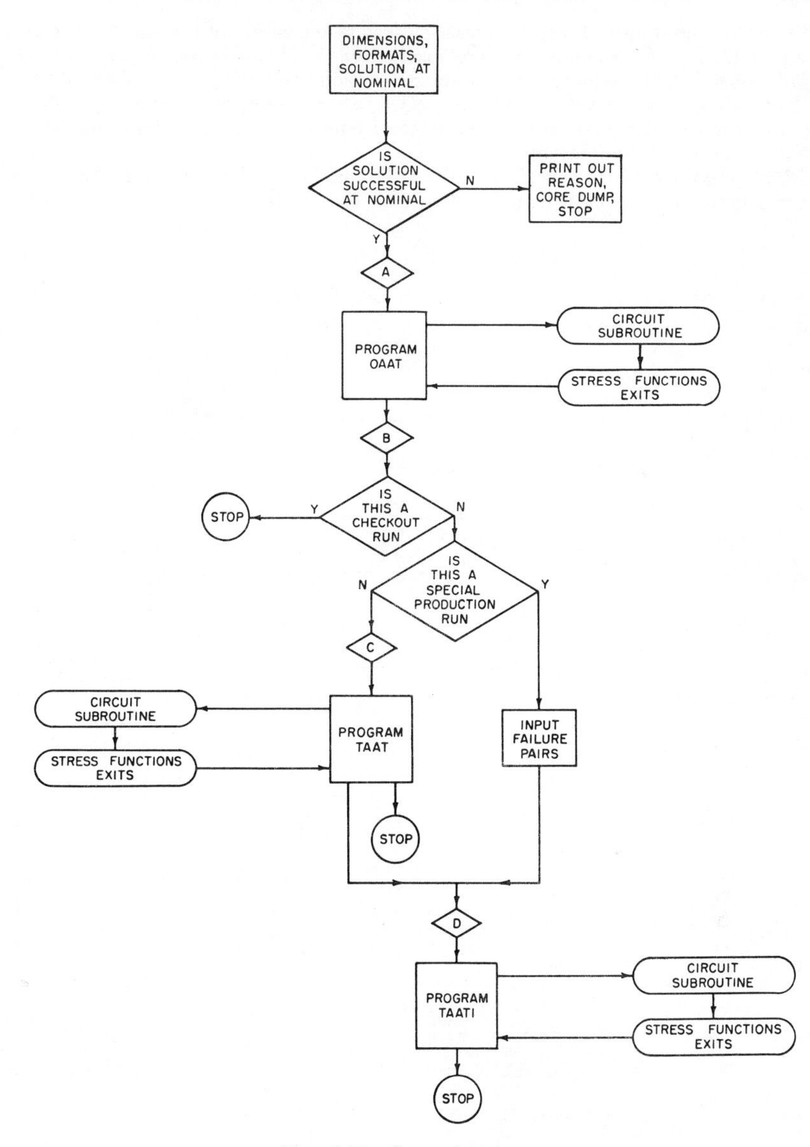

FIG. 5.21 General program.

Parameter-variation logic is available in a standard deck which is never rewritten and which systematically assigns and modifies the values of all the parameters either one or two at a time so that no more than two parameters have a value different from nominal at any one time. For each of these sets of parameter values the circuit-equation subroutine (used by all three programs) solves the circuit equations and evaluates the results against all the specified circuit-failure criteria. In addition, such functions as voltage stress and power dissipation are calculated to determine the combination of parameter values resulting in their maximum values. A general program-flow chart (Fig. 5.21) illustrates the relative position of the three parameter variation blocks and their connecting instructions. (The blocks labeled STRESS FUNCTION, EXITS represent one set of instructions used by all three programs.)

5.11a Program OAAT. Program OAAT, a one-at-a-time parameter-variation logic in which all parameters are held at their nominal value except the one being varied, performs in the following manner:

1. The parameter is set at its low limit: If success is indicated on the printout, the program proceeds to item 3; item 2 is utilized if the circuit fails.

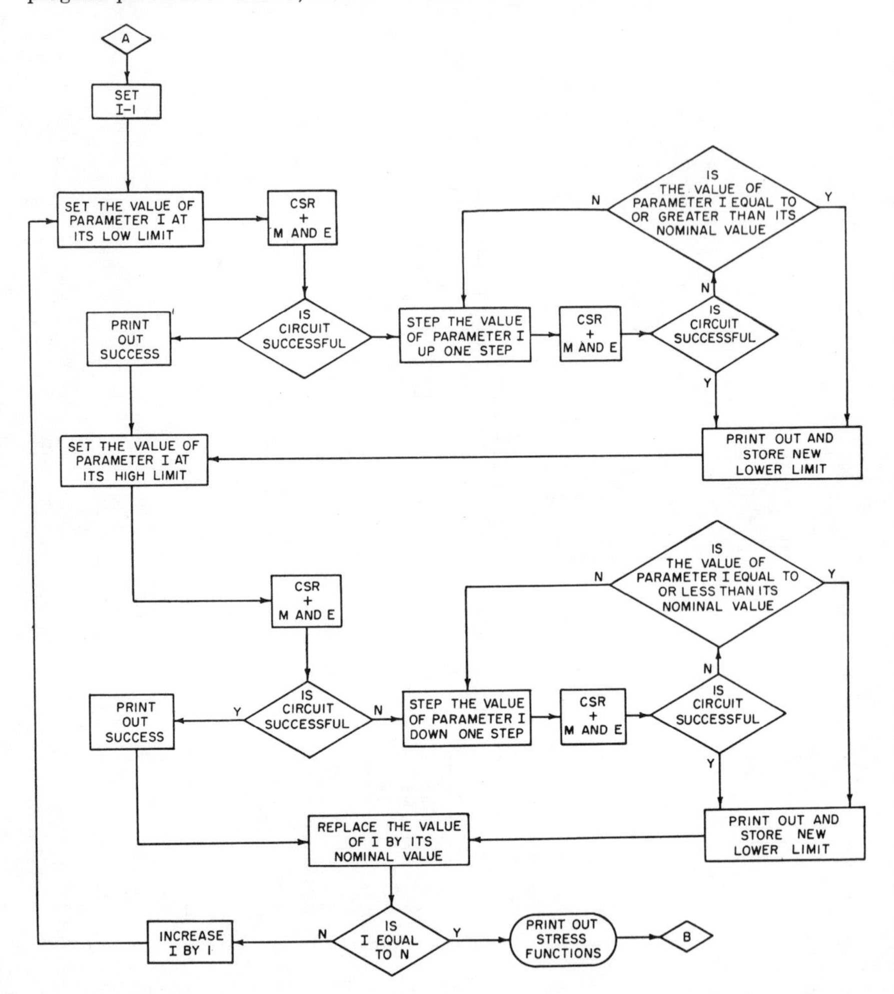

FIG. 5.22 Program OAAT. *Note.* Circuit subroutine (CSR), maximums (M), and exits (E).

2. The parameter is stepped from its low limit to its nominal value until the first success point is achieved, at which time the computer prints out the parameter value.

3. The parameter is set at its high limit. If success is indicated on the printout, the program proceeds to item 5; if the circuit fails, the program advances to item 4.

4. The parameter value of the first success point is printed out as the parameter is stepped from its high limit toward nominal.

5. This procedure will be repeated with the next highest parameter if all parameters have not been varied; the program will stop when all parameters have been varied if this is a checkout run; or program TAAT or program TAATI, depending upon the input data, will be initiated if this is a production run.

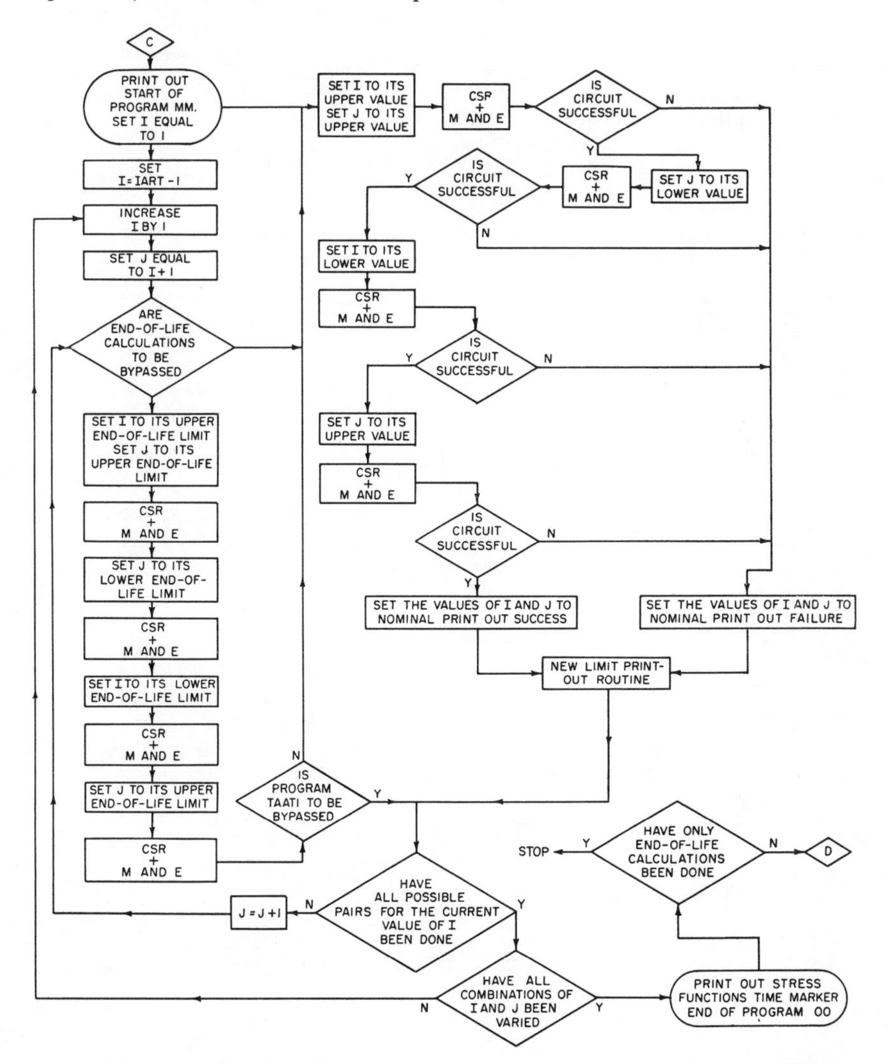

Fig. 5.23 Program TAAT. *Note.* Circuit subroutine (CSR), maximums (M), and exits (E).

In addition to printing out these new tolerance limits, program OAAT will store them for use in the following programs and thus produce a new set of tolerance limits on each parameter, some of which will be the same as the parameter-variation limits specified in the program input. A complete flow diagram for program OAAT is given in Fig. 5.22.

5.11b Program TAAT. (**Fig. 5.23**) Program TAAT, a two-at-a-time parameter-variation logic which will eliminate all straight vertical and horizontal failure lines from consideration in program TAATI, will also calculate the end-of-life stress functions for each pair of parameters, exiting directly to program TAATI (Fig. 5.21). Using the new limits from program OAAT if necessary, program TAAT tests each parameter pair at its four extreme points, failure at one or more of which identifies

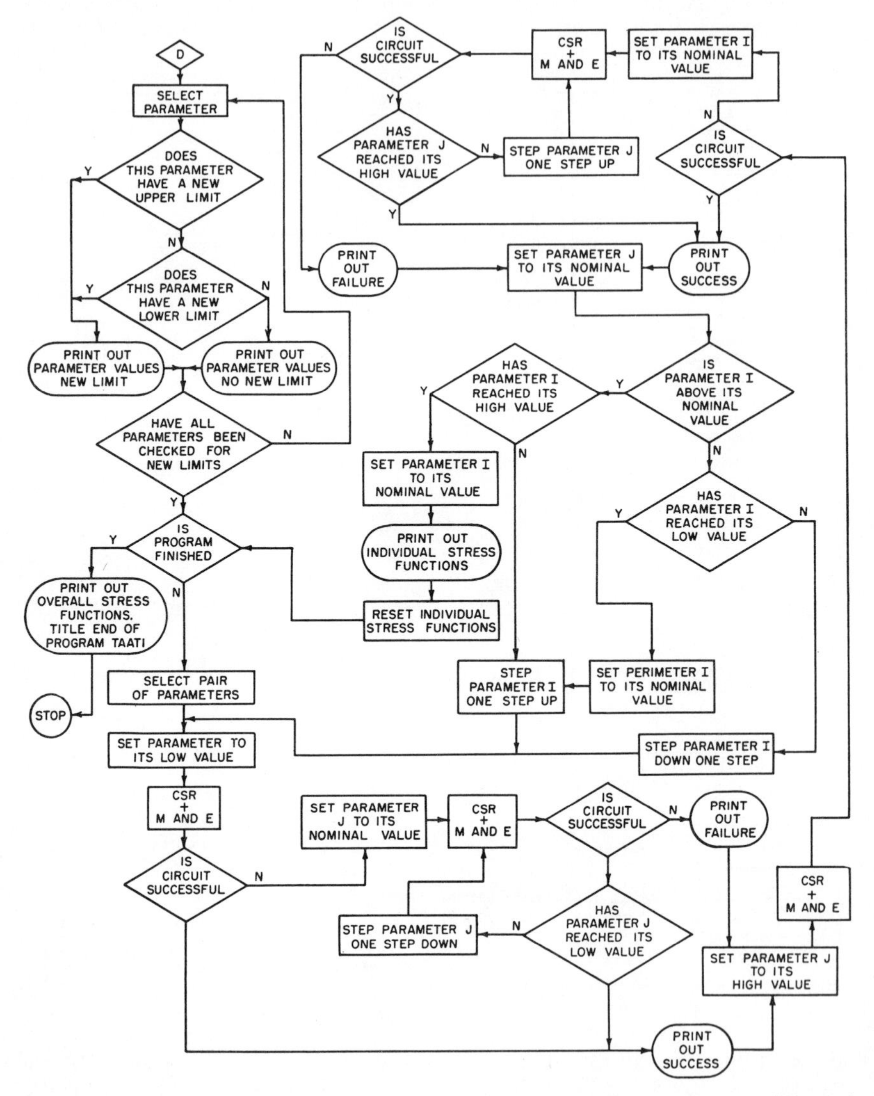

FIG. 5.24 Program TAATI. *Note.* Circuit subroutine (CSR), maximums (M), and exits (E).

the parameter as interdependent (success or failure as well as the type of failure will be indicated on the computer printout). Although the end-of-life stress functions also are calculated by using this method of parameter variation, the points to be used are determined by the upper end-of-life part-parameter drift data. Either the end-of-life calculations or the four-corner test routine may be bypassed.

5.11c Program TAATI. Program TAATI (Fig. 5.24) is a two-at-a-time variation of all interdependent parameters determined by program TAAT in which a circuit is solved for each pair of parameters (I and J), with J at both its low and high values and I at its nominal value (these values are established by the input data and/or program OAAT. Using this procedure, the logic varies J as before and steps the value of I down one step until I reaches its low value; at this point, I is placed one step above nominal and the process of varying J is begun. If a failure is encountered as J is varied, the logic returns J to nominal and steps it out toward the limit that failed until the failure point is reached. The computer then prints out the percentage of nominal value for both parameters and the type of failure. However, if no failures are encountered, the coordinates of the success points tested are indicated on the printout. When all parameter pairs have been varied, the computer will print out the maximum stress functions encountered during the parameter variation.

The circuit-equations subroutine, utilizing circuit equations written in matrix form, is a closed subroutine (fixed entry and exits) that solves the linear system of equations representing the equivalent circuit, tests the solutions against given circuit criteria, and evaluates the stress functions. Although it was found that a standard reduction routine gave the best results as far as time was concerned, each circuit subroutine has certain peculiarities which must be included. The following list describes in detail the flow diagram for the circuit subroutine (Fig. 5.25):

1. If the system of equations must be solved more than once for a given region A, that is, the matrix has more than one condition, the matrix coefficients and column vector elements must be evaluated, necessitating placement of the newly created matrix in a temporary storage region and setting ICD = 1. Move the present solutions and stress functions into the previous solutions and power stress functions and into storage location.

2. If this is the first time through the subroutine, transfer to block 3 (block 1 will not be zero). Proceed to block 4 if it is not the first time through.

3. The maximum and minimum selections are initiated, and the nominal matrix and vector columns are printed out.

4. The system of equations will be solved by reduction in this block.

5. Another test is conducted to determine if this is the first time through the subroutine.

6. If this is the first time through the subroutine, a check is performed to determine if the reduction routine was singular, overflowed, or was solved correctly.

7. The computer will print out MATRIX IS NOT SOLVABLE.

8, 9. The solutions are back-substituted into the equations to ensure satisfaction to a tolerance of ± 0.01.

10. If any of the equations are not satisfied, the computer will print out SOLUTIONS FAILED SUBSTITUTION CHECK as well as the proper solution.

11. Block 11 (which tests for condition 1) to block 15 are set for two conditions. The flow chart must be expanded if more than two conditions exist. However, these blocks are omitted if the system of equations has only one condition.

12. If tests determine that condition 1 is satisfied, transfer out of the condition testing area (blocks 11 to 15).

13. If condition 1 is not satisfied, set up the matrix for condition 2.

14. A printout indicates that condition 1 was not satisfied.

15. Tests determine if condition 2 is satisfied.

16. While a printout indicates if conditions 1 and 2 are both unsatisfied, proceed with the circuit tests.

17. As soon as one of these circuit tests (all of which are given a unique identification number) fails, set FAIL equal to the particular identification and transfer to the exit routine.

18. If all circuit tests are satisfied, calculate the new stress functions.

19. If end-of-life calculations must be made, circuit-failure criteria will be bypassed.

20. Exit location will be directed, depending upon whether or not the program is in end-of-life calculations.

Figure 5.26 presents three versions of a two-transistor, saturated amplifier circuit which is typically used to convert a sinusoidal input to a square-wave output, as well as assumed parts variations (typical for current military types when subjected to environmental stresses and prolonged periods of storage) and pertinent results of

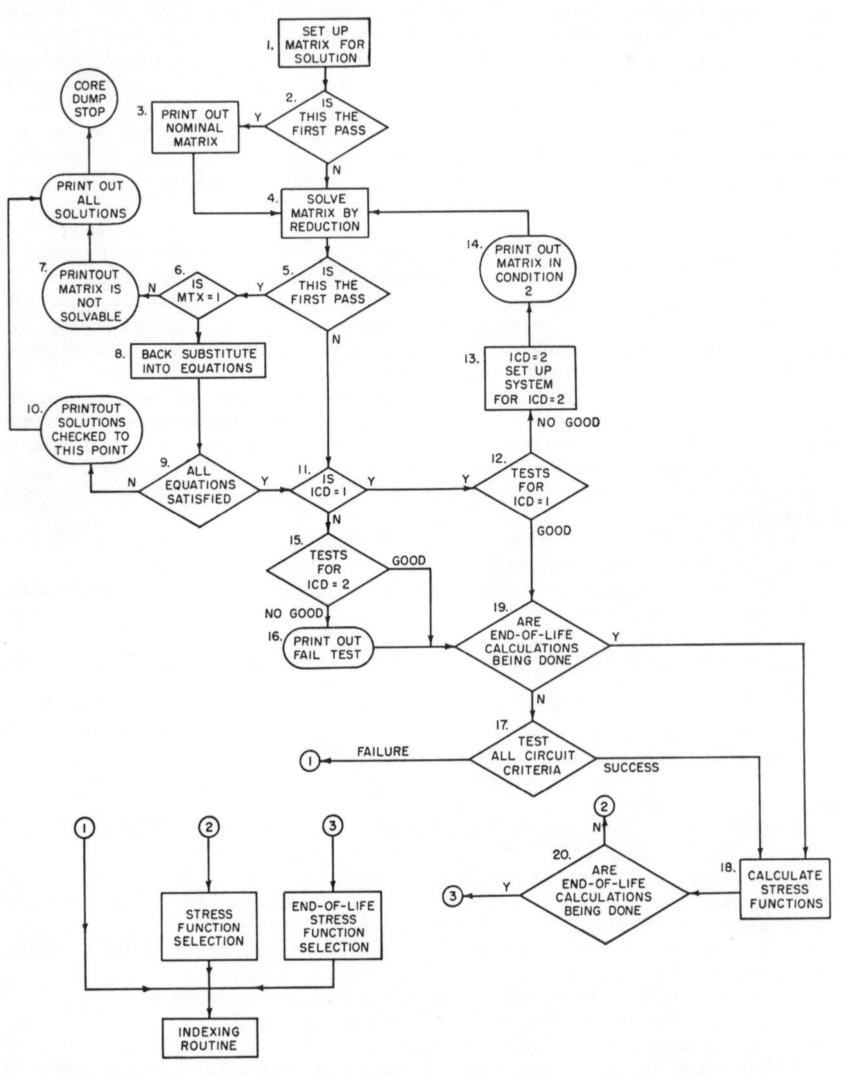

Fig. 5.25 Typical subroutine.

design calculations. Section *A* illustrates a design based entirely on nominal conditions and does not pass worst-case criteria; section *B* presents the same circuit configuration with part values appropriately adjusted to pass worst-case criteria (note that no parts were added); and the version illustrated in section *C* actually uses fewer parts and still passes worst-case criteria. Although use of the circuit of section *A* led to a number of failures because of marginal operation, the circuits of sections *B* and *C* eliminated the problem.

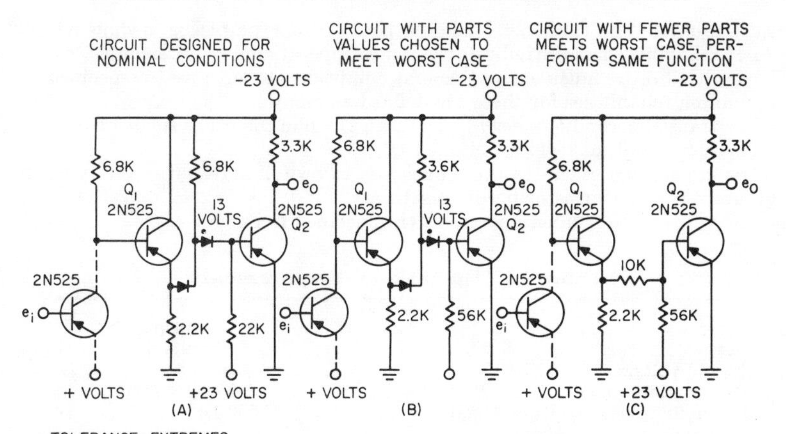

TOLERANCE EXTREMES OF PARTS PARAMETERS

PARAMETER	MINIMUM	MAXIMUM
+23 AND -23 VOLTS +1% POWER SUPPLIES	22.77	23.23
RESISTORS	+20% OF VALUE	-20% OF VALUE
13 VOLT ZENER DIODE	12 VOLTS	14 VOLTS
DIODE FORWARD VOLTAGE DROP	0	1 VOLT
TRANSISTORS 2N525		
h_{FE}	17	
I_{CO}	0	300μa
V_{BE}	0	1 VOLT

COMPUTED CIRCUIT PARAMETERS

PARAMETER	CIRCUIT (A) (NOMINAL CASE)	CIRCUIT (A) (WORST CASE)	CIRCUIT (B) (WORST CASE)	CIRCUIT (C) (WORST CASE)
I_C OF Q_2	6.96 MA	7.65 MA	7.65 MA	7.65 MA
I_B OF Q_2	0.33 MA	0.072 MA (*REVERSE)	0.66 MA	0.68 MA
REQUIRED MINIMUM h_{FE} OF Q_2	21	--	11.6	11.2
MAXIMUM ALLOWABLE I_{CO}	1.1 MA	--	0.313 MA	0.313 MA

*CIRCUIT FAILS TO OPERATE

FIG. 5.26 Two-transistor saturated amplifier circuit.

5.12 EXAMPLES OF DESIGN RELIABILITY DERIVATION

The object of the following discussion is to determine the design reliability of a standard transistorized, saturating, bistable multivibrator (Fig. 5.27). The specified operational requirements are listed in Table 5.5. Components and their character-

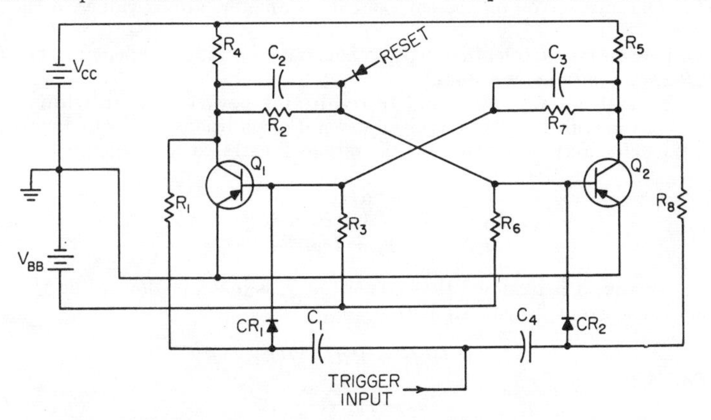

FIG. 5.27 Saturating bistable multivibrator.

istics are presented in Table 5.6. The purpose of the examples is to show with what degree of reliability the multivibrator will meet operational requirements.

The multivibrator must satisfy several conditions to operate as specified. The corresponding reliabilities for these are defined as follows:

1. One transistor conducts heavily (saturates) while the other is cut off (P_1).
2. The proper output level is obtained (P_2).
3. The trigger signal causes the transistors to switch states (the ON transistor is cut off and the OFF transistor becomes saturated (P_3).
4. The multivibrator must operate at the specified frequency (P_4).

Table 5.5 Operational Requirements

Input	+10-volt pulse ± 10%
Output	−10 volts d-c ± 10%
Supply voltages	±10-volt d-c ± 20%
Temperature range	−55 to +71°C
Maximum pulse-repetition frequency	500 kc

Table 5.6 Components

Q_1,Q_2	2N396 transistor, germanium
C_1,C_4	1,000-μμf capacitors
C_2,C_3	330-μμf capacitors
CR_1,CR_2	1N626 diode
R_1,R_8	10 kilohm ± 5% ½-watt composition resistors
R_2,R_7	12-kilohm ± 5% ½-watt composition resistors
R_3,R_6	47-kilohm ± 5% ½-watt composition resistors
R_4,R_5	22-kilohm ± 5% ½-watt composition resistors

The circuit operates as specified only when the conditions previously discussed are satisfied. Assuming that the conditions are independent of each other, the design reliability of the circuit P_D will be the product of the individual conditions reliabilities.

$$P_D = P_1P_2P_3P_4 \tag{5.1}$$

The following assumptions are to be made:

1. The component changes caused by temperature changes are not of a permanent nature.
2. The changes of component parameters caused by aging will be neglected.
3. The tolerances given on the component parameters are 3σ values of the normal distribution.
4. The percentage of tolerance will be constant, i.e., independent of the effect of temperatures on parameter values.

5.12a Equations for Meeting Individual Operation Conditions. Figure 5.28 presents the multivibrator circuit under static conditions. The following parameter relationships may be written for the circuit illustrated in the figure.

$$I_{E_2} = \alpha I_{E_2} + I_{B_2} \tag{5.2}$$

$$(I_1 + I_{CO_1})R_3 + I_1R_7 + (I_1 + \alpha I_{E_2})R_5 = V_{BB} + V_{CC} \tag{5.3}$$

$$I_{B_2}R_2 + (I_{B_2} + I_{CO_1})R_4 = V_{CC} \tag{5.4}$$

As can be seen, it is assumed that transistor Q_2 is in saturation, while Q_1 is cut off from the base current [Eq. (5.4)] of Q_2.

$$I_{B_2} = (V_{CC} - R_4I_{CO_1})/(R_2 + R_4) \tag{5.5}$$

From Eq. (5.2)

$$I_{E_2} = I_{B_2}/(1 - \alpha)$$

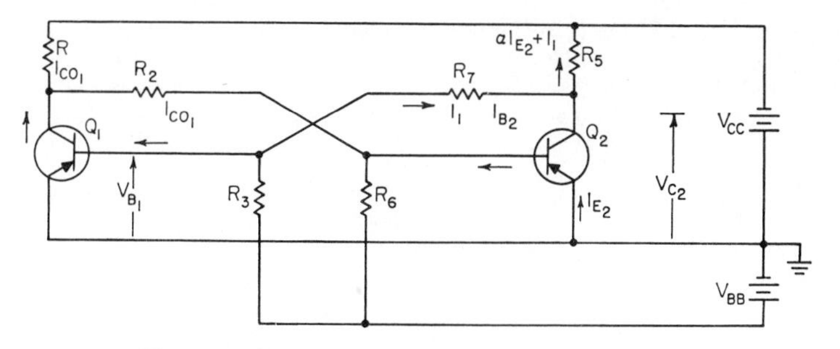

FIG. 5.28 Multivibrator circuit, static conditions.

and by substituting Eq. (5.5) for I_{B_2}

$$I_{E_2} = (V_{CC} - R_4 I_{CO_1})/(1 - \alpha)(R_2 + R_4) \tag{5.6}$$

On substituting Eq. (5.6) into Eq. (5.3) and solving for I_1,

$$(R_3 + R_5 + R_7)I_1 + R_3 I_{CO_1} + \frac{\alpha R_5}{1 - \alpha} \cdot \frac{V_{CC} - R_4 I_{CO_1}}{R_2 + R_4} = V_{BB} + V_{CC}$$

$$I_1 = \frac{V_{BB} + V_{CC}\left[1 - \left(\frac{\alpha}{1-\alpha}\right) \cdot \frac{R_5}{R_2 + R_4}\right] - \left[R_3 - \left(\frac{\alpha}{1-\alpha}\right)\frac{R_4 R_5}{R_2 + R_4}\right] I_{CO_1}}{R_3 + R_5 + R_7} \tag{5.7}$$

Condition 1. To ensure that Q_2 is in saturation, its collector voltage should be equal to the emitter voltage or positive with respect to it. Thus, $V_{C_2} \geq 0$, since the Q_2 emitter is at ground potential, and

$$V_{C_2} = (\alpha I_{E_2} + I_1)R_5 - V_{CC} \tag{5.8}$$

Substituting Eqs. (5.6) and (5.7) for I_{E_2} and I_1, respectively,

$$V_{C_2} = \frac{\alpha R_5}{1 - \alpha} \cdot \frac{V_{CC} - R_4 I_{CO_1}}{R_2 + R_4} + R_5 \left\{V_{BB} + V_{CC}\left[1 - \left(\frac{\alpha}{1-\alpha}\right) \cdot \frac{R_5}{R_2 + R_4}\right]\right.$$
$$\left. + \frac{- \left[R_3 - \left(\frac{\alpha}{1-\alpha}\right) \cdot \frac{R_4 R_5}{R_2 + R_4}\right] I_{CO_1}\}}{R_3 + R_5 + R_7} - V_{CC}\right.$$

$$V_{C_2} = \frac{\alpha}{1 - \alpha}\left[\frac{R_5(R_3 + R_7)}{(R_2 + R_4)(R_3 + R_5 + R_7)}(V_{CC} - R_4 I_{CO_1})\right]$$
$$+ \frac{R_5}{R_3 + R_5 + R_7} V_{BB} - \frac{R_3 + R_7}{R_3 + R_5 + R_7} V_{CC} - \frac{R_3 R_5}{R_3 + R_5 + R_7} I_{CO_1} \geq 0$$

From the equations shown previously,

$$\frac{\alpha}{1 - \alpha} = \beta \geq \frac{R_2 + R_4}{R_5} \cdot \frac{V_{CC} - \frac{R_5}{R_3 + R_7} V_{BB} + \frac{R_3 R_5}{R_3 + R_7} I_{CO_1}}{V_{CC} - R_4 I_{CO_1}} \tag{5.9}$$

For transistor Q_1 to be off, its base should be positive with respect to its emitter. Thus,

$$V_{B_1} > 0 \tag{5.10}$$

since Q_1 emitter is at ground potential. But

$$V_{B_1} = V_{BB} - (I_1 + I_{CO_1})R_3 \tag{5.11}$$

Substituting Eq. (5.7) for I_1 in the preceding equation,

$$V_{B_1} = V_{BB} - I_{CO_1}R_3 - R_3 \cdot \left\{ \frac{V_{BB} + V_{CC}\left[1 - \left(\frac{\alpha}{1-\alpha}\right)\frac{R_5}{R_2 + R_4}\right] - \left[R_3 - \left(\frac{\alpha}{1-\alpha}\right)\frac{R_4R_5}{R_2+R_4}\right] I_{CO_1}}{R_3 + R_5 + R_7} \right\} \tag{5.12}$$

$$V_{B_1} = \frac{R_5 + R_7}{R_3 + R_5 + R_7} V_{BB} - \frac{(R_5 + R_7)R_3}{R_3 + R_5 + R_7} I_{CO_1} - \frac{R_3}{R_3 + R_5 + R_7} V_{CC} + \frac{\alpha}{1-\alpha} \cdot \frac{R_5R_3}{R_2 + R_4} \cdot \frac{V_{CC} - R_4I_{CO_1}}{R_3 + R_5 + R_7} > 0 \tag{5.13}$$

From the preceding equation,

$$\frac{\alpha}{1-\alpha} = \beta > \frac{R_2 + R_4}{R_5} \cdot \frac{V_{CC} - \dfrac{R_5 + R_7}{R_3} V_{BB} + (R_5 + R_7)I_{CO_1}}{V_{CC} - R_4I_{CO_1}} \tag{5.14}$$

It may be seen from Eqs. (5.9) and (5.14) that the right-hand side of Eq. (5.9) is larger than the right-hand side of Eq. (5.14); therefore, Eq. (5.9) must be satisfied to meet condition 1.

Condition 2. To satisfy operating condition 2, the output voltage should be within the specified limits.

$$E_{out\,max} > V_{CC} - R_5(I_{B_1} + I_{CO_2}) > E_{out\,min} \tag{5.15}$$

when Q_2 is off.

Since

$$I_{B_1} = (V_{CC} - R_5I_{CO_2})/(R_5 + R_7)$$

when Q_1 is in saturation and Q_2 is cut off,

$$E_{out\,max} > V_{CC} - R_5V_{CC}/(R_5 + R_7) + R_5^2/(R_5 + R_7) - R_5I_{CO_2} > E_{out\,min}$$
$$E_{out\,max} > [R_7/(R_5 + R_7)](V_{CC} - R_5I_{CO_2}) > E_{out\,min} \tag{5.16}$$

The relationship in Eq. (5.16) must be met to satisfy condition 2.

Condition 3. The requirement imposed on the input pulse is the removal of the fixed change from a saturated transistor. This change is supplied by the trigger capacitor which, in turn, must be recharged by the input pulse during the remaining time in the cycle.

The charge to be removed is approximately

$$Q \cong i_{CM}/W_{\alpha b} \tag{5.17}$$

where i_{CM} is the maximum collector current of the saturated transistor and $W_{\alpha b}$ is the angular cutoff frequency of the transistor.

The trigger capacitor charges

$$q_C = C_4V_{C_4} = C_4E_T(1 - e^{-t/C_4R_8}) \tag{5.18}$$

where E_T is the trigger amplitude, t is the duration of trigger pulse, and $\frac{1}{2}T$ is specified (1 μsec). Therefore,

$$i_{CM}/W_{\alpha b} < C_4E_T(1 - e^{-T/2C_4R_8}) \tag{5.19}$$

will have to be satisfied to meet condition 3.

Condition 4. The maximum frequency of operation for a transistor multivibrator is approximately

$$f_{max} \cong f_{\alpha b}/\beta^{1/2} \tag{5.20}$$

where $f_{\alpha b}$ is the cutoff frequency in the common-base configuration and β is the common-emitter short-circuit current-amplification factor (low frequency). This is the transistor requirement for external circuitry. Since C_2 and C_4 must discharge in one-half the period of the specified repetition rate,

$$C_2R_2 \text{ and } C_3R_7 < \tfrac{1}{2}T = 1\ \mu\text{sec} \tag{5.21}$$

Equations (5.20) and (5.21) must be satisfied to meet condition 4.

5.12b Calculations. *Condition* 1. Substituting normal values of the parameters in Eq. (5.9),

$$\beta \geq \left[\frac{12{,}000 + 22{,}000}{22{,}000}\right]\left[\frac{10 - \dfrac{22{,}000}{12{,}000 + 47{,}000} \times 10 + \dfrac{47 \times 22 \times 10^6}{10^3(12 + 47)} 2.5 \times 10^{-6}}{10 - 22{,}000 \times 2.5 \times 10^{-6}}\right]$$

$$\beta > \frac{34}{22} \cdot \frac{10 - 220/59 + (2.5 \times 22 \times 47)/(59 \times 10^3)}{10 - 0.22 \times 0.25} \tag{5.22}$$

$$\beta > 1.55 \cdot [(10 - 3.73 + 0.044)/(10 - 0.055)]$$

$$\beta > 0.9.$$

Since min β given by the manufacturer is 15 (much greater than 9), the worst-condition calculations will be made. If the inequality condition remains satisfied, P_1 will be unity.

The following condition is the worst case at room temperature.

$$\beta > \left[\frac{10^3 \times 1.05 \times 34}{10^3 \times 0.95 \times 22}\right] \times \frac{\left\{10 \times 1.2 - \dfrac{22 \times 10^3 \times 0.95}{59 \times 10^3 \times 1.05} \times 10 \times 0.8 + \left[\dfrac{47 \times 22 \times (1.05)^2 \times 10^6}{10^3(12 \times 0.95 + 47 \times 1.05)} \times 6 \times 10^{-6}\right]\right\}}{10 \times 1.2 - 22 \times 1.05 \times 6 \times 10^{-6} \times 10^3}$$

$$> 1.55 \times 1.11 \times \frac{12 - 3.73 \times 0.9 \times 0.8 + \dfrac{1.11 \times 10.30}{61 \times 1000} \times 6}{12 - 0.132 \times 1.05} = 1.37 \tag{5.23}$$

Thus $\beta = 15(>1.37)$ is satisfied for room temperature.

The following condition presents β at $+71°\text{C}$.

$$I_{CO} = 120\ \mu\text{a}$$
$$\beta_{71°\text{C}} \cong 15 \times 1.1 \cong 16.5 \tag{5.24}$$

Variations in resistors are $+24$, -15 per cent.

$$\beta > [1.55 \times 1.11 \times 1.24] \times \frac{\left[12 - 3.73 \times 0.9 \times 0.8 \times \dfrac{0.85}{1.24}\right] + \left[\dfrac{1.11 \times 10.30 \times 1.24 \times 10^6}{10^3(12 \times 0.8 + 47 \times 1.29)} \times 120 \times 10^{-6}\right]}{12 - 22 \times 1.29 \times 120 \times 10^3 \times 10^{-6}}$$

$$\beta > 2.13 \times [(12 - 1.84 + 2.44)/(12 - 3.4)] = 3.4 \tag{5.25}$$

Therefore $\beta_{71°\text{C}} = 16.5(>3.4)$ satisfies the high-temperature worst case.

The following condition is presented for the worst case at $-55°\text{C}$.

$$I_{CO} = 0 \qquad \beta_{-55°\text{C}} = 15 \times 0.165 = 9.7 \qquad \beta = 2.13 \times [(12 - 1.84)/12] = 1.8$$

Therefore, if $\beta_{-55°\text{C}} = 9.7$ (which is greater than 1.8), the worst case is satisfied at $-55°\text{C}$. Thus, the condition 1 inequality holds for the worst-case parameter variations, and the reliability for this case is $P_4 = 1$.

Condition 2. Substituting normal values of parameters in Eq. (5.16)

$$\begin{aligned} E_{out\,max} &> [12{,}000/(22{,}000 + 12{,}000)](10 - 22 \times 10^3 \times 2.5 \times 10^{-6}) > E_{out\,min} \\ &> 3.53(10 - 0.055) > E_{out\,min} \\ &> 3.5 > E_{out\,min} \end{aligned} \tag{5.26}$$

Condition 3. To satisfy condition 3, Eq. (5.19) must hold. Thus, substituting nominal values,

$$\frac{10/22{,}000}{2\pi \times 8 \times 10^6} < 1{,}000 \times 10^{-12} \times 10(1 - e^{-10^{-6}/(10^3 \times 10^{-12} \times 10 \times 10^3)})$$

$$\begin{aligned} 10^{-9}/110 = 9 \times 10^{-12} &< 10^{-8}(1 - 0.9) \\ 9 &< 1{,}000 \end{aligned} \tag{5.27}$$

This inequality is satisfied by two orders of magnitudes and therefore for the worst-case parameter variations. Thus,

$$P_3 = 1$$

Condition 4. Equations (5.20) and (5.21) must be satisfied. Substituting nominal values

$$f_{spec} < 8 \times 10^6/95^{1/2} = 0.822 \text{ mc} \tag{5.28}$$

Inserting values for worst condition

$$f_{spec} < 5 \text{ mc}/(150 \times 1.2)^2 = 0.373 \text{ mc} \tag{5.29}$$

f_{spec} must equal 0.5 mc; therefore, worst case will not be satisfied, and a per cent success will have to be calculated.

$$\begin{aligned} f_{\alpha b} &= 8 \pm 3 \text{ mc} \qquad \sigma_{f\alpha b} = 1 \text{ mc} \\ \beta &= 95 \pm 85 \qquad \sigma_\beta = 85/3 = 28.3 \end{aligned}$$

$$\begin{aligned} \sigma_{f\,max} &= \sqrt{(\partial f_{max}/\partial f_{\alpha b})^2 \sigma_{f\alpha b}{}^2 + (\partial f_{max}/\partial \beta)^2 \sigma_\beta{}^2} \\ &= \sqrt{1/95 + 8^2/(4 \times 95^3) \times 28.3^2} = \sqrt{0.0105 + 0.015} = \sqrt{0.0255} \\ &= 0.16 \text{ mc} \end{aligned} \tag{5.30}$$

$$(f_{max_{nom}} - f_{spec})/\sigma_{f\,max} = (0.822 - 0.5)/0.16 = 2 \tag{5.31}$$

Therefore, 2σ deviation from nominal will satisfy the inequality. Reliability will be the area under the normal curve to the right of the -2σ limit.

$$\begin{aligned} P_{41} &= A_{n2\sigma} + (1 - A_{n2\sigma})/2 \\ &= 0.9545 + 0.0455/2 = 0.9545 + 0.02275 \\ &= 0.9772 \end{aligned} \tag{5.32}$$

Also C_2R_2 and $C_3R_7 < \frac{1}{2}T(1\ \mu\text{sec})$. Nominal values

$$330 \times 12 \times 10^3 \times 10^{-12} = 3.95\ \mu\text{sec}$$

This inequality cannot be satisfied; therefore,

$$P_{42} = 0 \qquad \text{and} \qquad P_4 = P_{41} \times P_{42} = 0.9768 \times 0 = 0 \tag{5.33}$$

The design reliability of this multivibrator is

$$P_1 = P_1 \times P_2 \times P_3 \times P_4 = 1 \times 0 \times 1 \times 0 = 0 \tag{5.34}$$

5.13 CONCLUSIONS

It has been shown that circuit-drift failures can be eliminated by worst-case design procedures but that a price is paid for this immunity in the form of increased system complexity, increased component stresses, and increased power demand. Consideration of the entire problem leads to the conclusion that decreasing the probability of circuit-drift failures (by increasing the tolerance margin of the circuit) tends to increase the probability of component catastrophic failures and, consequently, an optimum component-tolerance design point exists for maximum system reliability. The optimum tolerance margin depends upon the specific system and, generally, varies inversely with the number of components comprising the system. Thus, to maintain a specified system reliability in the face of increasing system complexity, it is necessary to assume a decreasing component-parameter spread (tighter tolerances) and a decreasing component catastrophic failure rate. Both these requirements may be relaxed if some form of redundancy is introduced to overcome the inevitable occurrence of catastrophic failures.

Inherent circuit reliability is achieved when circuit performance failure is caused by random catastrophic part failures. Part-parameter variations are inevitable and must be adequately considered; i.e., each circuit must provide an acceptable output if no part parameter violates the wide range of values such a parameter assumes over the useful part life in a specified environment. The objective of an analytical program is to ensure that each circuit evidences this essential characteristic. Part-parameter incompatibility failures should not be observed when using these techniques. Inherent circuit reliability should be realized physically.

The increasingly stringent and demanding requirements placed on today's circuits for military applications require such new designs and techniques of circuit analysis as high-speed digital computers to ascertain, in integrated form, the quantitative reliability while considering high functional requirements, light weight, and minimum volume. The output of these computers provides an objective, detailed analysis of the circuit which allows for design improvement and includes information on the part parameters, drift, and stability per unit time required for a circuit to perform within specified limits, visual graphs depicting the interdependence of part parameters as related to circuit performance, and data on the voltage and power stresses imposed on the parts in the circuit as voltage and part-parameter values change.

A check of the characteristics of the required circuit design parameters against other parameters, i.e., size and weight vs. voltage and wattage of various piece parts, may be made to ascertain any deficits per allocated requirements. Further refinement of these techniques allows the determination of the design reliability P_d and provides optimization of nominal part values for specific circuit configurations.

Bibliography

ARINC, "Evaluation and Prediction of Circuit Performance by Statistical Techniques," *ARINC Monograph* 5, ARINC Research Corporation, February, 1958.

Connor, J. A., "Prediction of Reliability," *Proc. Sixth Natl. Symp. Reliability Quality Control*, 1960, p. 134.

Dreste, F., "Circuit Design Concept for High Reliability," *Proc. Sixth Natl. Symp. Reliability Quality Control*, 1960.

Earles, D. R., "Reliability Growth Prediction During the Initial Design Analysis," *Proc. Seventh Natl. Symp. Reliability Quality Control*, 1961.

Feyerherm, M. P., "Basic Reliability Considerations in Electronics," *Proc. Fifth Natl. Symp. Reliability Quality Control*, 1959, p. 119.

Hellerman, L., and M. P. Rocite, "Reliability Techniques for Electronic Circuit Design," *IRE Trans. Reliability Quality Control*, vol. PGROQC-14, September, 1958.

Kaufman, M. I., and R. A. Kaufman, "Predicting Reliability," *Machine Design*, p. 178, August, 1960.

Reliability Engineering Group, "Circuit Analysis Techniques and System Analysis Techniques," vol. 1, North American Aviation, Inc., Los Angeles, 1963.

Reliability Engineering Group, "Computer Analysis Techniques," North American Aviation, Inc., Los Angeles, 1963.

Stevens, M., "Statistical Aspects of Systems Analysis," *Proc. Seventh Natl. Symp. Reliability Quality Control*, 1961.

Section 6

ACCESSION AND ORGANIZATION OF RELIABILITY DATA

MARTIN BARBE

IDEP OFFICE, RELIABILITY DEPARTMENT, AEROSPACE CORPORATION
EL SEGUNDO, CALIFORNIA

Since the Interservice Data Exchange Program (IDEP) was authorized by the U.S. Air Force, Army, and Navy, Mr. Barbe has had the responsibility for initiating and operating the program. As a member of the interservice committee which established the criteria for the program, Mr. Barbe was a principal contributor to the design of the system and did a great part of the research into methodology, data needs, contractor cooperation, and organization necessary for the program to be a success. He brought to this job four prior years of experience as Titan Program Reliability Coordinator (Ramo-Wooldridge/Space Technology Laboratories). During that period he designed and developed the concept of a reliability handbook for use by the several contractors involved in the Titan Program.

Mr. Barbe holds an M.E. degree from New York University and is the author of several technical papers on aircraft production methods, metal stitching, reliability, production environmental testing, and information transmittal and retrieval. During World War II, he was a member of the Central and Eastern War Production Councils. He was Chief Methods Engineer at Curtiss-Wright/Buffalo, and he has held various management and quality control positions with Lockheed, Whittaker Valve, Tevco Electronics, and in his own precision electronic and hydraulic parts manufacturing company.

ACCESSION AND ORGANIZATION OF RELIABILITY DATA

Martin Barbe

CONTENTS

6.1 RATIONALE AND SCOPE

Any reliability activity is exposed to a huge and rapidly growing volume of information with potential bearing on its problems. Much of the balance of this handbook treats of the manipulation, interpretation, evaluation, and utilization of this information. However, such actions are made difficult or impossible without the ability to retrieve quickly the significant facts from large masses of material varying from raw data to completed reports. The engineer's traditional system of individual desk-drawer filing rapidly becomes unwieldy, and search calls upon personal discrimination, experience, and memory. Presented herein are some basic rules to assist the engineer's technical skills during the planning and inception of data-storage projects. The engineer can then concentrate on data *interpretation*, while the search function can be taken over by clerical help or be partially automated.

6.1.1 Material Considered to Be Reliability Data. As necessary structuring of a rather amorphous topical area, the reliability data to be discussed in this section are considered to fall in one of the following categories:

Category of data	*Subdivisions*
A. Human-generated	Reliability program plans Environmental exposure requirements; system and lower levels Hardware performance specifications and test plans Mathematical/statistical, theory, concepts, and tables of functions
B. Tests results; stimulated use—controlled conditions	Exploratory R&D tests-hardware, materials, or principles Partial, initial, and qualification tests (including requalification) Comparative source/type evaluations Acceptance tests (including production environmental tests) Reliability tests
C. Tests results; actual use—partially controlled conditions	Field failure/success data Flight-test-failure incidents Flight-test-success (attribute) data

Category *A* will usually involve narrative-type documents whose information is generally handled as a single unit and is not quite as subject to normalization and reduction to automated manipulation, comparison, or retrieval as data in categories *B* and *C*.

Type *C* data frequently incorporate uncertain peripheral conditions; their interpretation is discussed in Sec. 9.

6.1.2 Data Not Specifically Considered in This Section. A reliability department may make use of data from many sources, but where the processing of such information is normally the line responsibility of some other department and any reliability actions would only utilize a tabulation, trend chart, or other summary prepared by the responsible department, such data will *not* be discussed in detail herein. Since corporate organization assignments vary, arbitrary decisions are necessary on precisely what constitutes reliability data. Thus receiving-department inspection data and in-process rejection reports are usually tabulated and handled by the quality control department and will not be discussed herein; the principles involved in field-failure-report handling can be applied with a slight shift in emphasis. The extraction and updating of failure rates from performance data are also a special problem; it is discussed in Sec. 7.

As another example, cost and productivity figures often enter into a reliability survey, but their handling is another specialized field not considered herein. Likewise, the vast mass of telemetered variable flight data leading to performance and accuracy conclusions is considered as reliability data only after its reduction to a final success/failure data input.

Since Sec. 9 deals with failure-reporting inputs and outputs, this section will discuss only the mass handling of this material while being machine-processed as a part of the general bank of reliability data.

6.2 INFORMATION-RETRIEVAL TERMINOLOGY AND INDEXING THEORY

Information retrieval (IR) people have invented many new terms and adopted specialized meanings for familiar terms. These terms will recur throughout this section, making a somewhat narrative combination of theory and glossary desirable at this early point.

Some approaches that are seldom practical for use with reliability data will be mentioned. Their identification, however, serves to orient the position of the applicable methods which are detailed in the overall information-retrieval field. Classification theory will be treated only as applied to retrieval systems known to be in actual use; many very interesting concepts are currently too abstract for handbook inclusion. (See Refs. 1 and 2 for some theoretical background.)

6.2.1 Major Indexing Methods. The term "retrieval" implies that the material or documents have been arranged or tagged in some manner to allow relatively rapid selection of the desired data. All such file structuring falls under the following types of *indexing:*

A. Normal. Traditional library practice. Use of one card (or one space on magnetic tape, etc.) for each *document.* Cards can be arranged (alphabetically) by title (author's), author's name, etc.

Sometimes the document subject(s) are interpreted *by an indexer* as fitting under appropriate subject headings from a preestablished framework such as the Dewey Decimal System, Universal Decimal System, Library of Congress System, AEC Subject Headings, etc. Such filing is still "normal," because there is one document entered per card or space.

B. Inverted. Newer system, claimed more adaptable to manual manipulations which can replace complex machine searchers to some degree. Use of one card (or tape space, etc.) for each *characteristic* by which a classification or search might be made, listing all documents (or document numbers) to which this *characteristic* applies on the one card (or space), usually in some organized sequence or pattern.

Classification control by means of a prearranged thesaurus of terms (such as the ASTIA "descriptors") is almost mandatory to the use of an inverted file system, although open-ended inverted systems have been advocated in theory. It is usually more difficult to update or purge inverted files.

6.2.2 Retrieval Terminology

A. Dictionary (Thesaurus). Any ordered listing of all the categories or terms (key words, descriptors) which anyone is *allowed* to use in cataloging a document under a particular IR system. It limits and defines a "completely structured universe," but it should allow for additions and changes. (Tables 6.1 and 6.2 constitute "dictionaries" specialized to reliability data.)

B. Sequential Search. This term is applied to any simple alphabetical or numerical search. When the material has been carefully cataloged, such searches can be productive with a minimum of complexity or equipment.

C. Coordinate-search Principles. A search utilizing a (coordinate) filing/retrieval system is designed to locate the "address" of the document(s) which deal(s) with any specified number of a desired list of properties. These properties are selected from the source list (thesaurus) of descriptive terms which was used by the originator or indexer. "Coordinate" implies those documents located at an imaginary junction of these metaphorical "descriptive axes." Some examples of coordinate-expressed inquiries would be requests for any reports indexed under specified (or a stipulated number of) the following (typical) words or phrases from the thesaurus:

Example 1 Part Information
Resistors
Variable
Encapsulated
Qualified to MIL XYZ
Established reliability
Example 2 Materials Information
Greenland
Dated after 1960
Cosmic rays
Northern hemisphere
Tetrafluoride
Friction, bearings
Life tests, plastic
Radiation effects, plastic

The system can be designed for "more-than" type queries, such as "successful operation at 250,000 ft altitude or more," or ranges, such as "between 0.90 and 1.2 watts." Variations can be introduced, with increasing computer program complexity until the term "coordinate" hardly applies.

D. Coordinate Search by Descriptors. In Calvin Moers' particular adaptation of a coordinate system (Zatacoding, Ref. 2, page 3, 122), each card represents one particular document and all the applicable descriptors (or their code) selected from the dictionary are marked on the card, indicated by a punched card, or indicated by a punched hole in a coded location. Usable card area tends to limit the number of terms in a dictionary for this system.

E. Coordinate Search by Uniterms. M. Taube's coordinate system involves an inverted card file, i.e., one card for each *property* (unit term) specified and an indicating punch mark, hole, etc. for each document in the file applicable to or using that term. The number of documents *per deck* is limited to the punch locations per card, but it can reach 40,000 with the largest Termatrex cards. The system capacity can be increased by additional decks or special variations of coding. (Incidentally, the word "descriptor" as used by ASTIA, now DDC, is closer to M. Taube's "uniterms" than to Moers' original "descriptors.") The coordinate principle has been mechanized through COMAC, TERMATREX, MATREX, etc. (discussed under Sec. 6.8, Hardware).

F. Permutational Indexing. This refers to indexing (usually a catalog or printout) according to a succession of aspects of the titles or subjects. In the key word in context (KWIC) permutational system, each *significant* word of the title is successively indexed on. Thus, a report title "*D*eterioration of *T*hin-wall *N*ylon parts from *G*amma *R*adiation" would be listed five times in the proper alphabetical sequence under D, G, N, R, and T. An example of KWIC indexing, reproduced from a 1962 *ASTIA (DDC) Title Index,* follows:

RMANCE AND INSTALLATION	BROCURE.z VIPER A.S.V. 12 PERFO	27	AD-278908
SIDERATION OF METHYLENE	BROMIDE AS A TOWING TANK FLUID.z CON	09	AD-279087
-DIMETHYLALLYMAGNESIUM	BROMIDE.z NUCLEAR MAGNETIC RESONANCE	25	AD-279290
0 DEGREES BEAM FROM THE	BROOKHAVEN ALTERNATING GRADIENT SYNC	20	AD-280790
INS RAISED BY SELECTIVE	BROTHER-SISTER MATINGS IN OUR LABORA	16	AD-279947
TAXA OF BENTHIC GREEN,	BROWN AND RED ALGAE. z NEW	16	AD-279697
SERIAL NO 6 APPLICABLE	BRUNSWICK CORP. REPORT. z GRUMMAN AIR	08	AD-281476

Note: z emphasizes the end of the title

G. (Columnar) Transposition. The changing of an original sequence of sort by assigning initial (highest) sort priority to different properties. Since the property being sorted on first usually is listed in the left-hand column, any change transposes the position of the columns or fields to the new order of sort. For example, a printout might initially be indexed by (1) author's last name, alphabetical, then (2) author's company affiliation, and (3) first significant word of title. A transposition of (1) and (2) as the initial sorting field would then result in grouping together all documents issued from a given company and *then* by author, instead of grouping all of one author's works together. The physical location of the columns can either be changed or remain constant; the important factor is the priority (hierarchy) assigned to each column in the sorting.

6.2.3 Terms Defining Physical Distinctions in the Indexing Media. Any indexing system which requires or allows a search or sort by other means than manually leafing through pages or cards is called a "manipulative" index system. Manipulative systems can become increasingly complex in nature until the distinction between manual and automated systems becomes rather arbitrary.

Automated systems will be discussed in Secs. 6.7 and 6.8. However, many successful desk-top systems can handle the appropriate level of mechanization for the volume and uniformity of data involved. These generally can be divided into the following.

A. Optical Coincidence. (Peephole, Batten, Peek-a-boo) Cards. These cards manually facilitate search on an inverted file system in which each card represents one of the topics, characteristics, descriptors, or other property by which the system has been designed to search. Various designs have 500 to 10,000 hole positions per card, each of which identifies a document, part, etc.

The cards representing the topics, characteristics, and descriptors for the particular search are removed from the general file. Light showing through all the cards at any hole position(s) indicates that those documents (parts) fulfill all the characteristics being searched for. A separate card position-to-acquisition number or title-cross-index list is necessary to identify or procure the actual document.

B. Edge Notch. This system uses a peripheral row of holes (one, two, or three deep) which are notched out to indicate dates, sizes, or other descriptors. This allows manual sort by separating cards which are allowed to fall off wire pins through those holes opened by notching. Combining this manipulation with punch-out or optical coincidence of second or third rows of holes can allow quite a complex set of search criteria with very simple equipment.

6.2.4 Terms Defining the Function of the System. The major function(s) of a manual or automated system are described herein by some combination of the terms locate, furnish, and compute/select. Definitions are given in Sec. 6.8.1.

6.2.5 Terms Relating to Document Numbers. The nature of the document *number* plays an important part in describing the above functions; distinctions in such numbers are:

A. Address Number (or Address). A number or code enabling a particular machine or system to locate the physical document immediately. Shelf numbers do this in some libraries. In some automated systems the address is an identifying number used internally by the machine and need never be known to the searcher.

B. Acquisition (Sequence, Serial, Boxcar, Accession) Number. Document identification which has no subject or positional coding significance; usually assigned in the order the document is received or cataloged.

C. Mnemonic Number. Classification or specific document identification which includes symbols or letters suggesting the topic of the document/classification, for example, LGR/L for landing gear-left.

D. Polyvalent Notation. Notation in which *each* digit signifies a code denoting a parameter or characteristic; often used in standard part numbers.

6.3 INTERNAL DATA SOURCES

Positive controls are necessary for a reliability department to assure flow of all relevant data into their files. Periodic check by reliability personnel is *not* satisfactory; formalized document routing to reliability and/or sign-off provisions are required.

The following areas often generate information vital to reliability control and should be periodically monitored to establish that no new data sources are bypassing the reliability files.

A. Purchasing/Subcontracting Department. Look for major subcontracts involving test requirements and individual tests subcontracted directly at project engineering request.

B. Library Acquisition Lists and Document Control Inventories. A large company has much valuable data generated from "one-time only" sources; libraries can serve as checkpoints which often turn up these occasional inputs.

C. Contracts Department. Often plans, proposals, or performances bearing on reliability are forwarded to the customer without the reliability department being notified. Screening or receipt of all documents is not proposed; only a positive check-off arrangement *within* contracts department to assure transmittal of relevant data.

D. Field-service Engineering. Too often this department is isolated physically from design-reliability engineering and runs its own failure analyses and quick fixes on customer-accepted assemblies. Full copies of all issuances should go to reliability department.

E. Environmental Test Laboratory. Laboratories usually compile schedules, plans, status reports, etc. on a regular basis. The need for reliability department to utilize these is determined by the degree of centralization of the test-control function and its integration within the reliability organization.

6.3.1 Assuring Information on Test Activity. All test reports on everything from resistor burn-in to complete system operation cannot usually be digested or utilized by a reliability department, but knowledge of their existence and immediate access to the desired details of such data must be available. The degree to which these are then perused in detail and reduced into the common data bank can be at the discretion of the reliability department and can be tailored to fit the immediate urgencies. However, there are usually several points at which test labor can be spent *in-house*, and often policy approval of all such activities does not go through some central point, or even through one point for each project.

An extra copy of all test authorizations as they are initially approved, before extensive planning, fixture, and part commitment, should be routed through some control point which we shall arbitrarily identify as the "test-control center." (See Sec. 8.) If such an identified function does not exist, the job of keeping tabs on test data is more difficult. (For proper experience retention at a permanent data center, it is essential that maximum limits be set on undocumented informal lab checks, such as "not over 4 hours," "not over one man-day," and "not over $100.")

6.3.2 Tests Performed in Independent Laboratories. Some companies have a routing through purchasing for outside testing which avoids most of the in-house controls. It is necessary to verify that a copy of such authorization, as mentioned in Sec. 6.3.1, reaches the reliability department in *all* cases.

6.3.3 Parts Testing by Subcontractors and Third-tier Contractors. This is probably the most difficult area to monitor, and it is one in which the prime contractor's engineering department often takes the position that it has contracted for talent to do a complete job and does not feel it appropriate to become involved in the component-parts selection details. In other cases the prime contractor's engineers may desire to be conscientious on such points, but a *contractural* requirement to submit detailed listing and justification of nonstandard parts applications is not always written or enforced—especially as it applies to *prior* approval of parts-test plans and expenditures.

When parts-test plans are submitted, they may not be cleared through an identified company central control point, but in any event the reliability department must assure that positive requirements for its information copies are contracted and enforced. All the above points apply, with some increased complexity, to the testing done by third-tier contractors, especially when required and funded by the prime contractor.

6.3.4 Checking Receipt of Parts-test Data. The actions mentioned in Secs. 6.3.1 to 6.3.3 establish flow of data on the prediction of events to come; it is then equally necessary to cross-check on the accomplishment of these tests and to receive the completed document (or notice of its existence when the full text would not be desired). Establishment of such a flow of documents, or notification, becomes a check on the efficiency of the *prediction* system; while the prediction schedules perform the corresponding check to "close the loop" on completed material.

6.4 STRUCTURING OF DATA AT ORIGINAL SOURCE

Tabulating, storing, or analyzing data is made difficult or impossible unless the data are *structured* (i.e., placed on standard format, coded, descriptors identified, etc.) at the source to the degree of rigidity appropriate to the volume and anticipated complexity of search. Typical uniform conventions of nomenclature and coding arrangement are described in Secs. 6.6 and 6.7.

6.4.1 Failure Reports. Control on failure reports is described in Secs. 7 and 9; naturally, the degree of enforced standardization will vary with the volume of reports to be handled. A small quantity can reasonably be tabulated, and the trends analyzed and studied, by using manual methods and by working from the original narrative descriptions. As the quantity of reports grows, the necessity of conventional coding and "restricted English" terms increases if the information is to be handled on a mass basis. A computer search is possible only when each "field" or box (by which a search might be made) is restricted to a stipulated selection of terms or figures on the original report. The trends thus revealed naturally require subsequent engineering interpretation of significance.

Some failure-reporting systems have over 60 such structured boxes, allowing search to be made by any combination of these factors. Usually an "uncontrolled" narrative description is also included, but unless key words are selected from an established dictionary (see Sec. 6.2.1*B*), machine search is of no value and comparison or matching is impossible.

6.4.2 Test Reports. The summary format illustrated in Fig. 6.1 is basically designed for reports on component parts, although it is applicable to tests of assemblies of greater size and complexity. Specific adaptations can be made easily if the testing in question is consistently on a particular type of product. However, the principles of utilizing a standardized format to facilitate rapid interpretation, coding, and retrieval still apply. The common requirements for date, full identification of the part, tabular description of the tests and results, plus a narrative summary of conclusions, constitute a universal disciplinary framework to guide the reporting of any methodical testing.

6.4.3 Test-reporting Format and Content. MIL-STD-831 (28 August 1963) prescribes requirements for format and arrangement of required elements when reporting on a test to the military. However, the principles stated therein can also be adopted as a checklist for test-data reporting on civilian items as well.

6.4.4 When to Require Test Reports. There is no uniformly accepted criterion defining *when* a formal report is justified, since this will vary with the urgency, complexity, schedule, and budget of each project. The important point is to establish

REPORT SUMMARY SHEET

1. COMPONENT/PART NAME PER GENERIC CODE	2. PROGRAM OR WEAPON SYSTEM	3.	DAY	MO.	YR.
	5. ORIGINATOR'S REPORT NO.	TEST COMPL.			
4. ORIGINATOR'S REPORT TITLE		REPT. COMPL			
	6. TEST TYPE, ETC.				

7. THIS TEST (SUPERSEDES) (SUPPLEMENTS) REPORT NO:

8 ITEM	8A. PART TYPE, SIZE, RATING, LOT, ETC.	9. VENDOR	10. VENDOR PART NO.	11. IND./GOV. STD. NO.	12 TOTAL TESTED
1					
2					
3					
4					(OVER)

13. INTERNAL SPECS.ETC REQ'D TO UTILIZE REPT.	ENCL	SENT WITH REPORT NO.	14. MIL. SPECS./STDS. REFERENCED IN 15C
A			D
B			E
C			F

15A. B ITEM	TEST OR ENVIRONMENT	C PER SPEC	D SPEC. PARAGRAPH/ METHOD/CONDITION	E TEST LEVELS, DURATION AND OTHER DETAILS	F NO. TESTED	G NO. FAILED
					(OVER)	

16. SUMMARY OF REPORT, NATURE OF FAILURES AND CORRECTIVE ACTIONS TAKEN:

(OVER)

17. TESTED BEYOND VENDOR CATALOG SPECIFICATIONS YES	18. VENDOR INFORMED OF TEST RESULT BY: LETTER / CY OF REPT / ORAL	19. SIGNED	20. CONTRACTOR	SUBCONTRACTOR

21. REPT. NO:

REPRODUCTION OR DISPLAY OF THIS MATERIAL FOR SALES OR PUBLICITY PURPOSES IS PROHIBITED.

FIG. 6.1 Report summary format stipulated in MIL-STD-831 (front).

ground rules in advance on exactly what minimum man-hours or dollars will automatically require a report, unless other considerations override these limits. There will be exceptions to require documentation below this limit or omission above, but the procedure and authorization to make such exceptions should also be prescribed in writing, in advance. Such procedures should also specify an approval routing and schedule on completed reports, since information is perishable.

6.5 EXTERNAL DATA ACQUISITION AND INTERCHANGE

6.5.1 Current Data Systems and Activities. The number and constant variations in information centers and systems in the United States forbid mention of any

IDEP FORM NO. 18: 9-62

8 ITEM	8A. PART TYPE, SIZE, RATING, LOT, ETC.	9. VENDOR	10. VENDOR PART NO.	11. IND./GOV. STD. NO.	12. TOTAL TESTED
5					
6					
7					
8					

15A. B ITEM	TEST OR ENVIRONMENT	C PER SPEC	D SPEC. PARAGRAPH/ METHOD/CONDITION	E TEST LEVELS, DURATION AND OTHER DETAILS	F NO. TESTED	G NO. FAILED

16. SUMMARY OF REPORT, NATURE OF FAILURES AND CORRECTIVE ACTIONS TAKEN:

21. REPT. NO:

FIG. 6.1 Report summary format stipulated in MIL-STD-831 (back).

but a few of those most relevant to reliability. Further details can be found in Refs. 4 and 5, which list over 200 such activities. Persons interested in pursuing the general status of information exchanges and repositories in the United States further are referred to the material in Refs. 6 and 7.

AFELIS: Air Force Engineering and Logistics Information Systems, HQ-AFLC, Wright-Patterson Air Force Base, Dayton, Ohio. Commonly referred to as USAF Parts Data Bank. Planned to collect, store, and rapidly retrieve data (and respond to inquiries) on reliability and availability of parts fulfilling stated requirements. Initial service tests in late 1964. (Inquiries containing requests for reliability estimates will be referred to RADC for added response.)

AGED: Advisory Group on Electron Devices, 346 Broadway, New York City 10013 (formerly AGEP and AGET). Activity sponsored by Office of Director of Defense,

R&E, DOD; to propose, monitor, and report on state-of-the-art advancement projects on electron parts and tubes. Issues annual and interim survey reports and answers status inquiries from qualified requesters.

Air Force Materials Laboratory, Research and Technology Division (MAAE), Air Force Systems Command, Wright-Patterson Air Force Base, Dayton, Ohio. Generates, collects, and makes available data on the mechanical, physical, and thermophysical properties of materials. Reports and handbooks are published.

CAMESA: Canadian Military Electronic Standards Agency. Close Canadian parallel to DESA, except that a larger portion of testing is done by the agency itself. Qualification data issued on Canadian branches of United States firms, Canadian firms, and United States products.

Ceramic and Graphite Technological Evaluation Section, R&T Division, Wright-Patterson Air Force Base, Dayton, Ohio. Processes, analyzes, and disseminates technical information on ceramics and graphite.

CPIA: Chemical Propellant Information Agency. A new agency from the combination of LPIA and SPIA.

DASA: Transient Radiation Effects on Electronics Systems, Defense Atomic Support Agency. Collects from and distributes to some 50 to 60 government and contractor locations. Works with REIC (Radiation Effects Information Center of Battelle) for indexing/analyzing data. The specific program of major interest to reliability people would be the TREES, administered by the DASA Data Center, GE/Tempco Division, Santa Barbara, California.

DDC: Defense Documentation Center, Cameron Station, Building 5, 5010 Duke Street, Arlington, Virginia 22314 (formerly ASTIA). Master center for all technical data generated on government contracts, including R&D, systems data, and logistics. *Technical Abstracts Bulletin* (TAB) issued twice monthly; holders of Government contracts can obtain reports within their designated field of interest without charge. Some complex computerized search, using descriptors from *ASTIA Thesaurus*, is possible.

DESA and **DESC-E:** Defense Electronics Supply Agency, Arlington, Virginia. (The Electronic Parts Function formerly covered by ASESA is now under DESC-E, Gentile Air Force Base, Dayton, Ohio.) This agency, among its many other activities of less direct interest to reliability engineers, issues official qualified products lists showing companies which have demonstrated potential to comply with a given MIL specification.

DMIC: Defense Metals Information Center, Battelle Memorial Institute, Columbus, Ohio. Collects, processes, and disseminates technical information on structural metals and closely related aerospace materials—titanium, beryllium, magnesium, tungsten, molybdenum, columbium, etc.

DTIE: Division of Technical Information Extension, U.S. Atomic Energy Commission, P.O. Box 62, Oak Ridge, Tennessee 37831. The AEC counterpart of DDC. Semimonthly *Nuclear Science Abstracts* available from Superintendent of Documents, Government Printing Office, Washington, D.C. 20402.

ECRC: Electronic Components Reliability Center, Battelle Memorial Institute, Columbus, Ohio. For an annual fee, members receive tabular summaries of parts tests contributed by other members, plus privilege of purchasing special searches or consulting on parts problems.

EDDS: Electron Devices Data Service, Electron Devices Section, National Bureau of Standards, Washington, D.C. 20234. Information on characteristics, availability, and manufacturers of receiving and microwave tubes, semiconductors, etc., both United States and worldwide.

EPIC: Electronic Properties (of Materials) Information Center, Hughes Aircraft, Culver City, California. One of a group of seven information centers on materials sponsored by the USAF/AFSC/ASD Materials Laboratory. Each center concentrates on different aspects (see REIC, MPIC, PLASTEC).

FARADA: Failure Rate Data, U.S. Naval Ordnance Laboratory (Code 60), Corona, California. Conducts a program which compiles failure-rate data into their *Failure Rate Data Handbook*, available without charge to participants.

Groth Institute, Florida Atlantic University, Boca Raton, Florida. Collects, correlates, and interprets data on physical and chemical properties of crystals.

IDEP: Interservice Data Exchange Program. Army office: Army Missile Command, Huntsville, Alabama. Navy office: Officer-in-Charge (Code E-6) U.S. Naval Fleet Missile Analysis and Evaluation Group, Corona, California 91720. Air Force office: Aerospace Corporation, For: AFSSD, El Segundo, California. Government-funded program to exchange summary and full test of parts-tests reports on microfilm cards between some 150 participating contractors of the three services and NASA. Contact points for follow-on inquiries are identified at all participants.

INDEX: Inter-NASA Data Exchange, Headquarters, NASA, Washington, D.C. Detailed and justified *Qualified Parts Listing* for NASA use, giving failure rates and performance data, compiled from many sources including vendor data. This was a portion of NASA Preferred Parts and Sources Program; status and availability outside NASA indefinite at time this list was compiled.

LPIA: Liquid Propellant Information Agency, Applied Physics Laboratories of Johns Hopkins University, Silver Springs, Maryland. Government-funded, Navy-administered activity to gather performance data on liquid propellants, motors, and components and distribute abstracts on reports to qualified recipients.

MCRC: Mechanical Components Reliability Center, Battelle Memorial Institute, Columbus, Ohio. Counterpart to ECRC; deals with mechanical subject areas.

MPIC: Mechanical Properties of Materials Information Center, Balfour Engineering, Suttons Bay, Michigan. Prepares and distributes evaluated data on the strength of aerospace materials. Air Force/ASD sponsored.

Office of Standard Reference Data, National Bureau of Standards, Connecticut Avenue and Van Ness Street, N.W., Washington, D.C. 20234. Compiles and critically evaluates quantitative data on physical and chemical properties of substances. Developing a complete data center in this area.

NASA: Scientific and Technical Information Division, Code ATSS, Washington, D.C. 20546. Central services for NASA, their contractors, and other government agencies on reports/data dealing with aerospace topics. Performs searches and prepares bibliographies. Issues STAR semimonthly; abstracts technical articles (25,000 to 30,000 per year). Available also from Government Printing Office. NASA also contracts with *Research Triangle Abstracts and Technical Reviews,* approximately 55 abstracts issued per month, 1,355 items abstracted through June, 1964. Details available from NASA, Code KR.

National Referral Center for Science and Technology, Library of Congress, Washington, D.C. 20450. *Refers* inquiries to the location(s), information center, or individual most likely to be qualified to answer. Occasionally the answer itself is furnished.

NFSAIS: National Federation of Science Abstracting and Indexing Services. A forum of some twenty nonprofit and government agencies working to achieve more uniformity among the overlapping services. Sponsored by the National Science Foundation.

OTS: Office of Technical Services, U.S. Department of Commerce, Washington, D.C. 20230. Government center for collection and organization of technical reports resulting from government-financed research and for technical translations (available from Government Printing Office; see below). In cooperation with Science and Technical Division of Library of Congress, performs extensive literature searching on a fee basis. Issues several monthly and biweekly indexes available on subscription; full catalog of government publications and selected government publications, technical translations, research reports, etc. Address inquiries to Superintendent of Documents, Government Printing Office, Washington, D.C. 20402.

PLASTEC: Plastics Technical Evaluation Center, Picatinny Arsenal, Dover, New Jersey. Collects, exchanges, collates, develops, and evaluates technical data on plastics of interest to DOD, with emphasis on structural, electrical, electronic, packaging, and rocket/missile applications.

Prevention of Deterioration Center, National Academy of Sciences, 2101 Constitution Avenue, Washington, D.C. 20418. Information on natural and induced environmental effects on materials and equipment. Consulting services free to its sponsors, NASA and U.S. Army Biological Laboratories, and their contractors.

PRINCE: Parts Reliability Information Center, Code R-ASTR-TR, Marshall Space Flight Center, Huntsville, Alabama 35812. Computerized storage and periodic printout into an index of reliability test data pertinent to NASA requirements. Available to NASA and approved contractor personnel. Automatic advisement on new material to personnel qualifying for their *Field-of-Interest Register*.

RADC: Rome Air Development Center, Rome, New York. (Air Force) Reliability Central (EMERC), monitors and stores data on compliance with MIL-R-38100 established reliability specifications for parts. Also will gather, analyze, and respond to inquiries on other parts-reliability data. Initial stages in 1964.

Radio Systems Division, National Bureau of Standards, South Broadway, Boulder, Colorado 80301. Reliability and efficiency of radio-spectrum use and standards of measurement in radio systems; consulting on a no-charge or contract basis depending on the size of the task.

Reliability-value Engineering Branch, Engineering Division, Assistant Chief for Research, Development Test and Evaluation, Bureau of Naval Weapons, U.S. Department of Navy, Washington, D.C. Information on methods of assigning reliability objectives to electronic and mechanical equipment available to government agencies and qualified contractors.

REIC: Radiation Effects Information Center, Battelle Memorial Institute, Columbus, Ohio. Collects, analyzes, and distributes radiation-effects information on aerospace materials. ASD-funded. Literature searches, answers to technical questions, and data compilation on request.

SETE: Secretariat of Electronic Test Equipment, New York University, New York City. Government-funded, Air Force–administered agency compiling varied vendor claims into organized data and furnishing other information on test equipment.

Shock and Vibration Centralizing Agency, U.S. NRL Code 1020, Washington, D.C. 20390. Information on shock, vibration, pressure, temperature, and radiation.

SPIA: Solid Propellant Information Agency, Johns Hopkins University. Similar to LPIA except dealing with solid fuels, casings, and motor parts. Because of *relatively* smaller topical area, coverage can be more complete than LPIA.

STAR: See NASA.

TIO: Technical Information Office, Office of Naval Research, Washington, D.C. The TIO will direct properly detailed requests for information to the proper channels within the ONR for possible answers.

TPRC: Thermophysical Properties Research Center, Purdue University, School of Mechanical Engineering, Lafayette, Indiana. Research and literature surveys on thermophysical properties of all substances. Annually publishes *Retrieval Guide to Thermophysical Properties Research Literature.*

6.5.2 Abstracting Services. The activities listed in Sec. 6.5.1 are either directly or indirectly government-sponsored; there is also a growing number of industry association, semicommercial, and commercial activities which furnish abstracts of published articles. Examples of these, generally relevant to the reliability area, are the following.

Annual Review of Binary Alloys, ITT Research Institute, Chicago. Review and critique on literature pertinent to binary equilibria in metallic systems.

American Mathematical Society, 190 Hope Street, Providence 6, Rhode Island, 12,000 references and abstracts per year.

Applied Science and Technology Index, The H. W. Wilson Company, 950 University Avenue, New York, New York 10052. 77,000 references per year.

Cambridge Commercial Corporation, 283 Main Street, Cambridge, Massachusetts. Surveys solid-state physics, computers and aerospace sciences; journals and abstracts on microcards, etc., available on fee basis.

DATA, Inc., Derivation & Tabulation Association, 48 South Day Street, Orange, New Jersey. Microfilm and publications on semiconductors and microwave tubes on fee basis.

Engineering Societies Library, United Engineering Center, 345 West 47th Street,

New York, New York 10017. Reference service, literature searches, translations, specialized bibliographies. Issues *Engineering Index;* articles retrieved can be purchased as microfilm or reprint.

International Abstracts in Operations Research, Operations Research Society of America, Mount Royal and Guilford Avenues, Baltimore 2, Maryland. 800 abstracts per year.

International Journal of Abstracts, The International Statistical Institute, c/o Dr. William R. Buckland, General Editor, 22 Ryder Street, London SW. 1, England. Quarterly; 1,000 abstracts per year.

Operational Research Quarterly, Operations Research Society, London. Published by Pergamon Press, 122 East 55th Street, New York, New York 10022. Articles and abstracts; about 100 per year.

Quality Control and Applied Statistics Abstracts Service, Interscience Publishers, 440 Park Avenue South, New York, New York 10016.

Reliability Literature Survey, Air Force Flight Test Center (1963). Abstracts of some 1,000 (unclassified) articles and papers on reliability. Includes the ASTIA (now DDC) bibliography of reliability literature on file.

6.6 CLASSIFICATION AND STORAGE FOR MANUAL RETRIEVAL SYSTEMS

The universal classifications included in this section as examples are intended primarily for use in manual systems, since these simpler systems require less critical tailoring to specific needs.

6.6.1 Classification: Manual vs. Automated Requirements. Most manual systems can be placed on automated equipment; but because of the simplicity desired for manual handling, it is then seldom possible to make full use of the machine's capabilities for complex searches. However, forethought in the initial design of a manual system can anticipate automation capabilities to an extent which will allow better efficiency in any future changeover. Study of advantages, restrictions, and liberties allowed by automated systems and their respective hardware (Secs. 6.7 and 6.8) is advised, but preferably, the review of any proposed *manual* system should be made by a person versed in automation. Such review can save much rework in the event that mechanization becomes desirable later.

6.6.2 Classification Categories for General Reliability/QC Information. The breakdown shown in Table 6.1 is primarily intended for classifying general-information material (specifically in the QC/reliability area) rather than for handling a mass of detail data on any one particular topic. For example, under 874, Maintainability Reporting, Analysis and Evaluation, one would expect to find general theory plans, outlines, correspondence, and possibly periodic summary reports, but not the bulk of input data and detailed day-by-day records unless they were very small in number. A descriptive table such as Table 6.1, covering a broad topical area, establishes a manual storage and referral classification scheme (adaptable to a technical field), rather than a data-*manipulating* or processing retrieval system, although many terms could be adopted into a retrieval dictionary.

6.6.3 Terms for Use in Classifying Engineering Information. The rigidity with which limited specific terminology (restricted English) must be enforced is somewhat relaxed for manual systems, where some human judgment can be applied during search. However, the increasing trend to design all systems for potential automation is illustrated by the considerable effort to create an *Engineering Thesaurus,* by the Engineers Joint Council, 345 East 47th Street, New York, New York 10017.

This document defines preferred usage of some 13,000 technical words. There is such notation as "use . . ." in all cases where a preference in a choice of terms exists. "Broader terms," "narrower terms," and "related terms" are also listed to establish the hierarchal position of each word relative to related words in its subject area and to aid in narrowly defining the area of search to a librarian or file clerk. This publication was issued early in 1964, but since it covers *all* phases of engineering, it is necessary for any reliability department to supplement it with its specialized or newly added

Table 6.1 ASQC Literature Classification System*

The full document classification number defined by this system is in the form 000:00:000, where the first three digits define methodology, the next two describe function, and the last three identify the industry.

METHODOLOGY CLASSIFICATION

- 000: General
  - 010: Quality Control Definitions/Symbols
  - 090: Bibliographies
- 100: Statistical Process Control
  - 110: Control Charts
    - 111: Control Charts for Variables
    - 112: Control Charts for Attributes
  - 120: Process Control Requirements
    - 121: Specifications
    - 122: Tolerances
    - 123: Process Capability
  - 130: Frequency Distributions in Process Control
- 200: Sampling Principles and Plans
  - 210: Principles of Sampling
    - 211: Bulk Materials
    - 212: Discrete Units
    - 213: Classification of Characteristics
    - 214: Sample Selection
  - 220: Sampling Plans
    - 221: Selection/Comparison of Sampling Plans
    - 222: Attributes Plans
    - 223: Variables Plans
    - 224: Continuous Plans
  - 230: Censuses and Surveys
- 300: Management of Quality Control
  - 310: Initiation of Quality Control
  - 320: Training in Quality Control
    - 321: Tutorial Manuals and Texts
    - 322: Training Aids/Devices
    - 323: Training Programs
    - 324: Licensing/Skill Certification
  - 330: Organization for Quality Control
    - 331: Quality Control Personnel
  - 340: Administrative Techniques in Quality Control
    - 341: Records and Reports of Quality
    - 342: Standards and Procedures
    - 343: Incentive Plans
    - 344: Quality Indices
    - 345: Quality Auditing Systems
    - 346: Drawing Change Control
  - 350: Economics of Quality
    - 351: Customer-Vendor Relations
    - 352: Quality Standards
    - 353: Quality Cost Measurement
- 400: Mathematical Statistics and Probability Theory
  - 410: Theory of Estimation and Statistical Inference
    - 411: Point Estimation
    - 412: Confidence Intervals
    - 413: Hypothesis Testing
    - 414: Decision Theory
  - 420: Properties of Distribution Functions
    - 421: Normal Distribution
    - 422: Poisson Distribution
    - 423: Binomial Distributions
    - 424: Composite/Multivariate Distributions
    - 425: Fitting Distribution Functions
  - 430: Probability Theory
    - 431: Stochastic Processes
    - 432: Queuing Theory
    - 433: Baysian Methods
  - 440: Transformations
- 500: Experimentation and Correlation
  - 510: Tests of Significance and Confidence Intervals
    - 511: Significance Tests
    - 512: Confidence Intervals
    - 513: Statistical Analysis
  - 520: Design and Analysis of Experiments
    - 521: Planning the Experiment
    - 522: Experimental Designs
    - 523: Special Methods for Analysis of Data
    - 524: Analysis of Variance
  - 530: Correlation
    - 531: Simple Correlation
    - 532: Multiple Correlation
    - 533: Rank Correlation
    - 534: Covariance Analysis
  - 540: Curve Fitting
    - 541: Linear Regression
    - 542: Non-Linear Regression
    - 543: Multiple Regression

* Reproduced by courtesy of E F. Taylor and R. Jacobs.

Table 6.1 ASQC Literature Classification System (*Continued*)

544: Orthogonal Polynomials
545: Time Series
546: Goodness of Fit Tests
550: Short-Cut Methods of Analysis
551: Non-Parametric Tests
552: Graphical Analysis of Data
553: Tables
554: Charts
555: Nomographs
600: Managerial Applications of Quality Control
610: Operations Research Methods
611: Collection of Operational Data
612: Special Techniques and Their Application
613: Managerial Systems Analysis
614: Linear Programing
615: Dynamic Programing
616: Queuing Theory
620: Industrial Engineering Methods
621: Packaging and Shipping
622: Production Scheduling and Control
623: Work Measurement and Wage Plans
624: Inventory Schedule and Control
625: Facilities Replacement
630: Business Economics Methods
631: Measurement and Analysis
632: Forecasting/Estimating
633: Financial Policy
634: Sales Control
635: Index Numbers
640: Performance Measuring and Reporting (critical path)
700: Measurement and Control in Quality Control
710: Measurement of Quality Characteristics
711: Physical Properties
712: Dynamic Properties
713: Structural Properties
714: Chemical Properties
715: Atomic and Nuclear Properties
716: Aging and Deterioration
717: Error of Measurement
720: Process Control
730: Data Handling
731: Data Collection
732: Data Reduction
733: Data Processing
734: Data Storage
735: Data Retrieval
740: Automation
750: Sensory Measurements
751: Visual/Sight
752: Taste
753: Smell
754: Tactile/Touch
755: Audio/Sound
760: Inspection
761: Receiving Inspection
762: In-Process Inspection
763: Assembly Inspection
764: Final Inspection
765: Shipping Inspection
766: Field Inspection
767: Calibration and Standards
770: Test Engineering
771: Test Planning
772: Test Reporting
773: Test Equipment
774: Test Methods (Destructive)
775: Test Methods (Non-Destructive)
780: Environmental
781: Environmental Conditions
782: Environmental Effects
783: Environmental Measurement
784: Environmental Equipment
800: Reliability
801: Definitions and Semantics
802: Texts, Manuals, and Handbooks
810: Management of Reliability Function
811: Organization
812: Training and Indoctrination Programs
813: Program Implementation/Evaluation
814: Value Analysis
815: Specifications/Contracts/Requirements
816: Procurement Relations
817: Trade-Off Evaluations
820: Mathematical Theory of Reliability
821: Probability and Prediction Theory
822: Reliability Distribution Functions
823: Life Testing Theory
824: Estimating and Assessment
825: Apportionment
830: Design (Reliability Aspects)

Table 6.1 ASQC Literature Classification System (*Continued*)

- 831: System Reliability Analysis/Evaluation
- 832: Human Engineering
- 833: Part Selection, Analysis, Standardization, and Derating
- 834: Maintenance Engineering (see 870)
- 835: Configuration, Packaging Standardization
- 836: Design Reviews
- 837: Tolerance Analysis/Safety Margins
- 838: Design Redundancy

840: Methods of Reliability Analysis
- 841: Data Collection
- 842: Data Reduction
- 843: Data Processing
- 844: Failure Modes/Mechanisms/Analysis
- 845: Data Exchange Systems
- 846: Data Usage (Prediction, Spares, Etc.)

850: Demonstration/Measurement of Reliability
- 851: Testing Methods
- 852: Usage
- 853: Reporting, Analysis, and Evaluation

860: Field/Consumer Activity
- 861: Training and Operations
- 862: Maintenance
- 863: Logistics
- 864: Reporting/Feedback

870: Maintainability
- 871: Management
- 872: Theory
- 873: Design
- 874: Reporting, Analysis, and Evaluation
- 875: Demonstration
- 876: Reporting/Follow-Up

880: Availability
- 881: Management
- 882: Theory
- 883: Design
- 884: Reporting, Analysis, and Evaluation
- 885: Demonstration

FUNCTIONAL CLASSIFICATION

Code	Function	Description
:00:	General	
:10:	Management	Those functions which direct the activities of an organization
:20:	Production	Those functions dealing with a marketable product or service
:30:	Financial	Those functions dealing with monetary aspects of an organization
:40:	Procurement	Those functions relating to obtaining material for an organization
:50:	Sales	Those functions concerning marketing and servicing the product or service of an organization
:60:	Engineering	Those functions establishing the specific requirements and standards of actual or potential marketable product
:70:	Quality	Those functions assuring that product or service conforms to requirements and standards
:80:	Industrial relations	Those functions dealing with the personnel and community relations aspects of an organization
:90:	Management services	Those functions dealing with the systems and services employed by an organizational complex

INDUSTRY AND BUSINESS CLASSIFICATION†

:000 General or Non-Classifiable Establishments
- :000 General
- :099 Non-classifiable

:100 Agriculture, forestry, and fisheries
- :101 Commercial farms
- :102 Non-commercial farms
- :107 Agricultural services and hunting and trapping
- :108 Forestry
- :109 Fisheries

:200 Mining
- :210 Metal mining
- :211 Anthracite mining
- :212 Bituminous coal and lignite mining
- :213 Crude petroleum and natural gas
- :214 Mining and quarrying of nonmetallic minerals except fuels

:300 Contract construction

:400 Manufacturing
- :419 Ordnance and accessories
- :420 Food and kindred products
- :421 Tobacco manufactures
- :422 Textile mill products
- :423 Apparel and other finished products made from fabrics and similar materials
- :424 Lumber and wood products, except furniture
- :425 Furniture and fixtures
- :426 Paper and allied products
- :427 Printing, publishing, and allied industries
- :428 Chemicals and allied products

† Adapted from Bureau of Labor Statistics Standard Industrial Classification (SIC).

Table 6.1 ASQC Literature Classification System *(Continued)*

:429	Petroleum refining and related industries
:430	Rubber and miscellaneous plastics products
:431	Leather and leather products
:432	Stone, clay, and glass products
:433	Primary metal industries
:434	Fabricated metal products, except ordnance, machinery, and transportation equipment
:435	Machinery, except electrical
:436	Electrical machinery, equipment, and supplies
:437	Transportation equipment
:438	Professional, scientific, and controlling instruments; photographic and optical goods; watches and clocks
:439	Miscellaneous manufacturing industries
:500	Transportation, communication, electric, gas, and sanitary services
:540	Railroad transportation
:541	Local and suburban transit and interurban passenger transportation
:542	Motor freight transportation and warehousing
:544	Water transportation
:545	Transportation by air
:546	Pipe line transportation
:547	Transportation services
:548	Communication
:549	Electric, gas, and sanitary services
:600	Wholesale and retail trade
:650	Wholesale trade
:652	Retail trade—building materials, hardware, and farm equipment
:653	Retail trade—general merchandise
:654	Retail trade—food
:655	Automotive dealers and gasoline service stations
:656	Retail trade—apparel and accessories
:657	Retail trade—furniture, home furnishings, and equipment
:658	Retail trade—eating and drinking places
:659	Retail trade—miscellaneous retail stores
:700	Finance, insurance, and real estate
:760	Banking
:761	Credit agencies other than banks
:762	Security and commodity brokers, dealers, exchanges, and services
:763	Insurance carriers
:764	Insurance agents, brokers, and service
:765	Real estate
:766	Combinations of real estate, insurance, loans, law offices
:767	Holding and other investment companies
:800	Services
:870	Hotels, rooming houses, camps, and other lodging places
:872	Personal services
:873	Miscellaneous business services
:875	Automobile repair, automobile services, and garages
:876	Miscellaneous repair services
:878	Motion pictures
:879	Amusement and recreation services, except motion pictures
:880	Medical and other health services
:881	Legal services
:882	Educational services
:884	Museums, art galleries, botanical and zoological gardens
:886	Nonprofit membership organizations
:888	Private households
:889	Miscellaneous services
:900	Government
:991	Federal government
:992	State government
:993	Local government
:994	International government

concepts. However, the terms therein should form a framework for indexing technical articles at the time of publication, or data at the time of manual storage, and they can also form a framework on which to build the more rigid classification systems for machine use. (See also Sec. 6.5.1, DDC.)

6.6.4 Selection of Characteristics Groupings. Examples used in this paragraph will refer chiefly to hardware items, since they lend themselves to clearer illustrations as well as typify a major portion of reliability data. However, these principles can also be extended to the nonhardware type QC/reliability topics covered in Table 6.2 or to the organization of a group of terms as in the *Engineering Thesaurus* (Sec. 6.6.3), which are not necessarily identified by digits.

A. Major Classes. Selection of the major classifications is seldom an IR problem; a part such as shown in Table 6.2 is either a relay or a resistor, and engineering decisions need little help from the document list.

B. Subgroups Classifying Specific Properties. Below this, a part possesses many qualities each of which falls under a particular grouping of terms (often known as a dichotomy). Examples of more specific group headings might be Contact Arrangement, or Wattage Rating. This more specific type of group usually defines strictly mutually exclusive physical distinctions; a part cannot be both 1 in. and 1½ in. long.

C. Example of Correct Classifications

Major class	*Construction*	*Rating*
651	.20 = Carbon Film	.06 = Less than .1 Watt
Resistors	.40 = Composition	.08 = .10 →| .125 Watt*
	.75 = Thermal	.12 = .125 →| .25 Watt*

* Where *ranges* are shown in a table, the problem that arises is where to file the dividing-line item (0.125 watt). One approach is to use a symbol such as .10 →| .125 to indicate "including 0.10 and up to but not including 0.125," or .10 |→ .125, indicating "all values greater than the lower value and including the upper value." The latter interpretation has been used in Table 6.2.

D. Alphabetical Arrangement. Wherever possible, terms should be arranged in alphabetical sequence, even though numerical codes are employed. The advantages of this universally known system can also be further employed when it is possible to use mnemonic codes.

E. Subgroups Classifying General Properties. It may be more appropriate to select broader group headings such as Physical Characteristics, Type, and Usage, but these are more subject to the overlap problem.

The broader and more general groupings involve the problem of dealing with *functions* and *applications*. Such categorization should be avoided when possible, since it frequently involves discretion in filing data involving overlapping areas. When group headings which require such ambiguous subclassification terms are unavoidable, the preferred location should be used, and cross references ("see . . ." or "see also . . .") added at the other possible file locations. An example of the misuse of the "general" group heading follows:

Under a Valves—*Type* group heading might erroneously be shown such terms as Fuel, Check, and Flapper. These terms belong under three different groups, i.e.: Fuel under a Media Handled grouping of terms; Check under Function; and Flapper under Construction, since a particular check valve might or might not *also* be categorized as flapper, etc. The word "type" to identify a grouping is rather vague, and it can lead easily to this type of ambiguity.

Extreme familiarity with colloquial usages of terms may lead an expert to create an index quite difficult for the average user to interpret. Since it is difficult for the creator of a classification system to check his own work, a pilot run with an average user is suggested before final classification.

A general library-type filing approach, dealing with more intangible subjects, assumes that much overlap and multiple choice of equally suitable filing classifications are inevitable. In filing hardware data such is less often the case.

Table 6.2 Decimal Coding System for Part/Component Data IDEP-II Classifications

(Expandable to include larger assemblies)

Usage Example: Interpretation of Sample File Category 141.30.40.12

	Major Class	Filler*	Base*	Material* Conductor	Material* Plating
	141	.30	.40	.1	2
Boards, Printed Circuit	141				
Melamine		.30			
Glass Mat			.40		
Copper				.1	
Gold					2

* The heading and hierarchal sequence of these subclassification groups vary according to the characteristics needed for filling each type of part. They were selected for reliability/quality control–type searches, and they do not pretend to the completeness needed by purchasing or stock control for completely definitive coded standard part numbers.

General or Multitopic Documents/Data. Where none of the subclass categories apply to the report in question, the .00 category should be used to indicate "general," "miscellaneous," etc. This has slightly different meaning from "NOC - not otherwise classified." Also, use .00 to identify a report dealing with *more than one category* under the same subclass. This is a compromise with "ideal" filing preciseness to eliminate excessive cross references. For example, a report dealing with ¼-, ½ , and 1-watt resistors should be numbered under the wattage subclass as General, 651.65.07.<u>00</u>. However, a document should be cross-referenced under several categories of the subclass when these subclasses identify items of *quite* different nature (example: see Instruments) and when the item would be lost under General. After filling in all significant digits, always complete the nine-digit spaces with zeros.

Codings Which Are Repeated Identically. Several properties, such as pressure, frequency, or power rating, are applicable to a number of parts/components. In these cases, common subclassification descriptors are detailed at the end of the table, and all repetitive usages refer to this list,"(common)." Always classify data under the lowest applicable code.

Referenced MIL Specifications. The MIL Spec numbers following subclass numbers are for reference only; they indicate that the general area agrees closely with that covered by the specification but is not limited to parts which comply with that specification.

MAJOR CLASSIFICATIONS OF THE PART/COMPONENT GENERIC CODE

	Accelerometers (See 852)
025	Accumulators
027	Actuators
*051	Amplifiers (Electrical or Electronic)
*081	Antennas
082	Inactive (See 081)
085	Attaching, Methods & Materials
091	Audio Devices
101	Batteries, Non-Rechargeable
102	Batteries, Rechargeable
104	Bearings
115	Bellows
117	Brakes
121	Blowers & Fans
124	Inactive (See 511)
141	Boards. Printed Circuit
151	Inactive (See 152)
152	Capacitors, Fixed
161	Capacitors, Variable
	Choppers (See 601 Relays)
	Circuit Breakers (See 341)
170	Circuits, Evaluation of
181	Coils, Inductance, Fixed
182	Coils, Inductance, Adjustable
191	Computer & Recording Elements
201	Connectors, Electrical
232	Counters
241	Crystals
271	Delay Lines
301	Electron Tubes
303	Inactive (See 306)
*306	Environmental Simulation Equipment
307	Fasteners
321	Filters, Electrical
325	Filters, Non-Electrical
331	Finishes & Surface Treatments (Materials & Processes)
336	Fittings, Tubing & Hose
337	Fluids
338	Inactive (See 337)

* Standard packages or assemblies; contract-peculiar data will be filed by assembly part number.

Table 6.2 Decimal Coding System for Part/Component Data (*Continued*)

Code	Description
341	Fuses & Circuit Protective Devices
345	Gaskets & Seals
	Gears (See 511)
347	General Technical Data
358	Gyros
361	Hardware (Mechanical & Electro-Mechanical)
371	Inactive (See 361)
381	Inactive (See 361)
386	Heat Exchangers
391	Heaters, Electrical
404	Hose
411	Identification Devices & Methods
415	Ignition Parts & Explosives
427	Inactive (See 428)
428	Instruments
461	Light Sources, Electrical
484	Lubricants
491	Inactive (See 361)
501	Materials, Non-Metallic
502	Materials, Metallic
511	Mechanisms, Power Transmittal
520	Inactive (See 347)
525	Modular Packaging Techniques
531	Inactive (See 532)
532	Motors & Motor Generators
541	Mounts, Resilient (Shoch & Vibration Isolation)
543	Inactive
545	Optical Devices & Cameras
555	Oscillators & Signal Generators
*556	Parts Testing Techniques & Equipment
557	Power Supplies, Electrical, Rotating
*562	Power Supplies, Electrical, Non-Rotating
563	Propulsion Parts & Materials, Liquid Fuel Engines
565	Propulsion Parts & Materials, Solid Fuel Engines
575	Pumps & Hydraulic Motors
*587	Recorders (Assemblies Only)
588	Inactive (See 817)
590	Inactive (See 587)
	Receivers (See 817)
	Rectifiers (See 301 & 741-3)
*593	Refrigeration Devices
595	Regulators, Electrical
596	Inactive (See 597)
597	Regulators, Fluid
601	Relays
651	Resistors, Fixed
661	Resistors, Variable (Potentiometers or Rheostats)
675	Inactive
	Resolvers, Electrical (See 795)
741	Semiconductors, Diodes
742	Semiconductors, Transistors
743	Semiconductors, Special Types of
751	Inactive (See 741)
752	Inactive (See 742)
753	Inactive (See 743)
	Sensors (See Transducers 852)
761	Shields & Shielding
	Soldering (See 085)
771	Solenoids
781	Inactive (See 361)
791	Switches, Electrical
795	Synchros (Incl. Resolvers)
805	Tanks & Tank Parts
*807	Telemetering Components
	Thermistors (See 651)
811	Timers & Programmers
*817	Transceivers, Receivers & Transmitters
851	Inactive (See 852)
*852	Transducers
901	Inactive (See 902)
902	Transformers
915	Inactive (See 817)
	Transmitters (See 817)
920	Inactive (See 921)
921	Tubing, Casings & Sleeving
925	Valves
941	Waveguide & Microwave Plumbing
	Welding (See 085)
951	Wire, Cables & Harnesses

COMPLETE PART/COMPONENT CLASSIFICATION CODE

025 ACCUMULATORS

025.00 Service, Media Handled (Common)

025.00.00 Pressure Range (Common)

025.00.00.00 Construction

.10 Cylindrical - Pressure or Surge Regulator

.20 " - Sump

Table 6.2 Decimal Coding System for Part/Component Data *(Continued)*

025 ACCUMULATORS (Continued)
- .70 Spherical - Pressure or Surge Regulator
- .80 " - Sump

027 ACTUATORS
- 027.00 Service, Media Handled
  - .25 Electric - See Common for others
- 027.00.00 Pressure Range (Common)
- 027.00.00 Type
  - . 1 Lead Screw
  - . 2 Manual
  - . 3 Piston & Cylinder
  - . 4 Solenoid
- 027.00.00.00 Action
  - .10 Double Acting - Linear
  - .20 " " - Rotary
  - .60 Single Acting - Linear
  - .70 " " - Rotary

051 AMPLIFIERS (Electrical or Electronic)
- 051.00 Function
  - .10 Audio
  - .20 " Mixer
  - .25 Computer
  - .30 Frequency Converter
  - .35 Magnetic (Mag-Amp)
  - .40 Power
  - .50 Regulating, Voltage or Current
  - .60 Servo
  - .70 Video
  - .80 " Mixing
- 051.00.00 Frequency Range (Common)
- 051.00.00.00 Power Rating (Common)

081 ANTENNAS
- 081.00 Function
  - .10 Radar Automatic Tracking & Scanning
  - .20 " Fixed Position
  - .30 " Manually Steerable Only
  - .40 Radio, Command & Telemetering, Auto Scan & Track
  - .50 " , " " " , Fixed Position
  - .60 " , " " " , Manually Steerable Only
- 081.00.00 Frequency Range (Common)

085 ATTACHING, METHODS & MATERIALS

(Incl. Solder, Welding Rod, etc.) (For Fasteners, See 307)
- 085.00 Method (or Materials)
  - .10 Soldering (MIL-S-6872)
  - .20 Welding (For brazing, use .15)
  - .30 Binding } (Electrical junction only, See Class 501 for other)
  - .40 Wrapping }
  - .50 Bonding, Electrical (For chemical bond epoxy, use .45)
  - .60 Crimping
  - .75 Swaging
  - .80 Staking
  - .85 Taper Tab Pinning

091 AUDIO DEVICES
- 091.00 Function
  - .10 Audio Emitting
  - .20 Audio Sensing
  - .30 Audio Emitting &/or Sensing
- 091.00.00 Type
  - .10 Bell
  - .20 Buzzer
  - .30 Handset
  - .40 Headset
  - .50 Hearing Aid
  - .55 Horn
  - .60 Mircophone
  - .70 Pickup
  - .80 Siren
  - .90 Speaker
  - .95 Whistle

101 BATTERIES, NON-RECHARGEABLE
- 101.00 Type
  - .10 Dry Cell (For wet cell, use .80)
  - .20 Gas
  - .30 Nuclear (Radioactive)
  - .40 Standard Cell
  - .50 Thermal
- 101.00.00 Electrode Material
  - .65 Lead
  - .70 Magnesium-Silver Chloride
  - .73 Mercury
  - .75 Nickel-Cadmium
  - .80 Silver-Cadmium

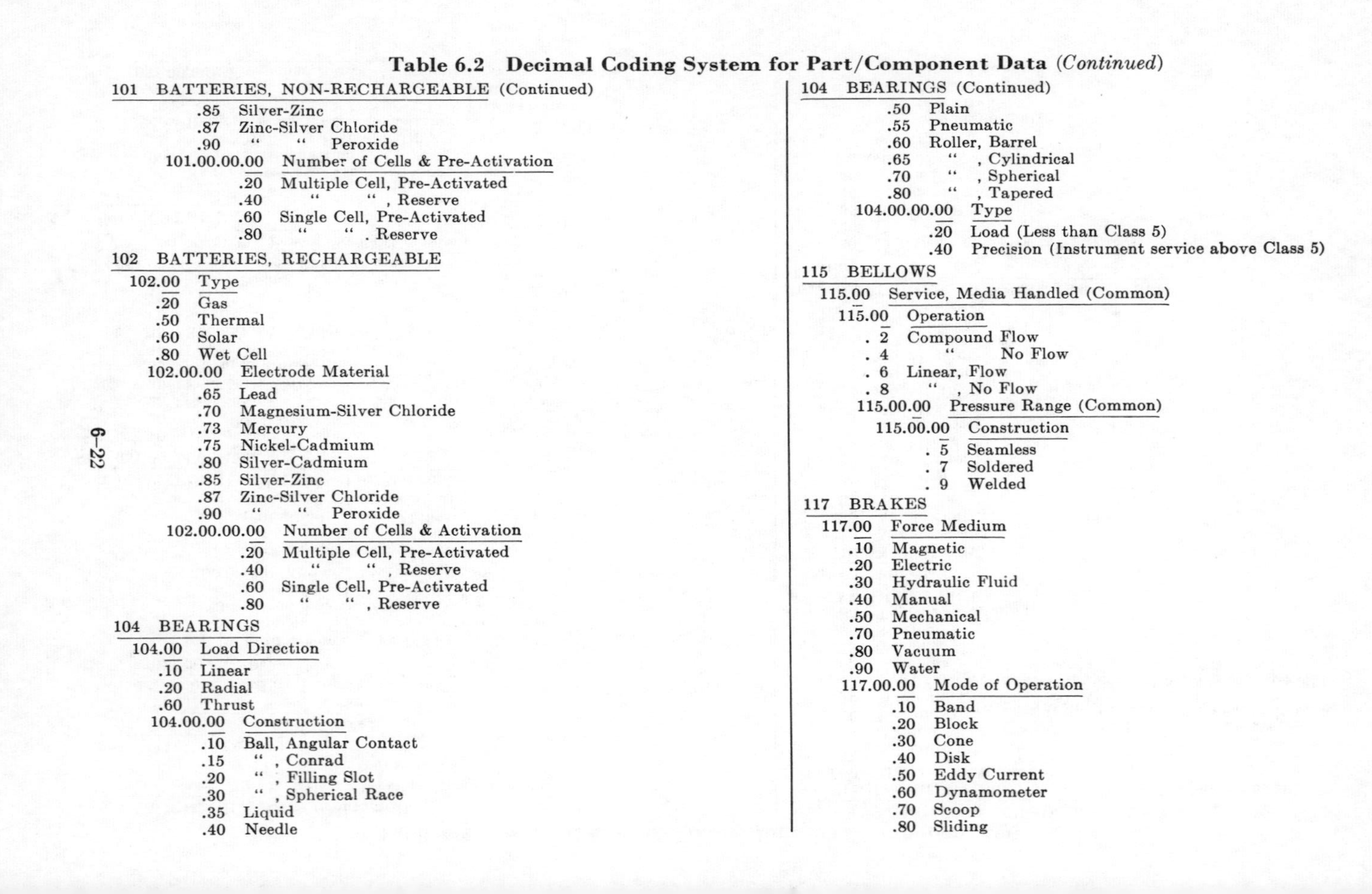

Table 6.2 Decimal Coding System for Part/Component Data *(Continued)*

101 BATTERIES, NON-RECHARGEABLE (Continued)

.85 Silver-Zinc
.87 Zinc-Silver Chloride
.90 " " Peroxide
101.00.00.00 Number of Cells & Pre-Activation
.20 Multiple Cell, Pre-Activated
.40 " " , Reserve
.60 Single Cell, Pre-Activated
.80 " " , Reserve

102 BATTERIES, RECHARGEABLE

102.00 Type
.20 Gas
.50 Thermal
.60 Solar
.80 Wet Cell
102.00.00 Electrode Material
.65 Lead
.70 Magnesium-Silver Chloride
.73 Mercury
.75 Nickel-Cadmium
.80 Silver-Cadmium
.85 Silver-Zinc
.87 Zinc-Silver Chloride
.90 " " Peroxide
102.00.00.00 Number of Cells & Activation
.20 Multiple Cell, Pre-Activated
.40 " " , Reserve
.60 Single Cell, Pre-Activated
.80 " " , Reserve

104 BEARINGS

104.00 Load Direction
.10 Linear
.20 Radial
.60 Thrust
104.00.00 Construction
.10 Ball, Angular Contact
.15 " , Conrad
.20 " , Filling Slot
.30 " , Spherical Race
.35 Liquid
.40 Needle

104 BEARINGS (Continued)

.50 Plain
.55 Pneumatic
.60 Roller, Barrel
.65 " , Cylindrical
.70 " , Spherical
.80 " , Tapered
104.00.00.00 Type
.20 Load (Less than Class 5)
.40 Precision (Instrument service above Class 5)

115 BELLOWS

115.00 Service, Media Handled (Common)
115.00 Operation
. 2 Compound Flow
. 4 " No Flow
. 6 Linear, Flow
. 8 " , No Flow
115.00.00 Pressure Range (Common)
115.00.00 Construction
. 5 Seamless
. 7 Soldered
. 9 Welded

117 BRAKES

117.00 Force Medium
.10 Magnetic
.20 Electric
.30 Hydraulic Fluid
.40 Manual
.50 Mechanical
.70 Pneumatic
.80 Vacuum
.90 Water
117.00.00 Mode of Operation
.10 Band
.20 Block
.30 Cone
.40 Disk
.50 Eddy Current
.60 Dynamometer
.70 Scoop
.80 Sliding

Table 6.2 Decimal Coding System for Part/Component Data *(Continued)*

117 BRAKES (Continued)

117.00.00.00 Material
- .10 Asbestos Fiber
- .30 Leather
- .40 Metal
- .50 Molded Fiber
- .60 Paper
- .70 Rubber
- .80 Wood

121 BLOWERS & FANS

121.00 Type
- .05 Axial
- .65 Radial

121.00.00 Power Requirement
- .05 AC
- .10 DC
- .15 Pneumatic

141 BOARDS, PRINTED CIRCUIT

(For Base Material only See 501)

141.00 Filler
- .10 Epoxy
- .20 Glass
- .30 Melamine
- .35 None
- .40 Phenolic
- .50 Polyester
- .60 Silicone

141.00.00 Base
- .10 Asbestos
- .20 Ceramic
- .30 Cotton Fabric
- .40 Glass Mat
- .45 " NOC
- .50 " Weave
- .60 Nylon
- .70 Paper

141.00.00.00 Conducting Material
- .10 Copper
- .20 Gold
- .30 Silver
- .40 Solder
- .50 Tin

141 BOARDS, PRINTED CIRCUIT (Continued)

141.00.00.00 Plating Material
- . 1 Copper
- . 2 Gold
- . 3 Silver
- . 4 Solder
- . 5 Tin

152 CAPACITORS, FIXED

152.00 Type
- .05 Air (or Gas)
- .10 Aluminum (MIL-C-62)
- .20 Ceramic (MIL-C-20 & 11015)
- .25 Film (Paper & Paper Plastic) (MIL-C-25, 14157, 26244, 27287)
- .30 Film Metalized (MIL-C-18312)
- .35 Glass & Vitreous Enamel (MIL-C-11272)
- .45 Mica (MIL-C-5, 10950, 10959)
- .75 Tantalum, Solid (MIL-C-26655)
- .76 " , Wet (MIL-C-3965)
- .77 " , Foil (MIL-C-3965)
- .85 Vacuum (See 741 for Varactors)

152.00.00 Enclosure
- .30 Dipped
- .40 Hermetically Sealed
- .50 Molded

152.00.00.00 Mounting
- .20 Chassis Mount
- .40 Lead Mount
- .60 Plug-in

161 CAPACITORS, VARIABLE

161.00 Description
- .05 Flat Plate Air Dielectric
- .20 Single Turn Ceramic Trimmer
- .25 Piston Type Tubular Trimmer, Ceramic Dielectric
- .35 " " " " , Glass Dielectric
- .45 Composition Mica
- .72 Piston Type Tubular Trimmer, Quartz Dielectric
- .75 Voltage Variable (For Semiconductors, See 741)

170 CIRCUITS, EVALUATION OF

170.00 Type
- .05 Amplifiers, Recurrent
- .06 " , Pulse

Table 6.2 Decimal Coding System for Part/Component Data *(Continued)*

170 CIRCUITS, EVALUATION OF (Continued)
- .10 Controller, Motor
- .11 " , Magnetic Modulating
- .15 Conversion, NOC
- .16 " , Heterodyning
- .20 Demodulator, Synchronous
- .21 " , Asynchronous
- .25 Inversion
- .30 Modulation
- .35 Oscillation
- .40 Power Output
- .45 Rectification
- .50 Superconduction
- .55 Switching
- .60 Timing
- .65 Triggers

181 COILS, INDUCTANCE, FIXED (MIL-C-15305)
- 181.00 Frequency Range (Common)
  - 181.00.00 Power Ratings (Common)
    - 181.00.00.00 Voltage Rating (Common)
      - 181.00.00.00 Core Material & Construction (Common)

182 COILS, INDUCTANCE, ADJUSTABLE
- 182.00 Frequency Range (Common)
  - 182.00.00 Power Ratings (Common)
    - 182.00.00.00 Voltage Rating (Common)
      - 182.00.00.00 Core Material & Construction (Common)

191 COMPUTER & RECORDING ELEMENTS

(For Delay Lines, See 271)
- 191.00 Type
  - .10 Crystal Memory
  - .15 Code Converter
  - .20 Logic Element (Inverters, "And," "Or," "Nor," etc.)
  - .40 Disc
  - .50 Drum
  - .70 Magnetic Core
  - .75 Tape
  - .80 Tube, Memory
  - .90 Magnetic Memory
  - .95 Magnetostrictive Memory
  - .98 Wire, Magnetic
  - 191.00.00 Frequency Range (Common)

201 CONNECTORS, ELECTRICAL

(For Fluid Fittings, See 336. For Terminations, Splices & "Quick Disconnects" for Single Conductor Wire Only, See 361)
- 201.00 Type
  - .20 Plug Only
  - .50 Receptacle Only
  - .85 Mated Pair (MIL-C-8384)
  - 201.00.00 Principal Application
    - .10 Flat Cable
    - .30 Hybrid RF & Power
    - .40 High Voltage (Over 240 volts)
    - .50 Low Voltage & Low Signal
    - .60 Printed Circuit
    - .70 RF (MIL-C-3608, 3650, 71, 3607, 3655)
    - .80 Umbilical
    - 201.00.00 Contacts
      - . 2 Captive
      - . 7 Removable, Solder Type
      - . 8 " , Non Solder Type
      - 201.00.00.00 Description
        - .20 AN or MS (MIL-C-5015)
        - .30 Miniature Round (MIL-C-26482, 26500)
        - .40 Phone Type (MIL-J-641)
        - .50 Micro Miniature Rectangular
        - .60 Miniature Rectangular (MIL-C-26518)
        - .70 Micro Miniature Round
        - 201.00.00.00 Significant Characteristic
          - . 1 High Temperature
          - . 2 Fireproof
          - . 3 Hermetically Sealed
          - . 4 Quick Disconnect
          - . 5 Waterproof

232 COUNTERS
- 232.00 Actuation
  - .10 Electrical
  - .25 Electronic
  - .30 Mechanical
  - 232.00.00 Function
    - .05 Angle
    - .07 Direction
    - .20 Frequency
    - .30 Impulse

Table 6.2 Decimal Coding System for Part/Component Data *(Continued)*

232 COUNTERS (Continued)

- .40 Photoelectric
- .50 Proportional
- .60 Pulse Rate
- .70 Radiation
- .80 Revolution Event
- .81 " Rate
- .90 Scintillation

241 CRYSTALS

241.00 Function
- .10 Band Pass Filtering
- .30 Frequency Control
- .60 Pickup
- .70 Time Delay

241.00.00 Material
- .20 Barium Titanate
- .40 Galena
- .50 Germanium
- .60 Quartz
- .80 Rochelle Salt
- .90 Tourmaline

271 DELAY LINES

(For Crystals only, See 241)

271.00 Type
- .10 Distributed Parameter
- .20 Lumped Parameter
- .30 Magnetostrictive
- .40 Network (Pulse Delay & Pulse Forming)
- .50 Variable (Continuous & Step)
- .60 Video

271.00 Regulation
- .1 Fixed
- .2 Variable

271.00.00 Delay Time (Microseconds)
- .10 0–.10 (Above zero, up to and including .10)
- .20 .10–1
- .30 1–10
- .40 10–100
- .50 100–1000
- .60 1000–10,000
- .70 Above 10,000

271 DELAY LINES (Continued)

271.00.00.00 Enclosure
- .10 Cased, Hermetically Sealed
- .20 " , Not Hermetically Sealed
- .30 Encapsulated

301 ELECTRON TUBES

301.00 Type
- .01 Cathode Ray, Display
- .02 Control Thyratron
- .08 Indicator
- .10 Power-Electric
- .22 Photoelectric
- .55 Special Purpose
- .59 Commutative, Beam Switching
- .60 Microwave
- .61 " , TWT & BWT
- .62 Proximity Fuse Types
- .63 Radiac
- .64 Sensing, Displaceable Element
- .66 Geiger-Muller Counter
- .70 Magnetic Field, Controlled
- .71 " " , Intensity Measurement
- .72 Pressure Measurement

301.00.00 Construction
- .10 Diode, Single
- .20 " , Multiple
- .30 Triode, Single
- .40 " , Multiple
- .50 Tetrode, Single
- .60 " , Multiple
- .70 Pentode
- .80 Dual Purpose, Unlike Sections
- .90 Six or More Elements

301.00.00.00 General Function
- .10 Receiving
- .30 Rectifying
- .40 Special Purpose
- .50 Transmitting
- .70 Voltage Regulating

306 ENVIRONMENTAL SIMULATION EQUIPMENT

306.00 Environment Simulated
- .05 Acceleration

Table 6.2 Decimal Coding System for Part/Component Data (*Continued*)

306 ENVIRONMENTAL SIMULATION EQUIPMENT (Continued)
- .10 Altitude
- .12 Combined
- .15 Humidity
- .20 Moisture
- .40 Radiation
- .45 Salt Spray
- .50 Sand & Dust
- .55 Shock
- .60 Temperature Cycling
- .65 " , Humidity Cycling
- .70 " , High
- .75 " , Low
- .80 Vibration, less than 20,000 cps
- .85 " , Ultrasonic

307 FASTENERS
- 307.00 Type
  - .05 Blind
  - .10 Bolt (See 415.00.30.00 for explosive bolts)
  - .20 Clip or Clamp (Non-electrical)
  - .30 Eyelet
  - .40 Hook
  - .50 Insert (MIL-F-18240)
  - .60 Nut
  - .65 Panel & Cowel (MIL-F-5591, 25173)
  - .70 Rivet
  - .80 Screw
  - .90 Staple
- 307.00.00.00 Material
  - .10 Aluminum
  - .20 Beryllium Copper or Copper
  - .30 Brass
  - .40 Magnesium
  - .50 Non-Metallic, NOC
  - .60 Plastic (Nylon, Teflon, etc.)
  - .70 Steel, Cad Plated
  - .80 " , Unplated, NOC
  - .90 " , Corrosion Resistant

321 FILTERS, ELECTRICAL
(For Crystal only, See 241)
- 321.00 Frequency Range (Common)

321 FILTERS, ELECTRICAL (Continued)
- 321.00.00 Type
  - .10 Band Pass
  - .12 " Rejection
  - .15 Bridged T
  - .25 Equalizer
  - .40 High Pass
  - .50 Low Pass (including power supply filter)
  - .55 Notched T
  - .60 Pulse Shaping
  - .65 Quadrature Rejection
  - .70 Single Sideband
  - .90 Vestigial Sideband
  - .95 Wave Shaping
- 321.00.00.00 Adjustability
  - .20 Fixed
  - .50 Tunable

325 FILTERS, NON-ELECTRICAL
- 325.00 Service, Media Handled (Common)
- 325.00 Principle of Operation
  - . 1 Chemical
  - . 2 Centrifugal
  - . 3 Electrostatic
  - . 4 Gravitational
  - . 5 Mechanical
- 325.00.00 Pressure Range (Common)
- 325.00.00.00 Filtering Material
  - .10 Cloth
  - .20 Ceramic
  - .30 Earth
  - .40 Fiberglass
  - .50 Paper
  - .60 Sintered Metal
  - .70 Spaced Plates
  - .80 Wire Mesh

331 FINISHES & SURFACE TREATMENTS (Materials & Processes)
- 331.00 Type
  - .10 Anodic
  - .20 Etches & Removal Processes
  - .30 Cleaning & Degreasing
  - .40 Paints & Applied Coatings

Table 6.2 Decimal Coding System for Part/Component Data (*Continued*)

331 FINISHES & SURFACE TREATMENTS (Continued)
- .60 Plating & Integral Additive Coatings
- .80 Stains & Penetrants (Incl. Iridite)
- .90 Ultrasonic Cleaning

336 FITTINGS, TUBING & HOSE

336.00 General Classes
- .20 Flared Connections
- .40 Beaded Connections
- .60 Straight Tube Connections
- .70 Swivel Connections
- .80 Quick Disconnects
- .90 Umbilical Disconnect

336.00.00 Service, Media Handled (Common)

336.00.00.00 Pressure Range (Common)

337 FLUIDS

(For Rocket Fuels, See 563.70)

337.00 Composition or Base
- .02 Air
- .05 Alcohol
- .06 Ammonia
- .10 Ammonium Hydroxide
- .12 Argon
- .15 Aniline
- .20 Benzol
- .22 Bromine
- .24 Carbon Dioxide
- .25 Fuels, General
- .26 Chlorine
- .28 Fluorine
- .30 Glycerin
- .32 Helium
- .36 Hydrogen
- .38 Krypton
- .40 Mercury
- .50 Naphthalene
- .55 Oil, Animal
- .60 Oil, Mineral
- .65 Neon
- .70 Nitrogen
- .75 Oxygen
- .80 Oil, Vegetable
- .90 Cleaning Fluids, NOC
- .95 Xenon

337 FLUIDS (Continued)

337.00.00 State of Material Under Conditions of This Test
- .10 Frozen
- .30 Standard Temperature & Pressure
- .50 Gaseous

341 FUSES & CIRCUIT PROTECTIVE DEVICES

(See also Thermal & Magnetic Switches, 791.40 & 791.50. For Semiconductor Devices, See 741 thru 743).

341.00 Type
- .10 Circuit Breaker
- .20 Instrument (MIL-F-15160)
- .30 Link (Renewable)
- .40 Plug
- .60 Screw
- .70 Sealed Cartridge

345 GASKETS & SEALS

345.00 Shape
- .30 Flange
- .40 Labyrinth
- .50 O-Ring
- .60 Flat
- .70 Split Ring
- .80 Packing
- .90 Pressure Energizing (i.e., Tubular)

345.00.00 Material
- .05 Asbestos
- .10 " , Metallic
- .15 Cork
- .20 Cotton
- .25 Fabric
- .30 Fiber
- .35 Flax
- .40 Hemp
- .45 Leather
- .50 Metal
- .55 " & Plastic
- .60 " & Rubber
- .65 Paper, Plastic
- .67 R.F., Special Composition
- .70 Rubber, Natural
- .75 " , Synthetic
- .85 Teflon
- .90 Wool

Table 6.2 Decimal Coding System for Part/Component Data *(Continued)*

347 GENERAL TECHNICAL DATA

(Any category below to be used only for documents not classifiable under a specific major classification of part, material or process)

347.00 Type
- .10 Bibliographies, Lists, Indexes
- .15 Compatibility of Materials Studies
- .20 MTTF/MTBF Surveys & Handbook Data
- .23 Design Techniques, Applications Data
- .25 Reliability, Q.A. Procedural & Organizational Studies
- .30 Multi-Component Surveys
- .35 General Test Specs (Not applicable to one class; MIL-STD-202 Type)
- .40 Rel. Eval., Prediction & Other Statistical Techniques
- .50 Test Philosophy Papers (Test Planning, Test Sequences, Sample Selection, etc.)
- .55 Parts Procurement Specs. (Procedures & related documents not primarily associated with a specific part or process)
- .60 Environmental Studies (Specs & Papers Applicable to various part types; the principal topic being environment, not the parts or simulation equipment, except Radiation)
- .65 Radiation Studies
- .70 Manufacturing Processes & Techniques
- .75 Biological Studies
- .80 Plant Facilities Studies, Discussions, Surveys, etc.
- .90 Human Factors Studies, Papers, etc. (except Safety Engrg.)
- .95 Space Mechanics; Studies & Papers
- .97 Safety Engineering

358 GYROS

358.00 Type
- .10 Azimuth
- .30 Free (Position)
- .50 Integrating
- .70 Rate
- .90 Vertical

358.00.00 Construction
- .05 Hermetically Sealed
- .10 Non-Hermetically Sealed

358.00.00.00 Drive & Size
- .10 Miniature, Air Turbine
- .20 " , Electric Motor
- .30 " , Gas Turbine
- .40 Sub Miniature, Air Turbine

358 GYROS (Continued)
- .50 Sub Miniature, Electric Motor
- .60 " " , Gas Turbine

361 HARDWARE (Mechanical & Electro-Mechanical)

(Include only documents not classifiable under a specific item heading. For Fastening Hardware that requires special tooling, such as swaging, crimping, etc., See 085)

361.00 Type
- .04 Brushes
- .08 Bus Bars
- .10 Bushings
- .12 Contacts
- .14 Cord, Thread, Twine (MIL-T-713)
- .16 Clips & Holders (Lamp & Fuse) (MIL-F-19207, 21346, MIL-L-3661)
- .45 Dials & Knobs
- .50 Drawer Parts (Handles, Slides, etc.)
- .52 Grommets & Fairleads
- .54 Slip Rings
- .56 Sockets
- .58 Hinges
- .60 Keys & Pins
- .62 Latches
- .64 Magnets
- .67 Lighting Panels (MIL-P-7788)
- .70 Insulators (JAN-I-7, 8, 9, 21)
- .74 Shims, Spacers & Washers
- .77 Springs (MIL-S-6715)
- .80 Terminal Boards & Strips
- .87 Terminations, Wire (MIL-T-7928)

386 HEAT EXCHANGERS

386.00 Service, Media Handled (Common)

386.00.00 Pressure Range (Common)

386.00.00.00 Mode of Operation
- .10 Liquid to Liquid
- .20 Liquid to Gas
- .30 Gas to Gas
- .40 Gas to Liquid

391 HEATERS, ELECTRICAL

391.00 Type
- .20 Conduction
- .30 Convection
- .50 Radiation

Table 6.2 Decimal Coding System for Part/Component Data (*Continued*)

391 HEATERS, ELECTRICAL (Continued)
- 391.00.00 Construction
  - .10 Ceramic Supported Spiral Element
  - .20 Deposited Film (NESA Glass, Windshield)
  - .30 " Metal, Ceramic or Glass Base
  - .40 Spiral, Open Element
  - .50 Static, Strip, Resistive, Enclosed
  - .60 Tubular, Liquid Immersion
  - .70 Blankets
- 391.00.00.00 Power Rating (Common)

404 HOSE

(For Tubing & Casings, See 921)
- 404.00 Service, Media Handled (Common)
- 404.00.00 Pressure Range (Common)
- 404.00.00.00 Material
  - .20 Metal, Flexible
  - .40 " , Reinforced
  - .50 " , Wire, Reinforced
  - .60 Plastic
  - .70 Rubber, Natural
  - .80 " , Synthetic

411 IDENTIFICATION DEVICES & METHODS
- 411.00 Type
  - .10 Bands
  - .25 Decals
  - .30 Etch Marking
  - .40 Ink, Stamping & Inks
  - .50 Nameplate, NOC
  - .70 Photo Etching Materials & Processes
  - .80 Silkscreen Marking Materials & Processes

415 IGNITION PARTS & EXPLOSIVES
- 415.00 Basic Purpose
  - .05 Command Destruct
  - .10 Missile Release at Launch
  - .20 Rocket Ignition-Solid
  - .40 " " -Liquid
  - .60 Stage Separation
  - .80 Provide Impulse for other secondary actions
- 415.00.00 Type
  - .20 Arm-Safe Mechanism
  - .30 Explosive Bolt

415 IGNITION PARTS & EXPLOSIVES (Continued)
  - .40 Explosive Charge, NOC
  - .50 Prima-Cord
  - .60 Squib Pyrotechnics
  - .80 " Explosives
- 415.00.00.00 Initial Actuating Signal
  - .20 Barometric Pressure
  - .40 Electrical
  - .60 Flash
  - .80 Percussion
  - .85 Impact
  - .90 Acceleration

428 INSTRUMENTS

(For Sensing Elements excluding Readout, See **852.** For Elapsed Time Indicators, See 811. For Revolutions Indicators, See 232.)
- 428.00 Media Measured
  - .01 Acceleration (linear or angular; not vibration)
  - .03 Altitude
  - .04 Attitude
  - .05 Barometric Pressure
  - .07 Capacitance
  - .09 Current (MIL-M-3823, 1725, 10304, 16034)
  - .12 Depth
  - .14 Dimension, Linear (Thickness, Diameter, Length, etc.)
  - .15 Direction or Heading (Angular measurements)
  - .18 Distance
  - .20 Ductility
  - .24 Elasticity
  - .25 Electrical, NOC
  - .27 Frequency
  - .30 Hardness
  - .33 Humidity, Absolute
  - .36 Humidity, Relative
  - .39 Impedance
  - .40 Light Intensity
  - .43 Liquid Level
  - .45 Phase Angle
  - .48 Power, Electrical
  - .49 Power, Mechanical
  - .50 Pressure, Gas
  - .51 Pressure, Liquid
  - .55 Radioactivity
  - .60 Rate of Flow (Fluids)

Table 6.2 Decimal Coding System for Part/Component Data (*Continued*)

428 INSTRUMENTS (Continued)

.63 Rate of Turn
.65 Resistance
.73 Sound
.74 Temperature
.75 Torque
.77 Vacuum
.79 Velocity (Linear or Angular)
.80 Vibration (Includes amplitude & frequency; not acceleration alone)
.85 Voltage (MIL-M-3823, 17275, 10304, 16034)
.87 Volume
.90 Wave Form

428.00.00 Action

.30 Controlling
.40 Indicating & Measuring
.50 Measuring Only
.60 Processing
.80 Testing

428.00.00.00 Readout

.10 Audio
.20 Dial
.30 Digital
.40 Plotting
.50 Recording (Tape)
.60 " , NOC
.70 Visual, NOC

461 LIGHT SOURCES, ELECTRICAL

461.00 Type

.10 Arc
.20 Fluorescent, Ultraviolet
.25 " , Visible
.30 Neon
.35 Incandescent, Instrument Lighting & Pilot
.40 " , General
.50 Lamp Assemblies, NOC
.60 Mercury Vapor
.70 Molecular, Solid State
.80 Infrared (Black Bodies)
.85 Lasers, Masers

484 LUBRICANTS

484.00 Application

.10 Extreme Cold

484 LUBRICANTS (Continued)

.15 Extreme Heat
.20 High Loading
.30 Radiation

484.00.00 Significant Use Condition

.10 High Speed
.20 Low Drag
.30 Vacuum

484.00.00.00 Composition

.10 Animal Product
.20 Compounded
.30 Fish Product
.40 Graphite
.60 Molybdenum Disulphide
.70 Petroleum Product
.80 Silicone
.90 Vegetable Product

501 MATERIALS, NON-METALLIC

(Include only documents not classifiable under a specific item heading, such as an item of hardware, a coating, finish, etc.)

501.00 Principal Constituent

.01 Adhesive (MIL-A-298, 5090)
.12 Asphalt
.16 Cork
.20 Ceramic
.32 Encapsulating (Incl. potting compound)
.35 Desiccants
.36 Fabric
.37 Edible
.44 Fiber
.58 Glass
.60 " , Fiber
.62 " , Mica (Woven or Laminated)
.64 Grain, Alundum
.66 Impregnating Compounds (When not used as a finish)
.68 Mica
.70 " , Glass Bonded
.74 Oxide, Aluminum
.76 Paper
.78 Plastic, NOC
.80 Porcelain
.82 Elastic Products, Synthetic
.84 Rubber, Natural

Table 6.2 Decimal Coding System for Part/Component Data (*Continued*)

501 MATERIALS, NON-METALLIC (Continued)

- .86 Sapphire
- .88 Silicone Compounds
- .90 Tetrafluorides

501.00.00 Final Shape

- .10 Extrusion (L-P-349, 590)
- .15 Cloth
- .20 Foil
- .30 Pellets
- .40 Rod (MIL-P-17091, 79)
- .50 Sheet (MIL-P-77, 78, 80, 997, 3115, 5425, 15037, 15047, 17721, 18177)
- .60 Tape
- .90 Powder

502 MATERIALS, METALLIC

502.00 General Type

- .10 Ferrous, NOC
- .20 " , Corrosion Resistant
- .30 Non-Ferrous

502.00.00 Basic Material (or Alloy)

- .10 Aluminum
- .15 Beryllium
- .20 Bronze
- .30 Cobalt
- .35 Columbium
- .40 Copper
- .42 Gold
- .45 Iron
- .50 Magnesium
- .55 Molybdenum
- .60 Nickel
- .65 Platinum
- .68 Silver
- .70 Steel
- .72 Tantalum
- .75 Tin
- .80 Titanium
- .85 Tungsten
- .95 Zirconium

502.00.00.00 Method of Fabrication

- .10 Cast
- .20 Drawn

502 MATERIALS, METALLIC (Continued)

- .40 Extruded
- .50 Forged
- .80 Rolled

511 MECHANISMS, POWER TRANSMITTAL

(For Servo Mechanisms & Synchros, See 795)

511.00 Type

- .10 Shafts
- .20 Gears (Incl. Gear Trains)
- .40 Couplings (Incl. Universal Joints)
- .60 Belts
- .80 Clutches
- .90 Pulleys
- .92 Chains
- .95 Cams

515 MICROELECTRONIC CIRCUITS

515.00 Function

- .10 Multivibrator (All kinds)
- .20 NOR, NAND, GATE
- .30 Shift Register
- .40 Buffer
- .50 Amplifiers (All kinds)
- .60 Adder (Including half adder)
- .70 Pulse Shaper
- .80 Switch

515.00.00 Process

- .01 Diffused Elements (Active substrate)
- .02 Diffused and Deposited Elements (Active substrate)
- .03 Thin Film Deposited Elements (Inactive substrate)
- .04 Hybrid Circuits (Any combination of above with/without discrete elements in single package)

525 MODULAR PACKAGING TECHNIQUES

(To be used only for documents where the packaging technique & not the function of the module is the predominant factor)

525.00 Enclosure

- .10 Non-Repairable
- .50 Repairable

525.00.00 Service

- .20 Shock Resistance
- .40 Extreme Temperature
- .60 Vacuum
- .80 Vibration Resistance

Table 6.2 Decimal Coding System for Part/Component Data *(Continued)*

532 MOTORS & MOTOR GENERATORS

- 532.00 Size
  - .20 Miniature
  - .40 Small Fractional HP
  - .60 From 1/4 HP to 1 HP
  - .80 Over 1 HP
- 532.00 Construction
  - . 1 Synchronous
  - . 3 Hysteresis-Synchronous
  - . 5 Universal
  - . 7 Induction
  - . 8 Permanent Magnet
  - . 9 Servo Motor
- 532.00.00 Voltage Rating (Common)
- 532.00.00.00 Frequency Range (Common)

541 MOUNTS, RESILIENT (Shock & Vibration Isolation)

- 541.00 Design Application
  - .10 Combination Shock & Vibration
  - .20 Shock only
  - .30 Vibration only
- 541.00.00 Principal Material
  - .10 Combination Metal & Rubber
  - .20 Metal Spring
  - .30 Rubber or Synthetic Rubber
  - .50 Wire Mesh
- 541.00.00.00 Damping Method
  - .10 Frictional
  - .20 Hydraulic, Fixed
  - .30 " , Variable
  - .40 Pneumatic, Fixed
  - .50 " , Variable
  - .60 Undamped

545 OPTICAL DEVICES & CAMERAS

- 545.00 Type
  - .05 Cameras, Still & Motion
  - .06 Projectors
  - .07 Film Processing Devices
  - .10 Comparators
  - .20 Microscopes
  - .30 Periscopes

545 OPTICAL DEVICES & CAMERAS (Continued)

  - .50 Range Finders
  - .60 Sextants
  - .70 Telescopes
  - .80 Television Cameras
- 545.00.00 Optical Element Used
  - .10 Crystal
  - .20 Lens
  - .30 Prism
  - .40 Mirror
  - .50 Porro Prism

555 OSCILLATORS & SIGNAL GENERATORS

- 555.00 Frequency Range (Common)
- 555.00.00 Type
  - .10 Electronic
  - .30 Resonant Reed
  - .50 Tuning Fork

556 PARTS TESTING TECHNIQUES & EQUIPMENT

(This category is for complex production inspection and parts checkout devices & techniques not normally classifiable as "instruments." The sub-categories listed below are not intended to cover the subject, but are intended as a guide to some of the possible types of documents to be included.)

- 556.00 Type
  - .10 Ultrasonic Analysis
  - .20 X-Ray Analysis
  - .30 Infrared Analysis
  - .40 Magnetized Particles
  - .50 Radioactive Tracers
  - .60 Semiconductor Testing
  - .70 Dielectric Testing

557 POWER SUPPLIES, ELECTRICAL, ROTATING

- 557.00 Voltage Rating (Common)
- 557.00 Type
  - . 1 Alternators
  - . 2 Converters
  - . 3 Dynamotors
  - . 4 Generators
  - . 5 Inverters
- 557.00.00 Frequency Range (Common)
- 557.00.00.00 Power Rating (Common)

Table 6.2 Decimal Coding System for Part/Component Data *(Continued)*

562 POWER SUPPLIES, ELECTRICAL, NON-ROTATING
(For Batteries, See 101 & 102)
562.00 Power Rating (Common)
562.00.00 Output Frequency Range (Common)
562.00.00.00 Voltage Rating (Common)

563 PROPULSION PARTS & MATERIALS, LIQUID FUEL ENGINES
563.00 Type
.25 Injector
.35 Manifold
.45 " , Vernier, Exhaust
.55 Nozzle, Main
.60 " , Vernier or Swivel
.70 Propellants

565 PROPULSION PARTS & MATERIALS, SOLID FUEL ENGINES
565.00 Type
.20 Attach Fitting
.40 Engine Case, Shell
.60 Nozzle
.70 Propellants

575 PUMPS & HYDRAULIC MOTORS
575.00 Operating Principle
.10 Axial
.20 Centrifugal
.30 Diaphragm
.40 Gear
.50 Reciprocating
.60 Radial
.70 Vane Type, NOC
575.00.00 Service, Media Handled (Common)
575.00.00.00 Pressure Range (Common)

587 RECORDERS (Assemblies Only)
(For Recording Elements, See 191)
587.00 Type
.20 Disc, Grooved
.25 " , Magnetic
.30 Drum
.40 Facsimile
.45 Film, Photographic
.55 Tape, Multiple Channel, Magnetic
.60 " , Single Channel, Magnetic

587 RECORDERS (Assemblies Only) (Continued)
.70 Tape, Perforated
.80 " , Plastic Wrinkle (Thermally Deformed)
.90 Wire, Magnetic

593 REFRIGERATION DEVICES
593.00 Principle of Operation
.10 Chemical Absorption
.20 Gas Expansion
.30 Physical Absorption
.40 Thermo-Electric
593.00.00 Type
.10 Closed Cycle
.20 Open Ended

595 REGULATORS, ELECTRICAL
(For Semiconductors, See 741–743. For Gas Tube, See 301)
595.00 Regulated Quantity
.10 AC Voltage
.20 DC "
.30 Current
595.00.00 Principle of Operation
.10 Carbon Pile
.20 Electro-Mech
.30 Electronic
.50 Saturable Transformer
.60 Static
.70 Vibrating Contact

597 REGULATORS, FLUID
597.00 Property Regulated
.10 Flow
.60 Pressure
.80 Temperature
.90 Volume or Level
597.00 Construction
. 1 Ball
. 2 Cone
. 3 Diaphragm
. 4 Dome Loaded
. 5 Internal Loaded
. 6 Poppet
. 7 Sleeve

Table 6.2 Decimal Coding System for Part/Component Data *(Continued)*

597 REGULATORS, FLUID (Continued)
- 597.00.00 Service, Media Handled (Common)
  - 597.00.00.00 Pressure Range (Common)

601 RELAYS
(For Solid State Devices which function as Relays, See 741–743)
- 601.00 Contact Ratings (Resistive Load)
  - .10 Less than ½ amp
  - .20 ½–1
  - .30 1–5
  - .40 5–10
  - .50 10–20
  - .60 20–50
  - .70 Over 50
- 601.00 Construction
  - . 1 Enclosed (Not Hermetically Sealed)
  - . 4 Hermetically Sealed (MIL-R-25018, 5757)
  - . 7 Unenclosed
- 601.00.00 Coil Voltage
  - .10 1/10 Volt and under
  - .10 1/10 Volt to 1 Volt (Over .1 volt and up to and including 1 volt)
  - .30 1–30 VDC
  - .40 30–109 VDC
  - .50 109–240 VDC
  - .60 240–1000 VDC
  - .70 1–30 VAC
  - .80 30–109 VAC
  - .90 109–240 VAC
- 601.00.00 Type
  - . 1 Chopper
  - . 2 D'Arsonval
  - . 3 Frequency Sensitive (e.g., Resonant Reed)
  - . 4 Latching
  - . 5 Sensitive
  - . 6 Power
  - . 7 Stepping
  - . 8 Time Delay
  - . 9 Polarized
- 601.00.00.00 Contact Arrangement (Common)

651 RESISTORS, FIXED
- 651.00 Material or Type
  - .20 Carbon Film (MIL-R-10509)
  - .40 Composition (MIL-R-11)

651 RESISTORS, FIXED (Continued)
  - .50 Metal Film, Power (MIL-R-11804)
  - .55 " " , Precision (MIL-R-10509)
  - .65 Wirewound (MIL-R-93, 9444)
  - .66 " , Power (JAN-R-184 MIL-R-26, 18546)
  - .75 Thermal (Thermistor)
  - .80 Voltage (Varistor)
- 651.00.00 Enclosure
  - .03 Ceramic
  - .05 Glass
  - .07 Metal Encased
  - .08 Molded or Dipped
  - .09 Vitreous Enamel
- 651.00.00.00 Power Rating (Common)

661 RESISTORS, VARIABLE (Potentiometers or Rheostats)
- 661.00 Material
  - .20 Carbon Film
  - .40 Composition
  - .45 Metal Film
  - .60 Plastic, Conductive, NOC
  - .61 " " , Precision ≤ 1%
  - .70 Wirewound, NOC
  - .75 " , Precision ≤ 1%
- 661.00.00 External Adjustment
  - .30 Multiturn
  - .50 Single Turn
- 661.00.00 Output Function
  - . 1 Linear
  - . 2 Non-Linear, Non-Trigonometric
  - . 3 Trigonometric
  - . 4 Logarithmic
- 661.00.00.00 Power Rating (Common)

741 SEMICONDUCTORS, DIODES
- 741.00 Function
  - .10 General Purpose
  - .20 Fast Switching (Computer)
  - .30 Rectifier
  - .40 Zener (Not Voltage Reference)
  - .50 Voltage Reference (Not Zener)
  - .60 Photo Sensitive
  - .70 Control Rectifier (SCR)
  - .80 Tunnel

Table 6.2 Decimal Coding System for Part/Component Data *(Continued)*

741 SEMICONDUCTORS, DIODES (Continued)

- 741.00.00 Material
  - .10 Copper
  - .20 Germanium
  - .30 Selenium
  - .40 Silicon
  - .50 Gallium Arsenide

742 SEMICONDUCTORS, TRANSISTORS

(For Multiple Unit Transistors, See 743)

- 742.00 Function
  - .10 General Purpose (& less than 1 watt)
  - .20 Power (larger than 1 watt)
  - .30 Fast Switching ($F_t \geq 100$ mc.)
  - .40 Unijunction
  - .50 Field Effect
  - .60 Switching, NOC
  - .70 Oscillator
- 742.00.00 Materials & Polarity
  - .10 Germanium, NPN
  - .20 " , PNP
  - .30 Silicon, NPN
  - .40 " , PNP

743 SEMICONDUCTORS, SPECIAL TYPES OF

(Packaged multi-element units or solid state developments not classifiable under 741 or 742)

- 743.00 Type or Function
  - .10 Amplifier
  - .20 Bridge
  - .30 Circuit Breaker
  - .40 Flip-Flop
  - .50 Limiter
  - .80 Switch

761 SHIELDS & SHIELDING

- 761.00 Energy Shielded From
  - .10 Electromagnetic Radiation
  - .20 Magnetic Flux
  - .30 Nuclear Radiation
  - .40 Light
  - .50 Heat
- 761.00.00 Item Shielded
  - .10 Coil
  - .30 Electron Tube

761 SHIELDS & SHIELDING (Continued)

  - .80 Transformer
  - .90 Wire & Cable

771 SOLENOIDS

- 771.00 Motion & Electrical Power
  - .10 AC Linear
  - .20 AC Rotary
  - .30 DC Linear
  - .40 DC Rotary
- 771.00.00 Power Rating (Common)
- 771.00.00.00 Voltage Rating (Common)

791 SWITCHES, ELECTRICAL

- 791.00 Method of Actuation
  - .10 Fluid Pressure
  - .20 Gravity
  - .30 Inertia
  - .40 Temperature
  - .50 Magnetic
  - .70 Manual
  - .80 Mechanical, NOC
  - .90 Squib
- 791.00.00 Type
  - .05 Coaxial (MIL-S-3928)
  - .10 Crossbar
  - .15 Float
  - .20 Jack
  - .25 Key, Telegraph
  - .30 Knife
  - .35 Mercury
  - .40 Plunger
  - .50 Push Button (MIL-S-6743)
  - .55 Rotary (MIL-S-3786)
  - .65 Slide
  - .70 Telephone Lever
  - .75 Toggle (MIL-S-3950)
- 791.00.00.00 Contact Arrangement (Common)

795 SYNCHROS (Including Resolvers)

- 795.00 Function
  - .10 Control Differential
  - .15 " Resolvers
  - .20 " Transformer
  - .30 " Transmitter

Table 6.2 Decimal Coding System for Part/Component Data (*Continued*)

795 SYNCHROS (Including Resolvers) (Continued)

.60 Torque Differential
.70 " Receiver
.75 " Resolvers
.80 " Transmitter
795.00.00 Case Size
.10 BUORD Size
.20 Special Non BUORD
795.00.00.00 Output Voltage (Common)
795.00.00.00 Output Frequency
. 5 60 cps
. 6 400 cps
. 8 Other

805 TANKS & TANK PARTS

805.00 Service, Media Handled (Common)
805.00.00 Material & Use
.10 Fiberglass (Airborne)
.15 " (Ground or Shipboard)
.20 Metal (Airborne)
.25 " (Ground or Shipboard)
.30 Plastic (Airborne)
.35 " (Ground or Shipboard)
.40 Synthetic Rubber (Airborne)
.45 " " (Ground or Shipboard)
805.00.00.00 Pressure Range (Common)

807 TELEMETERING COMPONENTS

811 TIMERS & PROGRAMMERS

811.00 Type
.10 Event Sequence Programmer
.30 " " " (Auto Reset)
.50 " " Time Recorder
811.00.00 Method of Operation or Receiving Signal
.10 Inertia
.20 Photoelectric
.30 Clockwork (Escapement)
.40 Electrical (Synchronous)
.50 Electronic
811.00.00.00 Readout or Output
.10 Signal
.20 Binary

811 TIMERS & PROGRAMMERS (Continued)

.30 Digital
.40 Printout
.50 Plot
.60 Dial
.70 Audio

817 TRANSCEIVERS, RECEIVERS & TRANSMITTERS

817.00 Type
.10 Radar, Airborne
.20 " , Ground or Shipboard
.40 Radio & Communication, Airborne
.60 " " " , Ground
.80 " " " , Shipboard
817.00.00 Frequency Range (Common)
817.00.00.00 Power Rating (Common)

852 TRANSDUCERS

(This category includes only the sensing or actuation head portions of devices which change energy from one form to another. Sensing devices which include readout provisions are considered Instruments, and will be found under 428)

852.00 Media Sensed
.01 Acceleration (linear or angular; not vibration)
.03 Altitude
.04 Attitude
.05 Barometric Pressure
.07 Capacitance
.09 Current
.12 Depth
.14 Dimension, Linear (Thickness, Diameter, Length, Etc.)
.15 Direction or Heading (Angular measurement)
.18 Distance
.20 Ductility
.24 Elasticity
.27 Frequency
.30 Hardness
.33 Humidity, Absolute
.36 " , Relative
.39 Impedance
.40 Light Intensity
.43 Liquid Level
.45 Phase Angle
.48 Power, Electrical
.49 " , Mechanical

Table 6.2 Decimal Coding System for Part/Component Data *(Continued)*

852 TRANSDUCERS (Continued)

- .50 Pressure, Gas
- .51 " , Liquid
- .55 Radioactivity
- .60 Rate of Flow (Fluids)
- .63 " " Turn
- .65 Resistance
- .73 Sound
- .74 Temperature
- .75 Torque
- .77 Vacuum
- .79 Velocity (Linear & Angular)
- .80 Vibration (Includes amplitude & frequency; not acceleration alone)
- .85 Voltage
- .87 Volume
- .90 Wave Form

852.00.00 Output

- .09 Current
- .40 Light
- .48 Power, Electrical
- .49 " , Mechanical (Force)
- .73 Sound
- .74 Temperature
- .75 Torque
- .77 Vacuum
- .79 Velocity (Linear or Angular)
- .80 Vibration
- .85 Voltage

902 TRANSFORMERS

902.00 Type or Function

- .10 Carrier (Wide Frequency Applications)
- .15 Computer Applications, NOC
- .20 Driver
- .25 Hybrid
- .30 Input
- .35 Interstage
- .40 Modulation
- .45 Oscillator (Sine or Square Wave)
- .50 Output
- .55 Pulse, NOC
- .60 Power, Auto Transformer
- .62 " , Filament (only)
- .64 " , Isolation

902 TRANSFORMERS (Continued)

- .66 Power, General Purpose
- .68 " , NOC
- .70 Saturable Core
- .70 Saturable Core
- .75 Sonar Applications

902.00.00 Construction

- .10 Enclosed (Not Hermetically Sealed)
- .20 Hermetically Sealed
- .30 Unenclosed

902.00.00.00 Frequency Range (Common)

921 TUBING, CASINGS & SLEEVING

(For Hose, See 404)

921.00 Material

- .10 Fiber
- .15 Aluminum
- .20 Glass
- .25 Copper
- .30 Paper
- .35 Steel, NOC
- .40 Plastic
- .45 Steel, Corrosion Resistant
- .50 Porcelain
- .60 Vinyl

921.00.00 Construction

- .10 Flexible
- .20 Heavy Wall
- .30 Thin Wall
- .40 Standard Wall
- .50 Zipper Wall
- .60 Wrapped & Sintered (Bundyweld)

925 VALVES

925.00 Service, Media Handled (Common)

925.00.00 Pressure Range (Common)

925.00.00 Function

- . 1 Check
- . 2 Multi-function
- . 3 Relief
- . 4 Servo
- . 5 Shutoff

Table 6.2 Decimal Coding System for Part/Component Data (*Continued*)

925 VALVES (Continued)

. 6 3-way Selector
. 7 4-way Selector

925.00.00.00 Principle of Operation

.10 Ball
.20 Butterfly
.30 Flapper
.40 Poppet
.50 Sleeve
.60 Slide or Gate
.70 Spool
.80 Globe
.90 Needle

925.00.00.00 Actuation or Control

. 4 Manual
. 5 Motor
. 6 Pilot
. 7 Pressure
. 8 Pyrotechnic or Explosive
. 9 Solenoid

941 WAVEGUIDE & MICROWAVE PLUMBING

941.00 Shape or Function

.02 Attenuator
.10 Bend
.20 Cavity
.25 Circulator
.30 Coupler
.32 Detectors and Detector Mounts
.35 Divider
.40 Flange
.50 Horn
.55 Isolator
.60 Joint, Fixed
.70 " , Rotating
.80 Junction, NOC
.85 Mixer
.90 Phase Shifter
.95 Terminations and Loads

941.00.00 Construction

.10 Coaxial
.30 Flexible
.70 Rigid

941 WAVEGUIDE & MICROWAVE PLUMBING (Continued)

941.00.00.00 Frequency Range

.01 Broadband
.02 Below 0.7 Gc
.03 0.7 |→ 2.0 Gc
.04 2.0 |→ 4.0 Gc
.05 4.0 |→ 8.0 Gc
.06 8.0 |→ 12.4 Gc
.07 12.4 |→ 18.0 Gc
.08 18.0 |→ 26.5 Gc
.09 26.5 |→ 40.0 Gc
.10 40. → 63 Gc
.11 63. → 86 Gc
.12 86. → 110 Gc

951 WIRE, CABLES & HARNESSES

(For Magnetic Recording Wire, See 191)

951.00 Strand Material

.02 Aluminum
.04 " Alloy
.08 Beryllium Copper
.12 Chrome-Nickel Alloy
.14 Constantan
.16 Copper
.18 " Alloy & Bronze
.28 Gold
.34 Iron
.38 Magnesium
.40 " Alloy
.42 Manganese
.44 Molybdenum
.46 Monel
.50 Nickel
.58 Platinum
.64 Rhodium
.70 Silver
.74 Steel, NOC
.75 " , Corrosion Resistant
.76 " , Copper Clad
.84 Tantalum
.88 Titanium
.92 Tungsten
.96 Zirconium

Table 6.2 Decimal Coding System for Part/Component Data *(Continued)*

951 WIRE, CABLES & HARNESSES (Continued)

951.00.00 Major Insulation Material

.10 Ceramic
.20 Cotton
.30 Fiberglass
.40 Mica
.45 Nylon
.50 Paper
.55 Plastic, NOC
.60 Quartz
.65 Rubber
.70 Silk
.75 Varnish
.80 Vinyl
.85 Tetrafluoride (Teflon, Kel-F)
.90 Uninsulated

951.00.00.00 Construction

.10 Solid
.20 Multiple Strand
.30 " " -Cabled Conductors
.40 " " -Harnessed, Molded
.50 " " - " , Sleeved
.60 Printed, Flat Conductor
.70 RF

951.00.00.00 Shielding

. 1 Shielded
. 2 Unshielded

COMMON SUBCLASSIFICATION DESCRIPTORS

SERVICE, MEDIA HANDLED

.10 Cryogenic Fluids
.20 Fuel, Exotic
.30 Fuel, Hydrocarbon
.40 Gas, Hot
.50 Gas, NOC
.60 Hydraulic Fluid
.70 Oxidizers, NOC
.80 Pneumatic
.90 Oil

PRESSURE RANGE

.10 0–25 mmHg
.20 25–35 mmHg
.30 35–225 mmHg (70,000′–30,000′)

PRESSURE RANGE (Continued)

.40 225–760 mmHg (30,000′–Sea Level)
.50 S. L. only
.60 0–100 psig
.70 100–1000 psig
.80 1000–5000 psig
.90 Over 5000 psig

POWER RATING

.06 Less than .1 Watt
.08 .10–.125 Watt
.12 .125–.25 Watt
.14 .25–.50 Watt
.16 .50–1 Watt
.20 1–2 Watts
.30 2–10 Watts
.40 10–100 Watts
.50 100–1,000 Watts
.60 1–10 KW
.70 10–100 KW
.80 100–1,000 KW
.90 Over 1 Megawatt

CONTACT ARRANGEMENT

.05 SPST (NC)
.10 SPST (NO)
.12 SPDT
.20 DPST (NC)
.25 DPST (NO)
.30 DPDT
.40 3PST (NC)
.45 3PST (NO)
.50 3PDT
.60 4 PST (NC)
.65 4PST (NO)
.70 4PDT
.80 6PST (NC)
.85 6PST (NO)
.90 6PDT
.95 Multiple Pole

VOLTAGE RATING

.10 Under 1 MV
.20 1–10 MV
.30 10 MV-1 Volt
.40 1–30 Volts

Table 6.2 Decimal Coding System for Part/Component Data *(Continued)*

VOLTAGE RATING (Continued)				
.50	30–109 Volts			
.60	109–240 Volts			
.70	240–1,000 Volts			
.80	1–10 KV			
.90	Over 10 KV			
CORE MATERIAL & CONSTRUCTION				
. 1	Air, Encapsulated			
. 2	" , Hermetically Sealed			
. 3	" , NOC			
. 4	Diamagnetic, Encapsulated			
. 5	" , Hermetically Sealed			
. 6	" , NOC			
. 7	Ferromagnetic, Encapsulated			
. 8	" , Hermetically Sealed			
. 9	" , NOC			
FREQUENCY RANGE				
.02	D.C.			
FREQUENCY RANGE (Continued)				
.05	60 Cycles Only			
.06	400 Cycles Only			
.07	0–3 Kc	Audio		Freq.
.10	3–30 Kc	"		"
.20	30–300 Kc	Low		"
.30	300–3,000 Kc	Medium		"
.40	3–30 Mc	High		"
.50	30–300 Mc	Very	High	"
.60	300–3,000 Mc	Ultra	"	"
.70	3,000–30,000 Mc	Super	"	"
.80	Over 30,000 Mc	Extremely	"	"

Note:

1. NOC = Not Otherwise Classified
2. Range: The Range is defined to be greater than the lower value, up to and including the larger value. Sample: Voltage Rating 1–10 mv. Any voltage greater than 1 mv, up to and including 10 mv.

6.6.5 Classifications for Component/Part/Hardware/Materials. Table 6.2 is intended for use in classifying data on component *parts*, although sufficient room has been allowed to interfile data on larger assemblies. However, since major assembly breakdowns differ completely for each system and since the actual drawing numbers frequently constitute a convenient and more specific breakdown, drawing numbers often form a preferable filing basis for strictly *program-oriented* data. There is no relationship between the general headings of Table 6.2 and such drawing-number files, although the headings should be fairly compatible with standard parts breakdowns.

Because Table 6.2 was developed for missile/space use, it is heavily electronic-oriented; but since 99^3 breakdowns under *each* three-digit major classification are possible, it should be quite simple to utilize the general principles involved and by insertions and expansion, to cover the particular topical needs of any industry from ice cream to furniture.

Materials and processes are coded in only sufficient detail to illustrate a possible arrangement for such topics, which would be integrated into the parts classification scheme, but which could also be greatly expanded for M&P specialist's use. For example, a materials engineer would require much subdivision under 502.00.10 Aluminum.

6.6.6 Importance of Hierarchical Arrangement in Manual Retrieval Systems. The sequence of a group of terms in the generation breakdown of Table 6.2 automatically places each heading at a particular rank or position in accordance with the conclusions of the devisers of the table. When their decisions *coincide* with those of the user, a manual search is easy and productive. However, when the users selection of terms or the devisers' sequence *poorly* reflects anticipation of search criteria, the manual search becomes cumbersome. Although machine systems can search according to the eighth and ninth digits almost as quickly as by the first and second, the search-time differential is more significant when the search is manual. Therefore, when devising a manual system, the *relative* position in the file-and-search sequence should be given considerable study. For example, in Table 6.2, under 601 Relays, Contact Ratings is placed ahead of Type. If most searches are for an *x-amp* relay, this is good; but if most searches are for an *xyz-type* relay, the sequence should be changed.

6.7 STRUCTURING FOR STORAGE IN AUTOMATED SYSTEMS

As mentioned in Sec. 6.6, the distinction between automated- and manual-system requirements is usually one of degree; the computer or punch-card sorter, being completely literal-minded, is a much stricter taskmaster over ambiguous expressions, almost identical punctuation, variations in capitalization and spelling, random choice of synonyms, etc. To a computer "the U.S. Air Force" bears no relation to "the US Air Force" or "the United States Air Force," unless it is told in advance to accept these forms as identical in meaning. Likewise, P2V-2 and the P2V2 represent entirely different projects to a computer.

For this reason, extensive assignment of *numerical codes* is necessary, although at times it is only necessary for the codes to be known and used by the punch-card or tape creator.

Numerical codes, however, often serve to clarify and standardize the engineer's specific meaning in the process of supplying a uniform numeric representation. Often, the reduction of general terminology to a series of numerically identified codes reveals much fuzzy thinking and varying interpretations of the conditions implied by the words. For example, the phrases "temperature shock," "thermal shock," "temperature cycling," or "hi and low temperature exposure" can be used loosely when accompanied by amplifying descriptive material. However, when a single numerical code must be selected to identify each particular exposure, the meaning of each phrase must be rigorously defined. The specialist, in programming for automation, cannot function properly unless the engineering specialist in reliability presents him with an organized input with consistency of meaning. In addition, uniformity in term usage is of immediate utility to assist any manual search or engineering decision when it is necessary to evaluate data from several sources comparatively.

At risk of some dissention from users of the more sophisticated data systems, it is emphasized that a system can retrieve (or manipulate) only by the criteria fed into it. Although extremely complex and generally impractical machine logic will enable a computer to go a little farther to remedy loose input expressions, this ingenuity and time are usually better spent on planning and enforcing the use of precise input terminology and codification.

6.7.1 Glossary of Terms for Environmental Exposures. The terms considered in this paragraph and listed in Table 6.3 describe the external forces applied to a part either before, during, or after the measurement or assessment of the output, which will be described in Sec. 6.7.3. These environmental forces are considered to include such exposures as thermal, bacteriological, chemical, electrical, mechanical, nuclear, and electromagnetic.

The listing in Table 6.3 is suggested as a base for a coding which can be expanded to any desired degree of detail under any of the respective headings by adding letter suffixes, although in most cases the detail breakdowns shown will more than suffice. The environmental tabulation shown in Fig. 6.2 illustrates the use of certain frequently used headings in a format designed to fit within a 132-column printout, allowing each field sufficient width to express the specification level or intensity of exposure in actual digits.

Many variations could be devised to "code" these levels in various bands or ranges. However, when the quantities to be expressed are actually numerical and ample space exists to express them without requiring recourse to a table look-up, it is preferable to employ the minimum amount of coding which is necessary to identify the respective fields and assure uniform interpretation.

In the case of those environments which are *not* numerically variable in a simple manner, it is more practical to utilize coding symbols plus a separate table. Although Salt Spray could be detailed as a series of intensity and duration values, it is more convenient to utilize the pregrouped combination of levels and conditions defined by the Military Specifications. Table 6.4, therefore, proposes coding which covers the more common specifications and illustrates a principle which could be expanded to cover other, similar situations.

Note that these specifications generally cover test environmental levels; to a lesser extent the test methods, procedures, sequences, measuring devices, and techniques; and in only a few cases the performance criteria for success/failure. The latter are more often expressed in a program-oriented requirement, and as such are not easily adapted to a general-purpose tabulation such as this. The specifications listed in Table 6.4 could be used to elaborate on almost *all* the environments tabulated in Fig. 6.2, but where digital values are more expressive, they have been chosen for printout. The format can be adapted to expanded specification definition on all environments, if desired, by using two lines of printout for each entry or by varying the environments selected, the columns assigned, etc. Further expansion could allow inclusion of test *results*, etc.

Table 6.5 is included to facilitate brief and uniform reference to specification source, where required. It can be expanded to include various categories of in-house documents, and it is suggested only as a prototype for adaptation to an individual, specialized automation program for some particular type of test data.

6.7.2 Single-symbol Environmental Coding. Table 6.3*A* has been included for those uses where only a general signpost toward existing data is desired or allowed by space considerations. The same words as in Table 6.3 have been used, but there is no correlation between the arrangements. This is because the mnemonic-term selection and alphabetical arrangement of the symbols in Table 6.3*A* allow a frequent user to remember most of this table quite easily, whereas the amount of detail in Table 6.3 renders memory relatively unimportant.

CAPACITORS, FIXED, TANTALUM, SOLID, HERMETICALLY SEALED, CHASSIS MOUNT		REFERENCED MIL SPECS/STDS: (A) MIL-C-3965B (B) MIL-STD-202B (C) (D) (E) (F) (G)	ANALYSIS NO. DATE: DAY / MO. / YR.

FIELD NO 1A 1B 2A 2B 3A 4A 4B 5A 6A 7A 7B 8A 8B 8C 9A 9B 10A 10B 10C 10D 10E 10F 10G 10H 10I 11A 12A 13A 14A 15A 16A 17A 18

CODING FOR FIELD 9B — TYPE(S) OF RADIATION: NUCLEAR: A - FAST, B - SLOW, C - THERMAL; ELECTROMAG: D - GAMMA, E - X-RAY; CHG PART: F - PROTON, G - ELECT., H - MISC.

OR LINE ENTRY	TEMPERATURE OPER/STOR. LO -°C	TEMPERATURE OPER/STOR. HI +°C	TEMPERATURE CYCLE/SHOCK LO -°C	TEMPERATURE CYCLE/SHOCK HI +°C	ALTITUDE FT X 10³	VIBRATION MECHANICAL SINUSOID MAX FREQ CPS	VIBRATION INTENSITY RMS G	VIBRATION RAND G²/CPS	VIBRATION ACOUSTIC DB	LIN ACCEL MAX INTENSITY G	LIN ACCEL DURATION SEC	SHOCK MAX INTENSITY G	SHOCK DURATION-MSEC	SHOCK TOTAL NO. BLOWS	RADIATION TOTAL ABS RADS x 10ⁿ	RADIATION TYPE - SEE REMARKS	LIFE TEST OPER TIME PER PC HR/CYCL	SAMPLE SIZE NO PC	FAIL PER 10⁶ PART HOURS	°C EXTERNAL CONTL TEMP	% PRESSURE	% CURRENT	% FREQ	% POWER	% VOLTS	SALT SPRAY	MOISTURE RESIST	HUMIDITY	SAND-DUST	IMMERSION/LEAK	RADIO INTERFER	EXPLOSION/BURST	TOTAL NO. PARTS TESTED	REMARKS REFER TO FIELD NO.	REMARKS
A			54C	74C	35	2,000	30								n		500H	23		+ 85					100			A					60	8C	RANDOM FALL 6 FT ON CONCRETE
B			18C	70C													2,000H	57							100								57	101	50% + 75% RATED VOLTAGE APPLIED
C	55	125																									A						15	12A	D-C LEAKAGE LESS 150% INITIAL
D	55	125																							116								15		
E	55	125																									A						24	12A	PARA 5.3.1 OF TEST PROCED #9
F	55	125																									A						24	12A	PARA 5.3.1 OF TEST PROCED #9
G			54C	37C																													17	2A2B	-8CY H1 TEMP INCREMENTS OF 15C
H	55	125			80					50	60	5	011	48			2,000H	12		+125							Z						15	12A	MIL-C-3965 F3-DC LEAKAGE CAPDF
I						2,000	60																										3	4B	VIB IN TWO AXES OF 35G AND 50G
J	55	125																									A						15		
K	55					2,000	15										100H	10		+ 65													25	1A	MIL-C-3965B
L	65		25C	85C																								A					10	1A	24 HOURS CYCLED 95% HUMIDITY

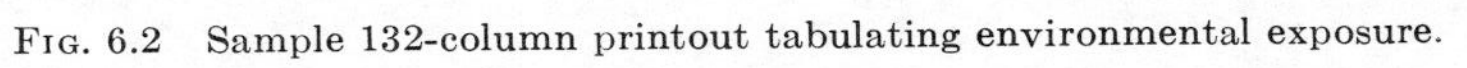

FIG. 6.2 Sample 132-column printout tabulating environmental exposure.

Table 6.3 Definitions for Environmental Exposure Terms

This table is generally designed to define the specified or controlled variable exposures. However, many of the terms under a grouping such as vibration could also be applied to observed outputs. This is not objectionable, since data comparison and fact retrieval require the normalizing of data under similar descriptions and identical units of measurement, and this requirement applies equally, whether the property considered is a controlled or an uncontrolled variable.

General Notes

1. † indicates environments cited by symbol in current NASA MSFC index (Listing).
2. * indicates environments which tentatively would be selected as most significant for printing out into a page tabulating the environmental exposures in parts test reports. Column headings would be preprinted; levels would be shown in digits or specs; combined environments would be indicated; *no* pass/fail data would be included; a 132-col single-line printout would be used.
3. All oscillating values, such as g and AC voltage, will be stated (in normalized data cards/sheets) as rms (root mean square) values. Values in previous reports citing peak or mean will be converted to rms before recording on cards, tab sheets, tape, etc.
4. All directional testing (vibration, shock, etc.) will be assumed to use similar input values for all axes tested. If this should vary (usually for larger components only), the axis indicator will be repeated and the new set of values shown for the *y* and/or *z* axes.
5. Identifying numbers and letters have been assigned to *stabilize term meanings* only; other codes or coding systems could be assigned without invalidating the definitions below.
6. "Operating" and "nonoperating" conditions have been identified only in the few cases where such seems a frequent question. These terms do not seem to require glossary attention. When data-reduction procedure is being discussed, codes could easily be proposed for O and N.O. before, during and after the exposure.
7. The "fixed" environments (10 to 17) can either be described by yes or no or detailed by utilizing code symbols similar to those suggested in Sec. 6.7.1.

Collective or basic environment	Condition, measured quantity, or detail environment	Measured units
1. Operating life	*A Time of operation per part at conditions of temperature, load pressure, etc. stated below	hours
	*B Actuations at stated load (for valves, solenoids, switches, relays, etc.)	cycles
	C Temperature, external, controlled (high)	°C
	D Temperature, external, controlled (low)	°C
	Conditions During Operating Life:	
	*E Pressure; specified applied and controlled internal or external pressure	% (rated)
	*F Current; specified electrical load	% (rated)
	*G Frequency; specified frequency input	% (rated)
	*H Power; specified power input (watts, hp, etc.)	% (rated)
	*I Voltage; specified voltage input	% (rated)
	J Flow	% (rated)
	K Mechanical; applied stress (tension, compression)	lb
	L Stress; applied torsion or bending load	in.-oz
	M Stress; applied torsion or bending load	in.-lb
	N Overload: operation at higher than rated output in watts, as in a generator, alternator, amplifier, etc.	% (rated)

Table 6.3 Definitions for Environmental Exposure Terms (*Continued*)

Note on Category 1. Accelerated life tests (step-stress increases) are not sharply distinguishable from *operating* tests under any above-normal environment. However, where the basic intent is to determine *wear-out* time (or cycles) under the *normal* mechanism of failure at *normal* ratings—but to determine this more rapidly by increase of stresses—these are herein considered variations of the life test.

When the external temperature, radiation, etc. is increased for the purpose of verifying the possibility of introducing a *new* mode of failure, then such test is entered under that specific environment, and not as a "life test at % of rating."

Otherwise, *above-normal* stress of *every* type, such as "overrated altitude," and "overrated vibration," would have to be repeated under Life. We have separately noted "(K) — Overload," the only exception.

Other environments where the *usual* checkout is an *overnormal* operating condition have not been so repeated, although possibly some other popular overtests could be added.

Collective or basic environment		Condition, measured quantity, or detail environment	Measured units
2. Mechanical vibration, sinusoidal	*A	Maximum frequency	cps
	*B	Intensity (see Note 4)	g
	C	Minimum frequency	cps
	D	Total duration of test; low-high-low	min
	E	Sweep cycle time; low to high	min
	F	Sweep cycle time; high to low	min
	G	Sweep cycle time; low-high-low (If this is given, E & F will not apply.)	min
	H	Resonance dwell 1:	
		time	min
		range	cps
		output intensity	g
	I	Resonance dwell 2:	
		time	min
		range	cps
		output intensity	g
	J	Resonance dwell, others	yes/no
	K	Max excursion at crossover (double amplitude)	in.
	L	Axis (axes) on which stress is applied (see Note 5)	x, y, or z
3. Mechanical vibration, random	*A	Maximum frequency	cps
	B	Minimum frequency	cps
	C	Total test time	min
	D	Power spectral density, mean	g^2/cps
	E	Maximum level	g^2/cps
	F	Frequency of maximum level	cps
	G	Axis (axes) on which stress is applied	x, y, or z
4. Acoustic excitation	A	Maximum frequency	cps
	B	Minimum frequency	cps
	C	Maximum sound pressure level (SPL), referred to 0.0002 dynes/cm^2	db
	D	Octave band at which maximum SPL occurs	cps
	E	Overall SPL	db
	F	Roll-off	db/octave
5. Shock	A	Pulse shape: (*a*) sawtooth, (*b*) ½ sin, (*c*) square, (*d*) other	*a*, *b*, *c*, or *d*
	*B	Maximum-intensity peak acceleration during shock	g
	*C	Duration of 5B	ms

Table 6.3 Definitions for Environmental Exposure Terms *(Continued)*

Collective or basic environment		Condition, measured quantity, or detail environment	Measured units
5. *(Continued)*	D	Rise time: time interval, zero to peak acceleration	ms
	E	Decay time: time interval from peak to zero acceleration	ms
	*F	Number of blows	no.
	G	Height of drop	in.
	H	Axis (axes) along which shock was applied	x, y, or z
6. Acceleration, linear	A	Maximum	g
	B	Duration at maximum g	sec
	C	Method of applying force; assumed centrifugal unless noted as linear (sled)	L/NL
7. Acceleration, angular	A	At specified rpm or rpa, acceleration	rad/sec^2

The following heading covers temperature, shock, cycling, soak, storage, etc., by fully defining the conditions without concern for the "label" attached.

8. Temperature (controlled, external)	*A	High	+°C
	B	Dwell time at high	min
N = C + D + E {	C	Interval to cool to ambient	min
	D	Dwell time at ambient (25°C unless otherwise specified)	min
	E	Interval to cool to low	min
	*F	Low	−°C
	G	Dwell time at low	min
P = H + I + J {	H	Rise interval to ambient	min
	I	Dwell time at ambient (25°C U.O.S.)	min
	J	Rise interval to high	min
	K	Portion of cycle part was operating (A, B, C, D, E, F, G, H, I, J)	
	L	Number of repetitions of complete cycle	no.
	M	Operational state: assume operating unless noted	N.O.

Notes on Category 8

1. If high and low exposures are on different units, specify C, D, E, H, I, J, all as zero. If cycling is not controlled as to time intervals, enter dashes.

2. If one sample is given prolonged soak and another is cycled, note the environment of the steady-temperature sample(s) first. The other sample or samples should be described under a second 8 heading, designated 8*. A third type could be under 8**, etc.

3. A "temperature test" is defined, for purposes of this glossary, as exposure to temperatures substantially above or below normal operating ambient conditions, with intent to induce failure as a result of such exposure or to evaluate high- or low-temperature performance. Such tests are not, therefore, included under Life Test even though the part may be operated at abnormal temperatures until failure.

9. Altitude	*A	Height, barometric pressure in terms of altitude above sea level	K ft
	B	Vacuum, specified barometric pressure in terms of millimeters of mercury (used in measuring space environments)	mm Hg

Table 6.3 Definitions for Environmental Exposure Terms *(Continued)*

Collective or basic environment	Condition, measured quantity, or detail environment	Measured units
10. Corrosion (salt spray)	†A Exposure to attack by oxidation and other deleterious effects of corrosive atmosphere	Spec exposure (Note 8)
	B Time of exposure	hours
11. Moisture, humidity	†A Continuous exposure to high relative humidity at 65–70°C	Spec (Note 8)
12. Sand and dust	†A Exposure to effects of concentrated sand and dust at ambient pressure and ambient temperature to 71°C	Spec (Note 8)
13. Fungus	†A Exposure to effects of fungi spores	Spec (Note 8)
14. Explosive atmosphere*	†A Operation within explosive mixture to determine whether or not item or equipment will cause ignition or to determine its flame-arresting or explosion-containing properties. (This is *not* to determine the part's ability to resist an external explosion.)	Spec (Note 8)
15. Fluid leakage	†A *Immersion* in liquid to determine integrity of construction and susceptibility to *inward* leakage. (Often called seal test.)	Spec (Note 8)
	B Fluid leakage *outwards*, from internal pressure	Spec (Note 8)
	C Pressure imposed. (A or B; not both will be imposed)	psi
16. Radio interference*	Susceptibility to electromagnetic interference sometimes called radio noise, r-f interference, hash, static, jamming, etc. (This is not a test to determine if part *causes* interference with the operation of other units.)	Spec (Note 8)
	A Susceptibility to conducted r-f noise	Spec (Note 8)
	B Susceptibility to radiated r-f noise	Spec (Note 8)
17. Radio-noise emission	A Causing objectionable *conducted* r-f interference to other units	Spec (Note 8)
	B Causing objectionable *radiated* r-f interference to other units	Spec (Note 8)
18. Pressure	†A Exposure to (usually) internal pressure, as in a tank, accumulator, or cylinder, to determine structural integrity	Spec †
	B Type of test, defined as: (*a*) Operating (*b*) Proof (maximum spec requirement; not destructive) (*c*) Burst (destructive test to failure) (*d*) Other	*a*, *b*, *c*, *d*

Table 6.3 Definitions for Environmental Exposure Terms *(Continued)*

Collective or basic environment		Condition, measured quantity, or detail environment	Measured units
18. *(Continued)*	C	Specific pressure imposed	psi (If not lumped under H or I)
19. Radiation, exposure to	A	Fast-nuclear	neutrons/cm² sec
	B	Slow-nuclear	neutrons/cm² sec
	C	Thermal-nuclear	neutrons/cm² sec
	D	Gamma	ergs/g/hr
	E	X-ray	ergs/g/hr
	F	Proton	
	G	Electron	
	*H	Total accumulated dosage (flux)	(particles/cm²) $\times 10^h$
	I	Total absorbed dosage	rads $\times 10^h$
	J	Duration of exposure	min
	K	Energy of particle	electron volts (mev $\times 10^h$)

Note on Category 19. A to G express general types of exposure, and for a quick-look printout only the most significant one would be indicated, by letter. Intensity (H) would only be quoted, for this predominant exposure, as two digits and the exponent h. For small parts the particles/cm² seems to be a preferable unit; for larger or living items the unit of absorbed damage effect, rads, might be used.

It would be quite possible to have a requirement to express several types, durations, and intensities for a series of tests on a single part. The above breakdown would not cover such cases fully, but it can be so expanded subsequently.

Collective or basic environment		Condition, measured quantity, or detail environment	Measured units
20. High-voltage tests	A	Dielectric strength (dielectric withstanding voltage): exposure to increasing high potential until insulation breakdown	kv
	B	Insulation resistance: measurement of specified leakage current under high potential, expressing performance in ohms	megohms
	C	Current leakage: measurement under specified high voltage	μa
	D	Overvoltage operation, considerably over rating for considerable time (over 5 min)	% rating
	E	Surge voltage or momentary overload	% rated
	F	Surge voltage, number of applications	cycles
	G	Surge voltage, duration of each application	sec
	H	Temperature at which test was conducted	°C
21. Terminal-stress tests	A	*Lead pull:* application of tension to leads or connections. Result is often termed "terminal strength" or "lead strength." (Enter a value for pounds pull instead of quoting a spec reference.)	lb
	B	*Lead bend* (tension applied during bends)	lb
	C	Number of 90° bends	no.

6.7.3 Glossary of Terms Expressing Performance

A. Terms Considered Output Parameters. This paragraph discusses words (Table 6.6) describing the *results of measurements taken,* as opposed to the *conditions* to which the item was exposed (Table 6.3). This distinction is not always clear-cut; and it is not strictly necessary that it be so, since the assignment of a code and definition

Table 6.3A Single-symbol Coding for Measurement, Test, or Environmental Exposure

The listed symbols represent an extreme simplification, and this condensation required the exercise of judgment in selecting the "related" environments to group under each symbol. Naturally, individual users will wish to make adjustments. For this purpose, the digits 3 to 9 may be used.

Symbol	*Measurements, tests, or exposures indicated by symbol*
A	Acceleration: angular (centrifugal) or linear
B	Bending of leads, pulling of leads, terminal stress/strength
C	Corrosion (salt spray)
D	Dust and sand
E	Explosive atmosphere: ability to operate within, and/or ability to withstand damage from external explosion
F	Fungus, exposure to
G	Sunshine, exposure to
H	Height (altitude—sea level to space)
I	Immersion: fluid leakage inward or leakage outward (seal test)
J	Magnetic field, effect of exposure to
K	Flammability
L	Life test: under normal load or overload (endurance)
M	Moisture resistance: (humidity) and/or precipitation (rain), vapor bath
N	Noise (acoustic)
O	"Operating" (used, where significance and space allow, to indicate that part was operating during the environment shown immediately preceding)
P	Pressure (internal or external above atmospheric): proof or burst
Q	High-voltage tests: dielectric withstanding strength, insulation resistance, current leakage, surge voltage, "hi-pot," dissipation factor, overload
R	Radio interference: emission of objectionable EM or r-f and/or sensitivity to EM or r-f emissions
S	Shock (mechanical)
T	Temperature: high, steady storage. T/U indicates temperature cycling or shock
U	Temperature: low, steady storage. T/U indicates temperature cycling or shock
V	Vibration: sinusoidal, random, or combined
W	Weight, measurement of
X	X-ray and/or any other particle radiation
Y	Initial parameters, measured or visually checked before exposures
Z	Parameters, measured or visually checked after exposures
$	(dollar sign) Solderability and/or effect of soldering
*	(asterisk) Failure analysis
.	(period) "End of environmental coding. Additional information follows."
SS	(after the period) Indicates sample size
&	(ampersand) Used in the final column when there are several other environmental exposures which space does not allow printout
/	(slash) Combined or associated environmental exposures will be joined by an intermediate slash (*X/Y* or occasionally *X/Y/Z*)
---	(dashes) Environmental data not given or not applicable

locates and identifies the property being referred to. Nothing is gained by debating, for example, whether the term "solderability" primarily refers to the act of exposing a part to molten solder or the subsequent satisfactory performance of the soldered connection; the important requirement is that "solderability" is always detailed in the same identified terms and location.

Since the outputs of larger assemblies are expressed in a great variety of specialized and detailed units, it is more practical to illustrate storage/retrieval principles by the use of *parts* performance parameters. Those shown in Table 6.6 illustrate typical requirements needed for describing *performance* of parts; the principles are adaptable to cover nonaerospace types. It is contemplated that each user would expand such a table in his specific interest area to cover all test criteria used including physical measurements, visual inspections, and electrical, mechanical, or chemical outputs or responses.

The primary requirement of any system is that a specific code number is assigned to each measured output property to assure that the file searcher and the using engineer (as well as any machines involved) interpret the term exactly as the test conductor intended and can refer to a single detailed exposition when in doubt.

Grouping terms by the *type of part being measured* is in most cases impractical, since the majority of terms apply equally to many different parts. For example, "resistance," in ohms, is applicable to resistors, relay coils, heating elements, etc. "Impedance" applies to antennas, speakers, coaxial cables, and amplifiers.

Table 6.4 Two-character Coding for Test Specification and Level

Second Code Character Expressing Condition, Level, or Type of Test. The letter or digit following the specification symbol indicates the actual "condition." No coding is used

Example: BC = MIL-STD-202, Cond. C
F3 = MIL-E-5272, Procedure 3

Code	*Specification*
A	FED-STD-151
B	MIL-STD-202
C	FED-STD-406
D	MIL-STD-750
E	MIL-E-4970
F	MIL-E-5272
G	MIL-E-5400
H	MIL-T-5422
J	MIL-I-6051
K	MIL-I-6181
L	MIL-E-8189
M	MIL-I-16400
N	MIL-I-16910
P	MIL-S-19500
Q	MIL-T-21200
R	MIL-I-26600
S	Test performed per internal (contractor) spec or program spec requirement. Spec to be cited in notes section of tabulation card.
T	Test performed per procedure detailed in test report; no spec cited.
U	Test performed, procedure detailed per MIL (or other general usage). Spec not included in above list. Spec to be cited in notes section of tabulation card.
Y	(Yes) Test performed. No other detail available.

There are exceptions when the same term acquires *different meanings* when applied to different parts. In these cases the units of measurement are usually different although the basic term may remain the same, and such terms especially should be defined and coded to preserve this distinction. Examples are:

Impedance, electrical; measured in ohms
Impedance, acoustical; expressed as a ratio of the pressure to the volume displacement at a given surface in a sound-transmitting medium
Noise, electronic; measured in microvolts per volt
Noise, acoustical; measured in decibels

The same type of differentiation will be needed when terms vary in the nature of measurement or extreme differences in magnitude. Examples could be:

Voltage, rms
Voltage, dc
Rotational speed; revolutions per second (gyro)
Rotational speed; revolutions per hour (recording drum)

Table 6.5 Specification Origin Identification Code*

There is no government requirement to use these codes, but since most procurement or engineering data reference some specification requirements, some uniform coding is useful in storing this information.

Agency or type of document	Abbreviation	Code
Military Standardization		
Aeronautical Material Specification	AMS	01
Army-Navy Standard	AN STD	02
Book of Ordnance Engineering Standards	Bk of ORD ENG STD	03
Federal	FED	04
Joint Army-Navy	JAN	05
Joint Communication-Electronic Nomenclature System	JCENS	06
Military Specification	MIL	07
Military Standard (book form)	MIL-STD	08
Military Standard (sheet form)	MS	09
National Aircraft Standard	NAS	10
Federal Specifications	FED SPEC	11
Federal Standards	FED STD	12
Military Handbooks	MIL HDBK	13
Air Force–Navy–Aeronautical Specifications	AN SPEC	14
Air Force–Navy–Aeronautical Bulletins	ANA BUL	15
Air Force–Navy–Aeronautical Design Standards	AND	16
Air Force Specifications	AF SPEC	17
Air Force Specification Bulletin	AF BUL	18
Qualified Products Lists	QPL	19
Other Military		
Aeronautical Standards Group	ASG	26
Air Force–Navy–Aeronautical Standards	ANA	27
Armed Service Electro Standards Agency	ASESA	28
Army	ARMY	29
Atomic Energy Commission	AEC	30
Bureau of Aeronautics	BUAER	31
Bureau of Ordnance	BUORD	32
Bureau of Ships	BUSHIPS	33
Bureau of Weapons	BUWEPS	34
Federal Aviation Agency	FAA	35
Naval Aircraft Factory	NAF	36
Navy	NAVY	37
Ordnance Corps	ORD	38
Signal Corps	SIG	39
U.S. Air Force	USAF	40
Other Departmental Documents		41
Commercial		
International Electrotechnical Commission	IEC	49
Industry Documents (National Aircraft Stds)	NAS	50
American Association of Railroads	AAR	51
American Iron and Steel Institute	AISI	52
American Society for Testing Materials	ASTM	53
American Standards Association	ASA	54
Electronic Industries Association	EIA	55
Joint Electron Device Engineering Council	JEDEC	56
Joint Electron Tube Engineering Council	JETEC	57
National Electrical Manufacturers Association	NEMA	58
Radio Electronic Television Manufacturers Association†	RETMA	59
Radio Manufacturers Association†	RMA	60
Society of Automotive Engineers	SAE	61
Underwriters Laboratories, Inc	UL	62
Aerospace Industries Association	AIA	63
National Microfilm Association	NMA	64

* Adapted from a Logistics Item Identification Guide issued by Defense Electronics Supply Center, August, 1963.

† Now superseded by EIA, which also has jurisdiction over JEDEC and JETEC.

Multipliers by powers of 10 or floating-decimal-point notation could allow for magnitude variations, but Table 6.6 is designed to typify a system easily readable by humans.

B. Descriptive Terms Not Coded under Outputs. Many part descriptions, for *stock control* identification purposes, use additional amplifying terms, which seem to apply primarily to usage and thus could be considered from one viewpoint as "output" terms. However, when these are used as *fixed physical descriptions defining the part,* they are not included herein as outputs.

Examples might be diameter, length, power dissipation and rating (not actual measured wattage output), winding design, and enclosure type. A "½-watt resistor" designates a rating of dissipation for a particular type or size, not a measured output.

Table 6.6 Output Performance Coding Examples

Note. This is a sample compilation illustrating an approach to handling this type of classification. Some terms border on being "environments," but none *duplicate* terms in Table 6.3. Code numbers and alphabetization are only typical examples and are not for permanent use. When this table was compiled, several agencies (DSA, AFLC Parts Data Bank, DOD/EDS-0016) were working on similar glossaries, but none had been proved sufficiently to be regarded as a national standard at that time.

Code	Term	Definition (measure of)	Units, symbols	Usually applicable spec
615	Resistance change, humidity	Measure of change in resistance with humidity exposure	$\pm\%\ \Delta R$	MIL-STD-202 MIL-R-10683, Para.4.6 .9
620	Resistance change, immersion	Measure of change in resistance resulting from immersion in fluid	$\pm\%\ \Delta R$	MIL-STD-202B
630	Resistance change, low-temperature operation	Effect of operating load at low ambient temperatures	$\pm\%\ \Delta R$	MIL-R-11D Para.4.6 .7
645	Resistance change, mechanical shock	Effect of shock	$\pm\%\ \Delta R$	MIL-R-11D Para.4.6 .1
650	Resistance change, momentary overload	Change in resistance under specified short-time overload in power (or voltage, not exceeding rating)	$\pm\%\ \Delta R$	MIL-R-11D Para.4.6 .10
655	Resistance change, salt spray	Measure of resistance change due to exposure to corrosion-producing environment (salt spray)	$\pm\%\ \Delta R$	MIL-STD-202
660	Resistance change, temperature	Effect of a series of gradual changes in ambient temperature	$\pm\%\ \Delta R$	MIL-R-11D Para.4.6

Likewise, arbitrarily fixed criteria for test conditions or acceptability, which are not measured parameters, would be omitted. Examples are "ambient-temperature range at which power rating applies," and "resistance tolerance" (which is often used as an identifying physical description).

6.7.4 Coding Digital Values for Machine Manipulation. The preceding paragraphs treated of input-data structuring, where the only desire is to rapidly pull out of storage the exact values or exposures as recorded. However, it is sometimes desired to obtain the minimum, maximum, mean, scatter, per cent, or number of items beyond 3σ from a group of results in different reports. Where data are numerically expressed in identical units and identified consistently with "labels" as illustrated in Tables 6.3 and 6.5, the computer programmer can easily arrange to perform such comparisons between separate reports. The possibilities of such requirements must be thought out in advance, because they will control the strictness of design and enforcement of coding and measurement-unit usage. Variations of data-manipulation requirements will require variations of the codings in Table 6.3 and 6.5, but the essential requirement for a standardized coding remains.

When it is decided to enter on the cards or tape any value (such as the max) other than the standards already established and coded, select an appropriate new identifying code, define exactly what value variation or function this applies to, and tag all such values with this code. This assures that the keypunch operator can give uniform instructions to the computer and the programmers can arrange to retrieve facts or manipulate data as anticipated.

6.7.5 Floating-point Notation. When the occasion requires storage of values which vary considerably in magnitude, the use of a standardized mathematical shorthand is quite convenient. The figures to be entered in the computer are expressed in the form 5EXXXX · · · X, which expresses the value $(X.XXX \cdots X)^{50+E}$.

The "XXX · · · X" expresses any desired number of significant digits, with the decimal point always understood to occur after the *first* digit. 5*E* expresses the exponent to which this base number must be raised to equal the actual value. A handy convention assigns the value of 50 to the exponent zero; thus an exponent of 10^3 would be coded as 53, one of 10^{-3} would be 47, etc. This is to eliminate the need for plus or minus signs. Complete examples are:

Value	*Multiplier*	*Floating-point code number*
13.6	1.36×10^1	51136
0.136	1.36×10^{-1}	49136
0.001367	1.367×10^{-3}	471367
13600	1.36×10^4	54136

6.7.6 Symbols for Vendor Names. *The Federal Supply Code for* (United States and Canada) *Manufacturers, Cataloging Handbook H*4 (H4-1, name to code; H4-2, code to name) identifies many thousands of firms and branches by means of five-digit codes. These handbooks are issued by the Defense Logistics Supply Center, DSA, Department C6-C, Battle Creek, Michigan 49016. They are very useful for procurement purposes, since they give company and plant location. They are updated frequently, but the rapid growth and shift of industry are inevitably ahead of the revisions. Reliability departments are urged to obtain these books for use in identifying suppliers of items for which reports are recorded.

6.7.7 Federal and Specialized Commodity Codes. All products purchased by the government (which includes practically everything manufactured, mined, or grown) are classified in the *Federal Supply Classification H2 Handbooks.* These handbooks, issued by the same agency as the H4, classify products in some 80 two-digit Commodity Area Groups. Examples: 10 = Weapons, 14 = Guided Missiles, 20 = Ships and Marine Equipment, 22 = Railway Equipment, 59 = Electrical and Electronic Components, 88 = Live Animals.

*Handbook H*2-2 further breaks the commodity groups into four-digit classes, FSC Numbers. Examples: Class 1005 = Guns through 30mm; Class 1010 = Guns, over 30mm to 75mm. Under each four-digit class are listed one to several pages of non-coded alphabetical item names and the FSC under which each name will be found. Volume H2-3 lists some 700 pages of these names alphabetically under the basic names. Examples: Under Brushes there are some 150 types listed, from Acid Swabbing = FSC 7920 through Motor = FSC 5977 and Mouthpiece, Band Instrument = FSC 7720.

There is a necessity to further identify *specific* items, and this is done by the H6 Handbooks on *Federal Item Identification Guides for Supply Cataloging.* Volume H6-1 contains:

Section A, an alphabetical list of names, followed by a Description Pattern Number and an Item Name Code, both five-digit numbers, plus a short description of the item(s) covered.

Section B, a numeric index of Description Patterns and Name Codes.

Section C, Abbreviations and Symbols.

Volume H6-2 contains the Descriptions Patterns, in 10 segments (A to J) arranged by (two-digit, H2) Commodity Area groupings. Volume H6-3 contains the Reference Drawings, arranged in similar segments. The problems of updating such a system in a rapidly changing area such as electronics are enormous, and the compromise necessary between various viewpoints on categorization will never group together in one area all information desired by some special-interest group such as reliability/QC engineers. The Federal terminology is a valuable reference, and some industries may be able to adapt entire sections of the *Handbooks* to their filing needs. As an alternative, Table 6.2 is offered as a hardware information-classification system tailored in complexity to a reliability engineer's probable needs. (See also Sec. 6.6.4.)

6.8 ELEMENTS OF RETRIEVAL HARDWARE

Since this field is explosively expanding, this section will only describe equipment by basic type, although where unique principles are involved, it is impossible to avoid defining a particular manufacturer's machine, whether or not his name is mentioned.

This treatment will be oriented to the *storage media* used, although data systems often involve conversion through several media. As in computers, there are various "standard" components and peripheral equipment functions which different manufacturers have emphasized or combined to suit special system requirements. Such combinations will not be discussed herein, and the topic will be limited to exclude the computer technology field on one side and the offset-printing field on the other.

6.8.1 Classification of Retrieval-equipment Components. Although many devices combine or cut across these gross categories, the following basic distinctions will be used in distinguishing between the general classes of retrieval equipment which:

1. *Locate* the address (specific file or machine location), report number, list of documents, or sources of data which (probably) answer a query. Many EAM and EDP systems which have been converted to library or data searching stop at this stage.
2. *Furnish* a facsimile of the actual document, excerpt, or abstract desired (when given the address).
3. *Compute and select* (manipulate and select) the data from a storage of facts rather than documents, displaying or printing this out in a form differing from the input.

6.8.2 Microfilm Cameras and Practices

A. Rotary. High-speed, usually semi- or fully automatic feed. For high-quantity, less-critical photographing of items of similar size. (Examples: failure reports, canceled checks.) Volume can run up to several thousand an hour, with both sides of a document photographed simultaneously.

B. Planetary. Lower speed; individually placed, lighted, and exposed frames. For critical items or where contrast or quality of original is variable or poor. Speed in low hundreds per hour. Usually used when MIL Specification quality is desired.

C. Step and Repeat. Special-application camera which exposes a series of images in selected locations on a precut sheet of film. Used for placing a number of pages on a microfiche card (4 by 6 in., for example). Can also cover a large original by making a number of exposures in a patchwork pattern. (See Sec. 6.8.4*B* for microfiche application details.)

D. Photographic Specifications for Microfilming. Microfilming of engineering documents (35 mm) is covered in MIL-M-9868, and reduction ratios are prescribed as 16×, 20×, 24× and 30× for stipulated sizes of drawings. Reduction ratio for documents normally varies from 8× to 30×; in special equipment, up to 60×. Anything over 24× involves extra care, fine originals, and expense. The input, reproduction equipment, user skill, etc., must all be considered in designing a system around the proper reduction ratio. Specifications MIL-C-9878, MIL-STD-804, and MIL-C-9877 also stipulate microfilm standards, largely relating to aperture cards. (See Sec. 6.8.4*B*.) For microfiche the National Microfilm Association has issued a Specification NMA-M-1-1963, which has been adopted into the NASA/STAR system.

6.8.3 Microfilm Image-storing Materials

A. Silver Halide. (Referred to in the trade as "silver original.") Emulsion-coated cellulose acetate film negative. The usual media for original image exposure; a high-contrast specialization of normal camera film. The image is reversed with each generation of reproduction—from positive to negative to positive, etc.

B. Diazotype. A cellulose triacetate base into which the emulsion is bled, forming a surface layer of about 0.0005 in. It is therefore more scratch-resistant than silver. Developed in a single dry process by ammonia vapor; slightly lower in quality than silver but very commonly used for copies. Reproduces without reversal (positive to positive), available in roll or sheet in thicknesses of 0.002 to 0.010 in.

C. Thermal, Light-scattering Image (Kalvar). Light-sensitive grains imbedded in polyester (Mylar) base; exposed by ultraviolet and developed by heat. Fine grain; good contrast and film strength. A newer material; processing not as firmly established. Normally reversal reproduction; direct-image option available.

6.8.4 Microfilm Image-storing Media

A. Roll Film. In continuous strip is most popularly 16 or 35 mm but is occasionally used in 8, 70 or 105 mm. Microfilm is normally not perforated. Major advantages of roll film are:

1. Inviolate file.
2. Rapid and simple transport for automatic sequential sort.
3. Less expensive quantity production and duplication of film by the entire roll.

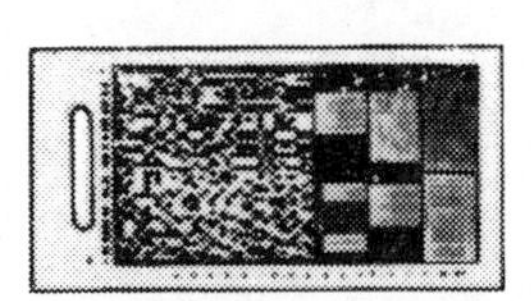

FIG. 6.3 Examples of microfilm chip.

FIG. 6.4 Microfilm–magnetic chip combination.

Some disadvantages are:

1. File is less flexible for adding or updating.
2. Single strip does not allow interest-area-oriented distribution of portions of original text.
3. Use by one individual ties up an entire roll for period of his search and study.
4. Search must generally proceed from start of a roll and cannot random-access direct to a particular image.

B. Cut Microfilm (Microfiche). Popular sizes are 3 by 5 in. and 4 by 6 in. A nonreduced title or identification is usually affixed or photographically produced at the top of the card, or film is joined to summary card (as in IDEP).

Microfiche can be produced by a step-and-repeat camera or by assembling from roll-film negative. Recently introduced equipment produces fiche masters from roll film by using an automatic taping/assembly machine; updating by this method is easier than reshooting the entire original with step-and-repeat.

C. Film Chips. Special case of microfiche where roll film has automatically been cut to a very precise size to give additional file-handling speed and flexibility. (See Fig. 6.3 and Sec. 6.8.6*A*.)

D. Graphic-EDP Combinations (Figure 6.4). A special adaptation of optical and magnetic-tape techniques uses polyester-film cards coated with ferric oxide and coded for magnetic "reading" over 70 per cent of the area, but carrying a photographic image on the balance. This combines complex search and structured data storage with the actual document facsimile, but it is presently a more expensive system than others described herein.

E. Film and Cardstock. Combinations are many, but by far the most common is the "aperture card" (Fig. 6.5). This uses a standard-size EAM card carrying one or more chips of microfilm affixed within windows in the card by glueing, ultrasonic bonding, or transparent retainers. These cards can be machine-sorted without damage if certain machine alterations are made, and they are widely used with the single D aperture (35 mm by 1¾ in.) for engineering drawings. Another advantage is the availability of many semi- and fully automatic aperture-card camera/developers and card-to-card duplicators.

F. Micro-opaque Cards. Similar to a high-contrast gloss photograph; offer a very scratch resistant surface, combined with ability to carry directly readable copy or to be offset-printed in large quantities. (Available in gummed-label form also.) Disadvantage is the higher illumination needed for readers and printers.

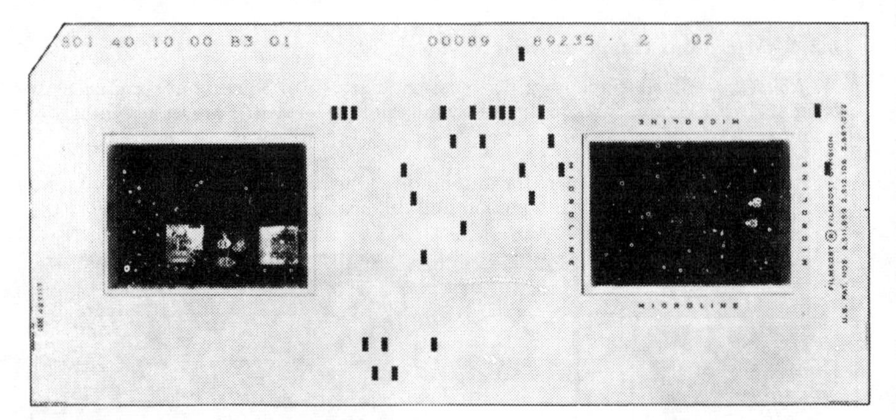

Fig. 6.5 EAM Aperture Card.

6.8.5 Viewing and Original-sized Copy-making Machines

A. Viewers. Vary from hand-held magnifying frames at $10 to $20 to large-screen projection devices at several thousand. Some are designed for roll film only and as a result are very efficient, utilizing a snap-in cassette or cartridge. As viewers become more general purpose, accommodating both rolls and the various size microfiche types, the efficiency in handling any one type naturally decreases.

B. Viewer-Printers. These devices are a relatively new development in compact desk-top machines allowing unskilled operation. They range in cost from about $700 up. Use is made of either (1) stabilized combined developer-fixer bath with conventional photographic paper; (2) electrolytic conductive paper, passing over a charged sponge after exposure, or (3) electrostatic charging of the plain or ZnO-coated paper, either directly or through a selenium drum, dusting with carbon powder and toner, and heat-fusing the image (if coated paper is not used). This last type has several larger automatic variations which are quite out of the desk-top range. In the direction of the least-expensive first-cost equipment, a one- or two-bath developing unit can be used in connection with a projection viewer, similarly to photographic enlarging.

C. Film-to-film Duplicators. These devices also have emerged somewhat from the photo laboratory, particularly in the specialized aperture-card duplicators. Roll to roll is still a process requiring more expensive equipment, although in all areas the dry-processing (Kalvar) film is making point-of-use duplication more practical.

D. Cathode-ray-tube Devices. This method cuts across the computer and microfilm fields, utilizing a magnetic-tape input to create a display of the desired document on the face of a CRT. This is not a duplicate image of the original, but rather a facsimile by means of letter outlines formed through a type font on a raster in the tube.

The momentary display is then either copied on microfilm or printed out as a full-size hard copy. A preselected form outline can also be impressed in the same image by using a transparency of the form flashed through a half-silvered prism. This method is efficient when rapid selection by complex search from a large number of similar documents (such as failure reports) is desired and when digested *information* in standard format, rather than an exact *image* of the original, is acceptable.

6.8.6 Search and Retrieval Equipment for Microfilm

A. Card and Film-chip Manipulation. Equipment for EAM cards is too generally familiar to warrant discussion, but there are several mechanical aids to creating and searching the edge-notched cards which facilitate large-volume use. In the peephole field there are precision drilling devices and frames which permit storage of many times the information content previously possible. Specialized retrieval machines are able to sort the film chips at very high speed by reading out by photocell the coded designations which have been photographically recorded in the film at the time of microfilming. Some handle the chips in cartridges or magazines and transport them individually by using suction drums; another uses a strip containing a number of images and selects the strip and images automatically from a carousel containing many thousands of strips.

B. Tape and Film Devices. As above, the normal punched- and magnetic-tape devices will be mentioned only as inputs which are used in some cases to control the coding or retrieval of the document-storage media. One 35-mm-roll storage device controls the black or white lighting of the binary-coded squares by means of punched tape. Simpler roll-film devices use punch cards, and one merely uses a manual keyboard to record a 10-digit number on the light panel, which is photographed with the page.

The binary coding of tiny black-and-white squares in the film is read through pinpoint photocells as the film runs by. This concept can be expanded to any degree of detail desired by adding more coding pages per image page. Both 16- and 35-mm systems are available.

The simplest roll-film search utilizes "streak" coding, in which lines across the viewer frame apparently move vertically in relation to numbered levels at the edge of the frame as the roll is run through the viewer. Search is visual.

6.8.7 Data-transmittal Equipment. Electronic transmittal of facsimiles by wire or microwave and sending data via telemetry are commonly visualized in connection with information retrieval, but several specialized remote devices may not be so generally known. Available viewers can scan (via closed-circuit TV) a microfilm frame at a central file and can even remotely control magnification or concentrate view on an enlarged portion of the field. Scanning of documents via a small, portable TV camera is being done for immediate microfilming from the TV receptor. Querying of a central fact repository from up to 50 remote teletype stations is being instituted in at least one large information system.

Scanners which can interpret several type fonts and reduce the letters to standard magnetic-tape coding allow much quicker transmittal or can utilize a much narrower bandwidth. Such facsimile transmittal is simpler and therefore more immediately adaptable to many scattered receiving stations than direct image transmittal, although the initial reduction would require more effort.

Although the field of reliability data retrieval could probably not *by itself* justify creation of any of these systems, it may well utilize portions of the hookups for other purposes.

6.8.8 Multiple-copy Production. Almost every microfilm viewer-printer device will also produce short-run masters which can be run directly on multilith presses for quantities of 50 to 300 or so. However, be sure to consider the delay and expense of changing paper, speeds, and solutions before comparing these more exotic systems with the cost of copying a relatively small quantity from a full-size page either by utilizing an ordinary office duplicator or by employing deposited-graphite devices or the more conventional direct photographic plate-making methods.

6.8.9 User Attitude toward Microfilm Operations. A short treatment such as this can give only a very sketchy summary of equipment status in the complex and expanding field of data retrieval. However, enough has been described to indicate why data system designers cannot arbitrarily centralize viewing and reproduction (and sometimes storage/retrieval) solely for device operating efficiency. Newer equipment is certain to increase specialization, both of the compact desk-top devices and the massive and expensive high-production units, and each usage will require individual consideration of the balance between centralization and convenience.

6.9 UTILIZATION OF STORED DATA

6.9.1 Establishment of a Reliability Data Center. Any sizable reliability activity can justify the maintenance of an area allocated to reliability data and operated by assigned and trained personnel. The physical and organizational location of such files within the functional organization interested in their day-to-day use encourages maintaining compatibility with current technical requirements and nomenclature. If such data are handled by a remote library-type function, dead material is more likely to accumulate and clog the current needs. The design of facilities for a reliability data center should include ample table space and a microfilm viewer, where applicable, in a subdued-light area.

6.9.2 Encouraging and Monitoring Data Utilization. A data center's unit efficiency depends on a proper volume of usage; and when it is contemplated that a portion of such usage will originate outside the reliability section, it is necessary to conduct a continuing plantwide educational program on the requirements to contribute into the center and the probable benefits from consulting with the center. Properly organized "experience retention" files have been called an integral part of the corporate assets, and a broad motivational effort can be based on this viewpoint.

Devices to monitor utilization of individual files can consist of voluntary log sheets which the users fill out with comments on utility, or they can be more formal sheets given out with each package of data and on which a return-acknowledgement copy detailing usage is requested. Methods will vary with the nature and volume of usage of each file, but consideration should be given to logging activity to the appropriate degree of detail for both planning and justification purposes.

6.9.3 Impact of Data-management Techniques on Reliability. The contents of this section were selected to outline the potentials of current and evolving information-retrieval techniques for assisting the overall effectiveness of a reliability/quality control activity. However, unless the reliability data center (or equivalent) is organized as a technical function, it remains—or deteriorates into—an inactive-file maintenance operation. Utilizing its potential requires a capable and progressive staff, plus management support.

References

1. Vickery, B. D., *On Retrieval System Theory*, Butterworth & Co. (Publishers), Ltd., London, 1961.
2. Vickery, B. D., *Classification and Indexing in Science*, Academic Press Inc., New York, 1959.
3. "Current Research and Development in Scientific Documentation," *NSF*-61-29, National Science Foundation, Washington, D.C. (No. 8, May, 1961)
4. "Non-conventional Technical Information Systems in Current Use," National Science Foundation, Washington, D.C. (No. 2, March, 1950)
5. *Directory of R&D Information Systems*, OAR-1, DCS/Plans & Operations Research, USAF, Washington 25, D.C. (ASTIA cat. 262958)
6. *Interagency Coordination of Information, Hearings of the Senate Government Operations Committee, Parts* 1 *and* 2, Government Printing Office, Washington, D.C., 1963. (Sept. 21, 1962)
7. *Science, Government & Information: A Report of the President's Science Advisory Council, The White House*, Government Printing Office, Washington, D.C., 1963. (Jan. 10, 1963)

8. *Guide to Microreproduction Equipment*, National Microfilm Association, P.O. Box 386, Annapolis, Md. Edited by Hubbard W. Ballou; issued 1959; supplements issued annually. Also issues *Annual NMA Symposium Proceedings* and *Monthly National Microfilm News*.

Bibliography

Cataloging Guides

Hargrove, C. E., *Subject Headings Used by the U.S./AEC Technical Information Service*, 3d ed., rev., Atomic Energy Commission, Washington, D.C., 1960.

Perry, James, and Allen Kent, *Tools for Machine Literature Searching*, Interscience Publishers, Inc., New York, 1958.

Stratton, F. E., *Corporate Author Entries Used by the U.S./AEC Technical Information Service in Cataloguing Reports*, 4th ed., rev., Atomic Energy Commission, Washington, D.C., 1959.

Indexing and Retrieval Theory and Mechanics

Cleverdon, Cyril W., "Report on the Testing and Analysis of an Investigation into the Comparative Efficiency of Indexing Systems," ASLIB Cranfield Research Project sponsored by the NFS, Cranfield, England, College of Aeronautics, 1962.

Frank, O., *Modern Documentation and Information Practices*, International Federation for Documentation, The Hague, Netherlands, 1961.

Ranganathan, S. R., *Prolegomena to Library Classification*, Chaucer House, London, 1957.

Scheele, M., and J. E. Holstrom, *Punch Card Methods in Research and Documentation*, Interscience Publishers, Inc., New York, 1961.

Shera, J. H., Allen Kent, and J. W. Perry, *Advances in Documentation and Library Science*, vol. II, *Information Systems in Documentation*, Interscience Publishers, Inc., New York, 1961.

———, *Information Retrieval and Machine Translation*, vol. III, *Information Systems in Documentation*, Interscience Publishers, Inc., New York, 1961.

Symposia Transactions and Collected Papers

Boaz, M., *Modern Trends in Documentation* (Symposium at University of Southern California), Pergamon Press, New York, 1959.

Hattery, L. H., and E. M. McCormick, *Information Retrieval Management*, American Data Processing, Inc., Detroit, 1962.

———, *Machine Indexing; Progress and Problems*, Center for Technology and Administration, The American University, Washington, D.C.

———, *Proc. Third Natl. Reliability Quality Control Symp.* Washington, D.C., 1957; *Proc. Sixth Natl. Reliability Quality Control Symp.*, Washington, D.C., 1960.

Listings of Abstract Compilations, Data Sources, Bibliographies

Fry, B. M., and F. E. Mohrhardt, *A Guide to Information Sources in Space Science and Technology*, John Wiley & Sons, Inc., New York, 1963. (Library of Congress Card 63-19662)

"A Guide to the World's Abstracting and Indexing Services in Science and Technology," *Rept.* 702, *Natl. Field Sci.*, Abstracting and Indexing Services, Washington, D.C., 1963.

National Information Center, hearings before an ad hoc subcommittee of the House of Representatives Committee on Education and Labor (first session on HR 1946), Government Printing Office, 1963. (21-226 0)

Government Surveys and Reports

Committee on Government Operations, U.S. Senate, 87th Congress, *Coordination of Information on Current Federal R&D Projects in the Field of Electronics*, Government Printing Office, Washington, D.C., 1961. (69547)

A. D. Little, Inc., *Organization and Documentation*, report for the National Science Foundation, July, 1963. (PB181548A)

Section 7

THE COFEC RELIABILITY DATA SYSTEM

LAWRENCE J. FOGEL

PRESIDENT, DECISION SCIENCE, INC.
SAN DIEGO, CALIFORNIA

Lawrence J. Fogel received the B.E.E. from New York University, the M.S. from Rutgers University, and, after doing doctorate work at the Polytechnic Institute of Brooklyn, Rutgers, and M.I.T., the Ph.D. in engineering from UCLA.

After holding a number of positions in engineering systems design and evaluation, Dr. Fogel joined General Dynamics/Convair as a design specialist concerned with cockpit instrumentation on the F-106. He performed research concerning anticipatory and antivertigo displays. For three years he was head of the reliability group in the engineering department, then accepted an invitation to serve as Special Assistant to the Director of Research of NSF. While in Washington, he prepared numerous technical papers and the report "Investing in Scientific Progress," which was sent to Congress and the governors of all the states. On returning to General Dynamics/Astronautics, he assumed the position of Senior Staff Scientist, serving as a consultant to management and directing research in the field of artificial intelligence and command control. In 1965 he organized Decision Science, Inc.

Dr. Fogel has published some forty technical papers and a book, Biotechnology: Concepts and Applications. He has been granted five patents, is a senior member of IEEE and AIAA, and holds a life membership in the New York Academy of Sciences.

AARON FOGEL

RETIRED, RELIABILITY CONTROL DEPARTMENT
GENERAL DYNAMICS/ASTRONAUTICS
SAN DIEGO, CALIFORNIA

Aaron Fogel completed a program of study, including business administration, accounting, and electrical engineering, at Pace and Pace Institute, New York, and at Pratt Institute, Brooklyn, New York. Since 1958 he has completed various courses in mathematics and reliability engineering in the Graduate School of the University of California.

Mr. Fogel held several positions for the Tasco Products Company, of which he was General Manager from 1942 to 1947. Until 1956 he owned and operated companies engaged in producing equipment under patents he held. In 1956 he joined General Dynamics as a research engineer and later contributed to the design of flight-control instrumentation and the cockpit configuration for the F-106 interceptor. In 1957 he became a member of the reliability engineering group in the capacity of mathematical statistician. Until his recent retirement, he was a member of the reliability control department of Astronautics, where he operated the COFEC System, which he devised for the automatic processing and storage of failure data and the corresponding corrective actions. He was primarily concerned with the Centaur and Atlas weapon systems. He is presently serving as consultant in the field of reliability data analysis. He is a member of the American Statistical Association, the Human Factors Society, the ASQC, and the American Ordnance Association.

THE COFEC RELIABILITY DATA SYSTEM

Lawrence J. Fogel and Aaron Fogel

CONTENTS

7.1 INTRODUCTION

The primary purpose of any reliability data system is the enhancement of product reliability. In order to achieve this end, such a system must serve the design engineer and personnel at all levels of management.

Specifically, the reliability data system must respond to the designer by providing a complete failure history on product items in some simple easy-to-interpret form. This history must include an indication of the specific mode of each failure, the cause of each failure, and an explicit record of the effectiveness of all previous corrective actions. It is this last factor which permits successive modifications and redesign with assurance that the product reliability is improved through the elimination of individual failure modes. The COFEC reliability data system fulfills this need; thus the name COFEC (cause of failure, effect, and corrective action).

As indicated above, a reliability data system must serve all levels of management. It must provide the manager with an up-to-date overview of the discrepancy activities in such a manner as to allow immediate assessment of the situation and directed inquiry into prevailing difficulties. Thus, there is need for a quantitative status report, a report which summarizes the number of significant failures which have occurred at each facility, on each end product (final assembly product) and on each system during the previous time interval. With such a record in hand, the manager can focus upon those areas which most deserve his attention; he can raise specific questions and determine whether previously identified difficulties have been overcome. At the same time, lower levels of supervision also require similar quantitative reports of the discrepancy activity, but at a greater level of detail, classified by system, by end product, by subsystem, by assembly, and by part in the system over which they maintain cognizance. The COFEC reliability data system can provide just such information in a readily usable form.

Throughout the life history of the product, effort must be devoted to vendor rating, selection, and performance monitoring. This activity can take place on a firm foundation only if the reliability data system provides an up-to-date record of the number and kind of failures which are attributable to the products of each vendor. It is important to develop quantitative evidence on the level of cooperation of each vendor and to measure the effectiveness with which each vendor carries out the requested corrective actions. Here again the COFEC reliability data system provides the essential information in a form suitable for these purposes. An additional by-product of this system is numeric data on parts removal which can form a basis for the ordering of logistic support for the end product and the estimation of spare-parts requirements.

In summary, there is need for a reliability data system which provides the designer with a complete up-to-date history of the failures, their cause, and the effectiveness of previous corrective actions and which provides all levels of management with a survey of current discrepancy activities classified by facility, by end product, by system, by vendor, and so forth. To prove worthwhile, such a system must operate economically. That is, it must be capable of expansion or contraction in keeping with the reliability activity of the product; it must be suitable for the use of automatic information-processing techniques if, and when, the volume of data warrants such expenditure; it should provide an institutional capability which crosses organizational lines and covers a variety of projects so that benefit may be derived from use of the largest possible sample of observable data; it must be organized in such a way that the value of the system is not degraded through the unexpected loss of key personnel; it must provide rapid response to all requests for specific information and at the same time maintain a permanent record of all of the original discrepancy data.

7.2 ORGANIZATION AND PRESENTATION OF DATA FOR THE DESIGN ENGINEER

For the sake of the following discussion, let the term "discrepancy" be defined as any cause for the removal of an item, including any noted deviation from the specification, the instruction of a technical order, the expiration of permitted operational time, and so forth, or any irregularity which might cause a failure in the performance of the

DISCREPANCY REPORT NO. 41603

DATE DAY MONTH YEAR	REPORTING ACTIVITY	END PRODUCT TYPE, SERIAL NO.	SYSTEM
PART NO.	SERIAL NO.	NAME OF PART	MANUFACTURER
NEXT ASS'Y PART NO.	SERIAL NO.	NAME OF ASS'Y	MANUFACTURER
REPLACEMENT PART NO.	SERIAL NO.	OPERATING TIME DAYS HOURS MINUTES CYCLES	COFEC CODE
DESCRIPTION OF DISCREPANCY			
		REPORTED BY	TELEPHONE NO.

Fig. 7.1 A discrepancy report form.

item or any other portion of the end product even though this condition may be discovered prior to installation of the item. Discrepancies may arise in many areas. For example, discrepancies may be found in receiving inspection, during production, in final checkout, in production testing, and in operations at off-site bases.

All discrepancies must be brought to the attention of a reliability data center. Collection of these data is facilitated through the use of a discrepancy report form such as that shown in Fig. 7.1.[1] This form should include the following essential information: the discrepancy report serial number (numbered in advance), the date of the

[1] The forms and procedures described in the following discussion are typical and do not necessarily correspond to those in use.

discrepancy, the identification number of the end product, the part number of that discrepant part, nomenclature of that part, number of the next assembly (that assembly which immediately includes the discrepant part), serial number of that next assembly, name of the manufacturer of the discrepant part, part number assigned by that manufacturer, the number of hours or cycles of operation on the part prior to its removal (if available), the replacement part number; serial number of that part, name of the manufacturer of that part, and the serial number that manufacturer had assigned to the replacement part. In addition, it is imperative to include a careful description of the deficiency if a failure occurred; the action which was taken to correct the difficulty, the particular test or maintenance action which was being performed at the time, any measurements which were taken, and an opinion of the possible cause of failure if the reporting individual has the capability for making such a statement. The signature and telephone number of the reporting individual must be included so that he can be contacted if further information is required. If a nonfailure removal is being reported, reference to the instruction for removal should be included.

The COFEC Data Center is staffed by engineers who have received specialized training in reliability and are conversant with the function of components and assemblies of a particular type or system. To gain benefit from this specialization, the responsibility for record keeping and control is assigned by system, such as electrical, hydraulic, or pneumatic, and in some cases by the particular group of parts the function of which is utilized in more than one system. For example, a transducer may function in both the telemetry and the hydraulic systems. The specific assignment of responsibility should be based on the background, experience, and preference of the available personnel. Of course, clerical help is provided to allow these personnel to devote their attention to technical details and the control of corrective actions.

An up-to-date history on each part number and its interaction, if any, with other assemblies is maintained in the COFEC Data Center. Each received discrepancy report is assigned to the reliability engineer who maintains cognizance over the discrepant part. He checks the discrepancy report for errors and omissions and then enters it into his record. If a file folder exists for this part number, it will contain a history to which the new data are added. If, however, no file folder exists on this part number, a file is prepared for that part number and the discrepancy report data are entered as the initial record of activity. Although these records may become voluminous, every failure that has an impact on the reliability of the end product warrants investigation and corrective action. The cost of such activity is always small in comparison to the cost associated with future repetitive failures. Each folder contains the discrepancy history of a part number recorded in two forms: A descriptive record such as that shown in Fig. 7.2 and a control record such as that shown in Fig. 7.3.

The COFEC system employs a unique method of coding. Six digits are used to identify a precise description of the failure mode, its cause, and the corrective action and to indicate whether or not previous corrective actions have proved to be effective. This code is established first by grouping the parts by functional system or by the function of the part (this in correspondence with the assignment of responsibilities to the available personnel). All possible modes of failure for each group of parts are then listed, this being the result of a cooperative effort between the cognizant design engineer and the reliability engineer. In some cases a record of a recent design review may be available. Such a record would already contain a suitable list of potential modes of failure. If such a record is not available and time does not permit preparation of a list, then the failure modes are to be numbered in the sequence in which they are reported. This list of failure modes is called a master code list. Each master code list is identified by a prefix indicating the particular group of parts it serves—T for transducer, H for hydraulic, and so forth. Of course, the number of different master code lists depends upon the complexity of the end product.

There may be difficulty in recognizing and describing the cause of failure. For each failure there is a cause at some level of organization, and each cause may be traced, in theory at least, to another level below it. Primary concern is at that level of cause at which corrective actions may be taken. There is, of course, no objection to indicating still lower levels of cause in order to increase the scope of the available information, but these added data should remain of secondary interest.

The six digit positions of the COFEC code have specific meaning. The first two digit positions are used for identifying the failure mode. The second two digit posi-

COFEC — DESCRIPTIVE RECORD OF CODE TO IDENTIFY FAILURE MODE AND CORRECTIVE ACTION

PART NO. xxx			NAME TRANSDUCER / SYSTEM xx INSTRUMENTATION	
DATE	FAILURE MODE & CAUSE	MODE, CAUSE & CORRECTIVE ACTION	REMARKS	
1963				1
1 5	02 01		Static error tolerance exceeded +5.42% - 0.92%	2
1 20		02 01 01	FAR No. xxx Cause: Drive link ball out of wiper arm ball	3
			socket. Corr. Action: Vendor requested to investigate.	4
2 15		02 01 [01]	Vendor response. Socket has been enlarged and potting has been	5
			added for strength. Pull test 150 grams.	6
1 12	02 00		+1.94%-1.64%	7
1 18		02 00 [01]	FAR No. xxx Failure not confirmed. Investigation reveals	8
			incorrect test instrument calibration. Corr. Action: Test	9
			Procedure No. xxx Section xxx revised.	10
2 4	02 02		+0.00%-2.29%	11
2 20		02 02 01	FAR No. xxx Cause: Loss of reference pressure due to rough	12
			evacuation hole seat and ball. Corr. Action: Vendor requested	13
			to examine machining process and to consider use of bronze	14
			ball instead of steel ball.	15
3 12		02 02 [01]	Vendor response. Honing of seat improved. Will use bronze	16
			ball. Effective S/N xxx	17
2 15	02 03		+0.00%-17.87%	18
2 25		02 03 01	FAR No. xxx Cause: Bourdon tube leaking at the braze joint.	19
			Examination revealed voids and insufficient bond. Corr.	20
			Action: Vendor requested to review brazing process.	21
3 10		02 03 [01]	Vendor response. Brazing process has been revised. See file	22
			for detail. Effective S/N 123	23
4 12	02 03 [01]		+0.00%-12.43%	24
			S/N 144 made after corrective action of 3-10	25
4 19		02 03 02	FAR No. xxx Cause: Bourdon tube leaks at braze joint. This is	26
			same failure made as reported in FAR No. xxx 2-25-63. Corr.	27
			Action: Vendor advised that previous corrective action has	28
			proven ineffective and therefore more effort must be devoted	29
			to the solution of the brazing problem.	30
5 12		02 03 [02]	Vendor response. Brazed section of transducer has been rede-	31
			signed. New brazing alloy now used. Effective S/N xxx.	32
5 1	09 01		Spiking was exhibited at vibration frequencies between 20 and	33
			1200 cps.	34
5 15		09 01 01	FAR No. xxx Cause: Wiper arm bounce due to low wiper arm	35
			tension. Corr. Action: Vendor requested to consider increase	36

FIG. 7.2 A typical descriptive record.

tions specify the cause of failure, and the last two digit positions indicate the corrective action. Thus there can be as many as 99 different failure modes, 99 different causes of each failure mode, and 99 different corrective actions for each failure mode and cause of that mode. Note that this coding technique offers certain important advantages over the conventional codes which provide for the choice of a descriptor from a given vocabulary. In many cases these descriptor codes are ambiguous or so general that their use is of little value as the starting point for the design of a corrective action. Even though the reference vocabulary may be lengthy, it probably will not contain

COFEC — DESCRIPTIVE RECORD OF CODE TO IDENTIFY FAILURE MODE AND CORRECTIVE ACTION

PART NO. xxx | NAME TRANSDUCER | SYSTEM xx INSTRUMENTATION

DATE	FAILURE MODE & CAUSE	MODE, CAUSE & CORRECTIVE ACTION	REMARKS	
			in tension and its consequences.	1
6 2		09 01 01	Vendor response. Transducer has been redesigned. Vendor	2
			Change Proposal VCP No. xxx New unit will be identified by	3
			prefix A. Production to start 7-14-63.	4
5 10	03 00		Transducer was found to have unsteady output about 6% below	5
			what was expected. Measurement H33P	6
5 15		03 00 01	FAR No. xxx Failure not confirmed. The reported failure is	7
			due to external cause, incorrect hydraulic accumulator	8
			prechange. Procedure No. xxx changed to require 2500 psig.	9
6 28	03 01		Output is erratic.	10
7 5		03 01 01	FAR No. xxx Cause: Discontinuities caused by wiper riding over	11
			foreign particles. Corr. Action: Vendor requested to examine	12
			cleaning process.	13
7 12		03 01 01	Vendor response. New cleaning procedure and inspection insti-	14
			tuted. See file for detail. Effective S/N xxx	15
8 14	03 01 01		Output erratic at 1.3 psia.	16
8 20		03 01 02	FAR No. xxx Cause: Piece of lint on winding caused dis-	17
			continuity when wiper arm reached that point. Corr. Action:	18
			Vendor requested to re examine process in view of previous	19
			ineffective action.	20
10 25		03 01 02	Vendor response. Vendor is now using ultrasonic cleaning.	21
			Effective S/N xxx.	22
11 12	12		Output non-linear. 2% greater than spec.	23
11 14		12 03 00	FAR No. xxx Cause: Upper pivot of accelerometer shaft had	24
			broken loose due to prior assembly. Corr. Action: None.	25
			Production complete. New purchase not contemplated.	26
				27
				28
				29
				30
				31
				32
				33
				34
				35
				36

FIG. 7.2 A typical descriptive record. (*Continued.*)

a suitable description for a mode of failure which may be peculiar to the particular part in question. For example, "Frequency drifted below lower tolerance after warm-up period" will not be found in the usual vocabularies. This is especially true when an attempt is made to code the cause of failure, because that is peculiar to the product.

COFEC-CONTROL RECORD OF CORRECTIVE ACTIONS

PART NO. XXX				NAME TRANSDUCER								SYSTEM XX INSTRUMENTATION		
DATE 1963	F & CD NO.	DISCR. REPT. NO.	FAIL MODE	FAILURE ANALYSIS REPORT				NEXT ASS'Y NO.	DASH NO.	ACT.	FAIL CLASS.	REMARKS	TIME TO FAILURE	END PRODUCT NO.
				NUMBER	DATE	CODE	EFF.							
1-5	XX	XX	0201	XX	2-15	0201 [01]	S/N XX	XX	X	X	X		XX	XX
1-12	XX	XX	0200	XX	1-18	0200 [01]	S/N XX	XX	X	X	X		XX	XX
2-4	XX	XX	0202	XX	3-12	0202 [01]	S/N XX	XX	X	X	X		XX	XX
2-15	XX	XX	0203	XX	3-10	0203 [01]	S/N XX	XX	X	X	X		XX	XX
4-12	XX	144	0203 [01]	XX	5-12	0203 [02]	S/N XX	XX	X	X	X	S/N 144 after S/N 123, see entry 2-15	XX	XX
5-1	XX	XX	0901	XX	6-2	0901 [01]	S/N XX	XX	X	X	X		XX	XX
5-10	XX	XX	0300	XX	5-15	0300 [01]	S/N XX	XX	X	X	X		XX	XX
6-28	XX	XX	0301	XX	7-12	0301 [01]	S/N XX	XX	X	X	X		XX	XX
8-4	XX	184	0301 [01]	XX	10-25	0301 [02]	S/N XX	XX	X	X	X	S/N 184 after S/N 175, see entry 6-28	XX	XX
11-2	XX	XX	1203	XX	11-14	1203 00		XX	X	X	X		XX	XX

FIG. 7.3 A typical control record.

In contrast, the COFEC coding permits a description of any required length for the mode of failure, for the cause of failure, and for the corrective action, so that the pecularities of the type of part are reflected in the numerically referenced descriptions which "tell the whole story." This coding technique, in principle, is unlimited in its scope. The advantage results from a fundamental difference of procedure. In the conventional techniques the vocabularies are established for reference and personnel are required to fit the failure to the available descriptions, whereas in the COFEC coding system the description is constructed to fit the particular situation. To

```
MASTER  CODE  LIST   for    Transducers

                    prefix     T

01  Physical Discrepancy
02  Static Error Out of Tolerance, Receiving & Inspection Test
03  Erratic Output
.
.
.
11  Resistance Measurement Out of Tolerance
.
.
.
16  Leaking
.
.
.
```

FIG. 7.4 First page of a master code list.

```
MASTER  CODE  LIST (continued)

02          Static error out of tolerance, receiving & inspection test
02 01       Drive link ball out of wiper arm ball socket
02 01 [01]  Vendor XYZ Mfg. Co. has enlarged socket and has added potting
            compound for strength
.
.
.
02 03       Leak atBourdon tube braze joint
02 03 [01]  Vendor XYZ Mfg. Co. operators are trained at -------
            Brazing School effective Jan.12,63. serial no.xxx
02 03 [02]  Vendor XYZ Mfg. Co. now using improved leak test
            effective Mar.3,63.
.
.
.
02 03 [01]  Vendor ABCD Mfg. Co. has revised brazing process
            effective serial no.xxx
02 03 [02]  Vendor ABCD Mfg. Co.  has redesigned brazed section
            effective serial no. 152
```

FIG. 7.5 Second page of a master code list.

illustrate, a master code list for transducer failures is shown in Fig. 7.4. This is the first page of a group of pages and shows only the failure modes. Each of these modes has a separate page on which the experienced cause of failures and related corrective actions are entered, as shown in Fig. 7.5. Further uses of the COFEC coding technique for analysis and search of records will be described below.

A discrepancy report is received. The reliability engineer compares the reported discrepancy with the past history. The discrepancy report furnishes information descriptive of the failure mode. Reference is made to the master code list in order to determine whether or not this failure mode already appears. If so, the two digits which correspond to that failure mode are entered in column 2 of Fig. 7.7. If the appropriate failure mode does not appear, an adequate description is entered into the master code list and this description is designated by the next number. This new number is then entered into column 2 of Fig. 7.7 together with a description, amplified if required. Measurements are also included if they are pertinent. The extent of this descriptive entry is, of course, a matter of personal judgment. The code for the failure mode is entered in the space provided on the original discrepancy report. This report will reside in a permanent file so that it can be referred to in the future if desired. Control over the sequence of serial numbers must be maintained so that all are accounted for. The code number and description for the cause of failure and the corrective action will be entered on Fig. 7.7 when they become known. In a similar

FAILURE ANALYSIS REPORT

FAR NUMBER	PAGE	OF
XXX	1	X

PART NAME	ANALYSIS PERFORMED AT	DATE OF ANALYSIS		DISCREPANCY REPORT
PRESSURE TRANSDUCER	IN-HOUSE	Jan.12,1963		XXX
	MANUFACTURER OR VENDOR	PART NUMBER		
	XYZ Mfg. Co.	XXX		
	SERIAL NUMBER	GD/A PART NUMBER	NEXT ASSEMBLY	
	XXX	XXX	XXX	

History
1. This transducer reportedly failed on Dec.28,1962 at ------ because of open circuit. Measurement Number XXX
2. This transducer functions in the pressure programming system to indicate and record ....
3. This transducer contains a double metal bellows actuating a .....
4. COFEC record shows that this failure mode was reported 3 times within the last 6 months.

Analysis
1. Functional testing confirmed the reported failure .....
2. The transducer case was removed as shown ....

•

•

Conclusion
1. Pressure transducer failed resulting in ....
2. This is attributed to .... fatigue ....

Corrective Action
1. Vendor states " Our standard written procedure now calls for ...

Fig. 7.6 A typical failure-analysis report.

manner, entry of information concerning the discrepancy report is made on Fig. 7.8 as required. Severity classification of the failure is entered if indicated.

The use of two forms, Figs. 7.7 and 7.8, is deliberate in that it separates the numerical data from the descriptive matter, thus facilitating comprehension. Original entries are always in pencil in order to allow erasure if later information should indicate that the change in the failure-mode number or description should be made from a general one, such as "no output," to a more specific one, such as "open circuit." When a request is received for the reliability history of a part, Fig. 7.2 is typed as illustrated. This report shows the descriptive information in a manner which brings together the failure mode, cause of failure, and corrective action, regardless of the dates of their occurrence.

At least two carbon copies of the discrepancy report must be made by the initiator. The original is sent to the reliability data center. The first carbon copy remains at the point of origin, and the second carbon copy accompanies the failed part if that item is submitted for failure analysis. The use of different-color paper for these forms is recommended.

Let it be assumed that a certain part has been failure-analyzed. A failure-analysis report (Fig. 7.6) is received by the reliability data center. This report indicates the

COFEC — DESCRIPTIVE RECORD OF CODE TO IDENTIFY FAILURE MODE AND CORRECTIVE ACTION

PART NO. xxx	NAME TRANSDUCER	SYSTEM xx INSTRUMENTATION

DATE	FAILURE MODE & CAUSE	MODE, CAUSE & CORRECTIVE ACTION	REMARKS	
			(a) at 1-5-63	
1963				1
1-5	02		Static error tolerance exceeded +5.42% -0.92% to p1 l8	2
			(b) at 1-12-63	
1-5	02		Static error tolerance exceeded +5.42% -0.92% to p1 l8	2
1-12	02		+1.94% -1.64%	3
			(c) at 1-18-63	
1-5	02		Static error tolerance exceeded +5.42% -0.92% to p1 l8	2
1-12	02 00		+1.94% -1.64%	3
1-18	→	02 00 01	FAR No. xxx Failure not confirmed.	4
			Investigation reveals incorrect test	5
			instrument calibration. Corrective Action:	6
			Test Procedure No. xxx Sec. xxx revised.	7
			(d) at 1-20-63	
1-5	02 01		Static error tolerance exceeded +5.42% -0.92% to p1 l8	2
1-12	02 00		+1.94% -1.64%	3
1-18		02 00 01	FAR No. xxx Failure not confirmed.	4
			Investigation reveals incorrect test	5
			instrument calibration. Corrective Action:	6
			Test Procedure No. xxx Sec. xxx revised.	7
1-20	→	02 01 01	from p1 l2 FAR No. xxx Cause: Drive link ball	8
			out of wiper arm ball socket. Corrective	9
			Action: Vendor requested to	10
			investigate. to p1 l13	11
			(e) at 2-4-63 preceding entries as above	
2-4	02		" +0.00% -2.29% to p1 l18	12
			(f) at 2-15-63 preceding entries as above	
2-15	→	02 01 01	from p1 l11. Vendor response, socket has	13
			been enlarged and potting has been	14
			added for strength. Pull test 150 grams.	15
			Effective S/N xxx.	16
2-15	02		" +0.00% -17.87% to p1 l24	17

(arrows indicate new entries)

FIG. 7.7 A typical descriptive record at different points in time.

cause of failure and a suggested corrective action. The file on this part is now updated by the cognizant reliability engineer. Suppose that a month later this corrective action has been implemented. Notice of this implementation is also received by the reliability data center, and again the file is updated. Reference is made to the master code list. If the particular cause of failure has been experienced previously for the same failure mode, then the master code list will show the second pair of digits that is to be added to the previous entry on Fig. 7.7, which now will have four digits in the

(g) at 2-20-63
preceding entries as above

2-4	02 02		+0.00% -2.29% to p1 l18	12
2-15		02 01 [01]	from p1 l11 Vendor response. socket has	13
			been enlarged and potting has been	14
			added for strength. Pull test 150 grams.	15
			Effective S/N xxx.	16
2-15	02		+0.00% -17.87% to p1 l24	17
2-20		→ 02 02 01	from p1 l12 FAR No. xxx. Cause: Loss of reference	18
			pressure due to rough evacuation hole	19
			seat and ball. Corrective Action: Vendor	20
			requested to examine machining	21
			process and to consider use of bronze	22
			ball instead of steel ball.	23

(h) at 2-25-63
preceding entries as above

2-15	02 03		+0.00% -17.87% to p1 l24	17
2-20		02 02 01	from p1 l12 FAR No. xxx. Cause: Loss of reference	18
			pressure due to rough evacuation hole	19
			seat and ball. Corrective Action: Vendor	20
			requested to examine machining	21
			process and to consider use of bronze	22
			ball instead of steel ball.	23
2-25		→ 02 03 01	from p1 l17 FAR No. xxx Cause: Bourdon tube	24
			leaking at the braze joint. Examination	25
			revealed voids and insufficient bond.	26
			Corrective Action: Vendor requested to	27
			review brazing process.	28

(i) at 3-10-63
preceding entries as above

2-25		02 03 01	from p1 l17 FAR No. xxx Cause: Bourdon tube	24
			leaking at the braze joint. Examination	25
			revealed voids and insufficient bond.	26
			Corrective Action Vendor requested to	27
			review brazing process.	28
3-10		→ 02 03 [01]	Vendor response. Brazing process has	29
			been revised. see file for detail.	30
			Effective S/N 123.	31

FIG. 7.7 A typical descriptive record at different points in time. (*Continued.*)

second column. If the cause of failure is a new one for this failure mode, then the next two digits are to be assigned and then entered in the second column.

The corrective-action entry is made in a similar manner. If the cause and the corrective action have not both been experienced previously, then the four digits are assigned and described in the master code. If the cause has been experienced previously but the corrective action is new, the next number is assigned for the last pair of digits. After the master code list has been updated, if necessary, the entries are also made on Fig. 7.8. A period of time usually elapses between the request for corrective action and its implementation. This is reflected in the entries described below.

			(j) at 3-12-63 preceding entries as above	
3-12		→ 02 02 [01]	from p1 l23 Vendor response Honing of	32
			seat improved. Will use bronze ball	33
			(k) at 4-12-63 preceding entries as above	
4-12	↓ 02		+0.00% -12.43%	35
			S/N 144 Unit was made after corrective action 3-10-63 to p2 l1	36
			(l) at 4-19-63 (start of page 2)	
4-12	↓ ↓ 02 03 [01]		+0.00% -12.43%	35
			S/N 144 Unit was made after corrective action 3-10-63 to p2 l1	36
4-19		→ 02 03 02	from p1 l36 FAR No. xxx Cause; Bourdon tube	1
			leak at braze joint. This is the same failure	2
			mode as reported in FAR No xxx 2-25-63.	3
			Corrective Action: Vendor advised that	4
			previous action has proven ineffective and	5
			therefore more effort must be devoted to the	6
			solution of the brazing problem to p2 l13	7
			(m) at 5-1-63	
4-19		02 03 02	from p1 l36 FAR No. xxx Cause; Bourdon tube	1
			leak at braze joint. This is the same failure	2
			mode as reported in FAR No xxx 2-25-63	3
			Corrective Action: Vendor advised that	4
			previous action has proven ineffective and	5
			therefore more effort must be devoted to the	6
			solution of the brazing problem to p2 l13	7
5-1	09 ←		Spiking was exhibited at vibration frequencies	8
			between 20 and 1200 cps. to p2 l17	9
			(n) at 5-10-63 preceding entries as above	
5-10	↓ 03		Transducer was found to have unsteady	10
			output about 6% below what was expected.	11
			Measurement H33P	12
			(o) at 5-12-63 preceding entries as above	
5-12		→ 02 03 [02]	from p2 l7 Vendor response. Brazed section	13
			of transducer has been redesigned. New	14
			brazing alloy now used	15
			Effective S/N. xxx.	16

FIG. 7.7 A typical descriptive record at different points in time. (*Continued.*)

It is important that this time lapse be indicated so that the exact status can be reported if a request for part history should be received. When implementation is made, the last two digits are *not* changed; but this implementation is indicated by placing a marker (a square enclosure) around the last two digits. This serves an additional important purpose as indicated in the following illustrative example.

To show how there is a gradual buildup of the item history, Figs. 7.7 and 7.8 are developed by successive entries. These figures are shown in segments as they would appear at particular points in time, Figs. 7.7*a* to 7.7*y* and Figs. 7.8*a* to 7.8*y*. On January 5, a static-error tolerance was exceeded (Figs. 7.7*a* and 7.8*a*). On January 12,

			(p) at 5-15-63 preceding entries as above	
5-1	09 01		Spiking was exhibited at vibration frequencies	8
			between 20 and 1200 cps. to p2 l17	9
5-10	03 00		Transducer was found to have unsteady	10
			output about 6% below what was expected.	11
			Measurement H33P	12
5-12		02 03 [02]	from p2 l7 Vendor response. Brazed section	13
			of transducer has been redesigned. New	14
			brazing alloy now used.	15
			Effective S/N xxx.	16
5-15		→ 09 01 01	from p2 l9 FAR No. xxx Cause: Wiper arm	17
			bounce due to low wiper arm tension.	18
			Corrective Action: Vendor requested to	19
			consider increase in tension and its	20
			consequences to p2 l27	21
5-15		→ 03 00 [01]	from p2 l12 FAR No. xxx Failure not	22
			confirmed. The reported failure is due	23
			to external cause; incorrect hydraulic	24
			accumulator precharge. Procedure No. xxx	25
			changed to require 2500 psig.	26
			(q) at 6-2-63 preceding entries as above	
5-15		09 01 01	from p2 l9 FAR No. xxx Cause: Wiper arm	17
			bounce due to low wiper arm tension.	18
			Corrective Action: Vendor requested to	19
			consider increase in tension and its	20
			consequences to p2 l27	21
5-15		03 00 [01]	from p2 l12 FAR No. xxx Failure not	22
			confirmed. The reported failure is due	23
			to external cause; incorrect hydraulic	24
			accumulator precharge. Procedure No. xxx	25
			changed to require 2500 psig.	26
6-2		→ 09 01 [01]	from p2 l21 Vendor response. Transducer	27
			has been redesigned. Vendor Change	28
			Proposal VCP No. xxx. New units	29
			will be identified by prefix A.	30
			Production to start 7-14-63	31
			(r) at 6-28-63 preceding entries as above	
6-28	03		output is erratic	32

Fig. 7.7 A typical descriptive record at different points in time. (*Continued.*)

another static-error tolerance was reported (Figs. 7.7*b* and 7.8*b*). On January 18, a failure-analysis report pertaining to the second entry was received, indicating that the failure was not confirmed. The cause for reported failure was incorrect test-instrument calibration; the corrective action was revision of test procedure. Figures

7.7*c* and 7.8*c* show the second pair of COFEC code digits, 00, indicating failure not confirmed. In addition, the last two digits are enclosed by a marker, previously referred to, indicating that the recommended corrective action has been implemented.

(s) at 7-5-63				
preceding entries as above				
6-28	03 01		output is erratic	32
7-5		→ 03 01 01	FAR No. xxx Cause. Discontinuities caused by wiper	33
			riding over foreign particles. Corrective Action: Vendor	34
			requested to examine cleaning process to p3 l1	35
(t) at 7-12-63 (start of page 3)				
7-12		→ 03 01 [01]	from p2 l36 Vendor response. New cleaning	1
			procedure and inspection instituted. see	2
			file for detail. Effective S/N 175	3
(u) at 8-14-63				
7-12		03 01 [01]	from p2 l36 Vendor response. New cleaning	1
			procedure and inspection instituted. see	2
			file for detail. Effective S/N 175	3
8-14	03 01 [01]	←	output erratic at 1.3 psia S/N 184 after 7-12-63	4
(v) at 8-20-63				
7-12		03 01 [01]	from p2 l36 Vendor response. New cleaning	1
			procedure and inspection instituted. see	2
			file for detail. Effective S/N 175	3
8-14	03 01 [01]		output erratic at 1.3 psia S/N 184 after 7-12-63	4
8-20		→ 03 01 02	FAR No. xxx Cause: Piece of lint on winding	5
			caused discontinuity when wiper arm	6
			reached that point. Corrective Action:	7
			Vendor requested to re-examine process	8
			in view of previous ineffective action.	9
(w) at 10-25-63				
7-12		03 01 [01]	from p2 l36 Vendor response. New cleaning	1
			procedure and inspection instituted. see	2
			file for detail. Effective S/N 175	3
8-14	03 01 [01]		output erratic at 1.3 psia S/N 184 after 7-12-63	4
8-20		03 01 02	FAR No. xxx Cause: Piece of lint on winding	5
			caused discontinuity when wiper arm	6
			reached that point. Corrective Action:	7
			Vendor requested to re-examine process	8
			in view of previous ineffective action.	9
10-25		→ 03 01 [02]	Vendor response. Vendor is now using	10
			ultrasonic cleaning. Effective S/N 204	11

FIG. 7.7 A typical descriptive record at different points in time. (*Continued.*)

On January 20, a failure-analysis report for the failure recorded on January 5 was received. This cause of failure for static error is designated by 01. The corrective action requested is indicated by 01 in the last two digit positions (Figs. 7.7*d* and 7.8*d*). If and when the corrective action is implemented, it will be indicated by enclosing these last two digits as shown in Figs. 7.7*f* and 7.8*f* (entered on February 15).

On February 20, the failure-analysis report indicating the cause of the failure reported on February 4 was received. Since this was a new cause, it required another identification code, in this case 02. The request for corrective action is the first one for this cause and is therefore indicated by 01 in the last two digit positions (Figs. 7.7*g* and 7.8*g*).

On April 12, a static-error failure report was received on transducer serial number 144 (Figs. 7.7*k* and 7.8*k*). On April 19 the failure-analysis report stated that this serial number had incorporated within it the corrective action which was entered on March 10. This stated that the brazing process had been revised. The failure occurred, therefore, despite the corrective action. This is indicated prominently on Figs. 7.7*l* and 7.8*l*, in which the failure is now coded as 0203$\boxed{01}$. *The whole story of the failure, the cause of failure, the corrective action, and, in this case, the ineffectiveness of the corrective action, is clearly portrayed in a simple, easy-to-understand manner.* A review of the failure history shows not only the activity of the part, the number of ways in which the product had failed, and the number of attempts which had been made to correct for the causes of failure but also the number of times in which the technical knowledge or process control was unable to cope with the reliability problem.

(x) at 11-2-63
preceding entries as above

11-2	12		Output non-linear. 2% greater than spec.	12

(y) at 11-14-63
preceding entries as above

11-2	12 03		Output non-linear. 2% greater than spec.	12
11-14	→	12 03 00	FAR No. xxx Cause: Upper pivot of accelerometer	13
			shaft had broken due to poor assembly.	14
			Corrective Action: None. Production complete	15
			New purchase not contemplated.	16

FIG. 7.7 A typical descriptive record at different points in time. (*Continued.*)

For the design engineer this is of paramount importance, since it directs his thinking in new product design and development. For management this indicates the capability of the producer of the product and its influence on cost and production schedule.

The entry on April 19, Figs. 7.7*l* and 7.8*l*, is coded 020302, which indicates that the failure mode and cause of failure were the same, but that a second request for corrective action had been made. On May 12, the implementation of the second request for corrective action regarding the brazing problem was recorded as 0203$\boxed{02}$ (Figs. 7.7*o* and 7.8*o*).

On November 2, a report was received of a "nonlinear output" (failure mode 12). On November 14, the failure-analysis report indicating the cause was received; the cause was identified as 03. The report stated that corrective action would not be taken because production was complete and new purchase was not contemplated. This is indicated by 00 in the last two digit positions. Note that "no corrective action" is always indicated by 00. Because successive entries pertaining to a particular discrepancy report are not consecutive, a means for relating these entries must be provided. The first entry that is made usually carries only failure-mode information. During the time period that elapses from an initial entry to the time of receipt of the failure analysis, there may be other reports of discrepancies which are entered chronologically; thus the second entry pertaining to the failure cannot be made on the following line. It is therefore advisable to relate entries on the same item by cross-referencing page and line numbers, as indicated in Fig. 7.7.

COFEC-CONTROL RECORD OF CORRECTIVE ACTIONS

PART NO. X X X | NAME TRANSDUCER | SYSTEM XX INSTRUMENTATION

DATE	F & CD NO.	DISCR. REPT. NO.	FAIL MODE	FAILURE ANALYSIS REPORT				NEXT ASS'Y NO.	DASH NO.	ACT.	FAIL CLASS.	REMARKS	TIME TO FAILURE	END PRODUCT NO.
				NUMBER	DATE	CODE	EFF.							
1963														
												(a) at 1-5-63		
1-5	XXX	XXX	02						XX	XX	X		XX	XX
												(b) at 1-12-63		
1-5	XXX	XXX	02						XX	XX	X		XX	XX
1-12	XXX	XXX	02						XX	XX	X		XX	XX
												(c) at 1-18-63		
1-5	XXX	XXX	02						XX	XX	X		XX	XX
1-12	XXX	XXX	↓0200	XXX	1-18	↓0200[01]	S/N XXX		XX	XX	X		XX	XX
												(d) at 1-20-63		
1-5	XXX	XXX	↓0201	XXX	1-20	↓020101			XX	XX	X		XX	XX
1-12	XXX	XXX	0200	XXX	1-18	0200[01]	S/N XXX		XX	XX	X		XX	XX
												(e) at 2-4-63		
1-5	XXX	XXX	0201	XXX	1-20	020101			XX	XX	X		XX	XX
1-12	XXX	XXX	0200	XXX	1-18	0200[01]	S/N XXX		XX	XX	X		XX	XX
2-4	XXX	XXX	↓02	XXX					XX	XX	X		XX	XX
												(f) at 2-15-63		
1-5	XXX	XXX	0201	XXX	2-15	↓0201[01]	↙S/N XXX		XX	XX	X		XX	XX
1-12	XXX	XXX	0200	XXX	1-18	0200[01]	S/N XXX		XX	XX	X		XX	XX
2-4	XXX	XXX	02						XX	XX	X		XX	XX
2-15	XXX	XXX	↓02						XX	XX	X		XX	XX

(arrows indicate new entries)

FIG. 7.8 A typical control record at different points in time.

(g) at 2-20-63

1-5	xxx	xxx	0201	xxx	2-15	0201 01	S/N xxx		XX	XX	X		XX	XX
1-12	xxx	xxx	0200	xxx	1-18	0200 01	S/N xxx		XX	XX	X		XX	XX
2-4	xxx	xxx	0202	xxx	2-20	020201			XX	XX	X		XX	XX
2-15	xxx	xxx	02						XX	XX	X		XX	XX

(h) at 2-25-63

2-15	xxx	xxx	0203	xxx	2-25	020301			XX	XX	X		XX	XX

(i) at 3-10-63

2-15	xxx	110	0203	xxx	3-10	0203 01	S/N 123		XX	XX	X		XX	XX

(j) at 3-12-63

2-4	xxx	xxx	0202	xxx	3-12	0202 01	S/N xxx		XX	XX	X		XX	XX
2-15	xxx	110	0203	xxx	3-10	0203 01	S/N 123		XX	XX	X		XX	XX

(k) at 4-12-63

4-12	xxx	144	02						XX	XX	X	S/N 144 after S/N 123 see entry 2-15	XX	XX

(l) at 4-19-63

4-12	xxx	144	0203 01	xxx	4-19	020302			XX	XX	X	S/N 144 after S/N 123 see entry 2-15	XX	XX

(m) at 5-1-63

5-1	xxx	xxx	09						XX	XX	X		XX	XX

(n) at 5-10-63

5-10	xxx	xxx	03						XX	XX	X		XX	XX

Fig. 7.8 A typical control record at different points in time. (*Continued.*)

(o) at 5-12-63														
4-12	xxx	144	0203[01]	xxx	5-12	0203[02]	S/N 152		xx	xx	x	S/N 144 after S/N 123 see entry 2-15	xx	xx
5-1	xxx	xxx	09						xx	xx	x		xx	xx
5-10	xxx	xxx	03						xx	xx	x		xx	xx
(p) at 5-15-63														
5-1	xxx	xxx	0901	xxx	5-15	090101			xx	xx	x		xx	xx
5-10	xxx	xxx	0300	xxx	5-15	0300[01]	5-20		xx	xx	x		xx	xx
(q) at 6-2-63														
5-1	xxx	xxx	0901	xxx	6-2	0901[01]	S/N Axx		xx	xx	x		xx	xx
5-10	xxx	xxx	0300	xxx	5-15	0300[01]	5-20		xx	xx	x		xx	xx
(r) at 6-28-63														
6-28	xxx	xxx	03	xxx					xx	xx	x		xx	xx
(s) at 7-5-63														
6-28	xxx	xxx	0301	xxx	7-5	030101			xx	xx	x		xx	xx
(t) at 7-12-63														
6-28	xxx	124	0301	xxx	7-12	0301[01]	S/N 175		xx	xx	x		xx	xx
(u) at 8-14-63														
8-14	xxx	184	0301[01]						xx	xx	x	S/N 184 after S/N 175 see entry 6-28	xx	xx
(v) at 8-20-63														
8-14	xxx	184	0301[01]	xxx	8-20	030102			xx	xx	x	S/N 184 after S/N 175 see entry 6-28	xx	xx
(w) at 10-25-63														
8-14	xxx	184	0301[01]	xxx	10-25	0301[02]	10-15		xx	xx	x	S/N 184 after S/N 175 see entry 6-28	xx	xx
(x) at 11-2-63														
11-2	xxx	xxx	12						xx	xx	x		xx	xx
(y) at 11-12-63														
11-2	xxx	xxx	1203	xxx	11-14	120300			xx	xx	x		xx	xx

FIG. 7.8 A typical control record at different points in time. (*Continued.*)

Figure 7.8 is confined primarily to numerical data which, in this form, are particularly useful for reference to supporting information if such data pertaining to any particular entry on Fig. 7.7 should be required. For example, the design engineer may note that spiking failure, 09, is reported on May 1 (Fig. 7.7*m*). He may want to know at what activity the failure occurred, on what end-product number, time to failure, and so forth; this related information is shown in the last entry of Fig. 7.8*m*. Further, a review of the serial-number column reveals whether or not the same part is reported after rework.

This compressed form of data presentation not only records supporting data, but, because of its arrangement, displays the entire history of failures, causes of failure, and corrective actions in coded form. With the master code list at hand the overall descriptive picture becomes apparent. After such an overall review the design engineer may wish to examine certain details more closely. He can do so by reference

FAILURE MODE ACTIVITY

TRANSDUCER PART NO. XXX	01	02 00 00	02 04 [01]	·	·	·	·	·	08 04 [02]	·	·	·	·	·	·	12 02 [01]	·	·	·	·	·	·	14 01 00	·	·	16 02 [02]	TOTAL
– 3	2	4	1	·	·	·	·	·	1	·	·	·	·	·	·	1	·	·	·	·	·	·	2	·	·	1	79
– 5	3		2	·	·	·	·	·		·	·	·	·	·	·	3	·	·	·	·	·	·	1	·	·	1	4
– 7		3	1	·	·	·	·	·	1	·	·	·	·	·	·	1	·	·	·	·	·	·	4	·	·	1	15
– 8	1	1	1	·	·	·	·	·	2	·	·	·	·	·	·	1	·	·	·	·	·	·	1	·	·	1	13
·	·	·	·	·	·	·	·	·	·	·	·	·	·	·	·	·	·	·	·	·	·	·	·	·	·	·	·
·	·	·	·	·	·	·	·	·	·	·	·	·	·	·	·	·	·	·	·	·	·	·	·	·	·	·	·
·	·	·	·	·	·	·	·	·	·	·	·	·	·	·	·	·	·	·	·	·	·	·	·	·	·	·	·
·	·	·	·	·	·	·	·	·	·	·	·	·	·	·	·	·	·	·	·	·	·	·	·	·	·	·	·
·	·	·	·	·	·	·	·	·	·	·	·	·	·	·	·	·	·	·	·	·	·	·	·	·	·	·	·
–25	1	1		·	·	·	·	·	1	·	·	·	·	·	·		·	·	·	·	·	·		·	·	·	12
–29			1	·	·	·	·	·	·	·	·	·	·	·	·		·	·	·	·	·	·	1	·	·	·	3
–30	2	1		·	·	·	·	·	1	·	·	·	·	·	·	1	·	·	·	·	·	·		·	·	·	11
TOTAL	12	15	11	·	·	·	·	·	8	·	·	·	·	·	·	15	·	·	·	·	·	·	22	·	·	·	144
RESPONSIBILITY FOR FAILURE																											
NOT CONFIRMED	1	1	2	·	·	·	·	·	·	·	·	·	·	·	·	1	·	·	·	·	·	·	2	·	·	·	·
NOT CHARGEABLE TO VENDOR	·	·	4	·	·	·	·	·	·	·	·	·	·	·	·	·	·	·	·	·	·	·	·	·	·	·	·
CHARGEABLE TO VENDOR	2	1	·	·	·	·	·	·	1	·	·	·	·	·	·	2	·	·	·	·	·	·	·	·	·	·	·
NOT ANALYZED TO THIS DATE	·	·	·	·	·	·	·	·	2	·	·	·	·	·	·	·	·	·	·	·	·	·	2	·	·	·	·
TOTAL	12	15	11	·	·	·	·	·	8	·	·	·	·	·	·	15	·	·	·	·	·	·	22	·	·	·	144

FIG. 7.9 A typical analysis chart for a stated period.

to the related entry on Fig. 7.7. From a control standpoint it is of extreme importance to know the length of the time period between failure and implementation of corrective action. These data are shown respectively in columns 1 and 6 of Fig. 7.8. At any time a glance reveals corrective actions that remain to be taken, those that have been requested but not yet implemented, those that have been implemented, and those that have been implemented but have proved to be ineffective and, finally, the total number of failures and whether or not they are repetitive.

If a part is at the frontier of the state of the art and is purchased in considerable quantity, it is reasonable to expect that the discrepancies will be numerous until the process has come under "control." In that event first reference should be made to Fig. 7.8 in order to grasp the overall picture. In many cases it will be found advantageous to prepare a chart such as that shown in Fig. 7.9. This indicates how 144 discrepancy reports can be examined by failure mode, and/or by cause of failure, and by dash number or type of part, where applicable. A similar analysis can be made for a group of parts, such as transducers, oscillators, and hydraulic valves, or by vendor for individual or comparative study. These data can be recast in terms of responsibility for failure such as improper handling, inadvertent destruction, and use under environment beyond that called for in the specification. It goes without saying that such failures should not be charged against the vendor.

Figure 7.10 offers a more detailed analysis of a complex assembly to illustrate further the procedure and the depth to which analyses can be carried out. The figure shows not only how the failure occurred on the assembly but also what components are

TELEPAK ASSEMBLY
PART NO. XXX

COMPONENT NAME AND PART NO.	FAILURE CODE	DASH NO.				FAILURE MODE TOTAL	SUBTOTAL	COMPONENT TOTAL
		-3	-5	-8	-10			
OSCILLATOR								
PART NO. XXX-4								
	04 02	2			1	3		
	08 07	1	2	1		4		
SUBTOTAL		3	2	1	1		7	
PART NO. XXX-1								
	04 02 [01]	2				2		
	14 05		1	2		3		
SUBTOTAL		2	1	2			5	
TOTAL		5	3	3	1			12
AMPLIFIER								
PART NO. XXX-36								
	01 02	3	1			4		
	03 03		1	3		4		
SUBTOTAL		3	2	3			8	
TOTAL								8
TRANSMITTER								
PART NO. XXX-10								
	05 02	3			2	5		
	05 03	1		1		2		
	05 03 [01]	1		1		2		
SUBTOTAL		5		2	2		9	
TOTAL								9
GRAND TOTAL		24	16	14	25			79

FIG. 7.10 A typical analysis chart for complex assembly.

responsible for the overall failure and the causes and corrective actions taken to correct these conditions.

In monitoring the development of a product it is important to know the progress which is being made from a reliability standpoint. This centers around not only the failure modes and their relative frequency, but more particularly whether or not they are diminishing with time.[1] This trend is clearly indicated by a chart such as is shown in Fig. 7.11.

It is intended that quality control problems be included in the COFEC reliability data system. If the size of the organization and activity level warrant, it is advisable

[1] The relative frequency is based upon a comparison of the number of failures to the number of items tested.

to establish a separate COFEC data center for quality control. In order to identify a clear distinction between domain of cognizance of the data centers, it is suggested that the in-house manufacturing operation be viewed as an entity similar to a vendor. All problems which occur prior to and including final test are to be considered quality control problems. When a part passes final assembly test, it is to be considered as equivalent to a purchased part. Any discrepancies which occur after final test are recorded as reliability discrepancies. The vendor's quality control problems do not have any influence on reliability unless they develop into a failure after the product has been completed. The QC-COFEC files should preferably be maintained by class of part. For example, all reports on circuit boards should be placed in one file. Here it is not the part itself which is of interest; rather, it is the assembly processes which are used in the manufacture of that part that are of interest.

As indicated above, the COFEC reliability data system is designed to answer the questions of the design engineer. He may inquire about the frequency of a particular failure mode and cause, not for an individual part, but for a group or class of parts. Such data may be retrieved from the COFEC file by scanning the permanent records of all the discrepancy reports for the particular numerical code, thereby selecting only those discrepancy reports which contain the desired information. Such a search of the discrepancy reports can be performed either automatically or manually. Manual

TRANSDUCER GROUP A PART NOS. XXX, XXX, XXX.

FAILURE MODES

1962	01	02	03	·	·	·	·	08	·	·	·	12	·	·	·	16	·	18	TOTAL
JAN	20	10	3	·	·	·	·	7	·	·	·	2	·	·	·	1	·	9	·
FEB	15	12	2	·	·	·	·	8	·	·	·		·	·	·	3	·	3	·
MAR	6	16		·	·	·	·	6	·	·	·	2	·	·	·		·	1.	·
APR	9	16	1	·	·	·	·	2	·	·	·	1	·	·	·	2	·		·
·	·	·		·	·	·	·	·	·	·	·	·	·	·	·		·	·	·
·	·	·		·	·	·	·	·	·	·	·	·	·	·	·		·		·
·																			
·																			

FIG. 7.11 A typical trend-analysis chart.

search can be facilitated by making an extra carbon copy or by reproducing all discrepancy reports as they are received and maintaining a separate file by part category. The number of discrepancy reports which must be examined is greatly reduced by such classification.

7.3 PRESENTATION OF DATA FOR EXECUTIVE ACTION

The COFEC system provides information to management and supervision in three different forms, each designed to satisfy a different need. The first, shown in Fig. 7.12, is a quantitative report which indicates how many discrepancies occurred during the time period covered, how many of these occurred at each severity level (if so classified), how many of these occurred on each end product, how many occurred in each system, and so forth. In the illustration the 0 in the end-product column indicates that the part had not yet been assigned to a particular end product. This report may be issued weekly or at some other suitable time interval. It allows week-to-week comparison by end product, by system, by activity, and so forth.

A second report is the descriptive report by end product shown in Fig. 7.13. It also may be provided weekly and is arranged by each end product, by system, and by part number. This periodic report allows top management to follow the weekly progress of the end product in terms of what parts failed on what end product, in what system, and so forth.

The third report is also descriptive but is arranged by system and by part number, as shown in Fig. 7.14. This report is designed for the purpose of maintaining a weekly flow of information to the head of the engineering design group of the discrepancy activity under his cognizance. It includes the same information as in Fig. 7.13 but is recast by system and part number.

QUANTITATIVE STATUS REPORT

WEEKLY END PRODUCT DISCREPANCY SUMMARY

DATE

		FAILURE CLASSIFICATION				
END PRODUCT	SYSTEM	A	B	C	OTHER	TOTAL
0-	04			2	13	15
0-	06				3	3
0-	08			1		1
0-	10				1	1
0-	14			1		1
0-	18				3	3
0-	24		2	2	2	6
		**	2**	6**	22**	30**
5	06				1	1
		**	**	**	1**	1**
6	08			1		1
6	20			1		1
6	24		1	2		3
		**	1**	4**	**	5**
11	10			1		1
		**	**	1**	**	1**
12	24			2		2
		**	**	2**	**	2**
13	04		1	1		2
		**	1**	1**	**	2**
14	04	2				2
		2**	**	**	**	2**
15	04	1				1
15	24		1			1
		1**	1**	**	**	2**
17	14				1	1
17	20			1		1
17	24		2		1	3
		**	2**	1**	2**	5**
18	06			1		1
		**	**	1**	**	1**

FIG. 7.12 A typical quantitative status report.

To complete the management presentation, three additional reports that are of identical format are prepared monthly and show cumulative data. These records present the entire history of all products to the end of the preceding month.

The COFEC reliability data system is also of value for the purpose of vendor rating. In arriving at a score for each vendor, several attributes are weighted. One of these is quality of product, which is indicated by the failure activity level. As mentioned

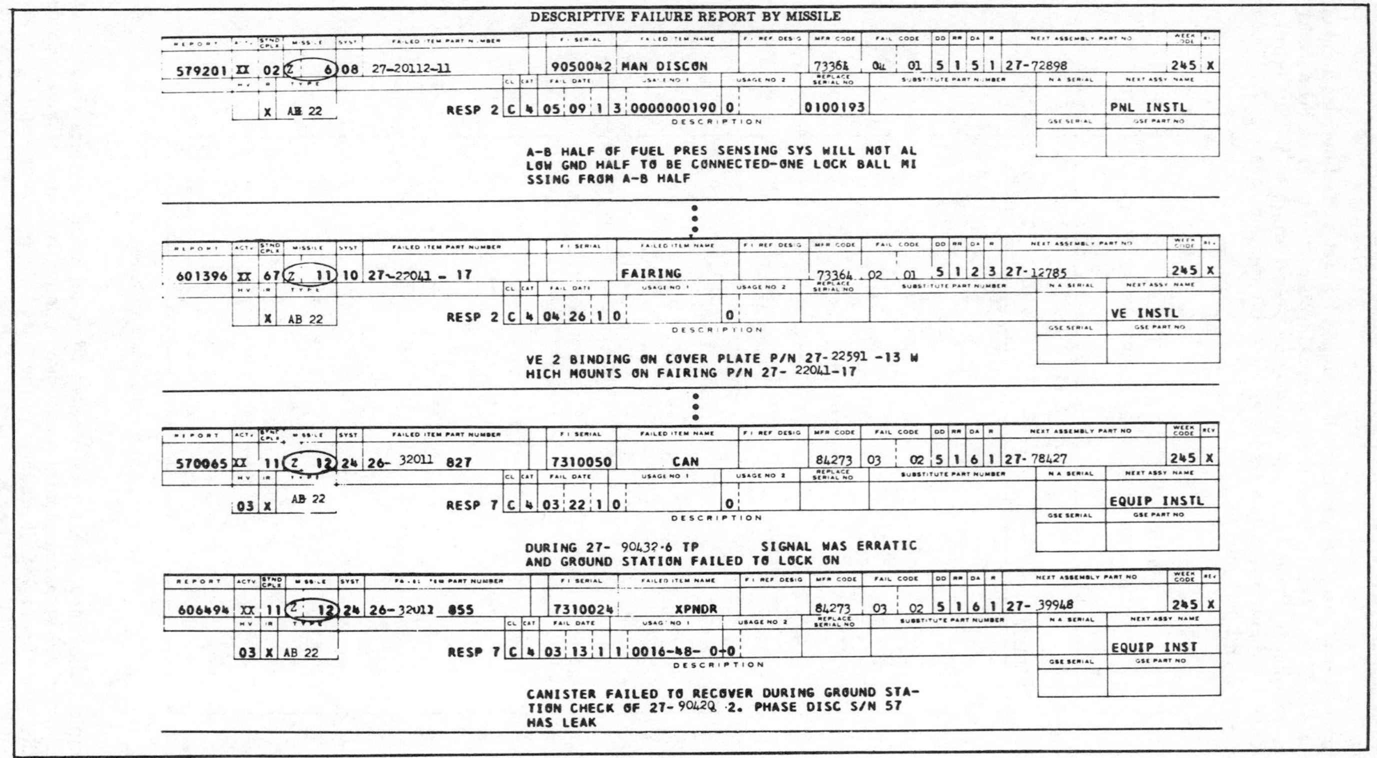

DESCRIPTIVE FAILURE REPORT BY MISSILE

REPORT	ACTV	STND CPLX	MISSILE	SYST	FAILED ITEM PART NUMBER	F I SERIAL	FAILED ITEM NAME	F I REF DESIG	MFR CODE	FAIL CODE	DD	RR	DA	R	NEXT ASSEMBLY PART NO	WEEK CODE	REV
579201	XX	02	Z 6	08	27-20112-11	9050042	MAN DISCON		73364	04 01	5	1	5	1	27-72898	245	X

HV	IR	TYPE		CL	CAT	FAIL DATE	USAGE NO 1	USAGE NO 2	REPLACE SERIAL NO	SUBSTITUTE PART NUMBER	N A SERIAL	NEXT ASSY NAME
	X	AB 22	RESP 2	C	4	05 09 1 3	0000000190	0	0100193			PNL INSTL

DESCRIPTION

A-B HALF OF FUEL PRES SENSING SYS WILL NOT AL
LOW GND HALF TO BE CONNECTED-ONE LOCK BALL MI
SSING FROM A-B HALF

GSE SERIAL	GSE PART NO

⋮

REPORT	ACTV	STND CPLX	MISSILE	SYST	FAILED ITEM PART NUMBER	F I SERIAL	FAILED ITEM NAME	F I REF DESIG	MFR CODE	FAIL CODE	DD	RR	DA	R	NEXT ASSEMBLY PART NO	WEEK CODE	REV
601396	XX	67	Z 11	10	27-22041 - 17		FAIRING		73364	02 01	5	1	2	3	27-12785	245	X

HV	IR	TYPE		CL	CAT	FAIL DATE	USAGE NO 1	USAGE NO 2	REPLACE SERIAL NO	SUBSTITUTE PART NUMBER	N A SERIAL	NEXT ASSY NAME
	X	AB 22	RESP 2	C	4	04 26 1 0		0				VE INSTL

DESCRIPTION

VE 2 BINDING ON COVER PLATE P/N 27-22591 -13 W
HICH MOUNTS ON FAIRING P/N 27- 22041-17

GSE SERIAL	GSE PART NO

⋮

REPORT	ACTV	STND CPLX	MISSILE	SYST	FAILED ITEM PART NUMBER	F I SERIAL	FAILED ITEM NAME	F I REF DESIG	MFR CODE	FAIL CODE	DD	RR	DA	R	NEXT ASSEMBLY PART NO	WEEK CODE	REV
570065	XX	11	Z 12	24	26- 32011 827	7310050	CAN		84273	03 02	5	1	6	1	27- 78427	245	X

HV	IR	TYPE		CL	CAT	FAIL DATE	USAGE NO 1	USAGE NO 2	REPLACE SERIAL NO	SUBSTITUTE PART NUMBER	N A SERIAL	NEXT ASSY NAME
03	X	AB 22	RESP 7	C	4	03 22 1 0		0				EQUIP INSTL

DESCRIPTION

DURING 27- 90432-6 TP SIGNAL WAS ERRATIC
AND GROUND STATION FAILED TO LOCK ON

GSE SERIAL	GSE PART NO

REPORT	ACTV	STND CPLX	MISSILE	SYST	FAILED ITEM PART NUMBER	F I SERIAL	FAILED ITEM NAME	F I REF DESIG	MFR CODE	FAIL CODE	DD	RR	DA	R	NEXT ASSEMBLY PART NO	WEEK CODE	REV
606494	XX	11	Z 12	24	26- 32012 855	7310024	XPNDR		84273	03 02	5	1	6	1	27- 39948	245	X

HV	IR	TYPE		CL	CAT	FAIL DATE	USAGE NO 1	USAGE NO 2	REPLACE SERIAL NO	SUBSTITUTE PART NUMBER	N A SERIAL	NEXT ASSY NAME
03	X	AB 22	RESP 7	C	4	03 13 1 1	0016-48- 0+0					EQUIP INST

DESCRIPTION

CANISTER FAILED TO RECOVER DURING GROUND STA-
TION CHECK OF 27- 90420 -2. PHASE DISC S/N 57
HAS LEAK

GSE SERIAL	GSE PART NO

The above numbers and descriptions are fictitious and have been chosen for illustrative purpose only

FIG. 7.13 A typical descriptive report by end product.

DESCRIPTIVE FAILURE REPORT BY SYSTEM

REPORT	ACTV	STND CPLX	MISSILE	SYST	FAILED ITEM PART NUMBER	FI SERIAL	FAILED ITEM NAME	FI REF DESIG	MFR CODE	FAIL CODE	DD	RR	DA	R	NEXT ASSEMBLY PART NO	WEEK CODE	REV
588979	XX		Z 13	(04)	27-42122 - 9	0110999	SERVO CYL ASS		64578	04 00	5	1	7	1	27-32762	245	X

HV	IR	TYPE	RESP	CL	CAT	FAIL DATE	USAGE NO 1	USAGE NO 2	REPLACE SERIAL NO	SUBSTITUTE PART NUMBER	N A SERIAL	NEXT ASSY NAME
08	X	AB 22	RESP 6	C	4	03 02 1	0	0				SRV CYL INST

GSE SERIAL	GSE PART NO

DESCRIPTION

B1 PITCH SERVO CYL HUNG UP DURING OPERATION S
0952 A/P POLARITY CHECK-PROBLEM IN SERVO AMPL

REPORT	ACTV	STND CPLX	MISSILE	SYST	FAILED ITEM PART NUMBER	FI SERIAL	FAILED ITEM NAME	FI REF DESIG	MFR CODE	FAIL CODE	DD	RR	DA	R	NEXT ASSEMBLY PART NO	WEEK CODE	REV
589085	XX		Z 13	(04)	27- 30210- 9	9050227	ACT ASSY		64578	02 01	5	1	7	1	27-48619	245	X

HV	IR	TYPE	RESP	CL	CAT	FAIL DATE	USAGE NO 1	USAGE NO 2	REPLACE SERIAL NO	SUBSTITUTE PART NUMBER	N A SERIAL	NEXT ASSY NAME
08	X	AB 22	RESP 1	B	4	02 02 0	0	0				SERVO INSTL

GSE SERIAL	GSE PART NO

DESCRIPTION

B-1 PITCH ACT WILL NOT ADJ TO SPEC PER B/P-LO
W LIMIT CALL OUT IS 12.917+HIGH 12.957-LOW CL
OSEST ADJ 12.914+HIGH 12.990

⋮

REPORT	ACTV	STND CPLX	MISSILE	SYST	FAILED ITEM PART NUMBER	FI SERIAL	FAILED ITEM NAME	FI REF DESIG	MFR CODE	FAIL CODE	DD	RR	DA	R	NEXT ASSEMBLY PART NO	WEEK CODE	REV
638783	XX	11	Z 17	(14)	27- 29111- 857		HARN		43789	01 01	5	1	8	1	27-68138	245	X

HV	IR	TYPE	RESP	CL	CAT	FAIL DATE	USAGE NO 1	USAGE NO 2	REPLACE SERIAL NO	SUBSTITUTE PART NUMBER	N A SERIAL	NEXT ASSY NAME
	X	AB 22				05 23 1	0	0				HARN INSTL

GSE SERIAL	GSE PART NO

DESCRIPTION

COAX CABLE Q154A IS BROKEN AT ELECT CONN 2Q1U
SP23 THIS WIRF RUNS FROM PERM SPLICE AT STA 0

⋮

REPORT	ACTV	STND CPLX	MISSILE	SYST	FAILED ITEM PART NUMBER	FI SERIAL	FAILED ITEM NAME	FI REF DESIG	MFR CODE	FAIL CODE	DD	RR	DA	R	NEXT ASSEMBLY PART NO	WEEK CODE	REV
638711	XX	11	Z 17	(20)	27-67439 -807		PROBE ASSY		43789	03 02	5	1	6	3	27-55614 5	245	X

HV	IR	TYPE	RESP	CL	CAT	FAIL DATE	USAGE NO 1	USAGE NO 2	REPLACE SERIAL NO	SUBSTITUTE PART NUMBER	N A SERIAL	NEXT ASSY NAME
04	X	AB 22	RESP 7	C	4	06 07 1	0	0				PROBE INSTL

GSE SERIAL	GSE PART NO

DESCRIPTION

FUEL 95 PCT SECONDARY PROBE INOPERATIVE-

The above numbers and descriptions are fictitious and have been chosen for illustrative purpose only.

FIG. 7.14 A typical descriptive activity report by system.

above, some of these failures are not the responsibility of the vendor and should not be charged to him. The COFEC file clearly indicates those failures which *are* and those which are *not* his responsibility. When a failure is first reported, it is charged to the vendor. If analysis shows that the failure was caused by factors beyond the vendor's control, then this charge should be voided so as to reflect the true situation.

Another attribute of great importance is the vendor's level of cooperation, this being measured in terms of the speed with which he implements corrective actions and their effectiveness. Such information is also revealed clearly by the COFEC files. It is fruitful to notify the vendor that nonresponsibility failures are not charged to him.

Design-review procedures have proved to be of distinct value in the enhancement of product reliability. The history of the part as revealed by the COFEC data serves as a firm basis for the detailed discussion of the potential failure modes in relation to the state of the art. From this it becomes possible to estimate the likelihood of reliability problems.

The time or cycles to failure is indicated on Fig. 7.3. It is important to understand that it is statistically improper to use these data for the computation of mean time between failure (MTBF), the reason being that this is a record of only those parts which have expired. Nowhere is there available a record of the complete population's potential performance at any particular time. The value of the time-to-failure data rests in that they can be compared to the minimum time to failure, which is often stipulated in specifications. These data are also of use for determining the time to removal as a preventive-maintenance measure.

Many of the spare-parts provisioning programs are based upon the assumption of a constant failure rate. The validity of this assumption can, in most cases, be disproved by reference to the time-to-failure records. Generally, new products show erratic behavior and demonstrate a varying failure rate. Even standard products show the effects of age, reflected in a failure rate which varies as a function of time. It is better to base the spares provisioning program on the replacement-part frequency, which is averaged over a constantly expanding time interval as the history accrues. This kind of information is directly obtainable from the COFEC system (Fig. 7.3).

7.4 CONCLUSION

The purpose of gathering failure data is to obviate future failures. All too often, the details and workings of failure data and analysis systems become the primary concern, and the essential purpose of the system is lost in a maze of good intentions relating to the use of forms, extensive coding procedures, sophisticated computer programming, dissemination of printouts, and the storage of the original data. To prove effective, a reliability data analysis system must in itself be as simple as possible so as not to jeopardize its function by its own complexity. The COFEC reliability data system has avoided these difficulties through the use of a unique digital coding which allows compact representation of large and complex descriptions in a manner which facilitates their manipulation and by anticipating the kind of questions which will arise so that answers are readily available.

In more than two years of successful operation at General Dynamics/Astronautics, the COFEC system has proved itself to be effective over a variety of large-scale end products and, at the same time, economical in its operation. The recognized requirements for a reliability data system have been fulfilled.

Bibliography

Fogel, A., "A Data Reduction and Failure Correction System (COFEC)," *IRE Trans. Reliability Quality Control*, vol. RQC-11, no. 3, pp. 56–70.

Fogel, L. J., "Biotechnology: Concepts and Applications," pp. 748–760, Prentice-Hall, Inc., Englewood Cliffs, N.J., 1963.

"Reliability Program Provisions for Research and Development Contracts," *Lewis Reliability Publ. Le RC-REL-1A*, Lewis Research Center, NASA, Nov. 20, 1962.

Section 8

TESTING PROGRAMS

ROBERT W. SMILEY

COMMANDER, USN

OFFICER-IN-CHARGE, POLARIS MISSILE FACILITY PACIFIC

BREMERTON, WASHINGTON

Prior to his present assignment as Officer-in-Charge of the Polaris Missile Facility Pacific, Commander Smiley was at BuWepsRep (Special Projects Office) at Sunnyvale, California, for six years. At that station, he conceived and directed the Navy and contractor quality control, inspection/test, and reliability programs for the design and manufacture of the Polaris missile, including the QC/reliability program at the major subcontractors and at the Polaris Depot at Charleston, South Carolina. He was responsible for organizing and chairing the Polaris—Minuteman—Pershing Rocket Motor Non-destructive Test Committee, which developed and proofed all of the nondestructive test techniques presently in use on the large solid-propellant rocket motors for ballistic missiles. He formed the Joint Ballistic Missile QC/Reliability Group with the Directors of Quality Control for the Air Force and Army Ballistic Missile Programs. From this group came the initiation of the Interservice Data Exchange Program (IDEP).

Commander Smiley served during World War II in the Ordnance Department, New York Navy Shipyard, and aboard the USS Kearsarge. After five years of consulting-engineering work and as representative for plumbing, heating, and air-conditioning manufacturers, he was recalled for the Korean conflict. This led to three years at the Philadelphia Naval Shipyard as Project Officer on TERRIER missile handling-equipment design and two years at the Naval Inspector of Ordnance Office, Pomona, California. There he served as Quality Control Officer during the development and production of the TERRIER missile. He is a Fellow of the American Society for Quality Control, and he has written a large number of articles on reliability programming and testing and the management of reliability programs.

TESTING PROGRAMS

Robert W. Smiley

CONTENTS

8.1 GENERAL

The overall test program for a product can be considered to be the most important single phase of a well-planned and executed reliability program, requiring the largest expenditure of reliability/quality funds and manpower. The test program provides the proof of all the theoretical calculations, both reliability and design. It provides the vital inputs on which the designer bases his design and subsequent redesign or design refinement. It is the source of almost all meaningful data from the inception of the project throughout the entire life of the hardware, the springboard for corrective action on design, process, and use, and the only sound basis on which logistics planning can proceed to ensure that the necessary parts and maintenance capability are available to support the equipment in actual use. It provides project management with the most vital information on the technical progress and problems, both real and incipient, of the project.

The importance of a complete, integrated, planned, documented, and vigorously prosecuted test program cannot be overemphasized, and it is essential that the most qualified personnel available be assigned to all phases of it. It is unfortunate that project management frequently loses sight of its importance and that the most capable personnel are often assigned to other functions such as design or production. Such projects are frequently in trouble because of unexpected hardware failures and inability to pinpoint and correct the causes of the failures. A highly qualified test group working in a well-conceived test program and with top-management backing in the form of funding, space, and equipment can preclude most of these project troubles.

8.2 KINDS OF TESTING

Since a comprehensive reliability test program encompasses all tests on the hardware from inception of the project through the final use and disposition of hardware, it follows that the test program includes many kinds of tests. Intelligent planning of an overall test program, then, requires an understanding of the kinds of tests that are available in order that optimum choices can be made. In the following paragraphs we shall subdivide tests into five categories by different factors. The reader will understand that any given test will fall into one of the subdivisions of each of these categories and that the final selection of a test includes a determination of each subdivision.

A common failing in the planning of test programs is the failure to consider all of these categories; too often the assumption is made that the approach used in the past or in other projects is the optimum one for the project being planned. Such a slipshod and haphazard approach to test planning results in an incomplete analysis being made of all the test regimes available and the selection of test techniques less than optimum for the purpose intended. Thus an expensive simulated environmental test requiring

CHECKLIST FOR TEST/INSPECTION CLASSIFICATION

TEST NAME________ PART NAME________
REQUESTED BY________ DEPT____ PART NUMBER________
TEST START DATE________ TEST SPAN________ COMPLETION________
CHARGES: Engineering________ Testing________

1. Test Type (underline one)
   - I. EVALUATION (Research, Feasibility, Evaluation, Qualification)
   - II. SIMULATED USE
   - III. QUALITY (a. Production, ambient or environmental; b. Installation; c. Field maintenance)
   - IV. RELIABILITY (a. Peripheral; b. Life; c. Accelerated Life; d. Service Life Evaluation; e. Surveillance
   - V. SPECIAL (Consumer research; Investigative)
2. Destructive or Nondestructive?________ Type________
3. Ambient or environmental?________ (Name environments)
4. Actual or simulated conditions?________
5. In-house or outside________ Where?________
6. Level: Parts, subassembly, assembly, subsystem, system?
7. Customer controlled?________ Extent?________
8. Test Equipment: Standard or Special?________ Design by whom?________
9. Test Conductor from what organization?________
10. Test Procedures prepared by whom?________ Basic Requirements outlined where?________
11. Reports submitted to whom?________ Copies?________
12. Extent of Inspection coverage________ Customer Inspection?________

FIG. 8.1 Checklist of test/inspection classifications.

new environmental equipment may be specified merely because the company has usually tested in a laboratory, whereas the factors of use of the particular new product being considered may dictate that actual environment, despite the drawback usually associated with uncontrollable natural environment, may be the most optimum selection. Use of a checklist like Fig. 8.1 will not only facilitate consideration of the available choices but ensure that none are overlooked.

8.2a Destructive vs. Nondestructive Testing. Simply speaking, a destructive test is one that will leave the tested hardware unfit for further use, whereas a nondestructive test is one that will not. In most cases, as with tests of explosives, this simple definition will suffice. However, in some rather rare instances the hardware may still be usable for limited purposes, as with a complete design or production qualification test which leaves the hardware unfit for delivery to a customer but perfectly good for testing to failure to determine failure modes. Hence it is important that the possible or potential "further use" be examined early in deciding on the exact elements of any test program so that a trade-off can be made whenever it is economically feasible.

Other factors being equal, economically it is always desirable to utilize nondestructive testing instead of destructive, provided the net cost to the program is not adversely affected because more nondestructive tests are required to achieve the same purpose as might be achieved with a small number of destructive tests. Furthermore, nondestructive testing leaves the test sample in condition to permit meaningful failure diagnosis, enhancing considerably the potential value of the test.

On the other hand, the indirectness of many nondestructive tests makes their selection questionable. An example is the use of vibration levels less than actual service conditions, where the extrapolation of results to the service condition is complex and not straightforward. In some applications, where considerable vibration data could be attained during an R&D program to correlate destructive to nondestructive vibration levels, the nondestructive test may be chosen for a production acceptance or assessment test, and both categories of testing programs may be chosen for the R&D phase to gather the necessary correlating data. An overall correlation program of this type is essential to developing meaningful nondestructive production inspection programs involving radiography, magnetic particle, ultrasonics, microwave, and the like. Unfortunately, such nondestructive production inspections are frequently programmed without the important prerequisite correlation work being performed during the R&D phase, and there results, therefore, no usable and proven set of accept/reject criteria against which the inspector or engineer can evaluate the results of the nondestructive tests on the actual production hardware.

The extremely high unit cost of some products makes nondestructive testing mandatory in all phases of a project. Notable in this group of products are the large solid-propellant motors used in ballistic missiles and satelites, where the cost of a single stage may approximate a half million dollars or more. Since project economics dictates that only a relatively small number of these may be fired, it is essential that everything conceivable be known about the motor to be fired. Generally, the bulk of this information comes from an extensive nondestructive examination of all the parts and systems of the motor. Metal parts and various subassemblies, as well as the final motor, are completely X-rayed and checked with ultrasonics and other techniques for case and liner bonding and weld and metal integrity. Samples of propellant, liner, their ingredients, and other chemicals are thoroughly analyzed in chemical laboratories. Complete variables data are recorded on not only product measurements but on all of the details of the process and the environment in which the process was carried out. These data are then minutely examined and analyzed to establish correlation with the firing results so that meaningful limits can be set on the eventual production process variables, as well as on the results of those nondestructive examinations that are selected for production use. In these projects the use of nondestructive testing is mandatory for both production and field maintenance testing, as well as for in-service life evaluation and surveillance test programs.

Most ordinary production testing is nondestructive, but in high-reliability programs these tests are backed up with destructive tests performed on samples drawn at regular intervals from the production line. With such a project, it is frequently possible to test only the critical parameters for every unit of a product, leaving for the sample production-assessment program those less critical parameters which can be safely sampled. Thus an economy of testing is achieved by combining in the sample testing both noncritical nondestructive tests and the destructive tests. Such a technique works well with complex functional electronic, hydraulic, and mechanical assemblies where test cost is a considerable portion of the cost of the product.

The use of vibration as an environment for nondestructive production testing is controversial, with some reliability personnel believing that such a regime is in fact degrading, particularly to closely machined or critical mechanical parts such as gyros.

The author does not concur with this contention, but believes that the routine use of vibration in in-process and acceptance production testing results in shaking out poor workmanship and incipient service failures before the hardware leaves the factory. The same thing can be said for the use of an elevated-temperature environment for certain critical electronic components, particularly in those designs where weight and space considerations have made underdesign of circuits a necessity and only the best components can be tolerated. The elevated-temperature technique has been used successfully to separate out low-grade semiconductors, tubes, resistors, capacitors, transformers, and potted assemblies. The temperature selected generally does not exceed the actual operating temperature of the part in its most severe application environment.

8.2b Ambient vs. Environmental Testing. Ambient testing is usually considered to include that testing performed under existing static conditions found in the laboratory or on the factory floor, while environmental testing includes all testing in which the specimen is subjected to some nonambient condition. However, some testing performed under actual-use conditions at existing environmental conditions, particularly when the locale is deliberately chosen to provide extremes in temperature, vibration, humidity, dust, etc., is also considered to be environmental. The discussion that follows will so consider such actual-use tests.

Ambient tests are usually used for production testing, largely because of their simplicity and economy. (They may run one-tenth to one-hundredth the cost of an environmental test.) To be useful in high-reliability production projects, it is essential that they be developed in the R&D phase, in conjunction with environmental tests, to determine their validity for separating out material which will not function in the actual environments that will be encountered by the hardware after delivery. Therefore, the production ambient tests should be planned at the beginning of a program, and each piece of R&D hardware destined for environmental test should be tested to the proposed production ambient test before it is tested environmentally. Subsequent failures in the environmental test will usually require revising the production ambient test.

With this ambient/environmental relationship in mind, it should be apparent that the ambient test is only a substitute, dictated by economy, for an environmental test, unless the actual-use conditions for the hardware are approximately equivalent to the factory ambient, as in the case of household appliances or military equipment such as computers destined for use in a protected environment. The comparative economy of ambient tests, however, makes them the most widespread of the two, and when properly correlated with expected performance in use environment, they provide a very high degree of assurance that the hardware will properly function in use. Because of their relative simplicity, they can be and are used at all levels of assembly, and it is not difficult to achieve a high degree of compatibility among tests at various levels. (See Sec. 8.7*a*.)

Environmental testing is necessary in a high-reliability project to determine in absolute terms the performance of the hardware in actual use. Most environmental test programs attempt to subject specimens to the expected extremes of use environments; however, in a rigorous reliability project it is also necessary to perform peripheral testing beyond the extremes actually expected in order to determine the design safety margin, as well as to detect incipient weaknesses. As a general rule, such overtesting should exceed the design limits by a significant margin, say, 10 to 15 per cent of the spread from ambient to the design environmental limit, in order to be sure that the effect of the environment is separated from random variations in the product. The high cost of environmental testing, plus the inevitable interaction of environments (e.g., elevated internal temperature is always present in tests of electronic assemblies), makes mandatory the use of statistical design of experiments to ensure the maximum return for each testing dollar. (See Sec. 8.3*a*.)

A well-planned environmental test program will also include a comprehensive ambient test program preceding the environmental testing, as indicated above, to provide the maximum correlation between the two and thereby permit economical ambient testing to be the work-horse program. Environmental tests are usually used

(1) during the R&D phase for evaluating the design at several stages of maturity for its ability to operate satisfactorily in use environments, (2) at the beginning of and during the production phase to determine whether the production process is first capable of producing and then subsequently continuing to produce hardware operable in the use environment, and (3) during the use phase to evaluate the rate of degradation of the ability of the hardware to continue to operate in the use environment. These tests are all further described in Sec. 8.2*e*.

8.2c Actual Conditions vs. Simulated (Laboratory). Environmental testing can be classified according to the method by which the environmental exposure is created, i.e., naturally in the actual-use environment or artificially in a laboratory. Consideration of these subdivisions is worthwhile, since the cost of testing and the usefulness of the data accumulated may vary markedly between them. The decision to choose one over the other is complex and is based on many factors, a few of which follow:

Size of Parts. Exposure of very large masses to some environments in a laboratory is very often difficult to achieve, particularly to the dynamic environments such as shock, vibration, and sustained acceleration or to rapid changes in atmospheric (or other) pressure and temperature. With this combination of factors, the decision to utilize natural environments is almost dictated, and hardware is commonly subjected to such tests as rough-road or transportation tests in the shipping container; flight tests of major packages and segments of large missiles, space vehicles, or airplanes; and proving ground, desert, and arctic tests of automobiles and other vehicles. It is possible to build laboratory environmental equipment to provide these and other difficult exposures for very large assemblies, but generally the cost is prohibitive unless other factors demand that the laboratory facilities be provided.

Nature of the Parts. Very complex functional parts whose performance must be checked during the exposure are difficult to test under many actual-use conditions because of the necessity for large, complicated test consoles to provide functional stimuli and output-measuring capability. This limitation is particularly applicable to flight testing of missiles, where the permissible payload severely limits the amount of data which can be transmitted by telemetry to ground stations. With this kind of parts it is also frequently necessary that the performance of the parts be checked at both high and low input-stimulus level, a condition that is generally impossible to achieve when the part is installed in its operating condition in the system. The operation of other kinds of parts, such as pyrotechnics and liquid or solid rockets, cannot be adequately measured in flight conditions, and simulated environmental testing is mandatory.

Frequency of Testing. If tests are to be performed frequently or continuously, as with acceptance testing under vibration in a production test program or when a test program is set up to test periodic samples from a production line environmentally, the decision to utilize natural environment or establish a laboratory capability is largely a question of economics, being based on the relative total costs of the two choices. However, time is sometimes an overriding factor, since the time required to attain a full laboratory capability may exceed the production run time or the time required to perform each test in actual-use condition may inject too much lag in the data acquisition. Within large, well-established concerns, laboratory facilities appropriate to the class of material being developed and produced are usually available, and repetitive tests are frequently tailored to the laboratory capability, with only a very limited portion of the test program performed in actual-use conditions.

Complexity of Instrumentation. The complexity of the instrumentation required, which is a function not only of the complexity of the part but of the extent of the data required, is a major factor in the choice of laboratory over actual-use conditions. If only a few attributes need to be sampled or if considerable error in data can be tolerated, it is entirely possible to utilize portable recorders or to telemeter data to a complex ground station and gather all the necessary data in an actual-use test. On the other hand, a requirement for large amounts of data or for very accurate data cannot usually be met except in the laboratory, where a wide array of very sensitive equipment is available.

Complexity of the Test. It follows from the discussion under nature of the part and complexity of instrumentation that the complexity of the test, which influences the instrumentation as well as the exposure, is an important factor. For example, test requirements which include extremes of temperature on the same specimen can generally be met more economically and more rapidly in the laboratory than in actual-use testing, unless an airplane or other airborne vehicle can be used as a test vehicle to transport the specimen rapidly from burning desert heat to subzero temperatures. Such airborne exposure would be expensive, however, unless the test vehicle is being used for other purposes at the same time, which would then increase the difficulty of scheduling the test.

Accessibility of Natural Environments. It is entirely feasible for an automotive designer to find within the confines of the United States the extremes of environment in which the product will live. Subzero and desert temperatures are readily and inexpensively available, as are extremes of altitude, road conditions, dust, rain, and other precipitation, and all in the multiplicity of condition combinations needed. On the other hand, the manufacturer of large solid-propellant rockets does not have a natural high-altitude (low-pressure) environment at his fingertips except by very expensive, short-duration flights. Paint manufacturers find the use of paint patch panels in the blazing California sun, or on heavily traveled freeways, to be the practical answer to providing the desired environment. In each instance, the natural environment should be examined to determine the relative difficulty not only of transporting the specimen to it but also of controlling the environment to attain the extremes desired within a time scale of the test program, of measuring the environment encountered, and of assessing anomalies or failures in the test with the degree of lack of control encountered.

Relative Costs. The relative costs of the two approaches are a major factor, and this includes not only the direct costs of performing the tests but the investment which the company already has in laboratory equipment, additional investment which may be required, and the utilization factor of equipment and personnel. A test which for other factors may be performed with somewhat greater facility in the field may be scheduled into the laboratory purely for the purpose of leveling out a saddle in the laboratory workload. Conversely, a test may be performed in the field, other factors being equal, merely because the only suitable laboratories are in a peak-load condition.

Relative Time. The relative time to achieve a required test capability or to perform a required test is frequently the overriding factor in the decision process. If the development of a product must be completed by a specified date, then data-acquisition milestones must be met and the decision becomes purely one of finding the environment, wherever it may exist, and accepting compromises of the other factors. New complex laboratory facilities can take six months to a year to design, build, and equip, a time scale which is frequently too long for tightly scheduled projects. On the other hand, production test results must be made available for the accept/reject decision within a relatively short time after the specimen is produced, and a tightly scheduled test program must be established. This makes almost mandatory the use of in-house laboratory facilities, where the environment is under complete control.

8.2d Levels of Tests. A fourth convenient way to classify testing is by the level of assembly. Tests can be performed at all levels, but for practical purposes the levels generally chosen are parts, subassemblies, assemblies, subsystems, and systems. Two opposing functions of each class of test operate in check and balance to require some testing at each level, and they are the principal factors dictating the selection of particular attributes to be tested at a specific level. Thus, for example, in production test programs it is desirable to test every attribute as soon as possible after it is created to preclude further investment in nonconforming hardware. This is the true quality control function. Opposing this function, however, is the necessity for testing attributes at the last possible time before the attribute is covered up to ensure that nothing in the production process has degraded the attribute. This is generally called acceptance testing, and is the customer's assurance that the product being delivered meets the functional requirements.

These two factors work in opposition, and the resulting compromise is the actual production test plan, which generally requires some testing at all levels. Similarly, in design evaluation or reliability test programs, the need to know the performance under varying environments of each design unit down to the piece part and the inability to make such a determination solely from tests of the higher assembly levels will tend to force the level of testing downward. Opposing this and tending to force the testing to higher levels are the need to know the performance of the end product, including spares, and the inability to determine from assembly-level tests the interaction of the various subassemblies that make up final assemblies. These forces, acting in balance and contained within the overall testing budget, will dictate the final evaluation/reliability test programs.

Of particular importance in considering testing at various levels of assembly is the necessity for integrating the testing performed at the different levels. In most industrial organizations, specialization of engineers along product lines coupled with specialization vertically along functional lines has resulted in considerable compartmentalization of personnel responsible for specifying the details of various aspects of reliability test programs. Thus the overall engineering evaluation test program is frequently under the cognizance of the design organization, while the tests at parts, subassembly, and system levels within this group of tests are respectively assigned to parts application, subsystem design, and system integration or the project office.

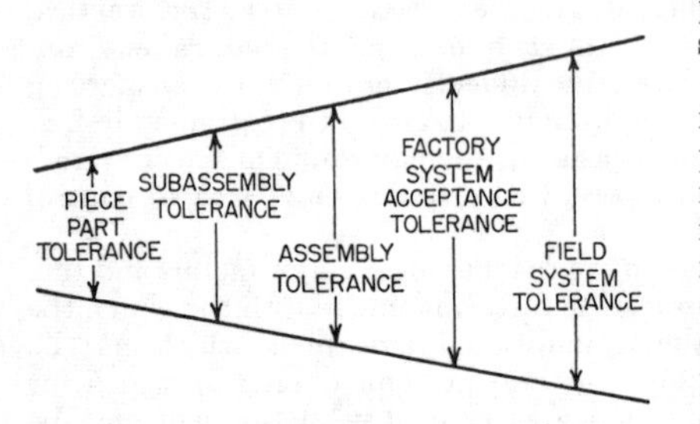

FIG. 8.2 Tolerance funneling.

As a result of this dual compartmentalization, the details of environmental exposure, attributes to be tested, test equipment and measuring techniques used, and input conditions are frequently incompatible, and the resulting data cannot be correlated. Usually when these uncoordinated groups are responsible for preparing a test program, it is necessary for another group, such as reliability, to review and approve all test plans and procedures before they are released in order to achieve the compatibility necessary to permit optimum utilization of the data. In a better organizational arrangement the whole job of test planning is the responsibility of reliability, but the work is based on test requests or test assignments from the design department.

The same principle holds for production testing (including that performed at field assembly or installation points), which should be completely planned in detail by a single quality or reliability organization. Another facet of the problem of integrating tests, however, needs to be considered with production testing, particularly with electronic or hydraulic functional hardware. Many functional attributes in this kind of hardware drift with time, handling, or functional cycling. If the acceptance limits on these attributes are set identically at successively higher levels of test, there will be a measurable percentage of hardware with attributes just inside the limits at one level of test which will drift outside the limits in the next test and be rejected back to the lower level for rework. To preclude the resulting circulation of hardware in a properly integrated series of successive-level tests, the tolerances of a single attribute are established in a funnel arrangement, with the tightest tolerance at the lowest level of assembly (Fig. 8.2). This funneling of tolerances is treated in more detail in Sec. 8.3*b*.

8.2e Tests by Purpose. When one suggests that a test program is needed, the first question is generally "What kind of test?," meaning a test for what purpose. It is natural to think of testing in terms of the intended purpose for which it is being run, since this is the usual departure point for all of the planning, funding, assignment of responsibility, and use of the resulting data. In a comprehensive test program associated with a high-reliability project, it is convenient to consider the many purposes for which tests are conducted in five large groups, with each subdivided into specific classes of test. Here these five groups are arbitrarily named as follows: I, evaluation; II, simulated use; III, quality; IV, reliability; and V, special.

Group I, Evaluation Tests. Although it can be said that in a larger sense all testing is evaluating a specimen against a set of explicit criteria, in modern projects a discrete series of well-defined testing programs is recognized as being run for the purpose of evaluating a design and/or a production process against the performance requirements for the item being tested. As used here, the term is limited to those classes of environmental tests performed during the preproduction phase of a project. Four such classes are generally recognized: research, feasibility, evaluation, and qualification.

RESEARCH. Research tests are the earliest tests performed during the conduct of a complete design/production project, and they are usually distinguished from the other tests in Group I by the facts that (1) the specimens are breadboard, (2) the capabilities of several design approaches are being compared, and (3) the tests are investigative or probing in nature. The designer generally has complete control over the testing and seeks data on which he can decide on the most nearly optimum of several possible design solutions. The tests will include both ambient and environmental types, and they will generally be performed only on those portions of the design for which the solution is not straightforward or for which the operating conditions are unfamiliar to the designer. Research tests are very informal, with procedures generally being cut and try, and they are performed under constant and close supervision by the designer, who generally retains the data in the form of design notebooks.

FEASIBILITY. The second series of tests in the development phase of a project are feasibility tests. Having selected a particular design solution from several possible solutions, the designer goes through a series of design refinements, advanced breadboard specimens of each being tested against the requirements for the design. These tests, as the research series, are generallv under the complete management of the designer, who will make on-the-spot changes to the specimen in a series of steps aimed at improving the ability of his design to meet the functional and reliability requirements. The data, which the designer retains in his design notebooks, are for his own use. It is not unusual for a designer to discover during the feasibility tests that his initial selection of a particular design solution was erroneous and that he must select another of the possible design solutions which were tried out in the research tests. It is against this not-uncommon eventuality that experienced designers carefully retain all research-test data.

EVALUATION. When the designer is satisfied that his design is completed and that to the best of his ability within the constraints of time and funding he has met his design requirements, he releases the design for the manufacture of enough R&D hardware to permit a complete series of tests to be performed to evaluate the design against those requirements. These tests are frequently the responsibility of another organizational entity within the company, usually reliability, to ensure that an unbiased evaluation is attained. The tests are comprehensive, testing all known requirements that can be simulated in the laboratory. The performance of the hardware is checked, with extremes of input stimuli, during and after exposure to all of the known expected environments. Failures are carefully diagnosed to determine whether the design is at fault (and therefore requires change and a rerun of all or part of the evaluation test) or whether the parts, workmanship, or manufacturing process was in error (in which case the hardware is corrected without design change, but again generally requires rerun of all or part of the test).

In a well-disciplined project, satisfactory completion of the evaluation testing signals the release of the design to prototype manufacture. The evaluation test program is a controlled test program, with formal procedures prepared jointly by the design and reliability groups. In a very mature project, inspectors will check the test setup, the recording of data, and the calibration of test equipment and ensure that complete control is maintained over the test program.

In customer-funded development projects, however, the customer will not generally participate in the evaluation test program, since this test series tests only the design and not the proposed production tooling and process. (Exceptions are sometimes taken to this generalization when the customer decides to waive the formal qualification program because of schedule pressure and to release the design and manufacturing process based on the ability of the R&D manufacturer to produce quality hardware.)

NOMENCLATURE AND GENERAL REQUIREMENTS FOR ENVIRONMENTAL TEST PROGRAM, CUSTOMER-CONTROLLED PROJECT

	RESEARCH TESTS	FEASIBILITY TESTS	ENGINEERING EVALUATION TESTS	DESIGN-QUALIFICATION OR PREPRODUCTION TESTS	REQUALIFICATION OR PRODUCTION-ASSESSMENT TESTS
OBJECTIVE	To determine functional capabilities of parts, components, assemblies, and packages for possible application into end-use hardware. To establish basic design.	To establish confidence on development parts, components, and packages prior to use on any field- or flight-test hardware. Includes items to be used in development program that are not intended for use in a tactical program.	To provide test data on a selected list of functional hardware and subcomponents that will satisfy engineering evaluation requirements. Acceptable data may be presented to customer in lieu of formal preproduction tests.	To demonstrate by a number of ground tests that parts, components, assemblies, and packages used in equipment system can function in their environments as specified in applicable specification and drawings. Successful completion of these tests will provide test data to qualify item.	To demonstrate by test that manufacturer continues to produce acceptable production hardware in strict accordance with design and environmental requirements.
SCOPE OF TESTS	To be determined by responsible designer.	To requirements of development specification. Requirements may be reduced to a minimum of those environmental attributes required for field- or flight-test program.	Tests to be performed to all requirements specified in development specification as defined for preproduction tests and to satisfy requirements of integrated reliability-test program.	Tests to be performed to all requirements specified in formal specification as defined for preproduction tests.	Functional and limited environmental tests as defined in applicable formal specification.
HARDWARE	Preprototype, vendor off-the-shelf items, and breadboard-type hardware.	Development hardware manufactured to engineering drawings.	Development hardware manufactured on development tooling to released development designs. Pilot-line or production tooling will be used if available.	Pilot-line or production hardware.	Production hardware.
ENGINEERING DOCUMENTATION	Hardware: vendor catalogue or lab sketches; test to engineering-memo requirements.	Hardware: R&D drawings, test to development specification and test assignment.	Hardware: released R&D drawings. Tests: development specification (released), integrated reliability-test program group summary and test assignment.	Hardware: released production drawings. Test requirements: formal specification and test assignment.	Hardware: released production drawings. Test requirements: formal specifications and test procedures.
CONFIGURATION CONTROL	Design group to maintain record of changes to hardware or test conditions.	Design control by responsible design group. All changes must be fully documented.	All design changes must be documented and approved by change screening board, the program director to be notified of approved changes.	All design changes must be documented and approved by change screening board, the program director to be notified of all approved changes. All changes must be approved by customer.	Customer-change control. Design changes require customer approval.

TEST PROCEDURE OR DOCUMENT	As defined by test assignment or as directed by responsible design engineer.	Test procedure per test facility format. Detail test assignments may be used in lieu of a test procedure.	A detailed test procedure is required.	A detailed test procedure is required.	Production-assessment-test procedure is required.
TEST-EQUIPMENT DOCUMENTATION	Test-equipment documentation not required unless requested by responsible design engineer.	Sketch documentation sufficient to assure repeatability of tests.	All test equipment must be completely documented and proofed.	All test equipment must be completely documented, and specialized test equipment proofed.	All test equipment must be fully documented and proofed.
QA INSPECTION PARTICIPATION	Inspection participation not required unless requested by responsible design engineer.	QA inspection surveillance. Source inspectors to witness tests performed at subcontractor facilities.	All tests to be witnessed by a QA inspector.	All tests, in-house or at vendors, to be witnessed by a QA inspector.	All tests performed by QA engineers and technicians.
TEST REPORTS	Interoffice notebook and flash report and/or engineering report.	Interoffice notebook and a final report and flash reports. Final report is to be prepared, maintained, and distributed by test agency.	A test report containing all test data, change sheets, failure reports, summary, and conclusion must be prepared. The test specimens must be completely identified in the report.	A test report containing all test-data change sheets, failure reports, summary, and conclusion must be prepared. Test specimen must be completely identified in report. Interim test report must be submitted to customer.	Data sheets from test procedures and test summary, including analysis of results, failure diagnosis, and corrective action.
CUSTOMER PARTICIPATION	Customer participation is not required.	Customer participation is not required.	Review and approve satisfactory engineering evaluation-test reports submitted in lieu of preproduction-test reports.	Review and approve preproduction test procedures and reports. Maintain surveillance over preproduction tests as desired.	Review and approve PAT program plan, PAT procedures, and surveillance of all tests. Review and approve test reports.
SCHEDULE	To be established by responsible design engineer.	Tests must be completed and final reports written prior to flight or field test of equipment on which item will be used.	Tests to be performed in lieu of preproduction tests to be completed in sufficient time to allow any retests to be performed on pilot-line and/or production hardware.	Test reports on pilot-line hardware must be approved by customer prior to ship date of first production article. Test reports on production hardware must be approved by customer prior to a negotiated ship date subsequent to first production unit.	Sampling rate as established in PAT program plan. Test reports approved by customer to permit release of hardware for production usage.

FIG. 8.3 Chart of principal differences in environmental tests.

Since the design is highly, though perhaps not completely, mature at this point, the data are normally used not only for signaling design release but as an important input into the reliability data pool. Evaluation test hardware is frequently given to the reliability organization for further reliability testing, i.e., testing to failure, peripheral testing, etc.

DESIGN QUALIFICATION AND PREPRODUCTION. In high-reliability projects, once the design has been proved out in a design evaluation test series (and provided associated field evaluation tests discussed in Group II are satisfactory), there remains only one more series of tests to release the production effort. These are the design qualification and preproduction tests. These tests are essentially a complete rerun of the design evaluation test, but in this second series the specimens are built on the production tooling, with production methods, processes, inspection, and procedures; the parts are procured to production specifications and inspected by production inspection methods.

If the project is customer-financed, the customer will usually monitor the test program completely, reserving the right to approve deviations from the plan or procedures, and will require that his representative formally waive any failures occurring which do not lead to corrective action on the design or production process. The specimens will be subjected to the full range of expected environments, with extremes of (high and low) input stimuli and the performance of the hardware checked before, during, and after exposure. It can be expected, in a complex-item project, that this series of tests will reveal that changes unwittingly introduced in the transition from R&D manufacture to production have degraded the ability of the hardware to perform satisfactorily and the production processes will require some correction. These corrections frequently require rerunning all or a part of the preproduction test. A demanding customer will insist upon approval of this decision.

In a similar fashion, it may be discovered that either changes creep into the production design disclosure or the two different shops (R&D and production) will interpret the design disclosure differently. The design-qualification hardware may differ significantly from the hardware which had previously passed the engineering evaluation tests, and corrections will be required, with some degree of test rerun.

These contingencies must be planned for, since they affect the time in which a program can be run. In missile and space projects, it is normal to schedule from 5 to 6 months for the design-qualification program, sometimes longer if the design is completely new or if there are not a considerable maturity and discipline in the manufacturing, design, and inspection organizations. This time scale is so long in many projects that the customer will allow production to start before the completion of the formal qualification program, with the attendant risk that parts may have to be reworked. This tendency should be resisted by the reliability group, since under these circumstances there will be considerable additional pressure from manufacturing and design groups to waive any failures or degradation found in the qualification test program.

Group II, Simulated-use Tests. Although the series of evaluation tests provide most of the data by which the design and the production process can be evaluated, there will remain an uncertainty that the product will function properly in its operating environment if the laboratory tests are not supplemented by testing and demonstration in the actual-use environment. Laboratory testing is designed to control environments and inputs so that cause-and-effect relationships can be accurately determined. This very control, however, with its unnatural orderliness and discipline, injects an element of unrealism into the testing program. Furthermore, in laboratory testing it is exceedingly difficult to provide a completely random combination of environments. And last, but by no means least, there is always in a laboratory test program the element of doubt that the actual-use conditions have been completely or properly simulated, besides the fact that there are usually some environments or operating conditions that cannot be feasibly or economically duplicated in a laboratory.

For these reasons a parallel program of use testing is established along with the evaluation test program. Such use testing takes the form of road testing of automobiles or flight testing of airplanes and space vehicles. These use tests provide the important evaluation of the response of the hardware to field conditions, and the response sometimes differs surprisingly from the one envisioned by the engineers. However, the tests are not without drawback. In particular, since they are of necessity performed under less-controlled conditions than can be achieved in a laboratory, the results are frequently difficult to assess, as well as difficult to reproduce.

It is frequently necessary to back up field use testing with special diagnostic laboratory tests in an attempt to reproduce use-test failures and permit intensive failure analysis, diagnosis, and corrective action. This is especially true with missile or space-vehicle flight tests, where the data from the test are severely limited to those available from telemetry and where the test hardware is destroyed and therefore unavailable for post mortem analysis. To overcome this serious limitation on flight-test analysis in some missile projects, it has been necessary to devise special recovery techniques, including parachute landings, or to design the test vehicles to include reflyable characteristics.

It must be recognized that field use testing is very expensive; literally millions of dollars are invested in missile ranges such as those at Cape Kennedy, White Sands, or the Naval Ordnance Test Station, China Lake, California; in a similar fashion, the General Motors Proving Ground, Milford, Michigan, also represents a multimillion dollar investment. These testing sites, and others like them, can be considered to be extensions of the normal laboratory, and every attempt is made to standardize or control the conditions of the testing. The tests must be performed in a manner that will not compromise the intent of field use testing, which is to try the product under the random conditions that are to be encountered in actual service use.

Group III, Quality Tests. In any high-reliability project it is essential that every step in the production process, in its broadest sense, be checked by many tests and inspections. In this broadest sense "production process" includes not only the actual factory production of the delivered product but the production of R&D and test hardware, the field installation or dealer assembly process, and the periodic use, maintenance, and overhaul process. While it is true that the inherent reliability of an item is set or limited by the inherent reliability of the design, it is also true that almost everything that happens after the design is released to production tends to degrade that inherent design reliability, and the quality test program provides the check and balance to prevent the degradation from becoming intolerable.

Quality tests and inspections take many forms, and all of the classifications discussed in Secs. 8.2*a* to 8.2*d* are utilized. When the tests are destructive, as they are when the condition of manufacture is such that the effect of variations in all of the product variables cannot be determined accurately except by destructive analysis of the parts, sampling must be utilized. If many identical items are to be produced, sampling can be performed in accordance with MIL-STD-105, "Sampling Procedures and Tables for Inspection by Attributes," and the consumer/producer risks (or AQL) can be chosen to ensure that the precise reliability level desired in the product is passed by the test program.

However, the problem is infinitely more difficult with low-production projects, since there is no statistically valid "sampling rate" for very small lots. This situation is common with the solid-propellant rockets and pyrotechnics for large ballistic missile and space projects, where production for a year may total from ten to a hundred. In such situations, sampling rates may vary from one for one to one for ten, with the exact rate established being dependent upon (1) the tolerable degree of risk, (2) the probability of uniformity of successive units of product, (3) the potential effect of failure of the item, (4) the ability of the production process to produce sufficient hardware for both delivery and test within the time scale available, and (5) the cost of the hardware and the test.

Variable sampling rates are sometimes used in long-run low-production projects, with, for example, a rate of one for one utilized for the first 10, one for four for the next 25, and one for ten for the balance of the production program. Such a variable sampling rate is also frequently used for expensive nondestructive testing. The decision to reduce sampling in this manner is predicated upon demonstration that the production process is capable of producing a uniform product of the desired quality level.

The risk, however, of having to recall or scrap unacceptable material increases directly as the sampling rate decreases, and in many projects the delivery schedule or cost of each unit of product precludes taking much risk beyond the one-for-four or one-for-five level.

For ease in discussion, it is convenient to subdivide Group III into three subgroups:

Group IIIa, Production (Both R&D and Production). In this subgroup are all those routine tests and inspections performed to determine that the hardware is being produced in accordance with the requirements of the drawings, specifications, and other applicable engineering and shop requirements. These tests are performed on every item of hardware (or on samples drawn from every lot), including the hardware produced purely for Groups I, II, IV, and V testing as described elsewhere in this section. Although the press of schedules, particularly those of flight tests of missiles, may seem to warrant waiving the requirement for acceptance or quality testing of R&D hardware, reliability personnel must resist this tendency; for it is *essential* in a high-reliability project that the *exact* configuration of the hardware tested be known in order to permit meaningful interpretation of both success and failure data. The validity of either result of a flight test is in serious question if all of the parameters in the item flight-tested have not been measured in the factory and thus are not known to be identical (or any variation known) to the item to be produced and delivered for service use.

A second basic reason that quality tests are performed during the R&D program is to determine the practical permissible limits of variability in the product. In complex functional material, limits on the many parameters involved in the design can be and usually are set theoretically by designers based on parametric studies and computations. However, when parts operate together in the complex interrelationship of an end-product configuration, the theoretical limits are frequently either too wide for optimum functioning of the parts or too tight for economical manufacture. To determine the optimum balance of limits, it is essential that accurate variables data be gathered during the R&D manufacturing and that variations in each parameter be correlated against variations in performance of the end product. This testing for the purpose of gathering variables data is a most important part of the R&D production test program.

There is a third basic reason that quality tests are performed on the R&D hardware. Quality or acceptance testing and inspection of high-reliability products is difficult, and the test equipment used is itself frequently more complex than the product being tested. The purpose of this acceptance/quality testing is not only for acceptance but also for assuring that a consistent product, uniformly meeting the engineering requirements, is produced and for weeding out for repair all items that fail to conform. It is essential that the items delivered for service use be identical with those which successfully passed the R&D tests. This identicality can be assumed only if the same quality tests are used in both the R&D and production phases, since differences in tests and inspections will yield different output populations.

Furthermore, since quality testing and test equipment are complex, they too must be developed during the R&D phase of the project in much the same way that the product is developed and in parallel with it. Before a project is ready for the production phase, not only must the design and the production process have been proved capable of producing a product able to perform the required function but the quality test program must have been proved capable of accepting only those items which will and of rejecting only those items which will not so perform. This latter proof is one of the important objectives of the R&D phase of a high-reliability project.

Production quality tests (Group III*a*) are composed of two general categories: ambient and environmental. In the ambient category are those tests performed (1) at receiving to check the quality of incoming parts and materials, (2) in-process to check the quality of the in-plant manufacturing process, and (3) at acceptance or delivery to check the final overall quality of the product as shipped. There is usually an additional test or inspection point after crating and packing to ensure that the protective packaging used is adequate to preclude damage or degradation to the product in transportation to the customer or consumer. For some classes of product, like electronic or precision mechanical assemblies, additional "tests" either separate from or combined with these ambient tests are performed on the product to deliberately exercise or "burn in" the product. The aim of these burn-in tests, which are not really tests but planned operating time, is to weed out infant-mortality failures.

All these ambient tests are generally performed on every unit of product or, if the production rate is high enough, on a sample drawn from every lot of material. The tests are generally performed in the same factory environment which was encountered in producing the item. The tests are performed by hourly inspectors or technicians who report organizationally to the manager of quality assurance and/or reliability. However, when the testing is performed by the inspection group on R&D hardware, the authority to make the final decision to accept or reject nonconforming material is frequently delegated to the engineering manager.

An important decision which must be made for each item and all material procured is whether the final quality testing of the procured material will be performed at the source of supply or at receiving. Although costly duplication of test equipment can be obviated by having the tests run at the vendor's plant, this advantage is frequently offset by a number of distinct disadvantages. Costwise it is generally necessary in a high-reliability project for the buyer to provide a source representative to witness tests. However, unless this representative is in residence at the plant, the value of the witness is somewhat dubious, since the pinch of economy frequently dictates that each representative be responsible for a number of plants and the scheduling difficulty frequently requires that he waive the witnessing. Under these conditions, source "inspection" rapidly deteriorates into a routine paper-stamping. (The real danger in this situation is that a false sense of security is generated in the assembly facility, where it is assumed that the source inspector's stamp on shipping documents is a guarantee of the required quality level of the procured material.)

A second disadvantage is that test equipment may not be available in-plant for failure diagnosis or repetitive testing of procured material that fails in in-process or final quality testing of the product. This leads to the highly unsatisfactory situation in which all such material must be returned to the vendor for failure verification and diagnosis, a situation that is costly in both dollars and time. A third disadvantage is that complete and accurate data may be difficult to obtain from vendor testing, which makes difficult the task of predicting or assessing the reliability of the item. Furthermore, without test equipment in-plant, it is difficult to perform investigative tests when the reliability of the item is in question. For these reasons vendor's test equipment is frequently duplicated at receiving, particularly on the critical, complex items of hardware, and the tests are performed at both the vendor's plant and at receiving.

In the environmental category of production quality tests are those performed during the production run to assess how well the production process continues to produce material capable of meeting the environmental requirements for the hardware. These tests are commonly called production assessment or requalification tests, and they are performed at several levels of assembly in a rigorous reliability test program. The testing can be either destructive or degrading to the product or not, and the decision is generally a balance between economics and the criticality of the parts involved. A good balance in this decision can be achieved if the lower-level (piece-part) tests are destructive, with the test samples discarded after use, and the higher-level tests nondestructive. With this system, the higher-level parts can be used in production and delivered.

Sometimes a better balance can be achieved if all the tests are run rigorously without regard to degradation of the samples and the samples are then given to reliability for peripheral testing or testing to destruction for failure-mode or other reliability studies. The sampling rate for production-assessment tests varies quite widely, but it is generally so set that a test is run on each item about once a month. If the production run is long, the interval is frequently gradually stretched out to two months provided the process proves to be stable and well controlled. The parameters tested and the environments utilized are the more critical ones tested in the qualification or preproduction test program, and particularly those which will most readily reveal degradation in the production process. In a typical regime these environments will include shock, vibration, temperature, and humidity. In the more sophisticated programs, time-to-time failure testing is included in the production-assessment test programs, particularly at piece-part level, with lot acceptance made conditional on the parts meeting specified minimums.

The qualification and requalification programs will be planned simultaneously in a well-conceived and integrated environmental test program, and the test equipment will be designed with the capability for both programs. Although there are some differences in the requirements—qualification testing requires the testing of more parameters in more environments, while requalification testing demands more efficiency in testing because of its repetitive nature and its impact on delivery schedules—a single set of test equipment can do both jobs, with a considerable savings in test-equipment cost to the project. Such an approach, furthermore, permits maximum correlation between qualification and requalification testing results, and it can be of considerable importance in detecting, diagnosing, and correcting shifts in production processes from those existing at the time of release to production. The same set of test equipment can also be used for evaluating material sampled from field stocks, as described under Group IV, Reliability Tests.

Group IIIb, Installation and Checkout Tests (Dealer/Field Level). In a high-reliability project it is not enough that a complete regime of ambient and environmental quality tests and inspections be performed at the factory; for the attained reliability of the population is as subject to degradation and deterioration from errors during assembly, installation, and handling in the field as from errors during manufacture. The conditions in the field are generally less controlled than in the factory and the personnel less experienced, so that the probability of error is therefore commensurately increased. Installation and checkout tests are used to detect these field errors, as well as to ensure that no inadvertent anomalies develop from functional interreactions at the system level. The field checkout tests are generally limited to the final-assembly levels and to examination of those parameters which are created or directly affected by the field assembly or installation process, though some additional assurance of quality can be attained by a repetition at field "receiving" level of the factory subsystem tests.

There are usually several levels of testing, comparable to the in-process testing in the factory. For complex installed systems the levels may include continuity testing of cabling, smoke tests or gross power-on tests to ensure that proper hook-up of cables was made to equipments, subsystem functional tests, and, finally, overall systems tests. For noninstalled equipments, such as missiles, the field-test plan will include tests at different levels of assembly, with a final overall missile systems test of the completed assembly. The tests, which are ordinarily ambient, since environmental equipment is generally not available in the field, are performed on every unit of the product.

It is most important that the field tests be compatible with the factory testing, and for this reason they should be planned by the same group that plans the factory tests and the test equipment should be designed by the same group that designs the factory test equipment. (The consequence of violation of this basic principle is constant circulation of hardware between the field and the factory and an overabundance of field failures which are unverifiable at the factory.) Compatibility of field testing also requires application of the parameter funnel principle discussed in Sec. 8.3*b*. Field checkout testing, like factory production testing, must be developed concurrently with the R&D program and utilized on the field installation and checkout of the R&D and preproduction models. Only in this manner will the validity of the ability of the field checkout test regime to weed out errors in tactical (or production) assembly or installation be assured before the project progresses into the production phase.

Group IIIc, Field Maintenance or Periodic Retest. Since the reliability of most equipments degrades with time and that of all equipments with use, it is essential that a production test program include planned retests at periodic intervals after the initial field assembly or installation so this degradation will be detected before equipments fail and corrective action can be taken. These retests should be performed at appropriate intervals on every item in the field. The interval is determined by the expected rate of degradation; and as with other tests, it may be variable. With the typical "bathtub" failure curve, the retest interval will be relatively short at the beginning of the use phase, when a high rate of infant mortality can be expected, and again at the end of the use phase when wearout occurs, but it will be relatively long in the intervening period. Thus in a typical item with a 5-year life, testing during the first and last year could be monthly, with quarterly or semiannual testing in the 3-year period between. The test regime will generally duplicate the final systems test described in Group III*b*, although lower-level tests may be necessary if the system-level test is too gross to measure all of the important parameters which affect performance.

Heavier emphasis is necessary in the periodic field maintenance tests than in the field installation tests on inspections, including not only visual but such nondestructive inspections as radiograph, magnetic-particle, or ultrasonic, particularly with high-reliability mobile equipment like airplanes and Army combat vehicles. In these equipments the structural integrity is critical, the normal usage results in wear and exposure to environmental stresses which degrade it. Thus periodic inspection for detection of reduction in cross-sectional area or of cracks, fatigue crystallization of metal, or for service damage is highly important.

In one segment of industry, commercial-airplane maintenance, this periodic reinspection has been developed to a highly sophisticated regime which ensures that all critical parts are examined at frequent intervals but without unduly increasing equipment down time. This regime, known as progressive inspection, is characterized by the examination of one or a few critical parts at each routine 100-hour inspection in contrast to a special examination of all critical parts at once in special 1,000-hour out-of-service inspections. The progressive-inspection technique can also be utilized on continuously operating functional equipments if redundant modules have been built into the equipment to permit one set of modules to function while the redundant set is being tested.

Group IV, Reliability Tests. Although all testing contributes data for reliability calculations and hence could be considered in a larger sense to be reliability testing, there are specific tests which are performed for no other purpose than to gather these data. These are the tests referred to in this section, and for purposes of this discussion they have been grouped into peripheral testing, life testing, accelerated life testing, service-life evaluation testing, and surveillance testing. The data from reliability testing are used to determine mean time or cycles to and between failure, to calculate or verify attained reliability, to establish storage and operating life limits on critically age-sensitive parts (and from both of these come the depth requirements for spare parts), and to determine modes of failure.

Reliability tests are performed at all stages of the project and on all levels of assembly. They are performed both in ambient and environmental conditions, and they include both destructive and nondestructive tests, inspections, and examinations. They may also include some actual-use tests, although they are usually confined to the laboratory to ensure control of input conditions. The hardware for reliability testing is often first used in one of the other testing programs, particularly when the purpose of the reliability test is to determine modes of failure or MTF/MTBF, since other test objectives can be met simultaneously and economically while the hardware is accumulating much of the necessary test or operating time.

Group IVa, Peripheral Testing. In almost all of the Group I, II, and III testing the parts are subjected to environments and input conditions which simulate as nearly as possible the actual range of use conditions, and a successful test is one in which the part functions properly in these conditions. From such testing, however, it is not possible to determine how much margin of safety has been designed and built into the product, since the part has not been stressed to functional yield. It is useful in predicting reliability of a population from data gathered on a limited sample to test the parts to environments and input conditions which are more rigorous than the expected service conditions by a substantial enough margin that failures can clearly be attributed to the peripheral conditions. As discussed previously in Sec. 8.2*a*, this margin should be at least 10 to 15 per cent of the spread from ambient to the limit of the service environment.

Peripheral testing is useful on all classes of material, but it is almost mandatory on solid propellants and pyrotechnics, where degradation with time may be a complex chemical reaction not fully understood or predictable. Peripheral testing is generally performed only on the top assembly levels, though occasionally the important functional field spares may be included in the program if the storage conditions of the spares are considered to be critical. Peripheral testing is usually performed in the laboratory, although, for the reasons discussed in Sec. 8.2*c*, it may be performed in actual conditions when the peripheral conditions are difficult or impossible to achieve in the laboratory.

Peripheral testing is almost always destructive, since it is highly desirable to continue to increase the severity of the environment until failure is achieved. If failure is not achieved in a selected environment, it is usually more advantageous to dissect the hardware or to subject it to life testing in normal environment to determine the degrading effect in terms of MTF of the specific level of over-environment selected, rather than to try to subject the same sample to the next more stringent environment in an effort to induce failure. This second course of action, if it does induce failure, frustrates analysis, since the cumulative effort of the two levels of environment cannot be separated and no conclusions on the effect of the higher environment can be drawn.

As a general rule, "used" hardware from other test programs is not suitable for peripheral testing, since the exposure effect at normal service limits cannot easily be eliminated from the peripheral defect, and meaningful conclusions cannot be drawn. It is useful to perform the peripheral tests at several increments of over-environments to permit establishment of gradient curves showing the relationship of degree of overstress to amount of part degradation. Such curves help in predicting reliability. Furthermore, in even the best-disciplined projects, tactical hardware will from time to time be inadvertently exposed to temperatures, shocks, vibrations, and other environments beyond the design requirements, and such curves are invaluable in determining whether service hardware so exposed to over-environments in service use is still usable, either with or without certain limiting restrictions. Peripheral test results are also useful in opening up expensive restrictions on the environments which the hardware is allowed to see, frequently resulting in lower logistics costs in the way of cheaper containers, less expensive air conditioning, less costly transportation or storage requirements, etc. Conversely, however, peripheral testing also frequently discloses unsatisfactory margins of safety in the design and may dictate more stringent handling, storage, and use restrictions than have been planned.

Group IVb, Life Testing. Reliability prediction and reliability assessment are vitally concerned with the determination of the mean time (or cycles) to and between failures, since this number is basic in reliability calculations. The number can be computed directly from the data gathered from the life test program, where tests are performed not only on samples of completed assemblies but on spares and piece parts as well. The tests are generally performed in the laboratory on test equipment which, for economy of testing cost, is designed to operate continuously or cycle the hardware automatically. The operation is interrupted at regular intervals, and functional tests or nondestructive inspections are made to find out whether there has been any degradation of the operability of the part with time or cycles of operation. Generally, the most severe expected service environments are chosen and a number of samples are utilized in a statistical design of experiments which permit the interpretation of results to include a determination of the effect of individual environments on the MTF. A control functional test is usually performed at time zero.

Life testing is slow and expensive and may take six months to a year to complete. In some situations, where real time is the same as operating time, the test program may take years; typical of these are tests of paint, where the actual service conditions are exposure to outdoor weather, or of submarine cable and equipment, where the actual service condition is exposure to ocean depths. In these situations it is essential that the life-testing program be instituted on the earliest production prototypes, so that field failures of service equipment delivered at a later time can be predicted prior to occurrence or that corrective action on the design or production process can be instituted before production actually begins. Sections 4 and 5 discuss the mathematics of predicting and assessing the reliability of a part from MTF; from this discussion it will be readily apparent that the life tests must be run for either up to several hundred times the expected life or to failure if data showing that a reliability of more than 0.99 has been achieved to a confidence of 0.95 or better are required.

With functional hardware, the life-test operation is interrupted periodically with an ambient functional test which determines the condition of the part, the hardware being returned to the test conditions after each functional test. With some other classes of hardware, such as pyrotechnics, solid propellants, and structural members, the "functional" test is inherently destructive. For these classes, it is necessary to provide a large sample, with a portion of these pulled out at periodic intervals and either fired or tested to yield. The hardware costs for life testing this class of materials is very high, and it is not unusual for no functional or yield tests to be performed until the hardware reaches two-thirds or three-fourths of the expected life. Such a decision, made in the interest of economy, does entail a risk that early catastrophic degradation with time will occur with the service hardware, but the decision is frequently dictated by project funding limitations, and is a better decision than one to reduce the sample size by shortening the total test span and eliminating testing at the expected life limit.

With complex functional assemblies, it is to be expected that some individual components will fail randomly as the life test progresses. These failures should be recorded and carefully diagnosed, and the failed parts should be replaced so that the same basic sample assembly can continue in the life test. The MTF of the individual piece parts as seen in the assembly-level life tests should be compared for statistical significance against the MTF of the same piece parts as determined by data from the piece-part life tests. When the failure occurs significantly earlier in the assembly level test, an application problem is generally indicated; it can be confirmed with continuation of the life test with a replacement part installed. Continuation of the test is mandatory to assure that other modes of failure, which would go undetected if the test were stopped at the first part failure, are revealed.

Frequent examinations of the samples during life testing is important to discover and correct such failures as loosened parts in vibration tests, embrittled materials in low-temperature tests, or softened plastics in high-temperature tests, since these failures can frequently result in secondary failures which would mask or make difficult the interpretation of the primary failure. In a 1,000-hour vibration test, for example, it is not unusual for the specimen to be examined every 50 or 100 hours. Frequent functional tests with variables data collection are also important to detect shifts in nominal values of the functional attributes at the earliest possible time. In this situation, in a stringent reliability program, the "failure" for calculation purposes should be considered to be the time the drift starts, rather than the time the attribute drifts out of specification limits.

Group IVc, Accelerated Life Testing. In a tightly compressed schedule, where R&D is hardly finished (or sometimes is not completely finished) before production starts, some assurance must be obtained relatively quickly that the hardware has an adequate life and that no gross weaknesses exist in the design that has been released on high-risk basis to production. Life tests are ordinarily too drawn out to provide such gross information quickly enough to permit design corrections to be made expeditiously. In these projects an accelerated life-test program is generally instituted. Accelerated life testing is also useful when spare parts must be manufactured concurrently with a short-run production program, as when the start-up time for a production rerun is long. In this second case, failure-rate data from ordinary life testing would not be available soon enough to make even rough order-of-magnitude determination of the range and depth of spares before the production run would be completed, and the accelerated life testing is mandatory. A third situation in which accelerated life testing is sometimes used results from the fact that there is a (statistical) distribution of the actual time to failure of all the items in the production lot. Some items in any lot will fail considerably earlier than the MTF predicted from the life testing and other testing programs.

If knowledge of the earliest possible time of failure (as opposed to the mean time of failure) is important, as with pyrotechnics and explosives, then the range of the variation in time of failure must be determined with comparatively high confidence. However, determination of the spread around the MTF is very slow with actual-time life testing, and the cost of keeping many samples in long-term storage is generally prohibitive. A comparatively large sample, however, can be subjected to accelerated life testing, and the results can be used to determine the range of variability in the time to failure with an adequate confidence. A fourth case when accelerated life testing is commonly used occurs if the product contains critical safety attributes. With pyrotechnics and solid propellants, certain modes of failure can be catastrophic, resulting in explosion and extreme danger to operating personnel and equipment. In these critical cases it is necessary to determine continuously the remaining life in the lot (see Service Life Evaluation, Group IV*d*). Samples are drawn from the service stocks at periodic intervals, generally every six months or yearly, and these "aged" samples are subjected to accelerated aging to determine whether some critical mode of failure is being approached.

Acceleration of aging may be achieved primarily by intensifying the environment to which the samples are subjected or by increasing the severity of the duty cycle or by a combination of the two. In the environmental approach, the environments are cycled from extreme to extreme in an attempt to reproduce in a short real time the degradation expected over the period of intended actual service life. The environments selected for cycling are a function of the type of hardware being tested. With homogeneous materials such as resistors, capacitors, structural members, pyrotechnics, solid propellants, plastics, and rubber materials, extremes of temperature are widely used, either at or somewhat beyond the expected (specification) service limits. When the expected mode of failure is the result of runaway chemical reaction, high temperature alone is often effective in attaining the desired acceleration. With certain classes of metals, extreme low temperature alternated with periods at ambient is appropriate.

When the expected mode of failure is not a material failure but a functional failure resulting from mechanical wear, the acceleration is usually attained by continuous operation of the device to obtain several years of expected cycling on the parts in a few days or weeks. If, however, the part is operated continuously in its normal service application, then this technique is of no avail and, instead, the input stress or loading is made more severe in an effort to increase the wear rate. With electronic equipment, ON-OFF cycling, with an ON duty cycle only long enough to warm up components, is an excellent way to accelerate aging and induce early failure.

The difficult part of accelerated life testing, particularly of the cycled or overstressed environment type, is not the determination of test conditions which will induce early failure, but the correlation of the failure data with the expected performance of the parts in their actual use and environment. A considerable amount of time and money is commonly spent in attempting to determine a statistically valid correlation between the two, but the attempt most frequently ends in failure. It is just not possible, within the practical constraints of ordinary budgets and time, to determine whether exposing, for instance, a solid-propellant motor to 125°F environment for 3 months is statistically equivalent to 3, 5, or 10 years of actual service conditions where the upper temperature is limited to 80°F.

Given enough time and enough testing samples, such valid correlation can be determined, but in the ordinary high-reliability program neither funding nor time will permit such a course of action. However, the lack of statistical validity does not obviate the need for, nor the advantages to be gained from, accelerated life testing. Intuitive but useful conclusions can be and are drawn from the results of very small samples subjected to accelerated life tests. In the rocket-motor test given as an example, a "successful" test with no failure would provide confidence that critical failures are not to be expected within the subsequent year for the tactical lot and no action would need to be taken to reactivate a deactivated production line or to increase the assurance quantities of spares being produced on an active line. Conversely, a critical failure would indicate to a prudent project management that such anticipatory action is required for insurance purposes.

Correlation is considerably simpler and more straightforward with accelerated duty-cycle testing. The expected number of cycles of actual service operation for a given unit of time can generally be estimated within rather practical limits of accuracy for most classes of mechanical and hydraulic functional equipment. Thus the number of landings an airplane will make in each 100 hours of operation or the number of times an automotive engine valve will open and close in each 100 miles of driving can be very closely estimated. The mean cycles-to-failure figures obtained from the accelerated duty-cycle testing of airplane landing-gear-operating equipment or of automotive valve springs can then be converted simply into expected mean time to failure of these parts in their actual service use. The same holds true to a somewhat lesser extent for duty cycling of electronic equipment, because with this class of material the failure is induced not only by the shock of power on–power off but also by an indeterminate effect resulting from the actual operation during the ON cycles. However, the extrapolation has enough validity to make the determinations discussed above.

Group IVd, Service-life Evaluation Testing. One problem facing top management of high-reliability projects (which is generally referred to the reliability group for an answer) is the determination of the amount of useful life left at any given time in equipments which have been delivered for service use. This knowledge is necessary to permit continuing intelligent evaluation of several aspects of the project and to make important decisions concerning them. Among these, perhaps the most important, particularly in a weapon project, is the decision that the tactical field or fleet stock either has sufficient remaining life that no replacement, refurbishing, or reworking action must be instituted or that it has not. The reasons for instituting such action may be that an intolerable degradation in performance has occurred or that the explosives have reached or are reaching a critical point where further degradation may result in explosive hazard.

If reliability is to provide management with an adequate estimate of the remaining life in the tactical population, particularly of the critical components, ordinary MTF calculations based on the general accumulation of data from early reliability tests and from field-failure data are not adequate. Specific tests are generally required to provide input data from tests on hardware drawn from existing stocks in field use so that the effect of the actual amount of field use and environmental exposure represented in the field stock can be evaluated. The service life evaluation (SLE) testing program is designed specifically to provide these up-to-date test data.

SLE testing is generally accelerated life testing, since the object of the testing is to provide management with immediate answers on the expected life remaining in the field population. The samples selected should be the oldest or those with the most use in order that the worst material condition can be detected. Functional hardware should be tested at ambient conditions both before and after being exposed to the accelerated-aging environment or cycling, and the results of these ambient tests should be compared with each other as well as with the original factory test data taken at the time the parts were delivered. This double comparison is necessary to plot degradation curves, which are prerequisite to predicting the probable remaining life in the population. For material which can only be tested destructively, e.g., pyrotechnics, two sample sets should be drawn with as nearly equal field time and environmental exposure as possible; one set is then tested as received for a control sample, and the other is tested after exposure to the accelerated-aging environment. As with functional hardware, the results of the two tests should be compared not only with each other but with the original factory-acceptance test data.

The same techniques for accelerating the aging, i.e., inducing early failure, are used with service-life evaluation testing as discussed in connection with Group IV*c*. Inasmuch as accelerated life testing is ordinarily performed early in the program, the correlation of that early testing can be verified or modified with the data from the first (control) ambient test of the service-life evaluation sample. Thus there will be a higher confidence in the ability to predict remaining life from the SLE testing than there was with the accelerated life testing program.

Group IVe, Surveillance Testing. The last test program in the reliability test group is surveillance testing. These tests, which are performed on samples drawn at regular intervals from the actual field-service stocks, consist of ambient tests and examinations performed on the samples at progressive levels of disassembly. The object of the testing is to discover evidence of failure or incipient failures in the hardware, including not only shifts in values of components in functional hardware but chemical deterioration of materials, fatigue cracks, corrosion, whiskers, hardening of O rings and seals, and any other unanticipated modes of failure.

The two characteristics differentiating surveillance testing from other kinds of reliability testing are the limitation of testing to ambient examinations and the complete disassembly of the specimens. Surveillance testing is not the same as service-life evaluation, even though both are performed on samples drawn from the service stocks, in that accelerated aging is performed on the SLE samples in an attempt to determine quantitatively the remaining life of the population, whereas the surveillance tests and examinations are aimed at finding specific existing or incipient deterioration. However, in some projects the SLE samples are completely disassembled at the end of the SLE testing, and in this way may provide data for the surveillance test program. These side-benefit data will not completely satisfy the goals of a surveillance program, since the effect of the accelerated-aging environments cannot be rigorously separated out, and a true picture of the service stock as it existed at the time the samples were drawn cannot be ascertained.

Surveillance testing of pyrotechnics, explosives, and rocket motors requires the withdrawal of twice as many samples as are needed for surveillance testing of functional and structural material, since one sample must be destroyed in a functional test in order to obtain parameter drift or shift data and another is required for dissection. The high unit cost of very large rocket motors generally precludes such double sampling, and in many instances project fund limitations may preclude any full-scale surveillance testing at all except for those examinations which are completely nondestructive (for example, radiographic examination of the quality of the propellant-to-liner or liner-to-case bonds).

Because of this economic limitation, in the large-rocket-motor projects it is not unusual for small samples of the propellant (gallon-carton size up through eighth-scale motors) to be cast and cured and put into relatively ambient storage for later surveillance examination and testing. These substitute tests are not completely satisfactory, however, because of the extreme difficulty of extrapolating full-scale functional degradation from the small-size-sample behavior. The program does provide some information on gross chemical degradation of the propellant, and for this reason it is generally considered worth the nominal cost.

The functional test regime selected for surveillance testing of functional hardware is generally the one utilized in the final factory test of the hardware. The variables data from the surveillance test on each sample are compared with the variables data from the factory acceptance test, with a particular purpose of determining whether any statistically significant shifts have occurred in any of the parameters. For this reason it is important that factory-test results keyed to package serial numbers be carefully retained and made available to the surveillance test agency. In a rigorous reliability project, when significant shifts are found in a single surveillance sample, additional samples are usually drawn from the service stock in an effort to determine the extent of the drift problem in the service stock.

The disassembly of structural and functional hardware must be performed with great care by skilled, experienced personnel in order that minute evidences of deterioration will not be destroyed in the disassembly process. A diligent search must be made for such signs as fluid-leakage paths or external traces, characteristic corrosion conditions, loose fasteners, stiff or worn mechanisms, loose connections, corroded connectors and couplers, dissimilar metal whiskers, and fatigue cracks. Any such conditions found, either by visual examination or with the aid of such nondestructive tests as black light or magnetic-particle examination, must be carefully photographed or otherwise recorded for further analysis and subsequent management consideration and decision on corrective action. It is also important that physical and chemical properties tests be made on representative samples of critical materials, especially O rings and gaskets, where failure in the form of premature hardening or cracking will result in second-order failure of supposedly protected parts.

Group V, Special Tests. Two classes of tests do not fall into the other four groups of tests described above. These are consumer research and investigative.

CONSUMER RESEARCH. All of the tests discussed above are performed by professional test personnel under conditions which are as controlled as possible and intended to simulate the actual-use conditions. Such controlled-condition testing is, of course, essential to ensure that a mass of analyzable data is available for making design and reliability decisions. However, controlled-condition testing is not without its drawbacks, since there is always an element of doubt that the conditions are truly representative of the actual-use conditions. In military weapon projects this uncertainty is small, because the use conditions themselves are disciplined and controlled and because the equipment must be designed to operate under the most severe combinations of conditions found anywhere that the equipment may be operated. However, with consumer goods the uncertainty is large, because civilian consumers are not disciplined and trained to follow certain set procedures in using equipment.

Furthermore, price competition precludes designing consumer goods to meet the stringent worldwide environmental conditions imposed on military weapons, and indeed the level of reliability of consumer goods must be carefully balanced against cost and price. Hence, in consumer-goods projects it is generally wise for a program of consumer testing to be established to provide the final proof of the acceptability, usability, and reliability of the design. It is surprising how often consumer testing reveals weaknesses in design, very severe use or environmental conditions, and consumer reactions not anticipated by even the most experienced and competent project management.

Many techniques, including random selection, have been successful in selecting the individuals who will be used to provide the consumer testing "laboratory." The important factor is to select individuals who will keep careful records, and it is also desirable that some be professional or scientific personnel who can follow explicit directions and that others be nonprofessional personnel to provide a mixture of consumer conditions and reactions. Samplings should be made in all of the representative geographic areas in which the product will be marketed as well as of the representative socioeconomic groups at which the marketing is aimed. The programs should be run for a period at least equal to the proposed warranty period and preferably twice that long. Written reports are desirable, but then should be backed up by periodic visits by field-service representatives.

Consumer test programs should be as carefully planned as the other programs outlined above. It is usually highly desirable to utilize prototype equipment so that consumer test data will be available to the project in time to introduce corrective design changes before the design is released for quantity production. The test samples should be returned to the factory for detailed disassembly and examination.

When time permits in a military project, particularly for the nonexpendable items, a similar test program will yield useful results in evaluation not only of the hardware design but of the operating and maintenance procedures as well. If such a program is conducted by a customer evaluation group, it can serve the secondary purpose of customer acceptance of the design. It is highly desirable that field engineering representatives of the design agency accompany the evaluation group, not only to ensure that the equipment is properly utilized but to ensure that there is a complete feedback to the factory of failure and difficulty reports.

INVESTIGATIVE. The last class of tests in a complete test regime is an important source of reliability data, but the tests cannot be planned in advance. Throughout the entire scope of any project there are many special tests performed to investigate specific problems or aspects of problems. These investigative tests include, for example, failure diagnoses of hardware which has failed in any of the test programs described above, production-process sampling tests (chemical and physical) for quality control or troubleshooting of production processes, functional tests and material chemical and physical tests to determine the adequacy of proposed substitutions of materials in the production program, and measurement-accuracy and -repeatability tests to assess the accuracy of production or other test equipment.

These tests are characterized by their ad hoc nature, but it is a frequent project weakness that the data resulting from them are not included in the overall reliability data bank. Such tests provide very useful background data against which reliability problems can be assessed; many provide information directly applicable to the reliability-prediction operation. It is therefore essential that a system that will ensure that at least a copy of all the test conditions and results is provided to the reliability organization be established. It is desirable from an economic point of view that reliability personnel be included in the test-planning function so that useful secondary data can be gathered simultaneously with the primary data desired. Furthermore, monitoring of the test by quality control inspectors will ensure that the validity of the data is acceptable and will preclude introduction of unknowns resulting from undisciplined testing.

8.3 TEST PLANNING

8.3a General Considerations. A reliability test program, like any other phase of a project, will be no better than the thoroughness and timeliness with which the program is preplanned. The five groups of tests described in Sec. 8.2*e* frequently represent millions of dollars of testing and test hardware and require the skillful coordination and scheduling of hardware, test facilities, and the work of experienced engineering and technical personnel. Without a thorough and timely job of planning, all of the necessary elements will not be available at the right time in the right quantity. This leads to the inevitable result that important tests will not be performed in time to permit intelligent decisions to be made from the resulting data, or the testing will be rushed and the test data will be incorrect or inadequate, resulting in erroneous decisions.

In most high-reliability projects, because of their compressed time scales, emphasis must be placed on starting the test planning very early in order that design of special test equipment can proceed (on a calculated-risk basis) and be ready in time for testing the first R&D hardware. Furthermore, the proposed test plan must be available to product designers so that the necessary test points can be designed into the product. In some instances a particularly high cost of a necessary test may dictate a change in the product design to one requiring less costly testing; the design change may be one of circuitry, material, or manufacturing process required. For example, a designer may specify a completely flaw-free sheet steel which would require excessive selective NDT, such as radiography, of large sheets of the material for acceptance inspection. Derating the material so that commercial grades of steel would be acceptable, perhaps at some permissible weight penalty, could measurably reduce the testing and overall cost without a sacrifice of necessary product strength.

In another common example, the very high cost of design evaluation and qualification testing of the product to unusual extremes of transportation environments may be the deciding factor in choosing to provide a protected transportation environment, such as heated containers, for the product rather than to design the product to withstand all of the extremes. Timeliness is important in the planning of production testing too. A frequent project weakness is to let the product and test-equipment designers work out the details of the acceptance or quality testing of the R&D hardware without reliability and quality assurance participation, with the result that the subsequent production test program is inherently dictated by these two design groups because the cost and length of time required to introduce tests desired by quality assurance and reliability preclude introducing them at all. (See Sec. 15.) In a well-managed project, reliability and/or quality assurance plan the R&D acceptance tests and inspections by utilizing initial inputs from the product design groups and adding the necessary quality control test points to them. This works out well only if the reliability group starts developing the acceptance-test plans concurrently with the inception of the product design effort.

It is not enough that the whole reliability test program be planned early in the project. It is equally important that the test planning be done in considerable detail and that it cover all elements of the test program. Only in this way will all of the testing be integrated and provide compatible and usable data. For example, details of the environments must be specified so that one group of tests will not be performed at 100°C, another at 90°C, and a third at 115°C, which, when it happens, precludes direct and meaningful comparison of test results. Furthermore, the test planning must be complete enough and in such a form that the overall test program can be studied with the intent of eliminating costly duplications or omissions.

Spread sheets for each item to be tested are very useful if they show the attributes to be tested and the test conditions (both inputs and environments), for they permit ready comparison and correction before the test procedures are prepared. Furthermore, the spread sheets will readily indicate where statistical design of experiments may be used to economic advantage. Proposed test schedules are also important, and they should be charted in a PERT network against such key project dates as points of design decision, R&D and production design releases, acceptance of first hardware, and qualification dates.

8.3b Planning the Program. *Organization.* A review of the purposes of the various testing programs which make up the overall reliability test program should make it clear that many organizational units have either a direct or an indirect interest in the conduct of the tests and the results. A common weakness in present industrial practice is to assign complete cognizance of the various tests to that organization which has the most immediate direct interest. Thus the design-oriented tests (feasibility, evaluation) are assigned to the engineering or development organization and the production environmental (requalification) tests are assigned to quality assurance. Such a division of responsibility almost guarantees that the test results from these many test programs will be incompatible and incapable of being combined into an overall reliability data bank.

In a better organizational arrangement the overall responsibility for planning all test programs and tests is assigned to the reliability organization, which ensures that the requirements of all of the individual groups, including design, quality assurance, and field service, are met. In this arrangement the organizational unit (reliability) most interested in having compatible data lending itself to population analysis will have the authority to set up the tests in a completely coordinated manner. One expeditious mechanism for reliability to use to provide this coordination is to chair an integrated reliability test-planning committee, which starts functioning at the initial phase of the project and continues to meet regularly to update the overall test program plan in accordance with changes in project direction. This committee should consist, at minimum, of responsible members of reliability, quality assurance, product design, test-equipment design, production planning, and test laboratories. (See Sec. 15.)

The test planning must not be thought of as a one-time process with an inflexible result, but instead should be regarded as a flexible selection of items and attributes to be tested which will vary with time and knowledge. In particular, testing should be considered primarily as a means of uncovering weaknesses or anomalies in either the design or the production process. As a project progresses, new weaknesses which will require new tests will be discovered; conversely, old weaknesses will be corrected, which will permit deletion of tests. Thus the test-planning committee (or committees if for convenience the work is subdivided into such functional subcommittees as testing, mechanical inspection, and NDT) is not an ad hoc committee, but a permanent part of the overall company committee structure. Inasmuch as the reliability organization is generally responsible for investigating failure modes and determining the corrective action for design and production weaknesses, it is appropriate for reliability to chair this committee.

As a general rule, the product designer will specify those items and attributes which need to be tested and inspected to ensure functional operability of the product; quality assurance will specify the additional items and attributes which must be tested for process and quality control; reliability will dictate the reliability-verification requirements, including items, attributes, and quantities; test-equipment designers will provide the limitations on testing; and the product designers will limit the number of test points and provide the allowable duty cycles. The committee, as a whole, will estimate costs and assess the overall balance between risk and cost.

Timing. The importance of starting the reliability test planning at the beginning of the project cannot be overemphasized. Unfortunately, in many projects, only those test programs directly associated with the development effort (research, feasibility, evaluation) are planned early in the project, and as a result many valuable data that might otherwise be collected on these early tests are lost. Included in these lost data are the basic correlation data comparing performance in controlled environments with performance in ambient conditions which are required to complete the production acceptance-test planning.

The whole reliability test program should be blocked out at the beginning, with best estimates made as to the parameters to be tested and the environments utilized for each individual program. Tentative decisions should be made on the various classifications of tests listed in Secs. 8.2*a* to 8.2*d* for each test. All these tentative test plans should be recorded in spread-sheet format, with the understanding that the planning will be revised at frequent intervals, perhaps monthly in the early stages and changing slowly to quarterly or semiannually by the time the project has reached the design-release stage. These changes in the program plan will result from many factors and events, including changes in hardware design concepts, changes in production techniques or in production location, changes in intended use of the hardware (which in themselves may result from changes in design), field-failure feedback information, or the discovery of new weaknesses and modes of failure.

The constraints of time and money will act as a check and balance against introduction of unnecessary changes to the preliminary planning, so that it will not be necessary or desirable to consider the preliminary planning to be frozen from the outset. Planning should, in fact, be considered to be quite flexible for most of the development phase of the project.

Planning Documents. A sample of the IRTP plan spread sheet is shown in Fig. 8.4. However, the integrated reliability test-program plan is itself incomplete as a document from which individual test procedures can be written. Although the plan specifies attributes (in words) to be tested, the environments to be utilized, and the general method of test, it does not contain the input stimuli, parameter values and tolerances, or test-equipment identification and errors. These supporting numbers are required for test-equipment designers to be able to design test equipment (and to compute or demonstrate the test-equipment error for each measurement) and write test-equipment calibration procedures and for procedure writers to be able to write test procedures.

The values for each parameter to be tested, including the tolerances and the input conditions with their tolerances, are conveniently set forth in a parameters document (Fig. 8.5) as well as in a parameter spread sheet (Fig. 8.6). These planning documents are necessary to ensure that the many facets of the overall test program are well integrated and that incompatibilities and inconsistencies do not occur, particularly in the larger organizations associated with modern weapons systems where the individual work of many different organizations is required to establish all of the details of the test program. Given a test program as outlined in the integrated reliability test-program plan sheet, with the parameters funneled as set forth in the parameters spread sheet and with test input conditions outlined in the individual parameters document sheets, the test-equipment designer can proceed with the design of his test equipment, the product designer can provide the necessary test leads or connections, production and production-scheduling personnel can schedule the necessary test hardware into the production plan, and inspection (QA) planners can plan for testing and inspection points, floor space, personnel, and standard equipment.

Although the integrated plans presented here have illustrated functional tests, the same kind of integrated preplanning is necessary for the special mechanical tests and inspections (including specialized nondestructive tests) which are required. However, because these kinds of attributes are not generally subject to deterioration with time or handling, no allowance is made in a funnel of tolerances; and wherever the attribute is to be measured, the same tolerances are used. For this reason there is not a need for a parameters spread sheet, although the integrated test plan and the individual parameters sheets are required to delineate the details of the desired test or inspection.

INTEGRATED RELIABILITY TEST PROGRAM PLAN

PART NAME: GYRO PACKAGE, FLIGHT CONTROL SUBSYSTEM: FLIGHT CONTROLS	PART NO. 1963822-C SPECIFICATION WS 1389B	TEST PLANS: PARA DOCUMENTS:	EET OD 22546 OD 22545	PPT OD 22548 OD 22547	VAT OD 22550 OD 22549	PAT OD 22552 OD 22551	FLD OD 22554 OD 22553	

ENVIRONMENT →		ACCEPTANCE					RANDOM VIBR.					SINUSOID. VIBR.					SHOCK					ACCELERATION					TRANS. TEMP.					FLIGHT TEMP.					HUMIDITY				
		LAB AMBIENT TEMP. 60–90°F HUMIDITY >90%					20–4,000 CPS 5.1 grms					FREQ. SWEEP 10–50 CPS 5 MINUTES/CPS 2 HOURS AT MAX RESONANCE MIL STD 167					50G PEAK 10 ±1.0 MSEC 300G PEAK 2 ± 0.5 MSEC					10G, 5 SEC MIN 3G, 60 SEC MIN					-40°F, 12 HOURS +150°F, 12 HOURS 4 CYCLES					120°F, 0°F 600°F FOR 6 MINUTES					MIL E 5272 120 HOURS MIN 5 CYCLES				
ENVIRONMENTAL EQUIPMENT TOLERANCE →		TEMP. ±5% HUMIDITY +10% -0%					±10%G					±2% OR 1 CPS					±10%					±5%					< 5%					< 5%					+10% - 0%				
TEST PROGRAM → / ↓ PARAMETER TESTED		EET	PPT	VAT	PAT	FLD	EET	PPT	VAT	PAT	FLD	EET	PPT	VAT	PAT	FLD	EET	PPT	VAT	PAT	FLD	EET	PPT	VAT	PAT	FLD	EET	PPT	VAT	PAT	FLD	EET	PPT	VAT	PAT	FLD	EET	PPT	VAT	PAT	FLD
EXAMINATION, NONOPERATING	VISUAL EXAMINATION OF MATERIALS DESIGN CONSTRUCTION DIMENSIONS WEIGHT COLOR AND FINISH IDENT AND WORKMAN	4	4	1	1	-																																			
LEAKAGE NONOPERATING	NO VISIBLE LEAKAGE	4	4	2	2	1	4	4	2	2		4	4	2	2	1	4	4		2		4	4		2		4	4		2		4	4		2		4	4		2	
POWER CONSUMPTION OPERATING	15.0 WATTS ± 2.5 WATTS AT AMB, ± 3.0 WATTS IN ENVIRONMENT	4	4	2	2	1	4	4	2	2		4	4	2	2	1	4	4		2		4	4		2		4	4		2		4	4		2		4	4		2	
INSULATION RESISTANCE	40 MEGOHM MIN	4	4	2	2																						4	4		2							4	4		2	
DAMPING RATIO OPERATING	SEE SHEET 3	12	12	6	6		6	6	3	3		6	6	3	3		9	4		4		9	4		4		9	4		3		9	4		3		4	4		2	
SENSITIVITY OPERATING	SEE SHEET 3	12	12	6	6	1	6	6	3	3		9	6	3	3		9	4		4		4	4		2		9	9		4		9	9		4		9	9		4	
NULL DRIFT OPERATING	MAX INCR - 3.6 MV (0-PK) PER G^2(RMS)						12	12		6		12	12		6																										
ELECTRICAL INTERFERENCE	PARA 3.24.6 WS 1389B	6	6				6	6				6	6				6	6				6	6				6	6				6	6				6	6			
STARTING CURRENT	250 MA MAX	4	4		2		4	4		2																	4	4		2											
AXIS ALIGNMENT OPERATING	OUTPUT OF EACH 2 AXIS ≤ 500 MV, AMBIENT OR ≤ 650 DURING ENVIRONMENT	12	12		2	1	6	6		2							12	12		2		12	12		2																

PAGE 1 OF 3

EET = ENGINEERING EVALUATION TEST
PPT = PREPRODUCTION TEST
} QUANTITY INDICATED IS TOTAL QUANTITY TESTED

VAT = VENDOR (FACTORY) ACCEPTANCE TEST
PAT = PRODUCTION ASSESSMENT TEST
FLD = FIELD TEST
} QUANTITY INDICATED IS SAMPLE SIZE PER PRODUCTION LOT

Fig. 8.4 Sample integrated reliability test-program sheet.

O.D. 22551 PART IV

REFERENCE:
TEST POINT # 201

PARAMETERS DOCUMENT SHEET

SHEET 1 OF 7 REV C

PART NAME GYRO PACKAGE, FLIGHT CONTROL			PART NUMBER 1963822-C
SUBSYSTEM Flight Control	DATE PREPARED 7-31-63	DATE RELEASED 9-15-63	RELEASE AUTHORITY JRA 52PD-20105C
PREPARED BY D.R. Keith (System analysis)	CHECKED BY P.P. Parish	RELIABILITY Fallon	
DESIGN CHECK & cty and EE Banghart	PROJECT OFFICE WA Steven		CUSTOMER APPROVAL J.E. Van Eav

TEST PLAN ITEM	ATTRIBUTE	PARAMETER	TOLERANCE
1	RANDOM VIBRATION		
	1.1 Input: 115 volts a-c rms ±0.5%, 400 ±2cps @ 70 ± 6°F		
	1.2 Load: Non-reactive impedance of 15K ± 300 ohms for pitch, yaw and roll outputs		
	1.3 The package shall be subjected to random vibration along each of its three mutually perpendicular axes for one minute. The level of vibration shall be 5.1 grams contained within the spectral density limits below:		
	1.3.1 Upper Limit Frequency (cps) — Envelope of Peaks (g^2/cps) 20 — 0.005 500 — 0.007 750 — 0.014 4000 — 0.025		
	1.3.2 Lower limit shall be one fourth of upper limit		
	1.4 Measure the in-phase null degradation of the demodulated output signal through a 0 to 4 cps filter (8 db/octave), of each gyro when vibrated in each axis (measure 0 to peak).		
	1.4.1 Pitch axis	1.7 mv 0-p/(gms)^2	max
	1.4.2 Yaw axis	1.7 mv 0-p/(gms)^2	max
	1.4.3 Roll axis	1.7 mv 0-p/(gms)^2	max
		(210 mv 0-p max, per axis)	
2	POWER CONSUMPTION		
	2.1 Input and load same as 1 above		
	2.2 Measure		
	2.2.1 Running power	15.0 watts	±2.5 watts
	2.2.2 Power factor	.960 ldg to .960 lag	
	2.2.3 Time to reach running power	50 sec	max
	2.2.4 With squelch circuit applied, measure running power and power factor	10.0 watts 1.0 to 0.7 lag	+1.8 watts -2.3 watts
3	INSULATION RESISTANCE		
	3.1 Apply 500 volts d-c minimum, between all isolated circuits and between circuits and case for 5 seconds minimum		
	3.2 Measure insulation resistance	40 megohms	min.
4	UNDAMPED NATURAL FREQUENCY AND DAMPING RATIO		
	4.1 Input and load per 1 above		
	4.2 Apply sinusoidal change of rate to produce 90° phase lag to pitch, yaw and roll rate gyros.		
	4.3 Measure the natural frequencies and damping ratios		

FIG. 8.5 Parameter document—sample page.

PARAMETERS SPREAD SHEET

Volume II, Section 2
Flight Control Subsystem
Gyro Package, Flight Control 1963822-C

Page 2-107
OD 22556

PERFORMANCE REQUIREMENTS			PARAMETERS							
ITEM	CHARACTERISTIC	CONDITIONS AND REMARKS	NOMINAL	In-Flight Tolerance	Submarine Tolerance	Tender Tolerance	Depot Tolerance	Factory Tolerance	Vendor Tolerance	Units of Measurement
GENERAL CONDITIONS										
1.0	External resistive load	(all output circuits) 15,000 ± 300 ohms								
ENGINEERING REQUIREMENTS										
2.0	Input									
2.1	Voltage	Single phase	115	±3%	±2.7%	±2.4%	±2%	±1.5%	±1.0%	volts a-c rms
2.2	Frequency		400	±0.8%	±2	±2	±2	±2	±2	cps
2.3	Harmonic content			2 max	2 max	2 max	2 max	2 max	2 max	per cent
3.0	Electrical									
3.1	Power consumption									
3.1.1	Starting current			125 max	125 max	125 max	125 max	125 max	125 max	milliamps a-c rms
3.1.1.1	Transient	Time to reach running power limits		50 max	50 max	50 max	50 max	50 max	50 max	seconds
3.1.2	Running power		15.0	±4.0	±3.8	±3.6	±3.3	±2.5	±2.0	watts
3.1.3	Running power factor	Leading		0.960 min	0.960 min	0.960 min	0.960 min	0.960 min	0.960 min	
		Lagging		0.960 min	0.960 min	0.960 min	0.960 min	0.960 min	0.960 min	
3.1.4	Squelch power		10.0	+1.8 -2.3	+1.8 -2.3	+1.8 -2.3	+1.8 -2.3	+1.8 -2.3	+1.8 -2.3	watts watts
3.1.5	Squelch power factor	Lagging only	1.0	-0.3 max	-0.3 max	-0.3 max	-0.3 max	-0.3 max	-0.3 max	
3.2	Insulation resistance	The resistance of insulation between isolated circuits and between circuits and the case with a potential of 500 volts d-c applied for not less than 5 seconds		40 min	40 min	40 min	40 min	40 min	40 min	megohms
3.3	Output frequency		36	±6.0	±5.94	±5.86	±5.75	±5.5	±5.25	cps
3.4	Damping ratio	At a natural frequency of 30 cps		0.91 max 0.49 min	0.90 max 0.50 min	0.89 max 0.51 min	0.88 max 0.52 min	0.85 max 0.55 min	0.82 max 0.58 min	
		At a natural frequency of 50 cps		1.32 max 0.38 min	1.31 max 0.39 min	1.30 max 0.40 min	1.27 max 0.42 min	1.23 max 0.45 min	1.19 max 0.48 min	
		(As the natural frequency increases from 30 to 50 cps the damping ratio limits shall increase linearly between the values given.)								

FIG. 8.6 Parameters-spread-sheet sample.

Selection of Attributes. The selection of attributes to be tested or inspected is a complex matter, with the final selection resulting from conflicting factors acting in check and balance. These factors include:

1. The need to demonstrate functionability under all use conditions
2. The need to demonstrate reliability
3. The cost of testing, including test and environment equipment
4. The time to perform each test
5. Equipment and personnel available to perform tests
6. Customer requirements
7. Need to assure spare-part interchangeability
8. Desire to provide optimum process and quality control and to assure repeatability of the production process
9. The required reliability of the part or system
10. The cost of and the cost of replacement of the part tested

It is obvious that except for the simplest of parts it is not possible to test all attributes of a part in every test regime, nor generally in any one regime. Consequently, it follows that any test plan represents a sampling of the attributes and that not only must attributes be sampled but so also must the items to be tested in each test regime. Since such a selection process is one almost entirely of judgment, the committee approach to test planning generally operates most efficiently to provide automatically the check and balance required to arrive at an optimum sampling.

To arrive at the final list of attributes to be tested, it is an inherent prerequisite that the importance of each attribute be classified. Although this classification is frequently informal, a better method is that first employed by the Navy and Army for ordnance material and known variously as the system of ordnance classification of defects, classification of defects, or classification of characteristics. In this system, each attribute is classified critical, major, or minor in accordance with its effect on coordination, life, interchangeability, function, and safety (CLIFS). (Further discussion of this system can be found in the Bureau of Ordnance document *Ordnance Standard* 78.) With such a system in operation, classification of attributes is standardized from project to project, as well as from item to item and from test program to test program within a project, and the many side advantages of classification can be realized throughout the whole life of a project. These latter include automatic designation of inspection and test sampling plans; automatic emphasis on important areas for such efforts as failure diagnosis, corrective action, inspection, and design-change control; and an automatic base for the quality incentive in incentive contracts. However, the direct benefit to inspection and test planning should be immediately apparent, since the otherwise subjective process of selecting attributes is reduced to objective application of an agreed-upon set of ground rules, largely eliminating the subjectivity that frequently accompanies this operation. (See Fig. 8.8.)

Selection of Input and Environmental Conditions. Concurrently with the selection of items and attributes to be tested, the input and environmental conditions associated with each test point should be specified, since the designation of an attribute to be tested is meaningless unless the conditions of test are also specified. As a general rule, input conditions should be at least as rigorous as those encountered by the part in its ordinary use. This will usually dictate that two or more sets of input conditions be provided: the "high" and "low" conditions which can be encountered as the part ordinarily functions. In some instances, as for example when a test of linearity, hysteresis, or sensitivity is required, other intermediate points will also be required to establish the performance curve.

In instances when the part is operating in a derated condition, it may be satisfactory to utilize a one-point nominal condition provided the allowable output tolerance is tightened by an arbitrary amount from that which would be allowed for a high-low input test. Since such a one-point approach always represents less than optimum assurance and is usually dictated by cost rather than quality considerations, it should be the exception rather than the rule. For Group IV reliability testing, where the

TEST PLAN SHEET

O.D. 225/52 PART I-C

PART NAME			PART NUMBER	TEST POINT	REV
GYRO PACKAGE, FLIGHT CONTROL			1963822-C	201	B

Specification	Test Procedure	Test Sta.	OCD POINT Yes X No	Page 1 of 4
WS 13898	22634	No. 52X220	TEST DATA REQD. Yes X No	

Prepared By	Checked By	Date Prepared	Release Authority
Johnson	P P Parish	5-13-63	JRA 52TP-22552
QA Engineering	**Design**	**Project Office**	**Customer Approval**
RA Hold	PD Stroux	WG Stevenson	JA Van Ess

OCD CAT	ITEM NO.	Attribute Tested and Test Condition
	1	RANDOM VIBRATION
		1.1 Input: 115 volts a-c rms ±0.5%, 400 ±2 cps @ 70 ± 6°F
		1.2 Load: Non-reactive impedance of 15K ± 300 ohms for pitch, yaw, and roll outputs
		1.3 The package shall be subjected to random vibration along each of its three mutually perpendicular axes for one minute. The level of vibration shall be 5.1 grams contained within the spectral density limits below:
		1.3.1 Upper limit:
		Frequency (cps) — Envelope of Peaks (g^2/cps)
		20 — 0.005
		500 — 0.007
		750 — 0.014
		4000 — 0.025
		1.3.2 Lower limit shall be one fourth of upper limit
MAJ		1.4 Measure the in-phase null degradation of the demodulated output signal through a 0 to 4 cps filter (8db/octave), of each gyro when vibrated in each axis (measure 0 to peak).
	2	POWER CONSUMPTION
		2.1 Input and load per 1 above
		2.2 Measure
MAJ		2.2.1 Running power and power factor; time to reach running power.
MIN		2.2.2 With squelch circuit applied, measure running power and power factor.
	3	INSULATION RESISTANCE
		3.1 Apply 500 volts d-c minimum between all isolated circuits and between circuits and case for 5 seconds minimum.
MAJ		3.2 Measure insulation resistance.
	4	UNDAMPED NATURAL FREQUENCY AND DAMPING RATIO
		4.1 Input and load per 1 above
		4.2 Apply sinusoidal change of rate to produce 90° phase lag to pitch, yaw, and roll rate gyros.
MAJ		4.3 Measure the natural frequencies and damping ratios of each of the following gyros:
		4.3.1 Pitch rate
		4.3.2 Yaw rate
		4.3.3 Roll rate
	5	SENSITIVITY, IN PHASE AND QUADRATURE AND HARMONIC COMPONENTS
		5.1 Input and loads per 1 above
		5.2 Apply the following rates to the pitch, yaw, and roll axes
		5.2.1 75°/second CW and CCW
		5.2.2 35°/second CW and CCW
		5.2.3 15°/second CW and CCW
		5.2.4 0°/second CW and CCW
		5.3 Measure the signal phase for each of the above rates
		5.4 Measure the in-phase voltage component (0° or 180°)
		5.5 Measure the out-of-phase voltage component (Quadrature and Harmonics).

FIG. 8.7 Test-plan-sheet sample.

margin of design safety is to be ascertained, the input conditions should be set at some value more stringent than that anticipated in actual operation.

Environmental exposure should be utilized whenever possible in a high-reliability test program. Ambient testing, except when the part will operate in an ambient or approximately ambient condition, is at best an approximation, and uncertainties

CODE IDENT 10001 BUWEPS OCD 1963822-C
ORDNANCE CLASSIFICATION OF DEFECTS
FOR
FLIGHT CONTROL GYRO PACKAGE

SHEET 1 of 4

Prepared by
Floyd Carpenter
ABC Manufacturing Co
29 January 196_

Approved
C. M. Smith
Special Projects Office
15 February 196_

A UNIT IS DEFECTIVE IF ANY OF THE REQUIREMENTS IN THE CLASSIFIED SECTIONS BELOW ARE NOT MET

NOTES:

A. This BuWeps (SPO) OCD covers only those defects which affect safety, function, life, interchangeability and coordination of items of Government prime procurement. It must be used in conjunction with applicable drawings and specifications, standard inspection instructions (MIL-Q-21549B) and the inspection cognizance letter.

B. All numbers are dimensions unless otherwise indicated. All dimensions are in inches unless otherwise indicated.

C. Change Directives A, 1 August 196_, B 28 December 196_ and C, 15 January 196_ are incorporated.

SECTION A--DIMENSIONAL REQUIREMENTS

	REQUIREMENTS	REFERENCE	METHOD OF INSPECTION
CRITICAL			
	None		
MAJOR			
101	0.250 ≠ 0.005 -0.000 dia of dowel pin hole		Gage 7422546
103	The 0.250 dia dowel pin hole shall be in true position within 0.005 dia	1963759G Sheet 3	Gage 7458649B
104	1.12 MAX distance from dowel pin hole to end of package (2 places)	1963759G Sheet 3	Gage 1870321
107	4.5 ≠ 0.2 pounds MAX, unit weight	1963822C Notes	SMI
SPECIAL REQUIREMENTS			
108	When packaged as a tactical spare, container must pass leak test per OD 21936A		Leak Test Console 2163455

SECTION B--FUNCTIONAL REQUIREMENTS

	The Functional Requirements will be found numbered in the margin of ATP-OD 22634 No Rev of 15 January 196_.		Section 3, OD 22634

SECTION C--MARKING AND FINAL VISUAL

151	Container is stencilled in a/c with Note 6, BUWEPS DWG 1963820K.		Visual
152	Humidity indicator less than 50% RH		Visual

Fig. 8.8 Sample classification of defects.

always exist in such testing. Unless the cost is prohibitive or the part has been designed with a limited duty cycle in its operating environment, the general rule should be established that all testing will be done under the use environment. The practical constraints of time, equipment, and money will in many instances mitigate against the rule, but starting with the rule makes it necessary that departures from it be justified, and the end result will be that a larger percentage of the testing will be performed under environments.

In particular, the vibration environment should be utilized wherever possible, primarily because vibration is one of the most economical and effective quality control tools available to the test engineer. The probability of detecting intermittent operation, loose and cracked parts, inadequate mounting or protection, poorly soldered joints, and a host of other common quality and workmanship defects is higher in this environment than in any other. From a practical standpoint, the part may be operating during the vibration exposure and hence little time is added to the overall test time or cost because of it. Some caution must be exercised in selecting the level of vibration, however, for with some delicate parts the exposure can be degrading if the service environment level or time limit is exceeded. However, the need for such care should not be accepted as an excuse for not using the environment. (Testing should include both random vibration to simulate the actual-use power spectra and sinusoidal vibration to permit meaningful failure diagnosis.) Shock, temperature, and humidity extremes are other common environments easily attainable and frequently used.

Some consideration should be given in planning an overall test program to the use of combined environments. The interaction of environments is difficult to assess; in a complex-item test program it is almost impossible to do so numerically, and empirical data are necessary. Hence, in the early phases, particularly in the R&D tests, at least one set of combined environment tests should be planned, coupled with a series of tests under single environments, so that adverse interactions may be noted. Obviously, not all environments can be combined at once (i.e., both high and low temperature cannot be combined in a single test), but usually a series of three or four tests will suffice. These tests include high and low humidity with high and low temperature, all combined with shock and vibration.

Interrelationship of Test Programs. A common weakness in test programs for high-reliability projects is to plan each of the separate test regimes individually. This weakness which is usually due to the assignment by top management of the responsibility for each regime to a separate organizational entity, is manifested by such division of the test responsibility as development evaluation tests to the engineering group, field evaluation tests to a field-test or consumer group, quality tests to the quality organization, and reliability tests to the reliability group. If management insists upon delegation of prime responsibility in this manner, its adverse affect upon the overall test program can be greatly minimized if an integrated test program is prepared by a test committee representing all of these groups. We have previously discussed some of the factors this committee should consider. Another is the interrelationship of each of the separate test regimes.

The first such interrelationship to be considered is that within groups, particularly Group I (Sec. 8.2*e*). Since all these tests are aimed at evaluation of the design, it is important that a considerable amount of consistency of testing be maintained for each one. This includes attributes to be tested, input and output conditions, test equipment used, environmental conditions, and data-recording techniques. The tests can and should vary in the degree of discipline and control, the numbers of samples tested, and the degree to which the test hardware resembles the final production design, but generally speaking the tests themselves should be as nearly alike as practicable to permit easy comparative evaluation and analysis of the test results as the hardware design progresses. In a similar fashion, the test regimes comprising Group III should be as alike as possible to permit comparative analyses of successes and failures at successive echelons of testing.

The second such interrelationship is that between groups. Thus Groups I and II should be established as complementary regimes which, taken in toto, provide a complete evaluation of the design. Generally speaking, tests of attributes in the simulated-use tests need not be duplicated in the laboratory when the only purpose is to evaluate the ability of the design to perform adequately in the use environment. However, since there is usually a good deal of uncertainty in the correlation between laboratory testing (and the selected levels of environments) and actual-use environments, it is usually worthwhile to duplicate as many tests and environments as possible in Groups I and II so that the laboratory testing which follows in Groups III and IV can be directly correlated to expected performance of the production hardware in the field or use situation. For instance, the transmission of vibration through the structure of a missile is generally assumed to be unity for the early Group I tests, and the vibration requirements for packages or subsections are set the same as for the overall missile. However, instrumented flight tests will gradually establish that both attenuation and amplification occur, and the vibration requirements for individual packages for the later Group I and for Group III and IV tests will be adjusted accordingly. This correlation can be achieved only if vibration is considered in both Group I and Group II testing.

A second important intergroup relationship is that between Groups III and IV. Group IV testing, if planned and performed without reference to data acquired in other test regimes, is extremely expensive and time consuming, and it may not produce meaningful results until long after any chance to use the information in corrective action on the design, production, or use phases has passed. Hence it is important that this group of tests be planned with constant awareness of the data which will be available from other tests, particularly those in Group III which will produce the greatest accumulation of data. Thus if life testing is considered in terms of mean time to failure, it is obvious that a considerable amount of pertinent information will come from the repetitive testing in the Group III production quality testing program, and it will be necessary to perform only a limited amount of testing to failure to round out the data bank for establishing the mean time to failure.

In a similar fashion, the Group III tests will provide adequate population data on success rate under expected environments, and it will be necessary in Group IV testing to provide only the additional tests beyond the expected environments to be able to assess the sensitivity of the parts to changes in environments. Similarly, the principal modes of failure will be ascertained in Group III tests, particularly those of a random nature whose determination would ordinarily require large lots of Group IV test hardware. This use of Group III test results for the purposes of reliability assessment is not without its cost, however, since such use will frequently decree that variables data must be gathered on many attributes in Group III testing for which attribute data would suffice if the purpose of the Group III testing were purely for determining whether the quality is within specified limits.

Test Procedures. No discussion of test planning would be complete without some words on the importance of the preparation and audited use of test procedures. Too many planned test programs are poorly executed because the performance of the tests is left to the skill and knowledge of the test technicians. It is important to realize that the necessity for detailed, formal, and controlled test procedures is a function not of the capability of the test organization, but of the level of reliability required of the product; the higher the reliability, the more detailed and controlled the procedures should be.

Test procedures should basically describe and control three separate and distinct areas of testing. The first of these is the calibration of the test equipment to be used against standards traceable to the National Bureau of Standards. Such calibration should be performed at the proposed interface between the test equipment and the hardware to be tested, i.e., at the test leads, and should include not only verifying the measuring or comparing part of the test equipment but also that part which establishes the input and the environmental conditions as well. For most optimum calibration, calibration should be performed only at the specific values which are to be measured or provided instead of calibrating the instruments or equipments involved throughout their entire range of values.

The second area to be controlled by procedures is that associated with proofing the test equipment with the hardware to be tested. This operation is essentially a demonstration that the test-equipment design does indeed provide its intended function when coupled with test hardware, and it is established to uncover unanticipated anomalies such as ground loops and variations in input conditions with variations in loading or in line voltages. This kind of proofing is particularly important the first time a given test-equipment design is used with a specific design of hardware, and it should be routine for all Group III testing.

The third area of testing to be controlled by procedures is the test program itself, and all the adjustments, hookups, and switch and button operations should be described in detail, including by diagrams. This portion should also contain detailed data sheets which will ensure that all the desired data, both input and output, are recorded and in the units desired. The data sheets should also include spaces for recording non-test information such as laboratory environmental conditions, date, precise hardware configuration, test operator and inspector identification, and other

ACCEPTANCE TEST PROCEDURE
ORDNANCE DATA (OD) 22634
Page 8

	6.3	Insulation Resistance
	6.3.1	Connect Cable 1607447 (ref para 3.3.2) to INSULATION TEST jack on test console and to unit under test.
	6.3.2	Actuate INSULATION MODE and MEGGER ANALOG switches
OCD 106	6.3.3	Actuate REF C switch and all pin switches except C sequentially. Actuate Megger foot switch after each pin selection. Record results on Test Data Sheet as instructed in paragraph 6.3.6.
OCD 107	6.3.4	Actuate REF D switch and all pin switches except D, E, F, and G sequentially. Actuate megger footswitch after each pin selection. Record results on Test Data Sheet as instructed in paragraph 6.3.6.
OCD 108	6.3.5	Actuate REF H and all pin switches except H and T sequentially. Actuate megger foot switch after each pin selection. Record results on Test Data Sheet as instructed in paragraph 6.3.6.
	6.3.6	Maximum indication on DVM shall be 1.198 volts (DVM will measure current: -1 volt = 10μA). Record results on Test Data Sheet.
	6.4	Natural Frequency and Damping Ratio.
	6.4.1	Pitch Gyro
	6.4.1.1	Mount the unit in the Holding Fixture (para 3.3.1). Mount the Holding Fixture securely on the Simulation Table (para 3.2.8) with face "C" down on the table (ref. Figure 2). Connect the circuit as shown in Figure 5.
	6.4.1.2	Actuate RATE, NORMAL LOAD, PITCH and GYRO ON switches. Allow 60 seconds for the gyros to spin up. Actuate COUNTER EXT. switch and adjust counter to read frequency times 10. Allow the counter 10 minutes to warm up.
	6.4.1.3	Position Servo Analyzer (part of 3.2.8) E2-DC, AC switch to the DC position. Position E1-DC, AC switch to AC position. Adjust gain on the amplifier to 5. Position Frequency Range switch to A position and adjust the Test Signal Amplitude control for approximately 30. Adjust the Frequency Control for 1 cps. Position E1/E2, Damping Ratio switch to E1/E2 position. Balance E1/E2 reading for 0 db by using appropriate db attentuators and assure that the Overload lamps are extinguished. Position E1/E2 Damping Ratio switch to the Damping Ratio position and zero Damping Ratio Meter.
	6.4.1.4	Position Frequency Range switch to C position and adjust the Frequency Control for 90° indicated phase shift as read on the Phase Angle Meter (assure Phase Angle Meter Sel. switch is in the A position).
OCD 109	6.4.1.5	Record the frequency as measured on counter (part of 3.2.1) as Natural Frequency on the Test Data Sheet. The Natural Frequency

FIG. 8.9*a* Sample test-procedure page and data sheet.

administrative data which will permit reconstruction of the test in the event of subsequent question. The data sheets should also include the accept/reject limits where testing is for the purpose of acceptance; these numbers are derived from the parameters-document tolerances from which the test-equipment error (see Sec. 8.4*d*)

Data Sheet Page 3 of 7

Test Data for ATP/OD 22634	Page 19
Report Number	

OCD No.	Procedure Para. No.	Function	Requirement	Actual Reading
107	6.3.4	Insulation Resistance		
		Ref D to Pin A	-1.20 volts max	________
		Ref D to Pin B	-1.20 volts max	________
		Ref D to Pin C	-1.20 volts max	________
		Ref D to Pin H	-1.20 volts max	________
		Ref D to Pin K	-1.20 volts max	________
		Ref D to Pin S	-1.20 volts max	________
		Ref D to Pin T	-1.20 volts max	________
108	6.3.5	Insulation Resistance		
		Ref H to Pin A	-1.20 volts max	________
		Ref H to Pin B	-1.20 volts max	________
		Ref H to Pin C	-1.20 volts max	________
		Ref H to Pin D	-1.20 volts max	________
		Ref H to Pin E	-1.20 volts max	________
		Ref H to Pin F	-1.20 volts max	________
		Ref H to Pin G	-1.20 volts max	________
		Ref H to Pin K	-1.20 volts max	________
		Ref H to Pin S	-1.20 volts max	________
109	6.4.1.5	Natural Frequency, Pitch Gyro	30 through 50 cps	________ cps
	6.4.1.5	Damping Ratio, Pitch Gyro (determine requirement from 6.4.1.5 and note it)	------	________
110	6.4.1.6	Natural Frequency, Yaw Gyro	30 through 50 cps	________ cps
	6.4.1.6	Damping Ratio, Yaw Gyro (determine requirement from 6.4.1.5 and note it)	------	________
111	6.4.1.7	Natural Frequency, Roll Gyro	30 through 50 cps	________
	6.4.1.7	Damping Ratio, Roll Gyro (determine requirement from 6.4.1.5 and note it)	------	________
112	6.5.4	Pitch: CW 75°/sec In-Phase VR	+0.7287 through +0.86713 mv max	________ mv
		CW 75°/sec Quad & Harmonic	680 mv max	________ mv
		CW 35°/sec In-Phase VR	+0.36857 through +0.43143 mv	________ mv

FIG. 8.9*b* Sample test-procedure page and data sheet.

has been subtracted. If the testing is not for the purpose of determining acceptability against a set of predetermined tolerance limits, the data sheet should list the test-equipment error for each data point and provide a place to record observed reading and to compute the actual reading including test-equipment error. (See Figs. 8.9*a* and 8.9*b*.)

Controls should be established over the test procedures to ensure that the original release and subsequent changes are authorized by an appropriate level of management and by interested activities, generally those represented on the test-planning committee. In addition, when the customer desires to control the project closely, his signature is also usually required. The degree of control is variable, depending upon the rigorousness of the discipline desired in the particular test regime. It will be noted that typical variations are outlined in Fig. 8.3.

8.4 TEST EQUIPMENT

8.4a General Considerations. The ultimate success or failure of a test program can well depend upon the care with which the test equipment is selected, designed, procured, and tested. The test equipment utilized will determine the accuracy of the measurements, the repeatability and usability of the results, and the cost of the test program, and these factors in combination frequently determine whether or not the test program is worthwhile. It is, therefore, very important that as much care be taken in choosing the test equipment as in planning the rest of the program. Furthermore, since test equipment is frequently more complex than the hardware being tested, it is equally important that the design of the special equipment be entrusted only to highly competent engineers with long and specific experience in designing such equipment.

The management of a test-equipment program should be approached with the same careful detailed planning as the management of the design and production of the production hardware, and, indeed, experience dictates that considerably more management attention is required on the scheduling aspects of the test-equipment program than on any other because of the complex nature of the equipment. Modern techniques of management, including PERT, are highly recommended as tools in this endeavor, and the program is generally complex enough to warrant establishing a test-equipment coordinating committee which should function from the earliest phase of the project until production is firmly established with proofed and approved test equipment. It is not uncommon for such a committee to be required to supervise design and development of new and unique test equipments, to delineate design requirements to ensure compatibility of test equipments used in different test regimes for the same hardware, and to establish schedules and then expedite and follow up attainment of them.

In its broadest sense, test equipment includes the equipment which provides the input stimuli to the hardware being tested, the measurement equipment which detects the output and either displays it on meters, compares it with known standards and displays the comparison, or prints out either the direct reading or the comparison, and the equipment which provides the environment to which the hardware is exposed during the test. The discussion that follows will include all these functions as part of the test equipment.

8.4b Comparative Features of Test Equipment. For ease of discussion it is convenient to consider several features of test equipment separately. These are purpose, type of control, calibration, and readout.

General Purpose vs. Special Purpose. Test equipment can usually be classified as either general or special purpose depending upon whether the equipment is usable respectively on one or more than one type of test article (although a hard-and-fast delineation is not always possible). In general, it can be said that general-purpose test equipment should be chosen whenever possible unless some feature of the test program dictates the use of special-purpose equipment. Among the factors which may so dictate are the following:

1. No general-purpose equipment is commercially available to make the test.
2. General-purpose equipment error is too large and consumes too much of the product tolerance.
3. General-purpose equipment setup time is too long considering the frequency with which the proposed test will be performed, and the general-purpose equipment utilization factor is too high to permit tying it up in a permanent setup.
4. Test time with general-purpose equipment is excessive, and the frequency of the test performance is high enough to warrant the cost of designing and building special equipment.

Usually, general-purpose equipment provides greater flexibility than special-purpose equipment but at a reduced efficiency of testing. In large projects the sheer numbers of instruments and test equipments that must be utilized make it possible to choose special test equipment whenever the economics of a particular test dictate, since there will, in any case, be a very large pool of standard equipment to provide all of the flexibility of test equipment desired. In small laboratories, however, the individual situation is not so easily settled, since quantities of standard equipment are limited and it is frequently decided to use standard equipment at an increased test cost over what might be realized with special test equipment just to build up the standard-instrument pool for future work.

Automated vs. Manual. Test equipment can also be classified in accordance with the manner in which it is controlled or programmed, i.e., either manually or automatically. As a general rule, manual control is cheaper in first cost and in maintenance, but it is usually more expensive to operate because it requires constant operator attention. Although in some instances an operator is required in attendance anyway (i.e., in development tests where each step of the test regime is determined by the results of the preceding step), in most repetitive testing the additional costs of automation can be recovered if the testing continues for a year or two. Perhaps more significant than the saving in direct operator cost is the saving resulting with automation from reduction in operator error in programming or recording data. Furthermore, automated test equipment normally provides more repeatable results, permitting easier understanding and diagnosis of test failures or anomalies (which represents additional cost savings), more uniform testing, and more consistent quality of product. Hence for repetitive testing, as in Groups III and IV, first consideration should be given to the use of automated test equipment, with manual equipment utilized only as the exception. Particular attention should be paid to automating the cyclic environmental equipment utilized in Group IV testing for accelerated aging and for vibration testing in both Groups III and IV, since substantial savings can be realized from an investment in test-equipment automation.

The form of the output from automated tests should be chosen to be compatible with the EAM equipment available in the plant to permit feeding test results immediately into computers for rapid data analysis. Equipment to provide either punched cards or tape is commercially available and both reliable and relatively inexpensive. Either can be combined with printout equipment to provide an immediate record of results for on-the-spot analysis. This equipment can also print out the accept/reject limits simultaneously with the observed values, and it can be programmed to identify an out-of-tolerance condition.

8.4c Standardization of Test Equipment. We have discussed in preceding sections the need for compatibility of testing not only between test programs but between test echelons. The latter is particularly true when the product is successively tested for the same attributes as it passes from vendor to purchaser or user; when it is built into complete units as in the assembly situation; or when it is tested in several different locations, as when these are multiple vendors, multiple field-assembly sites, or the same tests are used for acceptance testing, field evaluation testing, and repair-depot testing, each at a different location. In these situations it is most important that the testing and test conditions be as nearly identical as possible to preclude the introduction of testing errors and differences into the results of tests at the different locations.

Standardization of test procedures has previously been discussed, along with handling of parameters funnels. However, equally, if not more, significant is the potential contribution of errors and differences from test equipment utilized at the different locations. Ideally, the test equipment should be identical at all locations; when this is practically impossible, it is important that the differences not only be held to an absolute minimum but that they be known and evaluated carefully to determine the exact quantitative differences. There are several potential sources of error or difference. Among these probably the most important are those introduced by differences in test-equipment impedance and those introduced by the use of supposedly alternate and interchangeable instruments. The former can be eliminated by assigning the design of the test equipment to a single group and by instructing the design group to utilize the same testing techniques and circuits throughout the project. The latter are more difficult to control, since procurement policies generally require that competitive procedures be utilized for purchasing standard instruments and equipments, and this in turn leads to pressure to accept apparently or allegedly alternate and interchangeable instruments without complete comparative evaluation or analysis. Test personnel must constantly resist this pressure.

8.4d Test-equipment Error. Largely misunderstood is the subject of test-equipment error, and a general discussion is apparently in order here. All measuring equipment has an inherent error. If this simple fact were kept firmly in mind in the design, specification, and use of test equipment and in the establishment and interpretation of test parameters and observed values, there would be little confusion. Furthermore, the amount of the error is generally a significant portion of the allowable or desired tolerance on the parameter being tested, and considerable degradation of the product from the desired level of functionability and reliability can result from ignoring test-equipment error. Lastly, test-equipment error exists not only in the reading or sensing portion of the test equipment but in the portions which provide the inputs and environmental conditions as well.

One of the most misunderstood facts about test-equipment error is that associated with accuracy of meter movements. For example, a 5 per cent instrument with a 100-volt full-scale reading is accurate only to ± 5 volts *anywhere on the scale*, yet it is commonly misunderstood to mean accurate to ± 5 per cent of the reading! A bit of reasoning should make it obvious that meters or scales on a scale instrument should be so selected that the reading is as near to the maximum reading of the scale as possible to minimize the percentage of error. In this connection, it should be noted that this error normally includes errors introduced by variation in operators' sighting parallax.

Another common mistake is to split the test-equipment error around the limits of the parameter. Fortunately, this is seldom done with mechanical gauges, where the "gaugemakers' tolerance" (another name for test-equipment error) is taken completely from the parameter limits. Properly handled, for example, if the product limits are 90 and 100 volts and the test-equipment error is 2 volts, the product should be rejected if the meter reads more than 98 or less than 92 volts. Similarly, if an oven can be set to an accuracy of only 10° and the desired temperature for the test is 600°, then, to ensure that the product is tested properly, the oven should be set to 610°. Although this can mean that the product "sees" 620°, it is essential in a high-reliability program to test at least as rigorously as the specifications indicate the operating environment will be.

On complex special test equipment the total error should always be measured. The measurement is properly made at the interface between the test leads and the article to be tested (where it will include the error of the cabling and connectors), with the equipment and the article both energized and operating. The error can be computed in some cases, and this is useful during the test-equipment-design phase, but an actual measurement should be made to verify these calculations before the test equipment is released for use. In such calculations, and provided a verification is performed, it is generally permissible to utilize statistical techniques (summing the errors by root-sum-square) to combine the individual errors introduced by the many elements of the test equipment. Figure 8.10 is a sample of a test-equipment error-analysis report indicating one useful way of analyzing test-equipment errors and combining them for an overall estimate.

8.4e Test-equipment Calibration. Automation in comparatively recent years of complex weapons systems, especially missiles, spacecraft, and torpedoes, has tremendously increased the importance of calibration of test equipment. This arose because of the increased number and dispersal of the suppliers, manufacturers, and field activities that perform testing on all or part of these systems, and it became evident that test equipments utilized at these many different locations were not giving consistent and compatible results.

We have already discussed some of the factors which have contributed to testing

TEST EQUIPMENT ERROR ANALYSIS REPORT (TEEAR)

Equipment Title Flight Control Gyro Test Station 52X220	Parameter Document No 22551 Part IV	Prepared By *Mg*	Approved By MSL *TL Hamba*	Page 1 of 4
Equipment Drawing No 2334536	Test Point # 201 Date Prepared 8-26-63	Checked By *RC Hall*	Customer Approval *[signature]*	

Function			Measurement Device					
Title	Nominal Value	Tolerance (T_2)	Description	Accuracy (T_1) Full Scale	Accuracy (T_1) Measurement Point	T_2/T_1 Ratio	Error Formula	Final Readout Limit
Gyro output (pk-pk) Para 1.3	968 mv	Max	Ballantine 316S/2	±3% rdg	±28.8 mv	N/A	$(3\cdot10^{-2})(968) = 28.8$mv	939 mv
In-phase null degrad. Para 1.4	90 mv	Max	Keithley 151R Eqpt panel assy	3% rdg 3% F.S. Total	3 mv 7 mv 10 mv	N/A	$(3\cdot10^{-2})(90) = 2.7$mv + panel error	80 mv
Power consumption input Para 2.1 (115 vac input is adj. to desired value by setting Gertsch to .869565 and read output of Gertsch with DVM)	115 vac 400 cps 100 vac	± 1% +.00v, -2.00v	Gertsch Ratiotran NLS V35B, NLS125E Converter	.0058% 0.5% I.V. 0.05% FSx .3 (% 3rd harm. distort) IV	Negligible 500 mv 50 mv 300 mv		$(5\cdot10^{-2})(100\cdot10^{3}) = 500$ mv $(5\cdot10^{-3})(100\cdot10^{3}) = 50$ mv $(3\cdot10^{-2})(100\cdot10^{3}) = 300$ mv	
			TOTAL RSS	TOTAL ERROR	850 mv 585 mv	N/A	$\left[(500)^2(50)^2(300)^2\right]^{1/2} = 585$ mv	99.5 vac min 101.5 vac max
Running Power	15.0 watt	±2.5 W	Voltron #20.035	2% F.S.	±.40W	6.0	$(2\cdot10^{-2})(20W) = 0.4W$	15 ± 2.1 watt
Power factor	0.960	Max ld/lag	Voltron #20.036	- -	±0.02	N/A	Full Scale = 1; 2% = .02	±.980
Time to rch running pwr	50 sec	Max	Operator reaction	- -	±1 sec	N/A		49 sec
Squelch applied: Running power	10.0 W	±1.8W -2.3W	Voltron #20.035	2% F.S.	±.40W	4.50 5.75	$(2\cdot10^{-2})(20W) = .4W$	11.4 watt 8.1 watt
Power factor	.85 lag	±.15	Voltron #20.036	- -	±0.02	7.5	Full Scale = 1; 2% = .02	.85 ± .13
Insulation resistance Para 3.2 Para 3.2 (Msrd in terms of	500 vdc 40 megohm -1.0 vdc	min Max)	Wiley 5P-2 P.S. Wiley 5P-2 with analog output to NLS V35B DVM	±2%	±1.7 megohm corresponds to .042v	N/A N/A	$(4\cdot10^{2})(2\cdot10^{-2}) + 4\cdot10^{7} = 41.6$ megohm min $\frac{5\cdot10^{2}}{(41.6)10^{6}} = 12.0$ microamps max = -1.2 vdc max $-1.2[1-(5\cdot10^{-4})]$ - 1 digit = 1.198 v max = 41.7 megs	
Output Frequency Measurement (Para 4.3)	36	±5.5 cps	Micro Gee 64A Beckman 7350		No error ±0.5 cps	11/1	0.5	36 ± 5.0 cps
Damping Ratio @ 36 cps @ 50 cps	0.70 0.84	±0.15 ±0.39	Micro Gee 64A Data Log 204A Servo Analyzer	5% rdg	Table I for T/E error vs CPS	N/A		
Sensitivity, In-Phase & Quadrature and Harmonic 5.2.1 75°/second	8V	±650 mv	Genisco C181 Gertsch ACRB	.1% rdg .007% F.S. TOTAL	±8.0 mv ±0.7 mv ±8.7 mv	78:1		8000±641.3 mv

FIG. 8.10 Sample test-equipment error-analysis report sheet.

incompatibility; a major source has also been the lack of calibrations systems rigorously traceable to the standards of measurement held by the National Bureau of Standards. These incompatibilities have sometimes resulted in major catastrophic failures, as, for example, in missile projects, where flights have been lost because the flyable portion of the system was calibrated to a different set of standards than the ground controlling equipment. These differences were found to have resulted from breaks in the chains to NBS. Much work has been accomplished, particularly by Department of Defense activities, in establishing calibration systems and programs which preclude these differences.

Calibration is defined as the comparison of the indication of a measuring device with a known standard, the known standard itself being calibrated against more accurate standards in a series of rigorously controlled echelons up to a national standard held by NBS. It is most important that *all* test and measuring equipment utilized in a high-reliability project, including that used by R&D personnel, be calibrated against laboratory standards at specified intervals. These intervals should be set by reliability or quality assurance test-equipment personnel after thorough analysis of drift rates or susceptibility of instruments to inaccuracies resulting from handling.

A system of *mandatory* recall is usually required to ensure satisfactory operation of the calibration program, since almost all test operators are loath to release instruments for calibration. Mandatory recall, as used in this sense, means that the calibration laboratory is directed by project management to remove physically an instrument, when due, from the test floor to the laboratory for calibration. (To make such an operation practical, it is usually necessary for a loan pool of calibrated instruments to be available for replacement of instruments removed.)

Most large industrial firms and government activities now have their own calibration laboratories, and it is therefore necessary only for project test personnel to ensure that the laboratory be advised early in the project of any new (or more difficult) measurement requirements so that additional standards can be procured and calibration procedures prepared and proofed out in time to support these new measurements.

In smaller companies and activities, however, calibration laboratories are not usually considered economically feasible, and it is necessary for such activities to utilize the services of commercial calibration laboratories. They should do so with extreme care, however, for many commercial companies which represent themselves as capable of performing instrument calibration do not qualify as true calibration laboratories, but are in reality only instrument repair shops. Although some of these shops do have a limited number of standards, thorough investigation has often revealed that the "standards" have not been compared with more accurate standards traceable to NBS (or if they have been, the interval to the last calibration is excessive) and that the range of measurements supported by the standards held is far less than the range of instruments the shop is equipped to repair. When no suitable commercial calibration laboratory is available in a local area, arrangements can usually be made with the nearest Department of Defense industrial facility to perform instrument (or standards) calibration on a time-and-material basis.

Readers are referred to the *Standards Laboratory Information Manual*, published by the Department of the Navy, Bureau of Naval Weapons Representative, Pomona, California, for a complete treatise on calibration and standards laboratories. This manual includes recommended recalibration frequencies for a large list of specific commercial measuring equipment. There are three basic kinds of calibration: (1) calibration of individual instruments such as meters, gauges, and power supplies, (2) calibration of systems of complex test or environmental equipment, and (3) calibration of standards.

Individual instruments are generally transported to a calibration laboratory for calibration at intervals ranging from perhaps one to three months. A sticker is applied to the instrument to indicate the date calibrated and the date next due, and floor-inspection personnel are charged with the responsibility of ensuring that all instruments in use in their area of activity are in calibration. If a loan pool of instruments is available, the loan instruments should be maintained in an in-calibration condition and be cycled into the laboratory for recalibration at the same intervals as instruments in use on the floor. Where individual instruments are part of a complex test setup which is in constant use, it is economically feasible to calibrate only that one of several scales on an instrument which is used in the setup, *provided* the instrument is carefully and prominently marked as being usable only on that particular scale.

With large fixed instruments, or with delicate moving-coil instruments used for highly accurate measurements, it may be necessary for calibration personnel to carry fixed standards from the calibration laboratories to the instrument to perform the calibration, instead of bringing the instrument into the laboratory. When this is necessary, every effort should be made to provide as optimum working conditions as possible for the calibration to ensure maximum accuracy.

Calibration of complex measuring and environmental systems is usually performed in-place, although some consoles may be transportable to the calibration laboratory. It is generally economically feasible to dispense with the calibration of the individual meters and instruments in these complex setups (provided they are calibrated before the first installation). It is most important that the calibration be performed at the test leads or at the point of application of the environment so that the errors introduced by cabling and switching or by the input fixtures of environmental equipment will be detected in the calibration process. For control purposes, doors, panels, and removable instruments and equipments should be sealed and break-of-inspection procedures established to ensure that any changes or tampering will be detected and recalibration performed. The recalibration interval of complex equipment must be established by careful analysis of drift data, but it can usually be set initially to correspond with the shortest recalibration interval of any standard equipment installed in the system. Detailed procedures, including before and after data sheets, are essential to successful calibration of such equipment, and audit responsibilities should be assigned to an unbiased inspection group to ensure that the procedures are meticulously followed.

Calibration of standards consists of two types: (1) cross-checking of like standards held by a laboratory and (2) comparison of the standard with a standard of higher accuracy traceable to a comparison with a standard held by the National Bureau of Standards. Cross-checking is relatively easy and inexpensive, and its use sharply reduces the frequency with which the standards must be referred to a higher-accuracy standard. Certification of standards is common practice, and inspection should be charged with the responsibility of auditing all the standards in the chain from the local-level standard back to NBS to ensure that floor-level measurements are indeed within the specified accuracy.

Because widespread calibration started with metrology (i.e., mechanical measurement of length) and because in that measurement area it was easy to achieve, the minmum ratio of accuracy at each echelon of calibration was established at 10:1. Thus a set of working-level micrometers, with an accuracy of ± 0.0001 in., was calibrated against or compared with a set of gauge blocks with an accuracy of ± 0.00001, and these blocks in turn with another set of blocks of an accuracy of ± 0.000001, etc. As calibration of other measurement areas became more widespread, it became apparent that the accuracy demanded in many of the floor-level instruments was too nearly the same as the accuracy of the comparable national standard and that there was no room for two or three echelons of standards at the 10:1 ratio.

Since the National Bureau of Standards is incapable of providing mass-calibration service (which precluded sending large quantities of floor-level instruments directly to NBS), studies were instituted to determine the effect of reducing the optimum ratio to something less than the traditional 10:1. The result of these analyses indicated that a ratio of 4:1 provides an economical balance between cost of calibration and cost of scrapped material resulting from inaccuracy of measurement. Further studies have shown that the accuracy of a floor-level instrument relative to the corresponding national standard is most influenced by the lowest-level calibration performed and that 4:1 need be adhered to only at that level with but a slight degradation in the floor-level measurement accuracy.

No discussion of test-equipment calibration would be complete without some words on the necessity for the preparation and audited use of detailed calibration procedures for both commercial standard measuring equipment and special test equipment. Calibration is normally performed by highly skilled technicians in the class of personnel most inclined to believe that they possess the skill necessary to make the use of written instructions unnecessary. The errors and mistakes which result from this overconfident attitude are frequently catastrophic, particularly since other project personnel assume that the calibrations have been properly performed.

It is generally mandatory that auditing responsibilities be assigned to another disinterested group to ensure that detailed procedures are prepared and to enforce the use of them by calibration laboratory personnel. These procedures should be prepared in considerable detail and with extensive use of hookup diagrams, and they should include data sheets which require recording both the as-received and as-adjusted readings and setting forth the instrument errors. (The as-received readings are essential to permit studies of drift rates for the purpose of adjusting recalibration periods to suit individual usage conditions and specific drift factors.) Calibration procedures prepared by instrument manufacturers are not always in enough detail to provide this kind of discipline, and it is frequently necessary to rewrite them.

8.4f Test-equipment Ruggedization. Ruggedization of test equipment is not generally considered to be a matter of concern to the reliability department (except those personnel interested in the reliability of the test equipment itself). Instead, it is considered to be a matter of maintenance costs. Because of this misconception, inadequate test equipment is frequently carried over without review from the early R&D testing phase into the production and field-usage phases. Two important things happen as a result, both of them degrading to the product reliability. The most obvious is that inaccurate measurements or environmental exposures occur and introduce errors in the data on product performance or adjustment. The second, less obvious but perhaps more serious, is that the inadequate test equipment is incapable of performing continuously and consistently, with the result that project management loses faith in the ability of the test equipment and applies pressure to reliability and quality assurance to delete tests to permit achieving project schedules.

The effect of deletion of tests in favor of schedules is obvious. In the interest of product reliability, therefore, it is important that considerable attention be paid to ruggedizing the test equipment. In such work care must be taken that the substitution for production use of the more reliable measuring or environmental equipment does not introduce undetected changes in the tests, since such changes will preclude direct comparison of the R&D test data with data accumulated subsequent to the ruggedization. A convenient check on this problem can be made by scheduling compatibility runs, i.e., testing the same hardware on both the old and the new test equipment and carefully comparing the results before the new test equipment is released for production use.

In an overly cost-conscious project there will be considerable resistance from project management to the ruggedization of test equipment, as well as to the preparation of detailed construction drawings for the equipment manufacture, but these resistances must be overcome by reliability personnel. To minimize the ruggedizing cost, it is highly desirable that the reliability group review the initial R&D test-equipment designs to ensure that as much MIL-Spec quality material is specified as possible and that the original drawings are as detailed and complete as possible. Although this approach will increase the cost of the R&D test equipment, it will result in considerable overall net savings to the project by reducing the amount of redesign and rework effort for ruggedization and by improving the overall quality of both the R&D and the production testing efforts.

8.5 TEST FACILITIES

One of the more difficult decisions facing test-management personnel is the make-or-buy for the many kinds of testing which comprise the reliability test program. The straightforward economic factors involved are not particularly difficult to assess by common methods of cost accounting and analysis, and further discussion is not warranted here. However, discussion is in order on some of the many intangibles which are considerably more difficult to analyze and which frequently override or, at the very least, modify the economic considerations.

It can be stated that as a general rule it is good management practice to perform as much testing in-house as possible. First, the practice ensures getting the maximum utilization of capital investment as well as of the incumbent work force. Schedules are more readily monitored and a far greater flexibility is realized. The latter is a particularly important point for the R&D test program, in which test forecasts—being dependent almost entirely upon the success of a particular product design—are fluid at best. Control of test conditions, discipline of test operations, and liaison between the laboratory and other organizational elements are more easily established and maintained with in-house testing, all usually resulting in a higher quality of testing. In-house testing further precludes the necessity for training "foreign" personnel in the company techniques, standards, methods, and systems of technical and financial management, resulting in less lost time, fewer repeat tests, and a more satisfactory operation overall. Last, the company capability and capacity for performing tests are generally improved as the in-house test load broadens and increases, putting the company in a better competitive position for obtaining new projects.

There are balancing factors which must be considered, however, and which frequently dictate the "buy" decision. Time may be an overriding consideration. The in-house laboratory or test bay may not have the technical capability, personnel skills, or available capacity to perform some part or all of a required test, and the time to obtain them may be longer than the project schedules will permit. Or if there is time, the required test may be very special in nature with little likelihood of being repeated, so that it is not economically feasible to establish an in-house capability for it. This is particularly true when the cost of establishing it is comparatively high. Sometimes a rigorous customer may decide that the testing should be performed in an outside laboratory to provide an unbiased check on project performance, or he may have his own laboratories and decide that it is in his best financial interest to perform the testing there.

Particularly in R&D testing there may be very sharp peaks and valleys in the scheduled or actual testing load, and it may be more economical for the company to buy commercial-laboratory time to carry the peaks. In a multidivision company, sister divisions may have very low utilization factors and corporate management may dictate the interdivision transfer of a fixed percentage of testing. There is a distinct merit in maintaining a "mobilization base" of competent outside laboratories who are familiar with the company's way of doing business and are available on short notice to take care of unplanned emergency requirements.

In terms of groups as defined in Sec. 8.2*e*, Group IV testing and the formal tests from Group I are well adapted in most instances to being contracted to commercial laboratories. This is particularly true of the repetitive life and accelerated life tests in both groups. Research and feasibility tests from Group I should, whenever possible, be performed in-house, and it is generally prudent not to schedule the in-house laboratories to capacity in order to leave a substantial cushion of unplanned laboratory time available for emergency tests in these categories. Group III (quality) tests, because of their very close association with the production effort, should also be performed in-house, particularly the ambient tests. Requalification tests can be split, with those tests requiring environments not available in-house being farmed out to suppliers.

It is not surprising that managers of in-house test laboratories will usually decry the acceptability of vendors' testing, but in fact quite acceptable test programs at comparable or competitive prices can be let out on contract if they are performed by carefully selected vendors who have established good working relations over a long period of time. For most high-reliability programs, a time-and-material contract is much preferable to one of fixed price, inasmuch as the latter tends to destroy the flexibility which is required to keep a test program current with product changes or difficulties.

With either type of contract, it is most advantageous for the reliability department to station a representative at each vendor's laboratory not only to provide close liaison and direction but also to monitor and audit the vendor's performance. All of the company's in-house disciplines, including calibration of all test equipment, preparation, release, and change control of detailed procedures and data sheets, operator certification, and independent inspection coverage, should be required of the vendor and enforced. It may be desirable to have vendors' procedures reviewed, approved, released, and controlled in-house.

One often-overlooked source of outside laboratory support is the capability which exists in many universities and government (particularly Department of Defense, and more especially the Departments of the Army and Navy) activities. Although these activities will generally accept only time-and-material contracts, experience has shown that their costs are well estimated as well as very competitive with industrial laboratories. They offer the advantage of completely unbiased testing, and many have specialized capabilities and skills, developed to support pure research or defense purposes, that are uneconomical to establish or maintain in commercial laboratories which depend upon high utilization of a limited number of general-purpose equipments to maintain a profitable position. While it is true that higher-priority government work may at any time interfere with the projected schedules without recompense, nevertheless in practice the directors of these laboratories are well enough informed of their work-load forecast that they can and do maintain the promised deliveries.

One word of caution in regard to the university laboratories is in order. It is advisable to ascertain the qualifications of the specific personnel the university will assign to the particular tests in question, since the universities want to utilize as much undergraduate help as possible to ensure that the laboratories provide maximum student training. This can be acceptable only if an adequate number of fully qualified supervisors closely direct the undergraduate work. A one-for-one ratio is quite good.

8.6 CONDUCTING THE TESTING

As with all elements of industrial management, where so much value of a particular phase of a project can be lost if the planning and make-ready work is not followed by skillful execution, a well-planned test program, although backed up by properly designed and calibrated test equipment and detailed procedures and manned by highly trained technicians, can be relatively worthless if the tests are not conducted in a disciplined manner. Managers must actively manage the testing to preclude slipshod testing. Test personnel must be constantly impressed with the importance of doing a high-quality job of testing, and a meaningful audit system must be established to assess this quality constantly and pinpoint unacceptable practices.

The same quality control techniques that are applied to product design and product manufacture must be applied to the test program; the watchword must be "painstaking attention to details in a system of checks and balances." It is also important that the test programs meet their schedules and commitments to other organizations, or else project top management will devise ways of moving ahead without the reliability test program and despite the obvious resulting jeopardy to product quality and reliability. Last, the whole task must be done in an efficient manner to conserve project dollars for other important aspects of the project.

Thus, in a word, it can be said that the managers of the reliability test program must accept their responsibilities to the project as full-fledged members of the management team. In this they are recognizing that although they must remain independent and unbiased in their technical judgment of the quality of the product they are testing, they must at the same time work in a spirit of cooperation; or their technical judgment will be buried under a wave of resentment from planning, design, and production groups.

The discussion that follows will not attempt to cover the entire field of test-program management, since the normal principles of management hold true for this endeavor as much as for any other, but will instead pinpoint the common areas of weakness as they affect the quality of the testing. The areas covered are planning, work-loads, and progressing; data recording; station proofing; recording variations; interpreting results; inspection coverage; controls; failure diagnosis; and organization. (See also Sec. 16.)

8.6a Planning, Work-load, and Progressing. The overall reliability test program is, in a high-reliability project, a tremendously expensive operation—costing millions of dollars in some projects—to which as much as 35 per cent of the total project budget may be allocated. Such a substantial portion of a project not only deserves but demands an organized approach to the planning and progressing of the effort. Because of the magnitude of the task and its great variety of efforts and tests, all of the modern techniques of management control, including PERT and line-of-balance, should be employed, and top project management must be kept appraised of the status at all times.

It is penny-wise management to skimp on head count and budget allocation for this phase of the test program, since the test program is such an important and integral part of the overall project. Inasmuch as the work is actually performed by a very wide range of organizational elements, and not generally under a single department manager, it is wise to perform the administrative planning in a committee comprised of scheduling representatives of the many organizational elements involved. Firm commitments should be required from each for such functions as preparation of overall plans, preparation of detailed procedures, design of test equipment, procurement of test and environmental equipment, ordering and delivery of test hardware, proofing of test stations, start and completion of testing, preparation of test reports, and analysis of data. Milestone charts should be prepared, and a progressing and expediting group should be assigned specifically to the test program to monitor constantly the performance against the commitments. Management attention should be directed to the soft spots so that decisive, timely action can be taken. Testing which cannot produce results in time to be used in making design or production decisions might better not have been performed, no matter how valid and revealing the results.

Considerable attention should be paid in the planning and make-ready phase of the test program to progressing the preparation and release of supporting information. Included in this information are the product drawings and specifications, for it is in these documents that the basic criteria to be tested are defined. Furthermore, these engineering definitions must be available in time to permit the manufacturing department to produce the hardware to be tested. While a mature project organization will establish planning and progressing groups in the design department, it is essential that the test-program management independently follow the progress of the preparation and release of the engineering paper and constantly revise the test-program planning to compensate for the inevitable slippages in design release.

Work-around techniques are almost always necessary. Notable among these are the preparation of test procedures, design of test equipment, and the manufacture of test hardware all from preliminary unreleased product-design definition. It is recognized that this approach is high-risk costwise, but it is necessary in tightly scheduled projects to ensure that the test program will keep pace with the rest of the project.

The end goal of the planning, progressing, and expediting function must be to ensure that *everything* that is needed to perform a scheduled test is simultaneously available to the test group at the right time. This axiom seems so obvious as not to need to be stated, yet over and over again scheduling progressing groups lose sight of the "for the want of the nail" principle and fail to provide all of the essential material, personnel, equipment, paper, or budget. The need for completeness in this function cannot be stressed enough, for experience has shown that tests are repeatedly delayed for the want of but one of a long list of prerequisites. It is for this reason that there should be a cohesive identifiable group whose only function is to support the test programs. Dividing the responsibilities among specialists reporting to different supervisors (design, hardware, test equipment, facilities, etc.) has proved to be an inadequate approach.

We have noted previously that the reliability group (or the quality assurance group) should be given complete responsibility for the final technical content of the overall reliability test program in order to achieve test results which are coordinated, compatible, and of maximum usefulness. In a similar manner the reliability group should be given the complete responsibility for all aspects of planning, budgeting, and executing the overall reliability test program in order to achieve maximum efficiency of testing and the maximum amount of useful data on a schedule compatible with project schedules. The reason is basically twofold:

1. Responsibility should be vested in a single organization because the test program is composed of a myriad of interlocking requirements, including technical, budget, schedule, data, test equipment, and test hardware, all of which must be coordinated in an optimum balance of trade-offs and compromises. Each of the aspects mentioned must be carefully analyzed across the entire test program, and decisions must be made with due consideration of the results desired of the complete reliability test program taken as a whole. Splitting the responsibility among several organizations will not permit optimum decision making.

2. Responsibility should be vested in reliability (or quality assurance), because testing should be performed by an unbiased, disinterested third party having no direct or proprietary interest in either the design or production effort. Even the most objective or quality-oriented design or production manager will be under pressure by his operating division to "wish away" testing programs or test results which are in effect critical of their efforts. On the other hand, the reliability manager must in turn recognize that when he accepts the overall responsibility for the test program, he also accepts responsibilities to the company and to the project beyond those of pure quality/reliability, and he must be reasonable and understanding of the design and production problems. If he is not, both he and the test program he manages will lose their effectiveness.

8.6b Proofing Stations. Testing today is far more complex and difficult than it once was, primarily because the increased complexity of consumer goods, weapons, and space systems has been accompanied by a greater need for more assurance that the higher quality and reliability demanded of these goods have been met. This need for more assurance is the result, in consumer and industrial goods, of customer demands for better warranties and, in weapons and space systems, of the widespread automation which precludes operator compensation for equipment failures and forces reliability requirements to almost impossible levels.

These trends have made it necessary to institute testing at all levels of manufacture and of a much higher percentage of the functional attributes which are possible to test. The requirement for faultless operation of the equipment under more stringent environmental conditions than ever before has led to more extensive environmental testing during the development, manufacture, and field-use phases. The continued trend toward optimization of design, with more complex functional equipment, greater density of packaging, higher stress applications of parts and materials with corresponding reduction in safety factors, and more sophisticated operating features, has resulted in the requirement for more accurate testing because of the smaller allowable product-attribute tolerances. In response to these more difficult requirements placed on him, the test-equipment designer has been forced to design considerably more complicated and yet more accurate equipment.

To further complicate the test-equipment design problem, compressed project schedules have become the normal way of life, so that it is no longer possible for the test-equipment designer to wait for firm product designs before he starts his equipment design. The need for more, and more stringent, testing has grown in the R&D phases too, and the test-equipment development program has had to move nimbly along in parallel rather than in series with the product development program, requiring the test-equipment designer to try to second-guess the development of the product design so that he can have test equipment available to test the R&D hardware as soon as it is built.

All of this has resulted in immature test equipment being delivered to the various reliability test programs—immature in the sense that there is no time for shakedown of the test equipment with prototype product hardware prior to using the equipment in the actual testing for which it is intended. This situation has made it imperative that a program of test-equipment proofing be established as the first part of any test program. This proofing consists of a careful in-place test of the test equipment—mated with prototype product hardware, if available, or with actual hardware, as a last resort—to ensure that the test equipment actually tests all of the attributes it is supposed to test in the manner and to the accuracy specified.

This proofing is a series of tests which start with a smoke test (power applied to the test equipment but no product hardware attached), progress through activation of discrete sections of the test equipment without and then with prototype hardware, and finally a complete run of the test with full final configuration. During this operation, the test and calibration procedures are also proofed out, with corrections made as the proofing progresses, so that a final proofed-out copy is available when equipment is released for actual test purposes. Although this whole procedure appears to be a cumbersome, unnecessary precaution, it is amazing to anyone not intimately connected with test programs how many bugs and goofs are found in a proofing operation of this kind. Hardly a page of the test or calibration procedure goes through a proofing operation without extensive red-lining of errors, many of which would result in permanent damage to product hardware or in incorrect or inaccurate testing, and there are few sections of the test equipment or cabling that do not require some repair, rework, or redesign.

8.6c Recording Data. As it is folly in a high-reliability program to turn inspectors loose with no more instruction than "inspect to blueprint," so it is equal folly to turn test technicians loose on a test with no more instruction than to "record all data." As we must engineer specific and detailed inspection instructions, so we must also engineer detailed and specific data-recording instructions. When automated test equipment, including automatic printout of test results, is used, at least a portion of this task is done. Not all, however, because much "peripheral" information that is not handled by the printout equipment should be recorded. Included in these data are the administrative information concerning the part (its serial number, exact engineering configuration, date received) and the test (date tested, operator's name, inspector's name, revision letter of the procedure being used) and the technical data not included in the readout (test-equipment serial and configuration, date of last calibration, and laboratory ambient conditions including temperature, humidity, and line voltage). When manual test equipment is used, not only these data but all of the test results as well must be recorded. It is most important that the data be recorded in a consistent manner and in consistent and understood units of measurement and that there be a means of checking at the time of the test that all of the data that are needed are recorded.

It is not enough to expect that a skilled laboratory test technician or inspector will know what to record. For every test there must be a preprinted recording sheet or book made up by the test engineer with clearly defined spaces indicating the data to be recorded and the manner in which they are to be identified. It is very advantageous if the services of a human-factors engineer are utilized in the preparation of these data sheets to reduce the possibility of error by optimizing the probability of understanding on the part of the test technician. (See Sec. 12.) Thus the spaces should be well grouped so that, for example, the pretest data are all gathered and recorded in a single place at the beginning of the test-data section. Recording spaces should be sized to accommodate the amount and kind of data desired, and whenever possible, multiple choices should be preprinted so that the test operator need only check the appropriate statement. Units of measurement should be preprinted in large letters to indicate to the technician the desired figures, and the blank spaces should be quickly and easily identifiable to permit both rapid recording and easy recheck by inspection.

It is poor economy to slough off the job of preparing complete and well-engineered data sheets; for such an approach will inevitably require rerun of tests at an overall increased cost, and time loss, to the project. It is still axiomatic that unless managers guard against mistakes being made, someone will make them. (See also Secs. 6 and 7.)

8.6d Recording Variations. No matter how well a particular test is planned, when the test is actually performed, there will be variations from the detailed requirements of the plans and procedures. These variations occur in every phase of the test.

Hardware will be of a slightly different configuration than planned; the exact test equipment specified will not be available and a substitution will be required; operators will make errors and push the wrong button or connect leads improperly; the laboratory ambient conditions will be outside the limits specified and it will be necessary to perform the test anyway; input environmental equipment will not be controlled quite accurately enough or will suddenly develop trouble and be incapable of meeting the entire spectrum specified; the test program will call for testing a quantity of four specimens, but only three will be available when the test must be started. The list is endless. A wise test manager will recognize and accept from the beginning that these variation conditions will occur and that he cannot legislate against them or hold up the testing until they are all corrected, and he will establish a procedure for both controlling and recording them. The controls will be discussed later in Sec. 8.6*g*; here we shall discuss recording them.

Since it is axiomatic that the variations will occur, it is important that the personnel who perform tests, the personnel who specify and manage test programs, and the personnel who interpret or utilize the results of the tests all recognize the inevitable. In every test report should be a section specifically earmarked for recording these variations. This section should be conspicuously placed at the beginning of the report, where all those who use the test report will be sure to see it. Likewise, a place in the test-data sheets for recording these test variations must be provided. Test operators should be taught that recording variations is not an admission of error so much as it is the means of informing the users of the test result that something other than the specified test conditions occurred in order to permit meaningful analysis and use of the data. The problem is one of attitude which must be instilled in all echelons of the test organization to preclude personnel trying to hide what appear to be mistakes. Personnel should be shown that hiding these inevitable variations is not unlike a timid housewife trying to hide her expenditures by not recording the checks she writes!

8.6e Interpreting Results. Much of the value of a test lies in the manner in which it is interpreted. If, for example, tests are run at periodic intervals on production hardware and no examination is made of the trends of the test data, one of the important side benefits of the testing, i.e., predicting that a process is going out of control, will be lost. Similarly, if test data from a design evaluation test are examined only by the designer who "wishes" that the test data will show that his design meets his design requirements satisfactorily, many of the anomalies in the test data will be overlooked intuitively, and timely corrective action that might have been taken before the design is released for production will not be made. Therefore, it is essential that a test-analysis program be established to ensure that the data are adequately utilized. This program should not only specify how the data will be analyzed and for what but also establish that an independent group will perform the analysis and interpretation. The independence should be in freedom from bias. If the tests are to demonstrate the quality of the design, then the interpretation should be performed by reliability, provided that group is free from direct pressure from the design manager. If the tests are to demonstrate the quality of the production process, then the interpretation should be performed by quality assurance, provided that group is free from restraint by production management.

Of importance equal to independence of the group who perform interpretation of test data is the attitude with which the group works. Test analysis should always be approached with the firm belief that every test will reveal some weaknesses in the test specimen and that the data must be carefully scrutinized to find them. Developing this attitude in analysis personnel is not an easy task, since it naturally causes them to do more work not only in writing up the weaknesses indicated but in later supporting them in the face of opposition from the biased groups that want to wish away the apparently detrimental results. Analysis personnel should be screened from this opposition as much as possible by reliability or quality assurance management.

Until the hardware is removed from service, all test results for a project should be carefully filed in such a manner that the results of individual tests will be available for comparative analysis. The results of early tests should be used as a base against which later tests are compared, and analysis personnel should be instructed to look for trends or shift in the center points in attribute histograms. Of particular value in detecting these trends are the repetitive tests performed on the same serial-number specimens (which will reveal aging effects) or the periodic tests performed on samples drawn from production (which will reveal the shifts in the production process). It is important that the analysis personnel understand that every sudden change in center value should be investigated until a satisfactory explanation is found; the common failing is for such personnel to assume that someone else knows about the shift and that there is "probably" a very rational reason or that the problem has already been corrected. These are extremely poor quality attitudes, and they must be corrected at every opportunity by reliability management.

8.6f Inspection Coverage. Testing is commonly performed by one of the divisions of the quality assurance or reliability departments. Since these departments normally provide the check-and-balance function on production and design, respectively, it is human to consider that the personnel in them are like policemen or drawing checkers, and ordinarily no check and balance is established on the test work that quality assurance or reliability performs. In most of the functions performed by these departments such a check and balance is indeed not required, since the production inspection prior to QA inspection, or the design drawing check prior to reliability drawing review, provides a duplication of effort. However, the test function is somewhat different. It is not a check and balance in itself; rather, it is an original effort of considerable complexity. In this respect it is akin to the production process or the design effort, and it should therefore be checked by an independent group. As a general rule such checking can best be provided by setting up inspection on the test program by a different group in QA or reliability than the test section.

Because of the level of work involved, the inspectors selected to provide the inspection coverage should be the most qualified inspectors available, and they should be especially trained for this work. Their function is primarily to provide assurance that the detailed requirements for the tests are scrupulously met, that the procedures are followed precisely or that any variations are properly recorded and witnessed, that test equipment is properly calibrated and proofed and then sealed and placed under break-of-inspection control, that test specimens are of the proper configuration and properly inspected before delivery to the test area, and that the specimens, too, are sealed and put under break-of-inspection control. The presence of a good inspector in a test laboratory, at a factory test station, or at a preflight test installation will do wonders to ensure that testing is performed in an orderly and well-disciplined manner; and even if he hangs no inspection tags, his mere presence is well worth the added cost. This is also particularly true when tests are performed in a commercial laboratory, where the presence of a company representative has a most salutary effect on the laboratory operation. In the latter case, the inspector can also act as company representative and provide technical liaison to the commercial laboratory.

The inspector assigned to testing work must be trained to insist upon scrupulous conformance to written requirements both technical, in the form of test and calibration procedures, and administrative, in the form of control procedures. He should never be given authority to waive or modify any of these requirements, since such authority tends to degrade his inspection position. The authority to grant such waivers or to make such modifications should be vested in the test director (with appropriate controls as discussed later), to whom the test operator appeals with the inspector's tag. It is only by such a rigid system that the errors in procedures will be brought to management attention and corrected. If the inspector is given authority to overlook them, corrective action will rarely result. The test director given the authority to approve waivers or modifications should be the individual held responsible for the complete conduct of the test.

8.6g Controlling the Test. Closely allied with the discussion of providing inspection coverage for the testing is that of controlling the testing. It is generally wise to establish a single individual who is responsible for all aspects of the test, including meeting schedules, providing accurate and meaningful results, seeing that results are compiled, analyzed, and reported, and making sure that hardware and test equipment are on hand. The director is basically a manager. However, it is not enough just to assign an individual to this post and then assume that the problem of testing is solved. It is also necessary to set up a system of controls which will ensure that the complete test program is performed in a disciplined manner.

In most large companies engaged in high-reliability projects a very complex system of control over drawings is established to assure that the exact engineering configuration of the parts to be built is well controlled. Unfortunately, this control system is not always extended to other engineering definitions, and particularly not to the detailed procedures that are used in calibrating test equipments and in performing tests. The author firmly believes this to be a serious omission in many high-reliability projects and one which results in many errors and anomalies in test data. There must be a system of control over the preparation, review, signature, and release of these procedures identical with that over drawings. There should be an equally tight system of control over procedure changes, waivers, and variations. In a high-reliability project, everything that can be defined should be. A common example of looseness is the notation "or equal" in test procedures, referring to the test equipment to be used. These words grant carte blanche authority to anyone, including the test operator, to decide what test equipment is equal to that specified. A much improved wording is "Substitutions of test equipment will be approved by the test director," or "Substitutions of test equipment will be approved by engineering." Only in this way will there be assurance that such problems as impedance matching will be considered at the engineering level.

From the preceding paragraph it is apparent that the controls established should first ensure that technical documents are properly released and statused and that changes to them are also released and approved by the same levels of personnel as the original release. In addition, since in a complex project many documents require modification when actually put into use, the control system should ensure not only that variations are controlled but also that the approved or allowed variations are recorded in the test reports to permit an analysis of otherwise unexplainable data anomalies.

The control system should provide a check and balance; thus, since the test director can well be under pressure to keep the test program moving to meet the schedules for which he is responsible, he alone should not have the authority to waive requirements, make modifications, or substitute test equipment. A check and balance by another individual not responsible for schedules should be set up to review and approve the test director's approvals. This can usually be an individual in the group that will use the data resulting from the test or, for production acceptance testing, the customer. In a customer-financed and -controlled project, a rigorous customer will want to approve all the original releases of test and calibration procedures, as well as changes or variations to them, so that his engineering force can be used as an effective check and balance.

In a project with tightly compressed schedules it may be necessary to provide the test director with authority to give interim approval to changes and variations on the spot, with the stipulation that these changes are subject to review by higher authority within some specified length of time like 24 hours. This is a high-risk method, however, since it will put reviewing personnel under undue pressure to approve anything the test director has already approved, and it should be permitted only when absolutely necessary. It should never be established as the normal method of operation, but always set up as a temporary expedient.

8.6h Failure Diagnosis and Verification. Best-ball golf is a game that should never be played in high-reliability testing, but unfortunately it is. (For those readers who do not understand golf, best-ball golf consists of permitting a player to hit three balls at each lie, select the best of the three shots, and then shoot three more from that lie.) In testing, the equivalent of best-ball golf consists of pushing the start button to rerun a test which has shown a red-light failure, perhaps several times, until a green-light go test result is obtained and then accepting that test as the valid one! As ridiculous as this sounds, it is all too common, particularly in these days of highly complex test equipment, when it is not uncommon to have a higher percentage of test-equipment failures for a given test station than there are product failures. In the author's experience there is only one way to stop the practice, and that is to insist upon failure verification and failure diagnosis of every indication of failure.

Failure verification and failure diagnosis are two related but not identical functions. Failure verification is the first step required, and it consists of ensuring that the indicated failure is indeed the result of something wrong with the test specimen and not the result of an operator error or test-equipment malfunction. In order to permit failure verification, it is generally necessary to require that at an indication of product failure the operator will make no more runs, but will instead call for a failure-verification team to examine the test setup. To this end it is also important that the test operator be instructed not to move, change, or touch any part of the test setup, since too frequently such actions will destroy the evidence and make verification impossible.

Once the failure has been verified as being a product failure, the second step in this related process is failure diagnosis, which consists of performing necessary additional quantitative tests or examinations to pinpoint the exact cause of the failure in the product. This action is necessary not only to permit the product to be repaired or reworked to meet its functional requirements but also to ensure that corrective action can be taken on the design or process to preclude recurrence.

It is important that the failure-verification and failure-diagnosis results be well documented and provided to data-analysis groups to permit collation and trend analysis. Only from such data analysis will it be possible to pinpoint specific pieces of poor test equipment, untrained or otherwise substandard test operators, poorly prepared test procedures, and other factors contributing to inefficient testing. Such data analysis is also necessary to find the processes which are drifting out of control or the parts supplier who is providing a large percentage of marginal parts.

It is generally not adequate to have the failure verification and diagnosis performed by the test operator or the mechanic who replaces parts or makes adjustments, since the investigation that comes from these individuals will be cut and try and the data will be relatively meaningless. The verification and diagnosis teams should be composed of engineers. (It may be satisfactory to use engineering technicians with long experience working with engineers if they have been specially trained to utilize the analytical approach and to prepare engineering-level reports of such investigations.) Although the use of such high-grade personnel makes the verification and diagnosis effort expensive, it must be remembered that the data from this effort are the beginning of the true quality control function, since they are the foundation on which corrective action is to be taken, and the high cost is well warranted. In advanced organizations utilizing this system, these engineers are in the highest-pay grades.

Where there is a program of repetitive testing of the same item over a rather large (several hundred minimum) quantity of complex parts, it is possible to effect some economy in the failure-verification and -diagnosis effort by establishing "standard diagnoses." In this situation, which can start with material being returned from the field, patterns of part failures can be detected and established. If the diagnosis of these particular repetitive failures can be reduced to a series of finite steps which can be described in a procedure in sufficient detail for a test technician to perform the steps in cookbook fashion, it may not be necessary for the engineering-level failure-diagnosis team to be called to analyze each one of these specific failures. This economy procedure should be well controlled, and it should be clearly indicated that a variation from the expected result will require calling the team to examine the particular specimen before the standard diagnostic test can continue.

As a general rule, this method should not be used on in-plant tests on hardware still being produced, since it is possible in this case to take corrective action on the process, the design, or the parts. Generally, the standard diagnosis system is capable only of providing information for making standard repairs on the discrepant material, not for correcting the process.

8.6i Organization for Testing. The foregoing discussion has hinted several times at certain organizational concepts that appear to be required for a successful high-reliability test program. However, the subject is important enough to warrant a separate discussion which will tie all of the suggestions together in one place.

Perhaps the most important factor in deciding which organization, and its makeup, will perform the many functions related to testing is independence. Since, as stated in the opening paragraph of this section, testing is basically a check and balance for the work of many groups in the industrial complex, it should be axiomatic that the decision as to what hardware and which attributes should be tested, what environments will be used, how the test program will be conducted, and how the results should be interpreted should be made by some group other than the one that did the original work which is being tested. Hence it appears that the test program should not be under either the manager of engineering or the manager of production, since engineering and production are the two "doing" groups whose output will be checked by the test program.

In modern industrial organizations the quality assurance or reliability department has this independence from production and engineering, with all the managers reporting to the same level of management in the company. Therefore, test programs are very often assigned to either the quality assurance or the reliability group. Since the trend today is toward combining the two functions into a single department, the whole reliability test program as described can be assigned to this combined department. However, when the two are not combined, production testing (Group III*a*) is generally assigned to quality assurance, and the balance of the testing to reliability. Such assignment may lead to uncoordinated testing unless special steps are taken to ensure that the two organizations work in committee fashion on all of the test planning.

If engineering and reliability have separate laboratories, it may be feasible to assign the R&D tests (Group I, research and feasibility, and Group II, flight or field evaluation) to engineering; again, when this is done, special care must be taken to see that reliability and engineering work together in committee fashion to ensure that the overall test programs are compatible and yield complete data banks capable of overall analysis. In small companies where there is no separate reliability organization (fortunately, a rarity for a high-reliability project!) the test programs that would be assigned to reliability in this breakdown can be assigned to a separate test group in engineering. However, the engineering manager must be extremely careful to preserve the independence of the test group from pressure from his design managers.

A second important axiom in establishing an organization for a high-reliability test program is that test technicians are not capable of preparing the detailed test and calibration procedures for the tests. This function is a true engineering one, and it should be assigned to a group separate from the test operators. Inspection and testing (where inspection is visual examination and mechanical measurement and testing is measurement of functional attributes) are sometimes separated, with inspection planning and procedures preparation by an engineering group in quality engineering and the similar functions for testing by a group in reliability engineering.

This split is also carried out in the planning committees, where two separate committees consider inspection and testing, respectively. This split permits grouping of different technical specialties under a single supervisor and does provide certain advantages, but it usually requires either that another group be set up to integrate the two planning efforts or that production planners integrate the two requirements in the routing-sheet preparation. The important point is that the planning and procedure writing should not be performed by line inspection or by test-laboratory-operating groups.

The basic planning of the test and inspection program should be a committee function, with the committee or committees chaired by reliability/quality assurance. Section 8.3*b* discussed in some detail the makeup of the planning committee(s). The program plans prepared by these committees are implemented by the detailed procedures prepared by the procedure-writing groups mentioned in a preceding paragraph. Other committee functions having a direct bearing on the reliability test program are the engineering release committee and the make-or-buy committee. The release group is the controlling body for the initial release of and subsequent changes to detailed requirements documents (plans, procedures, specifications, etc.) and fundamentally is a group whose function is to determine the schedule and effectivity of these documents.

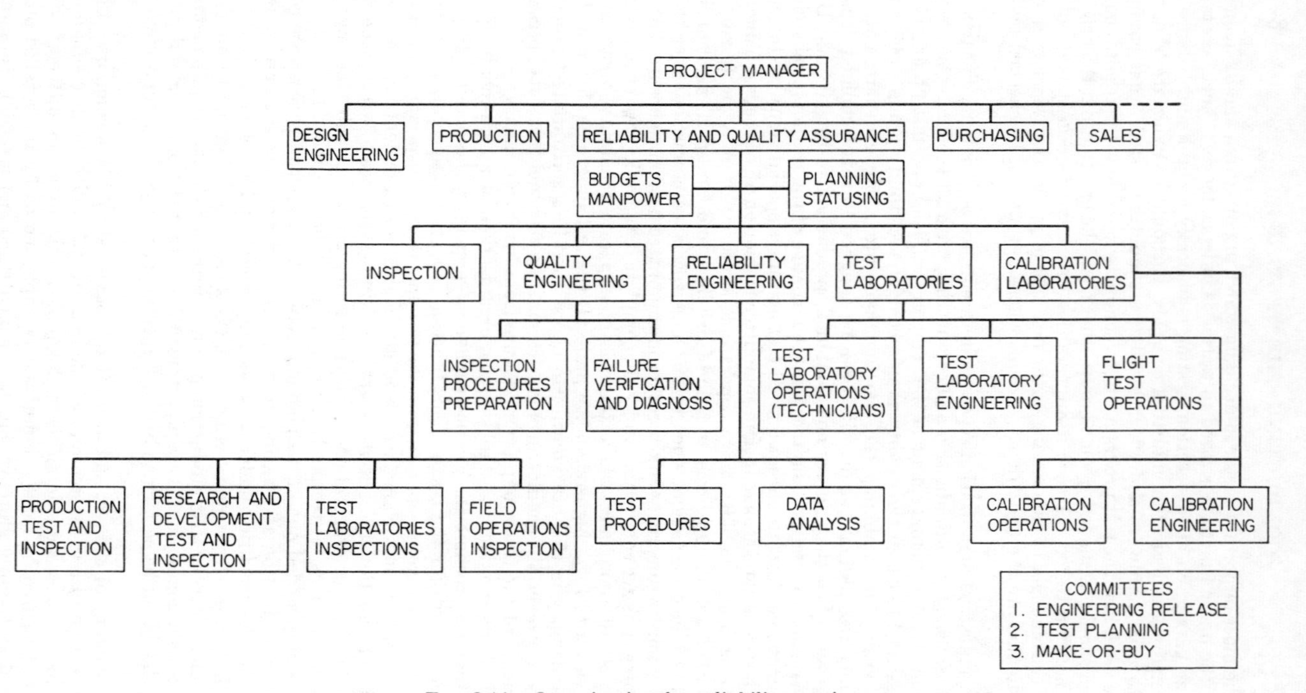

Fig. 8.11 Organization for reliability testing.

Each member of the group commits his parent organization to the schedules on which the group agrees. Thus the production member will assure that his organization can produce the necessary test specimens on the schedule agreed to; the inspection representative that his organization can provide the necessary inspection for not only the test-specimen manufacture but the test program as well; purchasing will commit to procurement of all purchased hardware as well as to procurement of any needed commercial test laboratory time; and so on with the other members, generally consisting of representatives of the interested divisions (i.e. the first echelon of organizations below the department level). The last committee having direct interest in the reliability test program is the make-or-buy committee, which will determine whether testing programs will be performed in-house or at commercial laboratories, as discussed in Sec. 8.5.

The complexity of the overall reliability test program and its sheer size in terms of man-hours and dollars make almost mandatory the establishment of special planning and statusing and of budget and manpower groups to ensure that the whole program will run smoothly. These are generally staff functions reporting to the reliability/quality assurance manager to keep him continually advised on the progress of the program and to provide schedule and budget coordination to the operating groups in the department.

All these concepts are represented in the suggested organization chart shown in Fig. 8.11, which is an actual organization chart of a large company which successfully manages overall reliability test programs of the type and complexity discussed in this section.

Section 9

MALFUNCTION AND FAILURE ANALYSIS

E. W. KIMBALL

RELIABILITY GROUP ENGINEER, MARTIN COMPANY, AEROSPACE DIVISION OF
MARTIN MARIETTA CORPORATION
ORLANDO, FLORIDA

Mr. Kimball first became associated with product improvement on military hardware when he joined the Signal Corps Engineering Laboratories in November, 1950. In those well-equipped facilities he became acquainted with both climatic and dynamic environmental testing on electronic equipment. After brief tours of duty with the New York Ordnance District and the Western Electric Company, he was transferred to the Rocketdyne Division of North American Aviation, where he was responsible for the reliability testing of Redstone and Atlas rocket-engine components. Later, at Aerojet General, he performed malfunction-analysis studies on the Titan propulsion system. At the Martin Company he headed the Pershing failure analysis and is now Reliability Group Engineer on the Sprint Project. Mr. Kimball has published numerous papers on failure analysis. He received the B.S. Degree from Columbia University.

Mr. Kimball has dedicated this section to Mr. R. C. Maloney, a pioneer in reliability research, who, with his wife and five children, was killed in an airplane crash caused by the same kind of malfunction he strove so hard to prevent.

MALFUNCTION AND FAILURE ANALYSIS

E. W. Kimball

CONTENTS

9.1 HISTORY AND PURPOSE

The importance of organized failure analysis was well recognized by the reliability founders, Robert Lusser and Dr. Leslie Ball. Many of the early symposium papers (1954–1956) extolled the benefits of conducting autopsies in a failed-parts morgue. However, in actual practice, it has taken many years to overcome management inertia and obtain adequate laboratories used specifically for determining the exact cause of failures. The problem is still partially unsolved; for R. S. Nelson, in his "Authority and Responsibility of the Central Reliability Group," July, 1962, reported that only 45 per cent of the aerospace contractors surveyed allowed their reliability groups to perform failure analysis. Some firms utilize quality and engineering laboratories for troubleshooting critical and repetitive failures, while others have no established procedures for analysis of failed hardware.

The reason for lack of interest in finding the cause of failures dates back at least as far as the Civil War. General U. S. Grant, in describing poor-quality hardware and the military contractor, resorted to poetry:

> It's not after glory that they pant,
> It's just the dollars and cents they want.

It is still all too common for a weapon-system or space-vehicle manager to state, "We have never had a failure—just a few malfunctions." In his desire to maintain his company's competitive position, he ignores the fact that any defect which can prevent successful accomplishment of a mission is a failure. Clearly, a need exists for reliability engineers to educate management on the advantages of product improvement. Prime consideration should be given to pointing out how company prestige can be raised by pinpointing the cause of failures and taking corrective action.

All contractors should have an unbiased reliability laboratory used specifically for analysis of all failed hardware. Design and quality engineers cannot be objective in their analyses, since their group may be responsible for the trouble. As a result, they tend to classify the great majority of failures as random and not likely to recur. Obviously, all failures have a cause, and the use of the word "random" is usually an excuse for not performing exhaustive troubleshooting. If there are no firm procedures for routing failed hardware to a reliability laboratory, many components are never disassembled, parts are not dissected, and causes for failure, which may have been easy to correct, are lost on the scrap heap.

Government agencies recognize the need for failure analysis, and appropriate instructions have been inserted in military reliability specifications. A typical paragraph reads as follows:

> All failures resulting from environmental, factory, field, and flight test should be analyzed by people in the reliability organization to determine the significance of the failures in a reliability sense. The analysis should be on both a statistical and engineering basis. An attempt should be made to determine causes of the failures and, where feasible, whether or not the deficiency was due to the design, quality control, or human factors.

The NASA Reliability Program Provisions (NPC 250-1), Section 3.7, state:

> The contractor and subcontractors shall employ a strictly controlled system for the reporting, analysis, correction and data feedback of all failures and malfunctions that occur throughout the fabrication, handling, test, checkout and operation of the space system. This system shall emphasize reporting and analysis of *all* failures and malfunctions that occur, regardless of their apparent magnitude, so that timely and appropriate evaluation, corrective and preventive action and follow-through can be accomplished.

The great bulk of reliability-improvement data comes from the laboratory analysis of environmental, in-house, and field failures. It is very difficult to ferret out the cause of flight-test failures, because the hardware is usually unavailable. It may be destroyed in the explosion triggered by the range safety officer, or it may be impossible to find on the ocean floor. In shallow water, missile assemblies bury into sand, ooze, or coral rock and are not easily removed. Sonar devices for locating hardware in deep water leave much to be desired. As a result, even with the best equipment, missile-residue recovery is rarely undertaken in depths below 300 ft. In troubleshooting hardware recovered from the ocean, problems arise because of salt-water corrosion. When flight tests are conducted over land, impact damage can also be a severe handicap. For these reasons the flight-test-analysis engineer must concentrate his investigation on telemetry data. There are never enough measurements taken, since limited space is available for instrumentation. Because of the difficulties mentioned, many in-flight failures are narrowed down to two or three possible causes, but the exact cause is unknown.

Since very little usable hardware is recovered from flight tests, failure analysis is performed primarily on malfunctions which occur during preflight testing. Prior to passing acceptance tests, in-house defects are predominantly of an obvious workmanship nature which do not require laboratory investigation. Environmental and field failures are more important, because they are the same kind which will occur in tactical operations unless corrective action is taken.

9.2 FAILURE RESPONSIBILITY

An objective decision is necessary in order to assign responsibility for failures properly. It is only natural for designers to attribute failures to poor workmanship, while the production people tend to blame design. Accordingly, these determinations should be made by reliability engineers or by a committee comprised of members from all divisions. A typical trouble-report committee is composed of ten members, two each from engineering, reliability, manufacturing, quality, and logistics. When a trouble report is returned from the field, the committee makes a preliminary decision of divisional responsibility. The report is then assigned to the appropriate division member for investigation. Meanwhile, the hardware has been sent to the laboratory for analysis. When the lab report is received, the investigator combines it with all the other information he has obtained and presents his case at the committee meeting.

If the committee is satisfied that the cause of failure has been determined and proper corrective action has been taken, the trouble report is closed out.

During an average month in the development of a weapon system, as many as 50,000 defects may be discovered during in-house inspection and testing. This compares with an average of 150 field failures per month. In-house defects include receiving inspection items charged to the vendors, cold-solder joints, improper dimensions, poor plating, and similar workmanship problems. The important thing to remember is that in-house malfunctions are caught before they get to the user. Even if the exact cause is never determined, these modes cannot endanger mission success. The field failures, on the other hand, are quite a different kind. These are the ones on which the reliability engineer should concentrate if his efforts are really to pay dividends. The accompanying table contains cumulative responsibility percentages obtained during a 3-year period of weapon development.

Responsibility	Per cent	
	In-house	Field
Manufacturing	57	15
Design	4	52
Improper operation	29	17
Handling	6	9
Unknown	4	7
Totals	100	100

9.2a Manufacturing. The best way to detect manufacturing variance may be 100 per cent inspection. To supplement this control, functional tests should be performed at the completion of each level of assembly. Automatic equipment for testing electronic parts is now available, and it is no longer necessary to use sampling plans to speed up the job.

Sampling plans can degrade reliability if they allow marginal parts to get into the system. Such a system may pass the final acceptance test but fail later on in the more stringent field environment. In addition, if a sample tells you to reject the lot and the hardware must be returned to the vendor, it then becomes difficult to maintain schedules unless a tremendous inventory is kept in stock. Tooling wear-out and faulty process control are major contributors to manufacturing variance. Difficulty in inspecting vendor's sealed units is the main reason why some poor-workmanship items get to the field. The most common types of poor workmanship are inability to hold tolerances, improper potting, faulty soldering, wiring errors, and contamination.

9.2b Design

"The achievement of reliability is of first importance, measuring and predicting are secondary. Adequate safety margins should be demonstrated in the laboratory prior to beginning an extensive flight test program."

—*George A. Henderson*

Before releasing a drawing, engineers should perform preliminary tests to determine whether their design will function properly under environmental stress. (See Sec. 8.2.) If they do this, the need for design reviews and troubleshooting is greatly reduced. By subjecting electronic breadboards and mechanical prototypes to environmental extremes, design weaknesses that a theoretical worst-case approach will overlook are uncovered. The engineering model shop can also build proposed electronic assemblies so that hot spots and ability to withstand vibration can be studied. The value of these methods is emphasized by the high failure percentage from vibration testing as shown in the accompanying table. These data were compiled from the

Environment	*Failures, per cent*
Shock and vibration	28.7
Low temperature	24.1
High temperature	21.3
Humidity	13.9
Altitude	4.2
Acceleration	3.2
Salt spray	1.9
Others	2.7
Total	100

results of over 3,000 development and preflight certification tests on major assemblies.

Since military characteristics stipulate that equipment should function through its intended life without wear-out, this type of field failure (<1 per cent) was included within design responsibility. Typical causes of design failures are electrical or mechanical tolerance buildup, transients, inadequate safety factors, and call-out of unreliable parts.

9.2c Improper Operation. Malfunctions listed under the improper-operation category include human errors by test engineers and technicians as well as design-change notices which were not incorporated. Mistakes in functional test procedures and inaccuracies in test equipment were also considered as improper operation. Even though test fixtures receive periodic calibration, they may drift out or become incompatible with other fixtures. As a result, some out-of-specification units will pass the test fixture and other assemblies will be returned as failed items when nothing is wrong. Common causes for improper-operation malfunctions are test-fixture drift, improper potentiometer settings, and parallax or interpolation errors in reading gauges. (See Sec. 8.4*e* and *f*.)

9.2d Handling. Handling malfunctions are usually attributed to abuse by assembly and operating personnel. During in-house testing and transportation, parts may be dropped or otherwise mistreated. When assemblies bind and cannot be moved in the field, technicians are often told to "get a bigger hammer." Of course, there are many examples of equipment which is not designed to withstand the normal handling experienced in military service. In that case reliability engineers must press for corrective action. Still a third cause of handling problems results from poorly designed shipping and storage containers.

9.2e Unknown. Failures are classified as unknown when two or more possible causes exist and the analysis lab cannot pinpoint the exact one. Flight-tested hardware is often recovered in such damaged condition that effective troubleshooting cannot be conducted.

During the classification process, it is important to eliminate from consideration all secondary malfunctions which occur as a by-product of a primary failure. This is necessary to prevent the data from being unfairly biased by problems which have no effect on reliability.

9.3 RECORDING FAILURE DATA

When establishing a reliability data center, management should concentrate on the storage of complete and accurate failure information. Reliability's raison d'etre is to improve the chances of success of every missile or space vehicle launched. This goal can best be achieved if all modes of failure are known and understood. However, the necessity for corrective action can be proved only when historical failure data are available as evidence. (See Secs. 6 and 7.)

Failure forms with space to record all necessary data must be provided to operating personnel. Technicians should be instructed to admit mistakes and improper operation in order that the data will present a true picture of unreliability. A major problem exists when failure reports are written by field-service and logistics personnel. Because data gathering is not their primary job, important information is often omitted. Logistics personnel are usually too busy ensuring the smooth flow of spare components and maintaining equipment to worry about obtaining accurate data. As a result, it has been found very desirable to assign reliability representatives to field locations. Their specific job is to secure complete information required by reliability. They maintain log books and can also determine if failures are caused by operator error or are secondary in nature. Of course, detailed troubleshooting is not performed in the field, since complete laboratory facilities are not available. In addition, flight-test schedules usually do not allow time for exhaustive failure analysis. The assembly causing the trouble must be quickly located and replaced so that the launch date can be met.

Almost every contractor and government agency has its own forms for recording failure data. Standardization attempts have been made, but it is very difficult for a single form to serve the needs of all users. The most common procedure is for contractors to use separate forms for in-house and the field.

9.3a In-house Forms. In-house forms should be relatively simple, since the majority of failures are discovered at the part level. These forms are used to record defects discovered during receiving inspection and manufacturing debugging or acceptance tests. A typical in-house form used by the Martin Company is shown in Fig. 9.1. Space is provided for important reliability information such as part name and number, circuit symbol, run time, and description of the defect or failure. The form has multiple copies which can be used for data-center storage, quality control, and reliability. The cardboard back copy is attached to the hardware when it is sent to the failure-analysis laboratory.

Another valuable aid to reliability failure analysis is a shop traveler or history card which follows all assemblies through manufacturing subsystem testing. In this way the number of tests performed can be obtained and cumulative success/test ratios can be computed. Serial numbers of subassemblies, test equipment used, and operator's numbers are also recorded on these forms. This type of information proves very valuable during a failure-analysis investigation.

9.3b Field Forms. During the R&D phase, it is desirable to use a field form which facilitates the reporting of the complete details of failures. Such data as the test measurements which indicated malfunction, printed-circuit board location within the chassis, and environmental conditions are invaluable to the failure-analysis engineer who must troubleshoot the equipment when it is returned to the contractor's plant. Martin Company's service trouble report (Fig. 9.2) is an example of a form which is very useful for product improvement. Fully one-third of the space can be used by the field engineer to record comments on the nature of the failure. In addition, blocks are provided to report the generation breakdown, circuit symbols, and run time on the hardware involved. Logistics procedures require that the failed hardware and service trouble report be turned in before spare replacements are issued. In this way, management is assured that all failures will be documented. The form is a reproduction master, so copies can readily be distributed to all interested groups.

Another form which has been used to advantage in the field is the quality reliability consumption report, Fig. 9.3. This form is modeled after the BMD-050 format of the Air Force ballistic missile programs. Its major advantage is that the data center can punch IBM cards directly from the source document. This method cuts down on the possibility of errors which occur when data are transferred from the source document to a transmittal form. The entire top row and blocks 13 and 19 of the consumption report are used for IBM codes.

A third type of field form, known as the failure and consumption report (Fig. 9.4), has been designed by the Tapco Division Reliability Office of Thompson Ramo Wooldridge. These forms are provided in a pad which can be kept in the back pocket of the field engineer. Instructions for filling out the form are printed on the covers of the pad. Engineers can thus document a failure immediately upon occurrence. This eliminates the practice of some field-service personnel who write trouble reports in their motel rooms after they have forgotten much important information.

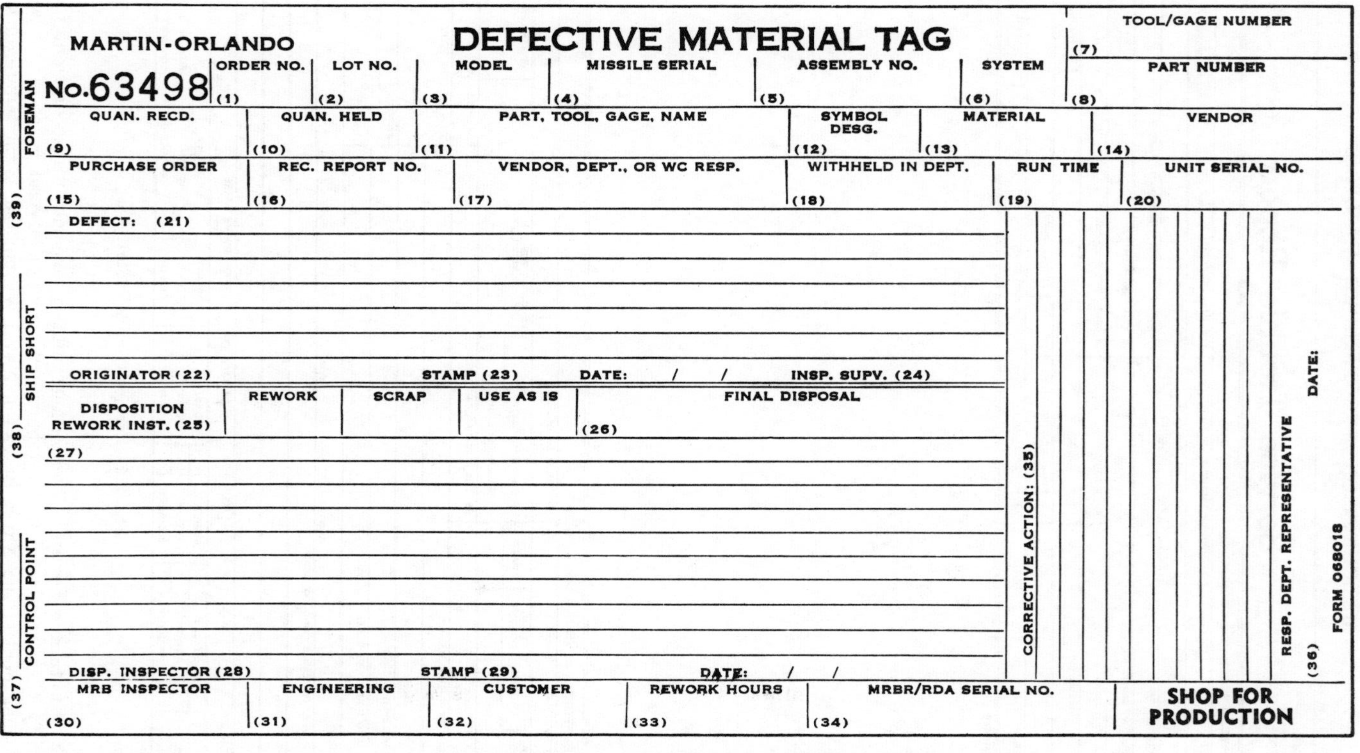

MARTIN-ORLANDO

DEFECTIVE MATERIAL TAG

No. 63498

ORDER NO. (1) | LOT NO. (2) | MODEL (3) | MISSILE SERIAL (4) | ASSEMBLY NO. (5) | SYSTEM (6) | TOOL/GAGE NUMBER (7) | PART NUMBER (8)

QUAN. RECD. (9) | QUAN. HELD (10) | PART, TOOL, GAGE, NAME (11) | SYMBOL DESG. (12) | MATERIAL (13) | VENDOR (14)

PURCHASE ORDER (15) | REC. REPORT NO. (16) | VENDOR, DEPT., OR WC RESP. (17) | WITHHELD IN DEPT. (18) | RUN TIME (19) | UNIT SERIAL NO. (20)

DEFECT: (21)

ORIGINATOR (22) STAMP (23) DATE: / / INSP. SUPV. (24)

DISPOSITION REWORK INST. (25) | REWORK | SCRAP | USE AS IS | (26) FINAL DISPOSAL

(27)

DISP. INSPECTOR (28) STAMP (29) DATE: / /

MRB INSPECTOR (30) | ENGINEERING (31) | CUSTOMER (32) | REWORK HOURS (33) | MRBR/RDA SERIAL NO. (34)

CORRECTIVE ACTION: (35)

RESP. DEPT. REPRESENTATIVE (36) DATE:

FORM 068018

SHOP FOR PRODUCTION

FOREMAN (39)

SHIP SHORT (38)

CONTROL POINT (37)

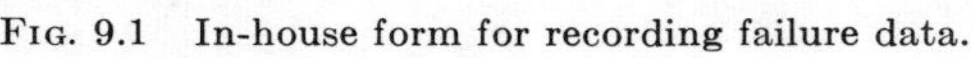

FIG. 9.1 In-house form for recording failure data.

FORM MAY61 2238
MARTIN ORLANDO

Service Trouble Report

ORIGINATORS BLOCK FOR COMPLETION

1 MODEL	2 DATE OF TROUBLE	3 SERIAL NO.	4 SAFETY PERFORMANCE
5 TOTAL HRS. ON PRODUCT	6 FAILED PARTS RETURNED YES ☐ NO ☐	7 SYSTEM NAME	
8 COMPONENT NAME		9 HOURS	10 COMPONENT PART NO.
11 COMPONENT SERIAL NO.		12 COMPONENT MFR.	
13 SUB ASSY. NAME	14 HOURS	15 SUB ASSY. PART NO.	
16 SUB ASSY. SERIAL NO.	17 PART NAME		18 HOURS
19 PART NUMBER	20 SYMBOL DESIGNATION	21 PART MFR.	
22 IMMED. ACTION	23 REPORTED BY	24 APPROVED BY	25 BASE / 27 WAYBILL NO.

26 COMMENTS

FIELD SUPPORT AND ENGRG. GROUP BLOCK

28 TYPE TROUBLE	29 CAUSE OF TROUBLE		
30 INDEX CODE NO. — SYSTEM / COMPONENT / PART	31 AIRBORNE EQUIP. / GND. SUPP. EQUIP.	32 DEPARTMENT RESPONSIBILITY	33 ITEM NO.
37 RECD. BY CCAS	38 DEL'D TO FA COMMITTEE		34 FILE CODE
39 RESULT OF FAILURE ANALY.			35 SUSPENSE DATE
40 CORREC. ACT. RECOMM.			36 STR.
41 DEL'D TO & REC'D BY CCAS			№ 24031

42 DEL'D TO SC ENG. LAB. MFG. ACCEP. BY QA

43 REC'D IN SERVICEABLE COND. BY CCAS

44 FINAL ACTION

SUBMITTED BY:	DATE	MAIL NO.	TELEPHONE	FOLLOW-UP DATES 1ST 2ND 3RD

ENG. ACTION COMMENT	QUAL. A ACTION COMMENT	MANUF. ACTION COMMENT	PUBS. ACTION COMMENT	SPARES ACTION COMMENT	SHIPPING ACTION COMMENT	ACTION COMMENT

Fig. 9.2 Service trouble report.

9.3c Customer Forms. The various military agencies utilize different forms to satisfy their requirements for reporting failures on several types of hardware. The Department of the Navy, Bureau of Naval Weapons, uses the failure, unsatisfactory, or removal report (Fig. 9.5) to document aeronautical-material deficiencies. Detailed instructions for preparation are attached to the report form to aid the originator in providing accurate data. Standardized codes, to describe the trouble,

QUALITY RELIABILITY CONSUMPTION REPORT

MARTIN
DEN 066124 (11-61)

01 L 34311
02 SYS. COMP. LINE
03 ASSY.
04 PART
05 FORM
06 FAILURE CAUSE
07 FAILURE CODE
08 DISPOSITION
09 TEST TYPE
10 MANUFACTURER CODE
11 FAILURE ANALYSIS REQ'D. 1 NO ☐ 2 YES ☐

12 ITEM PART NUMBER
13
14 ITEM SERIAL NUMBER
15 ITEM NAME
16 ITEM MANUFACTURER
17 INITIAL REPORT NUMBER

NOT TO BE COMPLETED BY OUTSIDE VENDOR OR RECEIVING INSPECTION

18 NEXT ASSEMBLY PART NUMBER
19
20 NEXT ASSEMBLY SERIAL NO.
21 MISSILE TYPE / MODEL / SERIES
22 MISSILE / GSE / GOE SERIAL NO
23 LOCATION CODE
24 FAILURE DATE DAY MONTH YEAR

TO BE COMPLETED BY OUTSIDE VENDOR AND/OR RECEIVING INSPECTION ONLY

25 PURCHASE ORD NUMBER
26 RECEIVING REPORT NUMBER
27 STOCK NUMBER
28 MATERIAL CODE
29 CONTRACT NUMBER

30-33 DESCRIPTION OF TROUBLE (INCLUDING TYPE OF TEST, MODE AND CONDITION OF FAILURE, ENVIRONMENTS AND MEASUREMENTS.)

34 CAUSE OF DISCREPANCY
35 CORRECTIVE ACTION
36 1 GFAE ☐ 2 ASSOCIATE CONTRACTOR ☐
37 QUAN. REJECTED
38 DEPT CHARGE
39 INSPECTOR OR ORIGINATOR
40 STAMP
41 REPLACEMENT PART REQ'D.
42 SHOP SUPERVISOR OR DEPT REP.
STAMP

43 DISPOSITION INSTRUCTIONS
44 MATERIAL REVIEW INSTR. REQUIRED ☐
45 RTV ☐
46 SCRAP ☐
47 COMPL. TO DRAWING ☐
48 MR CRIB ☐
49 QUALITY CONTROL SUPERVISOR
50 USE AS IS ☐
51 SCRAP ☐
52 REWORK ☐
53 QUALITY CONTROL MRB
54 ENGINEERING MRB
55 CUSTOMER MRB
56 ASSOCIATE CONTRACTOR OR CUSTOMER

FIG. 9.3 Quality reliability consumption report.

TAPCO GROUP RELIABILITY OFFICE—FAILURE AND CONSUMPTION REPORT

TPFC-2260 REV. 2 4/60 PRINTED IN U.S.A.

SYSTEM	1. TYPE, MODEL, SERIES	2. S/N	3. REPORTING FACILITY	4. TEST PHASE	5. REPORT NO. NO. 6479
UNIT	6. NAME (NOUN)	7. NUMBER	8. S/N ORIG. REPLACEMENT	9. MODEL DESIGNATION	10. INITIAL REPORT NO.
ASSEMBLY	11. NAME (NOUN)	12. NUMBER/MODEL	13. S/N ORIG. REPLACEMENT	14. MANUFACTURER	15. REF. DESIGNATION NO.
PART	16. NAME (NOUN)	17. NUMBER/MODEL	18. S/N ORIG. REPLACEMENT	19. MANUFACTURER	20. REF. DESIGNATION NO.

21. DATE AND TIME OF FAILURE
MONTH DAY YEAR TIME: (0-2400 HRS.)

22. HIGHEST LEVEL DISABLED .0 ☐ .1 ☐ .2 ☐ .3 ☐ .4 ☐ .5 ☐ .6 ☐ .7 ☐
NONE PART SUB-ASSY. ASSY. UNIT SUB-SYST. SYSTEM WEAPON
23. HIGHEST LEVEL REMOVED .0 ☐ .1 ☐ .2 ☐ .3 ☐ .4 ☐ .5 ☐ .6 ☐ .7 ☐

OPERATING PERIOD
24. UNIT ______ .1 ☐ SEC. .2 ☐ MIN. .3 ☐ HRS. .4 ☐ MOS. .5 ☐ CYCLES ______
25. ASSY. ______ .1 ☐ .2 ☐ .3 ☐ .4 ☐ .5 ☐ CYCLES ______

AT TIME OF FAILURE

26. LOCATION
.1 ☐ BENCH/CELL
.2 ☐ FINAL INSPECTION
.3 ☐ NORMAL USAGE ★
.4 ☐ IN TRANSIT.
.5 ☐ STORAGE

27. ACTION PERFORMED
.1 ☐ TESTING
.2 ☐ P.M./INSPECTION
.3 ☐ OVERHAUL
.4 ☐ NOTHING

28. OPERATING STATUS
.1 ☐ NORMAL
.2 ☐ ABNORMAL ★
.3 ☐ OFF

29. REASON FOR REPORT
.1 ☐ FAILED ITEM
.2 ☐ PREV. MAINTENANCE
.3 ☐ EQUIP. MODIFICATION
.4 ☐ FACILITY FAILURE
.5 ☐ DESIGN IMPROVEMENT
.6 ☐ PROCED. IMPROVEMENT

30. REPLACEMENT SOURCE
.1 ☐ STOCK ROOM
.2 ☐ BENCH STOCK
.3 ☐ OTHER FACILITY ★
.4 ☐ CANNABILIZED
.5 ☐ NONE AVAILABLE
.6 ☐ NONE REQUIRED

31. ACTION ON ASSEMBLY
.1 ☐ REPAIRED IN PLACE
.2 ☐ REPAIRED-REINSTAL'D.
.3 ☐ ADJUSTED
.4 ☐ REPLACED
.5 ☐ NONE OF ABOVE ★

32. DESCRIPTION AND APPARENT CAUSE OF TROUBLE
★ EXPLAIN

FS | FAT | FC

33. REPAIR TIME HRS. MIN.
.1 DIAGNOSIS
.2 OBTAIN REPL.
.3 REP. ACTION
.4 CHECKOUT

34. TEST EQUIPMENT USED

35. REPORTED BY/DATE

FIG. 9.4 Failure and consumption report.

Report Symbol BUWEPS 13070-3

1. Reporting Activity | 2. Report Ser. No. | 3. Date Of Trouble | 4. Installed In Aircraft/Arrest. Gear/Catapult/Support Equipment Model Buno or Ser. No. | 5. Aircraft Logbook Time

System, Set Equipment Or Engine: 6. Model Designation And Model No. | 7. Nomenclature | 8. Serial No. | 9. Time Meter Read./Logbook Time or Events (if applicable) Hour meter | Logbook hrs. | Starts | Landings

Unit, Component Accessory, Assembly Or Equipage: 10. Manufacturer's Part No. | 11. Nomenclature | 12. Serial No. | 13. Mfr's Code No. | 14. Contract No. | 15. Time Or Events Hrs. | Starts | Ldg's

Subassembly (Electronic) Or Primary Part Failure (Non-electronic): 16. Manufacturer's Part No. | 17. Nomenclature | 18. Serial No. | 19. Mfr's Code No. | 20. Location (if applicable)

Supply Identification Item(s) Returned: 21. Federal Stock Number | 22. (RM, MR copies only) | 23. Quantity | 24. (RM, MR copies only) | 25. (RM, MR copies only)

Reason For Report (Check one): 26. Removal Or Maintenance Action Required As A Result Of: 1 Failure/Suspected Failure Or malfunction 2 Damaged due To improper Maintenance/Operation/Test 3 Damaged or Defective On receipt 4 Damaged Accidentally 5 Scheduled/Directed Removal, high time Overage, excess To requirements | 27. Item overhauled by

DESCRIPTION OF TROUBLE

(If box 1, 2, 3, or 4 was checked in space 26, complete spaces 28 through 31. If box 5 was checked in space 26, leave spaces 28 through 31 blank.)

28. First Observed/Occurred During

1 Flight operations—Land based 2 Flight operations carrier based 3 Pre-flight 4 Daily 5 Conditional 6 Calendar 7 Overhaul/PAR 8 Shop maintenance bench test 9 Special directed inspection 10 Normal operation of support equip., catapults, arresting gear, mirror landing sys. only.

29. Symptoms—How Discovered Item

A Excessive vibration B High fuel consumption C High oil consumption D Incorrect display E Inoperative F Interference/Binding G Intermittent operation H Leakage I Low performance J Metal in oil K Noisy L None noticed M Out-of-balance N Overheating O Pressure out-of-limits P RPM out-of-limits Q Surging/Fluctuates R Temperature out-of-limits S Torque out-of-limits T Unstable operation U Visible defect V Other (Amplify)

30. Part Condition

007 Arced 780 Bent 135 Binding 429 Blistered/Peeled 070 Broken/Cracked 900 Burned/Burned out 120 Chafed/Galled 130 Changed value 910 Chipped/Nicked 999 Circuit defective 160 Connections defective 818 Contacts Burned/Pitted 170 Corroded 200 Dented 201 Distorted/Stretched 148 Eroded 250 Frayed/Torn 001 Gassy 381 Leaking 730 Loose 004 Low GM or emission 750 Missing 008 Noisy 450 Open 790 Out-of-adjustment 439 Plugged/Clogged 576 Ruptured/Split/Blown 935 Scored 585 Sheared 196 Shorted/Grounded 422 Soldering defect 660 Stripped 018 Tested OK—Did not work 389 Unknown (Cannot disassemble) 020 Worn—Excessively 099 Other (Amplify)

31. Cause Of Trouble

A Design deficiency B Faulty maintenance (Quality Control) C Faulty manufacturing (Quality Control) D Faulty overhaul (Quality control) E Faulty preservation/Packaging F Foreign object G Fluid contamination H Installation environment (Location in weapons sys.) I No failure-replaced to improve sys. performance J Operator technique/Adjustment K Other parts primary cause L Undetermined (Cannot disassemble) M Weather conditions N Wrong part installation O Other (Amplify)

32. DISPOSITION OR CORRECTIVE ACTION: Select appropriate code(s) from list below and enter in boxes at left to indicate disposition or corrective action taken with respect to each of the items entered in spaces 6, 10, and 16.

Space 6 / Space 10 / Space 16

Replaced And Returned To Supply — Code Reason: A Hold 90 days B Lack of repair facilities C Lack of repair parts D Lack of Tech. Pubs. E Lack of personnel F Beyond assigned maintenance level G Other—(Defective on receipt, high time, directed removal, excess to requirements, etc.)

Code Corrective Action: H Used as is I Adj./Realign./Serv./Repaired in place J Removed-Adj./Realign./Serv./Repaired-reinstalled K Removed-repaired-made RFI L Removed-tested Ok-made RFI M Removed-scrapped N Surveyed O Released for investigation and replaced (Indicate custody in space 35)

33. Maintainability Information — Hours | Tenths

Man-hours to locate trouble Space 10; Man-Hours to locate trouble Space 16; Man-hours to repair/replace/adjust; Actual time A/C was undergoing repair; Total time aircraft not flyable due to this malfunction

ACCESSIBILITY: S Satisfactory U Unsatisfactory (Amplify) SPECIAL: 1 Frequent trouble item 2 Can be installed wrong (Amplify)

34. Component/Assembly, Subassembly Replaced With: Mfr's Part No. | Serial No. | Mfr's Code No.

35. AMPLIFYING REMARKS (Furnish additional information concerning failure or corrective action not covered above. Do not merely repeat information checked above. Specify any severe operating conditions, such as hard landings, wheels-up landings, severe maneuvers, etc.)

36. Report Is: 0 FUR 1 AMPFUR 2 Urgent AMPFUR 3 Flight Safety AMPFUR 4 Follow up report | Signature | Rank/Rate | Date

Associated Parts Repaired Or Replaced (Do not list any item reported above)	37. Part No. (Non-electronic parts) Or Part Ref. Designator (Electronic parts)	38. Part Name, Tube Type, Semi-Conductor Type Or Description	39. Mfr's Code No.	40. Failure Code (From space 30)	41. Disposition (Code from space 32)	42. Activity Repaired By
						Signature
						Rank/Rate / Date

FAILURE, UNSATISFACTORY OR REMOVAL REPORT NAVWEPS FORM 13070/3 (10-62) (Mail this copy to NATSF) FUR

FIG. 9.5 Unsatisfactory or removal report.

facilitate mechanized data processing. IBM cards can be punched directly from the source document.

A similar form DD-787 (Fig. 9.6) is used by the Navy's Bureau of Ships to report failures of electronic equipment. The forms are provided in a pad which also contains instructions and data-processing codes (Fig. 9.7).

The Department of the Navy recognized that the increased complexity of weapon

ELECTRONIC EQUIPMENT FAILURE/REPLACEMENT REPORT DD–787 (PROPOSED) **REPORT BUSHIPS 10550–1**

1. DESIGNATION OF SHIP OR STATION

2. REPAIRED OR REPORTED BY

NAME	RATE	AFFILIATION
		1. ☐ U.S. NAVY 2. ☐ CONTRACTOR 3. ☐ CIVIL SERVICE

3. TYPE OF REPORT (CHECK ONE)

1. ☐ OPERATIONAL FAILURE
2. ☐ PREVENTIVE MAINTENANCE (POMSEE)
3. ☐ PREVENTIVE MAINTENANCE (NOT POMSEE)
4. ☐ STOCK DEFECTIVE
5. ☐ REPAIR OF REPLACEABLE UNIT OR PLUG-IN ASSEMBLY
6. ☐ OTHER

4. TIME FAIL. OCCURRED OR MAINT. BEGAN

MONTH	DAY	YEAR	TIME

5. TIME FAIL. CLEARED OR MAINT. COMPL.

MONTH	DAY	YEAR	TIME

EQUIPMENT

6. MODEL TYPE DESIGNATION

7. EQUIP. SERIAL NO.

8. CONTRACTOR (NAVY CODE OR COMPLETE NAME)

9. FIRST INDICATION OF TROUBLE (CHECK ONE)

1. ☐ INOPERATIVE
2. ☐ OUT OF TOLERANCE, LOW
3. ☐ OUT OF TOLERANCE, HIGH
4. ☐ INTERMITTENT OPERATION
5. ☐ UNSTABLE OPERATION
6. ☐ NOISE OR VIBRATION
7. ☐ OVERHEATING
8. ☐ VISUAL DEFECT
9. ☐ OTHER, EXPLAIN

10. OPERATIONAL CONDITION (CHECK ONE)

1. ☐ OUT OF SERVICE
2. ☐ OPERATING AT REDUCED CAPABILITY
3. ☐ UNAFFECTED

11. TIME METER READING

A. HIGH VOLTAGE

B. FILAMENT /ELAPSED

12. REPAIR TIME

MAN-HOURS	TENTHS

REPLACEMENT DATA

13. LOWEST DESIGNATED UNIT (U) OR SUB–ASSEMBLY (SA)	14. LOWEST DES. U/SA SERIAL NO.	15. REFERENCE DESIGNATION (V-101, C-14, R11, ETC.)	16. FEDERAL STOCK NUMBER	17. MFR. OF REMOVED ITEM	18. TYPE OF FAILURE	19. PRIMARY OR SECOND-ARY FAIL ?	20. CAUSE OF FAILURE	21. DISPOSITION OF REMOVED ITEM	22. REPL. AVAILABLE LOCALLY ?
						P ☐ S ☐			Y ☐ N ☐
						P ☐ S ☐			Y ☐ N ☐
						P ☐ S ☐			Y ☐ N ☐
						P ☐ S ☐			Y ☐ N ☐
						P ☐ S ☐			Y ☐ N ☐

23. REPAIR TIME FACTORS

CODE	DAYS	HOURS	TENTHS	CODE	DAYS	HOURS	TENTHS

24. REMARKS

(CONTINUE ON REVERSE SIDE IF NECESSARY)

SRA–1

Fig. 9.6 Form DD-787, Navy Bureau of Ships.

BLOCK 18 - TYPE OF FAILURE
QUICK REFERENCE LISTING OF MOST OFTEN USED FAILURE CODES
(IF PROPER CODE CANNOT BE FOUND, REFER TO ALPHABETICAL LISTING BELOW)

ELECTRON TUBES

CODE	TYPE OF FAILURE
002	AIR LEAK
007	ARCING, ARCED
960	BROKEN ENVELOPE
001	GASSY
380	LEAKAGE
004	LOW GM OR EMISSION
131	MARGINAL PART REPLACEMENT
009	MICROPHONIC
053	MISFIRES (THYRATRONS)
008	NOISY
003	OPEN FILAMENT
560	POOR REGULATION
011	SCREEN DEFECTS (CATHODE RAY)
005	SHORTED, INTERMITTENT
006	SHORTED, PERMANENT
018	TESTED OK, DID NOT WORK

TRANSISTORS AND SEMICONDUCTOR DIODES

CODE	TYPE OF FAILURE
741	ALPHA CUT-OFF LOW
744	BACK RESISTANCE LOW
739	BETA LOW
743	FALL TIME, EXCESSIVE
745	FORWARD RESISTANCE HIGH
742	Ico HIGH
737	OPEN, BASE-TO-COLLECTOR
735	OPEN, BASE-TO-EMITTER
156	POOR RECOVERY TIME
734	RISE TIME, EXCESSIVE
740	SATURATION RESISTANCE HIGH
738	SHORTED, BASE-TO-COLLECTOR
736	SHORTED, BASE-TO-EMITTER
731	SHORTED, COLLECTOR-TO-EMITTER
749	STORAGE TIME, EXCESSIVE

PLUG-IN ASSEMBLIES

CODE	TYPE OF FAILURE
035	DRIFTS
088	GAIN, LOW
094	GAIN, NONE
360	INTERMITTENT OPERATION
387	LOW PERFORMANCE
089	MODULATION, LOW
096	MODULATION, NONE
022	NO OSCILLATION
462	OUTPUT, LOW
255	OUTPUT, NONE
258	OVERHEATS
560	POOR REGULATION
097	RESPONSE, POOR
091	SENSITIVITY, LOW
680	UNSTABLE

OTHER COMMON TYPE OF FAILURE CODES

ELECTRICAL, ELECTRONIC

CODE	TYPE OF FAILURE
007	ARCING, ARCED
080	BURNED OUT
130	CHANGE OF VALUE
320	HIGH VOLTAGE BREAKDOWN
350	INSULATION BREAKDOWN
380	LEAKAGE
008	NOISY
450	OPEN
082	OPEN, INTERMITTENT
460	OPEN PRIMARY
451	OPEN ROTOR
470	OPEN SECONDARY
452	OPEN STATOR
453	OPEN WINDING
520	PITTED
005	SHORTED INTERMITTENT
006	SHORTED PERMANENT
620	SHORTED PRIMARY
612	SHORTED ROTOR
630	SHORTED SECONDARY
613	SHORTED STATOR

ELECTRO-MECHANICAL, MECHANICAL, CHEMICAL

CODE	TYPE OF FAILURE
710	BEARING FAILURE
780	BENT
040	BINDING, MECHANICAL
070	BROKEN
090	BRUSHES, IMPROPER TENSION
720	BRUSH FAILURE
150	CHATTERING
160	CONTACTS, CONNECTION DEFECTIVE
170	CORRODED
210	DETENT ACTION POOR
226	EXCESSIVE PLAY
567	HIGH CONTACT RESISTANCE
730	LOOSE
790	OUT OF ADJUSTMENT
570	RUSTY
770	SLIP RING OR COMMUTATOR FAILURE
164	SPEED INCORRECT
650	STICKY
945	STRUCTURAL FAILURE
020	WORN EXCESSIVELY

ALPHABETICAL LISTING

CODE	TYPE OF FAILURE
002	AIR LEAK
741	ALPHA CUT-OFF LOW
007	ARCING, ARCED
744	BACK RESISTANCE LOW
710	BEARING FAILURE
780	BENT
739	BETA LOW
060	BRITTLE
070	BROKEN
014	BROKEN BASE
960	BROKEN ENVELOPE
015	BROKEN GLASS
720	BRUSH FAILURE
090	BRUSHES, IMPROPER TENSION
080	BURNED OUT
120	CHAFED
130	CHANGE OF VALUE
140	CHARRED
150	CHATTERING
910	CHIPPED
180	CLOGGED
160	CONTACTS, CONNECTION DEFECTIVE
170	CORRODED
190	CRACKED
200	DENTED
210	DETENT ACTION POOR
230	DIRTY
035	DRIFTS
226	EXCESSIVE PLAY
743	FALL TIME, EXCESSIVE
240	FLAKING
745	FORWARD RESISTANCE HIGH
250	FRAYED
270	FROZEN
280	FUNGUS EFFECT
088	GAIN, LOW
094	GAIN, NONE
001	GASSY
290	GROOVED
300	GROUNDED
567	HIGH CONTACT RESISTANCE
320	HIGH VOLTAGE BREAKDOWN
742	Ico HIGH
340	INSTALLED IMPROPERLY
350	INSULATION BREAKDOWN
360	INTERMITTENT OPERATION
370	JAMMED
380	LEAKAGE
730	LOOSE
013	LOOSE BASE
012	LOOSE ELEMENTS
400	LOSS OF RESIDUAL MAGNETISM
004	LOW GM OR EMISSION
387	LOW PERFORMANCE
225	MANUFACTURER'S DEFECT (EXPLAIN)
131	MARGINAL PART REPLACEMENT
040	MECHANICAL BINDING
009	MICROPHONIC
053	MISFIRES (THYRATRONS)
750	MISSING
089	MODULATION, LOW
096	MODULATION, NONE
008	NOISY
022	NO OSCILLATION
920	NOT DETERMINED
737	OPEN, BASE-TO-COLLECTOR
735	OPEN, BASE-TO-EMITTER
450	OPEN, PERMANENT
003	OPEN FILAMENT
082	OPEN, INTERMITTENT
460	OPEN PRIMARY
451	OPEN ROTOR
470	OPEN SECONDARY
452	OPEN STATOR
453	OPEN WINDING
099	OTHER, EXPLAIN
161	OUTPUT INCORRECT
462	OUTPUT, LOW
255	OUTPUT, NONE
790	OUT OF ADJUSTMENT
258	OVERHEATS
928	PEELING
927	PINCHED
520	PITTED
010	POOR FOCUS
156	POOR RECOVERY TIME
560	POOR REGULATION
964	POOR SPECTRUM (MAGNETRON)
540	PUNCTURED
097	RESPONSE, NONE
734	RISE TIME, EXCESSIVE
570	RUSTY
740	SATURATION RESISTANCE HIGH
935	SCORED
011	SCREEN DEFECTS (CATHODE RAY)
091	SENSITIVITY, LOW
738	SHORTED, BASE-TO-COLLECTOR
736	SHORTED, BASE-TO-EMITTER
731	SHORTED, COLLECTOR-TO-EMITTER
005	SHORTED, INTERMITTENT
006	SHORTED, PERMANENT
620	SHORTED PRIMARY
612	SHORTED ROTOR
630	SHORTED SECONDARY
613	SHORTED STATOR
600	SHORTED TO CASE
610	SHORTED TO FRAME
615	SHORTED TO GROUND
640	SLIPPAGE
770	SLIP RING OR COMMUTATOR FAILURE
026	SOLDER JOINT DEFECTIVE
164	SPEED INCORRECT
650	STICKY
749	STORAGE TIME, EXCESSIVE
660	STRIPPED
945	STRUCTURAL FAILURE
018	TESTED OK, DID NOT WORK
947	TORN
965	TUNING DRIVE DEFECTIVE
670	UNBALANCED
680	UNSTABLE
690	VIBRATION EXCESSIVE
700	WEAK ELECTRICALLY
966	WINDOW SUCK-IN (MAGNETRON)
020	WORN EXCESSIVELY

Fig. 9.7 Data-processing codes.

A. JOB CONT NO. | B. PRI | C. TIME SPEC REQ | D. WK AREA | E. EST M/H | F. ORIG RPT NO. | G. REPORT NO. 26407-F | H. | I.

1. WEAPON TMS | 2. SERIAL NO. | 3. TIME | 4. WORK CENTER | 5. WORK ORDER NO. | 6. DAY MO YR | 7. WORK UNIT CODE

1A. AGE WUC | 2A. SERIAL NO. | 3A. TIME | 8. ACT TAKEN | 9. WHEN DIS | 10. HOW MAL | 11. UNITS | 12. LABOR HRS | 13. ASST WORK CEN

1B. ENG TM PSN | 2B. SER MOD YR-MFG SER NO. | 3B. TIME | 14. INST ENG TM PSN | 15. SER MOD YR-MFG SER NO. | 16. TIME

1C. ITEM FSC | 2C. PART NO. | 3C. SERIAL NO. | 17. INST ITEM PT NO. | 18. SERIAL NO. | 19. TIME

J. SYMBOL | K. DISCREPANCY | L. CORRECTIVE ACTION

CORRECTED BY-SIGNATURE & GRADE

DISCOVERED BY-SIGNATURE & GRADE | INSPECTED BY-SIG & GRADE | SUPERVISOR-SIG & GRADE

RECORDS ACTION

☐ UNCLEARED DISCREPANCY ☐ REPLACEMENT TIME CHANGE ITEM ☐ DATA TRANSCRIBED TO APPROP RECORDS

DATE TRANSCRIBED | TRANSCRIBED BY-SIGNATURE & GRADE

AFTO FORM JUL 61 211 PREVIOUS EDITIONS OF THIS FORM ARE OBSOLETE. MAINTENANCE DISCREPANCY/PRODUCTION CREDIT RECORD

Fig. 9.8 Maintenance discrepancy form, U.S. Air Force.

I ☐ WORK REQUEST ☐ SEPARATE EIR

1. DATE OF REQUEST *(Day - Month - Year)*
2. UNIT OR ORGANIZATION
STRAC ☐ CODE 1
3. WORK REQUEST NO.
17. JOB ORDER NUMBER

4. SERIAL NUMBER
5. REGISTRATION NUMBER

6. NOMENCLATURE AND MODEL
7. FSN

8. DATE MFG/LAST OVERHAUL
9. MANUFACTURER OR OVERHAUL ACTIVITY
10. ☐ HOURS ☐ MILES ☐ ROUNDS

11. FAILURE DETECTED DURING
☐ SCD MAINTENANCE CODE A
☐ INSPECTION/TEST CODE C
☐ HANDLING CODE B
☐ NORMAL OPERATION CODE D

12. FIRST INDICATION OF TROUBLE
A ☐ INOPERATIVE CODE 068
C ☐ LOW PERFORMANCE CODE 387
D ☐ NOISY CODE 008
F ☐ OUT OF ADJUSTMENT CODE 790
G ☐ OVERHEATING CODE 258
I ☐ OTHER CODE 099

13. DESCRIBE DEFICIENCIES OR SYMPTOMS ON THE BASIS OF COMPLETE CHECK OUT AND DIAGNOSTIC PROCEDURES IN THE EQUIPMENT TM. *(Do not prescribe repairs)*

14. SUBMITTED BY *(Signature)*
15. RECEIVED BY *(Signature)*
16. DATE RECEIVED *(Day - Month Year)*

III EQUIPMENT IMPROVEMENT RECOMMENDATION *(EIR)*

☐ EMERGENCY ☐ URGENT ☐ ROUTINE

29. ☐ NORMAL REPLACEMENT
30. FAILED OR DEFECTIVE PART FSN

31. RECOMMENDATION
a. ☐ IMPROVE DESIGN b. ☐ MODIFY c. ☐ REVISE PROCEDURE d. ☐ OTHER *(Specify)*

32. OPINION OR REMARKS *(Relate to block selected. Describe conditions under which failure occurred. Attach photos or sketches, if available.)*

33. SIGNATURE
34. ORGANIZATION
35. CONTROL NUMBER

DA FORM 2407, 1 APR 62

MAINTENANCE REQUEST
(TM 38-750)

FIG. 9.9 Equipment-improvement recommendation, U.S. Army.

systems required an improved system for rapid collection of data and analysis of failures. The forms shown here resulted from a detailed joint-requirements study by the Navy and industry representatives. Reports are completed for every repair action that involves a failure. They are also used when items are found defective upon removal from spare supplies. The forms are designed to provide the Navy with information to:

1. Correct deficiencies on equipment and parts.
2. Evaluate equipment reliability and maintainability.
3. Determine provisioning requirements.
4. Control stock of replaceable items.
5. Invoke contractual guarantees.
6. Control quality in current and subsequent production.

When the reports are received from the field, they are analyzed and the accumulated data are presented periodically in an *Information Bulletin*. Corrective action is initiated for all repetitive problems.

The Air Force counterpart failure form is the maintenance discrepancy/production credit record (Fig. 9.8). It also utilizes codes for data processing and resembles the layout of the BMD-050 reports.

The Department of the Army's equipment improvement recommendation (Fig. 9.9) is used to document field failures. The appropriate agency within the Army receives these reports from the field and then requests corrective action from the contractor.

It is difficult to achieve high reliability when field facilities are limited to reporting seriously repetitive failures. The same mode of failure may be occurring once or twice a month at each and every location, but no report will be made to the contractor because the situation is not understood. Also, some service personnel are reluctant to sign reports of this nature for fear that they may be wrong and that the equipment is not at fault. A third aspect of the problem is that the contractor cannot readily obtain failed hardware after a weapon becomes operational. All too often, military technicians will isolate trouble to a particular printed-circuit board, replace the board, and scrap the failed item. The exact cause of failure is never determined. The military services should consider a plan to return all failed hardware to contractors who maintain failure-analysis laboratories. Such a plan would cost very little and would improve reliability by orders of magnitude.

9.4 PROCESSING FAILURE DATA

In the past, many systems for processing failure data have proved ineffective. In trying to design a system to serve several organizations, compromises that did not adequately service any group resulted. Reliability requires a mechanized system which will accurately condense and store pertinent failure data as outlined in Sec. 9.3. Such a system must be capable of supplying special tab runs on a timely basis. This means a matter of minutes from the time a small report is requested until it is printed out. An example of this type of report might be a summary listing all the failures on a particular assembly which were attributed to design.

Experience has shown that it is sometimes faster to process special reports manually than by machine. Punched cards with the mode of failure printed out can be filed in assembly-number sequence. These cards can be removed from the file for immediate consultation. Thus, it is not necessary to wire boards and machine-sort a large deck. As a result, much time is saved in getting data to the user. This method can be used only for those types of special reports which are commonly requested. Otherwise, it would be necessary to have many files of duplicate cards sorted in various sequences.

Data storage on magnetic tape is another way to speed up the preparation of timely failure reports. This is especially true when the volume of data becomes very high. As is the case with most systems of data retrieval, more time is spent by man in preparation to use the machines than the machines themselves require to do the job. But machine time can also be a factor; for a tape-reading device must search over a 2,400-ft reel to retrieve reliability information character by character. It is probable that future reliability data will be stored on flexible magnetic cards for use in random-access computer equipment. These cards are kept in magazines which can be searched and read in a few seconds.

9.4a Data-condensation Problems. The biggest problem which arises when storing failure information is that the data may be misinterpreted by coding personnel. As was seen in Sec. 9.3, the problem can be alleviated by having test engineers place the codes on the failure-reporting form. The source document can then be used for keypunching, and human errors on transmittal forms are eliminated.

Transmittal forms are normally used if the reliability department reviews trouble reports prior to data storage. The reliability specialist is aware of the kind of information which should be stored to back up recommendations for corrective action. In addition, if some form blocks are left blank, the reliability coder can contact the field man to complete the report. An example of a transmittal form to condense failure data into machine language is shown in Fig. 9.10. Section 6 contains typical code classifications and identification symbols which are used for this purpose.

Even with the best data-condensation procedures, it is desirable for failure-analysis and design engineers to receive copies of trouble reports on hardware that is their responsibility. This is true because it is impossible to condense failure descriptions without losing important clues which aid in establishing the cause of failures and determining what kind of redesign may be necessary.

9.4b Data Summaries. Historical failure summaries prepared by electronic data-processing equipment serve to pinpoint repetitive problem areas. Martin Company's integrated failure report is an example of a comprehensive data summary

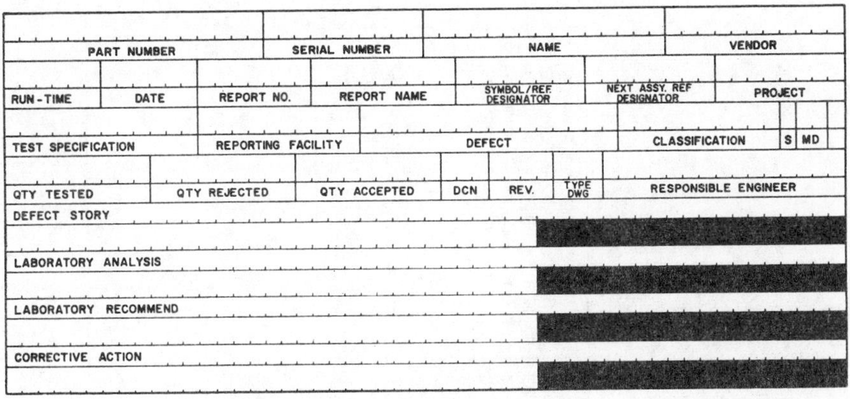

PART NUMBER | SERIAL NUMBER | NAME | VENDOR
RUN-TIME | DATE | REPORT NO. | REPORT NAME | SYMBOL/REF. DESIGNATOR | NEXT ASSY. REF DESIGNATOR | PROJECT
TEST SPECIFICATION | REPORTING FACILITY | DEFECT | CLASSIFICATION | S | MD
QTY TESTED | QTY REJECTED | QTY ACCEPTED | DCN | REV. | TYPE DWG | RESPONSIBLE ENGINEER
DEFECT STORY
LABORATORY ANALYSIS
LABORATORY RECOMMEND
CORRECTIVE ACTION

FIG. 9.10 Transmittal form.

which is distributed monthly to management, project, and design engineers. The report contains data on failure modes attributed to design which occur during subcontractor, in-house, environmental, and field tests. The data sources are shown on the flow chart, Fig. 9.11. If desired, it is also possible to produce similar summaries of poor workmanship or improper operation failures. Frequency of failure at the assembly level is shown as the ratio of failures divided by opportunities to fail or quantity tested. A listing of the "top 20" highest failure ratios for a moving monthly period as well as the number of the times the item has appeared on the top 20 is presented on the cover page of the report. The STR and DMT are the field and in-house reporting forms. The QDTR is the shop traveler history card.

9.4c Failure Graphs. Failure graphs can be used as a supplement to reliability measurements based on MTBF. The graphs can be prepared manually or mechanically by the use of an x-y plotter. Semilogarithmic paper gives meaning to the length of the lines between points, and percentage increase or decrease in failures can thus be readily detected. Moving periods of two or three months are used to smooth out trend lines and discount erratic fluctuations caused by factors which are not directly associated with unreliability. Three types of graphs have been found most useful: system failures, system-failure ratios, and cumulative percentage of total failures per system. None of these plots reflect varying system complexity, but they do offer an accurate means of comparing systems and determining where to concentrate cor-

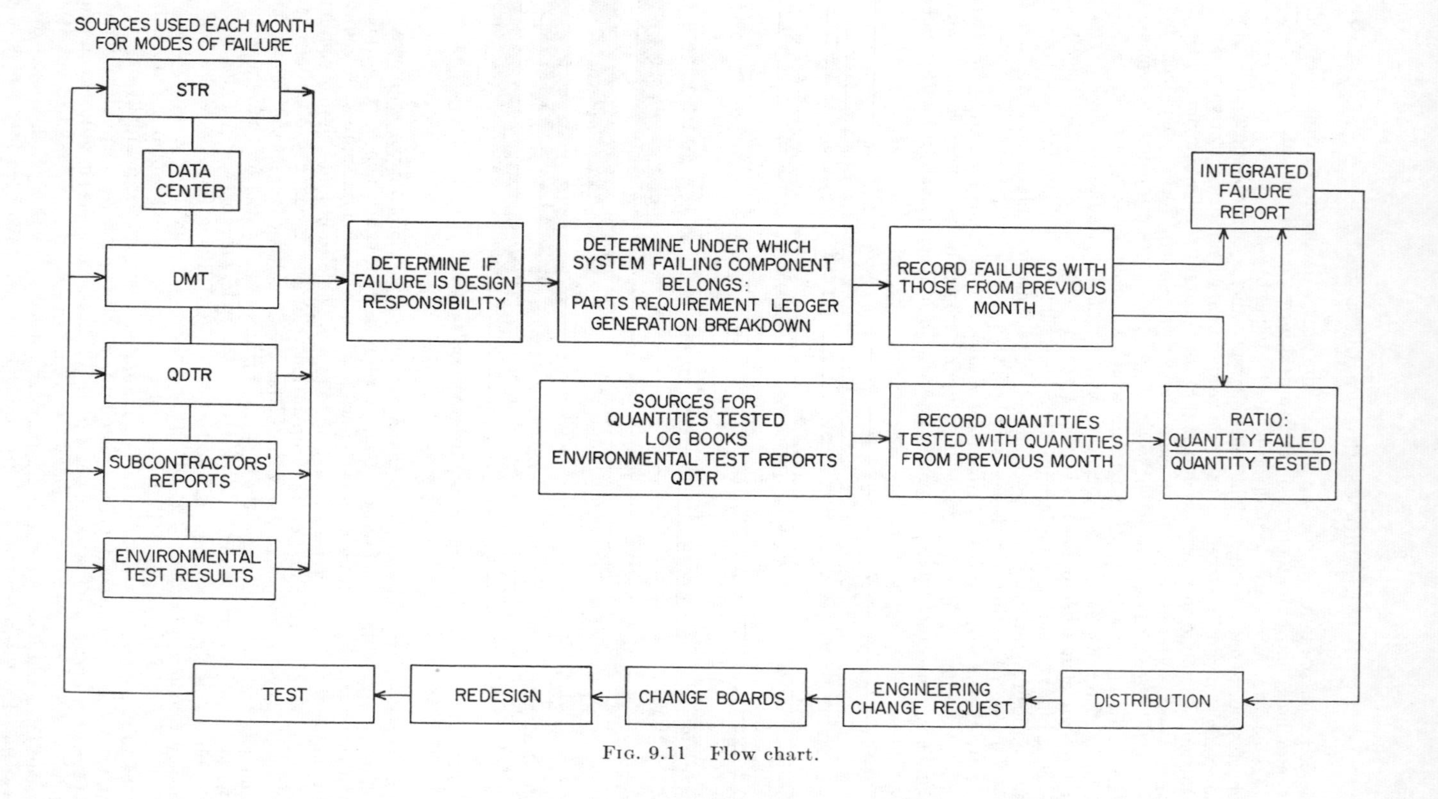

Fig. 9.11 Flow chart.

rective action. Failure ratios offer the most reliable estimate of trends, because quantity tested is taken into consideration.

9.5 ANALYSIS LABORATORIES

A laboratory used specifically for failure analysis is one of the most effective ways to achieve reliability improvement. If failed assemblies are sent to manufacturing for troubleshooting, the primary concern is what must be done to repair the unit. The manufacturing people have neither the time nor the inclination to perform exhaustive analysis to determine the exact cause of failure. It is much better to send failed hardware to a special laboratory staffed by inquisitive, research-minded engineers and technicians whose only reason for being is to find the reason why.

A properly equipped laboratory should contain at least one or more units of each of the following equipment:

Oscilloscope	Precision meters
Square-wave generator	Differential voltmeter
Power supply	Electronic counter
Recorders	Megohmeter
Phase meters	Microscope
LC bridge	Inspection light
Wheatstone bridge	Jewelers' tools
Transistor-curve tracer	Pyrometer
Multimeters	Signal generator
Oscillograph	Forceps
Curettes	Camera
Dielectric tester	Bone files

It is also valuable to maintain a library of functional test procedures and parts specifications.

The failure-analysis laboratory must be able to call upon the services of other laboratories to assist in performing special tests. Examples of laboratories which are commonly used are chemical, physical test, metallurgical, X-ray, hydraulics, pneumatics, dynamics, and environmental. The machine shop should also be available to aid in the disassembly of sealed units.

9.5a Problems. It is very important for the laboratory to be furnished with an accurate trouble report which details the test measurements recorded at the time of failure. When adequate spares are available, the complete assembly can be returned to the lab. When subassemblies are returned, they sometimes pass the unit-level functional test. In cases of this sort, the unit must be replaced in its original chassis to determine if there is a tolerance buildup which may cause the system to fail. Transient measurements of the inputs to a failed item are also useful in pinpointing the problem.

It is often necessary to observe strict procedures for accountability of hardware which has been accepted by the government. When failed equipment is returned from the field, it is placed in a bonded storeroom. Records are kept of all items which are sent to the failure-analysis laboratory. After the analysis has been completed, all parts are returned to field-service personnel, where they can be scrapped under the supervision of government inspectors. Normally, scrapped parts are sprayed with red paint and permanently stamped so there is no possibility that they will be redelivered to the contracting agency.

9.5b Failure Duplication. When the laboratory investigation cannot detect the cause of failure, it is desirable to purchase a new assembly and perform tests to duplicate the trouble. If the exact system inputs and failure environment cannot be determined, a worst-case approach should be used. The test can be thoroughly instrumented so that, when the failure under study reoccurs, it will be possible to pinpoint the trouble. The design engineer should be called in to witness these tests, since his knowledge may prove invaluable.

9.6 LABORATORY PROCEDURES

When a failed assembly is received by the analysis laboratory, pertinent data should be entered in a log book. Information normally recorded includes date of receipt, trouble-report number, assembly-drawing number, serial number, description of trouble, location of the facility where the failure occurred, and name of the engineer who will be responsible for the investigation. Two additional columns, for cause of failure and lab-report number, are filled out when the analysis is completed.

The first step in the analysis is to verify the failure. The functional test procedure is drawn from the library, and a test is performed to observe the symptoms stated on the trouble report. If the failure is verified, troubleshooting is begun. In the case of electronic assemblies it is often possible to localize the problem by consulting the schematic and analyzing the circuit. Many parts are secured to printed-circuit boards with an epoxy coating, and there are rarely enough test points to take the necessary voltage, current, and resistance measurements. In this case a soldering iron is used to loosen the epoxy and gain access to the part leads. This procedure is necessary, since no solvent that can quickly soften epoxy is available.

Potted units present quite a challenge, since the potting material must be removed prior to troubleshooting. The use of solvents is required, but from two to five days of repeated soaks and removal of partially softened potting may be needed before the operation is completed. Solvents which are employed for this purpose include trichlorethylene, methyl ethyl ketone, acetone, toluene, naphtha, phenol, ethylene dichloride, and acetic acid. Care must be taken not to use solvents which will damage or dissolve the parts.

When units are encased in sealed metal cans, it is always desirable to take X-ray photographs before starting disassembly. The reason for this procedure is to provide evidence that the failure existed prior to disassembly. If drawings are not available, X-rays are also valuable as an aid in avoiding damage to internal parts.

In the case of a high-cost warranty item the vendor's permission should be obtained prior to disassembly. Analysis should be performed at the prime contractor's facility so that system inputs to the failed assembly can be studied. The vendor is often requested to send a representative to assist in the troubleshooting.

If an intermittent condition is suspected, temperature and vibration tests are helpful in detecting the part responsible for the trouble. It is beneficial to construct a low-cost temperature chamber for this purpose, since environmental laboratory equipment may not be available on short notice.

9.6a Part Dissection. After assembly troubleshooting has isolated the part responsible for the trouble, dissection must be performed in order to pinpoint the exact mode of failure. Again, utmost care must be used to avoid destroying the evidence. X-ray photos should always be taken prior to proceeding with the analysis.

An electrochemical case remover (Fig. 9.12) is used to remove the tops of transistors. It is preferred over mechanical methods, because it does not subject the transistor to shock, heat, or vibration. Paint must be removed from the rim of the transistor case prior to placing the rim in the alligator clip. The transistor is centered above the hole in the metal plate and the beaker is filled with 20 per cent hydrochloric acid until the acid just touches the case. A positive d-c potential is then applied to the metal plate, and the negative lead is connected to the alligator clip. This device can cut through the top rim of transistor cases in five minutes, and no damage is done to the crystals. The metal plate and alligator clip must be gold-plated; otherwise, prolonged exposure to the acid would eventually corrode them.

After the junction is removed from the transistor, it is embedded in resin and hand-ground over water-cooled abrasive surfaces of varying roughness. The surface is then etched, and the grain structure is examined under a metallurgical microscope. Microphotographs of the junction cross section are compared with those of other malfunctions exhibiting similar electrical characteristics. If it appears that the failure is the result of manufacturing variance, the vendor is informed in order that he can resolve the situation.

Similar techniques are employed for dissecting other electronic parts. Jewelers' tools and a cutting wheel are very useful in opening relays, capacitors, and resistors.

9.6b Modes of Failure. Many thousands of failure modes have been uncovered during analysis of aerospace hardware. Samples of some of the more common modes are described in this section.

Transistors. Most transistor vendors choose their published maximum-voltage ratings conservatively. As a result, breakdown normally occurs at a voltage considerably higher than that specified. When breakdown does occur, the high local current densities can irreversibly alter the characteristics of the transistor. Test engineers should be instructed not to measure breakdown or punch-through values above the specified maximum ratings. Short-duration transients far in excess of the maximum rated voltage are usually applied under such conditions and are a common cause of failure. Transient breakdown is characteristic of ultrafast switching transistors which have a thin base region. The two other most prevalent failure modes are contamination and wafers which crack under shock and vibration. Improved manufacturing techniques and new methods for mounting the wafer on the header

FIG. 9.12 Electrochemical case remover.

have greatly reduced failures of this type. Most manufacturers have introduced hermetic-seal leak tests, and this has eliminated a serious problem.

Failure modes which occur less frequently are caused by poor bonding of contacts and inhomogeneous material. The laboratory sometimes detects current overloads occasioned by operator error. In such cases the internal leads are melted open with a ball of metal at the end. Thermal runaway is an assembly design problem associated with power transistors. It can be prevented by using adequate heat sinks.

Relays. Most cases of pitted and eroded relay contacts are caused by inductive loads. If protective diodes are not used, designers should specify relays with resistive contact ratings of at least twice the maximum anticipated inductive load. Use of parallel contacts to increase the current rating is undesirable because the contacts do not make and break simultaneously. If relays are not hermetically sealed, contact corrosion can also cause failure. Occasionally, coil opens or shorts occur, and they can usually be traced to poor manufacturing techniques. It is difficult to avoid small nicks and imperfections in wires of very small diameter. These irregularities constitute regions of relatively high resistance and are probable points of failure.

Transformers. Transformer opens occasionally occur because of poor bonding of the lead wire to the winding. Shorts within the winding can be caused by improper insulation or by overload attributed to human error. Normally, the overload mode can be detected because of the greater amount of melting and burning which is observed during dissection.

Diodes. Whisker-type diodes sometimes open or short because of off-center mounting of the whisker. The resulting spring tension can cause the whisker to break at the junction during a vibratory environment. Glass-type diodes often crack as the result of differential contraction of epoxy conformal coatings on printed-circuit boards.

Resistors and Potentiometers. High-temperature tests on printed-circuit boards have revealed intermittent opens on deposited-carbon and composition resistors using cemented leads. The problem is caused by the expansion of epoxy conformal coatings, which places undue stress on the leads. Ceramic resistors have cracked because of differential contraction between the ceramic and the epoxy. Wire-wound-resistor failures increase as the size of the wire decreases. This is due to manufacturing problems with fine wire similar to those which occur with miniature relay coils. The bulk of potentiometer failures are attributed to contamination on the resistance element. Occasionally, vibration will cause intermittent opens, but this mode has been virtually eliminated by improved design techniques.

Capacitors. Shorted wet-electrolyte tantalum slug capacitors have self-healing properties which can make it difficult to verify failures. Ion migration is a symptom of malfunction, so a chemical analysis should be performed on the electrolyte. Tantalum foil varies in color as a function of the formation voltage. Typical colors are shown in the accompanying table. It is often possible to determine if these capacitors

Approx. formation voltage	*Color*
5	Silver
8	Gold
13	Yellow gold
20	Dark brown
33	Light blue
40	Silver
67	Violet
100	Gold
133	Dark green
200	Light green

have been subjected to excess voltage by dissecting them and examining the color change of the foil. Polarized tantalums are very sensitive to voltage reversals. The failure-analysis engineer must, therefore, determine the peak voltage and repetition rate of any transients appearing across these capacitors.

Connectors. The biggest problems with connectors have been poor alignment and moisture. Guide pins for axial alignment of the receptacle and plug prior to mating are very helpful. New methods for environmental proofing to prevent moisture penetration have been developed.

Mechanical Problems. Corrosion, contamination, and leakage are major failure modes associated with mechanical systems. If adjacent dissimilar metals have not been anodized or passivated, galvanic action can easily occur in the presence of moisture. A breathing effect caused by changing temperature will draw humid atmospheres into internal areas which many designers believe to be inaccessible. When the failure-analysis engineer finds evidence of corrosion, he should determine the position of the dissimilar metals in the electrochemical or galvanic series and he should have the corrosion products analyzed to determine their chemical composition. This information is valuable in making recommendations for corrective action.

Contamination is usually caused by poor workmanship, although it can result from the machining action of moving parts. Seal leakage in high-speed rotating machinery has always been a difficult problem which is accentuated by the use of cryogenic liquids. Rather than design a positive seal, provisions are sometimes made to dispose of the leakage overboard. Rubber O rings used in both static and dynamic applications eventually loose their resilience. Failure-analysis engineers often test O rings with a durometer, because these measurements help back up suggestions to use metal-to-metal seals for systems kept in long-duration storage. Life-cycle tests are a useful way for the failure-analysis laboratory to uncover the effects of tolerance buildup in assemblies of moving parts. The increased viscosity of petroleum-base lubricants at low temperature is another common source of trouble. New methods for depositing dry lubricants on metal surfaces have in some cases eliminated this problem.

Fatigue in aircraft structures has become an important mode of failure owing to the longer operational life of commercial transports. This mode is also prevalent in military aircraft because of increasingly severe service conditions resulting from higher speeds and maneuverability. The use of exotic materials has intensified the problem, since the greater static strength of the materials is not necessarily accompanied by a parallel improvement in fatigue performance.

9.6c Case Histories

"After more than ten years of intense study of reliability problems, only a few failure analyses have revealed physical or chemical phenomena with which we were not familiar."

—*Lysle A. Wood*

There is nothing very difficult in pinpointing the exact cause of failure *if* the hardware is made available. The following case histories are typical of the work of a failure-analysis laboratory. More than forty additional interesting examples can be found in the failure-analysis references.

Example 9.1

A missile failed in flight, and a review of telemetry measurements revealed that pressure had been lost in the guidance section. The residue fell in the ocean to a depth of 1900 fathoms, so recovery was not attempted. The rate of pressure loss was such that it appeared that either the pressure regulator or the umbilical air-conditioning valve was responsible. New samples of both components were obtained by the failure-analysis laboratory and tests were designed to investigate the problem.

Three tests were performed on the air-conditioning valve. The first test simulated the launch environment by repeatedly engaging and disengaging a cable mast to sample valves. At the conclusion of each cycle, functional tests were performed and no leakage could be detected. In the second test, 30-mesh sand was placed in a valve and the piston was allowed to close. This test was repeated nine times, and again no leakage was observed. The third test evaluated the effects of drying on the piston lubricant. Valves which had been in long-duration storage were inspected and, although the surface of the lubricant was dry, the valves functioned properly. As a result of these tests it appeared unlikely that this valve was responsible for the missile failure.

The pressure regulator was used to dump gyro-bearing nitrogen to atmosphere in order to keep compartment pressure relatively constant during flight. Samples of the regulator were subjected to life-cycle and vibration tests. In addition, a pressure vs. flow check was made prior to life testing. This characteristic was measured with increasing and decreasing flow rates in order to determine hysteresis. The life-cycle test was performed with a regulator mounted in a 1-cu ft chamber. A calibrated orifice was used as the air inlet, and a vacuum pump was connected to the regulator to simulate in-flight atmospheric-pressure changes. The regulator was cycled 12,750 times. Two samples were mounted on a shaker so that vibration and functional tests could be conducted simultaneously. The regulators were subjected to frequency sweeps at 5, 10, and 20 *g*'s. At the conclusion of the tests, the data were analyzed and it was found that, although some pressure change had occurred, performance was still within specification limits.

The regulators were then disassembled, and it was observed that the hard anodized coating had been completely worn off the cylinder walls in the vicinity of the poppet stem pin. This pin, which is used to connect the poppet stem to the aneroid, is press-fitted into a hole in the poppet stem. Several additional regulators were removed from stock and disassembled. All pins extended outside the poppet stem—in one case by as much as four thousandths. Since the dimensions of the cylinder wall varied from sample to sample, it was possible that a maximum-length pin in a minimum-diameter cylinder could cause the poppet stem to stick, resulting in failure of the regulator. It was believed that this mode had been the probable cause of pressure loss during the flight test. The vendor relieved the stem diameter in the vicinity of the pin and the pin length was decreased. A more extensive functional test was put into effect, and no further failures were experienced.

Example 9.2

A missile launch countdown was aborted because the primary battery did not come up to voltage when it was activated. The one-shot battery was removed from the missile with fumes still spewing from the vent tube, and it was placed under load to effect discharge. The battery was then returned to the failure-analysis laboratory for troubleshooting. Continuity measurements were made on the connector to check the internal circuitry, and no defects were observed. The stainless-steel case was removed, and the battery was then carefully dissected by the manufacturer's representative who came to the laboratory to assist in the analysis. The specific gravity of the potassium hydroxide electrolyte was found to be 1.3007 at 25°C, well within the normal range. Spectrographic analysis of the electrolyte revealed the usual traces of aluminum, silicon, tin, lead, copper, and iron.

The two sumps which absorb excess electrolyte were inspected, and fluid was present in both. This indicated that all cells had been properly activated. This fact was further substantiated by dissection of the cells themselves. The gas generator and copper-tubing reservoir diaphragms indicated normal operation, and there was no evidence of electrolyte leakage. The battery-heater thermostats were painstakingly removed from the plastic-foam encapsulation and placed in a temperature chamber for testing. The tests determined that the thermostat which had been controlling the battery heater had a differential of about 15°F between open and close. This was in excess of the specification requirement of ± 5°F. A review of the launch-site records showed that the battery heater had come on just prior to the activation command. It was therefore concluded that the low battery voltage was caused by a marginal thermostat which allowed the battery temperature to decrease below the required level. Redesign of the thermostat by the vendor and 100 per cent functional testing during the battery assembly process succeeded in eliminating this mode of failure.

Example 9.3

A new type of exploding bridgewire system was incorporated into a missile. The system consisted of a high-energy single-pulse generator, a transmission line, and the exploding bridgewire. During testing of the missile, two problems were encountered. On several occasions, no pulse reached the bridgewire. On other occasions, the bridgewire did not vaporize, but merely melted. The defective systems were sent to the failure-analysis laboratory. A pulse generator was dissected, and electrical checks were performed on the parts. Figure 9.13 shows the pulse generator in a state of partial dissection. No defects were found, except that one electrolytic capacitor had a high leakage.

A detailed study of the circuit and procedures was performed. It was found that a very small reverse voltage was being placed across the above-mentioned polarized tantalum capacitor during missile system testing. To see if this small reverse voltage could be damaging the capacitor, a circuit that applied the reverse voltage to 10 samples was designed. After each hour, the capacitors were charged normally with an oscillograph across them to measure the charging rate. This process was continued for 10 hours. At that time, seven capacitors were found to be intermittently shorted. Three would not charge at all. Since the capacitor must be charged before the single-pulse generator can operate, it was concluded that this could have been the reason why no pulse had reached several of the bridgewires during testing of the missile.

In order to determine what could have caused the bridgewires to melt instead of vaporize, a breadboard version of the system was built. This system was fired with an oscilloscope connected to it so that the voltage waveforms could be seen at various points in the circuit at the time of firing. This revealed a reflected wave in the transmission line that caused excessive pulse energy to be dissipated in the cable rather than in the bridgewire. The use of a transmission line of the correct type resulted in greater energy being transmitted to the bridgewire. With the addition of a diode-resistor network to prevent reverse voltage from being seen by the capacitor and with the correct cable, no further failures occurred.

Fig. 9.13 Pulse generator in state of partial dissection.

9.7 LABORATORY REPORTS

Comprehensive reports of the laboratory analysis are prepared at the conclusion of each investigation. These reports should contain all the details of the test procedure and list the equipment used. Measurements should be provided in tabular form along with a schematic of the test setup. The past history of the failure mode is provided, together with a mission-effect analysis which classifies the failure as critical, major, or minor. Documentary photographs pinpointing the location and nature of the malfunction are always valuable. Conclusions and recommendations for corrective action are presented at the beginning for the benefit of managers who may not have time to read the entire report.

9.7a Report Distribution. Copies of the laboratory reports are distributed to the responsible design engineer and the members of the trouble-report committee. Additional copies are provided to project management and the customer, and to IDEP, if participating. (See Sec. 6.)

9.7b Vendor Failure-analysis Reports. Occasionally it is necessary to return assemblies to a vendor in order to take advantage of his specialized knowledge. Vendors have a very legitimate complaint in that they are rarely given sufficient information about the details of the failure to enable them to determine the exact cause. On the other hand, some vendors are very slow in returning reports, and they will not make an objective evaluation for fear of endangering their competitive rating. In order to maintain control of this operation, Martin Company places the following note on the purchase order:

Technical reports shall be prepared and forwarded same day return shipment is made. The report shall contain the following:

1. Specific cause of failure.
2. Contributory cause of failure, if any.
3. Recommended corrective action as to design, operating instruction, maintenance, etc.
4. Parts used to repair the item.
5. Any other data that will contribute to improving product reliability.

Receipt of this report is a condition precedent to the payment of your invoice.

9.8 CORRECTIVE ACTION

"Every failure must be regarded as significant until action has been taken to prevent its recurrence."

—*Dr. Leslie W. Ball*

Obviously there is no point in determining the cause of failures if the corrective-action loop is not closed. Perhaps the best method of accomplishing this task is to assign the job to a trouble-report committee. As was seen in Sec. 9.2, this committee is made up of members from all of the operating divisions. A trouble report cannot be closed out until all members are satisfied that proper corrective action has been taken. This results in a check-and-balance process which ensures that product improvement will be achieved.

It is beneficial to send field personnel copies of the final trouble reports which contain a description of corrective action. In this way, they are motivated to do a better job; for they can see the results of their efforts.

9.9 AIRCRAFT FAILURE ANALYSIS[1]

The Civil Aeronautics Board maintains a staff of accident-investigation engineers who are specialists in the fields of aircraft structures, power plants, systems, maintenance, and electronics. When an accident occurs, the closest CAB field office is notified immediately and arrangements are made for the security of the crash scene. The specialists proceed to the scene, where they are split up into several investigatory groups. Depending upon the nature of the accident, there may be one such group to question witnesses, another to inspect the aircraft structure, another to examine the power plants, etc.

Sometimes the work of the groups can be completed at the accident site, but more frequently tests or continued study of parts or components are carried on at the manufacturer's plant or some testing facility. Often it is necessary to move the power plants, instruments, and some of the system components immediately to more favorable locations for disassembly and study. The facilities of other Federal agencies such as the Federal Bureau of Investigation and the National Bureau of Standards can be used for specific tests or studies of parts and materials.

9.9a Examination of Airframe Wreckage. A thorough, detailed study of the airframe wreckage yields an amazing amount of information about the aircraft at the instant of ground contact and will sometimes give the full story of the accident. When the wreckage is widely scattered, only a small portion will be consumed by the ground fire which generally follows a crash. However, if an airplane contacts the ground at a steep angle with very little scattering of the wreckage, a large part of the aluminum, magnesium, and soft parts may be consumed by the fire, making the study much less rewarding.

Generally, one of the first things to do is to identify as many of the airplane sections, components, and parts as possible and to plot their exact position on a wreckage-distribution chart. Such a chart will often contain several hundred items, since the location of small items, interior furnishings, etc., may give clues to the exact order of breakup if there is a lengthy ground slide. The wreckage distribution chart should also show ground-impact marks and trees or objects struck or cleared just prior to

[1] Abstracted from a CAB paper entitled "Aircraft Accident Investigation Procedures and Techniques Employed by the Civil Aeronautics Board."

the point of initial ground contact. Construction of the chart is most easily accomplished by laying out a base line beginning at or near the point of initial ground contact and extending along the ground path as far as necessary. Measurements can then be taken along the base line and to the right or left with a minimum of equipment and trained personnel. If the wreckage is distributed over a small area, polar coordinates are generally more practical if the proper equipment is available. In addition to giving the members of the structures group a preliminary look at the overall wreckage, the chart allows a continuity check to be made to ascertain whether or not the entire aircraft is accounted for at the crash site.

Once the wreckage-distribution chart is prepared, the pieces can be moved as necessary to facilitate the study. It is frequently desirable to gather together all wing parts, fuselage parts, empennage parts, etc., so that related damage patterns can be identified. A complete or partial mock-up is sometimes necessary.

The examination of the airframe wreckage is just exactly that: a careful study of each piece of the aircraft to learn anything from it that can be learned. The pieces are examined for evidence of ground contact, in-flight fire (soot or molten-metal deposits), ground-fire damage, explosion, sabotage, corrosion, fatigue failures, in-flight collision with foreign objects (birds, wires, trees, etc.), midair collision (paint marks, abrasions, propeller cuts, etc.), aerodynamic flutter, foreign substances, and many other considerations. The torn edges of structure are examined for plastic yielding and paint cracking or other evidence that identifies the fracture as a "single overload type failure" and for evidence of fatigue. If one of the rare in-flight structural failures is suspected, the examination for fatigue must be very thorough. The direction in which mating pieces or sections of the aircraft break off relative to each other is determined to the extent possible. This is necessary to separate in-flight from ground-impact damage and to analyze the sequence of breakup.

All hydraulic cylinders and mechanical actuators are examined to determine their positions at the time of breakup so that the position of the flaps, landing gears, flight controls, trim tabs, etc., can be determined. All elements of the control system, including the mechanical portion and its supports, the boost units, and surfaces, are traced, and failures are examined to determine that the flight controls were or were not functional at the time of impact. This examination accounts for all balance weights and includes a close look at the hinge bearings, tab linkages, and other areas susceptible to damage from aerodynamic flutter.

The heaters, air-conditioning system, fire-extinguishing system, cargo, and other sources of toxic gases are examined for damage or evidence of malfunction. All external openings in the aircraft (doors, emergency exits, inspection openings, cargo hatches, cowl flaps, etc.) must be accounted for, even though extensive burnout may make this task extremely difficult.

The angle at which the aircraft contacted the ground and the attitude that it was in at contact are established by the examination of the structure and by the relative location of gouges and ground-impact marks shown on the wreckage-distribution chart. In addition to the considerations previously listed, all areas of the aircraft having known records of service difficulty are closely examined. All detail findings are accurately documented in writing and thoroughly photographed. The importance of good-detail photographic coverage cannot be overstressed.

9.9b Operations Group. An operations group is set up to develop all the facts relating to the history of the flight, the history of the crew, and the probable flight path. The history of flight includes flight planning and dispatching, weather en route, previous stops, radio contacts, operational weight, c.g. location at previous takeoff, and many other similar items. All of this evidence is carefully studied and evaluated to determine if any factor might have a bearing on the accident.

Another interesting phase which is normally handled by the operations group is radio-transmission analysis. Since most communications between aircraft and FAA radio facilities are recorded, the group makes an exhaustive study of this information. Occasionally, a crew member gets off part of an emergency message, but because of the excitement the message is not readable. Sometimes the message is garbled because of damaged equipment. In such cases, the message is transcribed onto tape and short groups of words are made into a continuous loop so that the message is repeated every few seconds. With this setup special sound equipment, capable of filtering out static and certain frequency bands, will generally help investigators to hear some words and associate sound or syllable patterns on others so that some intelligence can be obtained from the message.

A new device which should prove very valuable is a cockpit voice recorder which preserves all conversations on a continuous loop of magnetic tape. After 30 minutes, it starts to erase the old information and continues to record new conversations. The recorder shuts off on impact and the last half hour's conversations remain in a crash-proof housing to help investigators reconstruct the events leading to the accident.

All commercial aircraft operating above 25,000 ft are required to carry a flight recorder. This instrument records air speed, altitude, time, vertical acceleration, and heading information. Unfortunately, problems have arisen in the use of these recorders. In one instance, the aluminum foil on which the information is recorded was improperly installed and the resulting data were not conclusive. In two other cases, the recorder had run out of foil prior to the crash.

9.9c Witness Study. The main task of the witness group is to contact and interrogate all witnesses to the accident in order to uncover facts which will assist the other committees in their work. Statements made by witnesses soon after the accident are very often more accurate than those obtained later. Therefore, every attempt is made to secure witness statements as soon as possible. The technique used is to return the witnesses to the exact points from which they observed the aircraft and have them associate the aircraft's flight path with adjacent buildings, mountains, trees, etc. Then by measuring the related angles and distances it is often possible to determine the aircraft heading and altitude.

9.9d Radio and Electrical Equipment Study. The radio and electrical group checks the settings and indications of all instruments concerned with navigation of the aircraft. These data are recorded and evaluated, and, whenever possible, the instruments are given functional tests. As another routine procedure following an accident, the maintenance records of the electrical equipment are reviewed to uncover any previous difficulties which might reflect the degree of reliance placed on the equipment by the crew at the time of the accident.

When it is suspected that the failure or malfunctioning of an electrical component may have contributed to an accident, the various units are segregated and inspected for evidence of malfunction or failure. One of the first steps taken is to determine whether electrical power was available prior to impact. A technique that is often employed in this regard is the disassembly of rotary electrical machines such as inverters and motors and the examination of their armatures for evidence of rotational scoring. In cases of in-flight fire, particularly, suspicion is directed toward the electrical system until it can be eliminated as being a factor. Evidence of heat or burning in areas adjacent to electrical components or wiring is an indication of electrical system trouble. When the wiring is suspected, the terminal connections, routing, clamping, protection against chafing, and the capacity and condition of fuses and circuit breakers are all checked to determine whether they were contributing factors. Another technique used here is to lay out wire bundles and examine them for continuity of fire pattern in adjacent bundles.

9.9e Power-plant Study. In addition to the engine and propeller installations, the power-plant group is responsible for investigating all possible malfunctions or failures of related essential power-plant systems such as fuel, oil, and power-plant controls. A number of checks are made as soon as possible because they may direct attention to trouble areas. Propeller blade angle is one of the most important, since it is indicative of the power being developed by the engine. Blade angles at impact can usually be determined within an accuracy of $\pm 2°$ from a study of the impact marks found on blade shim plates, gears, or any part of the blade assembly that can be related to a nonmoving part.

The speed of the engine can be determined by study of the propeller speed control. Given the blade angle, speed of the aircraft, and propeller governor setting, the manufacturer of the engine can furnish a fairly reliable figure for the power being developed by the engine at impact. The speed of the aircraft can be calculated easily by applying the engine-propeller gear ratio to the governor setting to get propeller rpm and then multiplying the distance between propeller slashes in the ground by the number of blades by the rpm.

Other items that are checked early include all engine oil screens for free metal, spark plugs, and the interiors of the combustion chambers. This latter check is usually accomplished with a boroscope, which is indispensable for this purpose. As a matter of course, samples of fuel are obtained when possible for later analysis. Following the initial examination at the scene, if circumstances permit, all power-plant parts are removed to a shop for more extensive disassembly and examination. Sometimes the engines are in operable condition, and test-stand runs are conducted to establish their integrity. Review of the power-plant maintenance records and the aircraft logs is another task accomplished by the power-plant group.

9.9f Structural Mock-up. The use of a structural mock-up of specific sections of an aircraft is frequently very effective in finding an accident cause or in demonstrating beyond any doubt that a suspected occurrence did, in fact, happen. This is particularly true in the area of in-flight fires, midair collisions, and explosions.

Mock-up work is slow and tedious, but various procedures can be used to speed up the identification of the broken, battered, and crumpled parts. Part numbers can be found on many pieces, and these numbers can be checked against the parts catalog. In many instances, broken pieces can be matched with other identified items. In the beginning, wing parts are separated into one pile, fuselage in another, tail in still another, etc. The identification of parts and reconstruction of components are then started. Control-system parts and cable runs are generally laid out separately from the other reconstructions to facilitate study.

9.9g Accident Prevention. Aircraft failure analysis is still very much an art requiring the efforts of trained, experienced personnel with an inclination toward this work. Patience, perseverance, common sense, and logic combined with a thorough engineering background are the most effective tools available to the investigator. In recent years, aircraft have increased tremendously in range and weight, and this growth has brought with it greater design complexity. The higher operating speeds of these newer aircraft have generally resulted in more extensive breakup at impact. These factors present a terrific challenge to determine the cause of in-flight failures and to recommend measures for prevention of future accidents.

9.10 FUTURE IMPROVEMENTS

"The lessons learned on one system must be applied to improving reliability of future systems."

—Dr. Leslie W. Ball

When starting a new program, it is very valuable to prepare a malfunction-analysis prediction. Information is obtained from previous programs which incorporated similar hardware. The data should be presented in book form, and copies should be provided to each design group supervisor. The book should contain a tabular list of all potential modes of failure at the first level of assembly within each system. The causes and consequences of malfunctions are also indicated.

A rapid transition is taking place from the standard wired and printed-board circuitry to the integrated microcircuit. Microcircuits present problems for the failure-analysis laboratory, since all disassembly and troubleshooting must be done under a microscope. Microscopic inspection can sometimes detect the mechanism of failure, but this must usually be followed by microsectioning. New techniques such as infrared and electron-beam scanning will have to be developed to assist in this process.

Analysis of space-vehicle in-flight malfunctions will present a major challenge for the basic reason that it will be difficult to obtain the failed hardware. It would appear that analysis would be possible only if the failure were noncatastrophic or if emergency landings could be made on the closest terrain.

In summary, we have seen that laboratory troubleshooting is a straightforward operation which can be very effective when properly supported. If more contractors and government agencies would place emphasis on failure analysis, reliability in the aerospace industry would be greatly improved.

Acknowledgement. The technical comments and assistance provided by the Bureau of Safety, Civil Aeronautics Board, Mr. E. J. Nucci, Department of Defense, and Mr. R. R. Landers, Thompson Ramo Wooldridge, are greatly appreciated.

Bibliography

Auer, R. L., "Transistor Process Control through Failure Analysis," *Trans.* 1963 *Western Region Quality Control Conf.*, pp. 161–168.

Bevington, J. R., "Failure Analysis Program at the Component Level," *Semiconductor Reliability*, vol. 2, pp. 111–117.

Kimball, E., "Failure Analysis," *Proc. Eighth Natl. Symp. Reliability Quality Control*, pp. 117–128.

———, "Failure Analysis Laboratories," *Trans. Fifth Ann. West Coast Reliability Symp., ASQC*, 1964.

———, "Is Hindsight Becoming 20–20?" *Evaluation Eng.*, pp. 15–17, September-October, 1962.

———, "Product Improvement," *Proc. Sixth Natl. Conv. Mil. Electron.*, pp. 319–326.

———, "Weapon System Failure Analysis," *Quality Assurance*, pp. 40–44, November, 1962.

Landers, R. R., *Reliability and Product Assurance*, Prentice-Hall, Inc., Englewood Cliffs, N.J., 1963, pp. 418–427.

McWhorter, J. L., "Airplane Accident Investigation," *Tennessee Law Rev.*, vol. 28, no. 2, pp. 122–152.

Nelson, R. S., "Authority and Responsibility of the Central Reliability Group," master's thesis, San Fernando Valley State College, pp. 70–73.

Packard, C. C., "Reducing Reliability Risks through Failure Analysis," *Semiconductor Reliability*, vol. 2, pp. 91–99.

Southern, P. F., "A Practical Defect Analysis Program," *Semiconductor Reliability*, vol. 2, pp. 106–110.

Section 10

ENGINEERING DESIGN AND DEVELOPMENT FOR RELIABLE SYSTEMS

WALTER L. HURD, JR.

RELIABILITY AND QUALITY ENGINEERING DIVISION MANAGER
MISSILE SYSTEM, LOCKHEED MISSILE AND SPACE COMPANY
SUNNYVALE, CALIFORNIA

Mr. Hurd, a native of Montana, is a graduate of Morningside College, Sioux City, Iowa, and has taken post-graduate work at Hastings College, San Mateo College, and Stanford University.

Mr. Hurd served in the Army Air Force throughout World War II and is a Brigadier General in the Air Force Reserve. After the war, he became Vice-President and General Operations Manager for the Philippine Air Lines. Next, he became Quality Control Manager for the National Seal Division of Federal-Mogul-Bower Bearings, Inc. Since joining Lockheed Missile and Space Company, Mr. Hurd has been closely associated with the quality engineering and reliability programs for the Polaris Missile System.

He is a Fellow in the American Society for Quality Control and has served as chairman of the San Francisco–Bay Area section of that society. Mr. Hurd is also a Fellow of the American Association for the Advancement of Science and a member of the American Institute of Aeronautics and Astronautics. He served as a member of the Navy Special Projects Office's Committee on Product Quality which evaluated the reliability and quality program of the Fleet Ballistic Missile System in 1961. He served as a consultant to the Air Force as a member of the Weapon System Effectiveness Industry Advisory Committee (WSEIAC) and also as a member of the Council of Defense and Space Industries group which conducted a study of the National Aeronautics and Space Administration's quality program. Mr. Hurd has authored numerous technical papers and reports dealing with reliability and engineering design functions.

ENGINEERING DESIGN AND DEVELOPMENT FOR RELIABLE SYSTEMS

Walter L. Hurd, Jr.

CONTENTS

In this section we shall consider the primary role of the design organization in creating and proving designs with the required inherent reliability. Since the obvious is often overlooked and forgotten, let us start with the basic understanding that:

1. The designer creates the design and is responsible for all its characteristics including reliability.
2. Each hardware design has an inherent reliability potential.
3. The inherent reliability potential of the design is approached when hardware is produced to the design requirements, but the full inherent reliability is rarely achieved or maintained because production, handling, storage, and use all tend toward compromises which result in an achieved or actual reliability that is less than the inherent reliability.
4. Any complex product must start with a design containing a very high inherent reliability so that there will be a reasonable possibility of placing hardware with the desired or required actual reliability in the users' hands.
5. Reliability personnel must serve as an objective, independent check and balance on the designer, but reliability personnel must not be permitted to usurp the designer's responsibility for reliability.

Obviously, the best interests of design reliability are served by a strong, competent design function teamed with a strong, competent reliability function. If, however, an organization should be faced with the choice between a strong, competent design function and a weak reliability function *or* a weak design function and a strong, competent reliability function, then the better interests of design reliability would be served by the first condition.

The attainment of a high inherent reliability in a complex system can be very expensive in both time and money, although this expense is usually more than made up by decreased production and field problems in achieving and maintaining the required actual reliability. In fact, if the design does not possess a high inherent reliability, it may be impossible to reach the required actual reliability figure in the field, no matter how much time and money are poured into the effort.

Let us consider the costs of unreliability. A 10 per cent decrease in achieved reliability in an operational missile system would require at least 10 per cent more actual missiles to give the same degree of target coverage. These missiles would require additional launching sites (silos or submarines), checkout equipment, launching equipment, crews, and facilities. The additional personnel and equipment required would be both costly and time consuming.

The establishment of reliability goals thus becomes a matter of balancing the cost of reliability against the cost of unreliability. This is obviously a decision for the customer to make, since, in the end, he pays all costs. We shall in this section develop a logical approach to the methods for designing and developing reliable systems. In addition, we shall explore the methods, procedures, and practices involved in the design-reliability interface.

10.1 ROLE OF THE DESIGN FUNCTION

The importance of the design reliability function in a company is directly proportional to the importance of the design function. A strong, competent design function is essential in areas of rapid technological advances such as the aerospace area in general and the electronics area in particular. Within companies operating in these fields, the design organization is usually considered a line organization.

The design function is also of great importance to producers of consumer hard goods (such as automobiles, office equipment, and household appliances), to machine-tool producers, and in many other such areas. Design organizations are usually strong staff organizations in companies producing these products. Although of lesser importance in companies producing simple products or products of stable, proven design, the design function is an important one in all production industries.

As the rate and pace of technological advance increase, more and more individual companies and whole industries have been and are being forced to abandon complacency and to look to the future. Action, in such cases, usually starts with increased emphasis on the research and development function, with new design concepts as the output. The present trend is certainly toward increased emphasis on and importance of the design function. Some definitions are perhaps in order at this point:

Research. That scientific and/or engineering function of a company which attempts to develop new product or major new design approaches to existing product lines.

Development. That engineering function of a company which expands and completes a research concept. Sometimes development is paired with research as R&D, and in other situations it is paired with design engineering as project design or design development.

Design. That engineering function which applies established design approaches to the solution of specific problems. This is the function which produces the drawings and specifications for experimental and tactical hardware used in a specific program. In some companies the design function is known as product engineering. It is with the design function that the reliability function works most closely.

The design function within a company has certain responsibilities to the organization management. Working in the assigned product areas, the design function must create designs which are:

Functional. The design must be one which, when translated into hardware, will satisfactorily perform the functions for which it was designed.

Reliable. The design, when translated into hardware, must not only function when new, under favorable conditions, but also contain the capacity to continue to meet the full range of functional requirements over the required period of time throughout the specified range of environments. If maintenance is anticipated, the design must provide adequately for maintainability.

Producible. The design must be such that it can be economically produced by the production facilities and suppliers intended to be used.

Timely. The design disclosure must be completed and released within the established time schedule. The time schedule itself may be determined by contract, by model-change deadlines, or by the activities of competitors.

Competitive. The design must be salable. Factors involved in salability vary widely, but they include, in addition to those listed above, competitive cost, style, appearance, and many other factors.

Whenever possible, the designer is expected to use proven design techniques. When design objectives cannot be met by using the proven and familiar design methods, the designer is expected to adapt his methods, borrow design techniques from other industries, or use some of the new state-of-the-art materials and processes which are available. Since designers are usually, by disposition, creative it is often difficult for them to resist trying something new even though a technique of proven effectiveness and reliability is available. Designers are well known for their receptiveness to the efforts of parts and package suppliers' sales engineers, who are selling the outstanding merits of their new product. A major responsibility of the design management is the establishment of a system which makes it appreciably easier for a designer to use proven design than to use unproven design.

Since it is often impossible to meet all design objectives to the maximum desired extent, the designer is frequently required to trade off between his objectives. By requiring unusually tight tolerances or by specifying an exotic material, he may increase reliability at the expense of producibility. By not fully testing the ability of his design to function under the worst combinations of environment and aging, the designer may take a chance on lowered reliability so that he can release his design disclosure on schedule. Some of these compromises and trade-offs are nearly unavoidable, and it is the design function which has both the information and the responsibility to make the required decisions in these cases. However, both the fact that trade-offs have taken place and the reasons for the decisions made must be fully disclosed by the design function to the reliability function and to general management.

10.2 ROLE OF THE RELIABILITY FUNCTION

If a design is to have a high inherent reliability, provisions for reliability must be made within the design concept and must continue throughout development of the design to design completion. Since marginal reliability is often difficult to detect before the design effort is completed (this is specially true of complex, sophisticated designs), it is necessary that the reliability function work throughout the creation of the design as an independent, objective check and balance on the design function in respect to achievement of the design reliability goals.

There has been and is some confusion between feasibility and reliability during the early stages of design. Developing a design which will function more than once to establish the feasibility of a design concept is principally a design responsibility with only incidental assistance from the reliability function. While performing design analysis or design review (see Sec. 10.10), reliability personnel may discover and require correction of design features, errors, or omissions which affect feasibility, but reliability review of design is not primarily for this purpose. Reliability is concerned with the additional design requirements necessary to assure that a feasible design will continue to perform satisfactorily over the specified range of time, environments, and other operating conditions.

The reliability function works with the design function in several ways to achieve its objective. Reliability acts at various times as a helper, as a conscience, and as an inspector. As a helper, reliability performs certain analytical and statistical services for the design organization. These services include collection, analysis, and feedback of data on development hardware and on prior-model hardware production and testing.

Data functions are discussed in considerable detail in Secs. 6 and 7. In general, reliability helps design by predicting and measuring inherent reliability during the various design stages.

Reliability serves as a conscience to the design function by closely monitoring and reporting design progress toward the inherent reliability goal. In addition, all trade-offs affecting reliability are very closely examined. The techniques used in monitoring the design function are described in Sec. 10.9.

The reliability function gets its teeth when it inspects the design output and must approve before the design effort can progress. Some of the checkpoints for such enforcement of reliability requirements are the reliability approval of parts which can be used by the designer, reliability approval of actual usage of the approved parts, reliability approval of design reviews and, finally, reliability signature approval of the design-disclosure documents (drawings, specifications, and procedures).

The reliability function may also include coordination of design test programs, direction of independent reliability test programs, actual conduct of all testing, identification and establishment of control systems for portions of the design which have special age or operational limitations, preparation of reliability specifications applicable to suppliers, and the imposition of reliability requirements on suppliers through review and approval of procurement documents. These functions are discussed in more detail in Secs. 10.11 and 10.12.

10.3 WHEN AND WHY A RELIABILITY FUNCTION IS NEEDED

Generally speaking, the more difficult the design assignment, the larger the reliability effort required. The design problems encountered in a Telestar satellite, a major missile system, or a complex worldwide communications network require major design and reliability efforts. On the other hand, if the design is well within the existing state of the art, is simple, and has ample space, weight, and design-time allowances, then a relatively small reliability effort may be adequate.

It is evident that a major reliability effort (major in the sense of being a substantial fraction of the design effort) is required under the following circumstances:

1. In the design of any very complex equipment.
2. In the design of equipment with very high reliability requirements, particularly when the designer is working within severe space and weight limitations.
3. When the designer is operating under rigorous time constraints, particularly in time-compressed aerospace-type programs where production must start before the design is completed.

The general rule may be made that the more constraints on a designer and the tighter the constraints, the greater the required reliability program. One of the major reasons for this relationship is that, under the pressure of these constraints, the designer may intentionally or unintentionally neglect the reliability requirement. A strong reliability effort is needed both to assist and to check on the designer in matters affecting reliability.

In developing his designs the designer must work out compromises between his various requirements. The penalties for a designer not fully meeting his performance, schedule, cost, producibility, and other goals are much more immediate and certain than are the penalties for not fully meeting reliability goals *unless* a strong, independent reliability function is present to call immediate and forceful attention to any deficiencies in the design provisions for reliability. The presence of such a reliability function will help assure that full consideration is given to reliability requirements. It still may be necessary, under certain conditions, to trade off a reliability requirement for performance or for some other design consideration, but such a trade off must be made openly and with full knowledge of possible consequences.

Very few designers deliberately skimp on provisions for attaining the full required reliability in their designs. The dangers are, rather, those of oversight, lack of specific knowledge, and rationalization. Let us examine these areas in detail.

Oversight. These are the cases when the designer fails to take care of one of those innumerable details which make up the completed design. The designer knows that a special finish is required in a certain area but fails to call it out on his drawing. As a result of this omission, a general drawing note or specification becomes applicable. If this oversight is not caught, a substantial portion of the product may fail at some fraction of the required life.

Lack of specific knowledge. No designer can know everything about everything connected with his design, nor does he have the time to verify every detail. He does what he can, checks where he believes it necessary, and calls in the experts in certain highly specialized areas such as stress and thermal. The designer may, for example, specify a short-life rubber compound which was the best available for the purpose the last time he had need for such a part. In the meantime, superior materials that will function as well or better and that have a life well in excess of the design requirements may have become available.

Rationalization. This is a dangerous but wholly human practice. The designer is pressed for time; he honestly believes his design will meet all of the design requirements including the reliability requirements, but an additional series of tests should be run to be absolutely certain. To wait for test results will put the design behind schedule. It is easy for the designer to work himself into a frame of mind that the tests are not really necessary. This same practice of rationalization includes explaining away test failures with such comments as, "It was a testing error" or "It was an early design" or "The real environments will never be nearly that rigorous anyway."

When the consequences of unreliability are extremely dangerous or expensive in dollars or time or jeopardize national security, then reliability requirements become high. When high reliability is necessary, it cannot be left to good intentions or to chance. There must be an independent check and balance of every operation (including those operations which the reliability-quality function itself may perform, such as the writing of test procedures) and continued, everlastingly painstaking attention to details. No organization and no person can be considered so good or so omnipotent that its or his output need not receive a searching independent analysis.

10.4 NATURE OF DESIGN AND DEVELOPMENT TASKS

The words "design engineering" cover a wide range of activities with the common purpose of producing a design-disclosure package (drawings, specifications, procedures, and test-inspection plans) which contains the information required for competent production facilities to produce hardware which will meet or exceed all of the design objectives. It is not always realized that the end product of engineering design is the design-disclosure package itself. All other design efforts, no matter how difficult or spectacular, are simply steps in the production and proving of the design-disclosure package.

Depending upon the nature of the product, the use of the product, the demands of the customer, and the practices of the company, the design-disclosure package may be as simple as a sketch or as complex as a formal system of interlocking military-authenticated drawings, specifications, procedures, factory-test plans, factory-inspection plans, and field operation instructions with provisions for a formal feedback of field troubles, failures, and replacement information. The design-disclosure package is discussed in detail in Sec. 10.5.

To accomplish a design assignment, design management must receive or create clear-cut design objectives. These design objectives may be imposed by a customer or by the general management, or they may be developed within the design organization for submission to and acceptance (with or without modification) by general management. Regardless of origin, the design objectives must exist. Design management then subdivides the accomplishment of the objectives down within the design organization through formal or informal job assignments. The assignment of design work, allocation of resources to accomplish the work, coordination and integration of the efforts of the various design groups, and statusing and assessment of the design output against design objectives are the responsibility of design engineering management.

The individual designers, in working toward their assigned design objectives, draw upon their professional education and their experience, upon company-provided design policy and standards (such as design handbooks, material and process handbooks, reliability approved parts, and circuits handbooks), and upon the services of specialized design groups such as stress analysis, thermal analysis, and materials and processes groups. In many cases the designer will need to work with hardware in breadboard and developmental experimentation and testing. He must also plan semiformal and formal test programs to be carried out under ambient and specification-limit environmental conditions. (See Sec. 8.)

The designer's approach to achieving his assigned design objectives is basically the same whether he is designing a component part, a package, a subsystem, a system, or an integrated system. The difference in the design approach is only in the degree with which it is carried out. An understanding of the basic design actions after receiving the design objectives and assignment is of value to the reliability engineer. These actions are:

1. Create or select one or more design concepts which may achieve the design objective.
2. Analyze the feasibility of the various possible design concepts. This feasibility analysis may be made by use of experience (personal or acquired), by simulation and theoretical analysis, by breadboard experimentation and testing, by creating mock-ups, or by combinations of the foregoing.
3. Selection of a design concept, based upon analysis of possible choices and their respective likelihood of meeting all of the design objectives assigned to him. This selection usually requires the approval of the designer's supervision. This is a key reliability design review and analysis point.
4. Further apportionment, where necessary, of reliability requirements throughout the design down to and including the piece-part level.
5. Preparation of preliminary drawings and preliminary specifications (including, where necessary, dimensional and functional interface coordination and correlation requirements). Completion of these preliminary drawings and specifications is another key reliability design review and analysis point.
6. Release of preliminary drawings and specifications for production and procurement of development hardware to be used for feasibility and evaluation testing of the hardware alone and with other elements of the system.
7. Plan qualification test requirements and participate in planning production test and inspection requirements.
8. Participate in the direction of prototype and qualification testing, taking whatever corrective design action is found to be necessary.
9. Prepare the design portion of the complete, final design-disclosure package. It is at this point that a final reliability design review is accomplished.
10. Review and approve those portions of the design-disclosure package which are not created by the design function.
11. Release the completed design, after securing reliability and other required approvals, for factory use or for customer disposition as applicable.

The designer has several continuing tasks after the design is released. Two of these functions, design-configuration control and design-change control, are closely related. All design-change requests must be fully and carefully considered for impact on inherent reliability as well as for other impacts. As the design approaches completion, design-change control must come under the direct control of top management, because it is difficult to stop most design organizations from making changes. At some point a design freeze, after which no improvement-type change is permitted, is desirable. After the design freeze, only absolutely mandatory changes or very minor record changes (correction of errors) are usually permitted. (See Sec. 13.)

Design-configuration control relates to the control of requirements for a specific serial number, production block, or model type of hardware. If elements of the design are to be serialized when they are produced, these elements must be identified. If the drawing-change letter is considered a part of the part number, it must be so stated. Design-configuration control must supply the inspection-function listings of the as-designed part numbers along with the design requirements for record keeping of the as-built configuration. Any configuration departure of the as-built hardware from the as-designed hardware must be detected, recorded, and passed (as with any other hardware discrepancy) through material review for joint design-reliability (and customer, if required) disposition.

Thorough understanding and appreciation of the designer's problems and methods of solution are essential to the reliability engineer. In his actions to assure the required reliability the reliability engineer must be understanding of the designer's problems but unrelenting in insistence upon attainment of the reliability goals.

10.5 THE DESIGN-DISCLOSURE PACKAGE

Let us consider what makes up a complete design-disclosure package for a major complex weapon system. Such a system has a customer, has one to ten prime contractors who deal directly with the customer, has several major subcontractors who deal directly with each of the prime contractors, and finally, has several tiers of suppliers to both the prime contractors and the subcontractors. In general, the prime contractors and subcontractors supply both designs and hardware. The suppliers in

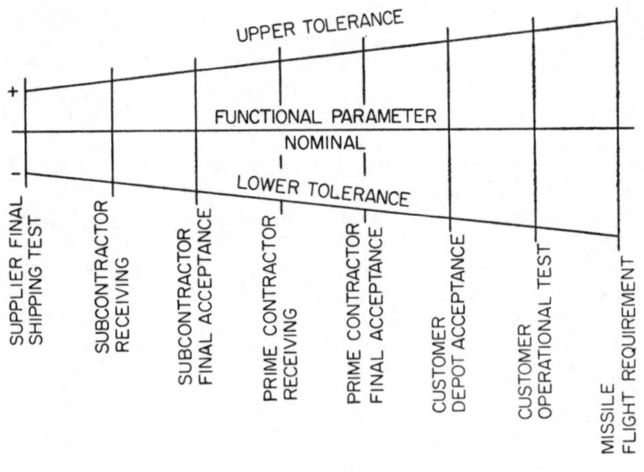

FIG. 10.1 Tolerance funnel.

the various tiers may supply hardware to their immediate customer's design, or they may supply hardware to their own design.

It is obvious that the design-disclosure package for any one major element of a weapon system has interfaces with the other major elements. Within each element, a tremendous number of interfaces exist. If the program is being carried out under accelerated conditions when design development and tactical production overlap to some degree, the coordination and correlation problems become of critical importance. The design-disclosure package is based upon the customer's operational and logistic requirements and will usually include the following in some form:

Basic specifications. A basic design specification which sets forth performance requirements, specifies environmental conditions, establishes reliability goals, and calls out basic logistic requirements.

Drawings. These include coordination drawings (interfaces between prime contractors), correlation drawings (interfaces between design elements within a prime contractor's responsibility), production drawings, procurement drawings, and drawings of special test equipment.

Parameters document(s). These documents detail the functional parameters with their tolerances starting at the operational-use end and working backward to the basic producer. Tolerances are tightened at each major step so that room is provided

EXAMPLE

OD XXXXX
PART IV

PARAMETERS DOCUMENT SHEET

REFERENCE TEST PLAN
TEST POINT NUMBER
XXX

(CONTINUATION SHEET)

SHEET 3 OF 5 REV.

TP ITEM NO	ATTRIBUTE	PARAMETER	TOL.
	3.3 Rocket Motor shall be safe to handle and shall be capable of operation following this test.	Shall be safe to handle & capable of operation (para.3, MIL-STD-353)	
4.	TEMPERATURE AND HUMIDITY TEST		
	4.1 Expose Rocket Motor to a 28 day Temperature and Humidity test in accordance with MIL-STD-354 as follows:		
	4.1.1 Subject Rocket Motor to two 14 day JAN Temperature and Humidity Cycles.		
	4.1.2 Cycle Rocket Motor between minus 20 degrees F and plus 145 degrees F. Temperature range with relative humidity of 95 percent.		
	4.2 Rocket Motors shall be safe to handle and shall be capable of operation following this test.	Shall be safe to handle & capable of operation (para. 3, MIL-STD-354)	
5.	PRESSURE TEST		
	5.1 Subject Rocket Motor to a cycle of pressure change and stabilization as follows:		
	5.1.1 Place Rocket Motor in a suitable pressure chamber.		
	5.1.2 Increase pressure from laboratory ambient to 65 PSIA at a rate of 2.1 psi per second.	2.1 psi/sec	±0.30 psi
	5.1.3 Maintain 65 PSIA pressure for not less than 75 minutes.		
	5.1.4 Reduce pressure from 65 PSIA to laboratory ambient at a rate of 21 psi per second	21 psi/sec	±1 psi
	5.2 Measure internal pressure rise of rocket motor.	Shall not exceed 0.5 psi during test	

Fig. 10.2 Sample of parameters-document sheet.

for some functional parameter drift or degradation with time and transportation. These adjusted tolerances are called "funnels of tolerance," with the small end of the funnel at the suppliers and the large end of the funnel at the operational users. Sometimes it is not possible to provide a normal slope to the funnel, but the slope should never be permitted to become negative (Fig. 10.1). Figure 10.2 shows one page of a set of parameter documents.

Specifications. These include internal-design specifications, procurement-design

2

Page from a Design Specification WS 2223A

3.5.3.1 Torque motor.- The torque motor shall operate when supplied with current from a driving amplifier as shown in figure 5. Excitation of the torque motor without pressure having applied to the assembly shall cause no deterioration or malfunction of the assembly during subsequent normal operation.

3.5.3.2 Resistance of motor coils.- The resistance of each motor coil shall be 1000 plus or minus 400 ohms.

3.5.3.3 Inductance of motor coils.- The inductance of each motor coil, when measured at the injector valve connector shall not be greater than 2.6 henries at 1000 plus or minus 100 cycles per second (cps).

3.5.3.4 Proof current.- Torque motor coils shall withstand, for a period of not less than one minute, a proof current of 43 milliamperes (ma) without deterioration or malfunction.

3.5.3.5 Dielectric strength.- The torque motor coils and the feedback transducer coils shall be capable of withstanding without damage, 500 volts of alternating current (ac) at commercial frequency, between the terminals of the coil and the body of the assembly for a period of not less than 10 seconds.

3.5.3.6 Insulation resistance.- Resistance of the insulation of the torque motor and feedback coils shall not be less than 10 megohms when a potential of 500 volts direct current (dc) is applied between the leads of the coil and the body of the assembly for a period of not less than one minute.

3.5.3.7 Continuity.- Electrical and shielding continuity shall be in accordance with the schematic shown on the applicable drawing in table I.

3.5.3.8 Polarity.- When the average current at pin F is greater than the current at pin H, fluid shall flow from the number 2 injector and the feedback signal from pin A shall be in-phase with respect to pin D. When the average current at pin H is greater than the current at pin F, fluid shall flow from the number 1 injector and the feedback signal from pin A shall be 180 degrees out-of-phase with respect to pin D.

3.5.3.9 Electrical inputs

3.4.3.9.1 Excitation voltage.- Feedback excitation voltage at the connector shall be 28.75 volts rms, plus or minus five percent with a frequency of 800 plus or minus four cps and a harmonic distortion of not greater than two percent.

3.5.3.9.2 Feedback power requirements.- The feedback power requirements of the assembly shall be 0.88 plus or minus 0.22 watt (real), and 2.34 plus or minus 0.51 reactive volt-amperes, lagging, when the feedback output is loaded with 10,000 plus or minus 300 ohms.

3.5.3.10 Electrical outputs.

3.5.3.10.1 Quadrature and harmonic distortion.- The quadrature and harmonic distortion of the feedback signal with both injectors closed shall be not greater than 0.53 percent of the specified excitation voltage. The quadrature and harmonic distortion, with each transducer operating singly at full scale output shall be not greater than 1.76 percent of the specified voltage.

5

FIG. 10.3 Sample page of a design specification.

specifications, material specifications, and production-process specifications. Design specifications must be compatible with the parameters document. Design specifications must also include a section on testing requirements. See Fig. 10.3 for a sample page from a design specification.

Test and inspection plans, factory, field, and formal test programs. The test plans call out the specific functional and other (pressure tests, etc.) test parameters to be tested.

TEST PLAN SHEET
(CONTINUATION SHEET)

OD XXXXX
PART I - B

TEST POINT	REV
XXX	B
PAGE 3 OF 9	

OCD CAT	ITEM NO.	ATTRIBUTE TESTED AND TEST CONDITION
	6.	RANDOM VIBRATION
		6.1 Vibration Environment (not to exceed 10 minutes total per axis): The assembly shall be subjected to random vibration along each of three mutually perpendicular axes for one minute per axis. The level of vibration shall be 13.7 grms from 20 to 2000 cps with the spectral density contained within the limits of 0.14 g^2/cps and 0.047 g^2/cps.
		6.2 Assembly shall be operated in a closed loop system conforming to Figure 1.
		6.3 Closed Loop Characteristics During Vibration
		6.3.1 Inputs:
		6.3.1.1 28.75 volts RMS alternating current at a frequency of 800 cycles per second to the feedback transducer primary coils.
		6.3.1.2 10,000 ± 300 ohms load across the feedback transducer output coils.
		6.3.1.3 Command input of a 100 percent amplitude modulated 800 cps carrier. Modulating signal shall be a triangular wave form at a frequency of 0.025 cps with a peak to peak voltage of 8 volts.
		6.3.1.4 Pressure of 750 psig at the assembly inlet port.
		6.3.2 Measure:
		6.3.2.1 Hysteresis
		6.3.2.2 Deadband
		6.3.2.3 Null current
		6.4 Feedback
		6.4.1 Inputs:
		6.4.1.1 28.75 volts RMS alternating current at a frequency of 800 cycles per second to the feedback transducer primary coils.
		6.4.1.2 10,000 ± 300 ohms load across the feedback transducer output coils.

FIG. 10.4 Example of a test-plan sheet.

Test plans must be in agreement with the parameters document and the specifications, the order of testing, the testing location, the test equipment to be used, and the test procedures to be used. The inspection plans call out the required minimum dimensional inspection and nondestructive testing required along with process control requirements. Inspection equipment and location are specified. See Fig. 10.4 for an example of a test plan. (See Sec. 8.)

Test procedures. These are the detailed, step-by-step instructions to the technicians or inspectors who will perform the functional test. Test-equipment capability is described in some detail, including accuracies. A special class of test procedures covers

Page 23 4

Page from Acceptance Test Procedure ATP-OD 22708

7.5.14 Adjust COMMAND INPUTS MANUAL control until VALVE CURRENT meter indicates a minimum current.

7.5.15 Position IN PHASE -180 switch to oposite position and adjust COMMAND INPUTS MANUAL control as necessary for a minimum indication on the VALVE CURRENT meter.

7.5.16 Position CONTROL PANEL VOLTAGE SELECTION switch to FEEDBACK. DVM shall indicate a zero voltage.

7.5.17 Position CONTROL PANEL FB switch to REVERSE. DVM shall indicate a zero voltage.

NOTE: If DVM should indicate a voltage in para. 7.5.16 or 7.5.17, it shall be necessary to repeat para. 7.5.14 and 7.5.15.

7.5.18 Position CONTROL PANEL switches as follows:

Switch	Position
IN PHASE-180	OFF
VOLTAGE SELECTION	OFF
COMMAND INPUT	STANDBY
FB	NORMAL

7.5.19 Position Freon Accumulator shut-off valve and REG SHUT OFF valve of Test Unit fully closed.

7.5.20 Position V3 MAIN PUMP BYPASS VALVE of Pump Unit to OPEN.

7.5.21 Position 2-way valve on Test Unit manifold fully open.

7.5.22 Position Freon Accumulator shut-off valve of Test Unit fully open and increase ACCUMULATOR PRECHG SUPPLY REGULATOR at a maximum rate of 500 psig/sec to apply a minimum of 1500 psig to valve inlet port for a minimum of one minute.

OCD 112 7.5.22.1 Verify no evidence of permanent deformation, failure or leakage, except at the vent port, and discharge ports, where leakage ia allowable. Record results on Test Data Sheet.

7.5.23 Position and adjust valves and regulators of Test Unit as follows:

Valve/Regulator	Position/Adjust
ACCUMULATOR PRECHG SUPPLY valve	Fully closed
Freon Accumulator shut-off	Fully closed
ACCUMULATOR PRECHG SUPPLY REGULATOR	Fully decreased
2-way valve on manifold	Closed

Fig. 10.5 Sample page from acceptance test procedure.

the proofing testing of the test-station test equipment. Test procedures must be compatible with the test plan, with the specifications, and with the test-equipment error analyses. Figure 10.5 shows a sample page from an acceptance test procedure. (See Secs. 8 and 15.)

Inspection instructions. These are the detailed instructions to the inspectors who will perform the minimum dimensional inspection and nondestructive testing called out in the inspection plan. These inspection instructions must agree with the inspec-

EXAMPLE 9

Page from Equipment Test Procedure ETP-OD 21681
Page 10

9.1.4 (continued)

RATIO BRIDGE AC POWER ON
800 CPS POWER SUPPLY LINE-ON
Visicorder POWER ON
LOW FREQUENCY FUNCTION GENERATOR POWER
COMPUTING DIGITAL INDICATOR POWER-ON

9.1.5 Position DIGITAL VOLTMETER Mode switch to AUTO.

9.1.6 Position AC-OFF switch of +35 VDC panel to AC. Observe that indicating light illuminates. Record results on Test Data Sheet.

9.1.6.1 Adjust +35 VDC panel voltage adjustment control for an indication of 35 ± 1 vdc on the panel meter.

9.1.6.2 Position CONTROL PANEL VOLTAGE SELECTION switch to +35 VDC and adjust Power Supply voltage adjustment control until DIGITAL VOLTMETER indicates +35 ± 0.5 vdc. Record results on Test Data Sheet.

9.1.7 Position AC-OFF switch to -30 VDC panel to AC. Observe that indicating light illuminates. Record results on Test Data Sheet.

9.1.7.1 Adjust -30 VDC panel voltage adjustment control for an indication of 30 ± 1 vdc on the panel meter.

9.1.7.2 Position CONTROL PANEL VOLTAGE SELECTION switch to -30 VDC and adjust Power Supply voltage adjustment control until DIGITAL VOLTMETER indicates -30 ± 0.3 vdc. Record results on Test Data Sheet.

9.1.8 Position CONTROL PANEL VOLTAGE SELECTION switch to +10 VDC. Verify that DIGITAL VOLTMETER indicates +10 ± 0.5 vdc. Record results on Test Data Sheet.

9.1.9 Position CONTROL PANEL VOLTAGE SELECTION switch to -10 VDC. Verify that DIGITAL VOLTMETER indicates -10 ± 0.5 vdc. Record results on Test Data Sheet.

9.1.10 Position VOLTAGE SELECTION switch to OFF.

9.1.11 Position DIGITAL VOLTMETER Function switch to AC.

9.1.12 Position OUTPUT-ON switch of 800 CPS POWER SUPPLY to ON.

9.1.13 Position VOLTAGE SELECTION switch to 28.75 VAC.

9.1.14 Adjust COARSE and FINE controls of 800 CPS POWER SUPPLY for an indication of 28.75 ± 0.25 vrms on DIGITAL VOLTMETER. Record results on Test Data Sheet.

Fig. 10.6 Sample page from equipment test procedure.

tion plan and the drawings. After a program is sufficiently mature, an indexed collection of inspection instructions can be used as the inspection plan.

Test-equipment error-analysis reports. These are, initially, theoretical analyses of the test equipment (to be used in carrying out the test plan) to establish the amount of applicable test-equipment error present. Test-procedure tolerances are specification tolerances adjusted by withholding the test-equipment error. This means that test-procedure tolerances are always tighter than test-specification tolerances (Fig.

OD XXXXX Sheet XX

TEST EQUIPMENT ERROR ANALYSIS SUMMARY SHEET

TEST STATION XXXXXX
TEST SET-UP TITLE IN-FLIGHT SAFETY CONVERTER, P/N XXXXXX AND XXXXXX

PREPARED BY: | APPROVED MSL
APPROVED | BUWEPS

TEST STATION XXXXXX
PAGE 4 OF 4 | REV
DATE

ATTRIBUTE			ERROR FACTORS						
TITLE	NOMINAL VALUE	TOLERANCE T_2	DESCRIPTION	ACCURACY (T_1) FULL SCALE	ACCURACY (T_1) MEASMT. POINT	ERROR FORMULA	RATIO T_2/T_1	MEASUREMENT SYSTEM ERROR	FINAL READOUT LIMIT
Voltage Drop Test	2.5 VDC	Max.	Scope Camera	N/A	± 3.0%	T_1 = ± 3.0% M.S.E. = $(V_{nom})(T_1)$ = ± 0.075 VDC	N/A	±0.075VDC	2.42 VDC max
	1.3 VDC	Max.	Scope Camera	N/A	± 3.9%	T_1 = 3.0% M.S.E. = $(V_{nom})(T_1)$ = ± 0.039 VDC	N/A	±0.039VDC	1.26 VDC max
Reset Time Output 3	60 usec	Max.	CMC Counter 226B	± 1 Digit	± 1.66%	T_1 = ± 1.66% M.S.E. = $(T_{nom})(T_1)$ = ± 1 usec	N/A	± 1 usec	59 usec max
Temperature Test (+77°F)	+77°F.	± 6.5%	CMC Counter 226B	N/A	± 0.25%	T_1 = ± 0.25% M.S.E. = $(T_{nom})(T_1)$ = ± 0.19°F.	26	±0.19°F.	+77±.19°F.
Temperature Test (-20°F)	-20°F.	± 25%	CMC Counter 226B	N/A	± 0.25%	T_1 = ± 0.25% M.S.E. = $(T_{nom})(T_1)$ = ± 0.05°F.	100	±0.05°F.	-20±5.05°F.
Temperature Test (+25°F)	+25°F.	± 20%	CMC Counter 226B	N/A	± 0.25%	T_1 = ± 0.25% M.S.E. = $(T_{nom})(T_1)$ = ± 0.06°F.	80	±0.06°F.	+25±5.06°F.
Temperature Test (+125°F)	+125°F.	± 8%	CMC Counter 226B	N/A	± 0.25%	T_1 = ± 0.25% M.S.E. = $(T_{nom})(T_1)$ = ± 0.31°F.	32	±0.31°F.	+125±6.56°F.
Temperature Test (+85°F)	+87.5°F.	± 2.8%	CMC Counter 226B	N/A	± 0.25%	T_1 = ± 0.25% M.S.E. = $(T_{nom})(T_1)$ = ± 0.21°F.	11	±0.21°F.	+87.5±2.71°F.
Temperature Test (+200°F)	+200°F.	± 5%	CMC Counter 226B	N/A	± 0.25%	T_1 = ± 0.25% M.S.E. = $(T_{nom})(T_1)$ = ± 0.50°F.	20	±0.50°F.	+200±10.5°F.
Dielectric Withstanding Voltage	500 vrms	Min.	Kilovolt Hi-Voltage Supply AC 5-2	± 2.0%	± 20%	T_1 = + 20% M.S.E. = $(V_{nom})(T_1)$ = ± 100 vrms	N/A	+100 vrms	600 vrms (min)
Insulation Resistance Test	500 VDC	Min.	Keithley 241 Power Supply	N/A	± 0.05%	T_1 = ± 0.05% M.S.E. = $(V_{nom})(T_1)$ = ± 0.25 VDC	N/A	+0.25VDC	500.25 VDC (min)
Insulation Resistance	100 Megohm	Min.	Keithley 515 Bridge	N/A	± 0.1%	T_1 = ± 0.1% M.S.E. = $(R_{nom})(T_1)$ = ± 0.1 Megohm	N/A	+0.1 Megohms	100.1 Megohm (min)

FIG. 10.7 Test-equipment error-analysis summary.

EXAMPLE 10

Page from End-to-End Calibration Procedure

SCP-OD 26345
Revision 1
12 September 1963

3.4.2.13.1

Console Load Resistor	Thermal Transfer Standard	AC Ammeter	Load Resistance	Tolerance
	______ Volts	____Amps	____Ohms	- 230 Ohms WIR

To obtain the value of the Load Resistance divide the voltage observed on the Terhmal Transfer Standard by the current observed on the AC Ammeter.

** 3.4.2.14 Observe the lead power factor as indicated by the Phase Angle Meter (Unit 1.2.1.3). Record on the Test Data Sheets.

3.4.2.14.1

Load Power Meter	Power Factor Meter	Tolerance
	____________	-.95 to + .95

NOTE: Paragraphs 3.4.2.15 through 3.4.2.24 are to be accomplished only if 3.4.2.13 is out of tolerance because of inadequate trimmer range.

3.4.2.15 The variable resistor under the Control Panel AC Milliameter has a capability of trimming the Console Load Resistance approximately 7.5%. In the event that this trimmer does not have a range that will permit setting the load resistance, in paragraph 3.4.2.13, to 230 ohms nominal, proceed as follows, and then accomplish paragraphs 3.4.2.13 and 3.4.2.14.

3.4.2.16 On the Control Panel (A6), adjust the Variable Resistor control under the Control Panel AC Milliammeter and labeled SET TO 500 MA, to the center of its range.

3.4.2.17 On the Power Amplifier (Unit 1.2.1.13), position the POWER switch to the 30 SEC WARMUP position.

Fig. 10.8 Example of end-to-end calibration procedure.

10.6). At the time of test-equipment proofing, the test-equipment error-analysis calculations are checked out against the test equipment itself and the error figure is corrected, if necessary. Any corrections must also be incorporated into the test procedures (Fig. 10.7). A test-equipment error funnel is usually created similarly to the funnel of tolerances described earlier. The purpose of the test-equipment error funnel (from supplier to field) is to detect situations in which a large test-equipment error may actually create an inverse funnel which may cause the rejection of a substantial amount of good hardware. A typical test-equipment error funnel is shown in Fig. 10.9.

An example of an integrated reliability test-program group review and recommendations relative to a test program for a part is given in Fig. 10.12.

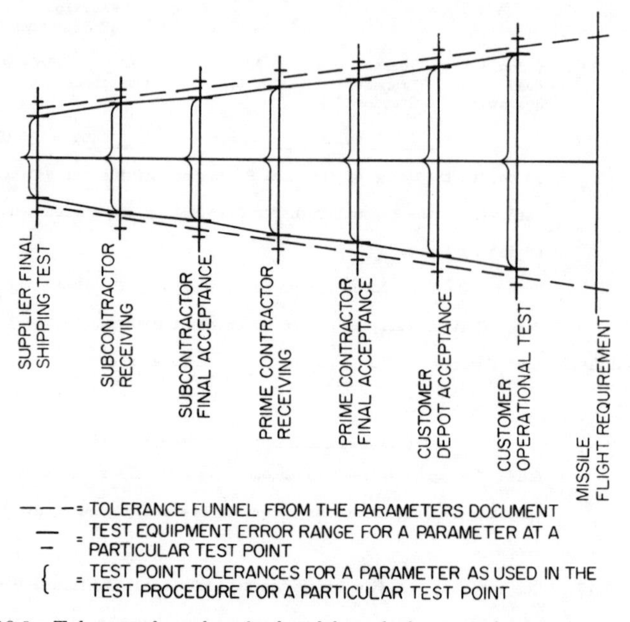

Fig. 10.9 Tolerance funnel and related funnel of test-equipment error range.

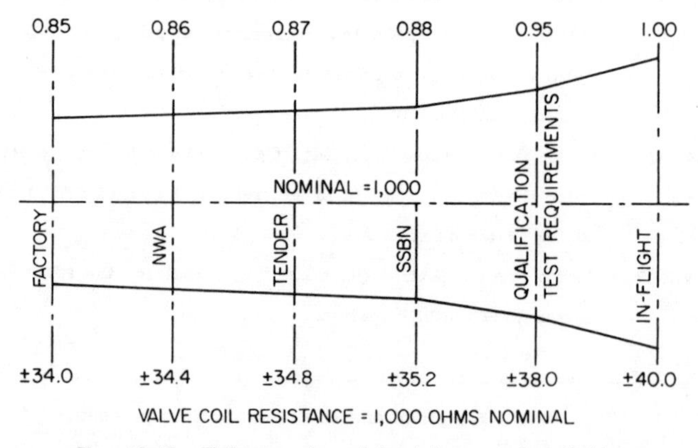

Fig. 10.10 Tolerance funnel for injector valve, fluid.

Most of the design-disclosure documentation is produced by the design function with review and approval by product assurance. The remainder of the design-disclosure documentation is produced by the product assurance (reliability and quality control) function with the concurrence of the design function. In this manner the check-and-balance feature is maintained. Creation and maintenance of a full design-

CLASSIFICATION OF DEFECTS

PART NAME	DWG. NO. 2363057	INSPCTN. POINT 13034	REV. 00
Valve, Injector Assy.	DWG. REV.	PAGE 1 OF 3	
PREP. BY / SQA SIG.		DATE	SAMPLING

	CLASS-IFIC-ATION	ITEM	ATTRIBUTES TO BE VERIFIED	REFERENCE	INSPECTION METHOD	TABLE	AQL PLAN
	Crit		None				
M	Maj	*101	$1.117 \pm {}^{.000}_{.002}$ dia. daturm - B	11D	Mics.	8A	1.5%
S		102	$1.006 \pm {}^{.000}_{.002}$ dia.	11D	Mics.	--	---
S		103	Datum Dia. - B - Shall be perpendicular to datum surface - M - within .001	11D	Indicator	--	---
M		*104	.330 ± .010 dim.	11D	Mics	8A	1.5%
S		105	$.250 \pm {}^{.005}_{.000}$ dim.	11D	Go-No Go Plug GA.	--	---
S		106	Keyway orientation	9C	Visual	--	---
S		107	1.000 ± .010 dim.	10C	Indicator	--	---
S		108	.187 dia. max.	10B	No Go Plug GA.	--	---
S		109	Datum Surface -M- shall be flat within .001.	8B	Indicator	--	---
S		110	.500 ± .010 dim.	8C & 5D	Mics.	--	---
S		111	.375 dia. max. (6) places	8A	Mics.	--	---
M		*112	.375 dia. max (6) places shall be in true position within .020 dia.	8A	Indicator 1870246	8A	1.5%
S		113	.468 ± .035 dim. (12) places	9C	Mics.	--	---
S		114	$\frac{1}{4}$ -28 UNF 3B THRD. (12) places	5D & 5A	Go-No Go THRD Plug GA	--	---

This Unit of Product is Defective if any of the Requirements Listed Herein are not Met.

Notes: A. This classification of Defect, originated by PA Inspection Engineering list those attributes which, when used in conjunction with the documents listed below, comprise the total inspection acceptance criteria:
1. Applicable Drawing and Specifications
2. Applicable PA & TS Policies & Procedures
3. Applicable Product Assurance Operating Instructions

B. Asterisk (*) before item number indicates an OCD attributes.

Remarks or Sampling Tables.

Fig. 10.11 Sample page from classification of defects.

disclosure package require a major design and product assurance effort. This is part of the total price for high design (inherent) reliability of a complex system.

With full design-disclosure documentation and with the test and inspection equipment invoked, it becomes possible for the inspection function to carry out its responsibility properly. The inspection responsibility is, of course, to determine whether or not the hardware has been built in conformance to the design in all essential respects. Hardware not in conformance to the design requirements cannot be accepted unless design, product assurance, and the customer are in agreement that reliability is not adversely affected.

IRTP GROUP RECOMMENDATION SUMMARY — SHEET 1 OF 5

PART INJECTOR VALVE — TEST SPAN 5 months — IRTP REPORT NO. 107 R-1

PART NO. 1963640 — RESP. DEPT. 58-63 — DATE REVIEWED 6-15-62

EFFECTIVITY A3Y 1 & Up — TYPE OF TEST EET/PPT/REL

☐ ENVIRONMENT ☒ DOCUMENTATION ☐ HARDWARE ☐ OTHER

DESCRIPTION	WAS (submitted)	IS (revised)
CHANGES TO IRTP TEST ASSESSMENT SHEETS REQUIRED BY WS 2223		
Environmental Level and Duration	1. Acceptance - Ambient Examination per DCD 6063038 and Para. 4.4, ERS 12259A.	1. Acceptance - Ambient Examination per Para. 4.6 WS 2223.
	2. Temperature - Trans. Cycle -20°F 6 hours, +125°F 4 hours. RH 15% Return to ambient.	2. Temperature - Trans. Cycle - Stabilize at Ambient RH controlled to prevent condensation. -20°F 6 hours +125°F 4 hours. Return to ambient.
	3. Pressure - Pre-Launch Depress to ambient within 20 sec. Rate decrease -21 psi/sec.	3. Pressure - Pre-Launch Return to ambient. Exponential pressure decay. init. rate 50 ± 10 psi/sec.
	4. Pressure 1st Separation ERS 12259A.	4. Pressure - 1st Separation WS 2223.
	5. Static Firing Tips of valves exposed, etc.	5. Static Firing - See Note 1. NOTE 1: Components from each vendor will be incorporated in complete MOD II TVC System for qualification under full scale static motor firings in accordance with LMSC 802501.
	6. Self conducted Elect. Interference	6. Delete in entirety.

Fig. 10.12 IRTP review group recommendation summary.

10.6 EFFECT OF RELIABILITY REQUIREMENTS UPON DESIGN ENGINEERING

Reliability requirements, goals, and objectives are established by the customer or by competitors' practices. Governmental entities such as the Department of Defense and the National Aeronautics and Space Agency are usually quite definite as to their reliability requirement, particularly on weapon systems, space systems, or other complex equipment. Large industrial customers such as those in the automotive and airline industries are also very definite in their reliability requirements upon their suppliers.

The reliability requirements for consumer goods and industrial products supplied to a number of small users are determined by the producer. Such a determination is based upon both customer reaction and the practices of competitors. It is certainly possible for a company to price itself out of the market by setting and maintaining unrealistically high reliability standards. Many more companies, however, have found themselves in trouble because they have had unacceptably low standards of reliability.

Reliability requirements, whether imposed or created, are not sacred and should be periodically challenged. This challenging of the need for and validity of any and all design requirements (including reliability) is the basic element of the value-engineering function. Removal during the early stages of a program of an unnecessary, unrealistic reliability requirement can save substantial time, effort, and money as well as increase the amount of resources which can be concentrated on solution of other design problems.

In a complex design it is necessary to break out the overall reliability requirement, goal, or objective into separate objectives for the elements of the design. This allocation or apportionment is usually made by the design integration function or is made for them by the reliability function. Design areas of great complexity or in which greatest state-of-the-art advances are required must be given the lowest possible reliability requirement, while design areas which are straightforward and use thoroughly known and tested design principles will have the most rigorous reliability requirements.

These apportioned reliability figures are frequently challenged by the responsible design supervision. As the design matures, design management usually makes some adjustments of the reliability requirements based upon relative progress in the various design areas. Specific reliability limitations may be imposed upon the designer as part of his design requirement. The designer may be prohibited from using, or enjoined against the avoidable use of, certain materials, parts, or practices such as:

1. Certain natural or synthetic elastomers with undesirable characteristics (e.g., neoprene).
2. Certain electronic parts with characteristics believed unsuitable for the particular design (e.g., relays, liquid capacitors).
3. Liquid rocket fuels for ballistic missiles to be carried in submarines.
4. The use of tin, zinc, and certain other metals in structural parts of electronic devices because these materials have the property of growing crystalline whiskers which may degrade reliability.

The designer may also be required by reliability to follow certain reliability practices, to use certain materials and parts, and to follow certain design practices such as:

1. Use solid-state electronic devices (transistors, diodes, etc.) rather than tubes.
2. Put in redundant circuitry at certain points.
3. Follow certain proven conservative design practices in buffering between dissimilar metals so as to assure against electrolysis under expected environments.
4. Follow certain conservative electrical bonding practices to control transient currents and ghost circuits.
5. Derate electronic parts a specified percentage to decrease transient failures and improve part life.

Numerical reliability requirements, goals, or objectives are a part of the basic design objectives imposed upon the design organization. In addition, reliability considerations may well result in specific design limitations (do's and don'ts) being imposed upon the designer. Both of these types of reliability requirement form a part of the design environment within which the designer must operate.

10.7 THE INTEGRATION OF FUNCTIONAL TESTING PROGRAMS

There are at least five general types of functional test programs during a development program with subtypes of each general type. These development-phase test programs are carried out at the component part, package, subsystem, and system level. They are:

Feasibility testing. This is the testing done by or for the designer to prove out his design concept or to enable him to choose the most promising concept from several possible design concepts. Feasibility testing is usually done with minimum formality.

Evaluation testing. This is the testing of early development hardware to the operating and environmental conditions for which it was designed. Written test procedures and test results are kept, but formality is low. Hardware, test equipment, and test procedures are modified as conditions require.

Qualification testing. This testing program is the formal proofing of the design against the design specifications. Corrective design action in the form of hardware redesign is taken when test results indicate the necessity for such design modification.

Reliability testing. The distinction between design testing and reliability testing is not a clear one. In general, however, design people are most concerned with meeting the performance specifications successfully and reliability people are interested in marginal condition and extended life testing. Reliability people test beyond the specification requirements by using more rigorous environments or by testing for a longer period (time or cycles). This overtesting is done to determine the performance margins and to induce actual failures so that modes of failure can be examined and possible corrective action can be studied.

Flight testing. This fifth testing category is a special one for aircraft, missiles, and spacecraft. While many flight environments can be closely simulated, it is rare that all of the actual environments and environmental combinations are known. Although successful completion of ground tests is very encouraging, flight-hardware design can be proved out only in actual launching and flight tests. In general, flight testing of unmanned vehicles is started very early in the development program to smoke out design weaknesses. Early flight failures are expected. A different philosophy must be used on manned-flight vehicles, when every possible method of ground testing must be used before attempting manned-flight test so that flight-test failures can be held to an absolute minimum.

It is obvious that the test requirements for the above-mentioned development test programs must be fully integrated to get maximum value from the various programs and to prevent program gaps, overlaps, and inconsistencies. While the need for and advantages of such test-planning integration are obvious, this integration is often very difficult in accelerated programs, when the various test programs overlap to a considerable degree.

The tracking and integration of the development test programs are one of the basic and earliest starting and long-continuing reliability efforts. This development test-integration function is carried right on into the planning of development and tactical production testing for quality control purposes, into the planning of periodic requalification (of production and suppliers) testing, of customer field testing, and of surveillance testing of hardware returned from the field.

In some companies it has been found desirable to have all development testing (except the earliest breadboard testing) actually performed by product assurance. Direction of the testing is a design function, but increased test objectivity is achieved by product assurance performance of the testing.

The subject of testing has been described in considerable detail in Sec. 8. The reader specifically interested in development and reliability testing is encouraged to read Sec. 8 carefully.

10.8 CUSTOMARY DESIGN PRACTICES IN VARIOUS SYSTEMS

Each of the design subdisciplines has some practices which are rather unique to it. Some of these design practices will be discussed briefly in this section.

10.8a Electronic Systems Design. Electronic systems design is the arrangement of more or less standard electronic parts into circuits which will accomplish the desired functions.

Since it is not practical for every circuit designer to be a specialist on all of the part types he may use, a special group of parts specialists is usually created. Within this parts specialists group are engineers who specialize in transistors, diodes, capacitors, relays, wire, potentiometers, and other electronic parts. These parts specialists are in touch with similar groups throughout industry through such functions as IDEP (Interservice Data Exchange Program), the Battelle Memorial Institute (BMI) Electronic Component Reliability Center, the BMI Mechanical Reliability Research Center, and the Guided Missiles Data Exchange Program (GMDEP). IDEP and GMDEP are directed by the Armed Forces; the BMI functions are industry-run. (See Secs. 5, 6, and 8.)

This group of parts specialists is usually a part of the reliability function. This function controls both the approval of parts and approval of the specific use of approved parts. Circuit designers are thus limited to the use of approved parts and must also secure approval for each specific application of each approved part. This application approval is to prevent the overstressing of approved parts by circuit designers. Overstressing is sometimes attempted by a designer in an effort to hold down package weight and size. When high reliability is required, design policy usually requires approximately 50 per cent derating of electronic parts with exceptions permitted only by special, high-level approval.

Electronics systems designers usually rely upon packaging specialists for design of the black-box structure and bracketry, upon thermal people for expert advice on heat-dissipation problems, and upon material and process specialists for advice on potting, soldering, welding, and other processes.

10.8b Structures Design. Designers of aircraft, missile, and satellite structures are not permitted (because of space and weight limitations) the massive overdesign used by structural designers of fixed structures or of surface mobile hardware. Aerospace designers work with fractional margins of safety and then rely heavily on destructive tests to confirm their design calculations. Such designers also rely on mechanical-parts specialists who are experts in bolts, rivets, screws, and other forms of mechanical fasteners as well as other common structural parts. The same type of parts control, to a lesser degree, is maintained on mechanical parts as is maintained on electronic parts as described above.

10.8c Hydraulic and Pneumatic Systems Design. Hydraulic and pneumatic system designers have several problems in common. They usually deal in high pressures and must design for safety from explosions. They are plagued with leakage problems and must always concern themselves with contamination problems. Such designers in aerospace work must carefully trade off their safety precautions against their space-weight limitations to get a design which will achieve its objectives but not endanger operating and maintenance personnel.

10.8d Mechanical Design. Functional mechanical design has a much longer history than electronic design, yet it is not nearly as standardized. There are few mechanical equivalents of the standard electronic parts and standard circuits available to the electronic systems designer. Much functional mechanical design is a modification or extrapolation of previously proved design. Again, if standard functional mechanical parts (such as bearings) are to be used, parts specialists are relied upon for approval or disapproval.

10.9 DESIGN METHODS FOR ATTAINING RELIABILITY

This section will discuss those practices a designer may follow to achieve reliability *which he would not necessarily follow if he were concerned only with getting his design to function to prove feasibility.* In other words, it will discuss the "extras" the designer does to make sure hardware built to his design will continue to function as it is supposed to within the full range of the specified environment or, put another way, will be "reliable."

Massive overdesign, when weight, space, and cost limitations permit, is one way of improving reliability. In structural design it means designing to hold 10,000 lb when the maximum specification is 1,000 lb. In electronic design it means derating electronic parts until they are being used to 10 or 15 per cent of their rated capabilities. In hydraulic or pneumatic designs it means designing a pressure vessel which will hold 1,000 psi when the design maximum specified pressure is 100 psi. This design technique is very effective when design constraints permit its use.

Simplicity and standardization are two useful approaches of the designer seeking high reliability. In general, the simpler the design, the higher the inherent reliability. A reduction in the quantity of parts or in the number of different parts used is a standard approach in trying to improve the inherent reliability of an electronic design. Standard parts and circuits have usually been thoroughly debugged and are less likely to develop unpleasant surprises for the designer. Since the usual designer is by nature a creative person, it requires real restraint for him to stay with simple designs and maximum use of standard parts and assemblies. While nearly any kind of Rube Goldberg device can be made to function, greater design complexity usually means lessened reliability.

Human-engineering considerations must be of concern to the designer. The designer should make it as difficult as possible to assemble or use his design incorrectly. When possible, cable lengths should be such that only the correct cable can reach the black-box connector. When the design is such that more than one cable can reach a black-box connector, the cable connectors should be of different sizes so that only the connector of the correct cable will fit. When a functional package is to be a spare, full maintainability consideration must be given to the problems of removal and replacement by ordinary people under field conditions. If the design is such that it is extremely difficult to remove and replace a spare part, the probability that the job will not be done properly becomes substantial. If the design is such that it is possible to drop an attaching bolt, nut, or screw into a vital or inaccessible area, the probability that this will happen (given a number of opportunities) becomes high. In either of these cases the design contains features which lower the inherent reliability of the total design. Human engineering and maintainability are discussed in detail in Secs. 11 and 12.

Production and field testability and inspectability of a design are of great importance to reliability. When a choice is possible, a designer of functional gear should create a design which can be subjected to full nondestructive, functional checkout. For example, a circuit breaker can be functionally checked, whereas such a check on a fuse is destructive. Here the testing advantage must be weighed against the probable higher reliability of the fuse. A solenoid actuator can be repeatedly functionally checked; an explosive actuator is a one-shot device. The ability to inspect important dimensions, surface finishes, and other nonfunctional attributes up to and past the assembly point where they are likely to be degraded is also a very important characteristic of a high-reliability design.

It should be noted that there are two strong schools of thought in respect to frequent operational checkout of complex weapon systems. The "wooden missile" school would like only very infrequent checks or no checks at all while the weapon is operational. The other group wants regular and frequent functional checks to make sure the missile is still operational. The first group believes that "to check out is to wear out" and that a weapon system's reliability is degraded appreciably at each checkout. In actual practice a compromise is generally effected to require relatively infrequent but regular abbreviated operational, functional checkouts.

When a design requires the use of a special process or requires unusual production capabilities, the drawings and specifications must clearly indicate this special requirement. In addition, design must also give production and quality control as much advance notice as possible so that the required equipment, installation, and personnel training can be arranged for with minimal adverse impact. An example is the design which must be produced in a clean room of a type which the production facility does not have.

Benign environments are usually much easier to design for than extreme environments. One way to increase reliability of a design is to protect it from extreme environments. This can be done by providing special temperature-controlled, shock-mitigated, dehumidified conditions during shipping, storage, and use. Although very expensive and not always attainable, modification of environments (usually by special environmental containers) is a "way out" when no reasonable design solution to achieve reliability throughout a full range of extreme environments can be found. The value-engineering technique of challenging the need for extreme environmental operation should be tried before going the modified-environment design route.

Redundancy is one of the important techniques used to achieve high design reliability, although it has some definite limitations. It is easy to see that a function which can be achieved by either of two (or more) independent functional routes has a reliability advantage over a function which can operate over only one route. Redundancy may be total or partial, i.e., all or only portions of a functional design may be duplicated. An automatic functional device is sometimes backed up with a manual device (where an operator is used). The standby generator in a hospital is there to improve electric power supply reliability in vital areas. The engines on a multi-engined boat or airplane provide a degree of redundancy and thereby may improve reliability. The decision to use a human crew in addition to fully automatic devices may be made to improve system reliability.

As mentioned in the first sentence of the preceding paragraph, redundancy as a design technique has some very definite limitations and cannot be considered as a cure-all. Here are some of those limitations:

1. Size and weight limitations may result in two duplicate circuits (or other functional devices) using smaller, less reliable components crowded into the available space. Under these conditions the reliability of the redundantized system may even be less than that of a single-circuit design.

2. Shock, vibration, or temperatures which will make one circuit or system inoperative will probably make both (redundant) systems inoperative. Under these design conditions, the design effort had better be concentrated on heat protection, vibration/shock mitigation, or other environmental protection for the single circuit or system.

3. What appears to be redundant may not really be so. For example, two pipes used to carry fluid in parallel (with necessary check valves, etc.) are fully redundant *only* if either pipe alone can carry the maximum fluid flow required. As another example, a twin-engined airplane which will not maintain altitude and flying speed on one engine is theoretically less reliable on a long overwater flight than a single-engined airplane. This is so because a twin-engined plane has twice the probability of engine failure that a single-engined plane has.

Another design technique to increase reliability is the deliberate decision not to design to meet the worst-on-worst operating conditions. Having so decided, the designer can use a simpler, less sophisticated design which will have a higher reliability under all but the worst possible combinations of use. The designer expects to gain reliability because of the statistically small probability of encountering the worst-on-worst environmental combination. This technique is sometimes known as performance reliability.

The final design technique to increase reliability is simply to give immediate, thorough design attention to alleged design discrepancies and inadequacies reported by design review, design analysis, design-change requests (from production inspection and liaison engineering), formal test program, and field-trouble and -failure reports and from failure diagnosis. Prompt, effective design corrective action on such discrepancies will substantially improve reliability.

To summarize, there are several design techniques which can be used to improve design reliability. These techniques must be carefully evaluated in each case against the design constraints affecting the particular design. Reliability-improvement design techniques include massive overdesign, derating of components, simplification, use of standard parts and assemblies, use of human-engineering and maintainability techniques, design provisions for test and inspection, call-out of special-process precautions, providing benign environments, judicious use of redundancy, use of performance-reliability techniques, and, finally, prompt effective design review of and corrective action on reported design discrepancies.

10.10 DESIGN-RELIABILITY ANALYSIS AND REVIEW

Design-reliability analysis is an embracive term which is used to cover many reliability functions. Among two of the most important of these functions are reliability-prediction analysis and reliability design review.

Reliability-prediction analysis is a function for assessing the potential inherent reliability in a design. It is attempted as soon as the possible design concepts appear. (See Sec. 5.) The analysis reports are updated as the design matures. The reliability-prediction analysis is the major reliability input to design and to design-review meetings.

A typical initial reliability-prediction analysis report on a functional electronics package design might contain the following sections:

Introduction. The introduction describes the unit physically and functionally. The use of the unit is explained, and a picture or sketch is included.

Summary of major conclusions and recommendations. This is a vital part of the report and should be emphasized by printing it on contrasting paper. It must be remembered that the purpose of reliability analysis is to identify design areas needing improvement and to propose those improvements so that corrective action will be taken.

Reliability block diagram. This diagram shows the function of the unit in the system as well as the major functions of the unit itself. Any circuit or functional redundancy will be shown. (The basic drawing prints and circuit diagrams should be included in the attachments or appendices.)

System-reliability estimation. The analysis will lead to a numerical estimate of the reliability of the unit by design and reliability personnel. The assumptions used are listed. These include required operating time, wear-out failure rates, aging characteristics, and nonstandard parts reliability. (The detailed analysis study itself should be an attachment.)

Component reliability. Sources of failure-rate data of the unit, the application of these data to the parts of the unit, and the assumptions used are contained in this section of the report. In addition, this section may contain information support, the failure rates used, reliability analysis of any special nonstandard part, a summary of mechanical and rotational stress analyses (if applicable), a description of the method used to determine the reliability of single-shot items (such as explosive devices), identification of all calendar-time- or operating-life-limited items along with references to provisions for their control, a statement on burn-in policy, and a statement on required procurement controls.

Failure-mode analysis. All primary failure modes are identified and described along with the effect of each on system performance. Statements will be made on design provisions present to prevent progressive failures, i.e., failures which, in turn, cause other failures.

Production-reliability analysis. This section of the report will describe special precautions and requirements necessary to maintain product reliability during production. This includes design-definition (documentation) review requirements, use of controlled production environmental requirements such as clean rooms, special process requirements, special testing and inspection requirements and limitations, and special handling/packaging requirements.

Maintainability analysis. This section should contain fault-detection and fault-correction information, accessibility of especially limited life items, suggested maintenance requirements, suggested service instructions, and logistic recommendations.

Conclusions and recommendations. This section should be a summary of all recommendations contained in other sections of the analysis report with a reference to the specific section paragraphs where the detailed information is contained. Detailed, specific recommendations for corrective action are also included.

As the design matures, so will the design-reliability-analysis reports. Later revisions of an analysis report will contain the design-review-meeting minutes and information on the resulting actions.

Design analysis is something less than an exact science, but techniques for analysis of electronic design have been quite well worked out. The mathematical and statistical techniques involved are described in detail in Secs. 4 and 5. In general, analysis of electronic design involves determining the number, kind, and applications of electronic parts, selecting (from handbooks or from test data) reliability numbers for the parts, assuming certain sets of environmental conditions, making allowances for derating of parts, making allowances for redundancy of circuits, and then calculating the inherent reliability of the design. On moderately to very complex designs the computations are usually done on a computer. While not an exact figure, the predicted reliability number resulting from such analysis does provide a rough guide to whether the design is anywhere near the required reliability. Design analyses of functional mechanical, hydraulic, and pneumatic designs are usually less exact. They are less exact because much less test experience is usually available on the parts used. Design analysis on structural designs is usually based upon estimation of safety factors followed by conversion of these safety factors into reliability numbers through use of a weighting system.

The reliability number predicted for a particular design as a result of reliability analysis is of special value in comparing alternative design concepts when the *relative* inherent reliability of the designs being compared is the major purpose of the analysis. The reliability-analysis report often is the only central source of complete, early design description with flow diagrams, schematics, operating theory, functional descriptions, predicted failure modes, and similar vital information. As such, it serves a valuable auxiliary communication and coordination function. The reliability-prediction-analysis report is, along with the design-disclosure information (drawings, specifications, and procedures) itself, a major input to the design review. A third basic input is a set of reliability design-review checklists completed by the designer.

Reliability design reviews are most successful when conducted within the design organization with the reliability engineer scheduling and setting up the meetings, taking the initiative, and publishing the minutes. Extensive premeeting preparations should be made by using reliability design-review checklists which are prepared and periodically revised by the reliability engineer. Whenever possible, design-review checklists should ask questions which require informative answers other than yes or no. A minimum of three (conceptual, interim, and final) design reviews is required on complex designs. Suppliers who furnish their own design work must also hold their own conceptual, interim, and final design reviews. The contractor's design and reliability personnel should participate in the conceptual and final design reviews.

The process of achieving high inherent reliability is cheaper and easier in some designs than in others. While nearly any design concept can be converted into a reliable design if enough money, time, and effort are expended, the relative ease (comparing two or more design approaches) with which reliability may be achieved can (and should) be recognized through design reviews. Conceptual design reviews have, of course, a potentially major impact on the design, with successive interim and final reviews having relatively less effect as the design becomes more fixed and less time is available for major changes.

Reliability design reviews should be combined, wherever possible, with other design reviews such as producibility and maintainability to minimize the demand on the designer's time and to resolve conflicting recommendations. The following are some of the design-review considerations. The areas marked with an asterisk may be accomplished separately as part of the reliability review of the basic design-requirements specification.

*Review of customer performance requirements
*Review of customer environmental requirements
Confirmation on use of approved parts in an approved manner
Circuit analysis and reliability prediction
Provisions for vibration and shock
Provisions for heat transfer
Correct use of dissimilar metals to minimize electrolytic action

Correct use of organic materials to minimize life limitations, while still meeting other design objectives

Search for conditions which might cause local hot spots

Search for potential sources of fluid or gas leakage

Minimizing the requirement for assembly adjustments and selective fit requirements in production

Provisions for testability and inspectability where required

Provisions for maintainability

Analysis of potential failure modes and their effects

The design department manager or his representative (senior staff engineer or supervisor) should normally chair the design-review meeting. As chairman, he makes the final decisions on suggestions and proposed changes. The design-review chairman is the only person in the meeting who has both the responsibility and authority to make design trade-offs when necessary. The design-review meetings are actually a device for measuring the performance of the designers, and the information gathered provides the chairman a better basis of fact for making the necessary trade-offs.

In practice, it is best for the reliability engineer to give the responsible designer the design-review checklist well in advance of the design-review meeting. The reliability engineer should go over the checklist with the designer to flush out any areas of concern. Ideally, these problem areas are resolved before the meeting. If the reliability personnel take exception to the design manager's decision on matters affecting reliability, the reliability recourse is through withholding approval or through the management ladder, depending upon the policies of the organization concerned.

The preparation required for a design-review meeting should usually result in most of the deficiencies being cleared up. Tests and analyses that might have been overlooked are made, and the designer will make more use of the various design specialists. In the final analysis, this preparation may, in itself, be the greatest single benefit of design review.

Properly done, design review can make a major contribution toward "getting the job done right the first time." Design review forms a counterforce against schedule pressures. Design time requirements are often underestimated. As deadlines approach there is considerable pressure for quick "intuitive" design solutions without the required searching analysis. A formal design-review program is a barrier to "quick and dirty" design solutions, since the designer knows he must face his supervision and the reliability personnel with the correct answers to a searching list of questions. With the design reviews known to be coming up, it is usually easier for the designer to do the design job thoroughly and correctly the first time.

To summarize, design-reliability analysis is a mathematical, analytical method of estimating or predicting the inherent reliability of a design by (in the case of electronic designs) assigning reliability numbers to the components and adjusting these figures for parts population, derating, redundancy, and other factors. (See Sec. 5.) Design-reliability analysis should be performed as soon as an interim design and schematics are available. The analysis should be revised every time a significant design change is made.

Reliability requests for corrective action are made when significant design discrepancies are found. The design-reliability analysis report forms a significant input to the design-reliability review meetings, which are a formal, scheduled series of meetings for the thorough, searching review of all aspects of each design with particular emphasis upon the reliability aspects. Design-review meetings are scheduled and conducted by the design organization with reliability engineers acting as instigators, recording secretaries, devil's advocates, and conscience. When necessary, design trade-offs are made at these meetings. Reliability personnel may refuse to approve or may react through management channels against unacceptable design reviews.

Design-reliability analysis and design-reliability review may be considered to be the "inspection and test" of the individual designs for the purpose of measuring and recording conformance to the basic design requirements and conformance to sound design principles and practices. In this sense, design-reliability analysis and design-reliability review are to the quality of design (inherent reliability) as inspection and test are to the quality (conformance to design requirements) of the hardware. Searching and competent design-reliability analysis and design-reliability review functions are essential in achieving a high inherent reliability in complex designs. High inherent reliability is a fundamental requisite for production of hardware which will still demonstrate high reliability under use conditions.

It is, unfortunately, far too easy for a design with a high inherent reliability to suffer a severe drop in reliability potential through inadequate control of design "gozinta" patterns, through sloppy change-letter control, and through drawings and specifications which contain ambiguous information and erroneous notes and call-outs or which omit essential requirements. The review of the design to make sure the designer's intent is clearly and completely communicated to the production and quality control people who need to use it is called design-definition review. This review, along with continual audit of the design-release, design-change, and design-configuration control systems, is essential in preventing reliability degradation in what is basically a highly reliable design.

10.11 OTHER RELIABILITY FUNCTIONS AFFECTING DESIGN

There are several additional reliability functions which affect the design function to some extent. These include:

Selection and designation of limited-calendar-life items. These are integral portions of the design which have an expected life that is less than that required of the assembly. Such limitations are usually caused by the use of organic materials (such as rubber or other elastomers) or of explosive chemicals. When feasible, design changes to remove the limitation should be made. If changes cannot be made, provisions to live with the limitation must be made. Such provisions include establishing a system of control and the individual limits of each limited-calendar-life part. Limited-calendar-life items must, of course, be made spares and included in provisioning plans.

Selection and designation of controlled items. These are functional parts which have operating life limits in time of operation, operational cycles, or power turn-ons that are less than the parts might encounter in actual use. Action is similar to that taken on limited-calendar-life items.

Surveillance. This is a program of detailed, searching analysis of material which has undergone extensive service. This service can be in amount of operation, in elapsed time in service, or both. The analysis is for the purpose of detecting signs of incipient failure as early as possible so that the maximum amount of time is available to forestall a crisis and to take both short- and long-range corrective action. Surveillance program results are fed back to design in the form of recommendations.

Design-definition review. Between the reliability design review on the one hand and the detailed drafting-practice review of the design checker on the other hand, there is need for another kind of reliability review which is for the purpose of making sure that the designer's intent is clearly and completely communicated to the technicians and inspectors who will use it. This is called a design-definition review. It is not enough that the designer's intent is clear to him or to another designer of comparable background. The design must be checked for clarity and completeness both within each design document and between related documents. The field of specification call-outs in particular and drawing notes in general is a fertile one for design-definition review.

Production reliability. This is a program wherein reliability engineers follow the design into the experimental shops and observe the translation of the design into manufacturing planning paper including shop orders, production drawings, and manufacturing-process specifications. The actual production process is then followed to observe what problems are encountered, what ambiguities or omissions exist, and what outright mistakes have been made in the design. Problems are called to design attention at once, and resolution is effected as promptly as possible. If the design organization has been concerned enough to perform the function described above, then the production-reliability function only monitors the adequacy and accuracy of their efforts.

Production-data feedback. Reliability serves as a key function in collecting selected variables, functional and dimensional data (including weights) from early production test and inspection. These data are then processed so that the various readings on each attribute are placed into a histogram along with specification limits, control limits, and a statement as to the normality of the population represented. Such data give the designer information of real value in adjusting his tolerances, if desirable, to better fit the production capability without degrading reliability and possibly while improving reliability. Tolerances may be tightened where the process is under control well within specifications so that other tolerances may be loosened where production process control is not so good. More information on reliability data is contained in Sec. 6.

Review and approve design changes. After the complete design is released, special reliability vigilance is required in the review of changes and in the selection of change effectivity. Design history is full of examples when changes that appeared to be perfectly desirable backfired in completely unexpected manners with serious degradation of both function and reliability. When reliability requirements are high and the consequences of unreliability are serious, the safe rule is not to tamper with a design of proven reliability unless it is absolutely necessary to do so. Even if a design change is acceptable, great care must be exercised in introducing it. The change must be so scheduled that necessary procurement can be effected, changes to test and inspection planning can be completed, test and inspection equipment changes can be completed, test and inspection instructions are modified, superseded material is dispositioned, and a myriad of other aspects are satisfactorily handled.

Request design corrective action on failures. During the development stage of a design program, reliability receives a great deal of data from factory inspection reports, from field failure and discrepancy reports, and from the actual diagnosis of failed hardware. As a result of analysis of these discrepancy data, requests to investigate and take design action to prevent recurrence of these discrepancies are directed to the design function. Prompt, sincere design action on these requests does a great deal to improve design reliability.

There are other reliability functions which affect design to some degree. Most of these are small efforts (such as the continuing sampling audit of design-release and -configuration control functions) although not unimportant. Those functions of greatest importance have been detailed in preceding paragraphs. All of these reliability functions which interface with design form part of the check-and-balance requirement and are part of the painstaking attention to detail required to attain and maintain high inherent reliability in a complex design.

10.12 RELIABILITY DEMONSTRATION, REPORTING, AND RESEARCH

This portion of this section will describe certain reliability functions which are only incidentally associated with the design function. In some cases the government and certain industrial customers have made a requirement for demonstration of reliability a part of the contract or purchase order. This has been talked about a great deal more often than it has actually been enforced. Cases when demonstrations of reliability have been required have usually involved functional piece parts or simple functional assemblies. If the object procured is very expensive or if the testing required to demonstrate reliability is expensive, it quickly becomes prohibitively costly to demonstrate any meaningful reliability figure with any significant degree of confidence. Where reliability demonstration is actually a procurement requirement, the reliability function takes a key part in negotiating the exact terms and conditions of the demonstration requirement. In addition, the reliability personnel of both the buyer and seller usually participate in the actual demonstration testing. Since a substantial portion of profit may be at stake, reliability-demonstration testing usually receives a great deal of corporate attention.

Starting with the inherent reliability estimate derived from the design analysis, the reliability function may wish to or may be required to track reliability growth as the development program matures. Such maturation of reliability requires both a statistical and an engineering approach. Results of ground ambient, ground environmental, and flight (if applicable) testing are evaluated to determine the extent of reliability growth. When failures have occurred and corrective action has been taken in the form of design change or other action, the evaluation of the adequacy of the corrective action and its impact upon reliability growth is one of the more difficult problems in evaluating and reporting on reliability growth. This subject is treated in greater detail in Sec. 5.

Reliability is a technical function with certain disciplines and technology of its own. These certainly do not stand still, and any substantial reliability function should devote some time and resources to research into improved methods for obtaining, maintaining, and measuring reliability. Some of the areas for such research include searching for improved statistical techniques, research into new and improved nondestructive testing methods and applications, research into improved functional test and inspection devices, research into improved calibration accuracies and better calibration-interval determinations, and research into the reliability potential and impact of new design techniques such as microminiaturization and molecular electronics. There are many additional areas where reliability research could be worthwhile. Part of such research includes keeping up with the reliability research activities of the professional organizations in the field such as the American Society for Quality Control and the Society for Non-destructive Test.

The foregoing paragraphs have described three of the principal activities wherein the reliability function is the principal prime mover with only minimal association with the design function. The importance and impact of all three of these areas of reliability effort appear to be on the increase.

10.13 ADMINISTRATION OF THE DESIGN AND DEVELOPMENT FUNCTION

There are two principal organizational forms for the design and development function. Either design and development are centralized for the entire company or, if the company is split into major production divisions, one or more of these major product divisions may have its own design and development function.

Figure 10.13 is a typical organization form with a centralized design and develop-

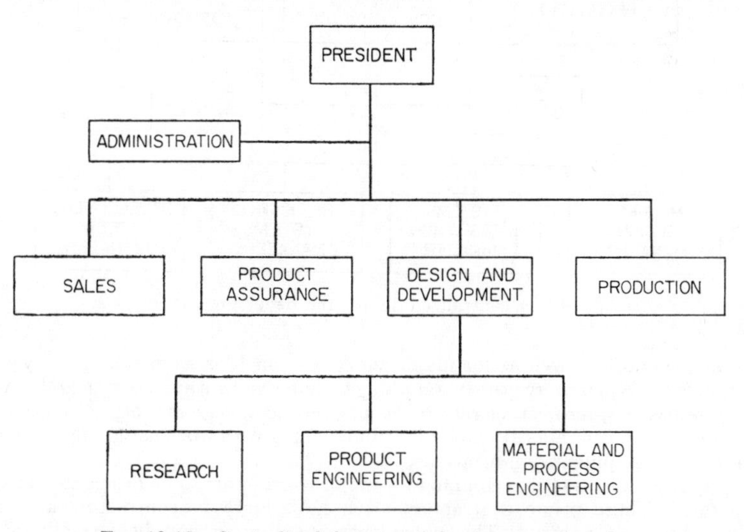

Fig. 10.13 Centralized design and development function.

ment function. Figure 10.14 is an example of a thoroughly decentralized design and development function in a major corporation.

If the organization also has a separate research function, the design and development function is closely allied with the research function. Often the design and development people will rely upon the research people for direct material and process

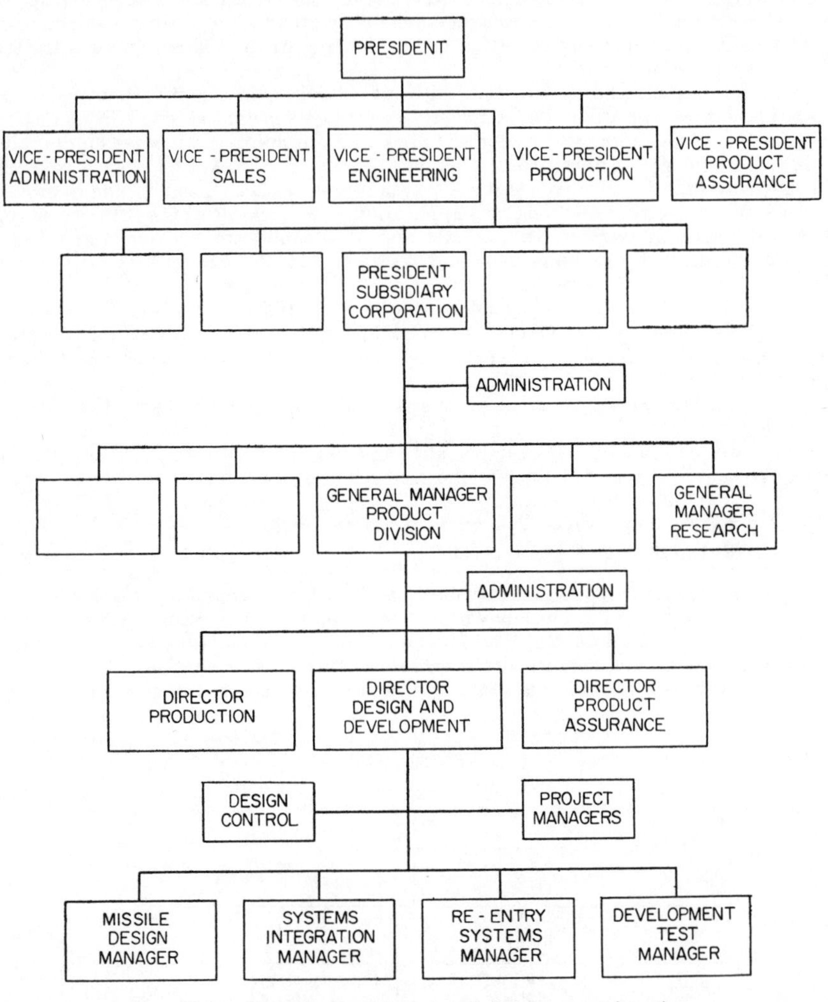

Fig. 10.14 Decentralized design and development function.

laboratory support as well as for major support in such areas as stress analysis and heat transfer. When there are several projects underway within a design and development organization, project engineers or project managers are often created with budget and manpower allocations to distribute, along with work assignments, to other design and development organizations.

In defense contracting it is common for major contractors to be designated as prime contractors. These prime contractors often have large subcontractors and major suppliers working for them. The subcontractors and major suppliers usually do their own design and development work. The prime contractor has the responsibility to make sure the subcontractors and major suppliers can and do perform their design and development function competently. In addition, the prime contractor's design and development organization has a review and approval responsibility over the design work performed by the subcontractors and suppliers. This relationship between prime and subordinate suppliers must, of necessity, be coordinated through their respective company contractual channels.

One of the major weaknesses of many design and development organizations is the lack of understanding of and a disdain for design documentation. It is not always realized that the end product of a design assignment is a package of proofed design-disclosure documents. The breadboarding, the feasibility testing, the circuit analysis, the stress analysis, the environmental testing, the mock-up, the production of development hardware, and the flight testing—all these are but means to achieve an end, and the end result is the proofed design-disclosure package. The common lack of understanding, among designers, of the importance of design disclosure causes much of the unreliability in complex development projects. As a result, much design and development work has been wasted or is of dubious value.

While it is understandable that a new design challenge will interest a designer more than "dotting all the i's and crossing all the t's" on the present design to properly complete it, a mature, competent design and development technical management will insist on completed design disclosure with rigorous design-change control.

10.14 ADMINISTRATION OF THE RELIABILITY FUNCTION

As described in Sec. 10.1, reliability came into being as a function to serve as a check on the quality of design. Most early reliability organizations were started by designers who had a quality control viewpoint in that they were aware of the consequences to product reliability of design errors and omissions. These early reliability organizations originated and grew within the design organizations.

As a parallel development, quality engineering developed rapidly as a function within many organizations, particularly if there was no effective reliability organization. The distinction as to whether certain functions are reliability functions or quality engineering functions is not and never has been clear. As a result, in many companies the activities have merged or come into conflict. At present, reliability is still a design-organization function in many companies, particularly where product reliability requirements are not extremely rigorous. It is sometimes difficult, though certainly not impossible, to maintain an independent and objective reliability attitude when reliability personnel are assigned within the design organizations.

Companies with substantial defense and NASA business with correspondingly rigorous reliability and quality requirements have tended to combine reliability and quality functions into a function called product assurance or quality assurance. When this is done, reliability engineering and quality engineering are usually also combined. A typical product assurance organization, including a detailed functional breakdown of the reliability and quality engineering function, is shown in Fig. 10.15. Section 16 deals with the subject of organizing for reliability and quality assurance in greater detail.

10.15 SECTION SUMMARY

It should be noted that the majority of this country's major missile and satellite launching failures have not been caused by the malfunction of some exotic device whose design pressed the state of the art. Rather, many of these failures have been caused by the failure of functional and structural parts of proven design and wide usage. In some case the parts were not made correctly, and in other cases there were human failures such as failure to torque and secure a fastener properly or failure to install an explosive device properly. No detail is too minor to cause a failure. High inherent and achieved reliability are, to a considerable degree, the product of painstaking attention to detail.

In this section we have reviewed the range of design responsibilities, which include the responsibility to provide for reliability requirements. We have developed the reasons why inherent reliability must be high and why reliability is needed as an independent check-and-balance function on design to assure that reliability requirements and considerations get their prompt and proper share of design attention. Further, we have explored the methods for designing reliable hardware and reviewed the methods, procedures, and practices used in achieving and assuring design reliability. Finally, we have reviewed the various types of design organizations and the various relationships between the design function and the design-reliability function.

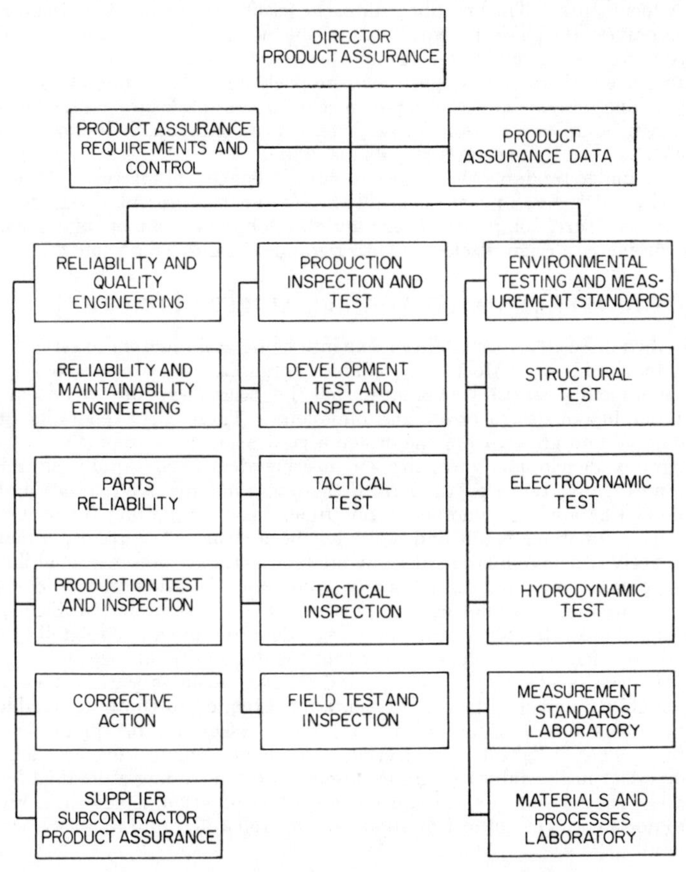

FIG. 10.15 Typical product assurance organization.

In summary, inherent reliability is the primary responsibility of the design organization with reliability serving as an independent check and balance on the design function, principally to make sure that the design function has given its reliability responsibility the painstaking, detailed attention which is necessary. In addition, reliability performs certain functions wherein its work is checked by design for the same reasons.

Bibliography

Hurd, Walter L., "Full Design Definition Is Essential to Missile Reliability," *Proc. Eighth Ann. Reliability Quality Control Natl. Symp.*, Washington, D.C., 1962.

———, "The Contribution of Design Analysis to Product Assurance," *Proc. Seventeenth Ann. Am. Soc. Quality Control Conv.*, Chicago, Ill., 1963.

Lubelsky, Benjamin L., "Reliability Techniques in Production," paper presented at Seventh Military-Industry Missile and Space Reliability Symposium, San Diego, Calif., 1962.

"Missile Systems Division Reliability Engineering Manual," *Lockheed Missiles and Space Co. Rept.* 801178, Aug. 1, 1963.

Smiley, Cmdr. Robert W., "Design Review and Field Feedback Trigger Product Improvement," *Trans. Joint AIAA, SAE, ASME Aerospace Reliability Maintainability Conf.*, Washington, D.C., 1963.

———, "Military Management of Missile Quality Control/Reliability Programs," *Proc. Ninth Natl. Symp. Reliability Quality Control*, San Francisco, Calif., 1963.

Section 11

CONSIDERATION OF MAINTAINABILITY IN RELIABILITY PROGRAMS

B. L. RETTERER

MANAGER, TECHNIQUES RESEARCH PROGRAM
ARINC RESEARCH CORPORATION, ANNAPOLIS, MARYLAND

Mr. Retterer earned a B.S. degree in electrical engineering from Ohio Northern University and did graduate work in the same field at the University of Cincinnati and Syracuse University.

He is currently associated with Arinc Research Corporation as Manager of Techniques Research Programs. In this capacity, research is directed for a broad range of technique developments in the product assurance field.

Formerly employed by RCA for thirteen years, Mr. Retterer was associated with the maintenance-engineering field for the entire period. This experience has included service as a field maintenance engineer and as a depot maintenance process engineer. Following this, Mr. Retterer engaged in reliability and maintainability studies conducted by the RCA Applied Research Activity. A major accomplishment was the performance of a three-year study for the U.S. Air Force on the development of maintainability-measurement and -prediction methods [Contract AF30(602)-2057].

Mr. Retterer is a member of the Institute of Electrical and Electronic Engineers, Armed Forces Communication and Electronic Association, and the American Society for Quality Control. He is a registered engineer and is the author of several reports and publications in the maintainability field which have appeared in technical journals and have been presented to national product assurance symposia.

CONSIDERATION OF MAINTAINABILITY IN RELIABILITY PROGRAMS

B. L. Retterer

CONTENTS

11.1 INTRODUCTION

Reliability as represented by failure rate establishes the frequency of maintenance activity. Maintainability, however, is concerned with the time, ease, cost, manpower, facilities, etc., required to restore a system to operational status.[1]

11.2 MAINTENANCE CONCEPTS

As stated, maintainability is a system characteristic concerning the facility with which a maintenance task may be accomplished. A maintenance task is an action or series of actions (manipulative or cognitive) required to preclude the occurrence of a failure or restore an equipment to satisfactory operating condition. Maintenance actions may include the following:

1. Assembly and disassembly
2. Inspecting, testing, and measuring (diagnosis and localization)
3. Removal and replacement (repair)
4. Checkout
5. Cleaning and lubrication
6. Securing materials (supply)
7. Preparation of reports
8. Contingency items
9. Administrative duties

Actions identified in (1) to (5) are considered productive, and time spent in their accomplishment is classified as active, whereas the remaining elements are nonproductive and are denoted as delay-time requirements. Within the active elements, item 2, inspecting, testing, and measuring, has been found to be the largest contri-

[1] Maintainability. The combined qualitative and quantitative characteristics of material design and installation which enable the accomplishment of operational objectives with minor expenditure including manpower, personnel skill, test equipment, technical data, and facilities under operational environmental conditions in which scheduled and unscheduled maintenance will be performed. Source MIL Std-778.

bution to active time for electronic systems investigated. Hence, during the system design, features of equipments which influence this element, such as test points and indicators, must be given careful consideration.

A maintenance task can result for two basic reasons defined as follows:

1. *Preventive maintenance.* That maintenance performed to keep a system or equipment in satisfactory operational condition by providing systematic inspection, detection, and correction of incipient failures before they occur or develop into major failures.

2. *Corrective maintenance.* That maintenance performed on a nonscheduled basis to restore equipment to a satisfactory condition by providing immediate correction of a failure which has caused degradation to the equipment below its specified performance.

Preventive (scheduled) and corrective (unscheduled) maintenance can be performed at several locations with respect to the system deployment, depending on the maintenance concept employed. Several concepts are identified as follows:

1. Repair in place.
2. Remove, repair, replace.
3. Remove, replace with spare, repair at base.
4. Remove, replace with spare, repair at factory.
5. Remove, replace with spare, discard defective package.

Each of these concepts may be further modified depending on the lowest unit of repair or replacement designated by the concept, which may include part, module (component), subassembly, black box, equipment, and/or redundant system. The choice of the appropriate concept and the unit of replacement is primarily one of economics, but in certain situations strategic implications must be considered. Factors of concern in the selection include failure rate, spares cost, inventory cost, transportation cost, maintenance-facilities requirements, system deployment, test-equipment requirements, and other factors which may influence cost or strategic implications.

No specific recommendation can be made concerning which concept forms the best approach, since each situation must be examined individually by relating the cost to the strategic factors. A typical trade-off situation is illustrated in Fig. 11.1. Here three alternate concepts *A*, *B*, and *C* were evaluated with respect to cost and maintenance time. For this illustration it was assumed that two constraints were imposed on the choice of concepts. These were (1) a maximum maintenance time acceptable to the using activity and (2) a maximum cost. From the plots made it will be observed that only concept *B* meets both conditions. Complete treatment of concept selection is considered beyond the scope of this section, but several studies which have been made concerning the topic are referenced.[1]

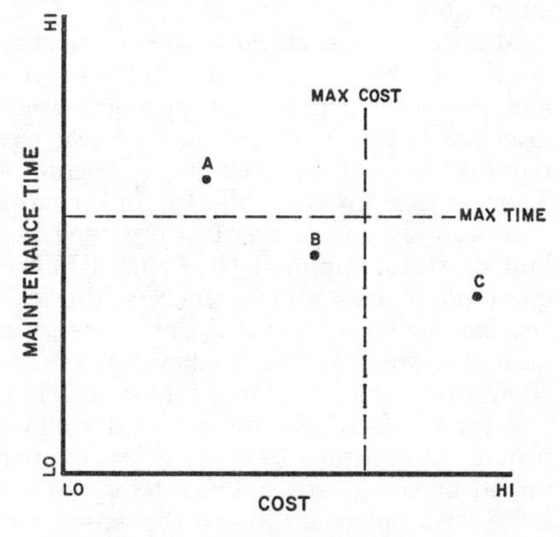

FIG. 11.1 Maintainability trade-offs.

As was previously stated, the diagnostic element is the largest contribution to active maintenance time and will be given further consideration here. Diagnosis which entails the isolation of the defective part, module, etc., in its classical form is characterized by a man-machine interface. The maintenance technician instilled with certain information concerning the equipment performance characteristics proceeds in a systematic manner, testing various elements of the equipment until the discrepant area is found. The technician is augmented in this process by technical

[1] "Micro-module Equipment Maintenance and Logistics Program (Final Report)," Contract DA-36-039-SC-85980, Radio Corporation of America, June, 1963.

manuals, test equipment, special tools, etc. This testing process can be accomplished under either dynamic or static, i.e., operating or nonoperating, conditions. Normally, it is more advantageous to diagnose under the dynamic situation, but certain types of tests require the static situation. Specifically, tests which require the application of external power source such as a resistance check necessitate a static environment. Operational requirements or restrictions may also dictate a static test situation.

System tests may be accomplished via closed- or open-loop modes. In the former the complete operation is monitored for faults. This method constitutes the ideal test mode, since a realistic operational condition is maintained. However, owing to operational restrictions or to the type of test required, it is often necessary to examine isolated functions of a system. For example, a neutralization test of a high-power r-f amplifier requires partial disabling of the system. Both closed- and open-loop testing entail checkout of major portions of a system and are contrasted against a third approach which encompasses the sequential examination of system-monitoring points. This technique, unless based on a logical progression (i.e., half split) to the faulty element, can represent a time-consuming process when applied to a highly complex system.

Consider for example, an equipment comprised of 5,000 serially connected parts. To test these sequentially without the logical progression obviously would require more than 2,500 tests 50 per cent of the time, assuming equal failure rates for all parts. By using the half-split method, the number of tests required is calculated as follows:

$$N = \frac{\log a}{\log 2} = \frac{\log 5{,}000}{\log 2} = 12.3, \text{ or } 13 \tag{11.1}$$

It follows that the sequential approach is feasible only where such tests can be automated.

Maintenance as accomplished in most systems today involves the human element to varying degrees. However, as the complexity of system continues to grow, the skills and senses of the man must be augmented by ancillary devices. Direct assists in wide usage include a range of manual test equipments (oscilloscope, VTVM, etc.) which require detailed knowledge concerning their use and application to the system. They possess great flexibility, and, comparatively speaking, they are low-cost items.

A second type of assist is provided in some recent equipment by the inclusion of built-in test equipment (BITE). BITE can be in the form of indicator lights, simple go-no-go meters and/or more elaborate signal-measuring devices. These provide the technician with status data concerning a number of key points within the operational system. This facility must be weighted against the additional equipment complexity and associated higher development and production cost.

A related form of built-in test device is the marginal-test facility. This test device provides the means to exercise critical parts and components across extremes of their operating conditions. The intent is to reveal items which may be changing characteristics with operation life to the extent that they are considered marginal. Owing to the complexity generally entailed and the limited benefits derived, this concept does not appear generally to offer sufficient return to warrant the investment required.

The final form of test device is the fully automated checkout system. Such devices vary considerably in their complexity depending upon the degree of automation and complexity of the system to be tested. Normally, such devices are separate from the system and contain the facility to make a large number of checks in a short time. These devices are capable of providing the necessary exercising stimuli plus the ability to measure and record the resultant values. Most operate in sequential mode, proceeding in a series of planned checks. The devices have the facility to stop when a discrepancy is found or to proceed and record all discrepancies, depending on their operational mode and/or programming. In the most highly developed form, the automated checkout device completely replaces the maintenance man in the diagnostic process. However, the technician is still required to monitor the process and accomplish required spares replacement.

The maintenance man is also dependent upon the support facilities available to him to accomplish maintenance. Such facilities include technical data, tools, special jigs, fixtures, and general work space. These facilities can vary widely, depending on the location of maintenance. This variance can range from minimal facilities at a remote operational site to an elaborate environment at a major repair depot. The more dependence that a particular design configuration places on such assists, the greater the probability that difficulty will be encountered in supporting the equipment when operated at an installation not having immediate access to supply channels. This is not to imply that such aids are not useful; but by diligent design evaluation, requirements for such support facilities can be reduced in many cases.

It has been established by previous studies that the maintainability design characteristics of an equipment have a definite correlation to maintenance-time requirements.[1] Relationships have been established for design-configuration attributes encompassing the following general area:

1. Facility with which entry may be made into the equipment (latches, fasteners, etc.)
2. Work space within the equipment, i.e., access
3. Diagnostic aids: test points, labeling, color coding, etc.
4. Maintenance instrumentation (fault indicating, built-in test equipment, etc.)
5. Packaging technique
6. Adjustment requirements (tuning, alignment, and other optimizing needs)
7. Safety requirements (provision to eliminate hazards, protective devices, etc.)

Design effort directed with full recognition of the impact of these features on maintenance requirements can substantially reduce maintenance-time requirements.

The maintenance man forms the final ingredient which must be considered in the overall maintenance concept. Attempts have been made to order maintenance capa-

Table 11.1 Technician Experience Factor

Months experience	*Experience factor*
5	0.31
10	0.44
15	0.61
20	0.74
25	0.91
30	1.05
35	1.14
40 or more	1.17

bility with a number of personnel characteristics.[2] Characteristics investigated include experience, proficiency measures, degree of training received, attitude, training scores, and related data. Of these, only experience has been found to provide a proven relationship. A series of ratings relating capability for different experience levels have been established; they are given in Table 11.1.[3] All other attempts to classify personnel by measurable attributes have met with only limited success.

Values given in Table 11.1 are normalizing factors used to relate the performance of technicians with varying experience levels to a base experience of 25 to 30 months. For example, a technician with 5 months experience who performs a task in 3 hours

[1] "Maintainability Engineering (Final Report)," Contract AF30(602)-2057, RCA Service Company, A Division of Radio Corporation of America, RADC-TDR-63-85, Feb. 5, 1963.

[2] *Ibid.*

[3] Federal Electric Corporation, "Advanced Concepts in the Development of a Numerical Maintainability Design Procedure for Shipboard Electronic Equipment and Systems," Bureau of Ships, Contract NObsr, Nov. 13, 1959.

would be expected to accomplish the task in 0.93 hours after he had acquired 25 to 30 months experience, that is, $3 \times 0.31 = 0.93$.

Equipment design should be guided by anthropometric data at all points where a man-machine interface will be presented.[1] In summary, the limitations and the capabilities of the operator and maintenance technician techniques should be carefully examined to assure proper division of responsibilities between man and machine.

11.3 MAINTENANCE PARAMETERS

Because of the many facets of maintainability, the development of a single all-encompassing figure of merit is not feasible at this time. Instead, a series of parameters are needed to describe these multiple maintainability characteristics. This point is further verified by reviewing the various consequences relating to maintenance performance. In a broad sense these include primarily (1) cost and (2) operational availability. These, in turn, may be related to lower-level measures such as man-hours, down time, and spares cost. The discussion presented here will be devoted to providing some detail concerning the more important maintainability-time parameters.

A problem associated with maintenance measurement stems from its dependence upon the design, personnel, and support factors. Of these, design is the only one in the operational environment which remains essentially constant, while personnel and the support environment are susceptible to continuous change. With these variable conditions prevailing, it is not possible to cite specific values; instead, the most probable estimate made must be accompanied by statements concerning the expected variation which the parameter may take. These estimates must be further conditioned by details concerning personnel and support environment associated with the maintenance requirements.

Time is probably the most important measure of system maintainability. Two time quantities to be considered will include system down time and technician time. (See Sec. 1.) Down time may be defined as the number of calendar hours that a system is not available for use, including both active and delay maintenance time. Active maintenance down time is defined as that time during which productive work is being done on the system from the time of recognition of the occurrence of a failure to the time the equipment is back in operation at its specified performance. Delay time relates the calendar time spent in administrative activities, excessive supply time such as off-base procurement, and other general areas which, although they preclude operation, cannot be considered productive toward task accomplishment. Down time is further described by the type of task, i.e., corrective or preventive.

Technician time is the man-minutes (hours) expended in the accomplishment of maintenance. This index may be further divided into active and delay technician time, which in turn is progressively expanded through the corrective and preventive categories.

Two mathematical terms that apply to time parameters are the mean and the 95th percentile, which are defined as follows:

Mean. The sum of a set of values divided by the number in the set.

95th percentile. A value which will encompass 95 per cent of all times under consideration. For example, if a value of 80 min were given, this would indicate that 95 per cent of the maintenance tasks would fall between 0 and 80 minutes.

Mean down time $\bar{M}_t$ includes both corrective- and preventive-maintenance contributions. Numerically, the relationship may be expressed as follows:

$$\bar{M}_t = \frac{F_c\bar{M}_{ct} + F_p\bar{M}_{pt}}{F_c + F_p} \tag{11.2}$$

where $\bar{M}_t$ = mean down time
F_p, F_c = number of preventive and corrective maintenance tasks per 1,000 hours
$\bar{M}_{ct}$ = mean corrective down time
$\bar{M}_{pt}$ = mean preventive down time

[1] *Ibid.* and J. L. Cooper et al., "Guide to Integrated System Design for Maintainability," Northrup Corporation, ASD-TR-61-424, October, 1961.

The mean down time for preventive maintenance may be approximated by an empirical expression showing the relation of $\bar{M}_{ct}$ to $\bar{M}_{pt}$:

$$\bar{M}_{pt} \cong \bar{M}_{ct}/1.4 \tag{11.3}$$

A more accurate estimate may be secured by applying the prediction criteria to a sample of preventive-maintenance tasks or, obviously, to direct measurement.

Since maintenance times have been observed to be distributed log-normally, the 95th percentile M_{max} can be derived by taking the logarithm for each of the values. The resultant distribution becomes normal, permitting utilization of the data in a normal manner. M_{max} is given by the equation

$$M_{max} = \text{antilog}\ (\overline{\log M_{ct}} + 1.645\sigma_{\log M_{ct}}) \tag{11.4}$$

$$\text{where } \overline{\log M_{ct}} = \frac{\sum_{i=1}^{N} \log M_{cti}}{N} = \text{mean of } \log M_{ct} \tag{11.5}$$

$$\sigma_{\log M_{ct}} = \sqrt{\frac{\sum_{i=1}^{N} (\log M_{cti})^2 - \left(\sum_{i=1}^{N} \log M_{cti}\right)^2 / N}{N - 1}} \tag{11.6}$$

N = number of tasks used in calculations

Two empirical expressions for derivation of the 95th percentile have been developed, and they respectively relate this quantity to mean down time and mean of the log (down time). They are

$$\log M_{max} \cong 1.5\ \overline{\log M_{ct}} \tag{11.7}$$

and

$$\log M_{max} = \overline{\log M_{ct}} + 0.5 \tag{11.8}$$

These expressions are useful when specific data are not available.

The mean may be tested in two ways summarized as follows:

1. Does the observed mean differ significantly from a stated value?
2. Does the mean *not* exceed a specified value?

Test 1 is appropriate when comparing the means developed from two sets of data, for example, predicted vs. observed field data. The t test forms an appropriate statistic for making this comparison and takes the following form:

$$t = \frac{\bar{X}_1 - \bar{X}_2}{S_p \sqrt{1/N_1 + 1/N_2}} \tag{11.9}$$

where $\bar{X}_1, \bar{X}_2$ = means of the two data sets
N_1, N_2 = respective sample sizes
S_p = pooled estimate of the variance

$$S_p = \sqrt{\frac{(N_1 - 1)S_1^2 + (N_2 - 1)S_2^2}{N_1 + N_2 - 2}} \tag{11.10}$$

where S_1^2, S_2^2 = respective variance for each data set

$$S^2 = \frac{\sum_{i=1}^{N} X_i^2 - \left(\sum_{i=1}^{N} X_i\right)^2 / N}{N - 1} \tag{11.11}$$

where X_i = task time in data set

To make the test, the hypothesis is formed that the means of the two data are the same. Calculation of the t value by using the above equations and referring to an appropriate table plotting t versus degrees of freedom in association with confidence levels will permit a decision regarding the hypothesis to be made. (See Sec. 4 for further details on Student's t test and Appendix A for values of t.)

The test 2 situation has its primary application to comparing a contractually specified mean to the value observed during maintenance testing. This test is concerned with determining if the computed mean can be considered statistically less than the specified value. The test situation is graphically illustrated in Fig. 11.2*a*, specification testing. Here the specified value has been depicted as the vertical line appearing to the right in the figure. The problem is to determine whether the mean or its possible range exceeds the specification. Merely comparing the observed mean to the desired value will not suffice, since the mean calculated from a sample is only an estimate of the true mean. However, it is possible to calculate the range in which the true mean may lie on the basis of the sample data. This range is illustrated in the figure by the distance between the upper and lower confidence limits. These limits are dependent upon the accuracy desired (confidence), sample size, and the data themselves. In this situation, the upper limit only is of concern, since the procuring agency must be assured that the true mean does not exceed the specified values. Since the upper limit marks the practical range of the true mean, it will be investigated further.

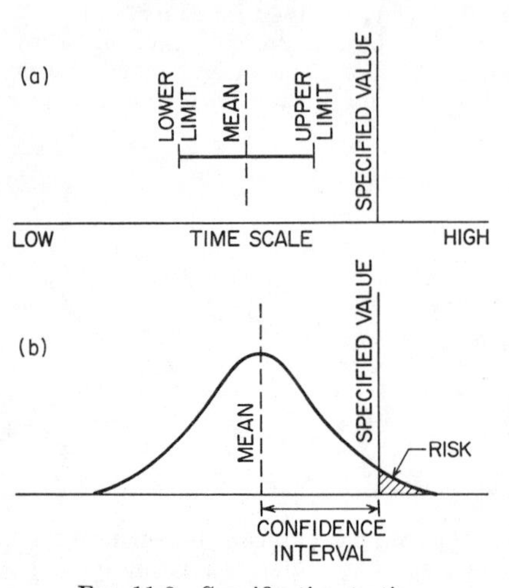

FIG. 11.2 Specification testing.

It has been found that the distribution formed by the sample mean is generally normal; hence, relationships appropriate to this distribution may be applied. The upper limit is thus given by the equation

$$UL = \bar{X} + Z\sigma_{\bar{X}} \tag{11.12}$$

where UL = upper limit
$\bar{X}$ = mean
Z = a constant to fix the confidence level
$\sigma_{\bar{X}}$ = standard deviation of the mean

$$\sigma_{\bar{X}} = \sigma/\sqrt{N} \tag{11.13}$$

where σ = standard deviation of sample data [see Eq. (11.11)]
N = sample size

The Z value is a constant determined by the contracting agency based on the level of confidence desired. Table 11.2 presents several typical values.

The confidence and risk columns may be interpreted by referring to Fig. 11.2*b*. In the figure the upper limit and the specified value have been so drawn that they coincide. In this situation the values for confidence and risk became exact. For example, using Z equals 1.282 with this coincidence prevailing means there is a 90 per cent chance that the true mean is less than the specified value. Correspondingly, there is a 10 per cent risk that the actual mean is greater than the desired level.

Choice of a testing level must always be a compromise. A high confidence level will require large sample size and more stringent maintainability achievement. Low confidence obviously increases the risk that the achieved maintainability will not meet imposed requirements. (See Sec. 4 for detailed discussion of statistical tests.)

Owing to the unique nature of the mean, the tests presented above are not directly applicable to the 95th percentile. Test of the 95th percentile will make use of its standard error, which is given by the equation

$$SE_{95} = 2.11\,\sigma/\sqrt{N} \qquad (11.14)$$

where SE_{95} = standard error for 95th percentile
σ = standard deviation of log M_t
N = sample size

On the basis of sample information, the calculated 95th percentile cannot be stated exactly, but the range can be determined in accordance with the expression

$$\log UL_{95} = \log M_{max} + Z(SE_{95}) \qquad (11.15)$$

where log UL_{95} = log upper limit for 95th percentile
log M_{max} = log 95th percentile
Z = a constant to fix the confidence level

A comparison of the upper-limit value calculated by using Eq. (11.15) with the logarithm of the specified 95th percentile will provide a means of determining compliance. The Z values may be obtained by consulting Table 11.2, which cross-references them to different confidence and risk levels. As in the case of the mean, compliance is proved if the upper limit is equal to or less than the specified 95th percentile.

Table 11.2 Confidence and Risk Factors

Z	Confidence, per cent	Risk, per cent
1.645	95	5
1.282	90	10
1.036	85	15
0.842	80	20
0.674	75	25

Comparison of the 95th percentile derived from two sets of data (prediction-observed) is provided by the statistic

$$SE_{95d} = 2.11\sqrt{\sigma_1^2/N_1 + \sigma_2^2/N_2} \qquad (11.16)$$

where SE_{95d} = standard error of the difference for 95th percentile
σ_1^2, σ_2^2 = respective variance of the log data
N_1, N_2 = respective sample sizes

The test is made by calculating the ratios of the difference between the 95th percentile and the standard error of the difference.

$$n_d = \left| \frac{\log M_{max\,1} - \log M_{max\,2}}{SE_{95d}} \right| \qquad (11.17)$$

If the value of n_d calculated is less than 2, the difference is not significant. Values of 2 to 3 are probably significant, and values over 3 are definitely significant.

The time parameter discussed may be related to higher-order measures descriptive of system or equipment capability, discussed in Sec. 1, System Effectiveness. System capability includes operational requirements such as reliability, performability, and maintainability, and they are expressed by availability, repairability, and operational readiness.

The availability of a system or equipment is the probability that it is operating satisfactorily at any point in time when used under stated conditions. "Availability" is a widely used term in both industry and the military. The availability may be stated in terms applicable to specifications (design) or operational use. These terms are intrinsic, operational, and use availability, and they are stated as follows:

1. Intrinsic availability A_i considers operate and active down time.

$$A_i = \frac{\text{operate time}}{\text{operate time} + \text{active down time}} \tag{11.18}$$

An alternative form is

$$A_i = \frac{\text{MTTM}}{\text{MTTM} + M_t} \tag{11.19}$$

where MTTM = mean time to maintenance
$\bar{M}_t$ = mean active down time

2. Operational availability A_o takes the same form as intrinsic availability A_i except that total down time is substituted for active time. It is as follows:

$$A_o = \frac{\text{operate time}}{\text{operate time} + \text{total down time}} \tag{11.20}$$

3. Use availability A_u considers off time in addition to total down time. It takes the form

$$A_u = \frac{\text{operate time} + \text{off time}}{\text{operate time} + \text{off time} + \text{total down time}} \tag{11.21}$$

The calculated availability value from this expression could be very high on an equipment operating for short periods and having a long off time.

In the use of the availability expressions, care should be taken when comparing values derived from one type of equipment with those obtained from another type of equipment. A clear statement should accompany each availability figure describing how the value was calculated and why it was used.

Repairability is the probability that when actual repair begins, the system will be repaired in a given period of time with a given manpower expenditure. This expression is useful when considering system operational readiness. This probability P_r is stated by the expression

$$P_r = \frac{1}{\sqrt{2\pi}} \int_{\infty}^{u} e \frac{-u^2}{2} \, du \tag{11.22}$$

where

$$u = (\log M_t - \overline{\log M_t})/\sigma \log M_t \tag{11.23}$$

The probability of repair P_r can be evaluated by using tables that develop the cumulative normal distribution. (See Sec. 4.) These tables are entered with values of u as calculated by Eq. (11.23).

Operational readiness may be defined as the probability that at any point in time a system or equipment is either operating satisfactorily or ready to be placed in operation on demand when used under stated conditions, including stated allowable warning time. Mathematically, this may be stated as

$$P_o = P_a P_s \tag{11.24}$$

where P_o = operational readiness
P_a = probability that system is operationally available
P_s = probability that system will operate satisfactorily for a time period t

The probability of availability P_a and probability of survival P_s may be related to repair and failure rate as follows:

$$P_a = u/(u + \lambda) \tag{11.25}$$

where u = mean repair rate
λ = mean failure rate

$$P_s = e^{-\lambda t} \tag{11.26}$$

where t = mission time
e = base of natural logarithms

The relationships stated are based on the premise that underlying distributions are exponential and that no administrative or other delay times are encountered in the repair process. (See Secs. 1 and 2 for other distributions.) Within a system, the same relationships would hold, provided the above conditions are met and the individual equipments are nonredundant. System operational readiness of each equipment is given by the equation

$$P_o = (P_1)(P_2) \cdots (P_n) = \prod_{i=1}^{n} P_i \tag{11.27}$$

By use of Eq. (11.27) in association with Eqs. (11.25) and (11.26), the overall system operational-readiness figure may be expressed in terms of individual equipment repair and failure rates. These expressions thus permit individual repair rates (maintainability goals) to be established for each equipment within the system through consideration of reliability and the overall system operational-readiness requirement. This modeling technique is a simplification of a more rigorous process which may be invoked, depending on the complexity of the system and on the number of factors to be considered.

11.4 MAINTAINABILITY AND THE SYSTEM LIFE CYCLE

Maintainability technology, to be effectively applied, must be considered throughout the equipment life cycle, i.e., design development, production, and operational use. The intent here is to discuss the more important aspects to be considered during each major phase.

11.4a Design. To meet overall system-specification requirements within a budgeted cost, it is often necessary to perform trade-offs among the major system parameters. Such trade-offs within the major parameters are also necessary to attain the specified levels for each parameter. In the case of maintainability a trade-off with reliability may be effected to achieve the desired availability. At the same time, however, mission requirements may dictate a minimum maintainability requirement below which a trade-off may not be made. In this situation trade-offs among the parameters of maintainability (design, personnel, and support) or among the components of the system may be necessary to achieve the required maintainability level. The following paragraphs give techniques for performing these trade-offs.

Since availability reflects two fundamental measures of system dependability, its use in analytically evaluating a system appears advantageous. To illustrate the use of availability for trade-offs, Fig. 11.3, weapon system A, was developed. Weapon system A is depicted as containing five subsystems for which the reliability and maintainability have been predicted. Table 11.3 summarizes the mean time between failure and the mean down time, showing the availability for each subsystem.

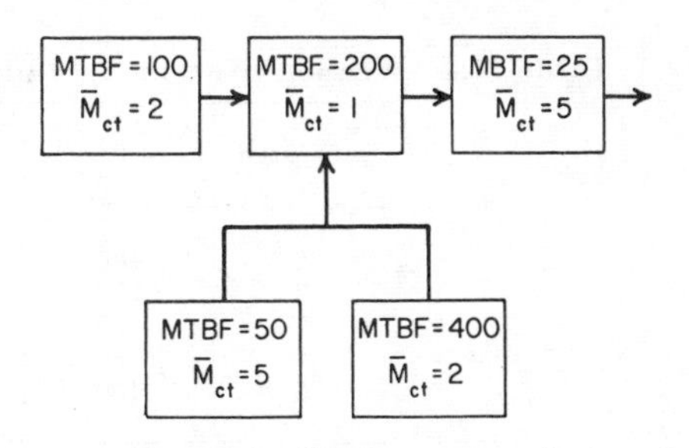

FIG. 11.3 Weapon system A.

The system availability is calculated by forming the product of the individual availabilities:

$$A_s = A_1A_2A_3A_4A_5 \tag{11.28}$$

Using this formula, the availability for weapon system A is

$$\begin{aligned} A_s &= (0.98039)(0.99502)(0.83333)(0.90909)(0.99502) \\ &= 0.73534 \end{aligned}$$

If the maintainability of an equipment has been improved to the state of the art or the budgetary limitations and still does not meet the specified level, a trade-off with reliability may be made to attain the desired availability. To illustrate this procedure, weapon system A as described above will be used. Examination of Table 11.3 shows that subsystem 3 has the lowest availability. Thus, it is reasonable to conclude that providing an alternate subsystem (redundant) will improve the total system availability. The availability for the redundant subsystem 3 can be calculated as follows:

$$A_{3r} = 1 - (1 - A_3)^2 \tag{11.29}$$

where A_{3r} = subsystem redundant availability
Substituting the value for A_3 gives

$$A_{3r} = 1 - (1 - 0.83333)^2 = 0.97222$$

Substituting the value for A_{3r} in the system availability equation gives

$$\begin{aligned} A_s &= (0.98039)(0.99502)(0.97222)(0.90909)(0.99503) \\ &= 0.85790 \end{aligned}$$

The introduction of redundancy for subsystem 3 has resulted in an increase of total system availability. This increase, however, was achieved with the added penalties of cost and complexity.

Table 11.3 Weapon System Availability

Subsystem	MTBF	$\bar{M}_t$	A
1	100	2	0.98039
2	200	1	0.99502
3	25	5	0.83333
4	50	5	0.90909
5	400	2	0.99502

Table 11.4 Weapon System A Availability (with Support Equipment)

Subsystem	MTBF	$\bar{M}_t$	A
1	100	1.0	0.99009
2	200	0.5	0.99751
3	25	2.5	0.90909
4	50	2.5	0.95238
5	400	1.0	0.99751

An alternative method for increasing availability is to increase the maintainability of the system through a trade-off between the design and support parameters. The basic method for performing this trade-off is to change the maintenance concept so that more of the burden is placed on the support parameter. As an illustration, assume that a sophisticated maintenance-checkout equipment is developed for weapon system A. This equipment reduces the maintainability requirements for the weapon system by one-half. Table 11.4 illustrates the availability achieved through the use of the checkout equipment. The weapon system availability achieved through this method is calculated as follows:

$$\begin{aligned} A_s &= (0.99009)(0.99751)(0.90909)(0.95238)(0.99751) \\ &= 0.85296 \end{aligned}$$

Again a substantial gain has been achieved but at a greater cost for the system. An additional degradation factor presented by use of the support equipment is its potential unavailability. This factor may be analytically treated to incorporate this degradation into the weapon system availability.

To select the best method for improving availability, the relative cost for each approach must be estimated for the examples given. It is assumed that each configuration possesses the same performance capability and that the development costs are as shown in Table 11.5. The data in Table 11.5 show that configuration II has the highest availability with the least increase in development cost.

Table 11.5 Weapon System Parameters

Configuration	Development cost	Availability
I. Basic system	$500,000	0.73524
II. Redundant system	$550,000	0.85790
III. Basic system plus support equipment	$560,000	0.85296

In summary, tabulation of system data as described in the above example permits detailed examination of alternative system configurations. Evaluation of the individual parameters against operational and other constraints permits the optimum configuration to be selected. In the example only two alternative configurations were considered. Actual use of this analytical approach for system optimization would entail examination of a broad range of alternatives.

It should be noted that the example is based on development costs and that the effect the proposed changes will have on support cost is not shown. When procurement is based on the cost for the life of a system, the total cost should be calculated for each alternative configuration and the selection should be made on that basis.

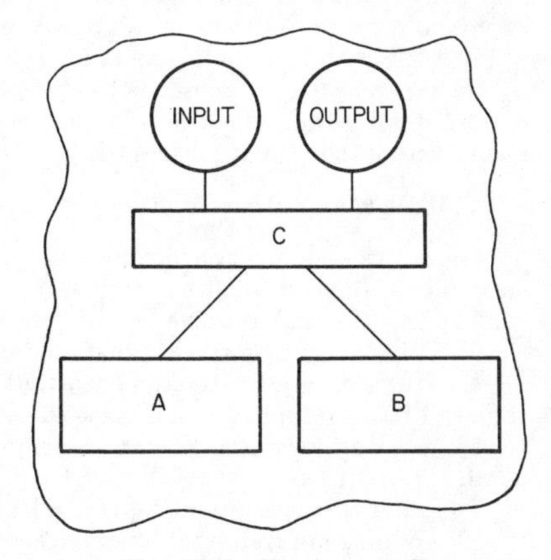

FIG. 11.4 Equipment *D*.

The procedure described establishes the required maintainability levels to be achieved considering the system requirements and the interface with reliability. Next, it is necessary to establish the design configuration and maintenance concept which will meet imposed maintainability requirements.

The system-maintainability goals must be apportioned to the three major parameters (design, personnel, and support). To accomplish this, a maintenance concept must be selected and a mathematical model describing this concept must be developed. Past experience with similar systems, together with the data given in this section, may then be used initially to apportion the overall goals to the major parameters.

The specification of maintainability for subsystems of a complex system presents a difficult problem. The distributions observed for down time have generally been log normal. This distribution does not permit direct addition of repair rates (reciprocal of mean down time), which is permissible with the comparable failure rate used in reliability technology. The following discussion reviews some of the problems inherent with combining or apportioning down time and presents a suggested approach.

Assume that an equipment was evaluated by subgroups; the next problem is how to combine these partial figures for an overall measure of maintainability. Consider, for example, the equipment pictorially represented in Fig. 11.4, equipment *D*. The portions of the diagram labeled A and B represent independent major functional units in equipment D. Section C integrates the outputs of A and B. It is desirable to evaluate each section, A, B, and C, individually and combine the results into a total equipment requirement.

Sections A and B are considered independent; hence, they will be evaluated as two separate equipments by following the method described in preceding sections. If the desired figure of merit is mean down time, this will yield $\bar{M}_{cta}$ and $\bar{M}_{ctb}$.

Section C is not independent: thus it requires a modified approach. Assume that sections A and B are instantly replaceable modules, and evaluate the maintenance of section C on this premise. Thus, section C will be evaluated by considering it to contain the two major replaceable units A and B. This analysis will provide the contribution of C to the total maintenance requirements of equipment D. However, the figure $\bar{M}_{ctC}$ for C was derived by considering A and B instantly replaceable, the contributions of which are not contained in the estimate. Since any maintenance of equipment D will begin at section C and proceed to either A or B, it is necessary to consider a method for combining the three figures derived. Numerically, the problem can be stated as maintenance of D equaling C plus A or B. The probability of A or B is dependent upon their expected rate of failure. Thus, mean down time for equipment D is expressed as follows:

$$M_{ctD} = M_{ctC} + \frac{\lambda A \bar{M}_{ctA} + \lambda B \bar{M}_{ctB}}{\lambda A + \lambda B} \tag{11.30}$$

where

$$\lambda A, \lambda B = \frac{1}{\text{MTBF}_{(A,B)}} = \text{failure rate} \tag{11.31}$$

Use of this technique requires strict adherence to stated assumptions, and it is highly dependent upon carrying out detailed maintainability analysis. Further, it is dependent upon the distribution of each subsystem being similar to that of the other. It is recognized that the model does not consider all ramifications of the problem, but serves to formulate the general approach to be used in its solution.

With the maintainability goals for the system and subsystem established, the next action necessary is to achieve the design which will meet requirements. A number of design guidelines which provide the designer with information to meet this end have been formulated. A typical set is presented here:

A. Maintenance Consideration for Electrical Designers

1. Are the maintenance and test-equipment requirements compatible with the concept established for the system?
2. Does the unit require special handling?
3. Can the unit be readily installed and connected to the system?
4. Are factory adjustments such that readjustment is not required when units are replaced in a system or when parts are replaced in the unit in the field?
5. What adjustments are necessary after a unit has been installed in the system?
6. Are adjustments capable of compensating for all possible tolerance buildups?
7. Is periodic alignment and/or adjustment recommended? How often?
8. Are all requirements for maintenance tests such that the specified time limitations can be met?
9. Has the number of factory adjustments been minimized?
10. Has the number of field adjustments been minimized?
11. Are interconnected circuits in the same package, thus providing minimal inputs and outputs at each maintenance level?
12. Is the interaction between adjustments and other circuit parameters minimized?
13. Is the design such that damage to the circuit cannot result from careless use of adjustment or combination of adjustments?
14. Are all adjustments and indicators of the center-zero type where possible?
15. Is periodic testing necessary? How often?
16. Are the test points adequate? Are they accessible in the installed condition?
17. What overhaul testing is required?

18. What specific test equipment is necessary?

19. Have factory and maintenance test-equipment requirements been minimized and coordinated with the requirements for other units?

20. What special techniques are required in the repair, replacement, or alignment of the unit?

21. Are parts, assemblies, and components so placed that there is sufficient space to use test probes, soldering iron, and other tools without difficulty? Are they so placed that structural members of units do not prevent access to them?

22. Are testing, alignment, and repair procedures such that a minimum of knowledge on the part of maintenance personnel is required? Can troubleshooting of an assembly take place without removing the assembly from a major component?

23. What special tools and/or test equipment are required?

24. Can every fault (degradation or catastrophic) which can possibly occur in the unit be detected by the use of the proposed test equipment and standard test procedures?

25. Have parts subject to early wear-out been identified? Have suitable preventive-maintenance schedules been established to control these parts?

26. Are the components having the highest failure rates readily accessible for replacement?

27. Are parts mounted directly on the mounting structure rather than being stacked one on another?

28. Are units and assemblies so mounted that replacement of one does not require removal of others?

29. Are limiting resistors used in test-point circuitry; i.e., is any component likely to fail if a test point is grounded?

30. Can panel lights be easily replaced? (Panel lights should not be wired in series.)

31. Have voltage dividers been provided for test points for circuits carrying more than 300 volts?

32. Will the circuit tolerate the use of a jumper cable during maintenance?

33. Are controls located where they can be seen and operated without disassembly or removal of any part of the installation?

34. Are related displays and controls on the same face of the equipment?

35. Are all units (and parts, if possible) labeled with full identifying data? Are parts stamped with relevant electrical characteristics information?

36. Are cables long enough to permit each functioning unit to be checked in a convenient place?

37. Are plugs and receptacles used for connecting cables to equipment units, rather than pigtailing to terminal blocks?

38. Are field-replaceable modules, parts, and subassemblies plug-in rather than soldered?

39. Are cable harnesses designed for fabrication as a unit in a shop?

40. Are cables routed to preclude pinching by doors, covers, etc.?

41. Is each pin on each plug identified?

42. Are plugs designed to preclude insertion in the wrong receptacle? Are plug-in boards keyed to prevent improper insertion?

B. Maintenance Considerations for Mechanical Designers

1. Are all items (parts and subassemblies) visually and physically accessible for assembly, wiring rework, and maintenance?

2. Are all test points accessible when the unit is properly installed?

3. Are all field adjustments accessible when the unit is properly installed?

4. Has sequential assembly which results in complicated sequential disassembly in order to make repairs and adjustments been avoided?

5. Is the design such that no unrealistic requirements for special facilities for maintenance, storage, or shipment are imposed?

6. Is the design such that no unnecessary requirements for a special maintenance environment (e.g., ground power carts, cooling, special primary power, etc.) are imposed?

7. Does the design provide for adequate protection of maintenance and test personnel against accidental injury?

8. Is each assembly self-supporting in the desirable position or positions for easy maintenance?

9. Can assemblies be laid on a bench in any position without damaging components?

C. Human-engineering Considerations for Operation and Maintenance (See Sec. 12)

1. Are visual indicators so mounted that operator can see scales, indexes, pointers, or numbers clearly? Are scale graduations, design or numerals and pointers, and scale progressions so presented that accurate reading is enhanced?

2. Do visual displays have adequate means for identifying an operative condition?

3. Have ambiguous information and complicated interpolations been eliminated from visual indicators to minimize reading errors?

4. Do controls work according to the expectation of the operator?

5. Do functionally related controls and displays maintain functional or physical compatibility, such as direction-of-motion relationships or proximity to each other?

6. Are controls so designed that the operator can get an adequate grip for turning, twisting, or pushing?

7. Does console design provide knee room, optimum writing surface, height, or optimum positions for controls and displays?

8. Do equipment design and arrangement allow space for several operators to work without interfering with each other?

9. Do arrangement and layouts stress the importance of balancing the workload, or do they force one hand to perform too many tasks while the other hand is idle?

10. Is the illumination designed with the specific task in mind, rather than with a general situation? (Many instruments are practically useless because of lack of illumination.)

11. Have extreme glare hazards such as brightly polished bezels, glossy enamel finishes, and highly reflective instrument covers been eliminated?

12. Are assemblies and parts so stacked that some have to be removed to repair or replace others, thus complicating maintenance?

13. Do fasteners for chassis and panels require special tools which hamper maintenance?

14. Do chassis door slides have means for holding the unit extended for servicing? Are the slides too loose, or do they bind?

15. Are handles provided, and are the chassis or units light enough to be moved without undue strain?

16. Is calibration indexing provided for maintenance-adjustment and calibration-adjustment controls? (Screwdriver adjustments are often too sensitive.)

17. Do the coding and symbols on equipments and instruction manuals coincide?

18. Is illumination provided for the maintenance technician?

Following the guidelines presented provides the proper design impetus toward meeting goal requirements. However, there is a further need to evaluate the maintainability characteristics of the system quantitatively during design to provide some assurance that the requirements will be met. Maintainability-prediction techniques permit this assessment to be made. Several techniques are available to fulfill this need.[1] A technique shown to be valid for ground electronic equipment will be described here. This technique establishes a relationship between the independent variables, design A, personnel B, support C, and the dependent variable, maintenance time T. Functionally stated:

$$\text{Time} = f(\text{design, personnel, and support}) \tag{11.32}$$

Through the use of regression analysis applied to actual maintenance data, the following empirical equation was established:

$$\text{Time} = \text{antilog}\,(3.54651 - 0.02512A - 0.03055B - 0.01093C) \tag{11.33}$$

[1] "Maintainability Engineering (Final Report)," *op. cit.*, and Federal Electric Corporation, *op. cit.*

In this equation the factors A, B, and C represent measures of the three variables which are evaluated for a specific maintenance task. Substituting the measured values into the equation provides an estimate of the expected maintenance-task time.

The measures of the independent variables are made by the use of checklists developed for each variable. Figure 11.5 illustrates a portion of the checklist developed for the design variable. Each question within the checklist is rated 0 to 4, depending on the condition of the design features for a specific maintenance task. Completing all checklist questions by using design information (drawings and related technical data) provides a complete task measurement. Repeated application of the checklists to a series of representative maintenance tasks permits the maintainability of an equipment to be estimated.

---

1. *a.* Access adequate for visual and manipulative tasks (electrical and mechanical)
   b. Access adequate for visual but not manipulative tasks
   c. Access adequate for manipulative but not visual tasks
   d. Access not adequate for visual or manipulative tasks ______

2. *a.* External latches and/or fasteners are captive, need no special tools, and require only a fraction of a turn for release
   b. External latches and/or fasteners meet two of above three criteria
   c. External latches and/or fasteners meet one of above three criteria ______

3. *a.* Task does not require use of test points
   b. Test points available for all needed tests
   c. Test points available for most needed tests
   d. Test points not available for most needed tests ______

4. *a.* Units or parts of plug-in nature
   b. Units or parts of plug-in nature and mechanically held
   c. Units of solder-in nature
   d. Units of solder-in nature and mechanically held ______

5. *a.* All parts labeled with full identifying information, and all identifying information clearly visible
   b. All parts labeled with full identifying information, but some information hidden
   c. All information visible, but some parts not fully identified
   d. Some information hidden, and some parts not fully identified ______

6. *a.* Task accomplishment does not require the use of external test equipment
   b. One piece of test equipment is needed
   c. Several pieces (two or three) of test equipment are needed
   d. Four or more items are required ______

FIG. 11.5 Design checklist (partial).

To illustrate the procedure, consider Table 11.6, which presents a sample calculation. Here, the first column identifies a number of tasks evaluated. The second column notes the part in the equipment assumed defective. The scores A, B, and C are the checklist scores determined for each task on the basis of design information. The last column is the expected maintenance time calculated from the scores by using the prediction equation (11.34). To facilitate the use of this equation, a nomograph which provides a graphical means of solution is given in Fig. 11.6.

The task-time data developed in the example cited can be used directly to determine equipment repair rate. Comparison of the predicted rates with the design goals will permit a judgment concerning the degree of maintainability achieved in the proposed design to be made. Should deficiencies be found, the technique may be repeated at appropriate intervals to determine the improvement affected by design changes.

11.4b Production. Current military maintainability specifications call for demonstration of system maintainability by the accomplishment of a series of typical maintenance tasks.[1] Such a procedure entails use of the equipment, test equipment, and technicians in a manner similar to the expected conditions in the operational environment. The maintenance situation is created by the introduction of failed parts into the equipment or similar actions creating a system malfunction. Measurement of maintenance-task time under those conditions provides a realistic evaluation

Table 11.6 Communication Equipment Maintainability Prediction

Task	Part	*A*	*B*	*C*	Time
1	0618	18	22	17	172.4
2	K101	44	26	20	26.8
3	T102	24	22	16	124.9
4	V101	40	22	20	44.8
5	V102	40	22	20	44.8
...	....	...	...	.....	....
47	V503	53	26	26	13.7
48	V504	35	21	17	69.2
49	V505	36	21	18	63.7
				Total	2,508.2
				Mean	51.2

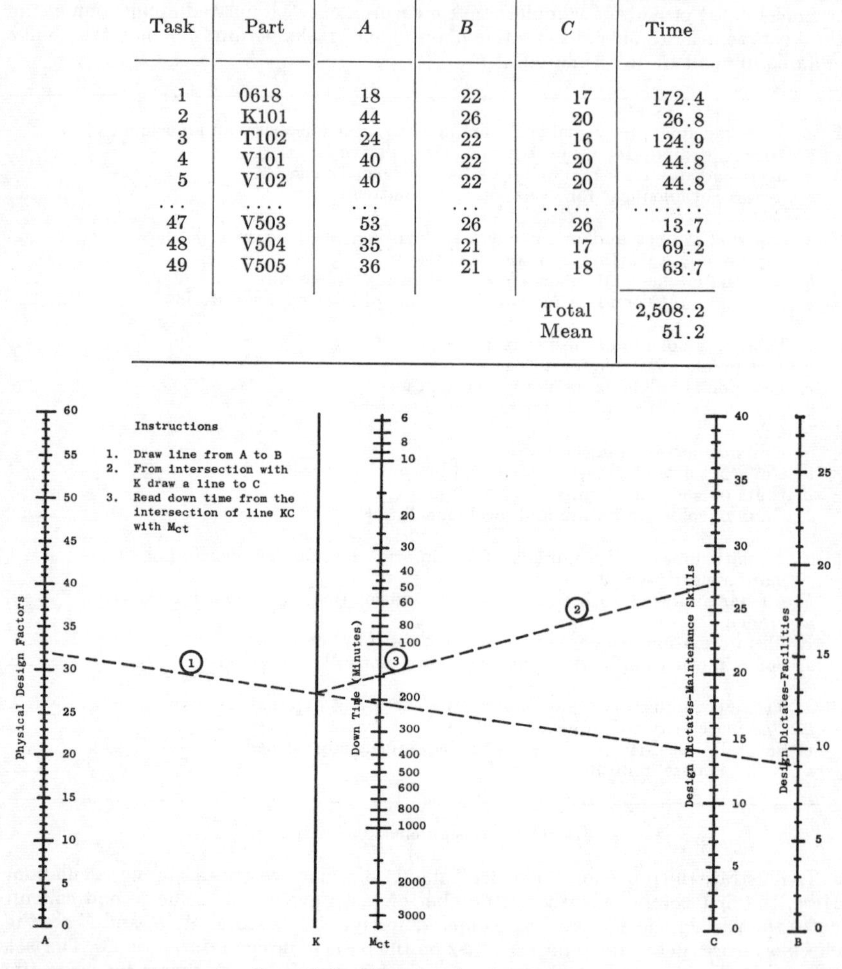

FIG. 11.6 Nomograph: down time.

of the expected maintainability to be achieved during operational use. The maintenance tasks observed in testing sequence are selected in accordance with the expected failure rate of the replacement units.

To illustrate this procedure, a typical calculation is given in Table 11.7. Here, the part types used in the equipment have been identified and the quantity used has

[1] Military Specification, "Maintainability Requirements for Aerospace Systems and Equipment," MIL-M-26512C (USAF), Mar. 23, 1962.

been noted. Reliability data provide the average part-failure rates. The product of the quantity used and the failure rate yields the expected number of failures per thousand hours. In this example, it was found by totaling the part failures that 6.391 equipment failures per thousand hours could be expected. This figure, used as a base, permits the percentage contribution of each part classification to be determined. These percentages in turn permit the apportionment of the desired sample, in this case 50. With knowledge of the part types to be tested, a random-selection process may be employed to identify the specific parts to be used.

Performance of the test sequence requires that careful attention be given to creating an environment similar to the expected use conditions. Aspects to be considered include technicians (training, experience, and general qualifications), test facilities (tools and test equipment), and technical data. Not only must consideration include the selection aptness, but control must be exercised during the testing to assure that these conditions are maintained. From this process, maintenance data which, when analyzed, constitute an evaluation of system maintainability are derived. As was

Table 11.7 Task Selection

Part class	Quantity	Avg. part failures, %/1,000 hr	No. expected failures/ 1,000 hr operation	Contrib. to total expected failures, %	No. failures for sample of 50	Actual failures
Blowers/motors	44	0.189	0.083	1.30	0.65	1
Capacitors	505	0.010	0.051	0.80	0.40	0
N-type diodes	19	2.983	0.567	8.87	4.44	4
Connectors	261	0.032	0.084	1.31	0.66	1
Relays	74	0.359	0.266	4.16	2.08	2
Coils	71	0.033	0.023	0.35	0.18	0
Resistors	1,517	0.015	0.228	3.57	1.79	2
Switch	176	0.045	0.079	1.24	0.62	1
Transformers	85	0.133	0.113	1.77	0.89	1
Tubes	301	1.567	4.717	73.81	36.91	37
Miscellaneous	101	0.178	0.180	2.82	1.41	1
Total	3,154		6.3910	100.00		50

noted, current specifications use the test data derived in this manner as a means of determining compliance with system-maintainability requirements.

11.4c. Operational Use. Essentially three types of data can be obtained during operational use; they are time, environment, and cost data. Time data result from measurements of equipment and personnel performance and are important to maintenance scheduling and capability determination. Data concerning the status of equipment, personnel, and support systems, as well as the natural environment, are necessary to isolate the factors affecting maintainability and to provide information for design improvement. "Cost data" refers to the costs associated with the maintenance and support of an equipment. Cost data are important to the determination of realistic specification requirements, and they form the basis for performing trade-offs between maintainability and other system parameters.

The least costly method of data collection is the documentation of equipment-maintenance activities by performing technical personnel. The principal disadvantage of this technique lies with the questionable accuracy obtainable. It is often difficult for personnel involved with maintenance to make impartial judgments, and the basic problems are not always apparent. (See Secs. 6 and 7.)

Another method of acquiring performance data is through the interview of personnel at operating sites. This survey may be conducted by personal interview or through questionnaires to be completed by the site personnel. Questionnaires and data forms must be developed to obtain the desired data, and the sites to be visited are selected to obtain a good cross section of the total population. If questionnaires are to be sent to maintenance personnel for their completion, instructions for form completion must be developed to describe the purpose of the questionnaire clearly. This data-collection technique is handicapped by usually not being capable of gathering time data.

Time-study technique entails employing an impartial observer to record times from the beginning to the end of a maintenance task. In addition to time, a description of the type of work accomplished during each element of the task may be recorded. These elements may be gross divisions or finite actions depending on the level of detail desired. It is also necessary to establish bounds as to what tasks and what elements will be measured so that the observer does not waste time gathering unnecessary data. If more than one technician is performing a task, each technician may be coded and the task elements may be recorded for each.

Work sampling is a technique whereby the activities of men and/or machines can be measured to within specified limits by sampling rather than by continuous observation. There are essentially five steps in performing a work-sampling study:

1. Determine categories for classification of the activities.
2. Determine the number of observations to be made for the degree of accuracy desired.
3. Develop randomized observation times.
4. Design the necessary forms.
5. Observe and record data.

Work sampling may be used to gather data on either maintenance-technician activities or on equipment status, and they may include:

1. Equipment maintenance
2. On-the-job training
3. Administrative duty
4. Nonproductive
5. Temporary duty

These activities may be subdivided to obtain the desired degree of detail. For example, equipment maintenance may be divided into corrective or preventive echelons or type of equipment (prime, ancilliary, test, etc.). Equipment activities would include operational, standby, out of service, and off.

To determine the number of observations required, it is first necessary to establish which of the categories will probably take the least amount of time and then estimate the per cent of the time this category will occur. The necessary sample size may then be determined by solving Eq. (11.34).

$$\text{Allowable error} = X_c \sqrt{pq/N} \tag{11.34}$$

where X_c = confidence interval in terms of standard errors
p = per cent of time activity occurs
$q = 1 - p$
N = number of observations

The allowable error is the amount of deviation between the sample and the actual per cent occurrence that can be tolerated. The confidence interval is the number of standard deviations for a normal distribution that would encompass the desired confidence level. (That is, ± 2 is approximately 95 per cent of all observations.) For 95 per cent confidence, Eq. (11.34) reduces to

$$N = 4pq/(\text{allowable error})^2 \tag{11.35}$$

The number of observations to be made per day is found by dividing the total sample by the number of days allotted to the study. (Sufficient time must be allotted so that the number of daily observations does not overtax the observer.) The times for the observations are selected randomly with a different set of times for each day.

Any valid method for randomizing the observations (selecting numbers from a hat or a table of random numbers) may be used. The times which are selected by this procedure are the times at which the observer will make his observations and record the activity in which the technician or equipment is engaged at that instant.

A field data-acquisition program must be carefully planned and controlled. The first step is to detail the prime objectives of the program and establish the limitations. It is then necessary to develop data-collection forms for recording the desired information. Depending on the data-collection technique to be used, either an observer training program or a field personnel indoctrination program must be developed. Finally, procedures for gathering data and for program control must be established.

Data forms must be carefully designed to ensure that all pertinent facts are obtained in accordance with established guidelines and limitations. It is important that forms be as simple and clear as possible to assure that they will not be misinterpreted or not be fully completed. The planned analysis methods should be considered so that data are gathered in a form that is easily used.

All personnel engaged in data collection must be trained to understand basic maintainability concepts and to derive and record the desired data accurately. If the data are to be obtained from the personnel performing equipment maintenance, the personnel must be indoctrinated in the use of the data-collection forms and be motivated to be complete and accurate in their reporting.

To implement a data-collection program, all reporting procedures and necessary scheduling should be completed in advance. The analysis procedures should be determined so that preliminary analysis may be performed as the data are received. Progress reporting and periodic checks should be used to determine the current status of the program. In addition, the initial reports should be reviewed for possible improvement in format or reporting procedures.

Final data analysis should be preceded by complete screening of all data to assure their aptness, accuracy, and completeness. Underlying distribution formed by the data should be ascertained to assure that maintenance parameters calculated are indeed valid. It is through this process that sound criteria may be formed and a true evaluation of operational maintenance conditions may be made by their use.

11.5 SUMMARY

The techniques described provide quantitative tools to assist in the achievement of system-maintainability goals through effective design planning, guidance, and evaluation. The mathematical modeling and apportionment procedure provides a means of translating overall system requirements into specific subsystem goals. Design guidance concerning the degree of maintainability achieved is provided by the prediction technique. Finally, specification compliance is demonstrated by the evaluation of the system through maintainability demonstration testing and field evaluation. Compositely, these techniques represent a major breakthrough toward establishing maintainability on a numerical basis.

The techniques are being employed on current programs, and from this experience it is expected that further refinement in the techniques will be achieved. Obviously, the work thus far accomplished is not complete. Although usable tools are now available, further investigation is required. Electronic systems have been given prime consideration in the technique development to date. Mechanical, hydraulic, and similar systems are, however, confronted with essentially the same maintenance problems. Procedure refinement is needed to increase both the accuracy and ease with which they may be applied.

More precise relationships are needed to equate specific design features to maintenance time and thereby permit greater guidance during the design phase. Relationships incorporating cost as a criterion, in addition to time, must be developed.

Progress made to date toward providing techniques for quantitative treatment of maintainability has been significant. A considerable challenge, however, remains: to encompass the complete scope of the maintainability problem within the domain of an explicit, well-defined technology.

Section 12

HUMAN FACTORS IN RELIABILITY

DAVID MEISTER

HEAD, SYSTEM RESEARCH SECTION
SYSTEMS EFFECTIVENESS DEPARTMENT, BUNKER-RAMO CORPORATION
CANOGA PARK, CALIFORNIA

Dr. David Meister, currently heading the System Research Section of the Systems Effectiveness Department of the Bunker-Ramo Corporation, was head of the Human Factors and Maintainability Group of General Dynamics/Astronautics, at San Diego, for eight years, where he was responsible for the human engineering of the Atlas Missile System. From 1954 to 1956 he was Assistant Project Director at the American Institute for Research, Pittsburgh, Pennsylvania, where he specialized in studies of ground electronic systems. From 1951 to 1954 he was at the Human Factors Division, Navy Electronics Laboratory, San Diego, performing research on methods of discriminating sonar stimuli.

Dr. Meister received his Ph.D. degree in psychology from the University of Kansas in 1951. He is senior author of Human Factors Evaluation in System Development, published by John Wiley & Sons, Inc., New York, in 1965. He is also the author of a number of publications on assessment and prediction of human error and reliability and system performance testing. He is a member of Sigma Xi, the Human Factors Society, and the Electronics Industries Association's committee on human factors.

HUMAN FACTORS IN RELIABILITY

David Meister

CONTENTS

12.1 INTRODUCTION

Human factors engineering (referred to henceforth in shorthand form as simply human factors) is the design of equipment, work environment, and work methods in accordance with human capabilities and limitations. Human factors specialists are therefore concerned primarily with (1) equipment design characteristics, (2) operating and maintenance procedures, (3) characteristics of the work environment (noise, temperature, work space, etc.), but also with (4) technical data, (5) communications, (6) logistics, and (7) system organization, because each of these interacts with and influences operator performance.

Human factors as a recognized discipline is of comparatively recent origin. It developed during and after World War II, when a tremendous increase in equipment complexity made it necessary to think in terms of tailoring equipment design to the operator. Prior to that time industrial engineers had attempted to modify the man and his work methods to fit the job. However, with weapon system complexity straining human capabilities to the limit, it was found that improvements in man-machine efficiency could be made more easily by modifying the machine rather than the man.

The expansion of human factors work since 1947 has been remarkable. Today there are approximately 2,000 people working in the area,[1] among them engineers, psychologists, physiologists, physicians, and anthropologists. Engineers provide the equipment orientation; psychologists, the behavioral concepts; physiologists and medical men, the biological background; and anthropologists, the anthropometric focus.

There is a natural relationship between the reliability and human factors disciplines. Both are concerned with predicting, measuring, and improving system performance, the reliability engineer through equipment, the human engineer through those equipment aspects that interact with and influence operator performance. In more than a few companies human factors personnel work in or for reliability-engineering

[1] J. A. Kraft, "A 1961 Compilation and Brief History of Human Factors Research in Business and Industry," *Human Factors*, vol. 3, pp. 253–283, 1961.

groups; in other companies reliability and design engineers are called upon to perform what is essentially human factors work. There are, moreover, compelling reasons for interrelating reliability and human factors:

1. All equipment systems consist of both equipment and personnel in interaction (hence, the phrase "man-machine system"). Consequently, to deal with total system reliability in terms analogous to Feigenbaum's "total quality control"[1] concept, engineers must consider both elements; we cannot speak of equipment or human reliability separately.

2. During the design phase, the reliability engineer monitors and reviews design. He must therefore know what human factors principles should be incorporated into design for maximum reliability. Even at present, human factors are not adequately considered in the design of equipment. This has resulted in complex systems that are difficult to operate and maintain and have a high malfunction rate and excessive downtime. Failure to apply human factors principles therefore directly results in lowered reliability.

3. The addition of any element to reliability estimates reduces that reliability unless the added element is invariably reliable. If the reliability engineer considers only equipment factors in his measurements, it means that he assumes the operator performance is optimal ($r = 1.00$). Since we know that the reliability of operator performance is obviously less than perfect, the reliability engineer must include that performance in his measurements. Otherwise, those measurements will be grossly inflated, *as they often are.*

The effect of the operator on reliability is shown by the high incidence of equipment failures resulting from human causes (human-initiated failures, or HIF). The frequency of HIF ranges from 20 to 80 per cent of all failures reported.[2] Reliability analysis of failure rates solely in terms of equipment failures therefore overlooks a major source of system unreliability.

4. Military specifications require the reliability program to consider human factors principles in system design. For example, MIL-R-27542A [3] requires that human engineering be one of the reliability-program parameters, and AFBM Exhibit 58-10[4] requires that equipment failures be analyzed for human causes (see also Bracha[5]).

To some reliability engineers the human is a foreign element operating according to rules vastly different from those regulating equipment functioning. Perhaps this is because human factors relationships are often expressed less precisely and quantitatively than are equipment relationships. There are two main reasons for this:

1. The operator is much more complex than any machine presently devised or contemplated. No machine can presently be built to duplicate higher human functions such as perception, recognition, and decision making.

2. The operator is inherently less stable than the machine; he is influenced by many more conditions. The operator's performance is affected by his physiological condition, by fatigue, the work environment (noise, for example), amount of learning, and incentives and rewards, among numerous other factors.

It is possible, however, to deal with the human operator as one does with the equipment component: in terms of inputs and outputs. This provides the reliability

[1] A. V. Feigenbaum, "The Management and Engineering Approach to Product Quality," *Proc. Ninth Natl. Symp. Reliability Quality Control*, 1963, pp. 1–5.

[2] J. I. Cooper, "Human-initiated Failures and Malfunction Reporting," *IRE, Trans. Human Factors Electron.*, vol. HFE, pp. 104–109, September, 1961.

Shapero, A., et al., "Human Engineering Testing and Malfunction Data Collection in Weapon System Test Programs," *Wright Air Develop. Div. Tech. Rept.* 60-36, February, 1960.

[3] "Reliability Program Requirements for Aerospace Systems, Subsystems and Equipment," United States Air Force, May 21, 1963.

[4] "Proposed Reliability Program for Ballistic Missile and Space Systems," Air Force Ballistic Missiles Division, June 15, 1959.

[5] V. J. Bracha, "The Air Force Reliability Program: MIL-R-27542," *Proc. Eighth Natl. Symp. Reliability Quality Control*, 1962, pp 17–23.

engineer and human factors specialist with a common descriptive language and permits them to apply the same mathematical treatment to both man and machine.[1]

12.2 HUMAN FACTORS THEORY

Before proceeding to a more detailed exposition of human factors principles, it is necessary to summarize the concepts on which these principles are based. Just as the reader could not properly understand how radar functions without some knowledge of electromagnetic phenomena, so he cannot understand human performance without becoming acquainted with underlying psychological concepts.

An understanding of operator behavior is based on three parameters: *stimulus-input, internal response,* and *output-response.* A stimulus-input S is any physical change in the environment which is perceived by the operator as a change. A flashing indicator light, the appearance of a radar pip, the failure of a machine to run after it has been activated, the sound of a factory whistle—all are stimuli.

The internal response O (for operator, because this response occurs within him) is the operator's perception and integration of the physical stimulus S. Remembering, deciding, and interpreting are all internal responses. The output-response R is the operator's physical reaction to O and also his response to S. Talking, throwing a switch, writing, and batting a ball are output-responses.

All behavior is a combination of these three elements: $S \rightarrow O \rightarrow R$. Complex behavior consists of many chains of S-O-R interwoven and proceeding concurrently. Each element in the S-O-R chain is dependent on the successful performance of the element preceding it. Human errors are made when any of the elements in the chain are broken:

1. A physical change in the environment is not perceived as S.
2. Several S cannot be discriminated by the operator.
3. S is perceived, but its meaning is misunderstood.
4. S is correctly understood, but the correct output-response is unknown.
5. The correct R to S is known, but it is beyond the operator's physical capabilities.
6. The correct R is within the operator's capability, but R is performed incorrectly or out of sequence.

The implications for equipment design are quite clear: in order for the operator to respond adequately, the stimulus must be perceivable by the operator and it must demand a response which the operator is capable of producing. Hence, equipment and task (stimulus) characteristics—not only controls and displays but the entire nature of the job to be performed—must be tailored to the operator's capabilities and limitations or he cannot handle them. Everyone is aware of human physiological limitations that absolutely bar performance; for example, high-frequency radar and sonar impulses must be converted to visible and audible frequencies; the human is incapable of detecting infrared or X-ray stimuli; we cannot expect him to lift (single-handedly) an equipment weighing 750 lb. However, many less obvious equipment characteristics may also present barriers to efficient performance.

How closely the equipment is designed to fit the operator will therefore determine how readily he can use it. Hence the design of equipment must take into account limitations of body size, body weight, and reaction times to environmental stimuli. Other environmental characteristics (noise, temperature, etc.) also act as stimuli and affect the operator's capability to respond. When such characteristics have a negative effect on operator performance, they must be modified or eliminated.

It is not enough, however, for the operator to make a response which he considers correct. He must receive some verification or feedback from the consequences of his response. If he were never to see the effects of his actions, the operator could never be sure that what he was doing was correct or not. Consequently, he would never learn the correct response, or else his response variability would be very high.

The consequences of the operator's responses affect the environment and are

[1] P. C. Berry and J. J. Wulff, "A Procedure for Predicting Reliability of Man-Machine Systems," *IRE Natl. Conv. Record*, pp. 112–120, March, 1960.

perceived by him in terms of some modification of his environment. This modification has associated with it certain feedback stimuli which are entirely analogous to equipment phenomena with the same name. For example, the consequence of firing a cannon shell is to destroy a target; the appearance of the destroyed target tells the gunner he has aimed his guns correctly. Feedback may be seen indirectly as a change in equipment displays (for example, an indicator light turns green, indicating that a circuit is energized); or the change may be perceived directly, as when the image of an approaching target grows larger.

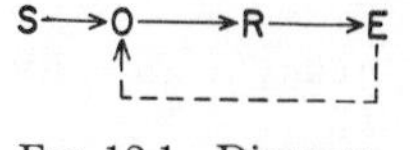

FIG. 12.1 Diagrammatic representation of feedback loop.

The correctness of feedback is determined by the operator's interpretation of the feedback stimuli. Certain stimuli have become associated with successful consequences; others with unsuccessful consequences. When the operator recognizes stimuli associated with successful or unsuccessful consequences, he knows enough to continue or to stop producing the same *R* to the same initiating *S*. This is shown in Fig. 12.1, where *E* represents the environment and the dashed arrow indicates the feedback stimuli.

In terms of reliable equipment design this means that equipment characteristics serve as both input and feedback stimuli to the operator. Equipment design must therefore be analyzed in terms of the *information* it provides to the operator about his environment. The goal is to supply maximum feedback information to the operator consistent with what he needs to know (without overloading his sensory channels). The designer must pursue his design in terms of the information requirements imposed by the task. He must then build that information into the equipment by providing required displays.

Unfortunately, even today the operator's need for information is often ignored, especially since our technological culture emphasizes automation ("make the equipment idiot-proof"). Recent studies,[1] however, suggest that more information rather than less is required.

12.3 MAN-MACHINE FUNCTION ALLOCATION

The starting point of all design is the allocation of functions to the equipment and/or the operator. The nature of the functions to be performed determines the characteristics of the equipment and the role of the operator.

Whether a given function should be performed by the equipment or the operator or by both jointly depends on which can perform that function more effectively in the operational situation. To determine this, a comparison must be made between the strong points of the machine and the strong points of the operator. These are shown in Table 12.1. It should be noted that the qualities described are *generic* qualities. Whether they hold for a particular equipment or task depends on the system's operating conditions. For example, it is obvious that machines possess incomparably more strength than humans; however, this consideration is irrelevant in a situation where the force that must be exerted is only 5 lb.

It is obvious from Table 12.1 that there is no simple answer to the question of whether the machine or the man should be assigned a particular function. One ground rule that has been suggested[2] is that if, during the planning of the system, the designer cannot specify the content of all the inputs and outputs of a function, that function should be assigned to the operator, since only man has the flexibility to make decisions about unexpected events.

So many situational factors affect human performance that the range of effective operator response may vary from zero to 0.9999 reliability. It is difficult, therefore, to develop generalizable relationships that cut across different types of systems and

[1] A. D. Swain and J. G. Wohl, "Factors Affecting Degree of Automation in Test and Checkout Equipment," Dunlap and Associates Report TR-60-36F, Stamford, Conn., March, 1961.

[2] A. Shapero et al., "A Method for Functions Analysis and Allocations (Final Report)," Contract AF 33(616)-6541. Stanford Research Institute, Menlo Park, Calif., August, 1961.

Table 12.1 Man-Machine Function-allocation Chart

Function	Men	Machines
Flexibility	Able to handle low-probability alternatives, unexpected events. High flexibility.	Flexibility limited, unexpected events cannot be adequately handled.
Ability to generalize	Able to recognize events as belonging to a given class, to abstract similarities, and to disregard conflicting characteristics.	Zero, or very limited abilities. Can react only to physical properties set up.
Ability to learn	Previous experience can be used to change subsequent behavior.	No change unless equipment configuration is changed.
Sensitivity	Sensitive to wide range of kinds of inputs. Can transduce many kinds of physical stimuli simultaneously. Small amounts of physical energy can be detected.	A given instrument is sensitive to only one kind of input. Thresholds not generally as low as human's. However, can sense energies beyond human spectrum.
Cost	Inexpensive and in good supply. Amount of training required depends on complexity of task.	Cost rises rapidly as complexity increases. Supply limited.
Weight	Light.	To reproduce human functions, would have to be immensely heavy.
Originality	Able to sense and report events and observations not of immediate concern.	No originality in discovering incidental intelligence or the relationships between functions.
Monitoring capacity	Poor.	Good.
Expendability	Nonexpendable, safety considerations have to be imposed.	Expendable.
Manipulability	Ability to perform precise manipulations required in maintenance.	To perform same tasks, machine would be extremely costly and complex.
Reliability	Subject to errors. However, reliability of manual equipment is higher than that of more complex automatic equipment because of smaller number of components.	Makes no errors, but reliability of automatic equipment is lower than that of manual equipment because of increased number of components.
Fluctuation in performance efficiency	Highly variable from day to day.	Minimal variability.
Reaction time	Relatively slow. Shortest human response about 200 msec.	Usually as fast as relay operating time, microsecond lags.
Physical force	Relatively weak unless aided by machinery.	Practically limitless in power.
Boredom or fatigue	A repetitive task results in error. Rest period required. 8 to 10 hour maximum efficient work expected.	Capable of many operations without decrement. Only physical limitations, such as heat, corrosion, or wearout, need be considered.
Environmental requirements	Can exist only in a very narrow band of environmental conditions. Physiological maintenance requirements are extensive.	Can tolerate many fluctuations in environment. Restricted only by design specifications.

Table 12.1 Man-Machine Function-allocation Chart *(Continued)*

Function	Men	Machines
Tracking ability	Comparatively poor tracking characteristics, although satisfactory in wide range of situations. Can change performance constants to produce best attainable system performance in any situation.	Good tracking characteristics. Considerable complexity needed, however, to track well in all conditions.
Channel capacity	Limited. Has maximum amount of information that can be handled at a given time.	Can be made arbitrarily large. Capacity limited only by design.
Overload operation	May be able to perform better in certain cases of overload than machines; can tolerate temporary overloads without complete disruption.	May break down completely if overloaded; with information-handling capacity fixed, overload leads to system disruption.
Survival	Interested in survival. Situations with possibility of danger are "stress-inducing," resulting in behavior degradation.	Not "conscious" of danger. No behavior decrement in the face of destruction.
Computational ability	Comparatively slow and poor computers.	Excellent and very rapid computers.
Memory storage	Poor short-term storage. Excellent long-term storage.	Excellent short-term storage. Long-term storage very expensive.
Deductive logic	Cannot always be expected to follow optimum strategy. The right premises sometimes lead to wrong conclusions.	Excellent in deductive logic, can exhaust all theorems from a set of postulates. Can store and use optimum strategy for high-probability situations.
Inductive logic	Can go from specific cases to general rules or laws.	No good at inductive logic.
Distraction	Easily distracted by competing stimuli.	Cannot be distracted by competing stimuli.

their missions. The trade-off considerations that we can apply to man-machine relationships are consequently largely qualitative.

Since the trade-off answer depends partly on the nature of the system and the job and partly on the way in which the question is asked, a human factors analysis of the system must be performed in each design situation. Cost, weight, hazard, the state of technology—all must be considered. Moreover, no single man-machine function will resolve a trade-off; resolution usually requires a consideration of multiple relationships. It is therefore impossible to recommend to the reader, in advance of analyzing the particular problem, that he should automate or use operators or some combination of the two and, if the latter, under what circumstances.

There are, however, some very definite advantages to using man in a system, particularly where his role is to act as the mission controller and to perform on-board maintenance. For example, Grodsky[1] has shown that the reliability of a space system with a maintenance man aboard is significantly greater than that of the same system with automatic maintenance.

[1] M. A. Grodsky, "Risk and Reliability," *Aerospace Eng.*, vol. 21, pp. 28–33, January, 1962. See also, same issue, P. M. Fitts, "Functions of Man in Complex Systems," pp. 34–39.

In performing trade-offs, the following hints may be useful:

1. Where the system mission contains major elements of uncertainty (unexpected events may occur and decisions may be required) the need for an operator is greatest.
2. Human error rate increases as a function of the constraints and demands imposed upon the operator. When task requirements tend to push the operator to his performance limits, automation of the affected functions may be required.
3. Human error rate is proportional to the number of series-interactive human links in the system. All other things being equal, human error rate is directly proportional to the length of tasks and procedures, the number of controls and displays to be operated, and the number of communications, decisions, and calculations required by the system. All this means is that the more operators in a system and the more they have to do, the greater the probability of unreliable performance. On the other hand, operator redundancy will, just as in equipment reliability, tend to increase the probability of errorless performance.
4. All other things being equal, experience suggests that the reliability of highly automated equipment, for example, an automated checkout equipment, readily degrades under operational conditions, so that it is wise to add supplementary or backup human functions to the system.

One way to decide whether a function allocation or trade-off is desirable is to determine the relative reliability of the equipment or system in terms of the two conditions represented (human or machine). The determination can be made by comparing the anticipated reliability of a human function with the anticipated reliability of the same function when automated.

Let us assume that we wish to compare the reliability of activating a missile firing circuit when preparations are complete by manually pushing the commit button vs. automatically activating the same circuit. (In the example below hypothetical values for equipment operation have been assumed.)

Example 12.1

Manual sequence

Prep. comp., circuit energized →
0.9999 ×
prep.-comp. lt. becomes green → operator perceives lt. →
0.9999 × 0.9998 ×
operator decides to fire → operator pushes commit button
0.9990 × 0.9997
= 0.9983

Automatic sequence

Prep. comp., circuit energized →
0.9999 ×
firing-circuit relay energized → relay energizes firing circuit
0.9999 × 0.9999
= 0.9997

It is obvious from a consideration of the relative reliability of the two sequences that the automatic means of firing the missile would be more reliable. This is, however, not the only factor to be considered. It is unlikely that the decision to automate would be based solely on relative reliabilities. Any trade-off must consider the consequences of the failure of system elements and should therefore be accompained by a failure modes and effects analysis, not only for the equipment but also for the human. For example, the operator has the capacity to withdraw his decision on instructions from higher command, even though the preparation-complete light is green. No such decision capability exists in the automatic loop, and to build it in would be inordinately expensive.

Obviously, a value judgment over and above relative failure rates is involved. The basic question which must be decided is what the designer gains or loses by introducing the man into his system. This is a decision which cannot be equated with failure rates and must be decided by examining all the possible modes of response required of the system.

12.4 HUMAN FACTORS IN DESIGN

The equipment designer often thinks of human factors in terms of principles or guide rules which can be used as a sort of checklist: if every rule is checked off, the design is satisfactory for the operator. The uncritical application of these principles to design will not, however, automatically produce a superior product. Design for operability and maintainability must proceed from a systematic analysis of the interaction between the requirements of the task and the limitations of the operator.

The designer often thinks of human engineering as being largely concerned with cockpit controls and displays, because human engineering has been commonly associated with aircraft design. But all manually operated and maintained equipment (the latter includes practically everything), whether it be ground equipment, facilities (brick and mortar), test and checkout equipment, even airborne equipment in unmanned vehicles—all have human factors implications.

The consideration of human-engineering principles must, moreover, start and proceed concurrently with design because that design usually goes through several preliminary configurations before a final one is decided upon. Each configuration may imply a decision which is crucial to operability and maintainability. The designer who waits until the final blueprint is about to be released before examining its human-engineering characteristics will find those characteristics frozen as solidly as if they were cast in concrete.

The application of human-engineering principles and the examination of design to ensure that those principles have been incorporated must stem from a human factors analysis of system requirements. The analysis proceeds through the following steps:

12.4a Allocation of Functions. The starting point of all design is the allocation of functions between equipment and operators which has been described in the preceding section. If the reliability engineer acts as human factors design consultant or critic, the questions he must ask the designer first are:

1. What functions are to be performed by the equipment and what ones by the operator?
2. What was the rationale (justification) for this allocation?
3. Under all the operational circumstances, is the allocation justified?

Occasionally the designer may not have explored all the details of this allocation. Therefore, it is necessary to determine what allocation was used as the basis of design and why.

12.4b Automation. Stemming directly from Sec. 12.4*a* is the question of automation. As a consequence of the technological culture in which he works, the designer has a tendency to automate whenever the state of the art permits him to, even when there is no requirement that he do so. As was pointed out earlier, excessive automation is often responsible for overly complex equipment with high malfunction rates and high down-time rates. The designer should be warned that where no significant loss in performance or danger to personnel or equipment will result from manual operations, the equipment ought *not* to be automated.

12.4c Operator Tasks. Also as a follow-on from Sec. 12.4*a*, the designer must know before he begins his design what operator tasks will have to be performed with his equipment. The reliability engineer should determine whether the designer knows what these tasks are and has incorporated their requirements into his design.

12.4d Operator Stress Characteristics. Certain requirements of the task tend to approach the limits of operator capabilities; when this happens, error inevitably arises. Both the designer and the reliability engineer should know what these characteristics are, so that they can look for them in the task requirements and attempt to eliminate them or design around them. Some of these characteristics are:

1. Operator must perform individual step(s) at high speed and/or at highly precise times (at −17:03:04 exactly, for example).

2. Individual step(s) in the operator task must be coordinated with another step(s) performed by another operator.
3. Operator must compare two or more displays rapidly.
4. Two or more displays are difficult to discriminate.
5. Changes in sequencing of equipment displays are very rapid.
6. The sequence of steps required to perform a task is overly long.
7. There is inadequate feedback to the operator to determine the correctness of his responses.
8. Decisions must be made on the basis of information from several sources.
9. The time required for decision making is extremely short.
10. Two or more controls must be operated simultaneously at high speed.
11. The nature and/or timing of stimulus inputs cannot be anticipated by the operator.
12. Prolonged monitoring by the operator is required.

12.4e Equipment Information. Earlier we spoke of the equipment as providing certain information required by the operator and maintenance man to do their jobs. This information is of three types:

1. *Feedback information.* For example, when the operator throws a switch to activate a circuit, a light should illuminate to indicate that the circuit has been activated (not just the relay, but the circuit).
2. *Preventive maintenance information.* The designer should analyze the preventive maintenance to be required of his equipment to determine whether any special displays will be needed if the maintenance man must calibrate or service the equipment.
3. *Malfunction symptom information.* Many human factors principles also relate to maintainability. These involve the maintenance man's response to the stimulus characteristics of the failed equipment. In order to perform maintenance properly, the maintenance man must be able to detect and diagnose a malfunction and then remove and replace the failed component. In the first case the problem for the designer is one of supplying enough information for detection and diagnosis; in the second case the problem is one of accessibility. (See Sec. 11.)

The designer can provide the information needed for detection and diagnosis of malfunctions by examining his design to determine which malfunctions are likely to occur and which will be most critical to the system. (Admittedly some of this information may not be easy to secure, but reliability failure modes and effects analysis will help to provide it.) For these malfunctions he can attempt to provide symptomatic displays which indicate that a malfunction has occurred or is in process of occurring.

12.4f Accessibility. Next to the problem of diagnosis the greatest maintenance problem is presented in securing access to the failed component to remove and replace it without damaging adjacent components. The designer should examine his drawing to determine whether sufficient hand space exists to grasp all major modules, whether access spaces are large enough and covers are fastened with only the required number of screws.

12.5 HUMAN FACTORS DESIGN PRINCIPLES

Immediately following are a series of human factors design principles which the reliability engineer should use in the process of reviewing designs. Much of this material has been taken from MIL-STD-803A-1.[1] Obviously, for reasons of space, it is impossible to supply all the relevant factors to be considered. The list of recommended readings at the end of the section will supply additional sources of material to which the interested reader can refer.

12.5a Control Panel Arrangement

1. Controls and displays should be arranged according to the procedure to be followed in operating the equipment; or, if that procedure is variable, controls and

[1] "Human Engineering Design Criteria for Aerospace Systems and Equipment, Part 1, Aerospace System Ground Equipment," Space Systems Division, Air Force Systems Command, Jan. 27, 1964.

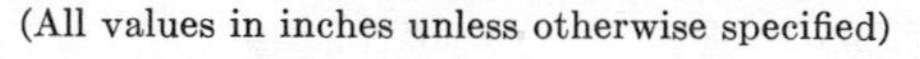

Table 12.2 Standard Console Dimensions
(All values in inches unless otherwise specified)

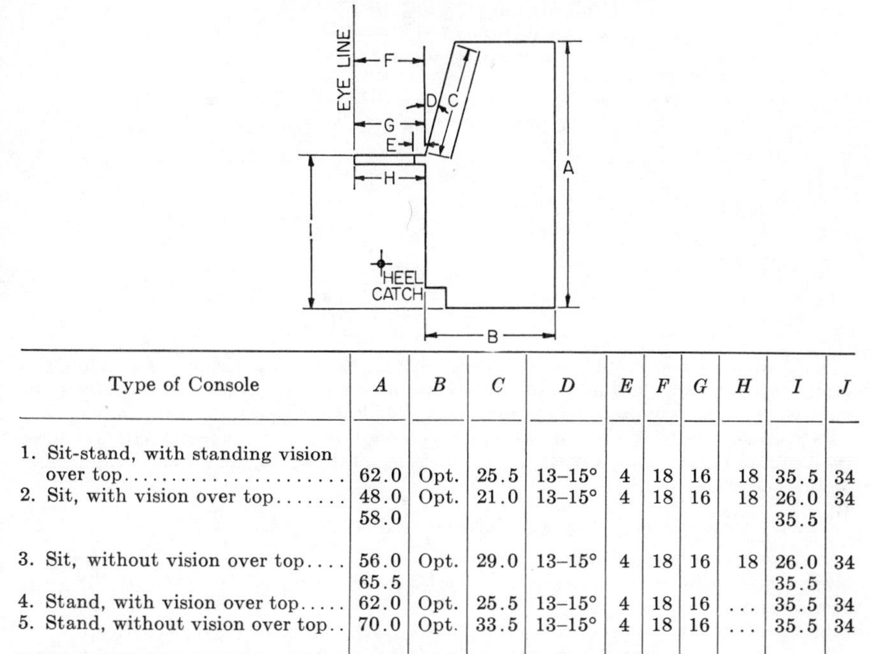

Type of Console	*A*	*B*	*C*	*D*	*E*	*F*	*G*	*H*	*I*	*J*
1. Sit-stand, with standing vision over top.	62.0	Opt.	25.5	13–15°	4	18	16	18	35.5	34
2. Sit, with vision over top.	48.0 58.0	Opt.	21.0	13–15°	4	18	16	18	26.0 35.5	34
3. Sit, without vision over top.	56.0 65.5	Opt.	29.0	13–15°	4	18	16	18	26.0 35.5	34
4. Stand, with vision over top.	62.0	Opt.	25.5	13–15°	4	18	16	...	35.5	34
5. Stand, without vision over top.	70.0	Opt.	33.5	13–15°	4	18	16	...	35.5	34

A. Maximum total console height from floor
B. Console depth at base
C. Minimum vertical dimension of console panel (excluding sills)
D. Console panel angle (from vertical, in degrees)
E. Minimum depth of pencil shelf
F. Distance from eye line to panel base
G. Minimum writing surface to edge of pencil shelf
H. Minimum knee clearance
I. Height from floor of writing surface and pencil shelf
J. Maximum console width (not shown)

displays which are operated together or which have an associated function should be located together.

2. Control-display organization should be such that visual displays occupy central panel areas and controls occupy peripheral areas.

3. Each control should be as close as possible to the display it controls.

4. The most important and frequently used controls should be located in the most favorable position for ease of reaching and right-hand operation.

5. Functionally similar or identical controls and displays should be located consistently from panel to panel.

6. Control-equipment consoles should be designed in accordance with standard console dimensions as shown in Table 12.2.

7. Controls and displays should be located according to Table 12.3.

8. The following color schemes for control equipment are recommended (all colors are from Federal Standard 595):

a. Console exterior, 24300, green
b. Console interior, 26622, gray
c. Console panels, 26492, gray
d. Lettering, 37038, black
e. Controls, 17038, black, or 26231, gray

Table 12.3 Location of Controls and Displays

Operation	Standing operator, distance above standing surface, in.		Sitting operator, distance above sitting surface, in.	
	Displays	Controls	Displays	Controls
Routine:				
Maximum......	72	72	44	35
Minimum.......	40	34	8	8
Precision:				
Maximum......	68	57	35	30
Minimum.......	50	34	16	8

9. Control panels should be laid out according to *module* design. A module is a grouping of controls and indicators based on the interrelationship of the controls and indicators in performing a system function. Module design locates functional groupings within a colored block. A dark-colored module contrasted against a light-colored background produces an association of controls and indicators, thus encouraging correct operation.

12.5b Control and Display Characteristics. *Controls*

1. The direction of control movement should, with practically no exceptions, always be to the right (clockwise) to increase and to the left (counterclockwise) to decrease.
2. For high precision over wide ranges of adjustments, use rotary controls.
3. Toggle switches should be used only for control functions requiring two discrete positions.
4. Guard critical switches against accidental activation.
5. Requirements for selecting controls are shown in Table 12.4.

Table 12.4 Requirements for Selecting Controls

	Push button	Toggle switch	Rotary control	Knobs
Requirements for selection	Momentary contact	Two discrete positions	Three or more detent positions	Accurate adjustment required
Desirable characteristics	Concave surface, positive indication of activation		No more than 24 positions	Size unimportant
Dimensions:				
Maximum...	Diam: none	Arm: 2 in.	Pointer length* Pointer width: 1 in.	Fingertip grasp knob: 4 in.
Minimum...	Diam: ½ in.	Arm: ½ in.	Pointer length: 1 in. Pointer width*	Fingertip grasp knob: ⅜ in.
Displacement	⅛ in.	30°		

* No limitation set by operator performance.

Displays

1. Only the information necessary for the operator to perform his task should be displayed, and then only to the degree of specificity and accuracy required for the job. Requirements for decoding, computing, or calculating that information should be kept to a minimum.

2. If a scale is used, it should start at zero. Graduations should progress by 1, 2, or 5 units or decimal multiples thereof. Numbers should increase in a clockwise, left-to-right, or from bottom-up direction. Whole numbers should be used in numbering major graduation marks, and there should be no more than nine intermediate marks and fewer if possible. All numbers should be oriented upright. The pointer should just meet but not overlap the shortest scale marking, and the pointer should be as close as possible to the dial face. Displays should be coded by color, size, location, or shape.

3. The following meanings should be assigned to colored indicators:

Red: system failure or malfunction
Flashing red: emergency conditions, personnel or equipment disaster
Amber: marginal condition, hazardous condition approaching
Green: system in tolerance, OK to go ahead
White: action or test in progress

4. Legend lights should be used in lieu of simple indicator lights.

12.5c Labels. Labels should primarily describe equipment functions and only secondarily their engineering characteristics. They should be as concise and as familiar as possible. They should be capitalized and oriented horizontally to be read from left to right, not vertically. Abbreviations should be used only when necessary. Futura or Airport Semibold type and Groton Extended engraving should be used for letters. Futura Medium or Tempo Bold type and Groton Condensed engraving should be used for numbers.

The following sizes of letters and numerals should be used:

Height, in.	*Viewing distance, in.*
0.09	20 or less
0.17	20–36
0.34	36–72
0.68	72–144
1.13	144–240

12.5d Illumination and Noise Levels. Both illumination and noise levels may have critical effects on operator performance. Equipment to be used in areas where maintenance demands the continuous presence of personnel, but where a minimum of voice communications is required, should not produce noise levels[1] in excess of those defined for condition *A*; see the accompanying table. Equipment to be used regularly by personnel in operational situations should not produce noise levels in excess of those

Table 12.5 Illumination Levels

Task conditions	Level, foot-candles	Type of illumination
Very difficult and prolonged visual tasks with objects of low brightness contrast; high speed and extreme accuracy required	100 or more	Supplementary type of lighting; special fixture such as desk lamp
Small detail, fair contrast, close work, speed not essential	50 or more	Supplementary type of lighting
Ready rooms, launch control facilities, prolonged reading, assembly	25 or more	Local lighting; ceiling fixture directly overhead
Passages, tunnels, stairways, occasional reading, washrooms, power plants	10 or more	General lighting
Emergency survival	5 or more	General or supplementary lighting

[1] Measured in decibels (db) at a distance 3 ft from the equipment.

defined for condition *B*. Equipment to be used in executive offices, conference rooms, etc., especially where telephone use is highly critical, should not produce noise levels in excess of those defined for condition *C*.

Condition	Noise levels, db, for octave bands, cps			
	20–75	75–150	150–300	300–600
A	100	89	82	76
B	79	68	59	52
C	76	64	55	48
Condition	600–1,200	1,200–2,400	2,400–4,800	4,800–10,000
A	73	70	68	67
B	48	45	43	42
C	43	40	38	37

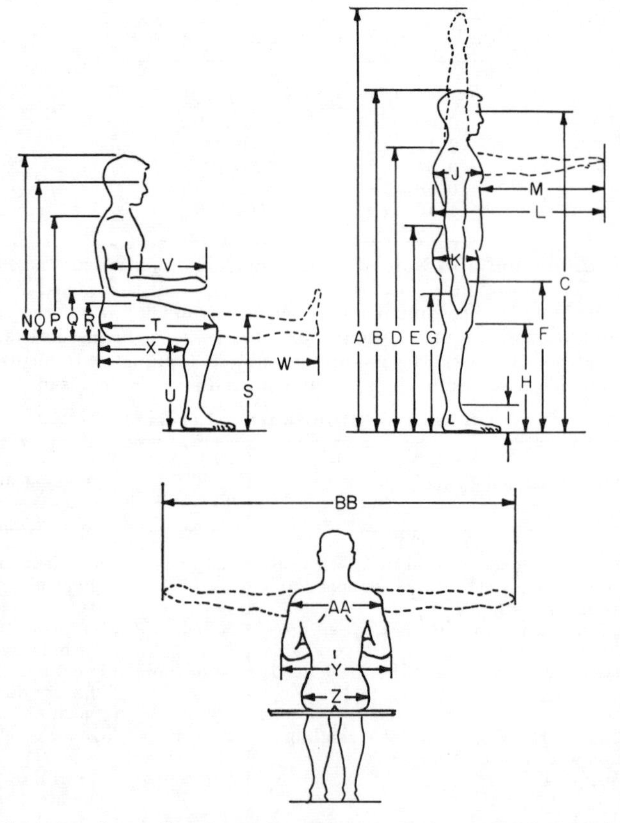

FIG. 12.2 Seated and standing body dimensions. (Refer to Table 12.6 for numerical values.)

Table 12.6 Standing and Seated Body Dimensions, in Inches

Parameter	5th Percentile	Mean	95th Percentile
Standing			
A. Vertical reach	77.0	83.6	90.3
B. Stature	65.2	69.1	73.1
C. Eye height	60.8	64.7	68.6
D. Shoulder height	52.8	56.5	60.2
E. Elbow height	40.6	43.5	46.4
F. Wrist height	31.0	33.5	36.1
G. Knuckle height	27.7	30.0	32.4
H. Kneecap height	18.4	20.2	21.9
I. Ankle height	4.9	5.6	6.8
J. Chest depth	8.0	9.1	10.4
K. Buttock depth	7.6	8.8	10.2
L. Functional reach	29.7	32.3	35.0
M. Depth of reach	23.0		
Seated			
N. Seated height	33.8	35.9	38.0
O. Eye height	29.4	31.5	33.5
P. Shoulder height	21.3	23.3	25.1
Q. Elbow rest height	7.4	9.1	10.8
R. Thigh clearance height	4.8	5.6	6.5
S. Knee height	20.1	21.7	23.3
T. Buttock knee length	21.9	23.6	25.4
U. Popliteal height	16.7	17.0	18.2
V. Forearm-hand length	17.6	18.9	20.2
W. Buttock-leg length	39.4	42.7	46.1
X. Buttocks to inside knee	17.7	18.9	20.1
Y. Elbow-to-elbow breadth	15.2	17.3	19.8
Z. Hip breadth	12.7	14.0	15.4
AA. Shoulder breadth	16.5	17.9	19.4
BB. Span	65.9	70.8	75.6

12.5e Location of Components

1. Whenever possible, equipment should be modularized so that the weight of removable units does not exceed 45 lb. Equipment should, if at all possible, be mounted in roll-out drawers or on racks.
2. All units designed to be removed and replaced should be provided with handles or other means of being carried; these should be located over the unit's center of gravity.
3. Parts should be mounted in an orderly array on a "two-dimensional" surface and not "stacked" one on the other.
4. Large units which are difficult to remove should be so mounted that they do not prevent convenient access to other parts.
5. Units should not be placed in recesses, behind or under stress members, floor boards, etc.
6. The design of equipment should be in accordance with the anthropometric standards described in Fig. 12.2 and Table 12.6.

12.5f Safety Requirements

1. Design should be such that commonly worked on parts, fasteners, or test points are not located near exposed terminals or moving parts.

Table 12.7 Clearance Dimensions, in Inches

Dimension	Minimum	Optimal	Arctic conditions
Squatting work space:			
A. Height	48	.....	51
B. Width	27	36	40
Optimum display area	...	27–43	
Optimum control area	...	19–34	
Stooping work space:			
C. Width	36	40	44
Optimum display area	...	32–48	
Optimum control area	...	24–39	
Kneeling work space:			
D. Width	42	48	50
E. Height	56	.....	59
F. Optimum work point	...	27	
Optimum display area	...	28–44	
Optimum control area	...	20–35	
Kneeling crawl space:			
G. Height	31	36	38
H. Length	59	.....	62
Prone work or crawl space:			
I. Height	17	20	24
J. Length	96		

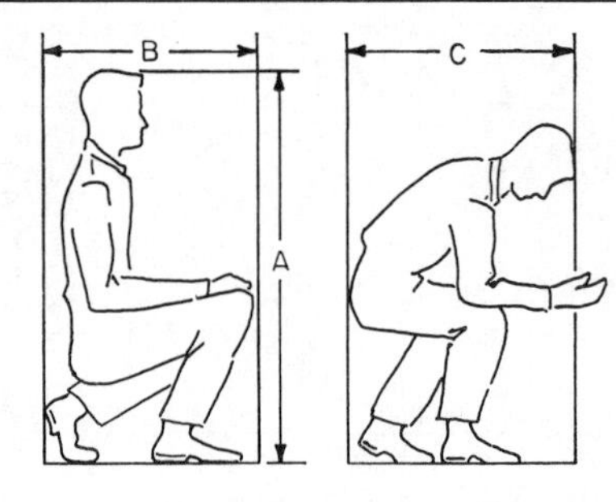

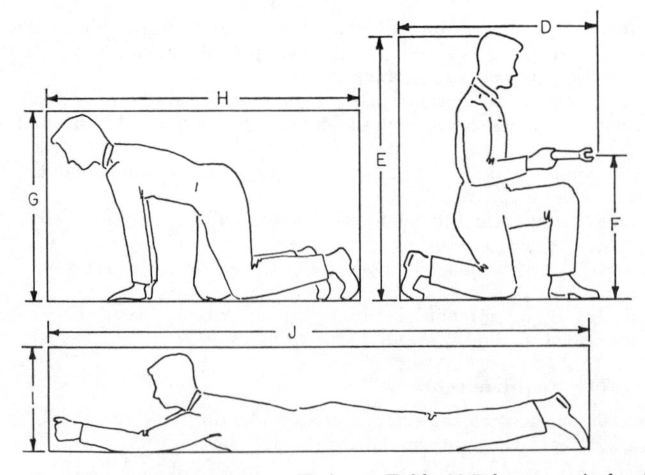

FIG. 12.3 Clearance dimensions. (Refer to Table 12.7 for numerical values.)

2. Guards or shields should be provided to prevent personnel from coming into contact with dangerous parts and to prevent devices from being inadvertently activated.

3. Vital or fragile equipment should be so located that they will not be used for handholds or footrests, walked on or rolled over by vehicles, or located near areas where heavy maintenance is performed.

4. Fail-safe design should be used wherever possible.

5. Tubes and plug-in items should be secured with clamps.

6. Adequate working space should be provided in accordance with standards of Fig. 12.3 and Table 12.7.

12.5g Connectors and Fasteners

1. Only the minimum number and type of fasteners required to secure equipment should be used. Fasteners should be few and large rather than numerous and small.

2. Locate fasteners so that they can be removed without first removing other units.

3. Connectors should be far enough apart that they can be grasped firmly for connecting and disconnecting (0.75 in. if bare fingers, 1.25 in. if bare hands or gloved fingers are to be used).

4. Cross-connection should be prevented by selecting:

a. Different sizes or types of connectors when these are used with similar, adjacent leads
b. Lines of different size
c. Special guide pins and keys
d. Different sizes of prongs and prong receptacles

12.5h Accessibility

1. Whenever possible, equipment should be left exposed for maintenance and major units and assemblies should be designed with removable housings. Accesses should be designed to avoid the necessity for removing components. Access spaces should be located on the same face of the equipment as related controls and displays and away from potential safety hazards. The bottom edge of a limited access should be no lower than 24 in. and the top edge no higher than 60 in. from the floor or work platform. (See Table 12.8.)

Table 12.8 Access Dimensions

Minimal Two-hand Access Openings	
Reaching with both hands to depth of 6 to 25 in.:	
Light clothing.......	5 in. high by 8 in. or three-fourth depth of reach*
Arctic clothing......	7 in. high by 6 in. plus three-fourth depth of reach
Reaching full arm's length (to shoulders) with both arms	Width = 19½ in. Height = 4 in.
Inserting box grasped by handles on the front	½ in. clearance around box, assuming adequate clearance around handles
Inserting box with hands on the sides:	
Light clothing.......	Width = box plus 4½ in. Height = 5 in. or ½ in. around box*
Arctic clothing......	Width = Box plus 7 in. Height = 8.5 in. or ½ in. around box*

Note. If hands will curl around bottom of box, allow an additional 1½ in. in height for light clothing and 3 in. for arctic clothing.

* Whichever is larger.

Table 12.8 Access Dimensions (*Continued*)

Minimal One-hand Access Openings, Width by Height	
Empty hand to wrist:	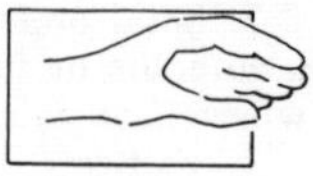
Bare hand, rolled	3.75 in. square or diam
Bare hand, flat	2.25 by 4.0 in. or 4.0 in. diam
Glove or mitten	4.0 by 6.0 in. or 6.0 in. diam
Arctic mitten	5.0 by 6.5 in. or 6.5 in. diam
Clenched hand, to wrist:	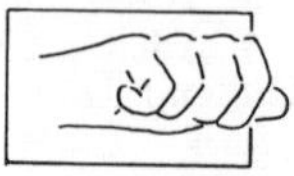
Bare hand	3.5 by 5.0 in. or 5.0 in. diam
Glove or mitten	4.5 by 6.0 in. or 6.0 in. diam
Arctic mitten	7.0 by 8.5 in. or 8.5 in. diam
Hand plus 1-in.-diam object, to wrist:	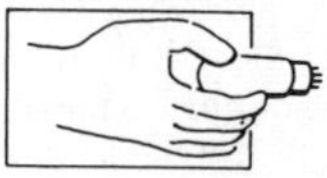
Bare hand	3.75 in. square or diam
Glove or mitten	6.0 in. square or diam
Arctic mitten	7.0 in. square or diam
Hand plus object over 1 in. in diam to wrist:	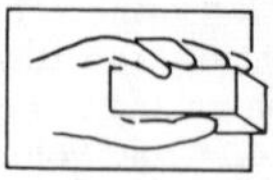
Bare hand	1.75 in. clearance around object
Glove or mitten	2.5 in. clearance around object
Arctic mitten	3.5 in. clearance around object
Arm to elbow:	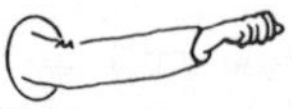
Light clothing	4.0 by 4.5 in. or 4.5 in. diam
Arctic clothing	7.0 in. square or diam
With object	Clearances as above
Arm to shoulder:	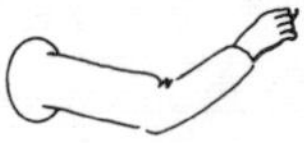
Light clothing	5.0 in. square or diam
Arctic clothing	8.5 in. square or diam
With object	Clearances as above
Minimal Finger Access to First Joint	
Push-button access:	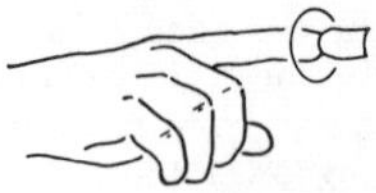
Bare hand	1.25 in. diam
Gloved hand	1.5 in. diam
Two-finger twist access:	
Bare hand	2.0 in. diam
Gloved hand	2.5 in. diam
Vacuum-tube insert (tube held as at right):	
Miniature tube	2.0 in. diam
Large tube	4.0 in. diam

2. Internal components should be packaged and mounted so that:

a. There is adequate access and wrenching space around fasteners.

b. Components to be serviced and repaired in position are between hip and shoulder height.

c. All large, heavy, or awkward units are so located that they can be slid or pulled out rather than lifted out (by mounting on sliding racks, in roll-out drawers, etc.) whenever practicable.

d. When it is necessary to place one unit behind or under another, the unit requiring more frequent maintenance should be most accessible.

e. Unshielded electron tubes should be 1.5 tube diameters apart.

f. Resistors and soldering points should be separated by at least 3/16 in.

g. Solder connectors, terminals, relays, etc. should be at least 1/4 in. apart.

h. The distance between items requiring insertion should be equal to the length of the longest related insert plus 1/2 in.

12.6 HUMAN FACTORS IN PRODUCTION

To understand why the reliability engineer should be concerned about human factors in production, two points must be made:

1. Anywhere from 20 to 50 per cent or more of all equipment failures result from human causes.
2. From 50 to 80 per cent of all human-initiated failures result from inadequate workmanship primarily in the factory but also in equipment installation in the field.

Since equipment failures resulting from poor workmanship seriously degrade system reliability, the higher reliability resulting from design improvements may well be completely nullified by inadequate production of that design. (See Sec. 13.) The factors in production that predispose to workmanship errors are:

1. Inadequate work space and poor work layout
2. Poor illumination, high temperature, and high noise level
3. Inadequate human-engineering design of machinery, hand tools, and checkout equipment
4. Inadequate methods of handling, transporting, storing, or inspecting equipment
5. Inadequate job-planning information and information transmission
6. Inadequate or unavailable operating instructions or blueprints
7. Too little or poor supervision
8. Inadequate selection, training, and/or motivation

The effect of these factors is to create a work situation favorable to the commission of production errors. That is, the probability of error increases as a function of inadequate production-system characteristics. For example, if a worker assembles a circuit incorrectly because his blueprint information is out of date or unclear, the fault lies with the system that has provided the incorrect or out-of-date or unclear information.

The kinds of errors which production workers typically make are probably familiar to all reliability engineers. These errors are of two main types:

1. Failure to fabricate or assemble the product to agree with design instructions: use of the wrong materials or parts, incorrect wiring of circuits, incorrect soldering or welding, forcing parts together, damaging parts with tools (denting, scoring, etc.), failing to include all required parts, cutting material to wrong dimensions.
2. Failure to check equipment to inspection requirements as reflected in rejection of correctly operating equipment or, even worse, acceptance of incorrectly operating equipment.

The second kind of error is responsible for much defective equipment in the hands of users: an estimated 25 per cent of all HIF reported from five missiles was ascribed to production error and therefore must have passed final acceptance inspection.

If we are correct in assuming that individual workmanship errors largely reflect production-system deficiencies, then it does little real good to attempt to eliminate the individual error (after it has occurred). The cost in time of tracing back on individual error or failure to its source (in one system found to be approximately eight hours per failure) is prohibitive, even when the number of failures is not excessive. Standard corrective-action processes in the factory often focus on correction of an on-going error situation but completely ignore the prevention of such situations. The reliability engineer may have more success by systematically evaluating the characteristics of the production process to discover where they are deficient and can be improved. In this connection there are a number of techniques which can be useful:

12.6a Review of Reports. Systematic review of all reports written on failed and reworked items in the factory to determine the percentage that are human-initiated and to pinpoint characteristic types of workmanship errors. Unfortunately, because of the lag in time between the occurrence of the failure or reject and the receipt of its report by the reliability engineer, and because the report characteristically does not pinpoint the factory department and individual responsible for the failure, failure reports are at best suggestive. Another difficulty is that failure-reporting personnel are reluctant to assign the responsibility for a failure to themselves or others, preferring instead to describe the failure as functional. At the same time, failure descriptions are usually vague and imprecise.

12.6b Process-flow Observation. Process-flow observation involves following the performance of a production operation as the item being manufactured passes through whatever stages of fabrication, assembly, and/or checkout are involved in the operation. During the period of observation, judgments are made by the reliability engineer concerning the various characteristics of the production process observed. These characteristics, which include the operating environment, personnel characteristics, work methods, logistics, equipment design, work-information channels, and supervision, can be studied by means of a checklist which is filled out by the engineer. (For a description of the checklist, see Meister.[1]) The engineer regulates his observation by the flow of the product through the assembly line and completes a new checklist for each major stage or function observed. Following an item through the production process ensures that every relevant aspect of that process is covered, as well as provides a valid sampling on which to base conclusions.

After the checklist observation is completed and analyzed, resulting recommendations should be submitted to production management. Efforts should be made to have production supervision handle their own problems. At the end of 30 to 60 days, the procedure should be repeated with the same department to see that changes in production characteristics have occurred as a result of implementing these recommendations. (See also Sec. 13.)

12.6c Product-improvement Training. Process-flow observation will, of course, be supplemented by interviews with individual workers, because the latter know most about their tasks. However, individual interviews are very costly in terms of time. A more effective technique developed at General Dynamics/Astronautics[2] is something called *Product-improvement Training* (*PIT*) which combines motivational training with the group interview for data-collection purposes.

PIT is a 4-hour, 1-hour-a-day class specifically set up for production personnel on a departmental basis. Each class is comprised of no more than 25 workers, with a desired mean of 15. Prior to beginning the class a 10-day check is made of the errors committed by the production department selected for training. Reject and rework reports and inspection "squawks" will provide this information. This forms the "before" part of the statistical evaluation of PIT. A comparison error check is made 30 days after training. The difference between the number of before and after errors measures the effect of PIT in terms of error reduction.

The class is basically a free-discussion group in which permissiveness is deliberately encouraged. The class is run democratically, avoiding as much as possible any suggestion of the traditional teacher-student relationship. In the first class session the mechanics of the class are explained. Attendees are encouraged to report problems they have encountered and working conditions which they feel are hampering efficient production. The 10-day-check errors are used as material to induce discussion, but in this discussion there is no imputation of blame. In successive sessions students bring up their own problems. These are recorded by a special reporter, and an investigation of any problem with substance is promised. Naturally, the anonymity of informants is preserved.

In order to ensure that production supervisers are favorable to the training sessions, the mechanisms are explained to them also. They are not permitted to attend the first three sessions, but they are invited to attend the last session and indicate their support of the program and review the problems needing solution.

For almost all production departments trained, error reduction as a function of PIT has been highly significant, ranging up to 60 per cent. However, it must be pointed out that the motivational effect is transitory; the training must be repeated, preferably at 3-month intervals. The value of PIT for reducing production error lies not only

[1] D. Meister, "Human Reliability Production Audit," *Proc. Ninth Natl. Symp. Reliability Quality Control*, 1963, pp. 184–192.

[2] D. Meister, "Motivational Training for Production Personnel," paper presented at Western Regional Conference, American Society for Quality Control, Seattle, Wash., 1962.

in its motivational aspects but at least as much in the uncovering of production problems that might not otherwise be revealed. These problems are brought to the attention of management and investigated further.

12.7 HUMAN FACTORS IN FIELD TESTING

Although human factors are not involved in typical environmental stress and life reliability tests, they are uniquely involved in field-performance tests.[1] That is because, if the field test is correctly carried out, it involves the total system; and this automatically involves human factors.

In field-performance tests it is possible for the first time to study the system as a whole and thus to verify that the human factors influencing performance have been correctly analyzed and applied to design during the design-development process.

It is unnecessary in the field test to develop a special test program for human factors purposes. Provided the engineering evaluation program includes the exercise of all system elements in a manner which approximates the operational use of that system, all that is required for human factors study is for the specialist to be in position to observe and record test data. Among the activities performed by the human factors specialist during testing, the following are most important:

1. Measurement of noise and illumination levels in various areas within test facilities to verify their adequacy for operator use. In many cases adequate lighting and noise baffling have not been provided.
2. Evaluation of the adequacy of the equipment installed in the testing area from the standpoint of the human-engineering principles discussed earlier. Until now all human-engineering evaluation has been performed with drawings; now it is possible to examine the equipment in its hardware form. The purpose of this evaluation is to determine whether the operational equipment is in fact adequate for operator use and, if it is not, what modifications must be made to improve it. This work requires systematic examination of the equipment by the human factors specialist using a standard human-engineering checklist.
3. Similar evaluation of physical test facilities and, in particular, equipment accessibility and safety factors.
4. Periodic surveys to discover potential user problems. Test personnel are interviewed to determine their attitudes concerning the adequacy of the human-engineering characteristics of the equipment they use.
5. Observation of major test operations to determine the causes of errors and delays in performance as a consequence of inadequate design, procedures, technical data, or training.
6. Investigation of repetitive human error and HIF problems reported during testing. Such investigations are similar to the periodic surveys described above, but they are directed to the solution of specific problems.
7. Measurement of operator performance reliability by observing tests and recording the amount of human error manifested in these tests.

The human factors engineer who participates in field-performance tests is largely problem-oriented; that is, he wishes to determine where a human factors problem area exists and to recommend solutions. In this respect he acts in a manner identical with that of the reliability engineer concerned about the mechanical and electrical functioning of the equipment to be tested.

12.8 MEASUREMENT AND PREDICTION OF HUMAN RELIABILITY

We define human reliability (HR) as the probability that a job or task will be successfully completed by personnel at any required stage in system operation within a required minimum time (if the time requirement exists). Note the similarity to the conventional definition of equipment reliability. Successful job performance is, how-

[1] "Personnel Subsystem Testing for Ballistic Missile and Space Systems," *AFBMD* 60-1, Air Force Ballistic Missile Division, Los Angeles, Calif., Apr. 22, 1960.

ever, not errorless performance; rather, it is performance without errors that have a significant effect on correct completion of the task.

The need for the reliability engineer to measure and predict HR has been touched upon before. Quantitative measurement of HR is required if HR is to be included with equipment reliability measurements to present a correct picture of total system reliability. What really do we want to know when we attempt to predict and measure HR? The things we wish to determine are:

1. In analyzing procedures we wish to anticipate the individual human error(s) that will be made in performing a single procedural step. For example, if the step says "adjust focus on scanner," what major errors will most probably be made in performing that step? If we know, we can design equipment or procedures to compensate for that error probability (provided that is economically feasible).

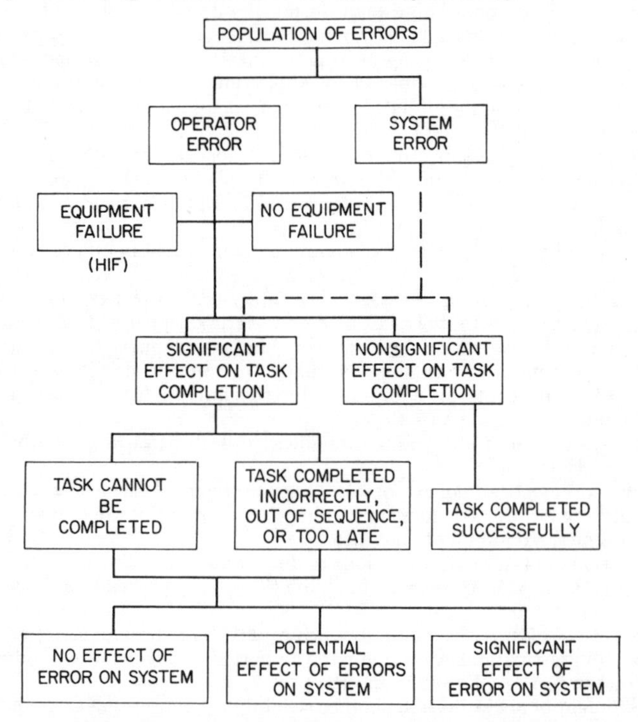

FIG. 12.4 Error taxonomy: type and effect of error.

2. For the same reason, we would like to predict the most significant and most frequent errors that may occur in operating and maintaining a particular equipment, subsystem, and system.

3. We would also like to know the frequency of HIF we can expect for that same equipment, subsystem, and system. This will tell us what to inspect for most carefully.

4. Not every error actually affects system performance, and many errors are reversible, once called to the operator's attention. Hence we must predict not only the probability that errors will be made but also which errors will have a significant effect on system performance and the probability that performance will be completed successfully by the operator.

The reliability engineer will undoubtedly object that reliability prediction and measurement techniques as he presently employs them do include the effect of human performance, since any measure of equipment operation must include the operator performance required by the operation. However, conventional MTBF reliability estimates often deliberately exclude HIF; moreover, even when they do include HIF data, these data do not account for all the effects of human error on the system. Many human errors do not affect the functioning of equipment but do affect the performance of the task.

We see from Fig. 12.4 that all human errors are not equivalent, just as all equipment failures are not equal in terms of their effect on the system. The population of errors can be differentiated in terms of source; in some cases errors are the responsibility of the individual operator, while in others they are a consequence of the way in which the system was developed and is being used.[1] The first type of error, which we call "operator error," occurs when there is a:

1. Failure to perform a required task or procedural step
2. Performance of a required task or procedural step incorrectly
3. Performance of a required task or procedural step out of sequence
4. Performance of a nonrequired task or procedural step

The distribution of individual operator errors tends to approximate a Gaussian (bell-shaped) distribution in which the mean of the distribution is a constant error and the standard deviation of the distribution is a variable error. Operator errors compound linearly in much the same way as do those of machines; hence they add up according to the squares of the standard deviations as shown by Eq. (12.1):

$$\sigma^2{}_{1+2} = \sigma_1{}^2 + \sigma_2{}^2 \tag{12.1}$$

where σ_{1+2} is the total error of the system, σ_1 is the first source of the error, and σ_2 is the second source. Chapanis[2] has shown that when two sources of error accumulate in the system, the overall error of the man-machine system can be greatly reduced by decreasing the size of the larger error, whereas attempting to reduce the size of the smaller source of error has little effect on the overall system error. Much of the rationale for human engineering lies in the fact that in the *operational* situation the larger source of error tends to be that of the man; hence greater reduction in error can be achieved by redesigning the system to reduce this source of error.

System error occurs when:

1. System elements (personnel, equipment, procedures, technical data, logistics, and communications) are nonavailable or inadequate.
2. Organizational procedures are inadequate.

System error might be suspected when, for example, an operator performs a task incorrectly because a checklist procedure gives him incorrect instruction. The fault then lies with the system. Among operator errors we must also differentiate between errors that affect equipment functioning (HIF) and those that do not (non-HIF failures). Errors causing HIF are nonreversible, because equipment, once failed or out of tolerance, usually cannot restore itself, whereas errors that do not result in equipment failure can often be reversed.

The error population must also be differentiated in terms of its effect on the system. As Brady[3] has pointed out, while some errors terminate the task unsuccessfully (have a significant effect on task completion), others do not. The latter may create only the risk that a system failure will occur, while still others merely delay the performance of the task.

Moreover, even if the effect of the error on the task is significant, this is still not to say that the effect of that error on the *system* will also be significant. There may be no effect or merely a potential (risk) effect on the system. This is because a significant

[1] D. Meister, "Individual and System Error in Complex Systems," paper presented at American Psychological Association Convention, St. Louis, Mo., 1962.

[2] A. Chapanis, W. R. Garner, and C. T. Morgan, *Applied Experimental Psychology*, John Wiley & Sons, Inc., New York, 1949.

[3] J. S. Brady, "Application of a Personnel Performance Metric," *Rept.* 6101-001-MU-000, Space Technology Laboratories, Los Angeles, Calif., Dec. 7, 1961.

effect on a task may "dampen out" in terms of the system because of subsequent task performances which cancel out the individual task effect.

It is impossible, therefore, to identify human error unequivocally with HR, because we must know what the effect of that error is before we can determine what its effect on HR is. A similar problem exists in determining how error-prone a task is, because the amount of error and the effects of error will vary as a function of the difficulty level of the task and the stress conditions under which a task is performed.

Despite "commonsense" observation, which is primarily attentive to errors rather than to correct responses, human performance is highly reliable once a task is learned. At least for reasonably simple tasks Klemmer[1] indicates that

> The average trained card punch operator . . . makes no more than four errors per 1,000 key strokes; skilled operators of bankproof machines go 5,000 key strokes per undetected error after a year of training. On simple or elemental tasks, with a little training error rates of better than 10^{-2} are obtainable; after a year of training the figure rises to 10^{-3}.

The error rate for HIF resulting from workmanship error is also relatively low. Rook[2] provides the sample error estimates, based on 23,000 production defects examined, listed in the accompanying table. How are these error rates, some of them

Error	*Error rate*
Transposition of wires	0.0006
Component omitted	0.00003
Component wired backwards	0.001
Wrong component used	0.0002
Component burned from soldering iron	0.001
Excess solder	0.0005
Solder joint omitted	0.00005
Wrong-valued component used	0.0002
Lead left unclipped	0.00003
Solder splash	0.001
Insufficient solder	0.002
Hole in solder	0.07

quite low, to be interpreted in view of what has been said earlier concerning the high incidence of HIF? As Rook[3] points out, errors occur frequently, but the variety of errors in any process is usually small. He reports, for example, that in the assembly of a complex radar fusing unit, fewer than thirty kinds of human error showed up, and only eight of these contributed materially to failure.

Reliability engineers are most concerned with the analysis of HIF. HIF's are analyzed by identifying the failure as human-initiated (i.e., as a failure *directly* caused by some human action) and classifying the HIF in one of a number of categories roughly describing the source of the failure. These categories are:

1. Human-engineering design error. The equipment was designed in such a manner that an operator error occurred, resulting in equipment failure. Example: requiring excessively precise manual calibration, so that the equipment cannot be kept in tolerance.

2. Fabrication error. Failure to construct the equipment per blueprint or engineering instructions. Examples: using wrong material, wiring a circuit backwards, poor soldering.

3. Inspection error. Accepting out-of-tolerance equipment or rejecting in-tolerance equipment.

[1] E. T. Klemmer, Personal communication.

[2] L. W. Rook, "Reduction of Human Error in Industrial Production," *Rept. SCTM* 93-62(14), Sandia Corporation, Albuquerque, N.M., June, 1962.

[3] L. W. Rook, "A Method for Evaluating the Human Error Contribution to System Degradation," presented as part of a symposium on human-error quantification at the sixth Annual Meeting of the Human Factors Society, New York, 1962.

4. Installation/maintenance error. Failure to install or repair equipment per blueprint or engineering instructions. Examples: reversing leads or wiring to wrong terminal points.

5. Operating error. Failure to operate equipment according to procedure. Examples: leaving out required step, adding unnecessary step, performing step out of sequence.

6. Handling error. Failure to transport, store, or treat equipment according to specified requirements. Examples: dropping, striking, or denting fragile equipment.

It will be noted that we do not include ordinary design inadequacies, although they are due to the designer's error, except when they directly result in an HIF. To do so would be to describe almost all failures as human-initiated and thus to make the HIF classification pointless.

The purpose of HIF analysis is to determine and correct the factors leading to the commission of the error responsible for the failure. However, as was pointed out earlier, the detailed investigation of individual failures requires an exorbitant length of time; so that, unless the number of failures being investigated is very small, HIF analysis must be statistically oriented in terms of gross percentages of HIF per subsystem or equipment, department, or location or in terms of determining the failure rate per type of error (as in Rook's listing). The difficulty of securing accurate HIF data is increased by omissions of pertinent data, overly terse failure descriptions, and the unwillingness of personnel to report failures as human-initiated.

Because HIF analysis is at a gross level, it can only point to a work area or a type of failure as presenting a problem to be solved. Corrective action to reduce the number of HIF consists of systematic investigations of the work area and work conditions which are conducive to these errors. Inasmuch as the greatest proportion of HIF seem to stem from the factory, the investigative methods described in Sec. 12.6 should be used.

One way to determine the impact of the HIF problem upon system reliability is to secure two measures of system reliability, one with all failures (including those human-initiated) included and the other without HIF. A comparison of the two resulting probability values will indicate how much reliability degradation results from HIF.

Van Buskirk and Huebner[1] suggest that HIF occur primarily in the early portions of development testing. Presumably they reach an early peak and decrease as development continues. Van Buskirk and Huebner suggest that upon the completion of development testing, human-initiated malfunctions should be reduced to an acceptable error rate and should be completely random in nature.

There is, of course, the problem of determining what an acceptable error rate is. Moreover, there is an implication here that after certain developmental factors inducing human error are presumably eliminated through redesign and testing, HIF result from the random variability of the operator manifesting itself as carelessness, temporary changes in motivation, etc. This agrees only partially with our experience.

For example, in one weapon system, measurements of HIF were made at successive intervals during development through operational use and indicated a striking stability. Over a period of approximately five years and a sample of some 35,000 failure reports the percentage of HIF ranged from 20 to 30 per cent. It is only fair to say that some observers would consider this percentage quite low.

Variations in percentages of HIF result from specific causes. If the total population of failure reports is gathered only from the production area, the percentage of HIF may rise as high as 70 to 80 per cent. If data gathered only from testing sites are analyzed, the percentage of HIF may drop to 10 per cent. Where a great deal of equipment is being installed, the percentage increases proportionately; where the only activity is routine checkout and periodic maintenance, the percentage of HIF decreases. The percentage of HIF increases, of course, with the amount of equipment being operated and the type of testing performed. Beyond such systematic fluctua-

[1] R. C. Van Buskirk and W. J. Huebner, "Human-initiated Malfunctions and System Performance Evaluation," *Rept. AMRL-TDR*-62-105, Behavioral Sciences Laboratories, Aerospace Medical Division, September, 1962.

tions, the month-to-month variations in percentage of HIF are not significant unless some situational factor has changed significantly.

Moreover, in the system mentioned above the error percentage was rarely random in the sense of having little or no relationship with the immediate work situation. For example, in this system fabrication, handling and inspection errors were produced mostly in the factory (as one would expect) and installation errors at the test site; handling errors occurred in both areas. Fabrication errors did not once occur during testing, while operator errors were rare in the factory. Apparently the type and cause of the error are not independent of the conditions under which the error is made.

A word of warning is advisable at this point. When different analysts categorize failure reports, significant variations in percentage of HIF may result if the analysts have not been provided with a common basis for differentiating the HIF. The method of securing the most reliable analysis is to have two or more skilled analysts review each failure report independently and categorize each report as HIF or non-HIF. After a comparison of the reviews, all failure reports on which analysts do not agree should be either considered as non-HIF or subjected to further investigation. This method will produce a lower percentage of HIF but one in which greater confidence can be placed.

Manifestly, every HIF reported is not of equal significance in terms of effect on system performance. The use of HIF data in reliability estimates may be objected to on the basis that the error, once made, is unlikely to be made again, and therefore reliability estimates based on such material cannot be used for predictive purposes. Underlying this, perhaps, is the idea that while units of equipment are highly alike—except for quality control variations, so that a failure of a given type tends to be representative of the same underlying functional process—personnel are quite variable. Under those circumstances, is it reasonable to suppose that errors will repeat themselves? In actuality, errors of a given type do tend to repeat themselves if the system factors responsible for them are not corrected.

There are two types of errors that may be of great significance to future system operations. The first is the extremely critical error responsible for a *catastrophic failure* of the system. Examples of this type of error are the cross-connection of two cables responsible for blowing up a power generator and throwing the wrong switch in a cockpit so that a jet engine flames out and the aircraft crashes. Catastrophic errors are symptomatic of some deficiency in the system configuration which, if uncorrected, will tend to cause a repetition of the situation.

The other type of error is *workmanship error* which may fail under the stress of operating conditions, such as a weak solder joint or weld or the use of improper material, that is, moderately out-of-tolerance conditions which are less likely to be detected. Such workmanship defects, which are likely to pass undetected through inspection and which could have a serious effect on the operational system, must be prevented. This is the reason for being concerned about inadequate workmanship.

Catastrophic HIF's tend to be infrequent, but many failures are reported as resulting from inadequate workmanship. The significance of these workmanship defects is not in the individual reported failure, because reported failures have, of course, been detected and corrected; instead the significance lies in the fact that the workmanship defects are symptomatic of an on-going situation which will produce *more* errors and defects of a given type.

HR can also be evaluated in terms of the probability of occurrence of operator error in subsequent events or trials.[1] The method by which this evaluation is made is identical with that by which equipment reliability is estimated. The equations that follow illustrate this identity. A probability estimate of an operator performing a procedural task correctly is given by the ratio r/n, where r is the number of successful task completions and n is the total number of task trials. However, r/n is only an estimate based on the particular data at hand and may not be interpreted as the actual

[1] The author is indebted to Mr. R. L. Mador of General Dynamics/Astronautics SLV Reliability Analysis Group for assistance in checking the probability concepts described in this section.

Table 12.9 Probability of Successful Task Completion (0.80 Confidence Level)

No. of Successes	Number of trials																			
	1	2	3	4	5	6	7	8	9	10	11	12	13	14	15	16	17	18	19	20
0	0.000	0.000	0.000	0.000	0.000	0.000	0.000	0.000	0.000	0.000	0.000	0.000	0.000	0.000	0.000	0.000	0.000	0.000	0.000	0.000
1	0.200	0.106	0.071	0.055	0.043	0.038	0.031	0.027	0.024	0.022	0.020	0.018	0.017	0.017	0.016	0.015	0.014	0.013	0.012	0.011
2		0.447	0.287	0.212	0.168	0.140	0.119	0.104	0.092	0.083	0.075	0.069	0.064	0.059	0.056	0.051	0.048	0.046	0.044	0.041
3			0.585	0.427	0.326	0.268	0.228	0.199	0.175	0.158	0.143	0.131	0.121	0.111	0.104	0.098	0.092	0.086	0.082	0.078
4				0.668	0.510	0.414	0.350	0.307	0.268	0.240	0.216	0.198	0.172	0.169	0.157	0.147	0.138	0.130	0.123	0.117
5					0.725	0.567	0.483	0.416	0.365	0.327	0.293	0.266	0.248	0.229	0.213	0.200	0.189	0.176	0.163	0.159
6						0.764	0.629	0.538	0.471	0.418	0.382	0.343	0.316	0.292	0.272	0.254	0.238	0.224	0.212	0.201
7							0.794	0.669	0.583	0.516	0.469	0.422	0.387	0.358	0.332	0.310	0.291	0.274	0.258	0.246
8								0.817	0.703	0.619	0.555	0.503	0.461	0.425	0.394	0.367	0.345	0.325	0.307	0.291
9									0.836	0.729	0.650	0.588	0.527	0.494	0.459	0.427	0.400	0.376	0.356	0.337
10										0.851	0.751	0.686	0.616	0.567	0.524	0.488	0.457	0.430	0.406	0.384
11											0.864	0.770	0.699	0.641	0.593	0.551	0.515	0.484	0.456	0.432
12												0.874	0.788	0.718	0.662	0.615	0.574	0.540	0.509	0.481
13													0.884	0.801	0.736	0.682	0.636	0.597	0.561	0.530
14														0.888	0.813	0.750	0.700	0.654	0.616	0.582
15															0.898	0.824	0.764	0.715	0.671	0.634
16																0.904	0.834	0.778	0.719	0.686
17																	0.910	0.842	0.788	0.742
18																		0.914	0.850	0.798
19																			0.919	0.858
20																				0.922

true probability P_T. If one desires to develop a confidence interval on P_T based on the r/n data, the procedure is indicated by the equation

$$\sum_{i=r}^{n} \binom{n}{i} p^i q^{n-i} = 1 - \alpha \tag{12.2}$$

where α = confidence that the interval p to 1 contains the true probability P_T
p = lower bound of P_T (true but unknown reliability)
$q = 1 - p$

See Sec. 4 and Lloyd and Lipow[1] (pages 209 to 216) for a more detailed discussion of the statistical methodology involved. Table 12.9 is based on this equation for various combinations of r and n to provide simple determinations of the lower bound probabilities, which are the entries in the table. The use of the table is illustrated by the following example.

Example 12.2

If the number of trials n is 15 and the number of successes r is 10, then

$$p = 10/15 = 2/3 = 0.67$$

and $p = 0.524$ at 80 per cent confidence level. The interval (0.524 to 1.00) contains the true but unknown probability with 80 per cent confidence.

Rook[2] has also developed a mathematical model for measuring error effects quantitatively, and in doing so he has closely followed methods used in reliability analysis. The mathematical form which had the widest applicability is

$$Q_i = 1 - (1 - F_i P_i)^{n_i} \tag{12.3}$$

where P_i is the probability that an operation will be so performed as to produce error i, F_i is the probability that failure will result if error i occurs, n_i is the number of similar operations in which error i can occur, and Q_i is the probability of failure as a result of error i. If two errors must combine to produce a failure,

$$P_i = P_1 P_2 \tag{12.4}$$

where P_1 and P_2 are the respective probabilities of the two errors.

The total probability of failure is given by

$$Q_T = 1 - \prod_{k=1}^{n} (1 - Q_k) \tag{12.5}$$

where Q_T is the probability that one or more failure conditions will result from one or more human errors in at least one of n classes or errors.

It should be noted that the kinds of error to which this model fits most closely are those described in inspection reports of equipment failures such as "omitted solder joint" and "bad weld," in other words, failures resulting from errors performed in repetitive manual production work. To apply the model, estimates of P_i are needed for various kinds of error and estimates of F_i for various kinds of hardware.

So far, the prediction of human reliability has been hampered by the paucity of human-error-rate data. Most descriptions of methods for predicting human reliability, such as the method of Williams,[3] assume that the necessary error-rate data already exist when, in fact, little is known in this vital area. All reliability predictions are based on historical failure data and judgments of the similarity between the system on which the data are based and the one for which the reliability analyst is predicting.

[1] D. K. Lloyd and M. Lipow, *Reliability: Management, Methods and Mathematics*, Prentice-Hall, Inc., Englewood Cliffs, N.J., 1962.

[2] *Ibid.*

[3] H. L. Williams, "Reliability Evaluation of the Human Component in Man-Machine Systems," *Elec. Manuf.*, pp. 78–82, April, 1958.

Table 12.10 Tables of Human Error Rate

Parameter	*Reliability*
Circular Scales	
Scale diameter, in.:	
1	0.9996
1.6 to 1.75	0.9997
2.75	0.9993
Scale style:	
Moving scale	0.9966
Moving pointer	0.9970
Color-coded	0.9999
Pointer style:	
Horizontal bar, 0 at base	0.9990
Triangle or vertical bar at base	0.9987
Distance between marks, in.:	
Less than 1/20	0.9975
More than 1/20 to 1/4	0.9986
More than 1/4 to 2	0.9996
Proportion of scale marks numbered:	
1:1	0.9999
1:5	0.9991
1:10	0.9980
Number of units shown on scale:	
50 to 100	0.9996
200	0.9984
400	0.9962
600	0.9952
Number of scales and arrangement:	
1 or 2 by 1	0.9999
2 by 2, 2 by 4, 4 by 4	0.9997
4 by 10, 6 by 4	0.9990
8 by 4, 9 by 5	0.9975
Scale increase:	
Right to left	0.9996
Left to right	0.9999
Counters	
Size (length), in.:	
1	0.9990
1 to 2	0.9998
3 and up	0.9995
Number of drums or digits	
1 to 3	0.9997
4 to 5	0.9993
7 and up	0.9985
Labeling	
Digit span:	
2	0.9998
3	0.9994
4 or 5	0.9992
6 or 7	0.9991
Words:	
1 or 2	0.9999
3 to 5	0.9995
6 to 11	0.9985
Size of printing (height), in.:	
1/5 or more	0.9997
1/8	0.9994

Table 12.10 Tables of Human Error Rate *(Continued)*

Parameter	*Reliability*
Lights	
Diameter, in.:	
Less than ¼	0.9995
¼ to ½	0.9997
½ to 1	0.9999
Number of lights on:	
1 or 2	0.9998
3 or 4	0.9975
5 to 7	0.9952
8 to 10	0.9946
Presentation:	
Intermittent (blinking)	0.9998
Continuous	0.9996
Linear Scales	
Size (length), in.:	
3	0.9997
6	6.9998
9	0.9996
Scale style:	
Moving pointer	0.9977
Moving scale	0.9970
Color-coded	0.9999
Scale direction:	
Horizontal	0.9998
Vertical	0.9995
Distance between scale marks, in.:	
1/10 or less	0.9975
1/10 to ¼	0.9992
¼ or more	0.9982
Number of units shown:	
50 to 100	0.9998
200	0.9988
400	0.9968
Scale increase:	
Left → right (or bottom → top)	0.9998
Right → left (or top → bottom)	0.9992
Proportion of scale marks numbered:	
1:1 or 1:2	0.9999
1:5	0.9995
1:10	0.9985
Scopes	
Number of range marks:	
1 or 2	0.9980
3 to 5	0.9997
6 to 10	0.9999
10 to 20	0.9990
20 and up	0.9983
Bearing-estimation method:	
Estimate (no aid)	0.9975
Use overlay	0.9990
Use cursor	0.9995
Scope size, in.:	
3	0.9990
4 and up	0.9999
Visual angle (from operator to scope face), deg:	
0 to 45	0.9999
45 to 80	0.9995
Target exposure time, sec:	
3	0.9990
5	0.9995
Over 5	0.9999

Table 12.10 Tables of Human Error Rate *(Continued)*

Parameter	*Reliability*
Semicircular Scales	
Size (radius), in.:	
½ to ¾	0.9996
¾ to 1	0.9997
1 to 2	0.9993
Scale style:	
Moving pointer	0.9981
Moving scale	0.9978
Color or zone code	0.9999
Scale arc length, deg:	
25	0.9937
50 to 100	0.9950
200	0.9964
Distance between scale marks, in.:	
Less than ½₀	0.9965
½₀ to less than ⅒	0.9933
⅒ to less than ½	0.9955
½ to less than 1	0.9969
1 to less than 2	0.9962
Proportion of scale marks numbered:	
1:1 or 1:2	0.9999
1:5	0.9995
1:10	0.9985
Scale increase:	
Left to right	0.9999
Right to left	0.9996
Cranks	
Diameter, in.:	
3	0.9970
4	0.9990
8	0.9975
12	0.9985
Control force, lb:	
Less than 5	0.9985
5 to 10	0.9972
Control/display movement direction:	
Direct	0.9992
Reverse	0.9975
Time lag between control turn and display movement, sec:	
Less than 1.5	0.9980
More than 2.0	0.9965
Control/display movement ratio, rev per in. of cursor movement:	
1:1	0.9975
2:1	0.9990
3:1	0.9992
Joy Stick	
Stick length, in.:	
6 to 9	0.9963
12 to 18	0.9967
21 to 27	0.9963
Extent of stick movement, deg:	
5 to 20	0.9981
30 to 40	0.9975
40 to 60	0.9960
Control resistance, lb:	
5 to 10	0.9999
10 to 30	0.9992

Table 12.10 Tables of Human Error Rate (*Continued*)

Parameter	*Reliability*
Joy Stick (*Continued*)	
Joy-stick support:	
Present	0.9990
Absent	0.9950
Time lag between control movement and display movement, sec:	
0.3	0.9967
0.6 to 1.5	0.9963
3.0	0.9957
Control/display movement ratio (distance):	
1:1 or 1:3	0.9936
1:4 or 1:6	0.9967
1:15	0.9950
1:30	0.9967
Control/display movement relationship (movement direction):	
Direct or positive	0.9998
Reversed	0.9970
Knobs	
Size (diam), in.:	
Less than ½	0.9995
½ to 3	0.9997
3 or more	0.9994
Resistance, oz:	
Light to moderate (4 or less)	0.9995
Heavy (6 to 16)	0.9998
Control/display movement relationship:	
Clockwise for increase	0.9999
Counterclockwise for increase	0.9995
Control/display distance ratio (scopes), in. of indicator movement per rotation:	
1 or less	0.9999
2 to 6	0.9997
6 or more	0.9996
Control/display distance ratio (meters), proportion of scale traversed by pointer per knob rotation:	
Less than ¼	0.9999
¼ to ½	0.9997
More than ½	0.9996
Knob grip:	
Knurled	0.9999
Smooth	0.9997
Lock mechanism:	
Present	0.9999
Absent	0.9996
Lever (Including Wrench or Pliers)	
Length:	
Long	0.9990
Short	0.9920
Plane of movement:	
Vertical	0.9992
Horizontal	0.9999
Control movement amplitude, deg:	
5 to 10	0.9964
10 to 20	0.9970
30 to 40	0.9975
40 to 60	0.9985
Control resistance, hand operation, lb:	
2 to 5	0.9999
10 to 20	0.9992

Table 12.10 Tables of Human Error Rate *(Continued)*

Parameter	*Reliability*
Lever (Including Wrench or Pliers) (*Continued*)	
Control resistance, arm operation, lb:	
2 to 5	0.9990
10 to 20	0.9999
20 to 30	0.9995
Direction of movement:	
Direct	0.9999
Reverse	0.9985
Control/display movement ratio (distance):	
1:1	0.9957
1:3	0.9970
1:6	0.9983
1:15	0.9975
1:30	0.9985
Push Buttons	
Size:	
Miniature	0.9995
½ in. or more	0.9999
Number of push buttons in a group:	
A. Single column or row:	
1 to 5	0.9997
6 to 10	0.9995
11 to 25	0.9990
B. Double column or row or rows and column:	
1 to 5	0.9997
6 to 10	0.9995
11 to 25	0.9990
C. Matrix:	
6 to 10	0.9995
11 to 25	0.9995
25 or more	0.9985
Number of push buttons pushed within group:	
2	0.9995
4	0.9991
8	0.9965
Distance between edges, in.:	
⅛ to ¼	0.9985
⅜ to ½	0.9993
½ or more	0.9998
Detent:	
Present	0.9998
Absent (switch returns)	0.9993
Rotary Selectors	
Size (diam), in.:	
1 to 3	0.9997
3 or more	0.9995
Number of positions:	
3 to 6	0.9997
6 to 12	0.9992
12 or more	0.9975
Distance between positions, deg:	
Less than 15	0.9975
15 to 30	0.9998
More than 30	0.9996
Indicator style:	
Dot	0.9995
Line	0.9996
Pointer	0.9999

Table 12.10 Tables of Human Error Rate (*Continued*)

Parameter	*Reliability*
Rotary Selectors (*Continued*)	
Distance between edges of adjacent switches, in.:	
½	0.9988
¾ to 1	0.9995
1 and more	0.9997
Toggle Switch	
Size	
Miniature	0.9997
Regular and large	0.9999
Number of positions:	
2	0.9999
3	0.9991
Direction of throw:	
Vertical	0.9999
Horizontal	0.9996
Angle of throw, deg:	
20	0.9997
40	0.9998
90	0.9999
Number of switches in group	
A. Single column or row:	
1 to 5	0.9998
6 to 10	0.9996
11 to 25	0.9990
B. Double column, double row:	
1 to 5	0.9998
6 to 10	0.9996
11 to 25	0.9992
C. matrix:	
6 to 25	0.9996
25 or more	0.9988
Distance between switch centers, in.:	
½ or less	0.9993
¾	0.9998
1 or more	0.9999
Connecting Cables	
Cable weight:	
Light (5 lb or less)	0.9997
Heavy (over 5 lb)	0.9992
Locking method:	
None	0.9987
Automatic	0.9990
Less than ¼ turn	0.9992
More than ¼ to less than 1 turn	0.9995
Clamp	0.9997
Pin or threaded	0.9999
Probes within connection:	
0, 1, or 2 or keyed	0.9998
3 or 4	0.9990
5 or more	0.9986
Disconnecting Cables	
Cable weight:	
Light (5 lb or less)	0.9999
Heavy (over 5 lb)	0.9997

Table 12.10 Tables of Human Error Rate (*Continued*)

Parameter	*Reliability*
Disconnecting Cables (*Continued*)	
Locking method:	
None	0.9999
Automatic	0.9999
Less than ¼ turn	0.9998
More than ¼ turn	0.9999
Clamp	0.9995
Pin	0.9997
Threaded	0.9999
Positioning Objects	
Weight of object, lb:	
25 or less	0.9998
30 to 70	0.9997
75 to 100	0.9993
More than 100	0.9991
With supporting equipment	0.9994
Object size, cu ft:	
3 or less	0.9998
4 to 8	0.9997
10 to 15	0.9992
More than 15	0.9989

Note. Assume a reliability of 0.9998 for speaking and writing and 0.9990 for all higher mental processes (recognition, decision making, etc.).

When the historical (error-rate) data do not exist, predictions are little better than educated guesses. This is another example of the similar problems faced by human factors and equipment reliability.

The point estimate prediction of human reliability is based, as is that of equipment reliability, on the product rule. After determining the probability estimate (in the manner just described) for the individual steps in a task and assuming their independence (at best a gratuitous assumption), the probabilities for the steps are combined by multiplication:

Example 12.3

step 1 = P_1 step 2 = P_2 step 3 = P_3

$P_{\text{Task 1}} = P_1 \times P_2 \times P_3 = P_{1,2,3}$

Likewise, the probabilities for individual steps of tasks 2, 3, . . . , n are combined to give us the probability of successful accomplishment of these tasks. It is possible by combining task probabilities in this way to develop predictions for the human reliability of an entire system operation, but the task is very tedious unless a computer is used. It must be remembered, however, that everything depends on the accuracy of the individual step probabilities; if these are incorrect, the entire calculation is dubious.

Up to this time probability values (error estimates) for human performance have been unavailable, but a recent attempt has been made by the American Institute for Research[1] to convert the data from 164 psychological laboratory studies of control and display utilization into human error rates which can be used for prediction purposes.

Table 12.10 presents the human-error failure rate (human reliability) associated with individual dimensions of controls and displays. This failure rate is the *proba-*

[1] S. Munger et al., "An Index of Electronic Equipment Operability (Data Store)," *Rept. AIR*-C43-1/62-RP(1), prepared by American Institute for Research, Pittsburgh, Pa., Jan. 31, 1962, for U.S. Army Signal Corps on Contract No. DA-36-039-SC-80555.

bility of correct use of a control or display having the indicated dimensions. Since the studies on which this table was developed were performed under laboratory conditions, the resultant error values themselves contain a built-in error and must be used *as a guide only*. The essential thing is that, regardless of the precise degree of their accuracy, for the first time human-performance error values are available to compare with equipment-performance reliability values.

Table 12.10 can be used in two ways:

1. As a guide to the selection of particular control-display hardware. It is assumed that the designer will wish to select those control-display features which give him the greatest probability of reliable use.

Since each control and display has a number of dimensions and each dimension is associated with a human reliability estimate, it is necessary to *combine individual dimension reliabilities* in order to secure an overall estimate of the degree of human reliability to be achieved with the particular control or display. This is accomplished by multiplying the individual dimensional reliabilities as shown in the following example:

Example 12.4

Component: Joy Stick

Component dimension	Value	Applicable reliability
Stick length................	6–9 in.	0.9963
Extent of stick movement...	30–40°	0.9975
Control resistance..........	5–10 lb	0.9999
Product.................		0.9937

The designer can then determine whether this particular combination of control dimensions gives him sufficient reliability or whether some other set of dimensions will be more satisfactory. Presumably the designer will pick the control with the combination of dimensions providing the highest reliability.

2. It is also possible to analyze written procedures to predict the probability of error occurrence in performing the procedure. For this analysis each procedural step should be analyzed individually and broken into its stimulus-input, internal-process, and response-output elements as shown in the following example:

Example 12.5

X light becomes green. Operator activates Y push button.

The stimulus element here would be the X light, which has the following characteristics:

Diameter, ¼ to ½ in.............	0.9997
Numbers of lights on, 3 or 4.......	0.9975
Presentation, continuous...........	0.9996
Product........................	0.9968

Note that in order to make a realistic estimate of the error probability associated with this light, it is necessary to know something about the equipment configuration in which the light will be read. The internal process, recognition of the light by the operator, is assumed to have a human reliability of 0.9990. We have arbitrarily assumed this value for all internal processes but this undoubtedly underestimates the reliability of most simple thought processes. The output-response here is the Y push button, which has the following characteristics:

Size, miniature	0.9995
Single column, 1-5	0.9997
Distance between edges, ⅜-½ in.	0.9993
Detent, absent	0.9998
Product	0.9983

Overall values for the input, internal process, and output are then multiplied together (using the product rule because these behavioral elements are in series) to form the human-reliability estimate for the step.

Example 12.6

Light (0.9968) × internal process (0.9990) × pushbutton (0.9983) = 0.9941.

A similar process must be performed for all the steps in the procedure. Thus, by progressively multiplying individual procedural elements together, one finally arrives at a total human reliability for a given procedure. It should be noted that this human reliability does not imply any equipment reliability. To determine the reliability of the entire system while performing a particular operation, it would be necessary to multiply reliability figures for human performance against estimated reliability figures for equipment performance.

To summarize, we have attempted to point out the essential identity of interest and necessary interrelationships between the reliability engineer and the human engineer. An increasing interaction between the two disciplines they represent cannot but redound to the advantage of each.

References

1. Air Force System Command, *AFSCM* 80-3, *Handbook of Instructions for Aerospace Personnel Subsystem Designers* (HIAPSD).

2. Air Force System Command, *AFSCM* 80-5, *Handbook of Instructions for Ground Equipment Designers* (HIGED).

3. Air Force System Command, *AFSCM* 80-6, *Handbook of Instructions for Aircraft Ground Support Equipment Designers* (HIAGSED).

4. Baker, C. A., and W. F. Grether, "Visual Presentation of Information," *WADC Tech. Rept.* 54-160, August, 1954.

5. Ely, J. H., R. M. Thomson, and J. Orlansky, "Layout of Workplaces," chap. V of *Joint Services Human Engineering Guide to Equipment Design, WADC Tech. Rept.* 56-171, November, 1957.

6. Ely, J. H., R. M. Thomson, and J. Orlansky, "Design of Controls," chap. VI of *Joint Services Human Engineering Guide to Equipment Design, WADC Tech. Rept.* 56-172, November, 1956.

7. Fogel, L. J., *Biotechnology*, Prentice-Hall, Inc., Englewood Cliffs, N.J., 1963.

8. Javitz, A. E. (ed.), "Human Engineering in Equipment Design," reprinted from *Electrical Manufacturing*, Gage Publishing Co., New York, 1952.

9. Lincoln, R. S., "Human Factors in Attainment of Reliability," *IRE Trans. Reliability Quality Control*, vol. RQC-9, pp. 97–103, April, 1960.

10. Majesty, M. S., "Personnel Subsystem Reliability for Aerospace Systems," *Proc. IAS Natl. Aerospace Systems Reliability Symp.*, 1962, pp. 199–204.

11. McCormick, E. J., *Human Factors Engineering*, 2d ed., McGraw-Hill Book Company, New York, 1964.

12. Meister, D., "The Problem of Human-initiated-failures," *Proc. Eighth Natl. Symp. Reliability Quality Control*, 1962, pp. 234–239.

13. Meister, D., "Methods of Predicting Human Reliability in Man-Machine Systems," *Human Factors*, vol. 6, no. 6, pp. 621–646, 1964.

14. Morgan, C. T., J. S. Cook, A. Chapanis, and M. W. Lund (eds.), *Human Engineering Guide to Equipment Design*, McGraw-Hill Book Company, New York, 1963.

15. Peters, G. A., and T. A. Hussman, "Human Factors in Systems Reliability," *Human Factors*, vol. 1, no. 2, pp. 38–50, 1959.

16. Rabideau, G. F., "Prediction of Personnel Subsystem Reliability Early in the System Development Cycle," *Proc. IAS Natl. Aerospace Systems Reliability Symp.*, 1962, pp. 191–198.

17. Rigby, L. V., J. I. Cooper, and W. A. Spickard, "Guide to Integrated System Design for Maintainability," *ASD Tech. Rept.* 61-424, October, 1961.

18. Van Cott, H. P., and J. W. Altman, "Procedures for Including Human Engineering Factors in the Development of Weapon Systems," *WADC Tech. Rept.* 56-488, October, 1956.

19. Willis, H. R., "The Human Error Problem," paper presented at American Psychological Association Meeting, St. Louis, Mo., September, 1962. (Martin-Denver *Rept. M*-62-76.)

20. Woodson, W. E., and D. W. Conover, *Human Engineering Guide for Equipment Designers*, 2d ed., University of California Press, Berkeley, Calif., 1964.

Section 13

RELIABILITY CONSIDERATIONS FOR PRODUCTION

JAMES A. MARSHIK

SENIOR QUALITY CONTROL ENGINEER, AERONAUTICAL DIVISION
MINNEAPOLIS-HONEYWELL, MINNEAPOLIS, MINNESOTA

Upon being deactivated as a Naval Reservist in the Korean War, Mr. Marshik joined Minneapolis-Honeywell, in the Inspection Department. Since then he has progressed through positions as inspector, inspection procedures writer, and quality control engineer to his present position. He has served as a quality control engineer in most of the departments of the Aeronautical Division, with special emphasis on the reliability considerations for gyros, electronic equipment, and electromechanical and hydraulic devices.

Mr. Marshik's most recently completed assignment was that of Lead Quality Control Engineer on Honeywell's Polaris Project, when he was responsible for all phases of quality and reliability of that highly successful program. He is now assigned as Senior Quality Control Engineer for the Apollo Command Module Stabilization and Control System Project.

He has long been an advocate of devoting ample consideration to reliability in the production phase of a project. He has presented a number of papers on this subject, including one, "Reliability Emphasis on Production Contracts," at the Eighth National Symposium on Reliability and Quality Control. He received his engineering education at the University of Minnesota and is a registered professional engineer in Minnesota.

RELIABILITY CONSIDERATIONS FOR PRODUCTION

James A. Marshik

CONTENTS

13.1 THE IMPORTANCE OF RELIABILITY EMPHASIS IN PRODUCTION

A number of the sections in this handbook deal with the systematic development of sound product design which possesses a high degree of inherent reliability. The benefits of these sophisticated approaches can be realized only if a successful transition from design to the production of hardware can be achieved. All persons concerned with the attainment of reliability goals should be ever aware that the fruit of their efforts can be realized only through the performance of the physical equipment. No matter how theoretically sound the design concept and the design development are, all is for nothing if the hardware is not fabricated in a manner that will produce reliable performance.

This section deals with the interface between product design and the production of physical product. It is written as a guide for design, production, quality control, and reliability engineers. The methods suggested are primarily aimed at the complex product, although modifications for less complex products are included. Utilization of a quality and reliability plan is highly recommended as a means of organizing the necessary considerations for successful incorporation of a design into production. Since a Q&R plan is a method of organizing other activities, it will be the first applied subject covered in the section.

It will be noted that the word "quality" and the word "reliability" are frequently used together in this section to emphasize the definite relationship between them. Product quality is generally defined as the condition of the product with respect to applicable specifications at a particular time of evaluation. Reliability, in general terms which do not conflict with the mathematical definition, is the capability of a product to continue to meet applicable requirements in usage.

In a sense, a reliability problem is really a quality problem of a special type. During product evaluation, a characteristic may test within specifications and yet bear a latent tendency toward gradual or sudden change; the essence of a reliability problem is this latent tendency toward change. Since this latent condition already exists in a characteristic when it is inspected, the desired quality is not really present at the time of acceptance. Unfortunately, a normal inspection is not capable of detecting this tendency. The application of "reliability," then, is a specialized form of "quality" which deals with defects of latent change—defects which are not visible to traditional quality assurance methods.

13.2 DEVELOPMENT OF A QUALITY AND RELIABILITY PLAN

All companies perform regular functions to incorporate a new design into production. The majority of these functions have a bearing on the ultimate quality and reliability of the product, and the successful performance of these functions involves many people and departments. Special efforts to organize the functions that affect product quality and reliability have proved so successful that application of this method is wholeheartedly recommended to firms not using an equivalent technique.

The organization of necessary functions into a quality and reliability plan *for a specific product* ensures awareness of necessary responsibilities and supplies the coordination required to direct all elements of the organization to the achievement of the goals. The Q&R plan does not necessarily involve the performance of more functions; rather, it serves to assure that procedures are provided and the *needed* functions are satisfactorily performed. The planning and documentation of a Q&R plan may, however, clarify the *need* for additional procedures.

The principal contributors to a Q&R plan are the design, production, quality control, and reliability-engineering groups. The developers of the plan are the quality control and reliability-engineering representatives. Input from the following departments is also utilized in Q&R plan development:

Program management, sales, or contracts (the department that deals directly with the customer)
Engineering test department
Support-equipment engineering
Production
Quality assurance or inspection (the department that performs product acceptance)
Instrumentation engineering (electrical measurement)
Gauge engineering (mechanical measurement)
Tool engineering

13.2a Considerations for a Quality and Reliability Plan. Development of the Q&R plan should involve these considerations:

Liaison with the customer
Reliability program for development and design
Customer-specifications review
Prototype test program
Engineering-specification review
Producibility and inspectability

Special processes
Potential quality and reliability problems
Instrumentation
Gauging
Required vendor data
Training
Sourcing and surveillance for purchased and subcontract items
Environmental testing
Failure reporting and analysis
Variable-data analysis
Training programs
Internal and customer quality and reliability reporting
Audits of quality assurance
Document audits
Nonconforming material procedures
Inspection plan
Configuration control
Lot control of parts
Classification of characteristics

The scope of the plan will vary somewhat depending on the requirements for the individual products. A Q&R plan document for a device which will become a component in a larger system can be prepared in memorandum format. In this instance the five- to ten-page plan covers the special quality and reliability procedures on the product type. A large portion of the quality and reliability programs for such a device are covered by standard company quality and reliability procedures.

As is frequently the case with complex products, numerous special Q&R requirements were encountered. In this instance an entire operating manual, including both the special and standard Q&R procedures, was deemed necessary. Figures 13.1 and 13.2 are excerpts from the manual, that illustrate special programs. Many special programs are directly responsive to customer requirements; others describe special operating techniques for optimizing quality and reliability.

The point of emphasis in this first subject is that planning is necessary for the introduction of a new design into production. Work done in the initial production phases affects, to a very large extent, the ultimate success or failure of the product as a reliable device. A Q&R plan is an efficient method of establishing and controlling the necessary planning functions.

13.3 CONCEPTS OF PRODUCIBILITY

"Producibility" is a quality of product design that enables the product to be yielded by a manufacturing process. The word "reproducibility" is helpful to our definition in that it is the capability to produce copies, to repeatedly duplicate an "original." It must be realized that the original to be duplicated is no single physical piece of hardware. Instead, the original is the somewhat abstract set of design specifications and drawings which contains permissible latitudes and tolerances representing a population rather than a single device. A design which can be manufactured with relative ease, at low cost, and with good quality is a highly producible design.

Normally, in human endeavor, hard work and accomplishment of the difficult pay bountiful benefits. Unfortunately, this is not true on the production line. The product characteristics on which the production organization works the hardest are the very characteristics that will produce the most scrap and rework and will be the most likely to produce failures in the field. When the field-failure history of a device is compared with the factory history of rejections, failures, scrap, and rework, a surprisingly close relationship, characteristic by characteristic, usually exists between the two.

High producibility is therefore not only of value in minimizing costs and assuring timely shipments of product; it is a major achievement in the quest for high reliability.

POLARIS MARK 2 GIMBAL ASSEMBLY **Honeywell**

DATA COLLECTION AND CONTROL SYSTEM

Significant quantities of both variable and attribute data are associated with each Polaris Gimbal Assembly. The data control system described in this section is designed to efficiently accomplish the necessary functions of data collection. The system also features specification control in the sense that it assures that the proper specification revision is utilized during system, sub assembly and major component tests.

The procedure supports the generation of Data Logs for each gimbal assembly. Data collected in this system is readily transferable to individual Data Logs. Provision is made for "special" data which is frequently required for engineering studies.

The term "worksheet" as used in this procedure shall include all internal forms which are designed for the purpose of recording Polaris functional or visual characteristics on both the final assembly and individual components. Many of these forms are computational in nature and the required basic computations are retained.

The Polaris worksheet program has been designed so that individual worksheets numerically correspond with individual OD's. Worksheets are prepared by Quality Control for both Production and Inspection. The procedure outlines the elements of control which are utilized.

VI - 5

FIG. 13.1 An excerpt from a plan describing a part of the data-collection program.

In general, when production difficulty and costs go down, field reliability goes up. This inverse relationship provides a strong incentive to strive for high producibility.

Attainment of this high producibility, however, requires attention to each of the product's characteristics and the relationship between those characteristics. Study should begin as early as the concept stage. As soon as the designer has isolated the possible approaches to attaining the product requirements, he should consider producibility. Past experience is the most concrete source of information. This expe-

POLARIS MARK 2 GIMBAL ASSEMBLY **Honeywell**

Procedure:

Specification Control Log

A Specification Control Log is maintained by Quality Control Engineering. Individual Log Sheets specify the current correct revision status of the OD and work sheet to be used for Polaris data recording purposes. A separate log sheet is maintained for every worksheet in use.

Revisions to the OD or to the worksheet must be entered into the appropriate log sheet. These entries shall be accompanied by the effectivity of the revision. Effectivity is entered in terms of system serial number for system tests. Effectivity of sub-assembly or component revisions will be entered by calendar date. The Specification Control Log provides a central reference of current revision status.

Change entries for OD's or Inspection worksheets must be accompanied by the signatures (initials) of the Systems Engineer and the Quality Control Engineer who have respective responsibilities for the affected OD and worksheet.

Changes to the Production worksheets must be accompanied by signatures (initials) of the cognizant Systems Engineer, Production Engineer and Quality Control Engineer.

Presence of all required signatures signifies acceptance of the

VI - 6

FIG. 13.2 Another excerpt from a plan describing a part of the data-collection program.

rience, to a certain extent, is within the designer's own knowledge. Consultation with an experienced production engineer will often aid the design engineer in his selection of the best design approach. Presence of documented "preferred engineering practices" is representative of the experience of many engineers and therefore should rank high as an indicator of producibility.

Production engineering should take an active part in influencing the preferred engineering practices. Production engineering best knows the equipment that the company has available for production use, and it has a better and more current knowledge of the capabilities of the processes it has developed. The application of production-engineering skill and knowledge to the concept and development stages of design is an important step toward producibility.

A relatively new practice of assigning a contingent of production engineers to a design operation has been applied on complex products. These same production engineers will ultimately lead the production-engineering effort when the product is ready for factory production. This practice provides a genuine incentive for the production engineers to influence the design along the lines of high producibility. The arrangement also provides a significant time benefit, since the production engineers are becoming familiar with the product and are developing production processes concurrently with the development of the design.

A thorough understanding between design engineer and production engineer can also eliminate the uneconomical design practice of applying extreme "safety factors" in tolerances. Traditionally, it is said, designers consider the tolerance actually required and then tighten it to make sure the production line will at least meet the original limits. The safety-factor approach may have some merit on easily produced characteristics, but it can be a serious drawback to producibility if the tightened limits strain the process capability.

The same understanding can make entirely feasible the practice of tolerancing on a root-sum-of-squares[1] approach or some modified version. Normally rss tolerancing is intended for economic purposes, but some advanced products rely on the method because they would otherwise exceed the state of the art in parts fabrication. Root-sum-of-squares tolerancing is particularly attractive in applications where a simple check ensures that a worst combination has not occurred.

Product inspectability should be included as an element of producibility. Capability to determine quickly and efficiently whether a characteristic is good or defective is of obvious value to quality and reliability. While experience is the best source of producibility information, certain product characteristics will not be represented by past experience. On these characteristics all available knowledge must be projected to the new application. Many of the characteristics in this category will involve final-device performance. Some determination can be made on prototype models, but it must be realized that these devices represent only *some* of the forms of the

[1] Characteristics of an assembly may be dependent on characteristics of a number of parts. These assembly characteristics are "traditionally" controlled by restricting the tolerances of parts so that the "worst possible" combination of parts would still produce a satisfactory assembly.

The probability of a worst combination is rather remote on two parts and becomes increasingly unlikely when the number of independent characteristics becomes greater. The root-(of the)-sum-of-squares method serves as a predictor of the actual limits of the independent variable.

An equivalent measure of process dispersion is assigned to each of the independent variables. Each of these values is squared, and all of the squared values are added together. The square root of this sum is the process dispersion of the dependent variable in the same dispersion measure. Statistically, the rss method is equivalent to the addition of variances. (See Sec. 4.)

When the actual process is unknown, the part tolerances are frequently used for estimating the probable dispersion of the dependent characteristics. For example,

$$A = (B \pm 1) + (C \pm 1) + (D \pm 1) + (E \pm 1)$$

$$\text{Limit}_A = \sqrt{1^2 + 1^2 + 1^2 + 1^2} = \sqrt{4} = \pm 2$$

The expected limit of A is therefore ± 2 rather than ± 4 yielded by "traditional" tolerancing. The method can, of course, be applied in reverse by stating and squaring a tolerance for A and assuring that the sum of squares of the tolerance of the independent variables does not exceed the A tolerance-squared figure.

The method, in theory, relies on a normal distribution but is very tolerant of nonnormal distributions in practice. The explanation here is intended only to familiarize the reader with the method rather than to recommend its application. There are pitfalls, and adoption of the method removes a safety factor that the designer has, perhaps unknowingly, applied to his designs in the past by traditional tolerancing.

design. Therefore, the findings determined during the initial production phase should be reviewed for producibility. Product changes in this stage are still possible with a minimum of cost and difficulty. For producibility, experience has been called in as a principal source of information. Next we shall see how experience can be gathered and how producibility can be measured.

13.4 PROCESS CAPABILITY: A MEASURE OF PRODUCIBILITY

Process capability is the inherent ability of a process to develop characteristics within tolerance limits. The ultimate term of measurement is the percentage of yield of acceptable product that can be obtained from the process when applicable tolerances and conditions are specified. Process capability information is normally considered to be the end product of a special study, but it need not be limited to this category. Our first discussion on this subject will cover process-capability information that is probably available in your plant even though no special effort has been made to obtain it.

A wealth of process-capability information can be obtained from inspection, scrap, and rework records. Normally, records of this type are associated with things that went wrong. In this application we are equally interested in what went right. If we have knowledge of what was produced and accepted in a particular period by a given process, and we also know what went wrong, we can calculate the percentage yield for the particular process. Frequently, the yield will already have been calculated as an element of cost control or as an inspection record.

Information that is gathered in this manner is not precisely process-capability information in the pure sense. It reflects more than the "inherent" capability of the process; it represents overall typical experience, which is probably better than "pure" process capability. Even though the data are of attribute nature, vast quantities of available data compensate for the lack of efficiency. Since applied-capability studies are time consuming and expensive, use of past rejection data offers a documentation of process capability which is economical both in time and money. A partial output from the study of past data when a *certain machine type* and *material* are specified as the process is shown in the accompanying table.

Outside diam, in inches, formed within 0.750 *in. of collet*	*Expected yield*
0.050–0.075 ±0.005	0.991
0.050–0.075 ±0.003	0.991
0.050–0.075 ±0.002	0.985
0.050–0.075 ±0.001	0.970
0.050–0.075 ±0.0005	0.940
Less than ±0.0005	Not recommended
0.075–0.125 ±0.005	0.998
0.075–0.125 ±0.003	0.998
0.075–0.125 ±0.002	0.992
0.075–0.125 ±0.001	0.980
0.075–0.125 ±0.0005	0.950
Less than ±0.0005	Not recommended

Special process-capability studies are warranted in the event of problems or special considerations. As an example, we shall consider an application on a subassembly. Design engineering determined that the thermal characteristics of a floated gyro could be improved if the power consumption of the gyro spin motor could be reduced. The spin motor process capability was required to determine a realistic reduction. The synchronous motor was originally designed for 3 watts maximum power consumption. Computation of the distribution of power readings on a large sample of motors indicated the distribution to be normal, with a mean of 2.6 watts and a standard deviation of 0.11 watts. This distribution was adapted to a yield-vs.-power tolerance graph which is simply a cumulative presentation of a normal curve (Fig. 13.3).

The graph displays the expected yield with any selected maximum power consumption tolerance. It shows that little difficulty would be expected in accepting motors at a 3-watt maximum level. If, on the other hand, the maximum were reduced to the desired level of 2.7 watts, the yield would slump to 82 per cent. The present process was not capable of economically producing to the tightened requirement. With the knowledge that the process itself had to be improved, the goal was achieved through an improved method of isolating laminations during bonding and an improved grinding technique to avoid metallic smear between laminations. Progress during the improvement was determined in the same manner as in the original study.

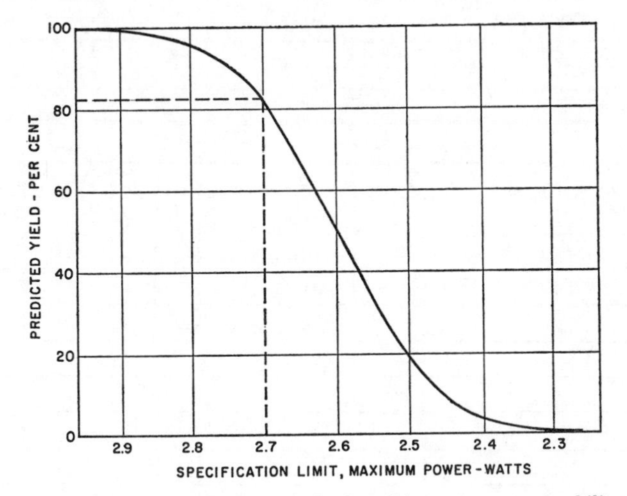

FIG. 13.3 One method of concisely describing a process capability.

Although Fig. 13.3 is characterized to a special problem, it has a shape typical of all normally distributed processes. It will be noted the curve is relatively flat at the top and bottom; a process capability can be pressed on the ends with not too great a loss in yield. Continued restriction of tolerances, however, moves the condition to the steep portion of the curve, where yields fall severely. The designer and production engineer should be aware of this natural condition.

13.5 AN AUTOMATED METHOD OF DETERMINING PROCESS CAPABILITY

Applied-process-capability studies are usually expensive, and in the past they were seldom utilized on other than problem characteristics. If digital computer facilities are available, however, much of the labor and cost can be eliminated. Vast programs of product-capability analysis that simply could not be considered if manual methods were to be used can be undertaken. Such an automated program, designed by the Honeywell Aeronautical Division, has been used with great success for a number of years. Called Automated Data Analysis, or ADA, this program is used here as an example of an automated capability study.

13.5a ADA Data Collection. In the ADA program a standardized form is used to collect all the information for the computer input (Fig. 13.4). Provision is made for 20 characteristics on each sheet. As many sheets as required may be used. Each characteristic is described in the space provided. Data are coded so each reading consists of three digits (000 to 999). The data, plus the necessary identification, use all 80 positions of a standard punched data card. Specification limits are coded in the same manner as the data and are recorded on the standard form. The data-type entry for the low specification limit is 7; the entry for the high specification limit is 9. The data type for actual data is 8. The identification at the top of the form aligns the data and specifications in the computer operation.

13.5b Raw-data Tabulation. The completed forms are keypunched on data cards and verified. A raw-data tabulation is made; it consists of an arranged listing of

ADA AUTOMATED DATA ANALYSIS — QUALITY ENGINEERING DEPARTMENT — DATA SHEET - Revised 9/10/58 RJP — 81-3312-020

WEEK		DEPARTMENT		GROUP		DEVICE		STATION		DATA TYPE	UNIT NUMBER (Last 3 Digits)			For Special Investigation				SHEET NO.	
3	6	1	5	1	1	2	7	0	2	8	1	3	7	0	0	0	1	0	1
1	2	3	4	5	6	7	8	9	10	11	12	13	14	15	16	17	18	19	20

E-27 D.C. Amplifier Module - Functional Test - Station 2

TEST NO.	DESCRIPTION AND/OR REMARKS		1 ST READING	2 ND READING	3 RD READING
1	Gain: Input (shorted)	Output ±0.015 v max.	0 0 4 (21 22 23)		
2	+ 1 mv	+ 0.095 - 0.105 v	0 9 9 (24 25 26)		
3	+ 2 mv	+ 0.190 - 0.210 v	1 9 8 (27 28 29)		
4	+ 3 mv	+ 0.285 - 0.315 v	3 0 1 (30 31 32)		
5	+ 5 mv	+ 0.475 - 0.525 v	5 0 7 (33 34 35)		
6	+ 10 mv	+ 0.95 - 1.05 v	1 1 8 (36 37 38)		
7	+ 15 mv	+ 1.42 - 1.58 v	1 5 3 (39 40 41)		
8	+ 30 mv	+ 2.80 - 3.20 v	3 1 8 (42 43 44)		
9	- 1 mv	- 0.095 - 0.105 v	0 9 7 (45 46 47)		
10	- 2 mv	- 0.190 - 0.210 v	1 9 4 (48 49 50)		
11	- 3 mv	- 0.285 - 0.315 v	2 9 8 (51 52 53)		
12	- 5 mv	- 0.475 - 0.525 v	5 1 0 (54 55 56)		
13	- 10 mv	- 0.95 - 1.05 v	1 0 2 (57 58 59)		
14	- 15 mv	- 1.42 - 1.58 v	1 5 1 (60 61 62)		
15	- 30 mv	- 2.80 - 3.20 v	3 0 9 (63 64 65)		
16	Total Current: Input shorted	0.010 ampere max	0 1 8 (66 67 68)		
17	+ 5 mv	0.012 ampere max	0 1 9 (69 70 71)		
18	+ 30 mv	0.020 ampere max	0 1 7 (72 73 74)		
19	- 5 mv	0.012 ampere max	0 0 9 (75 76 77)		
20	- 30 mv	0.020 ampere max	0 1 6 (78 79 80)		

FIG. 13.4 A typical input sheet for the automated data-analysis program.

all 80 punched digits presented as a horizontal string of numbers. Suitable columns on the tabulation runoff form (Fig. 13.5) identify the numbers in the same terms as the input form (Fig. 13.4). The data, as tabulated, are a suitable record of the measurements of all items of product considered. This operation compresses numerous data forms onto a single page of the raw-data tabulation. A capability that isolates each out-of-specification reading and presents the list of such readings as a separate output is also employed.

MINNEAPOLIS-HONEYWELL REGULATOR CO.

AERONAUTICAL DIVISION

D.A. RAW DATA

QUALITY CONTROL ENGINEERING

81-1141-011

	WEEK	DEPT	GROUP	DEVICE	STATION	DATA TYPE	UNIT	SPECIAL INVEST.	TYPE	SHEET	1	2	3	4	5	6	7	8	9	10	11	12	13	14	15	16	17	18	19	20
		16		1	2	1				1	625	575	001	625	575	001	625	575	001	410	410	410	650	650	650	650	160	160	160	001
		16		1	2	2				1	850	800	150	850	800	150	850	800	150	490	490	490	790	790	790	790	190	190	190	999
TEST=8	19	16	00	01	02	3	147	1115	0	1	810	750	060	770	730	040	810	740 840 ***	070	450	450	450	676	674	674	676	178	180	177	
	19	16		1	2	3	179	1115		1	790	730	060	770	730	040	810	710	100	451	451	451	664	672	670	674	178	178	178	
	19	16		1	2	3	185	1116		1	790	720	070	760	730	030	770	730	040	445	455	445	665	666	666	666	180	181	180	
	19	16		1	2	3	186	1115		1	770	690	080	780	720	060	760	720	040	448	448	448	677	677	677	677	177	180	177	
	19	16		1	2	3	192	1111		1	790	720	070	770	720	050	800	730	070	446	451	448	678	677	677	677	178	179	178	
	19	16		1	2	3	196	1115		1	760	720	040	755	715	040	775	700	075	449	449	449	667	674	668	675	177	177	177	
	19	16		1	2	3	224	1111		1	770	700	070	760	720	040	770	720	050	452	452	452	680	682	670	680	176	180	179	
	19	16		1	2	3	238	1115		1	780	710	070	800	720	080	780	710	070	454	454	454	679	685	673	679	176	178	176	
	19	16		1	2	3	242	1111		1	760	720	040	790	710	080	780	720	060	448	448	448	665	668	665	667	177	180	178	
	19	16		1	2	3	263	1111		1	790	710	080	750	710	040	780	720	060	452	452	452	675	675	671	678	179	180	180	
	19	16		1	2	3	284	1111		1	770	750	040	760	720	040	780	740	040	451	451	451	678	670	668	680	180	181	181	
	19	16		1	2	3	301	1115		1	790	700	090	790	720	070	800	710	090	454	454	454	679	679	677	679	177	177	176	
	19	16		1	2	3	318	1115		1	770	710	060	760	730	030	790	720	070	455	454	454	680	679	675	675	178	180	177	
	20	16		1	2	3	326	1115		1	780	710	070	800	740	060	780	720	060	452	452	452	679	683	681	680	178	180	177	
ST=11	20	16	00	01	02	3	329	1115	0	1	800	720	060	800	700	100	790	700	090	656	656 ***	656	693	680	680	681	178	179	177	
ST=10	20	16	00	01	02	3	329	1115	0	1	800	720	060	800	700	100	790	700	090	656 ***	656	656	693	680	680	681	178	179	177	
ST=12	20	16	00	01	02	3	329	1115	0	1	800	720	060	800	700	100	790	700	090	656	656	656 ***	693	680	680	681	178	179	177	
	19	16		1	2	3	336	1111		1	750	730	020	760	720	040	780	720	060	450	450	450	670	674	670	673	175	176	174	
		16		2	1	1				1	350	350	053	000	085	000	000	136	001	001	230	350	350	053	136	080	000	000	001	001
		16		2	1	2				1	550	550	071	090	115	020	080	196	999	999	999	550	550	071	196	100	020	080	999	999
	18	16		2	1	3	425	13		1	455	447	062	020	108	005	005	140	530	485	370	448	449	062	140	094	004	004	530	485
	20	16		2	1	3	57	1613		1	440	452	062	024	092	001	020	144	520	490	350	435	448	061	142	087	001	019	520	485
	20	16		2	1	3	159	1213		1	466	470	065	019	108	005	020	152	470	490	340	466	470	064	150	087	000	020	460	485
	20	16		2	1	3	299	1113		1	440	442	063	001	105	005	010	140	500	485	340	439	441	060	150	092	004	020	500	485
	18	16		2	1	3	317	1213		1	447	441	061	027	107	005	030	140	560	485	360	443	445	061	140	088	002	032	550	485
	18	16		2	1	3	334	1213		1	434	433	062	018	103	005	005	140	560	485	400	430	429	060	140	088	002	013	560	485
	20	16		2	1	3	337	1213		1	462	454	064	002	103	005	030	145	520	490	370	459	451	061	142	090	004	023	520	485
	20	16		2	1	3	362	1213		1	449	459	062	039	097	005	005	141	560	485	380	444	453	060	142	100	002	004	540	485
	20	16		2	1	3	364	1213		1	464	460	061	013	108	005	005	144	540	485	390	457	454	060	140	096	002	002	540	485
	18	16		2	1	3	369	1213		1	439	437	059	038	099	005	005	142	540	485	370	432	433	060	145	086	002	008	530	485

FIG. 13.5 A typical automated data-analysis raw-data tabulation.

13.5c ADA Computer Data Analysis. Most significant in the ADA program is the analysis of process capability. The raw-data cards plus an established program are fed into the digital computer. The data-analysis output is shown in Fig. 13.6. Each line of the tabulation represents a complete analysis of up to 100 tests on a single specification. The information, by column, includes:

Columns 1 to 3. Complete identification of the data immediately to the right. This includes the coded number of the device, the sheet number, and the specific test on that sheet.

Columns 4 to 6. The number N of observations available for the data identified in columns 1 to 3, the average $\bar{x}$ value of all of those data, and the process standard deviation σ associated with the data.

Columns 7 to 9. The specification information, including the low value of the specification, the high value, and the tolerance spread.

Columns 10 to 12. The three standard deviations, 3σ, limits of the process, and the total spread of six standard deviations, 6σ.

Columns 13 to 15. The same for four standard deviations, 4σ.

Column 16. Answers to six basic questions about the process. An affirmative answer is indicated by a 1, a negative answer by a blank. These questions are as follows:

1. Is $6\sigma \leq$ specification tolerance spread?
2. Is $8\sigma \leq$ specification tolerance spread?
3. Is $\bar{x} - 3\sigma \geq$ specification low limit?
4. Is $\bar{x} - 4\sigma \geq$ specification low limit?
5. Is $\bar{x} + 3\sigma \leq$ specification high limit?
6. Is $\bar{x} + 4\sigma \leq$ specification high limit?

Columns 17 and 18. Process shift; a 9 in column 17 indicates a shift to the left or low side of the specification nominal; a shift to the right or high side is represented by a blank space. Column 18 gives the percentage of the tolerance that the process average has shifted in the direction given.

The last set of 10 columns is a frequency distribution of the process data. The specification tolerance is divided into six equal cells, columns 3 to 8. Cells 2 and 9 each equal one-sixth of the tolerance immediately above and below the specification limits. Values more than one-sixth of the tolerance beyond the limits are in columns 1 and 10. The total in each column is the number of times the variable was within the range of the cell.

13.5d ADAgraph. An associated routine, using the output from the basic **ADA** program, provides a graphical display of the distribution of each variable covered in the ADA program. This display consists of two vertical reference lines defining, respectively, the lower and upper specification limits. The lower 3σ limit of the process, the mean $\bar{x}$, and the upper 3σ limit are printed as 1, 2, and 4, respectively, and they are positioned appropriately to the fixed specification limit lines. If two or more of the numbers are coincident, the sum of the items falling in that position indicates which items are coincident.

ADAgraph Output

		Lower spec limit				Upper spec limit	
Test 1			1		2 4		Well distributed
Test 2	1		2		3		Shifted to low side
Test 3	1			2		3	Process too wide
Test 4				7			Process extremely good—well centered and negligible variation

13.5e Normality of the Distribution. A program such as ADA relies on the normality of the process distribution. A recent innovation to ADA is an automated

H Aeronautical Division — AUTOMATED DATA ANALYSIS — QUALITY CONTROL ENGINEERING

IDENTIFICATION			PROCESS			SPECIFICATION			3σ PROCESS LIMITS			4σ PROCESS LIMITS			ANALYSIS			FREQUENCY DISTRIBUTION (LO SPEC. – NOM – HI SPEC.)									
DEVICE	SHEET	TEST	QTY. (n)	AVERAGE ($\bar{X}$)	STANDARD DEVIATION (σ)	LOW	HIGH	TOLERANCE	LOW	HIGH	6σ	LOW	HIGH	8σ	SIX QUESTIONS	SHIFT 9-	SHIFT % OF TOLERANCE	1	2	3	4	5	6	7	8	9	10
1	1	1	73	72986	4432	625	850	225	596	862	265	552	907	354		9	3			2	25	12	18	15	1		
1	1	2	73	65993	4089	575	800	225	537	782	245	496	823	327	1	9	12			6	35	8	21	3			
1	1	3	73	6993	3273	1	150	149	28	168	196	60	200	261		9	4			6	20	17	19	6	5		
1	1	4	73	73103	4361	625	850	225	600	861	261	556	905	348		9	3			2	27	9	18	17			
1	1	5	73	66336	4432	575	800	225	530	796	265	486	840	354	1	9	11			7	32	7	20	7			
1	1	6	73	6767	3304	1	150	149	31	166	198	64	199	264		9	5			4	26	16	18	3	6		
1	1	7	73	73041	4500	625	850	225	595	865	270	550	910	360		9	3			3	26	9	16	19			
1	1	8	73	65178	8695	575	800	225	390	912	521	303	999	695		9	16	1		9	31	6	22	4			
1	1	9	73	7041	3280	1	150	149	27	168	196	60	201	262		9	3			6	20	19	18	4	6		
1	1	10	73	45227	643	410	490	80	432	471	38	426	477	51	111111		3					31	35	6	1		
1	1	11	73	45210	589	410	490	80	434	469	35	428	475	47	111111		3					30	37	6			
1	1	12	73	45237	593	410	490	80	434	470	35	428	476	47	111111		3					27	40	6			
1	1	13	73	67927	755	650	790	140	656	701	45	649	709	60	111 11	9	29			11	60	2					
1	1	14	73	67968	735	650	790	140	657	701	44	650	709	58	111111	9	29			9	61	3					
1	1	15	73	67645	764	650	790	140	653	699	45	645	707	61	111 11	9	31			20	50	3					
1	1	16	73	67910	652	650	790	140	659	698	39	653	705	52	111111	9	29			9	62	2					
1	1	17	73	17737	98	160	190	30	174	180	5	173	181	7	111111		8						71	2			
1	1	18	73	17848	90	160	190	30	175	181	5	174	182	7	111111		12						62	11			
1	1	19	73	17756	153	160	190	30	172	182	9	171	183	12	111111		9						70	2	1		
1	1	20																									
2	1	1	99	44671	1234	350	550	200	409	483	74	397	496	98	111111	9	2					53	45	1			
2	1	2	98	45046	1199	350	550	200	414	486	71	402	498	95	111111							46	52				
2	1	3	99	6698	278	53	71	18	58	75	16	55	78	22	1 11		28					5	13	37	43		1
2	1	4	99	4602	4842		90	90	99	191	2.90	147	239	387			1			24	27	10	11	7	10	2	8
2	1	5	98	10470	745	85	115	30	82	127	44	74	134	59			16	1	1	3	5	11	16	36	24	1	
2	1	6	98	608	364		20	20	4	17	21	8	20	29	11	9	20			11	62		21	2	2		
2	1	7	98	1628	2327		80	80	53	86	139	76	109	186		9	30			60	24	1	9	2	1		1
2	1	8	98	14770	766	136	196	60	124	170	45	117	178	61	1 11	9	31			49	35	12	2				
2	1	9	98	51035	3624	1	999	998	401	619	217	365	655	289	111111		1					34	64				
2	1	10	98	47338	6176	1	999	998	268	658	370	226	720	494	111111	9	3				2	96					
2	1	11	89	35836	2858	230	999	769	272	444	171	244	472	228	111111	9	33			36	53						
2	1	12	86	44591	1318	350	550	200	406	485	79	393	498	105	111111	9	2					47	38	1			
2	1	13	86	44970	1264	350	550	200	411	487	75	399	500	101	111111	9						44	42				
2	1	14	86	6749	252	53	71	18	59	75	15	57	77	20	1 11		31					4	6	28	48		
2	1	15	86	14770	662	136	196	60	127	167	39	121	174	52	11 11	9	31			44	32	10					
2	1	16	86	8848	458	80	100	20	74	102	27	70	106	36		9	8			12	10	31	18	8	7		
2	1	17	86	459	247		20	20	2	12	14	5	14	19	11 11	9	27			22	30	32	2				
2	1	18	86	1737	1606		80	80	30	65	96	46	81	128	1	9	28			47	19	10	4	5	1		
2	1	19	86	49534	6150	1	999	998	310	679	369	249	741	492	111111	9				1		37	48				
2	1	20	86	48500		1	999	998	485	485		485	485		111111	9	2					86					
3	1	1	46	32107	398	311	329	18	309	333	23	305	336	31			6		1	1	3	9	18	8	6		
3	1	2	45	31320	331	302	320	18	303	323	19	299	326	26	1		12				2	6	20	9	8		
3	1	3	46	32130	526	311	329	18	305	337	31	300	342	42			7		1	4	2	10	6	16	6	1	
3	1	4	44	31355	462	302	320	18	299	327	27	295	332	36			14			3	1	6	12	9	11	2	
3	1	5	44	3541	2409	1	100	99	36	107	144	60	131	192		9	15			14	10	9	6	3	2		
3	1	6	46	676	152	1	10	9	2	11	9		12	12	1 1		14				5	4	21	16			
3	1	7	46	709	174	1	10	9	1	12	10		14	13	1		18				4	3	19	16	4		
3	1	8	42	1129	584	1	20	19	6	28	35	12	34	46			4				8	13	12	2	6		1
3	1	9	40	1183	602	1	20	19	6	29	36	12	35	48			7				2	22	6	3	5		2
3	1	10	44	1577	555	1	30	29		32	33	6	37	44			1			1	8	16	10	7	2		
3	1	11	44	1639	593	1	30	29	1	34	35	7	40	47			3			1	5	9	20	6	3		
3	1	12	44	1509	565	1	30	29	1	32	33	7	37	45		9	1			1	13	7	13	10			
3	1	13	44	1561	614	1	30	29	2	34	36	8	40	49							13	11	8	10	2		
3	1	14	41	1841	536	1	30	29	2	34	32	3	39	42	1		10				3	11	13	11	3		

5461

FIG. 13.6 The automated data-analysis output providing the analysis described in the text.

normality test. Consideration is being given to automated calculation of the third and fourth moments of the distribution as a further indication of the nature of the distribution. Usually a nonnormal distribution will appear wider than it really is (except in certain considerations of the fourth moment). Therefore, some "safety factor" on nonnormal distributions usually exists.

13.6 PROCESS CONTROL

In the broad sense, process control is the aggregation of all techniques required to produce products. Process capability, as previously discussed in this section, and failure analysis and corrective action, which will be covered later, are examples of long-term process controls. In the strict sense, however, process control is a tight, closed, feedback loop between the developed characteristics and the process that produced them. It consists of the measurement of characteristics at the site with provision for immediate adjustment of the process when need is indicated.

Process control is attractive from both a quality and an economic viewpoint. It serves to build quality into the product rather than to maintain a quality level by culling out defective items after they are produced. Inspections made after a product is produced cannot improve the quality of individual items. Such inspections can only upgrade the population quality by eliminating the worst or defective items. Process controls, on the other hand, have the advantage of sensing process deviation at a point where product quality can be improved by immediate adjustment of process elements.

The basic elements of an industrial process are specifications or requirements, equipment, materials, manpower, and written or unwritten techniques and methods which unite the first four elements for the specific application. Adequacy of individual elements and proper interactions between the elements yield a state of control. Loss or absence of control represents a deficiency in one or more of the elements or an improper relationship between the elements. In addition to having the process itself, a process control must:

1. Have the capability to sense a deviation or unsatisfactory state in the process.
2. Decide if anything should be done, and if so, what should be done to correct the condition.
3. Apply the necessary corrections to the process.

Process-control techniques vary widely depending on the particular application. On processes where control is critical or difficult to maintain, statistical charting methods are valuable. The same methods used on processes that are not extremely critical and are not particularly difficult to control would prove to be unnecessarily expensive. For an example of process control, we shall describe a technique that can be applied to an entire machine shop or assembly area. It is not a rigorous control, but it can be applied economically with a very favorable process improvement-to-cost ratio.

13.6a Personalized Quality Control as a Method of Process Control. Personalized Quality Control, or PQC, was developed by the Honeywell Aeronautical Division as an innovation to roving or patrol inspection. It is applicable to general workmanship as well as to characteristics that are directly controlled by specifications. The technique works as follows:

1. A chip or tag is identified and placed in a random-selection box for every process presently in progress. In most areas with a relatively fixed work force, the name of each operator on a separate chip serves to represent all activity in the area.
2. The inspector shakes up the box of chips and selects a chip at random. He will inspect work that has just been completed by the operator whose name appears on the chip.
3. The inspector performs a detailed inspection on the last specified number of items produced. Any discrepancies are brought to the attention of the production supervisor for corrective action.
4. A chart is maintained on each operator at his station.
5. The characteristics of the plan can be varied by adding chips to the random-selection box for operators who have poor records.

This process-control method is obviously directed at the process element manpower. It does, however, serve to isolate other deficient elements, because the operator will readily point out other factors which are contributing to his difficulty. The efficiency of the method is evaluated by the percentage of defective lots reported during a subsequent lot-by-lot final inspection. Admittedly, this technique is effective primarily because of its psychological aspect; nevertheless, it is effective. In numerous applications final-inspection yields were boosted to and maintained in the above 95 per cent yields region. The number of inspectors utilized for PQC work can be varied without changing the method.

13.6b Statistical Process-control Methods. Statistical process-control methods are extremely valuable where the nature of the process warrants the expense. Process controls of this type should be applied in the following typical instances:

1. When acceptability of the product can be determined only by accurate control of the process. Examples of such processes are spot welding and lead-wire welding.
2. When production volume is high and loss of control could represent a serious quality or economic loss.
3. On problem characteristics when the process capability is marginal.

The $\bar{x}$ and R (x-bar and R) chart is perhaps the most popular form of variable-characteristic process control. Other types of control are fraction-defective charts (p charts) and defects-per-unit charts (c charts).

13.7 PREVENTION OF THE DEGRADATION OF INHERENT PRODUCT RELIABILITY

For purposes of illustration, consider this simplified design-production relationship. Ideal or perfect production would yield a product that exactly bears the inherent reliability of the design. The ideal or perfect production process does not exist. Therefore, the reliability of produced product cannot be better—and, in fact, will be poorer—than the inherent reliability of the design. The inherent characteristics of the design are therefore the upper limit. The better the production process, the more closely approached are the inherent virtues of the product.

Strictly speaking, then, degradation of inherent reliability cannot be *entirely* prevented in production; it can only be minimized. On the positive side of the ledger, however, the inherent reliability can be improved through information obtained from the production operation. This serves both to raise the upper limit that could theoretically be reached and to raise the actual reliability of the product.

The gap between the inherent reliability and the actual reliability is very significant. Review of field-failure information will indicate that a large portion of the total failures are due to product fabrication—with no reflection on product design. The percentage of such failures frequently ranges between 40 and 85 per cent of the total failures. By itself, a figure of 60 per cent of the field failures assigned to fabrication without an indication of total failure rate is not necessarily an unsatisfactory level; it simply points out that production-caused failures are an extremely important consideration. There are therefore two objectives in our quest for the production of reliable product:

1. The inherent reliability of design must be achieved to the highest degree possible. This objective will be most difficult to attain. It will involve effective controls on the entire production process from the design requirements to the shipment of product. The major portion of this section is devoted to these controls.
2. The inherent reliability of the design and design process relationships must be optimized. This objective is achieved largely by experience in the initial production phase and results from planned reporting, corrective action, and feedback of information. Means for achieving the improvements are covered in Secs. 13.15 to 13.19.

13.8 CHANGE CONTROL OF ENGINEERING SPECIFICATIONS AND DRAWINGS

Engineering specifications and drawings are the primary source of information on both the nature of a product and the criteria for its acceptance. Changing these documents does not promote popularity among production personnel. Nevertheless, some changes to engineering documents are necessary, because they are the means for improvement. A realistic goal is to minimize rather than eliminate engineering changes. With the concession that changes are necessary, provision must be made for an orderly method of obtaining original issues and incorporating document changes into production.

The unhappy results of neglecting effective change-control procedures can range from financial loss when an incorrect product is subsequently discovered and rejected to serious reliability problems in the field. The basic requirements for effective change control are simple, but the change-control program needs extreme exactness at every step. The term "document" applies to drawings, specifications, engineering-change orders, and all similar documents generated by the engineering department. The following three requirements are essential elements of a change-control program.

1. Proper distribution of new documentation including changes
2. Removal of obsolete documentation
3. Assurance that effectivity is properly incorporated

While the first two requirements are distinct, the tasks are performed together, dealing as they do with providing the correct, and only the correct, information to the production and acceptance areas. The third requirement ensures that the information is converted into hardware at the effectivity assigned to the change or revision.

13.8a Methods of Distribution of Documents and Changes and Removal of Obsolete Material. Provision must be made that documents are distributed to all the "documentation areas" in which they are required and that these individual areas are thoroughly defined. Preprinted distribution lists are helpful in this instance.

The areas on distribution must maintain a rigorous, enforced discipline with respect to documentation. Limited-access files with checkout clerks and checkout control cards indicating both the location of a particular item of documentation and the date of checkout are desirable. Prints should be stamped in a manner that both identifies the rightful location of the document and shows that the item is on change control and checkout control. Only documents bearing this stamp may be used for acceptance purposes. The stamped impression on change-controlled documents permits the use of personal drawing or specification copies by engineers and others who require them. Ideally, copies of documents which are not on change control should be stamped "Not on Change Control—Not to Be Used for Product Acceptance." Documents on change control should never be marked up, and any other documents which have been marked should be clearly identified as such. Emphatic defacing with ink of all identifying *numbers* and *titles* of the document on every page is one method of identifying a marked-up document. The obliteration need not be so complete that the original information cannot be read.

The hazard presented by obsolete documentation in a production or inspection area is obvious. On receipt of new documentation, as a minimum, the clerk in charge of maintaining a documentation area should ensure that the equivalent old document is destroyed. In the event a document is checked out, the clerk should seek out the holder of the document and exchange the new for the old.

More stringent control can be achieved, at added expense, if all old documentation is routed to a central clearing area with a list of all items of official distribution. As new material is issued, the clearing area is informed of the old material to be received from each documentation area. Failure to receive an item in the allotted time results in action to recover the obsolete document. Serialization of individual documents adds a further control, but in most instances adequate control can be achieved without serialization.

An effective, economical compromise between the method requiring complete reliance on the individual distribution clerk and the clearing area is the use of a "distribution traveler" (Fig. 13.7). A distribution traveler is a two-copy carbon form that accompanies each package of one or more new documentation items. The form is numbered to identify the distribution package sequentially. It lists each item of new distribution that is contained in the package with a check-off or initial block in which the clerk indicates the processing of each item into the files. Parallel to the listing of each new item is a listing of each old item to be destroyed. Here, too, an initial box is provided to indicate disposal of the item. When the distribution package is processed, the clerk signs the form and routes one copy to a clearing area, which

DISTRIBUTION TRAVELER

To File Clerk: DT - ____________ Area ____________________

Return by: ____________________

This package contains new documentation which is to be filed in your area.

The superceded obsolete documentation shall be destroyed.

1) File each new item listed in the left-hand column. Initial the box aş filing is completed.
2) Remove and destroy the obsolete item which is listed directly to the right of the new item. Initial the box when the item is removed.
3) Retain one copy of this traveler in your file. Return the other completed copy to the Distribution Center, Mail Station 673, prior to the date indicated.

New Document	Rev	Filed (Initial)	Old Document	Rev	Destroyed (Initial)

FIG. 13.7 A distribution traveler for change control.

may be quality assurance in this application. She retains one copy as an area record. Clearing consists of simply assuring that boxes are checked off and that the forms are received within an allotted time after the distribution. While some expense is involved in initiating the form, checking off the boxes, and clearing, the method provides the following important benefits in addition to maintaining general order in the distribution process:

1. This is a psychological incentive for the documentation clerk to perform accurately. The clerk must give a positive indication that every element required to maintain an accurate file has been performed.

2. A break in the number sequence of the distribution travelers indicates that, somehow, a distribution package has been missed. The missed package may be replaced, and, while the changes in the missed package may now arrive slightly late, this fail-safe provision prevents long-term discrepancies and serves to detect distribution breakdown. The clearing group has further indication of missed distribution when the completed form fails to appear. This provides a double fail-safe against missing new documentation.

3. The collection of completed forms by the clearing group provides a centralized record of the volume of change activity. It is physical evidence to quality assurance that documentation areas are being properly maintained. (Supplementary auditing is desirable, however.) Also, government inspectors and customer representatives seem to be inclined to take a close look at the condition of specification files, and the records provided by this method should serve to assure them.

During the transition from one documentation revision to another, it may be necessary to retain the old document in the documentation area. This is particularly true when the effectivity of the revision applies somewhat later than the time when the change document is issued. In this instance, both the old and new revisions should be retained together, and a stamp stating "This document applies to———" should be applied to both documents with an appropriate description of the applicability of each. For example, the description on the old document might be: "This document applies to system serial number 1107 through 1291, inclusively." The description on the new document might be: "This document applies to system serial number 1292 and all higher numbers." The stamp impression identifying the document area and any special applicability should be applied where it will certainly be seen. Specific instructions for stamping and special handling can be included on the distribution traveler.

A central file, or storage area, for obsolete documents should be maintained for purposes of failure analysis and other reference requirements on older product configurations. These documents should be stamped "Obsolete Document—for Reference Only" or an equivalent. Normally, this file should be remote from the production area. In the event old-document revisions are maintained in a production area for repair of field returns, it is essential that the Obsolete Document stamp appear on the out-of-date item.

13.8b Assurance that Engineering-change Effectivity Is Properly Incorporated. There are several methods of indicating the effectivity of an engineering change. Effectivity may be stated by date, by serial number, or simply "after old-style parts are depleted." Many companies classify engineering changes into groups. Each classification reflects the nature and effectivity requirement of the change. For example, a typical system would use one of the following nine standard classifications for each engineering change.

Class 1. Mandatory on all devices. Hold shipment. Systems in the field shall be retro-fitted.

Class 2. Mandatory on all devices which have not been shipped. Hold shipments.

Class 3. Mandatory on all devices which have not passed final acceptance tests.

Class 4. Mandatory on all new production. Devices already started may use old style.

Class 5. New-style parts will be used within 30 days. Old-style parts will be scrapped in 30 days.

Class 6. New-style parts will be used when old-style parts are depleted. Old-style parts may not be used after use of new-style parts is begun.

Class 7. All new parts shall be made to this change order. Old-style parts may be used.

Class 8. Clarification or record change. Product not affected.

Class 9. Special disposition. Disposition must be explained on the change order.

The nine classifications provide a broad but necessary range of change-order actions.

A system such as this serves to denote relative importance of the change to concerned persons who know the code. The quality assurance group, of course, must enforce the denoted provisions of every change-order classification, regardless of its level. The inspector is able to ensure this only if the documentation he uses is accurate and up to date. It is good practice to have one documentation area for production personnel and one for quality assurance inspectors in the same area if usage activity warrants the duplication. In the event that only one set of files is maintained, it is best that quality assurance have jurisdiction over and control of the file. While perhaps any capable clerk could maintain the file, in some isolated instances the favored arrangement may advance the cause for quality and reliability. Also, the responsibility of the quality assurance supervisor to protect the "true word" presents a logical and favorable appearance to government inspectors and customer representatives.

13.9 CHANGE CONTROL ON OTHER DOCUMENTS

Controls similar to those for engineering specifications and drawings should be exercised on other documents which influence the quality and reliability of the product. These include:

1. Detailed test procedures
2. Detailed inspection instructions
3. Process and assembly instructions
4. Operating instructions and approval procedures
5. Purchase orders
6. Subcontracts
7. Measurement-equipment maintenance instructions
8. Measurement-equipment and tool specifications

Placing these documents in the same distribution program as engineering specifications and drawings should be considered. (See Secs. 7 and 9.)

13.10 CONFIGURATION-CONTROL METHODS

The need to associate the revision status of individual parts with the final device has been emphasized. In addition to this information, it is of great value that the final device be associated with the revision status of the tests by which the product was deemed acceptable, even though the tests do not alter the physical makeup of the device. There are several relatively simple methods of configuration control which are adequate for simple products and even more elaborate products such as electronic modules, gyros, accelerometers, motors, resolvers, and instruments. Configuration control on more complex products which would typically contain numerous components such as those mentioned will be dealt with separately.

Configuration control is rapidly becoming a rather routine customer requirement for complex equipment. In some instances a full configuration report is required with each item of product. Adaptation of the method described in this handbook for complex equipment should satisfy requirements of this type.

13.10a Simple Methods of Configuration Control. One relatively uncomplicated method of configuration control employs the marking of the date of manufacture on the product, either in code or in clear language. A log of engineering changes linked with the date of incorporation is maintained at the factory. Knowledge of the date of manufacture of a particular item allows the factory to ascertain the configuration of the item. Administrative effort on the log is negligible, but it must be maintained.

One shortcoming of the method is that every configuration change requires a date change. This is not a serious problem if dates are changed daily, but it is one if dates are changed weekly or monthly. Also, a customer reporting a failure is likely to give the device number and serial number but not the date or date code. Other methods are preferred for serialized components. The date-code method is very good for nonserialized high-volume components, since the marking is low in cost. Presence of the date, or a code recognizable as the date, is reassuring to a customer's reliability and quality control people. It implies that the vendor at least has the rudiments for isolating a problem to a particular group of product. Absence of a mark could mean that every item of the type will always remain suspect in the event of problems.

Another method related to the date-code method is simply to stamp the latest drawing revision letter on the item. Again, a log or copies of all drawing and specification revisions should be maintained. This, alone, is actually a cleaner solution to configuration control than the date code is. The date code, however, serves to isolate more easily problem periods of production in the event of trouble. A monthly date code with the revision letter represents a simple, three-digit solution to configuration control along with a capability to pin down problem production periods. An example of such a code is as follows:

Revision	Month	Year
A	January, A	1965, A
B	February, B	1966, B
C	March, C	1967, C

The code CBA means revision C built in February, 1965. Symbols could be used if letters or numbers could cause confusion with other data.

A degree of configuration control can be maintained by assigning a block, series, or group number as part of the device nameplate information. The block number is advanced only on significant changes. Again a log must be maintained. The effectivity of changes between block numbers cannot be determined exactly. However, the log can include all effective minor changes. When a block change is made, the minor changes may be stated to be in a "not later than block X" category. The problem and pitfall of this system are in deciding what is a "major" and what is a "minor" change.

Serial numbers represent an excellent medium of configuration control, again using the log method. Control by serial numbers links the particular device with factory data for purposes of failure analysis and measurement correlation. One complication with serial numbers is that many companies do not identify a device until it is ready for shipment. Nameplates are fastened immediately prior to shipment with serial numbers in sequential order. A primary objection to assignment of serial numbers at the beginning of assembly is that the customer may suspect that any device shipped out of normal sequence required considerable rework and repair. Control can be retained under these conditions, but at least an extra clerical step must be added. One method is to retain the factory data in a packet along with the device until the serial number is assigned. The packet is identified with the serial number and the link is established. However, if the packets get mixed prior to marking, confusion reigns. A more foolproof method is to assign each device a "factory serial number" stamped or etched in a remote position or a position eventually covered by the nameplate. All data are identified by the factory serial number, and a cross index with the serial number is prepared as nameplates are applied. This method has the additional, rather limited, merit of preserving device identification even if the nameplates are removed or switched.

Although the latter method would work on a complex device, it is best that such a device be assigned a permanent serial number at the beginning of assembly, with the number prominently and permanently displayed. This will minimize the risk of clerical mistakes.

13.10b Configuration Control of a Complex Product. For the discussion of configuration control on a complex device the following requirements are assumed:

1. Control is based on the product serial number.

2. A capability exists to specify the desired revision status of every part in the individual assembly. Provision is made to review and approve any changes required from the specified configuration.

3. The revision status of every part actually used is documented and supplied to the customer with the assembly. Parts are marked with a revision identification.

4. A number of critical components are identified by serial number. Backup test data on these items are available.

5. Acceptance tests are specified and identified by revision. Variable test data resulting from these tests are furnished.

6. All repair required in production is included.

Control of the revision status of individual parts presents the greatest difficulty because of the large volume of detail. The control can be accomplished manually, but a semiautomated program employing tabulation equipment will minimize the labor required. Prepared forms are valuable, and their use need not be limited to one particular product. The following steps are performed in a typical program:

1. *Preparation of a master deck of data cards.* A master configuration-control deck of data-processing cards is prepared to call out the configuration of the first assembly to be produced under the control system. All current drawings and specifications and the total parts list are reviewed, and the configuration is written on keypunch sheets. The data are punched into regular data-processing cards.

2. *Generation of configuration-control decks for each assembly.* Since the master deck was prepared from the first system, the deck for that system is simply a copy of the master deck with the addition of a serial-number card. On subsequent assemblies, the master deck is systematically modified as engineering changes become effective. Modification is accomplished very simply when effectivity is stated by serial number. On certain changes, old-style parts may be used until the supply is depleted. In that case the quantity of old-style parts on hand is considered in the plan. The changes to the master deck are usually very limited between assemblies, depending, of course, on the volume of change activity. The master deck is modified by removing an old card and replacing it with one that has the new information. More than one part number may appear on a single data card to reduce the volume of the decks. The modified master deck is duplicated, and a serial-number card is added to the duplicate, providing a deck for the assembly with the required modifications.

3. *Preparation and use of the configuration plan.* The deck of cards for a particular system is processed through a tabulation machine, yielding a configuration plan for the parts to be used in that particular assembly. A typical configuration plan with the title Accumulation Log is shown in Fig. 13.8. The "planned" and "actual" revision-status columns are evident in the form. The assembler installs a part with the "planned" revision letter and verifies it by writing the revision of the part used in the "actual" column.

If the assembler does not have a part of the correct revision status, a review must be made. This condition occasionally occurs if one or more old-style parts are damaged and scrapped. The plan calls for an old-style part but, because of the loss, only new-style parts are available. Normally, the new-style part will be found acceptable to use, but an engineering review must be made in each instance of change from the configuration plan. For example, two interdependent old-style parts may work well together and the two new-style parts may work even better. To mix an old-style and a new-style part, however, may produce results that are entirely unsatisfactory in either performance or reliability. In all cases the "actually used" part revision is entered in the "actual" column.

4. *Printout of the assembly configuration.* The configuration-plan form with the "actual" entries is suitable, in itself, as a record of a system configuration. Usually, a more concise format is desirable, particularly if the configuration is to be supplied to the customer. To provide this printout, the punched-card deck for the particular assembly is modified to incorporate the parts actually used. A tabulation run is made on a prepared form, yielding a summary as shown in Fig. 13.9. A complex device will require a number of pages for a complete printout. A prepared printout form can be omitted if an explanatory cover page is placed in front of the tab run pages.

#81-1137-012

POLARIS PROJECT ACCUMULATION SHEET

1203 MHA NO. 07 22 63	ASSEMBLY SERIAL NO.	ACCUMULATED BY ________
		CHECKED BY ________

*1.—INDICATES PART MAY BE GFE
2.—INDICATES PART REQUIRES SERIAL NO.
3.—INDICATES PART REQUIRES SERIAL NO. AND MAY BE GFE

●—XXX INDICATES SELECTIVE TABULATED PART NO. FILL IN DASH NO. USED AND QTY.
AR INDICATES INDEFINITE QTY. ENTER QTY USED.

	ASSY LAYOUT NO.	BUORD ASSY (REF)	QTY ● PER ASSY	MH LAYOUT PCO	BUORD PCO (REF) ●	BUORD PART NAME	BUORD PCO DWG REV PLND.	BUORD PCO DWG REV ACT.	SERIAL NO./GFE*	REMARKS
1	1979286	D1979286			P 1978054	GUIDE	0A		1	
2	1979286	D1979286			P 1978055	BUMPER	0C			
3	1979286	D1979286			P 1978056-2	SHIM	0C			
4	1979286	D1979286			P 1978057	GUIDE	0C		1	
5	1333069	D1980602			A 1979286	LIMIT STOP	0E			
6	1979286	D1979286			P 1979295	SPRING	0C			
7	1979286	D1979286			P 1979296	SPRING	0A			
8	1333069	D1980602			P 1979323	NUT	0B		1	
9	1979286	D1979286			P 1979352	NUT	0A		1	
10	1979286	D1979286			P 1979385	CAP GUIDE	0B			
11	1333069	D1980602			A 1980001-1	BRG MOUNT	0D		3	
12	1333069	D1980602			P 1980003	STUB SHAFT	0C		3	
13	1333069	D1980602			P 1980004-2	STUB SHAFT	0E		3	
14	1979286	D1979286			P 1980006-052	SCREW	0F			
15	1979286	D1979286			P 1980043	BRACKET	0E			
16	1333069	D1980602			P 1980226	BRG GIMBAL	0C			
17	1333063	D1980602			A 1980242	TAPE GUIDE	0N			
18	1333069	D1980602			P 1980626-1	PITCH GIMBAL	0J		3	
19	1979286	D1979286			P 2076841	RING				
20										

FIG. 13.8 A page from an accumulation log which calls out the configuration for a complex assembly.

13.10c Adaptations of the Configuration-control Method. Adaptations of the method described can tailor it to a particular product or a particular requirement. On some products, revision status may not be expressible as simply as the example in Fig. 13.9. Revision status on some parts could be "Revision L less Revisions D and

Honeywell • AERONAUTICAL DIVISION

=81-1212-002

DRAWING STATUS LIST

FOR ENCLOSED PARTS

POLARIS GIMBAL ASSEMBLY MHA SERIAL NO. 4001

PART NUMBER	NAME	DWG. REV.	PART NUMBER	NAME	DWG. REV.
2104105	CLIP WIRE	B	2104164	RETAINER BRG	A
2104167	NUT,PLAIN	C	2104186	COVER ASSY	F
2104193	CUSHION	D	2104193	CUSHION	D
2104199	CLAMP ASSY	B	2104224	SLEEVE AIR OUT	D
2104231	FLANGE AIR IN	D	2104233	STOP 2 MID AX	D
2104235	STOP RH MID AX	D	2167416	TEX SLEEVING	
2168286	CAP PROTECTIVE		2236910	SCREW	A
2295245-2	SEAL LEAD		2317670	SCREW	C
2317671	SCREW	C	2317672	SCREW	C
2317675	CONNECTOR	D	2317675	CONNECTOR	D
2317675	CONNECTOR	D	2317675	CONNECTOR	D
2317678	TERM STUD		2317680	CONNECTOR	D
2317681	CONNECTOR	E	2317681	CONNECTOR	E
2317681	CONNECTOR	E	2317682	CONNECTOR	D
2317682	CONNECTOR	D	2317701	PIN,STRAIGT	
2317708	FAN VANEAXIAL	B	2317710	HSG BEARING	E
2317710	HSG BEARING	E	2317711	GIMBAL ASSY	E
2317715	STAB MEM ASSY	C	2317716	IGA AZI AXIS	G
2317717	IGA AZI AXIS	F	2317718	IGA MID AXIS	H
2317719	IGA MID AXIS	J	2317720	IGA PIT AXIS	G
2317721	IGA PIT AXIS	G	2317723	SLIP RING 32	B
2317723	SLIP RING 32	B	2317723	SLIP RING 32	B
2317723-1	SLIP RING 32	B	2317723-2	SLIP RING 32	B
2317724	SLIP RING 72	B	2317726	STUB SHAFT	F
2317727	STUB SHAFT	C	2317727	STUB SHAFT	C
2317727	STUB SHAFT	C	2317728	HSG BEARING	C
2317733	HTR,TAP MNT	D	2317736G3	J PIP ASSY	A
2317736G4	SV PIP ASSY	A	2317739	STOP ASY MID	B
2317740	STOP ASY MID	B	2317742-2	PLATE IDENT	D
2317743-2	PLATE IDENT	D	2317744-2	PLATE IDENT	H
2317747	STUD	A	2317748	NUT	C
2317755	CONNECTOR	E	2317755	CONNECTOR	E
2317755-1	CONNECTOR	E	2317755-2	CONNECTOR	E
2317756	CONNECTOR	D	2317756-1	CONNECTOR	D
2317756-2	CONNECTOR	D	2317764-REF	BAL WEIGHT	C
2317765-REF	BAL WEIGHT	D	2317785	BAL WEIGHT	B
2317785-REF	BAL WEIGHT	B	2317788	CLAMP RING	A
2317788	CLAMP RING	A	2317792	RESOLVER	C
2317796	WASHER,CLAMP		2318710	SILASTIC	A
2318787	RESISTOR		2318983	RESISTOR	
2318983-REF	RESISTOR		2318984	RESISTOR	B
2318984	RESISTOR	B	2318984-REF	RESISTOR	B
2318987	SV PIP QUAD NK	K	2318987	J PIP QUAD NK	K
2401550-GFE	PIP	B	2401550-GFE	PIP	B
2401571-ALT	PIP		2401571-ALT	PIP	
2401779	SV PIPA TQ GEN	K	2401779	J PIPA TQ GEN	K
2401809	ANAL EX TELE	F	2401862	PIGA WHEEL SUP	G
2401942	ENC DR SHAPER	M	2401976	ENC BIAS AGC	M
2401978	XFMR MATCH PR		2402023	SV PIP PREAMP	K

FIG. 13.9 A page from a final configuration-control output describing the "actual" configuration of a specific assembly.

J" or "Revision M plus E.O.'s 69392 and 69460." More space is then required for printout. The program can be changed to allow the printout of one part per line or the printout in the long direction of the normal 8½- by 11-in. page. Provision can be made to print out additional data immediately below the basic entry in the event only some parts require a lengthy description.

In some instances it may be desirable to require that all parts accepted by material review board (MRB) actions be identified by MRB number. If configuration control is also required, MRB numbers can be included as supplementary information. A suitable code on the data card will allow sorting and an independent run on MRB'ed parts alone. The assembler would be required to transcribe all MRB identification supplied with the parts into the "actual" column.

Some other items which could be recorded are the part-manufacturer's name or selected measurements of certain parts. Although this information is normally handled in other ways, an elaborate configuration-control printout may serve a number of purposes in some product applications.

In some applications the basic system described may be considered to contain so much volume that changes are hidden. This problem can be solved easily: Simply print out the unchanged master list for every final assembly. Add an appendix for each system, stating essentially that "the configuration of this assembly is the same as the master list (Device Serial Number 1) except for." Then list the present status of all part changes. When the except-for list becomes too long, make a new master list and identify the new master list with the first device that contains exactly that configuration. This modification sets up certain devices as "milestones" and contains only the pertinent changes on a unit-by-unit basis.

13.10d Configuration Control Applied to Final Calibration and Test. Complex-equipment specifications are usually written in a form which, for purposes of configuration control, can be divided into two basic groups. The first group, which contains the greatest volume of information and which can be subdivided much more completely than will be done at this time, is the collection of all specifications needed to produce the physical assembly of the final device. The second group of specifications apply after the physical assembly is made. This second group describes how the assembly should finally be calibrated, how it should be tested, and how it should perform.

The configuration of the physical assembly is defined adequately by a printout of the revision letters of all the parts and, perhaps, even the revision letter of the final assembly. This satisfies control and status of the first group of specifications. The second group of specifications is normally a rather elaborate collection of specifications in text form, containing "knob-turning" detail. The same document is frequently used for calibration and test. An individual document covers a logical portion of the final-device test. Each document is maintained on its own revision status. A suitable control for such an arrangement is entitled Specification Control Sheet (Fig. 13.10), which positively identifies the proper specification for a particular system. The control sheet also provides a control on the proper collection of test data on the device. It should be noted that effective change control with indicated effectivity should automatically provide the correct specification for use. The control sheet provides a simplified backup to assure the required effectivity is met. (See Sec. 8.)

There are two schools of thought on how final-device specifications should be handled on complex assemblies. One approach, involving the separate updating of individual documents, has been discussed. Another approach is to make an overall test manual with a single revision letter. The first method has the disadvantage that the final configuration must be described in a list rather than by a single letter or number, but it has the change-control advantage of being small enough to allow the old specification to be destroyed and the new specification to be distributed in complete form. Its revision history is pertinent to the particular test.

The advantages and disadvantages of the first method are reversed in the single-document method. Note that the advantages of each can be carried over to the other, except for the favorable change-control feature of the separate-document approach. It would be impractical to throw away a massive manual every time a change was incorporated. Page-by-page upgrading of a manual, however, has a great potential for producing errors. It first relies on perfect distribution of new pages. Next, the exchanging of new pages for old is not particularly exhilarating, and the new page occasionally gets thrown away instead of the old page—and the old page still looks official. Replacing an entire section of the manual would be a satisfactory compromise. While this approach may consume quite a lot of paper, *not* having to tie up a group of engineers while they figure out what went wrong and *not* having to rerun an elaborate test will pay for a good-sized stack of paper.

An overall revision status on the separate-document approach is obtained by advancing the overall revision status every time an individual document is changed.

F-33 Inertial Platform
Specification Control Sheet

System Serial Number ____________

Test	ATP	Rev	Production Data Sheet	Inspection Data Sheet
1. Isolation Amplifier Test	12-1			
2. Azimuth Gimbal Alignment	12-2			
3. Stops and Limit Switches	12-3			
4. Gimbal Orthogonality	12-4			
5. Gimbal Balance	12-5			
6. Resolver Alignments	12-6			
7. Gimbal Position Indicator	12-7			
8. Precision Resolver Alignment	12-8			
9. Accelerometer Scale Factor and Bias	12-9			
10. Gyro Drift	12-10			
11. Environmental	12-11			
12. Accelerometer Wobble	12-12			
13. Accelerometer Encoder Tests	12-13			
14. Gimbal Torque	12-14			
15. Bearing Preload	12-15			
16. Start-up Tests	12-16			
17. Gyro-Amplifier Gain Products	12-17			
18. Accelerometer-Amplifier Gain Products	12-18			
19. Isolation Test	12-19			
20. Phasing Test	12-20			

The indicated Acceptance Test Procedures and Data sheets shall be used for test of this platform assembly.

____
Quality Control Engineering

FIG. 13.10 A specification-control sheet describing the correct test specifications and data sheets to be used on a serialized complex device.

13.11 LOT CONTROL OF PIECE PARTS

Lot control of piece parts has been developed to improve reliability analysis and directed corrective action on high-reliability-requirement devices. In a sense, lot control is a next step beyond configuration control as previously described in this section. In addition to providing revision identification, lot control, in its fullest sense, provides traceability of each component part back to the part-manufacturer's particular production run and, ideally, back to the material used by the manufacturer in that run. Although traceability to the serial number of the steam shovel that dug the ore for the material in the part will probably never be necessary, a high level of traceability is the mark and purpose of lot control.

The principal concentration of interest in and application of lot control is centered in the manufacture of complex electronic equipment. Equipment of this type is normally produced in separable subassemblies or modules. Modular construction is convenient for the application of lot control but is not essential.

Lot control works this way: When a vendor ships a lot of parts, included is a certificate identifying the manufacturing run of parts and other pertinent data required by the equipment manufacturer. The equipment manufacturer inspects the parts and assigns a lot number to the accepted lot. Certification provided by the vendor must be filed and identified by the same number. Thereafter, the equipment manufacturer must take special care to assure that the lot remains segregated and the identification number stays with the lot. A traveler form (Fig. 13.11) is provided for each subassembly or module to be built with lot-controlled parts. The module is assigned a serial number at the beginning of assembly, and that serial number is also indicated on the traveler.

The traveler and module must remain together all through production and test. (An appropriate container is useful in assuring "togetherness.") As each lot-controlled part is installed into the assembly, the assembler writes the lot number in a space provided on the traveler. The space is preidentified by circuit symbol, part number, and component value. Some manufacturers find it convenient to make up kits of parts for a particular module. The traveler can then be filled out as the kit is made up.

Frequent audits of stocking methods and accuracy in recording lot numbers should be provided by the quality assurance group. When the module is completed, all traveler spaces should be filled with lot numbers. Sequential arrangement of preidentified spaces, in order of assembly, aids in prompt detection of missing numbers. After acceptance, the module is identified by its serial number only and is subsequently controlled by normal configuration control. The only link with the lot numbers of the parts is now the traveler, which is filed at the factory. The customer may require a copy of each module traveler. A capability now exists to trace any part back to the information supplied by the vendor on his certification, although considerable digging may be required.

Wide use of a limited number of component types is very common because of the numerous reliability advantages inherent in such an arrangement. Lots are normally large, and it is very likely that parts of one lot will be assembled into many different module types. Semiautomated procedures can again be used to great advantage for prompt location of all parts from a suspect lot. To utilize this program, the data on every traveler must be punched into data cards. Also, the configuration of each major device must be punched with all module identifications and serial numbers. The traveler data cards are sorted for the suspect lot number, thereby isolating the modules that contain parts of the particular lot. The major-device cards are then sorted for the defective-module types and their serial numbers. The isolated cards are arranged and a tab run is made; this provides a printout of every major device affected, the type and serial number of each module affected, and the circuit symbol of each position containing a part of the subject lot.

If the major device is stationary, such as general-purpose computer, a creative programmer can so design the program that the printout is, in itself, suitable copy for immediate TWX transmittal to all affected service-engineering areas.

Perhaps the greatest potential difficulty in and expense of lot control lie in maintenance of communication and coordination with component vendors. The control and the information desired must be clearly spelled out to the vendor; it should form part of the purchase agreement. If the vendor produces high-reliability components as a major line, he can perhaps already supply the information needed for lot control. If his major line is components for entertainment electronics, some difficulty may be involved, particularly if the parts ordered are of a type that would be run the same as

LOT CONTROL TRAVELER 649682 Resistor Capacitor Network Revision _____ Lot Number _____							Serial Number 649682
		S/N	Lot Number	Rev	Symbol	Operator	Inspection Verification and Date
Circuit board	623679	--			TB1		
0.05 μf capacitor	624563-10	--			C1		
0.05 μf capacitor	624563-10	--			C2		
0.05 μf capacitor	624563-10	--			C3		
0.05 μf capacitor	624563-10	--			C4		
1.1K resistor	626593-11	--			R2		
1.3K resistor	626593-13	--			R3		
5.2K resistor	626593-52	--			R1		
0.01 μf capacitor	624563-2	--			C6		
0.01 μf capacitor	624563-2	--			C5		
0.01 μf capacitor	624563-2	--			C7		
5.0K resistor	626593-50	--			R4		
6.0K resistor	626593-60	--			R5		
1.1K resistor	626593-11	--			R6		
epoxy	2640-1	--			--		

FIG. 13.11 A lot control traveler for a module.

an entertainment component. This facet of lot control is part of the procurement of reliable supplies. (See Sec. 14 for details on part procurement.)

In summary, lot control is a relatively new program featuring high traceability of every part in a subsystem or major assembly. In-plant costs of maintaining lot control are moderate and are incurred primarily in the extra care taken in stocking parts, the extra time required by the assembler, and the time required to enforce the program through quality assurance audits. The biggest problem is the extraction of the desired information from component vendors, a problem which will be alleviated as lot control becomes commonplace. Vendors will then be able to supply the desired information as normal practice or at a moderate certification fee.

Lot control is particularly useful on products with large numbers of parts. In some applications lot control has included serialization of each component. Serialization provides a link between each component and its test data which could prove valuable in a failure analysis involving characteristic drift. Most high-reliability-component vendors can provide serialization of components; some can supply punched data cards for each serialed part with pertinent test data. Extra cost is usually involved, however.

13.12 CONTROL OF MEASUREMENT AND TEST EQUIPMENT

A large portion of the total number of product characteristics must be evaluated by measurement. The measurements performed on a product accomplish one or more of the following determinations.

1. Acceptability of the characteristic as compared with the specifications.
2. Repeatability of the product characteristic as an indication of characteristic stability and reliability. This is accomplished not by single measurement, but, rather, by an entire correlated test program.
3. A long-term input of the capability of the production process to produce the characteristic (process capability).
4. Immediate input for process control.
5. Characteristic data required for later operations or field use.

The measurement and test equipment which evaluate product must be maintained at an accuracy standard compatible with the requirements of the measured characteristics. For the production of high-quality equipment, a formal program of measurement-equipment control is absolutely necessary. (See Secs. 8 and 10.) In certain instances the tooling that develops the characteristic is used as the acceptance criterion when it would be impractical or even essentially impossible to provide gauging or instrumentation to measure the characteristic after it is developed. When tools are so used, they must be checked and controlled just as though they were gauges or measuring instruments.

Metrology facilities for evaluation, maintenance, and calibration are normally divided into two areas. A tool and gauge inspection laboratory specializes in mechanical measurements; an instrumentation inspection laboratory specializes in electrical measurements. This division is made because of specialization of personnel and management in each field. When two such labs are maintained, they are frequently coupled at higher management levels under a chief quality engineer or a chief inspector.

Although the measuring techniques for mechanical and electrical measurement equipment are different, the elements of control are essentially the same. The following steps are required:

1. When a piece of measurement equipment is presented for use, whether it has been purchased, subcontracted, or built in the company's own toolroom, the laboratory should thoroughly evaluate it against its specifications. The instrumentation-engineering group, or another engineering group selected as an impartial evaluator, should document the evaluation plan. Evaluation programs involve repeated tests of the equipment over a period of time, rather than a one-time inspection.
2. Immediately after the equipment is accepted by the laboratory as adequate for use, a permanent serial number should be applied.
3. A checking interval for the equipment should then be established. Experience,

particularly bad experience, should be utilized to modify checking intervals. A single out-of-calibration gauge or instrument should be treated as a serious matter. Ample effort to investigate and correct the problem is essential.

4. The new equipment should be incorporated into the calibration-control program to ensure that checking intervals and records are maintained by having a sticker on the equipment to indicate when the next calibration is due. Equipment should not be used after the due date indicated on the sticker.

5. The location of measurement equipment should be known at all times. Check-out control should be provided for small items.

6. Periodic checks are made according to the checking interval provided by the calibration program. Specifications and check procedures should be made available for each type of equipment. The nature of any instrument shift should be thoroughly investigated and the instrument corrected. Determination of product status after the discovery of measurement error is a difficult business, and the possibility of its occurrence exists in any organization.

In the development of a quality assurance plan, every characteristic should be considered before the fact with respect to the consequences to product status in the event of measurement-equipment error. Measurement-equipment checking intervals and data-recording systems should be influenced by this consideration. Use of "floor" standards for daily or weekly checks should be considered if it is determined that the status of product acceptability would be difficult to determine. It cannot be emphasized too strongly that incorrect measurement equipment is a serious matter which requires review and corrective action in each instance. Product acceptability must always be considered. *Simply recalibrating the equipment is not adequate action.*

13.13 TOOL CONTROL

Tool control should be conducted similarly to measurement-equipment control with some special considerations:

1. New tools should be inspected when they are received.

2. Provision should be made for a tool tryout when new, special tooling is developed. This may consist of a special evaluation run, or it can be defined as the first-run pieces of the initial production run.

3. After the tool-tryout period, the tools should be reinspected for any abnormal change encountered after operation. A tool is usually subjected to much greater stress than a gauge, and the tool tryout has definite merit.

4. Tooling should be individually identified except for small expendable items such as drill bits and taps.

5. Tools should be checked and certified periodically on long production runs. The period should be defined in terms of time or number of parts run. Tools should be inspected after sharpening. Consideration should be given to the relating of tool sharpening and tool inspection with process-control methods.

6. Tools should be checked before each job in a shop producing primarily short-run items.

A particular shop should establish tool-inspection requirements according to its mode of operation. Frequently, a shop will have a continual production of some items and a small-lot production of other items. In that event, two tool-control procedures are desirable.

13.14 RAW-MATERIAL CONTROL

Raw-material control consists of three basic steps:

1. Assurance that the proper material has been received and that it complies with applicable requirements.

2. Provision for the identification of all material to prevent its use in an incorrect application.

3. Proper handling and storage of material to preserve its quality and observation of expiration dates on perishable material.

Raw material can best be controlled by use of a material specification. The specification should completely describe the material and call out specific controls for characteristics. Certain material characteristics are more important than others, and a classification should be associated with each material characteristic to inform quality assurance in the level of control required. A representative classification system for raw materials follows:

Classification of Raw-material Characteristics

Class A. All material characteristics in this classification shall be tested in the company laboratory. Quantitative results shall be placed on file and identified by the material lot. At least three random samples of the lot shall be analyzed, unless the specific sample size is stated in the material specification.

Class B. All material characteristics in this classification shall have quantitative test results supplied by the vendor. The vendor shall certify the test report. In the absence of vendor test reports the material may be accepted only through class A testing.

Class C. All material characteristics in this classification must be certified by the vendor as meeting the applicable portions of the material specification. Absence of such certification will require class A testing for acceptance, except that quantitative records need not be retained.

Class D. All material characteristics in this classification may be accepted by visual observation. No testing or certification is required.

Such a material classification system should have provision for occasional audits of class B and C characteristics by the company laboratory.

Identification should also be covered in the material specification. Strict control of stock is important; use of an incorrect material has obvious cost, quality, and reliability hazards. The specification should have provision for controlling the expiration dates on perishable materials such as epoxies, rubber parts, and cements. The expiration date should be clearly marked on each container. The storage environment for perishable materials should be specified.

Material identification should be preserved as long as possible in the production area. Bar stock should be fed into a machine in such a way that the color code is preserved until the bar is consumed. Small containers of fluxes, varnishes, and cleaning agents should be identified with the material specification number at all times. Quality assurance procedures should be utilized to tie together the general steps required to meet material specifications.

13.15 THE PRODUCTION LINE AS A SOURCE OF RELIABILITY INFORMATION

In the discussion of producibility it was mentioned that field failures and production-line problems are closely related. This relationship was used to emphasize the importance of providing a producible design. The same field-production relationship can also be used as a source of reliability information. Its scope is expanded from the previous discussion of producibility in that the entire manufacturing process, as well as the design, is included for consideration.

Even though prototype models are fabricated prior to production as subjects for reliability testing, the information gathered during the initial production phase has two important advantages. First, the devices produced in production are production devices. They are built to approved procedures, specifications, and drawings. The model-shop practice of "filing to fit" no longer prevails. Production devices are the devices that will be used in the field application. Differences between prototype and production models may not be clearly obvious, but the likelihood that truly comparable devices will be produced under model-shop and production conditions is a remote one. (See Sec. 8 for production evaluation tests.)

The second advantage of reliability information obtained from production models is the larger volume of assemblies available for consideration. A concept developed under "producibility" was production that involves the repeated copy of an "original." This original is never demonstrated in a single physical product. Because of allowable variations of each characteristic, the original, described by engineering specifications, is an infinite population of possible assemblies. Reliability weaknesses in the whole population of devices may not be evident in a limited number of pilot models, even though more detailed tests and more severe tests can be made. The greater representation of product in the initial production phases offers a substantial advantage in allowing a closer estimate of the inherent characteristics of the entire population.

Information available for reliability and producibility purposes is separable into two basic types. The first is the selective indication that something is out of specification or something has gone wrong. Information of this type typically appears on forms with titles such as Failure Report or Defect Report. Since this information is supplied only when a problem is encountered, the problem indication stands out from the bulk of product information. Selective information or "defect and failure data," because of its natural isolation of problems and its ability to describe any conceivable problem, should be the primary source of reliability information. Failure and defect data should be gathered in every production organization. Elaborate preplanning is not required. If failure and defect reporting is carefully controlled for completeness, it tells, through elimination, what is going right as well as what is going wrong. (See Secs. 6 and 7.)

The second type of information is variable data, the quantitative recording of measurements on selected product characteristics. Preplanning is required for variable data, and entries are best made on a prepared data sheet with a suitable blank for each desired reading. Variable data are not a substitute for defect and failure data; rather they are a refinement, applicable to variable characteristics only, that significantly expands the scope of the information obtainable.

13.16 FAILURE-REPORTING SYSTEM SUITABLE FOR COMPLEX ASSEMBLIES

Failure and defect reporting requires a systematic procedure which ensures that every instance of failure is reported and that the proper information about the failure is included. The reporting system discussed has been used successfully for both in-plant and field-failure reporting.

The input of this failure-reporting system is a four-part carbon form which is concisely titled Failure Report (Fig. 13.12). Three copies are information copies, and the fourth is the action copy. The action copy is printed on card stock with provision for information on the reverse side (Fig. 13.13). The failure-reporting system includes *all* failures within a particular production area whether they occur during a production operation, a production calibration, or an inspection test. All personnel in the production area must report every failure they discover. The following information is provided on each report:

1. Identification of the system, device, subassembly, and component involved in the failure, with the serial numbers and schematic symbols where applicable.
2. Description of the failure, including quantitative measurements and a reference entry indicating the specification requirements on the particular characteristic.
3. Description of the stage of production at which the failure was detected, including accumulated operatimg time. The relationship of the failure to environmental testing is also indicated.

The condition is further defined as the device is disassembled for correction. The repairman indicates:

1. The isolated cause of failure after teardown
2. A further identification of the parts contributing to failure (when such identification could not be made prior to actual teardown)

At this point, the color-coded copies of the form are separated and distributed to the reliability group, the production group, and the quality control engineering group.

SUB ASSEMBLY PART NUMBER (LAST SIX NUMBERS) | FAILURE DETECTED AT

	NAME	PART NUMBER	T.T. OR SERIAL NO	OPERATING HOURS
SYSTEM				
DEVICE				
SUB ASSY.				

DATE BUILT
DEVICE
SUB ASSY.
DATE REJECTED

FAILURE DETECTED AT
1 RECEIVING INSPECTION ☐
2 PRODUCTION TEST ☐
3 INSPECTION ACCEPTANCE ☐
4 PRE ENVIRONMENTAL ☐
5 POST ENVIRONMENTAL ☐
6 ENVIRONMENTAL ☐
7 FLIGHT TEST ☐
8 OTHER ( ) ☐

14178

ENVIRONMENT: AT FAILURE PREVIOUS

DESCRIPTION OF FAILURE

REPORTED BY | DEPT. OR AREA

REPAIR ACTION & CAUSE OF FAILURE

REPAIRED BY | DATE

PARTS HELD FOR ANALYSIS ☐ YES ☐ NO | PART CONDITION
WHERE HELD OR SENT

FAILED PART NAME	PART NUMBER	CIRCUIT SYMBOL	VENDOR (IF KNOWN)	REPAIR AREA

COMPONENT PART NUMBER (LAST SIX NUMBERS)

DEVICE | SYSTEM SERIAL NUMBER | YEAR

Fig. 13.12 Front side of a failure report.

PART TYPE | CUSTOMER COMPLAINT | HELD FOR ANAL.

CAUSE OF FAILURE CATEGORY
☐ 1. SPECIFICATIONS ☐ 3. PROCESSES ☐ 5. MISAPPLICATION ☐ 7. MISHANDLING
☐ 2. DESIGN ☐ 4. WORKMANSHIP ☐ 6. ACCIDENTAL OVERLOAD ☐ 8. DEFECTIVE MATERIAL

DISPOSITION:
1. NON-CONFIRMED ☐
2. RECALIBRATION ☐
3. REPAIR ☐
4. MRB ☐
5. REPLACE ☐

ANALYSIS RESPONSIBILITY:
1. PRODUCTION ☐
2. VENDORS ☐
3. DESIGN ☐
4. MH OTHER AREAS ☐
5. PROD. ENGINEER ☐
6. QUALITY ☐
7. INSTRUMENTATION ☐
8. OTHER ☐

CUSTOMER COMPLAINT:

☐ 1. CONFIRMED ☐ 2 NON-CONFIRMED ☐ 3. NONE

PART TYPE FAILURE CODE
1. ACCELEROMETERS ☐
2. CAPACITOR ☐
3. CHOPPER ☐
4. CONNECTOR ☐
5 DIODE ☐
6. GYRO ☐
7. HEATER ☐
8. MOTOR-VG ASSY. ☐
9. POTENTIOMETER ☐
10. RELAYS ☐
11. RESISTOR ☐
12. SWITCH ☐
13. TRANSFORMER ☐
14. TRANSISTOR ☐
15. OTHER ☐

CORRECTIVE ACTION:

SIGNATURE | DEPT.

DISPOSITION | CAUSE OF FAILURE | ANALYSIS RESPONSIBILITY

Fig. 13.13 Reverse side of a failure report.

and the corrective action employed or to be employed. Action requests are directed to specific group supervisors, where required. Resolved problems should also be included as a part of the report.

The greatest value of such a committee is its fact-finding function rather than any major decision-making function. Individual groups have supervisors who should make the decisions when presented with adequate facts. Committees represent an efficient method of exchanging information, but, because of divided responsibility, committees, in general, are not effective decision-making bodies.

13.18 DATA COLLECTION AND CONTROL

Complex products normally require the collection of significant quantities of variable and attribute data. This subsection describes a representative data-control system for a complex product. Many of the ideas are applicable to all products when a need exists to "know exactly what happened."

The data-control system will be tied into the configuration-control system described in Sec. 13.10. It is imperative that the correct specification be used to test the product and that appropriate data sheets be provided to record the data. The specification-control sheet (Fig. 13.10) calls out which specification should be used and the data sheets applicable for the test; this is actually a part of configuration control.

On receipt of a specification-control sheet, a "product log" is prepared for a serialized final assembly. The final assembly is just being started as the log with blank data pages is presented for use. The log is so arranged that comparable data recorded by production and by inspection are side by side for comparison as the log is opened to the particular test. This arrangement is effective in cursory data analysis for shifts in characteristics. Normally, test results are such that the inspector cannot be influenced by the production readings.

Data pages are color-coded for easy discrimination between production and inspection tests and to identify rerun data. A typical color code is:

Production initial test	Green
Production rerun test	Yellow
Inspection initial test	White
Inspection rerun test	Red

If a rerun test should be required, the data sheet with the unusable data is marked Void with a special rubber stamp. All reruns are recorded on an appropriately color-coded data sheet.

On completion of device tests, all data from subassemblies in the final device are filed with the product log. Reproduction of data may be necessary if the customer requires a log with the product. If the data are to be keypunched for analysis, a suitable data form should be used.

Test reruns were mentioned in this section, and an awareness of the reliability hazards encountered in indiscriminate reruns of failed tests is needed. If a product fails a test, the production organization may request a rerun. If the product subsequently passes the test, the first test is considered in error and the product is accepted. The obvious hazard here is that the first test was, in fact, correct and the product characteristic is changing. A false intuitive—but dangerous—solution is to rerun the test many more times successfully and then decide that the first test was in error.

The solution to the problem of the failed test is not to rerun the product; rather, it is to improve the testing method to the degree that confidence can be given test results. If a discrepancy in the test *can be isolated*, rerunning should be permitted. *Otherwise, the product should be considered defective*, and reruns should be made only to isolate the source of trouble.

13.19 PROGRESS CURVES

It is well to keep track of the progress of an organization in quality and reliability improvements. This is best accomplished by establishing certain statistics as criteria of progress. Goals should be set for improvement, and the statistic should be measured and graphed at weekly or monthly intervals. Statistics should be based on volume of product or volume of work. The following statistics are typical indicators of quality and reliability progress.

1. In-plant reliability failures per unit built
2. Total failures per unit built
3. Defects per unit (or per 100 units if defect rate is low)
4. Module replacements per unit built
5. Field failures per quantity in the field
6. Quantity of MRB's per unit built (nonconforming-material waivers)
7. Visual defects per unit built
8. Mean time between failures (for in-plant and field)
9. Retest required per unit built
10. Scrap and salvage costs per unit built

In general, all of these consist of a criterion of progress as a numerator and a "standardizing" element as a divisor. A standard should form a part of every progress statistic to compensate for changing volume. For example, a 10 per cent increase in defect quantities would be an improvement if product volume were increased by 30 per cent. The statistic "defects per unit built" makes the necessary compensation.

Progress curves should be considered to show trends and absolute progress toward goals. If care is used in selecting the statistic, the effectiveness of improvement programs and normal learning is measurable. Trends should be steep toward improvement early in the program and gradually flatten in an exponential manner as improvements saturate ability for further improvement.

13.20 EMPLOYEE CAPABILITY AND ATTITUDE FOR RELIABLE PRODUCT PRODUCTION

By its existence as a latent rather than a visible or measurable hazard, the reliability problem is elusive and difficult to detect. A great advantage can be gained in combatting reliability problems if every person in a production area is conscious of potential reliability troubles and is constantly looking for these conditions while performing his normal work. To produce reliable products, reliability-consciousness must be thoroughly implanted in a production organization. Development of a group's pride in its ability to produce a reliable product will amplify the effectiveness of all production-phase reliability-improvement efforts.

Satisfactory performance of a production worker, with respect to reliability, rests in his ability to do his work and in his attitude toward his job. Attitude of the worker is as important as ability, and attitude is strongly influenced by the importance placed on reliability by his supervisors. An environment of strict discipline coupled with fairness and recognition for fine work is conducive to the achievement of a reliable product.

What should be expected of a production worker with respect to reliability can be summed up in two basic requirements:

1. He must perform his specified job skillfully in strict adherence to the instructions provided. He should suggest improvements, but he should not change methods without approval.

2. He must report detected failures and defective conditions in accordance with specified reporting procedures. He should report irregularities to his supervision when standard reporting does not apply.

The two requirements may at first seem too simple; they are very inclusive, however, when their total scope is considered. The employee should be made aware that he is personally responsible for his performance against these requirements.

13.20a Development of Reliability Consciousness. Development of reliability-consciousness is essentially parallel to the redevelopment of craftmanship. Reliability-consciousness embodies both responsibility and pride in work performed. The following suggestions should be considered in developing it.

1. *The worker should be told what product he is making and why it is to be made.* It is difficult for a person to take pride in his work if he doesn't even know what he is making.

2. *Provision should be made to identify the work with the person who produced it.* Signing off a traveler or data sheet reminds the worker of his responsibility to perform skillfully. An element of pride is developed in signing off work that has been done well. The worker knows that if the work is not done correctly, it will be reflected back to him. This is as it should be; people expect to be disciplined if they do not perform properly. Stringent control is expected by people and is conducive to good morale. It is *not* detrimental to morale.

3. *Disciplinary action should not be applied to workers who report discrepancies they have accidentally generated.* The employee should be told that reliability of the product depends on him as well as on the inspectors and testers. If he accidentally generates a discrepancy, he should immediately report the condition. There are, of course, many discrepancies that could be covered up by a worker and that would result in a reliability weakness in the product. When a worker reports discrepancies but generates an unacceptable number of them, something obviously must be done. Management should recognize the honesty of the worker and his valuable attitude toward reliability. He should receive additional training in the particular job, or he should be assigned to a job he can do effectively.

4. *The worker should be told if he is doing a good job.* A person wants to know how he appears to his supervisor. Failure to receive any indication generates indifference to the job.

5. *Production and inspection should have the same quality standards.* It is important that clear lines of agreement regarding quality standards exist on supervisory levels. Agreement on standards should be conducted at a supervisory level, and a united decision should be presented to production workers and inspectors.

6. *Supervision should be fair in dealing with personnel.* Time standards should be reasonable. Employees should not be criticized unjustly.

7. *The employee should be made to know that he is a member of the team.* His skills and his favorable attitude are essential to the production of reliable product. Management relies on him and he should know it.

13.20b Employee Training. On all skill levels, the most directly applied training for a particular job is on the job. The instructor for this form of training is usually the assistant foreman or the group leader, who should be qualified to train and should be allotted sufficient time to do it adequately. Prerequisites for on-the-job training should include basic skills in the primary functions of the job. For example, a trainee for a job requiring the assembly of electronic components should already be skilled in soldering techniques and general assembly methods. On-the-job training should adapt these skills to the particular assembly operation. Training aids, such as visual and audio presentations, are extremely helpful in adapting the worker to the particular job.

Classroom or group training is a preferred method of developing basic skills and is also helpful in presenting an overall picture to the worker. As mentioned, a great advantage can be gained if all persons are constantly looking for reliability weaknesses. If they are to look for them, it is necessary that they have a basic knowledge of the product on which they are working. Classroom training can provide this necessary familiarization. Refreshers in mathematics as a basic skill are always valuable when calculations are required for the generation of product. Need for general training can frequently be determined by weak areas in the group, such as frequent mathematical errors and high rejection rates. The topics covered in a typical training program for an operating inspection group follows. The curriculum includes a mixture of basic skills and applied information and is aimed at developing a favorable attitude as well as knowledge:

Session I	The D-68 Computer—What It Is, How It Is Used
Session II	Design Requirements of the D-68 Computer
Session III	The D-68 Inspection Plan and Its Relation to Production Processes
Session IV	The Effect of Workmanship Errors
Session V	Identifying Product Acceptability

Session VI	A Production Supervisor's Comments on Quality and Reliability
Session VII	Reliability of the D-68
Session VIII	Mathematics Refresher I
Session IX	Mathematics Refresher II
Session X	Application of Mathematics to D-68 Acceptance
Session XI	Understanding the D-68 Test Console
Session XII	Soldering for Reliability I
Session XIII	Soldering for Reliability II
Session XIV	Handling Methods for D-68 Modules and Parts

References

1. Burr, Irving W., *Engineering Statistics and Quality Control*, McGraw-Hill Book Company, New York, 1953.

2. Cowden, Dudley J., *Statistical Methods in Quality Control*, Prentice-Hall, Inc., Englewood Cliffs, N.J., 1957.

3. Duncan, Acheson J., *Quality Control & Industrial Statistics*, Richard D. Irwin, Inc., Homewood, Ill., 1952.

4. Feigenbaum, A. V., *Total Quality Control*, McGraw-Hill Book Company, New York, 1961.

5. Grant, E. L., *Statistical Quality Control*, 3d ed., McGraw-Hill Book Company, New York, 1964.

6. Juran, J. M. (ed.), *Quality Control Handbook*, 2d ed., McGraw-Hill Book Company, New York, 1962.

7. Western Electric Company, Inc., *Statistical Quality Control Handbook*, Western Electric Company, Inc., 1956.

Section 14

RELIABILITY SPECIFICATION AND PROCUREMENT

E. JACK LANCASTER

DIRECTOR OF QUALITY ASSURANCE, KEARFOTT SYSTEMS DIVISION
GENERAL PRECISION, INC., WAYNE, NEW JERSEY

In addition to the above position, Mr. Lancaster is Assistant Manager of General Precision's Corporate Product Assurance Council. Before joining General Precision, he was Manager, Quality and Value Assurance for the Advanced Products Group, American Machine and Foundry Company. Prior to that, Mr. Lancaster spent many years with the Air Force, including the following: Assistant National Inspector, Air Matériel Command; Director Quality Assurance, U.S. Air Force in Europe; Chief, Quality Assurance, Air Force Ballistic Missile Division. He received the Air Force Meritorious Civilian Award in recognition of outstanding effort in behalf of quality assurance aspects of the Air Force Ballistic Missile and Space programs. He served as Chairman of the Joint Army–Navy–Air Force Coordination Group for Quality Assurance of ballistic missiles and space systems.

Mr. Lancaster studied mechanical engineering at the Pratt Institute, Brooklyn, New York, and business administration at Washington University, St. Louis, Missouri.

He was President of the American Society for Quality Control, and he is currently serving that organization as Chairman of the Board of Directors. He has written many articles and papers dealing with both technical and managerial subjects in reliability and quality control.

EMIL M. OLSEN

GENERAL MANAGER, AIR CRUISER DIVISION
THE GARRETT CORPORATION, BELMAR, NEW JERSEY

Mr. Olsen has been actively engaged in fields allied to reliability and quality control for over fifteen years. He has been with the Garrett Corporation a major part of this time, having joined the Air Research Division in 1957 as Senior Quality Control Engineer. In 1962 he became Corporate Director of Quality and Reliability, in which capacity he was responsible for policies, planning, and programming in the functional areas of reliability and quality control for the Garrett Corporation's seven manufacturing divisions and subsidiaries.

During World War II, Mr. Olsen served as a project test officer at the Air Force Proving Ground Command, Eglin Field. His career in industry began in 1946 when he joined Sprague Warner and Company in Chicago. He became affiliated with the AC Spark Plug Division of General Motors in 1948, where his activities were concerned with the administration of worldwide field service and quality control, including the technical training of customer personnel and the formulation of operational test procedures. In 1951, Mr. Olsen joined Lear, Inc., and five years later he became Manager of Lear-California's Quality Control and Inspection Division. He became General Manager of Air Cruiser in 1963.

Mr. Olsen was educated at Muskegon Junior College, Marquette University, and the University of Southern California.

RELIABILITY SPECIFICATION AND PROCUREMENT

E. Jack Lancaster and Emil M. Olsen[1]

CONTENTS

14.1 INTRODUCTION

The primary purpose of any specification is to provide an absolute, positive identification of what is wanted. Ideally, specifications should be simple and brief without sacrificing completeness. In addition, the perfect specification should be unequivocal, enforceable, and self-contained. Reliability, however, is not a product or function that can be described alone. It must be associated with the parameters of equipment design, manufacturing processes, and the operation aspects, including the personnel concerned with those functions.

Reliability has a tremendous economic impact and therefore has complex contractual and procurement implications. Many of the problems associated with specification of reliability requirements arise from contractual implications and should be considered in that context. The common-law prerequisites of a valid contract include such elements as identification of the parties, competence of the parties, explicit definition of the purposes of the contract, mutual understanding and agreement of the parties, legality, consideration, possibility of accomplishment of contract purposes, and time of performance and completion. Procurement and contracts are discussed in several subsections at the end of this section.

Of the above basic elements, the ones most frequently compromised in reliability specifications are those involving explicit definition of the purposes, mutual understanding and agreement, possibility of accomplishment, and time of completion.

Reliability is a relatively new discipline, and its techniques are therefore not

[1] Mr Lancaster wrote Secs. 14.1 to 14.6; Mr. Olsen wrote Secs. 14.7 to 14.10. Mr. Lancaster and the editor would like to recognize the valuable assistance of Mr. Griffith W. Lindsay, Aeronautical Systems Division (U.S.A.F.) Staff Reliability Engineer, for his assistance in the preparation of this section. Mr. Lindsay, a graduate of Carnegie Institute of Technology and formerly of Chrysler Corporation, Kurz-Kash, Inc., and Buckeye Iron and Brass Co., edited Mr. Lancaster's contribution and prepared several illustrations and examples for inclusion in this section. Mr. Lindsay has been prominent in Air Force reliability work and has presented a number of papers on the subject before learned groups.

universally understood. Specifications are frequently written by engineers who assume that, in writing a specification, they are communicating directly with another engineer who will understand what they want. In fact, they are usually telling a group of administrators (supported by lawyers and accountants) in their own organization what they should tell a similar group in another organization what they, in turn, should tell their engineer/manager.

Needless to say, these two communications are not the same. For that reason, many modern contracts not only describe the design requirements of the product or system but also state, by reference to standard specifications, explicit requirements for the entire reliability program including the processing, fabricating, assembly, storage, and shipping functions. Illustrations of this type of specification are MIL-STD-785 and MIL-R-27173.

14.2 RELIABILITY APPORTIONMENT

The basic criteria for most specifications for reliability stem from the operating or program requirements established by either you or your customer. Such requirements may be expressed as a period of time with no maintenance or a prescribed time for a continued level of operation. In any event, quantitative reliability requirements should be expressed either as a probability of mission success or mean time between failures. In the case of some missile systems such operating times may include considerations for standby operations (in a readiness condition), countdown, and flight. With the use of generic reliability values or by assessing the complexity of major subsystems or from past history on similar systems, the total weapon reliability goals and/or requirements are apportioned among the various major systems by using one of several mathematical techniques. Allocation of systems reliability into straight subsystem levels is somewhat fallacious unless the apportionment also considers the interface problems in various combinations of components and subsystems.

A classic example of a major weapon system reliability apportionment is given in the accompanying table. The government may award contracts for each of the major systems identified in the table, or it may elect to award the entire weapon system to a single contractor who, in turn, will subcontract for the major systems that he cannot produce. By joint agreement, the buyer and the supplier of these major systems will agree upon further apportionment among the subsystems and/or major components of the systems in much the manner outlined in the table.

System	MTBF (examples)	Failure rate per 1,000 hr (examples)	Reliability of countdown and flight (example)
First Year (or First 50 Weapons)			
1. Guidance	3,500	0.286	0.70
2. Reentry	100,000	0.100	0.86
3. Engine	770,000	0.013	0.90
4. Ground oper. equip.	28,500	0.035	0.985
5. Launch-facility equipment	64,000	0.016	0.984
6. Command and control	4,000	0.25	0.90
Second Year (or Weapons 51 to 100)			
1. Guidance	4,500	0.222	0.89
2. Reentry	1,300,000	0.077	0.94
3. Engine	900,000	0.011	0.97
4. Ground oper. equip.	35,000	0.029	0.99
5. Launch-facility equipment	75,000	0.013	0.99
6. Command and control	5,000	0.200	0.97

14.3 RELIABILITY-PROGRAM REQUIREMENTS

Accompanying the quantitative reliability requirements in military contracts for aerospace equipment is usually a requirement for an organized reliability program. The U.S. Air Force procurement agencies, for example, use Specification MIL-STD-785, Reliability Program Requirements for Systems, Subsystems and Equipment, for defining the overall reliability-program requirements. This specification requires the contractor to establish, maintain, and document a reliability-program plan in accordance with the provisions of the specification. The specification expresses the desired reliability concepts which encompass management as well as technical factors. It requires consideration for all phases of the life cycle. The document establishes the concept of hardware and operational reliability and provides for recognition of the interfaces and for coordination with related activities such as systems analysis, maintainability, value engineering, human engineering, and environmental control.

MIL-STD-785 requires that the system reliability objectives and minimum acceptable requirements expressed numerically in the prime contract be reflected in the major-subsystem requirements. It requires the establishment of specified control points prior to release of design for initial fabrication. Mathematical models, apportionment, and initial predictions are also to be included in the reliability-program plan. The reliability program must provide for preliminary and continuing study of quantitative reliability requirements throughout the program to provide for progressive refinement of the reliability analyses and validation of specified requirements for all planned phases or operational modes of the system.

14.3a Reliability-program Plan. The contractor is required to develop a detailed reliability-program plan. This plan is to describe exactly how he intends to assure conformance to the specified reliability requirements of the contract work statement and to MIL-STD-785. The specification requires that the contractor's program plan include his proposed demonstration method and the required or the contractor's recommended confidence levels. The program plan requires a description of the contractor's reliability organization and personnel and a definition of their responsibilities, functions, and authority. It requires detailed listings of specific tasks, man-loading of the tasks, and detailed procedures to implement and control the individual tasks. It requires the designation of milestones, definition of interrelationships, and estimation of times required for reliability-program activities and tasks. Further, the program plan must include a schedule for review of its status and progress.

14.3b Reliability-program Elements. The reliability-program plan, required to be submitted by the contractor to the procuring agency for approval, requires identification of specific program elements and a description of the contractor's intended activities with regard to each program element. The program elements identified in the specification are as follows:

Test requirements for development, qualification, and acceptance
Environmental requirements for equipment design and testing
Components part testing
Maximum preacceptance operation
The integration of government-furnished equipment in the system from a reliability point of view
Parts reliability control
Identification of critical items
Supplier and subcontractor reliability programs
Reliability training and indoctrination
Human engineering
Statistical methods
Effects of storage, shelf life, packaging, transportation, handling, and maintenance
Design reviews
Manufacturing control and standards
Failure-data collection and analysis and corrective action
Reliability-demonstration plans (general and specific)
Periodic and final reports

14.3c Product Assurance Requirements. The last few pages have been devoted to the general reliability-program requirements of military specifications. MIL-STD-785 was used as an example. For some major weapon systems, the system program office of the procuring agency will supplement the military specification with exhibits. One such exhibit may be entitled, Product Assurance Requirements (to include both reliability and maintainability). The purpose of this document is to set forth product assurance requirements for the system specifications, program plans, and individual equipment specifications. This document is also contractually binding on the contractor and requires the contractor to prepare and submit a Contractor Product Assurance Plan. This is to be an integrated plan covering the separate areas of reliability, quality assurance, and maintainability programs with defined program interfaces.

14.3d Reliability Requirements and Measurement. The exhibit also defines in detail the basic probability of success rationale based upon requirements of the tactical phases of the weapon system support and operation. In addition, the weapon system, missile system, and subsystem reliability requirements needed to support these tactical phases are specified. The exhibit includes requirements for probability of success criteria and associated reliability-measurement procedures. The criteria include standard procedures and ground rules for the establishment of reliability models, hardware block diagrams, and MTBF derivations. Included under reliability requirements and measurement criteria are detailed definitions of pertinent terms and a narrative description of the following requirements:

System reliability requirements
System numerical reliability requirements
Reliability measurement
System reliability model including model requirements and model analysis
Subsystem reliability models
Design prediction assessment
Achieved reliability estimates, including evolution assumptions and techniques, ground rules for reliability data, approved sources of data, screening of data, the statistical treatment, assumptions, derivation of MTBF estimates, reliability computation, and a description of how to use the system model

14.3e Product Assurance Testing. The Product Assurance Requirements Exhibit also includes a section on product assurance testing. The purpose of this section is to establish minimum requirements or controls at system level, subsystem level, and component level which must be translated in so far as possible to the module level of equipment. It includes requirements for all tests conducted during development to determine satisfactory failure rates, satisfactory environmental design, and those tests which may be indicative of failure during use or of maintenance capabilities. It also includes detailed requirements and explanations for:

1. Program-change procedures
2. System test program, including simulated environmental tests, system reliability tests, and system acceptance tests
3. Subsystem test program, including environmental qualification tests, flight-proofing tests, acceptance tests, and component test programs

14.3f Environmental Criteria. The Product Assurance Requirements Exhibit includes requirements for establishing and controlling system and subsystem environmental tests. The details of the environmental requirements and tests may be included in another voluminous exhibit reference in the Product Assurance Exhibit.

14.3g Design Assurance. This section of the Product Assurance Requirements Exhibit outlines a program which the associate or major subcontractors are required to provide for reliability review and evaluation of equipment design throughout the program. It requires that reliability reviews identify, through design analysis and systematic audit of design concepts, drawings, documentation, and all significant modes of failure. This reliability review is to be a supplementary review and is not intended to relieve the design engineers of their responsibility for formulating a design with the required inherent reliability. Such design reviews are to identify critical areas or components after performing the following minimum analysis:

Trade-off conclusions reached through reliability analyses
Use of parts beyond normal specifications
Environmental controls
Packaging and configuration analyses
Producibility
Procedures and/or equipment that will be required for manufacture or test
Results of tests, as applicable
Design-criteria changes since design initiation
Failure-mode analysis
Hazards and safety analysis
Tolerance analyses
Mechanical and thermal-stress analyses

A series of successive design-review meetings is required to assure that the designers are continually analyzing the product. The minimum number of such reviews is established at three levels: preliminary review, detailed review, and final review. A detailed list of considerations, elements, and parameters to be included in these reviews is listed in the Exhibit. Design-review checklists are required with numerous suggested elements listed for inclusion in the checklists.

14.3h Parts Reliability Program. This section of the Products Assurance Requirements Exhibit requires the development of a parts reliability program. It establishes a Parts Working Group for all associate or major subcontractors for the purpose of establishing policies, procedures, and control for a program-wide parts program. It also requires identification of preferred parts, critical parts, failure-rate information, and detailed supplementary information concerning parts used in the system. These data are exchanged among the various major subcontractors or associate contractors working on the reliability program for the weapon system.

14.3i Reliability Integration Procedure. In some very large missile or space-system projects with many major associate contractors, a Reliability Integration Procedure is sometimes prepared by the contractor responsible for the integration of the major subsystems. This is generally a voluminous document which describes the plans and tasks, their implementation, responsibilities, end products, and schedules essential to the overall integration of the weapon system reliability program. It defines procedures for the formal exchange of information deriving from the program-wide reliability effort. This document is sometimes prepared by the integrating contractor, approved by the procuring agency, and included as a contractual requirement for all the associate contractors, including the integrating contractor. This document implements and integrates the military specification in the contract and the Product Assurance Requirements Exhibit, or similar document.

14.3j Summary of Reliability-program Requirements. To summarize this example of a specification of reliability for a major weapon system, the major documents and their purposes are listed:

Contract Work Statement. This document describes the overall objectives and/or minimum acceptable requirements for the system and the major subsystems.

*MIL-STD-*785. This is an example of a military specification for a reliability program. This document describes *what* is necessary in a contractor's reliability program and further requires the contractor to develop a reliability-program plan for approval by the procuring agency.

Product Assurance Requirements Exhibit. This document instructs the contractor *how* to develop his reliability-program plan and includes detailed instructions for interface with maintainability, safety, value engineering, etc. It also establishes ground rules to assure consistency among the producers of the various major subsystems with respect to mathematical models, reliability measurement, etc.

Reliability Integration Procedure. This document prescribes the overall procedure for integrating the efforts of all major contractors associated with the weapon system.

It provides even more detail as to *how* to prepare the plan. It also establishes standard procedures and formats for reports, exchange of information, etc. Its stated purpose is to assure an orderly and coordinated evolution of effort throughout the entire weapon system program by establishing methods of communication that will assure integrated technical and administrative relationships.

These detailed program requirements usually agree with the customer's or the integrator's reliability concepts and internal programs and thereby provide the customer or integrator a more convenient opportunity to monitor, and in some cases, to dictate the contractor's reliability program. Thus, there can be a customer/supplier relationship in which the reliability program is a cooperative effort, and the adequacy of the program is judged by the conformance of the supplier to the broad, but clear, rules promulgated by the customer. If it is properly and carefully administered, such an arrangement can be beneficial to both, but under other circumstances it is difficult to hold the supplier solely responsible for the results. Usually when the supplier contracts for critical subassemblies or major subsystems with another company, the same general approach is passed on to his supplier.

14.4 RELIABILITY PARTS PROGRAMS

The specification of reliability for parts and components may be much simpler. It usually prescribes the required operation over time under certain environments, identifies the important characteristics, establishes the required tests, and schedules periodic monitoring points by the customer. If, however, the parts or components required exceed the "state of the art" or if they are extremely critical to the reliability of the overall system, a much more detailed specification is usually required. An example of such a specification in one of the Air Force "Hi-Rel" programs for a diode is given below. To avoid possible security violations, the numerical values used are fictitious:

The contractor shall execute a reliability improvement program with the following objectives:

a. The contractor shall produce a computer diode, a high conductance silicon diode, and a low leakage silicon diode in accordance with the specification appended hereto.

b. The contractor shall institute a program which will establish process specifications, complete production facilities, production controls, and personnel training programs which will meet the reliability requirements of 0.0003 per cent, 0.0015 per cent and 0.009 per cent per 1000 hours at 70 per cent confidence level in the order indicated for the types in paragraph a. above. This program will be designed to provide contractor's and buyer's management with detailed visibility of program status and performance including product homogeneity status and reliability growth rate.

c. The contractor shall institute a program to discover causes of component part discrepancies arising in materials, processes and controls that give rise to failures and the contractor will institute corrective action, as appropriate, so that the discrepancy and the cause are eliminated or minimized in order to meet the reliability objective.

The task headings for the complete work statement given below suffice to outline the *detail of input the contractor* laid on the component manufacturer.

Task 1. Program plan
Task 2. Production processes
Task 3. Production control
Task 4. Failure analysis
Task 5. Corrective action on production processes
Task 6. Corrective action on control
Task 7. Evaluation of corrective action
Task 8. Tests
Task 9. Contractor's program organization
Task 10. Training
Task 11. Test equipment
Task 12. Technical direction and monitoring
Task 13. Serilization and IBM format
Task 14. Handling and packaging of devices
Task 15. Documentation and reporting
Task 16. Facilities provided by contractor

This vendor was required to identify all documentation on the internal processes and the materials being used on the devices, and the production control specifications. Documentation required by every step is identified and scheduled in this manner.

An objective of the parts improvement program was "a disciplined production line." The Government and the prime contractor were able to deal with proprietary information problems. They wanted evidence that (1) every process and production step had been identified and documented, (2) written procedures existed for each step, and (3) the written procedures were being followed. Confidence that these objectives are being met can be obtained without disclosure of all detailed (perhaps proprietary) process specifications to the prime contractor.

Some demonstration of certain major units of the overall guidance system for a minimum number of hours without failure is required as a part of the acceptance procedure. However, the actual proof of the overall system capability will not be demonstrated until after the equipment is operating in the missile silos for years to come.

14.5 OTHER DOD AND NASA SPECIFICATIONS AND RELIABILITY DOCUMENTS

Many different reliability-program specifications are used by each of the military services and NASA. The various specifications usually reference many other military documents, some of which are written for specific weapon systems, equipment, components, and special environments and applications. Such documents are modified, expanded, canceled, and combined with other documents at such a rapid pace that it is almost impossible to list all of them at any one time with any degree of accuracy or completeness. As an example, however, a list of the many government documents that may be cited in contracts or otherwise referenced by reliability specifications in contracts is provided below. (*Note.* Because of the frequency of revisions to and/or modifications of these documents, only the basic specification numbers are identified.)

MIL-A-8866	Airplane Strength and Rigidity Reliability Requirements, Repeated Loads and Fatigue
MIL-R-19610	General Specifications for Reliability of Production Electronic Equipment
MIL-R-22732	Reliability Requirements for Shipboard and Ground Electronic Equipment
MIL-R-22973	General Specification for Reliability Index Determination for Avionic Equipment Models
MIL-R-23094	General Specification for Reliability Assurance for Production Acceptance of Avionic Equipment
MIL-R-26484	Reliability Requirements for Development of Electronic Subsystems for Equipment
MIL-R-26667	General Specification for Reliability and Longevity Requirements, Electronic Equipment
MIL-R-27173	Reliability Requirements for Electronic Ground Checkout Equipment
M-REL-M-131-62	Reliability Engineering Program Provisions for Space System Contractors
NASA NPC 250-1	Reliability Program Provisions for Space System Contractors
NASA Circular No. 293	Integration of Reliability Requirements into NASA Procurements
LeRC-REL-1	Reliability Program Provisions for Research and Development Contracts
WR-41 (BUWEPS)	Naval Weapons Requirements, Reliability Evaluation
NAVSHIPS 900193	Reliability Stress Analysis for Electronic Equipment
NAVSHIPS 93820	Handbook for Prediction of Shipboard and Shore Electronic Equipment Reliability
NAVSHIPS 94501	Bureau of Ships Reliability Design Handbook
NAVWEPS 16-1-519	Handbook Preferred Circuits—Naval Aeronautical Electronic Equipment
PB 181080	Reliability Analysis Data for Systems and Components Design Engineers
PB 131678	Reliability Stress Analysis for Electronic Equipment, TR-1100
TR-80	Techniques for Reliability Measurement and Prediction Based on Field Failure Data

TR-98	A Summary of Reliability Prediction and Measurement Guidelines for Shipboard Electronic Equipment
AD-DCEA	Reliability Requirements for Production Ground Electronic Equipment
AD 114274	(ASTIA) Reliability Factors for Ground Electronic Equipment
AD 131152	(ASTIA) Air Force Ground Electronic Equipment-reliability Improvement Program
AD 148556	(ASTIA) Philosophy and Guidelines—Prediction on Ground Electronic Equipment
AD 148801	(ASTIA) Methods of Field Data Acquisition, Reduction and Analysis
AD 148977	(ASTIA) Prediction and Measurement of Air Force Ground Electronic Reliability
MIL-HDBK-217	Reliability Stress and Failure Rate Data for Electronic Equipment
RADC 2623	Reliability Requirements for Ground Electronic Equipment
AR-705-25	Reliability Program for Material and Equipment
USAF BLTN 2629	Reliability Requirements for Ground Electronic Equipment
OP 400	General Instructions: Design, Manufacture and Inspection of Naval Ordnance Equipment
MIL-STD-721	Definitions for Reliability Engineering
MIL-STD-756	Procedures for Prediction and Reporting Prediction of Reliability of Weapon Systems
MIL-STD-781	Test Levels and Accept/Reject Criteria for Reliability of Nonexpendable Electronic Equipment
MIL-STD-785	Requirements for Reliability Program (for Systems and Equipments)
DOD H-108	Sampling Procedure and Table for Life and Reliability Testing

14.5a Special Military Reports and Other Considerations. In addition, there are many special reports, handbooks, sampling tables, and bibliographies which also may be included in contracts or referenced by specifications. Many of the contracts or specifications for reliability include numerous requirements or references to requirements in special areas related to reliability such as human factors (personnel subsystems), value engineering, quality control, systems engineering, maintainability, systems effectiveness, safety, and producibility. Many military standards for specific parts and/or components and for special tests, statistical predictions, packaging, cleanliness, and nondestructive tests may also be referenced in the reliability specifications. Specifications, handbooks, and other military documents specifying methods, procedures, and other instructions relating to the design aspects may be included in the contracts for reliability or referenced in reliability specifications. Many of these documents are developed by different government agencies and/or by different organizational components at various levels within a particular agency. *It is because of these numerous independent activities of special interest that much care must be taken to screen all of the documents specified or referenced in reliability contracts and documents to identify inconsistencies or actual conflicts that may exist.*

14.6 CONTRACTING FOR RELIABILITY

The remainder of this section is devoted to promoting an understanding of the specification for reliability from a contractual point of view, which may serve to reduce the dangers of negotiating void or voidable contracts. Knowledge of the purposes and techniques of development of reliability requirements is one factor necessary in such an understanding. While much of the discussion will involve contracts with government agencies, the basic principles are inferentially applicable elsewhere.

The effectiveness of a system is a function of the quality, capabilities, and reliability of a product. Optimum effectiveness in the attainment of an overall objective is the proper goal of all system planning, including that for hardware items. The word "optimum" involves cost/effectiveness trade-offs; it does not necessarily mean "maximum." The overall objective may be to maximize profit, to damage an enemy, or to reduce the effectiveness of enemy action. Needless to say, whatever it is, it should be kept firmly in mind throughout a program.

While there is nothing new about the foregoing concepts, it is only since World War II that large-scale attempts have been made to quantify them within the framework of a formal discipline. Since these attempts have received their greatest emphasis in the development of military systems, the following discussion will largely and unavoidably be so oriented.

The ratio of successes to trials is the measure of achieved operational reliability. Operational reliability itself is a prediction of this ratio. Capability involves such things as range, speed, and altitude.

14.6a System Effectiveness. Let us consider a weapon system with a maximum range of 3,000 miles. Its reliability is known to be 0.9 for striking an objective at that distance. Its capability approaches zero, however, if the objective is 4,000 miles distant, and its effectiveness for this use also approaches zero. Any increase in reliability would be of no value in these circumstances. We may be able, in the present state of the art, to develop a system with a range of 4,000 miles but with a reliability of 15 per cent. System effectiveness is low because of low reliability, but it is still far better than that of the previous example. It would be folly to compromise capability to permit a reliability increase in such a situation.

Conversely, if the most distant military objective is 3,000 miles away, effectiveness would not be improved by increasing the range capability, but it could be improved by increasing reliability. A decision to spend resources to improve reliability and therefore effectiveness must be based on such considerations as whether the reduction in the number of systems deployed or a reduction in maintenance cost would offset this expenditure.

The optimum level of effectiveness of a system thus must be set by evaluating the various means of attaining the overall objective with respect to their costs of attainment. In arriving at such an optimum figure, various cost/effectiveness trade-offs have been made. Within this effectiveness level, further refining trade-offs of reliability and capability are evaluated. By the time of contractual commitment to production, these values should be frozen. No change can then take place without compromising the validity of the decisions on which the overall program is based or formally reorienting the program. (See Sec. 1 for detailed discussion of system effectiveness.)

14.6b Acceptance Criteria. Another point to consider is the conditions specified for acceptance. This point is important because of the current trend toward requiring that a specified reliability be proved with "high confidence." Translated, this expression means: "If the actual reliability is the same as the specified reliability, the acceptance plan should be very likely to reject the product." Under these conditions the manufacturer, to be reasonably sure of acceptance, must make a product with reliability much better than specified.

Consider, as a specific illustration of this point, a specified MTBF of 100 hours to be proved with 90 per cent confidence. This requirement could be met by, for instance, an acceptance plan requiring a test of 390 unit-hours with not more than one failure. But to be 95 per cent sure of acceptance under such a plan, the manufacturer would have to attain an actual MTBF of about 1,115 hours—more than eleven times that specified.

Under such impractical conditions there are bound to be "strict requirements loosely enforced." As a few years of reliability experience have shown, this leads to many difficulties. The most serious of these is the direction of efforts toward the circumvention of requirements rather than the attainment of true reliability.

14.6c Operational Reliability Requirements. Ultimately, a system is a means for accomplishing a stated objective. It includes not only the primary but also the supporting equipment, as well as the necessary personnel and procedures. Since personnel and procedures are not procurable as part of a hardware contract, and since they grossly affect system operational reliability, contracts for hardware cannot logically contain full-scale requirements for a high order of operational reliability with demonstration a condition of acceptance.

To illustrate this, let us consider the case of a manufacturer of golf clubs. He is in no position to guarantee that customers using his clubs on any golf course and under any weather conditions will shoot par golf or better or that they will have a specific probability of shooting par. He can say and demonstrate that his products meet USGA specifications and that the inherent characteristics of the clubs, determined by design and manufacture, *permit* a skilled player to shoot par, barring bad luck or mistakes. He can further demonstrate, at a predetermined confidence level, by stressed tests that under certain definite conditions his clubs will last for so many simulated rounds of golf without random breakage or wear-out with a certain probability. This hardware reliability can be purchased by a customer and supplied by a manufacturer; however, the operational reliability of the golf clubs depends grossly on what the customer does with them.

Producers of complex missile and space systems have a somewhat different problem with respect to operational reliability. In many instances military contracts for complex systems express quantitative reliability requirements in terms of mean time between failures (MTBF). In fact, more than one MTBF may be specified. No matter how many MTBF's are specified, there should be a definition of what MTBF means. One of the pitfalls in current contracts is that a MTBF is required but what it stands for is not explained. Let us take an example. If a contract for a system calls out 5,000 hours MTBF, it is necessary to ask this series of questions, as a minimum:

1. Is this MTBF the design, minimum acceptable hardware, or operational requirement?
2. When is this MTBF to be attained?
3. What definition of failure is to be used in calculating this MTBF?

In most cases the MTBF will be an expression of the operational requirement or the hardware reliability needed to meet operational requirements. It will then be necessary for the manufacturer to establish his own design objective to provide assurance of meeting hardware reliability tests and assurance that, when the system undergoes field usage, it will meet the operational requirement. In other words, overdesign by a factor to allow for production and field degradation is necessary.

The second question above, regarding "when," is of paramount interest. The 5,000 hours MTBF may be attainable at some future date, but not in the first operational system. Usually, the best approach is to make this MTBF the average of a group of systems, for instance, the first squadron or wing of delivered systems for a military contract. If the MTBF is associated with a space, or similar, mission for which only one or possibly several systems at the most are to be delivered, then a different approach is used.

If, in order to meet the specified MTBF, component reliability-improvement programs are required, it may not be possible to bid accurately for such programs at the time of the basic contract negotiations, and provisions should be made in the contract for these programs to be proposed at a later time.

14.6d Customer-Supplier Relationships. The attainment of required hardware reliability by a manufacturer, when consideration is given to all the possibilities of design and manufacturing errors, can scarcely be fortuitous. It obviously must be the result of a carefully planned, meticulously controlled program. In general, three basic types of manufacturer-customer relationships exist:

1. The manufacturer develops, designs, advertises, and sells proprietary products. Owing to the degree of control he exercises over the entire process, he is able to provide limited guarantees to the customer of certain characteristics of his hardware when used under certain conditions. Golf clubs, automobiles, and refrigerators are examples.
2. The manufacturer is a custom supplier. In this case he manufactures items to customer specifications. His liability to the customer is normally limited to a warranty that articles offered for acceptance are in accordance with the design and quality agreed upon in the contract. Die castings, automobile parts, plastic moldings, and screw-machine specialties are examples. Some degree of customer surveillance, in the form of inspection of shipments and limited evaluation of the status of production tools and processes, is frequently encountered.
3. The third type of relationship is between government agencies and their contractors. This relationship is strictly controlled by statute and many implementing regulations. The government agency is inhibited from arbitrarily limiting bidders on

government material contracts to those it considers the most competent; and only with great difficulty, by assuming the burden of proof, can it reject a low bid in favor of a higher one. Consequently, the agency is frequently in the position of having to award a contract for complex equipment to organizations of unknown and highly questionable technical capability. Timely delivery of satisfactory items in many cases is vital to the continued existence of the nation and, therefore, the government agency is forced into the position of exercising a detailed surveillance of the contractor's effort in design and manufacturing. Based on the premise that all bidders must receive the same treatment, programs for achievement of contractual objectives are required as contractual elements with both highly competent and dubious contractors. This is sometimes a source of irritation to efficiently managed industrial organizations, especially when the amount of detail is excessive, duplicative, or inconsistent.

It should be pointed out, however, that in the area of minimum acceptance reliability, for which valid requirements demonstrable by tests have been contractually incorporated, the agency's concern with "how" does not limit the responsibility of the contractor for "what." The agency's how is usually a minimal requirement in the absence of which attainment of reliability requirements is most unlikely. There are innumerable instances when contractors have neglected the agency's how with the result that the what did not meet requirements, and when success was attained only by eventual compliance with the how. Even efficiently managed contractor organizations have experienced this.

14.6e Hardware Reliability. Hardware reliability is procurable contractually. Since hardware dimension and magnitude are dependent upon a test, no contractual requirement can be said to exist unless the precise means of demonstration accompany the number. A stated value of hardware reliability with demonstration methods to be determined later has no validity and merely postpones contractual agreement. Some general principles of hardware reliability testing may be stated:

1. The test stresses should at least include the maximum values foreseeable in operation; and the specifying activity should assure that test stresses, when applied to equipments of common technology, are standardized at the maximum values which good design and manufacture may reasonably be expected to withstand. This assures the earliest possible visibility of the most likely failure modes.
2. Since test time is usually a limiting factor, provision should be made for decisions at the earliest possible time.
3. The test must provide verification of the integrity of both design and production processes.
4. The test must be effective in the determination of hardware reliability as a valid element in the prediction of operational reliability.
5. The test disciplines and interpretations must be specified at the time of contractual action, not later.
6. Statistical aspects must provide for not only evaluation of test samples but for a sampling density adequate to permit legitimate decisions as to the reliability level of the population they represent.
7. Generally, the tests should be designed to provide the maximum amount of information in the minimum amount of test time. (See Secs. 3, 8, and 15.)

14.6f Test Considerations. Figure 14.1 illustrates some further test considerations. In this chart the abscissa, complexity, is increasing from the basic hardware building block, the nonrepairable part, to the overall system. The ordinate is some function of cost. It should be explained that these curves are not meticulously plotted from data but serve to illustrate a logical principle. The curve of required test time decreases exponentially with increasing complexity. It illustrates that to qualify a system for reliability by testing of all nonrepairable parts (whose MTBF's are of the order of hundreds of thousands of hours) would be prohibitively time consuming and expensive. An existing body of knowledge of part-failure rates under specific conditions is mandatory, with tests confined to a few special parts not subject to analytical evaluation.

The parts subject to analytical verification are those whose reliability is dependent on structural fatigue or wear conditions in a relatively simple stress-strength relationship. These characteristics are determined by design and normal control of geometry and metallurgy and may of themselves provide adequate assurance of homogeneity. Figures of merit approaching unity would result from a specific reliability test which, to provide statistically significant data, would be disproportionately expensive and time consuming. Generally, the facilities for tests of such simple elements are relatively inexpensive.

The curve of costs associated with facilitation increases exponentially to the overall system level, where, for a system including an air vehicle, for example, it would not be possible to create or operate facilities for measuring reliability under fully controlled conditions. However, flight testing can give essential information as to integration problems of system components and can also give a negative evaluation of reliability. Its importance should by no means be underestimated as a factor in reliability demonstration. However, the disciplines of flight testing for reliability demonstration should be much more stringent than the simple mission success/trials ratio. Any flight-test failure of a component should be classed as a relevant failure whether or

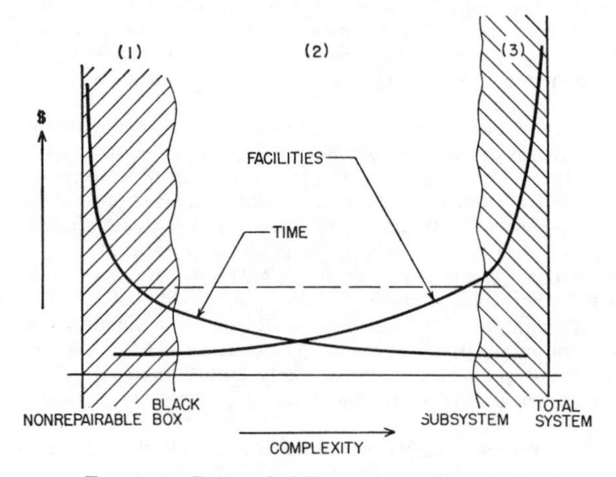

Fig. 14.1 Basic reliability test considerations.

not the component was needed on the test mission, and effective corrective action should be required. "Importance factors" are not properly applied in evaluating hardware reliability tests, although they are useful in predicting operational reliability.

The curves suggest that minimum cost and maximum value will be obtained from specific, controlled reliability tests in the area of subsystems, subassemblies, and/or components. Here, the magnitude of the reliability dimension should be moderate. The AGREE techniques for complex electronic equipment whose reliability cannot, practically, be adequately evaluated by analytical procedures alone fall in this category.

The sequential percentages of operational mission successes measure the achievement of operational reliability objectives. It is easily possible to distinguish from these an operational "growth curve" as personnel become knowledgeable in use and support of the system and as operating and supporting procedures become better defined. These improvements are sometimes due to the efforts of the user. When, in a production contract, hardware reliability requirements are correctly set by the procuring agency and a specific, adequate test procedure (with adequate funds and time) is provided for, it is most unlikely that a growth curve of hardware reliability will be encountered. There will be a growth in productivity and a reduction in the number of failed periodic tests during the course of a production program, but these should not be confused with a growth in reliability of accepted items. Realistic incentives for attainment of cost and schedule targets are difficult, if not impossible, to compute until contractual reliability requirements have been met.

In view of the limitations in accuracy of hardware reliability tests, which are oriented to a simple accept-reject criterion, and the obvious negative-exponential relationship between magnitudes of hardware reliability dimensions and their dollar value, only peculiar, nontechnical considerations could conceivably justify contractual incentive awards for superficially improved hardware reliability. Since increases in operational reliability, whose achievement is measured by the percentage of successful missions, are due mainly to efforts of the user, incentives could not logically be awarded on this basis.

Some hazards to be avoided in specification of, and contracting for, reliability include the following:

1. Confusion of operational reliability with contractual hardware reliability.
2. Very large reliability dimensions with requirements for demonstration at a high confidence level.
3. Inadequate funding of reliability efforts.
4. Summation of parts reliability as being equal to equipment reliability, neglecting the hazards of workmanship defects, imperfect process controls, and design defects.
5. Use of historical operational parts failure rates in predicting equipment reliability. It seems evident that only failure rates arrived at by a standardized test procedure could serve even as a point of departure for extrapolation or interpolation for prediction purposes. However, historical failure rates may be useful in predicting average spare-parts consumption on a technology-wide basis.
6. Assuming that reliability of any type is determined by design only and then confining efforts to the design area.
7. Entering into a contractual relationship involving incentives in which either the recipient of the incentive award is not responsible for results on which the award is based or in which the means of measuring performance are not unmistakably clear to both parties. Evaluation techniques must be relevant, accurate, and timely.

14.6g Incentive Contracting. Incentive contracts, however, are fast becoming the most common type of contract awarded by the government. Most of these contracts contain multiple incentive features on several parameters. Reliability is usually included in these parameters; whence the importance of considering the specification with the incentive.

The first item for consideration regarding a reliability incentive is how much it should be weighed with regard to the other incentive parameters such as cost, accuracy, and schedule. This is usually a matter for negotiation between the manufacturer and customer, but the manufacturer should recommend a weighting that best meets the prime parameters of the contract. (If, for example, the prime purpose of the contract is accuracy, then accuracy should bear the heaviest weight.)

At this point it should be clearly stated that demonstration tests prescribed by the specification and tests for purposes of determining incentive fee are not necessarily the same. In some cases there may be little or no resemblance. Discussed earlier was the fact that when exceedingly high MTBF's are involved, it may not be economically feasible to demonstrate reliability statistically.

The customer usually desires an early demonstration of reliability and tends to associate the incentive-fee provisions with this type of testing. He usually also desires that these early tests on early equipments demonstrate the final level of reliability. As stated by one reliability expert of considerable experience and renown, "The customer sometimes desires that the equipment take its high school final exam while it is still in the early grades of primary school." This is a matter that must be considered in the development of the specification.

Definition of a failure for reliability incentive purposes must be established and included in the contract. This may vary somewhat from the specification definition of failure, or in some cases it may be the same. It is important to remember that other parameters are being measured for incentive purposes, and there is a definite possibility of double jeopardy to the manufacturer. An example might be accuracy and reliability. If inertial navigation system instruments drift excessively, is this an accuracy discrepancy or a reliability failure? One certainly does not want to suffer a penalty in both cases. Even if definitions are more arbitrary than scientific, they must be spelled out in detail. It is often easier, and in the case of incentive contracts more prudent, to define what does *not* constitute a reliability failure. In this manner types of failures which do not affect MTBF can be excluded from the MTBF calculations and the incentive criteria. The producer should also be careful of how incentive provisions regarding reliability are affected by equipment whose use has been under the direction of the customer.

To keep pace with the rapidly changing reliability technology and techniques, the reader is encouraged to review the new reliability documents as they are issued by DOD and NASA. This is true for all aspects of reliability specifications, but it is especially important in the area of demonstration. It should be noted that the author has been somewhat critical of present demonstration requirements and practices in Sec. 14.6*b* Acceptance Criteria.

14.7 THE PROCUREMENT FUNCTION

Whenever one considers specifications for reliability, he must immediately consider the matter of procurement practices and contracts. The preceding subsections have examined the customer-supplier relationships as influenced by the specifications; the following subsections will examine the factors which affect the customer's procurement policies, practices, and procedures. Factors which have contributed to developments within the procurement function and potential problem areas, from the viewpoint of product reliability, will be considered along with recommended means for handling such areas to preclude degeneration of product reliability.

The evolution of procurement technique from that of elemental bartering to the sophisticated disciplines embraced in modern procurement was precipitated by many events. The Industrial Revolution of the early 1800s initiated a degree of interdependence among business enterprises previously nonexistent. The very products emerging from the phenomenon of the Industrial Revolution and subsequent technological development contributed to the shrinkage in time and distance which, in turn, gave added impetus to the interrelationship of the world business community. Events occurring during periods of time identified with World Wars I and II and the totally new factors which have had to be reckoned with since man first crossed the threshold of travel into space are all contributory to the development of progress in the field of procurement.

The question that arises at this point is what effects the evolution of procurement techniques and the greater interrelationship which now exists in the business community have had in relation to product reliability. The answer to this question might very well be that the predominent result has been the preciseness of product definition.

Procurement practices involving goods or services have always embraced considerations pertinent to reliability, either directly or indirectly, in addition to numerous others. Until recently, however, certain product characteristics were generalized and remained relatively undefined in the contract between the buyer and seller. Product reliability was lauded by both parties to a contract, but it was seldom made explicit by contract language. The very lack of definition in the area of product reliability between a buyer and supplier can have a significant impact on the reliability of the end product if from no other viewpoint than the volume of items procured by industries highly concerned with the reliability characteristic of product.

Conservative estimates relative to the proportion of dollars expended for procurement of supplies, components, subcontracted items, and services by industries engaged in development, design, and manufacture of hard goods total in excess of 40 per cent of the gross dollar volume of the business encompassed by those industries! Because of the dependence of all manufacturers on their suppliers and the monetary aspects attendant on the procurement function as reflected by the foregoing figure, it becomes evident that a great number of end items contain a disproportionate number of procured elements, compared to the number of elements produced under the complete control of the manufacturer. It follows, therefore, that the statistical probability of end-item degradation (including reliability) by procured elements increases as additional suppliers' items are utilized.

In view of the impact of procurement on reliability, and in light of the changes which have evolved through the years as regards the definition and quantitative requirements of current procurement documents and the attendant factors to be considered in a buyer-supplier relationship, it would be well to analyze this relationship relative to its effects on reliability of product. The following paragraphs will be concerned with such an analysis.

14.8 THE BUYER-SUPPLIER RELATIONSHIP AS REGARDS RELIABILITY

The foregoing discussion indicates that one of the significant developments which has occurred in the field of procurement has been an increasingly greater use of precise and quantitative terminology regarding reliability requirements in contract and purchase-order documents. This increased emphasis directed toward product reliability has had an impact in several areas of the buyer-supplier relationship.

From the broad perspective, this trend of events has created the need for better communication and coordination between the parties to a contract than previously existed. Additionally, and of paramount importance, areas of management decision are being affected by the definitized reliability requirements being written into contracts. Examples of such areas include management's concern with additional capital expenditures for test equipment, the influence of quantitative reliability requirements on established warranty policies, and decisions concerned with proprietary information and rights in data. This development has further had an effect on the activities attendant on contract performance. A number of considerations must be analyzed from the viewpoint of the manner in which a quantitative requirement with finite parameters would bear on the buyer-supplier relationship as compared with a generalized reliability requirement or one which would simply be a design goal.

The requirement with defined parameters will generally require considerably greater coordination than requirements of a general or a design-goal nature. The aerospace industry today is guided primarily by reliability requirements which have been documented and made mandatory by government agencies (including the several military agencies). Because government agencies do not normally include the capability for product design and manufacture, basic requirements concerned with reliability, in the form of governmental specifications, are made a part of the contract between a government agency and the prime contractor.

The determination as to which of the reliability requirements are to be imposed on the several tiers of suppliers concerned with total contract performance is within the purview of the prime contractor. In a large-scale program which eventually results in a product for a government agency (including military), tiers of contractors which may be identified as associates, major subcontractors, and second, third, and subsequent lower tiers of suppliers must be considered as well as the prime contractor. Section 14.3 reviewed the common reliability requirements in a contract, and Sec. 15 describes methods of verifying and/or demonstrating compliance with these requirements. Those portions of Secs. 14.3 and 14.6 concerned with prime contractors' responsibilities in this area should be reviewed in light of the variables existing in the interrelationship between various tiers of buyer-seller combinations.

A brief review of the usual arrangement by which a need is transformed into usable product is in order here. The agency instituting the basic requirement with a selected prime contractor will negotiate with that contractor as concerns the firm requirements of the ultimate item being procured. These requirements, if not already a part of a government-issued document, will become a specification by virtue of the requirements being embraced in a legal document such as a contract or purchase order. Subsequently, the prime contractor must determine the depth of penetration to which these finite requirements must be applied. Associate subcontractors and major subcontractors are, except in unusual circumstances, liable for conformance to all the basic contract requirements which have been imposed on the prime contractor by the ultimate using agency. The effects of the type of agreement existing between the prime contractor, associate contractor, and/or major subcontractors can, in certain respects, affect the manner in which such requirements are implemented.

The manner in which a prime contractor would handle surveillance and follow-up of finite requirements with a major subcontractor working under a cost-plus-fixed-fee contract would differ from the manner in which the same requirements would be monitored when a fixed-price incentive-type contract existed between the two parties. The cost-plus-fixed-fee contract allows minute and detailed direction of the seller by the buyer because, under this form of contract, the buyer is purchasing services, time, and material as opposed to conceptual design and implementation of that design into a finished product. When a proprietary-design manufacturer performs under a fixed-price incentive-type contract, he, as supplier, has recourse to parts of the Armed Services Procurement Regulations (ASPR) which provide him protection for proprietary-design portions of the product being supplied. Accordingly, in this kind of situation a program for control and reporting, excluding technological details of a proprietary nature, must be implemented.

In the case of the seller operating under the terms of a CPFF contract, the very nature of provisions contained in ASPR governing the execution of CPFF contracts requires complete disclosure of information by the seller at the discretion of the buyer. It can be seen, therefore, that the degree of surveillance and buyer participation in the supplier's activities are affected by the type of contract in a buyer-supplier relationship.

14.9 CONTRACT/PURCHASE DOCUMENTS PERTAINING TO RELIABILITY REQUIREMENTS

The normal course of development from concept to product realization embraces several general phases of documentation between the parties to a contract. Predicated on the need of a particular item/product, and having been allocated funding for obtaining the required items, the agency having a specific need submits invitations to bid (ITB) or requests for quotations (RFQ) to prospective suppliers. Reliability requirements pertaining to the ultimate item being procured, in addition to configuration, material, performance, and other requirements, should be included in RFQ and ITB documents. Standard terminology has been utilized in the work statements contained in invitations to bid and requests for proposals and is generally well understood within the industrial areas concerned with particular types of equipment. However, considerations relative to reliability requirements in the form of definitive statements have emerged only recently. It follows that specifications and terminology controlling reliability requirements have not had the advantage of long-term usage true of work statements and requirements utilized for defining product configuration and performance requirements. Consequently, both the buyer and seller should take the necessary steps to assure that there is a mutual understanding regarding reliability requirements outlined in ITB's and RFQ's.

The essentials of clauses contained in ITB's and RFQ's for requirements pertaining to reliability may be summarized as follows: Such statements must (1) provide clear definitions of all specific parameters, (2) explain and define the extent of the seller's obligation, (3) consider the type of contract existing between the buyer and supplier, (4) be compatible with warranty practices and policies concerned with information-disclosure policies, including the establishment of reliability warranties, and (5) outline the method of reporting all elements of cost concerned with the contractual obligation pertinent to reliability.

A number of companies in the aerospace industry utilize a document known as a basic agreement. Definition and interpretation of factors which are concerned in the business relationship are mutually agreed upon and incorporated in this document. The primary advantage to be gained is the elimination of a multiplicity of interpretations as regards the technical and legal aspects of a contract by having mutual agreements based upon considerations of such factors as the type of product involved, state of the art, production processes, warranty policies, rights in data, and other elements included. Agreements relative to the five elements outlined in the preceding paragraph should be included along with other factors in the basic agreement. Examples of clauses clarifying pertinent points in such a document may be obtained by reference to Sec. 15.

Infrequently, the performance of a contract will be implemented on the basis of the systems-management concept: the subcontractor is delegated full design and performance responsibility, with the buyer having responsibility for interface problems only. Under the systems-management concept, a minimum of coordination between buyer and seller would normally be necessary in regard to detail tasks that must be accomplished by the seller to meet subsystem or system requirements. A number of major contracts in the past few years have necessitated significant amounts of money for review and approval of reliability programs attendant on these contracts. The systems-management concept would eliminate the need for such costs and would require only a minimum amount of coordination concerned with detailed activity necessary in reliability programs. The essential factors concerned in a formal reliability program, such as allocations, predictions, hazard rate, and other stochastic processes, could be embraced in the systems-management concept and, as such, would be executed within one organization.

The broad scope of technological detail concerned with major programs, especially those which affect the national interest, makes mandatory the need to implement work on such contracts prior to definitization of all terms concerned with performance of the contract. Although the formal contract is not normally consummated until such time as definitization of the details is accomplished, actual performance can commence upon receipt of a pertinent letter contract. A letter contract merely defines that a portion of work is to be accomplished. Again, the importance of including in a letter contract reliability requirements which are pertinent to particular portions of a contract cannot be overemphasized. Specifications and standards affecting the task should likewise be included.

It has been pointed out earlier in this section that one of the results emanating from many years of development in the procurement processes is that of preciseness of product definition. The foregoing statements indicate that such definition should commence at the earliest possible time during contract performance and, accordingly, basic agreements, letter contracts, and purchase orders should, as applicable, define these areas. To assure full consideration of all facets concerned, invitations to bid and requests for proposals should likewise include as complete definition of the program as possible at the time of issue.

14.10 SUPPLIER SURVEILLANCE AND SOURCE CONTROL

14.10a Compatibility of Policies. The policies and procedures which guide a company's relationship with its suppliers to ensure that contractual requirements, including those pertinent to reliability, are satisfied must be compatible with the frame of reference of the particular procurement involved. Examples of some of the considerations which contribute to determining the frame of reference of a given procurement include, but are not limited to, the following: The state of the art as related to the item being procured; the function and application of this item; the dollar value of the procurement; the physical facilities of the supplier; and, with respect to product reliability, the criteria or form of objective evidence required to demonstrate realization of an acceptable level of reliability.

14.10b Government Monitoring. It has become both policy and practice for firms engaged in the manufacture of sophisticated items for government agencies to be guided in their operations by procedures specifically designed to promulgate product reliability. To provide common baselines for guiding contractors in the establishment of company policies and operating procedures which affect product reliability, the various agencies within the government have issued documents and specifications directed toward reliability in design, development, and production of equipments, subsystems, and systems. (See Sec. 14.5.) The very number of such documents precludes detailed discussion herein of specific subject matter covered. However, all areas of activity attendant on each phase of effort concerned with an ultimate product or service are dealt with to the end of obtaining optimum reliability.

Not the least of the areas of effort concerned with product reliability is that of supplier surveillance and source control. As has been previously mentioned, the depth of detail of procedures designed to guide the relationship between various tiers of contractors, subcontractors, and suppliers must be compatible with a given procurement and be based on the factors discussed above. In every procurement wherein product reliability is of importance, and more especially when product reliability is defined as a contractual obligation, there are certain basic areas which, with proper control, will greatly assist in realizing optimum product reliability. These areas may be summarily listed and briefly explained as follows:

A. General Policies. These documents would be concerned with general-policy considerations and would include instructions and directions applicable to all procurement activities notwithstanding the product involved. The language of general policies could be compared with boiler-plate language common in all purchase-order contracts and considered as standard-terms-and-conditions language. The following are examples of areas covered by this type of issuance: general instructions for the handling of government-furnished property, procedures for evaluating and/or auditing adherence of a company to its internal quality and reliability programs, and procedures and instructions for dealing with other than government agencies.

B. Administrative Control. This type of control would be concerned with procedures attendant on the recording of quality and reliability data, standard procedures for identification of product and/or special processes, the criteria and procedure for certifying personnel who accomplish special-process work, and policies and procedures concerned with the control of engineering drawings.

C. Product Controls. Procedures and policies concerned with product controls would include specific instructions and detailed acceptable criteria pertinent to various types of product such as raw castings, forgings, and rubber goods, as well as special instructions pertinent to sampling plans for a given product, the control of tooling and methods of tool repair, certification and identification, and product-reliability procedures specifically designed to evaluate the level of quality and reliability of a given product.

D. Outside Procurement Controls. Policies and procedures concerned with supplier relationships, including surveillance and control, are of prime importance to this particular section. To ensure that optimum product reliability is realized, basic principles in the following activity areas must be covered:

1. A definition of buyer responsibility and supplier responsibility
2. Instructions for evaluating supplier capabilities
3. Instructions for handling nonconforming material
4. Special instructions and procedures pertinent to testing and inspecting the item being procured
5. Instructions relative to the application or nonapplication of sampling
6. Any additional instructions peculiar to the particular procurement

To give an example of the type of instructional data which should be included in policies and procedures noted in (1) to (6), we shall, for a moment, discuss item 2, instructions for evaluating supplier capabilities. To provide a standard means of recording information during survey of a potential supplier and to allow for comparative evaluation of different facilities surveyed, many companies have formulated checklists. This survey form considers the facilities and the manner in which a potential supplier would handle each area of activity relevant to the procurement involved.

Areas which are relevant to supplier activity encompass the receipt and inspection of material, including records pertinent to such material; special-process capabilities;

the manner in which the supplier maintains information with regard to the repair and calibration of tooling and test equipment; the manner in which reliability controls are effected during product development; blueprint controls; and procedures exercised during final acceptance and shipping operation to preclude degradation of the reliability designed and manufactured into a given item. Additional data concerned with the fiscal and physical conditions pertinent to a supplier's facility would also be covered during a supplier survey, as would an evaluation of manufacturing methods, material practices, and contract administration policies and procedures.

14.10c Supplier Source Inspection. Major procurements normally include provisions which give the buyer the prerogative of examining a supplier's operation and performing product inspection and acceptance at source. Government specifications make government source inspection mandatory on many types of product. Government source inspection can be requested only by and for the government. Such a request must be contractually documented and approved by the cognizant government service handling a given contractor's facility. Current specifications relevant to a contractor's quality and reliability program specifically mandate the prime contractor as responsible for the performance of all areas concerned with the delivery of an item. Based on this, it has become common practice for all subcontract purchase-order contracts to include the right of buyer source inspection.

Accordingly, a thorough review of requisites of a purchase-order contract should be accomplished at the outset of contract performance to eliminate diverse thinking between the buyer and supplier. Of special benefit to both parties will be the establishment of clearly defined communication channels to handle any problems which might possibly arise during the performance of a contract. Although a myriad of policies and procedures are used in source control, all are in vain unless firm and mutual understandings on intercommunications are reached by the parties to a contract.

14.10d Control During Research and Development. The control of product quality and reliability during engineering research and development phases of a contract has, until recent years, been outside the purview of other than engineering personnel. The implementation of quality and reliability specifications by the government has imposed the requirements to effect production-type controls on activities primarily concerned with engineering research and development. It is recognizable that when constraints are placed upon the creativity so necessary during these early phases of conceptual work, the very quality of the engineering can suffer. It is therefore important that both parties to a contract recognize that special and minimum controls be used during engineering research and development phases of a contract.

Minimum controls are considered to embrace only those necessary to afford coordination of an intercompany and intracompany level. Further, such controls should consist of management-type controls as opposed to product-type controls. Management-type controls are those concerned with the fiscal and schedule areas and the procedures used in preparing for the transition of product-engineering research and development to follow-on phases of contract performance. It would be impossible to define specific procedures to be utilized during the engineering research and development phase of a contract in that the variables, such as product considerations, state of the art, type of contract, and facility size, would, in each case considered, be diverse. Accordingly, the need for full understanding is again emphasized.

14.10e Goals of Buyer/Supplier Relationships. The procurement of reliable components, subcontracted items, and services has been discussed from a general viewpoint of several considerations which must be reckoned with during the procurement process. Realization of optimum quality and reliability in procured items necessitates a good buyer/supplier relationship, a knowledge of the type of contract/purchase order documentation used in such an association, continuing surveillance and evaluation of product received from the supplier by the buyer, and feedback of findings to the supplier.

An attempt has been made to implant in the reader's mind the value of disciplines. A number of examples in the areas where such disciplines would be applicable has been included. No amount of policy, procedure, or instruction can supplant common sense.

Section 15

ACCEPTANCE TESTING

CLIFFORD M. RYERSON

SENIOR STAFF ENGINEER, HEAD, COMPONENTS ASSURANCE
HUGHES AIRCRAFT COMPANY, CULVER CITY, CALIFORNIA

Mr. Ryerson, after receiving the B.S. and M.S. degrees in physics from Stetson University and Duke University, began his career in the Physics Department of Duke University, where he designed and developed nuclear research equipment for the graduate school, including personally developing and managing the recording and analyzing equipment for three cosmic-ray-investigation expeditions.

He then joined the Naval Ordnance Laboratory, Washington, D.C. He was appointed general electronic consultant to the Research Division of the Laboratory. During this period, he received the Meritorious Civilian Service Award for his work in magnetic field instrumentation. He then worked for the Naval Gun Factory, and in 1952 joined RCA as a systems design engineer. Owing to his work in reliability, he was chosen in 1953 to be reliability coordinator for the plant and in 1954 moved to Camden, where he directed the central reliability program for the whole company.

During the last five years with RCA, Mr. Ryerson served as a management consultant to all the autonomous divisions of the company. During that time, he was loaned to the Department of Defense for assistance on the AGREE group and to the Naval Bureau of Weapons for assistance in organizing the BIMRAB Advisory Board. For this service he received the Admiral Coates Award in 1961.

From 1961 until January, 1964, Mr. Ryerson was engaged in private consulting work, providing Technical Audit Service to industry. He joined Hughes Aircraft Company in 1964, where he is responsible for the assurance of reliability, maintainability, quality, and value engineering in the Components Department.

ACCEPTANCE TESTING

Clifford M. Ryerson

CONTENTS

15.1 INTRODUCTION

Preceding sections have dealt with reliability theory and general test plans. This section deals with acceptance testing. Specific testing instructions are given, and enough discussion for assistance in interpreting the resultant test data is presented.

Most reliability tests are simple and easy to perform if the basic requirements are understood. One of the most common areas of misunderstanding is the figure of merit used to describe the results of the reliability tests. Reliability is defined in different ways by different people and confusion frequently results because a test related to a certain figure of merit is used for the wrong category of test.

The following paragraphs describe the various types of acceptance testing and explain the standardized figures of merit and related test plans involved with each. A few paragraphs are devoted to explaining the significance of confidence interval and confidence limits. Charts and graphs and specific application examples are presented.

15.2 TYPES OF ACCEPTANCE TESTING

15.2.1 General. One of the most common types of acceptance testing relates to the acceptance of parts or components prior to the manufacture of an assembly or subsystem or for use as replacement parts. A second major category of acceptance testing relates to instruments or equipments which are purchased as complete entities to perform specific functions. A third category of acceptance testing relates to large systems. The major factors related to acceptance testing in each of these three categories are summarized in Table 15.1.

15.2.2 Category I, Part Acceptance. The first category, relating to part acceptance, is indicated to have two divisions. These relate to the destructive testing of small lots of parts and the nondestructive testing of large lots. Type IA acceptance tests are usually performed on selected samples, whereas type IB tests are frequently performed on entire lots. Usually a test plan will require both types in order to provide for a complete evaluation of the critical reliability factors.

A characteristic of type IA testing is that accelerated stress conditions are frequently used. The type IB tests usually are performed at nominal operating conditions. A better plan, but one not always feasible, is to operate the IB tests at the maximum ratings for the parts. These factors are fully discussed in later paragraphs.

Table 15.1 Categories of Acceptance Testing

Category	Identity	Typical quantity involved	Common characteristic of test	Figures of merit	Abbreviation or symbol
IA	Component parts (samples)	Small sample	Destructive	Time to failure Stress to failure Per cent failure or success Failure rate	TTF STF %F, %S
IB	Component parts (lots)	Large lots	Nondestructive Screening	Acceptable performance Per cent reject Stability	a or r % rej
II	Equipment	Small quantity Typical of population	To evaluate degradation	Mean time between failure Mean time to first failure Time to wear-out Time to uniform replacement Mean time between replacement	MTBF MTFF (MTTF) L(longevity) LR MTBR
III	Systems	Single Unique service	Nondestructive Seldom completely simulates end-use conditions	Probability of success	R, P_s

15.2.3 Category II, Equipment Acceptance. Acceptance testing for equipments is nearly always nondestructive and is planned to evaluate performance under clearly specified operational conditions. Usually a power-on test that results in the destruction and subsequent repair of several parts or components is required. The usual purpose of this test is to determine if the failures are reasonable in number for the duration of the test according to the specification and if they result in degraded performance of the equipment after repair. A characteristic of category II acceptance tests is the number of test items (equipments) involved is usually only unity or a small number. The test item or items may, however, be withdrawn from a larger population about which considerable information might be known.

15.2.4 Category III, Systems. Acceptance testing for systems is usually performed on single unique systems. Great variation is inherent in system test plans, and considerable engineering effort is frequently required to develop appropriate procedures and techniques. A characteristic of this type of testing is that the size of large systems frequently prevents the systems being tested under conditions closely simulating actual end use. Subsystems are frequently tested independently, and the results are then combined to obtain a theoretical performance acceptance rating. A major exception to this is when missile systems are tested in space by actual firing. It is questionable if this type of testing can rightfully be considered a conventional acceptance test, and most manufacturers of such systems refuse to be bound by the results on the ground that too many parameters are unknown or uncontrolled as specified.

15.2.5 Reliability Acceptance Figures of Merit. The various categories of test employ different figures of merit in measuring acceptability. Confusion is created when an improper figure of merit is used. Most of the figures of merit used are interrelated mathematically. Refer to Secs. 4 and 8 for these relationships. The following paragraphs describe the various figures of merit and how they apply to acceptance testing.

A. Time to Failure (TTF). The time-to-failure measure is usually used in connection with a small sample of parts all of which are stressed at a specified level for a time long enough for the entire sample to fail. The average or mean of the times to fail is considered the TTF for the sample. The distribution around this mean of the times to fail is usually studied to obtain a knowledge of the part life under specified conditions. This reveals if the stress applied in the test is realistic for actual use of the part. Accelerated test stresses are sometimes used to study quickly the improvement in failure times achieved by changes in the product design, material, or process. See also Sec. 15.4.5 on step-stress accelerated testing. Do not confuse time to failure (TTF) with mean time to failure (MTTF) or mean time between failures (MTBF) as described later.

B. Stress to Failure (STF). The stress-to-failure technique was originally sponsored as an aid to the evaluation of weak links. One or more stress factors such as temperature or voltage is increased until an entire sample has failed. The mean value of the stress to failure (STF) and the distribution of failure events about this mean are studied to evaluate the part strength. The shortcoming in this approach is that few parts show simple correlation between the accelerated stresses and reliability at normal or derated stresses. The major use for STF is in the analysis of design safety factors to help determine the proper application of specific parts. See Sec. 8 for a discussion of this use.

C. Percentage Failure or Success (%F, %S). Percentages of failure and success are complementary figures of merit frequently used to specify the acceptability of one-shot devices. For example, an acceptance test using this figure of merit may be established to determine the usability of rifle ammunition. Since it is known that powder degrades with age and storage conditions, a sample may be taken at specified intervals of time and fired in a standard or typical rifle. Acceptability of the stored rounds for service use may be based on a fixed percentage of each sample performing as specified. (Per cent success = %*S*.)

This figure of merit is sometimes also applied to such items as electrical current-limiting fuses or even to one-shot systems such as guided missiles. The latter is an unfortunate choice, however, since the figure of percentage of failure cannot be determined in the same way and used in prediction for unique (single) one-shot systems. See instead the discussion under the heading Probability of Success (P_s, R) (Sec. 15.2.5*K*).

D. Failure Rate (λ). This figure of merit, failure rate, is generally applied to component parts. It denotes a time-based average rate of failure. Common levels of failure rate are in fractions of per cent of a lot for each 1,000 hours of operation. See Secs. 4 and 9 for detailed explanation of failure rate and its mathematical relations. Some people use failure rates measured as 10^6 part hours per failure, which is one order of magnitude smaller than the 10^5 equivalent of per cent per 1,000 hours. Still another failure-rate figure of merit sometimes used is 10^9 part hours per failure. Both these latter figures of merit have sometimes been called failure-rate "bits," but this terminology is generally discouraged because of the conflict caused when reliability data are computed and analyzed by using computer techniques.

E. Mean Time between Failures (MTBF). The figure of merit MTBF is commonly applied to equipments. It denotes the average time between failures which can be expected for an equipment during the portion of its life cycle in which the exponential failure "law" applies and during the period prior to the first longevity termination. See Secs. 3 and 4 for a discussion of life patterns and longevity. A danger in the use of this figure of merit (MTBF) is that it is frequently interpreted as a permanent inherent characteristic of an equipment, which it is not. Unless an equipment is designed, constructed, and known to have a "flat" portion of its life characteristic prior to a longevity termination, any MTBF measured will be valid only for the period during which it is measured. At some other time period another value of MTBF might apply. Even when the life characteristic for an equipment is known and the MTBF is measured, this value will be valid only during the flat portion of the life characteristic curve. The use of MTBF for equipment acceptance tests must be accompanied by other qualifying test requirements describing the life characteristics such as longevity.

The MTBF may be determined by testing several equipments through several failure incidents and averaging the total operating time for the number of failures. See the subsequent paragraphs with specific examples for determining specific quantities of equipments and test times.

F. Mean Time to First Failure (MTFF). Frequently an equipment or subsystem is to be installed in a use location which is not accessible for maintenance. The important figure of merit for this application is the average time the equipment can be expected to perform before it experiences its first failure. This figure of merit is identical with the term, sometimes used, mean time to failure (MTTF). This author prefers the use of MTFF to prevent confusing the time *to* first failure with the time *between* failures.

The MTFF can be found by testing a group of equipments a length of time required for each equipment to experience its first failure. It is obvious that the appropriate figure of merit used in the acceptance test plan will depend upon the particular application conditions described in the specifications. It is interesting to note that MTFF will not equal MTBF if there is an early debugging or infant-mortality problem.

G. Time to Wear-out (L) (Longevity). The longevity period is defined in the AGREE[1] report and other documents as the time to wear-out on the life characteristic curve. The termination of longevity is the time at which the failure-rate life characteristic curve becomes twice the value of the reciprocal of the acceptable MTBF. Beyond this point in time the MTBF will be different. A major overhaul at this point may or may not restore the condition of the equipment to the previous value of MTBF. Whether it does or not will depend on how closely all the component parts exhibit a one-horse-shay characteristic with the same duration.

H. Time to Uniform Replacement (LR). Complex equipments containing parts with many different life characteristics will exhibit a multiple series of level failure-rate plateaus following the longevity period. Multiple major overhauls will provide a "mix" of life distribution patterns which collectively can exhibit a constant failure-rate characteristic usually somewhat higher than the reciprocal of the original MTBF. The time in a complete life cycle of an equipment to establish this new constant failure rate is known as the LR.

I. Mean Time between Replacement (MTBR). The reciprocal of the average failure rate for an equipment during the period following its LR is its MTBR. The value of this will equal the value of the MTBF of the same equipment only if the design, construction, process, control, and materials used achieve identical life failure characteristics for all the parts. Major overhaul to the extent of complete replacement of the entire equipment (new replacement) having the same reliability as the original equipment is the simplest way to accomplish this. In practice the MTBR may range from one-half to one-quarter of the MTBF. These factors must be considered in developing the acceptance test plan.

J. Mean Time between Maintenance (MTBM). The mean time between maintenance is the average up time of a system or equipment between maintenance to remove or to prevent malfunction. This differs from MTBR in two respects. First, it may be based on a preventive-maintenance schedule during which no failures or malfunctions have occurred. Second, MTBM usually contains all available up time including nonoperating available time for a specific type of mission or service and is not based solely on power-on time.

K. Probability of Success (P_s, R). Probability of success P_s and system reliability R are usually synonymous; see Secs. 1, 4, and 8 for detailed explanations. The figure of merit P_s usually denotes that the acceptability of a system can be established by a prediction derived from a composite of theoretical analyses and computations based

[1] Advisory Group for Reliability of Electronic Equipment, DOD, June 4, 1957.

on qualified measurements. See Sec. 5 for a discussion on reliability prediction. In brief review

$$P_s = e^{-t/m} \tag{15.1}$$

where P_s = probability of survival
t = time of test
m = MTBF
e = Naperian base

L. Acceptable performance (a or r). Most acceptance test specifications define the conditions for acceptance or rejection on the basis of adequate or inadequate performance. Performance degradation beyond a certain specified acceptable limit represents, in these cases, as much of a failure as though function ceased entirely. Per cent degradation (% degrad) or instability about a norm ($\pm X$) is specified as the go-no-go acceptability limits a and r. The variations of key parameter values during a power-on test are frequently referred to as the Δ (delta) values. The numerical assessments of the Δ values are usually known as Δ measurements. See Sec. 15.4.6 on degradation testing.

15.3 CONFIDENCE INTERVAL

This subject has been discussed in other sections, but a special presentation is included here because of the importance of confidence to acceptance testing. Frequently, the concept is misunderstood to be a much more complicated statistical tool than it actually is. This presentation is deliberately developed in nonstatistical language so that the casual reader can quickly understand the important implications for acceptance testing. In general, a confidence interval is bounded by upper and lower confidence limits. Generally speaking, the broader the limits the higher the confidence that a particular group of events is enclosed. This is illustrated further in the following discussion.

15.3.1 Confidence Limits. To clarify the principle of confidence limits, consider the illustration of the time that a guest will arrive in town by train on a certain day. If all you know is the date of arrival, you can express 100 per cent confidence that he will arrive sometime between the limits 12:01 A.M. and 12:00 midnight. If you know that no night trains stop in town between the hours of 6:00 P.M. and 6:00 A.M., you can narrow your confidence interval to 100 per cent confidence that the time of arrival will be 12:00 noon ±6 hours. If you happen to know that all the train arrivals from his direction are in the morning, this again narrows your 100 per cent confidence interval. Your limits are then 6:00 A.M. and 12:00 noon. However, if this were all you knew about the arrival schedule, your confidence would be very low that he would arrive at any specific minute. Even if you knew the train number and the expected time of arrival, your 100 per cent confidence interval would have to be broad enough to allow for any possible exigency that might affect this particular event.

Suppose the train is scheduled to arrive at 11:00 A.M. You might investigate the record of this particular train and find that eight out of ten days, on the average, the 11:00 o'clock arrives within five minutes of 11:00. Your confidence would then be 80 per cent that the particular train would arrive at 11:00 A.M. ±5 min. Putting this another way, you would have an 80 per cent confidence that the exact time of arrival would be between 10:55 and 11:05 A.M. The 80 per cent confidence interval would be 10 min long and extend from the upper limit of 11:05 A.M. to the lower limit of 10:55 A.M. This, in statistical language, is described as a two-sided confidence interval, meaning that there are both upper and lower limits.

But suppose you want to make sure that the particular train is typical of those which arrive normally within the average confidence interval. You could check at the information window or with the stationmaster sometime before train time to see if this particular train is running on time at earlier stops. Twenty per cent of the trains normally arrive at times outside the 80 per cent confidence interval because of events which make them nontypical. This is the equivalent engineering action of evaluating a test result in terms of ancillary factors to determine mitigating circumstances or system interaction factors.

Suppose also that you are out of town on business and cannot get to the railroad station until a specific time. In that case you might want to know the confidence that the train will arrive some time after you do, so that you will be on hand to greet your guest. If you can arrive an hour or more ahead of the normal train time, your confidence will be almost 100 per cent that the train will arrive later than you do. However, as the two times of arrival approach coincidence, the confidence in your arriving first will approach 50 per cent. Under these conditions the variability in the train arrival is a major factor. This example illustrates a statistical approach described as a one-sided confidence determination or interval.

Both one-sided and two-sided confidence intervals are described in the following paragraphs. Figures 15.1 and 15.2 illustrate them.

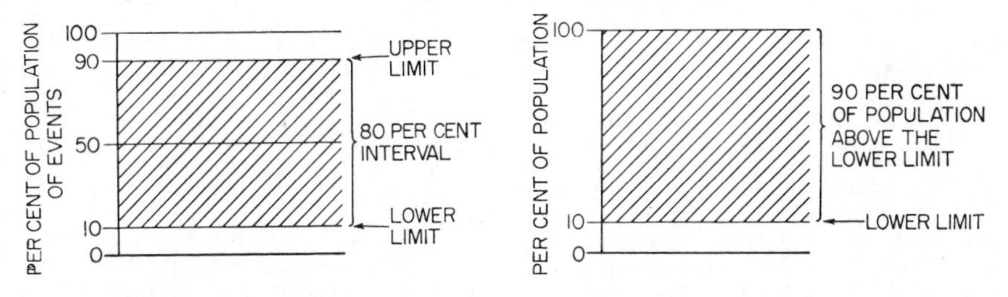

FIG. 15.1 Two-sided confidence interval.

FIG. 15.2 One-sided confidence interval.

15.3.2 Two-sided Confidence Interval. For reliability acceptance testing the two-sided confidence interval can be determined from the so-called two-tailed functions of the chi-square (χ^2) distribution. The χ^2 table is given in Table A.6.[1] Other values can be found on page 484 of Mood's *Theory of Statistics*, McGraw-Hill Book Company, New York, 1950. The upper and lower limits (UL and LL) can be found from the expressions:

$$\text{UL} = \frac{2f(\text{MTBF})}{\chi^2_{2f;1-(1-P)/2}} \tag{15.2}$$

$$\text{LL} = \frac{2f(\text{MTBF})}{\chi^2_{2f;(1-P)/2}} \tag{15.3}$$

where UL = upper confidence limit
LL = lower confidence limit
f = number of failures
P = confidence level

For example, suppose that five failures occurred during a test of 1,000 hours and it is desired to determine the 90 per cent confidence interval for the MTBF. The greatest likelihood estimate for the

$$\text{MTBF} = \frac{\text{time of test}}{\text{number of failures}} = \frac{t}{f} = m \tag{15.4}$$

Thus

$$m = 1{,}000/5 = 200 \text{ hours}$$

and from Eq. (15.2)

$$\text{UL} = \frac{(2)(5)(200)}{2_{@10,\,0.05}} = \frac{2{,}000}{3.94} = 508 \text{ hours} \tag{15.5a}$$

and from Eq. (15.3)

$$\text{LL} = \frac{(2)(5)(200)}{2_{@10,\,0.05}} = \frac{2{,}000}{18.3} = 109 \text{ hours} \tag{15.5b}$$

[1] Tables numbered A are part of Appendix A.

This result can be described as follows: The 90 per cent confidence interval for the MTBF demonstrated by this test extends from the upper limit of 508 hours to the lower limit of 109 hours. Or it can be stated that there is a 90 per cent confidence that the true value of MTBF lies between 109 and 508 hours.

Table 15.2 Multiplier K for Two-sided *MTBF* Estimate

Number of failures f	Upper limit K_U					Lower limit K_L				
	95%	90%	80%	70%	60%	60%	70%	80%	90%	95%
1	28.6	19.2	9.44	6.50	4.48	0.620	0.530	0.434	0.333	0.270
2	9.2	5.62	3.76	3.00	2.43	0.667	0.600	0.515	0.422	0.360
3	4.8	3.68	2.72	2.25	1.95	0.698	0.630	0.565	0.476	0.420
4	3.7	2.92	2.29	1.96	1.74	0.724	0.662	0.598	0.515	0.455
5	3.0	2.54	2.06	1.80	1.62	0.746	0.680	0.625	0.546	0.480
6	2.73	2.30	1.90	1.70	1.54	0.760	0.700	0.645	0.568	0.515
7	2.50	2.13	1.80	1.63	1.48	0.768	0.720	0.667	0.592	0.535
8	2.32	2.01	1.71	1.57	1.43	0.780	0.730	0.680	0.610	0.555
9	2.19	1.92	1.66	1.52	1.40	0.790	0.740	0.690	0.625	0.575
10	2.09	1.84	1.61	1.48	1.37	0.800	0.752	0.704	0.637	0.585
11	2.00	1.78	1.56	1.45	1.35	0.805	0.762	0.714	0.650	0.598
12	1.93	1.73	1.53	1.42	1.33	0.815	0.770	0.720	0.660	0.610
13	1.88	1.69	1.50	1.40	1.31	0.820	0.780	0.730	0.652	0.620
14	1.82	1.65	1.48	1.38	1.30	0.824	0.785	0.736	0.675	0.630
15	1.79	1.62	1.46	1.36	1.28	0.826	0.790	0.746	0.685	0.640
16	1.75	1.59	1.44	1.35	1.27	0.830	0.795	0.750	0.690	0.645
17	1.71	1.57	1.42	1.33	1.26	0.835	0.800	0.760	0.700	0.655
18	1.69	1.54	1.40	1.32	1.25	0.840	0.805	0.765	0.710	0.660
19	1.66	1.52	1.39	1.31	1.24	0.845	0.808	0.767	0.715	0.665
20	1.64	1.51	1.38	1.30	1.23	0.847	0.810	0.768	0.719	0.675
25	1.55	1.44	1.33	1.26	1.21	0.860	0.830	0.790	0.740	0.700
30	1.48	1.39	1.29	1.23	1.18	0.870	0.840	0.806	0.756	0.720
40	1.40	1.32	1.24	1.19	1.16	0.884	0.860	0.826	0.787	0.750
50	1.35	1.28	1.21	1.17	1.14	0.892	0.872	0.847	0.806	0.770
70	1.28	1.23	1.18	1.14	1.11	0.910	0.890	0.860	0.830	0.800
100	1.23	1.19	1.14	1.12	1.09	0.924	0.906	0.880	0.852	0.830
200	1.16	1.13	1.10	1.08	1.06	0.940	0.935	0.916	0.890	0.870
300	1.12	1.10	1.08	1.06	1.05	0.955	0.942	0.930	0.910	0.895
500	1.09	1.08	1.06	1.05	1.04	0.965	0.954	0.942	0.930	0.915

The solution to this problem can be simplified if the following transformations are utilized:

$$\text{UL} = K_U m \tag{15.6}$$

$$\text{LL} = K_L m \tag{15.7}$$

where

$$K_U = 2f/\chi^2_{2f;(1-P)/2} \tag{15.8}$$

The values of K have been worked out by the author for various values of f and are summarized in Table 15.2. A plot of the upper and lower limit K values for various confidence intervals is shown in Fig. 5.3.

15.3.3 One-sided Confidence Interval. For many reliability testing purposes a one-sided confidence interval is most appropriate. This can be illustrated by a situation in which a test is desired to demonstrate with a specific confidence that an equipment MTBF is greater than a minimum limit. This is sometimes described as a single-tailed function. It can be specified for the exponential distribution by the expression

$$\bar{m} = 2f\hat{m}/\chi^2_{2f;1-P} \tag{15.9}$$

where f = the number of failures
P = confidence
$\bar{m}$ = actual MTBF of the population
$\hat{m}$ = observed MTBF

and χ^2 is from the referenced χ^2 tables at the values of $2f$ and $1 - P$.

Suppose it is desired to demonstrate with a 90 per cent confidence that an equipment MTBF = $\bar{m} \geq 100$ hours. What must an observed $\hat{m}$ be if the test is terminated at the first failure?

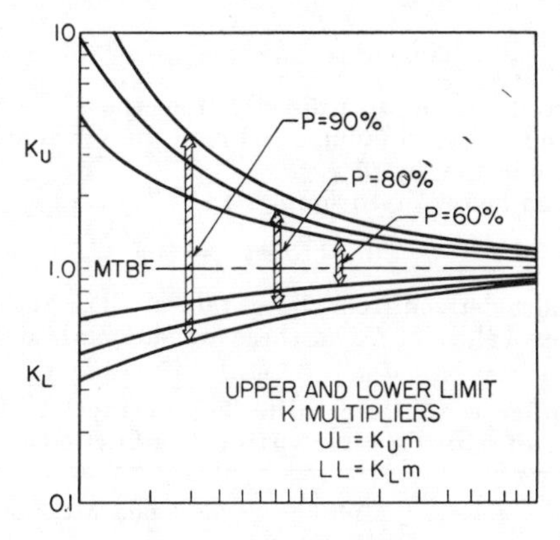

FIG. 15.3 Two-sided confidence plot.

Rearranging Eq. (15.9),

$$\begin{aligned} \text{Observed } \hat{m} &= \bar{m}\chi^2_{2f;1-P}/2f \\ &= (100)(4.605)/(2)(1) \\ &= 230 \text{ hours} \end{aligned} \tag{15.10}$$

In other words, if the equipment operates for 230 hours to the first failure, the test has demonstrated with a 90 per cent confidence that the inherent reliability is as good as or better than 100 hours MTBF.

But suppose the equipment fails before the end of the 230-hour interval. This does not prove that the MTBF is less than 100 hours. Remember that even if the actual MTBF is 100 hours, there is a 90 per cent chance that a failure will occur some time during the first 230 hours. So if a first failure occurs early, it is possible to continue the test to obtain the benefit of knowledge about subsequent failure events to adjust the confidence in the test results.

Suppose the test to demonstrate a required MTBF ($\bar{m}$) is continued to the second failure. How long must the test continue before the second failure occurs to demonstrate with a 90 per cent confidence that the MTBF ($\bar{m}$) is 100 hours or more? From Eq. (15.10),

$$\text{Observed } \hat{m} = \bar{m}\chi^2_{2f;1-P}/2f$$

Now $f = 2$ and χ^2 @ (4, 10%) = 7.779; therefore

$$\begin{aligned} \hat{m} &= (100)(7.779)/4 \\ &= 194.5 \text{ hours} \end{aligned}$$

But the observed MTBF ($\hat{m}$) is the time of test divided by the number of failures, i.e.,

$$\hat{m} = t/f \tag{15.10}$$

so the time of test $t = \hat{m}f$. Since $\hat{m} = 194.5$ and $f = 2$, the time of test to the second failure

$$t = (2)(194.5) = 389 \text{ hours} \tag{15.11}$$

A simplification of Eq. (15.10) can be arranged for finding the time of test directly as follows: From Eq. (15.10),

$$\hat{m} = \bar{m}\chi^2_{2f;1-P}/2f$$

but from Eq. (15.4), $\hat{m} = t/f$. Therefore,

$$\text{Time of test } t = \bar{m}f\chi^2_{2f;1-P}/2f$$

Canceling the number of failures f in numerator and denominator yields

$$\text{Time of test } t = \bar{m}\chi^2_{2f;1-P}/2 \tag{15.12}$$

Thus the time of test can be read directly from the χ^2 Table A.6. Enter the df column at 2 times the number of failures and read the χ^2 value in the $1 - P$ column. Divide this by 2 and multiply by $\bar{m}$.

Equation (15.12) can be rewritten simply as

$$\text{Time of test } t = \bar{m}w \tag{15.13}$$

where w is a multiplier derived from the χ^2 tables. The author has calculated the values of w for various values of f, and these are summarized in Table 15.3. Enter

Table 15.3 Multiplier w to Determine Test Time t to Demonstrate *MTBF* of m or Greater with a Confidence P

To number of failures f	Multiplier w for confidence P of						
	95%	90%	80%	75%	50%	20%	10%
0 to 1	2.99	2.30	1.61	1.38	0.69	0.223	0.105
2	4.74	3.89	2.99	2.69	1.68	0.824	0.532
3	6.29	5.32	4.28	3.92	2.67	1.530	1.102
4	7.75	6.70	5.51	5.10	3.67	2.297	1.745
5	9.15	8.00	6.72	6.25	4.67	3.089	2.432
6	10.51	9.25	7.90	7.40	5.65	3.903	3.152
7	11.84	10.60	9.07	8.55	6.65	4.733	3.895
8	13.15	11.70	10.23	9.70	7.65	5.576	4.656
9	14.43	13.00	11.38	10.80	8.65	6.428	5.432
10	15.70	14.20	12.52	11.90	9.65	7.289	6.221
11	16.96	15.40	13.65	13.00	10.65	8.160	7.020
12	18.20	16.60	14.77	14.10	11.65	9.031	7.829
13	19.44	17.80	15.90	15.20	12.65	9.91	8.646
14	20.67	18.90	17.01	16.30	13.65	10.794	9.469
15	21.88	20.10	18.12	17.40	14.65	11.682	10.299

$t = wm \qquad nt = w$

Example. To demonstrate $m \geq 100$ hours with 90% confidence

$t_{90,1} = 230$ hours to first failure
$t_{90,2} = 389$ hours to second failure
$t_{90,3} = 532$ hours to third failure
$t_{90,4} = 670$ hours to fourth failure

this chart under the confidence-desired column and read the multiplier for $\bar{m}$ along the number-of-failures row. From this it can be seen that to demonstrate a MTBF of 100 hours with a 90 per cent confidence, it should be possible to operate to 230 hours to the first failure or to 532 hours to the third failure, etc.

The method just described is based on the assumption that the exponential failure law applies during the test period. If this assumption cannot be made or if, for some reason, evidence indicates that a Poisson process in the time domain does not exist, then the use of a nonparametric test is required.

15.3.4 Nonparametric One-sided Test. It is sometimes necessary to determine something about reliability without making any assumptions about the failure distribution density in the time domain. For example, it may not be known whether or not the exponential failure law applies. In these cases a nonparametric test is required. Such a test can be based on the F distribution and the relationship

$$\hat{R}_{(t)} = 1/\{1 + [(f + 1)/(n - r)]F_{\alpha};\ \nu_2;\ \nu_1\} \tag{15.14}$$

where: $\hat{R}_{(t)}$ = reliability obtained during test of length t hours
f = number of failures observed
n = either number of items involved during the test to a single failure each or number of tests (or missions) performed by a single equipment
F = upper α percentage point of the F distribution with the corresponding degrees of freedom (see Table A.7)

Table 15.4 Nonparametric Test

n	Per cent reliability at 90% confidence for f equals											
	1	2	3	4	5	6	7	8	9	10	11	12
1	0											
2	10	0										
3	19	10	0									
4	26.5	18	8	0								
5	33.5	25.5	15	8	0							
6	40	32	21	14	8	0						
7	44	37	27	20	13	16	0					
8	48	41	33	25	18	11	5	0				
9	51	45	37	30	23	16	10	4	0			
10	54	48	41	34	28	21	15	18.5	3.5	0		
11	56	50	44	38	32	25.5	19	12	8	4	0	
12	58	53	47	41	35.5	29.5	22	16	11	8	3	0
15	64.5	59.5	54.5	45.5	44	39	33	27	22.5	18.5	14.5	10
18	69	65	60.5	56	51	47	42	37	32.5	28.5	24	19.5
20	72	68	64	60	55.5	51	46.5	42.5	38	34	30	26
24	76	73	69	65.5	62	58	55	51	47	43.5	40	36
28	79	76	72.5	69.5	66.5	63	60	57	53.5	50	47	44
30	80	77	74	71	68	65	62	59	56	53	50	47
35	82	79.5	76.7	74	71.3	68.5	66	64	60.6	58	55.1	52.5
40	84	81.5	79	76.6	74	71.5	69	66.7	64.5	62	59.5	57
50	87	85	83	81	79	77	75	73	71	69	67	65
60	89	87.5	86	84	82.2	80.6	78.8	77	75.5	73.6	72	70
70	90.6	89.2	87.8	86.1	84.5	83	81.4	79.8	78.5	77	75.5	74
80	91.6	90.4	89.1	87.7	86.2	84.9	83.5	82	81.9	79.5	78.2	76.9
90	92.3	91.2	90.1	89	87.8	86.5	85.3	84.1	83	81.8	80.6	79.5
100	93	92	91	90	89	88	87	86	85	84	83	82

Table 15.5 Nonparametric Test

n	Per cent reliability at 50% confidence for f equals											
	1	2	3	4	5	6	7	8	9	10	11	12
1	0											
2	38											
3	51.5	33.5										
4	60.5	44										
5	66	52	36									
6	70.5	58	44	31.5								
7	74	62.5	50	39	29							
8	76.6	66.5	55	45	36							
9	79	69.5	59	50.5	41	31.5						
10	80.8	72	63	55	46	37	29					
11	82.2	74	66	59	50	42	34					
12	83.5	76	68	62	54	46	38.2	30.2				
15	86.5	80	74	69	62	56	50	43	38	31.1		
18	88.5	83	78	73.6	68	63	57.5	52.5	47	41.6		
20	89.8	84.2	80	76	71	66.6	61.5	57.2	47	43	39	
22	90.8	85.6	81.5	78	73.6	69.5	65	61	57	52	47.5	44.2
24	91.5	86.6	83	79.5	75.6	72	67.7	64	60.2	56	52	48.5
26	92.1	87.4	84	80.8	77.3	74	70	66.8	63	59	55.1	52.2
30	93.2	89	86	83	80.2	77.2	74	71	67.5	64	61	58.5
35	94.2	90.2	87.7	85.3	83	80.5	77.5	75	72	69	66.5	64
40	95	91.3	89	87	85.2	83	80.2	78	75.2	73	70.3	68.5
50	96	93	91.2	89.5	88	86	84	82	80	78	76	74
60	96.8	94.2	93	91	90	88	86.5	84.8	83.2	81.5	80	78
70	97.2	95.2	94	92.3	91.2	89.6	88.2	87	85.5	84	82.6	81
80	97.7	96	94.8	93.3	92.4	91	89.7	88.5	87.3	86	84.8	83.3
90	97.8	96.5	95.5	94.2	93.2	92	91	90	88.8	87.8	86.5	85.1
100	98	97	96	95	94	93	92	91	90	89	88	87
200	99	98	97	96	95	94	93	92	91	90	89	88

The following statement can be made concerning this estimate of reliability: There is a probability of $1 - \alpha$ that the true reliability for t hours is equal to or greater than $\hat{R}_{(t)}$. It must be noted that the reliability is obtained directly from sample size and failure events without consideration of unknown parameters such as the MTBF or failure rate.

To use the F table for solving Eq. (15.14), the following convenient transformations are employed.

$$\text{Confidence} = 1 - \alpha \qquad \text{upper } \alpha \text{ percentage point} \tag{15.15}$$

$$\nu_2 = 2f + 2 \tag{15.16}$$

$$\nu_1 = 2n - 2f \tag{15.17}$$

For example, suppose an equipment is operated for 20 repeated identical missions with only 4 failures being observed. Apply a nonparametric test to establish an estimate for the equipment reliability in this service to a 90 per cent confidence level. From Eq. (15.14),

$$\begin{aligned} \hat{R}_t &= 1/[1 + \tfrac{5}{16}F(10\%)(10)(32)] \\ &= 1/[1 + (0.316)(2.15)] \\ &= 60\% \end{aligned} \tag{15.18}$$

This answer can be interpreted that you have a 90 per cent confidence that the reliability is greater than 60 per cent.

To expedite the transformations and eliminate the use of a variety of tables for this nonparametric test, the author has prepared a simplified chart, Table 15.4, for use at the 90 per cent confidence level and Table 15.5 for use at the 50 per cent confidence level. To use these charts, enter the number of failures f at the top, enter the number of items or test intervals n at the left, and read the reliability lower limit from the chart.

For example, suppose an equipment was operated for 60 repeated identical events and 3 failures were observed. To find the nonparametric reliability at 90 per cent confidence enter Table 15.4 at $n = 60$, $f = 3$ and read 86 per cent reliability. Or on Table 15.5 enter at $n = 60$, $f = 3$ and read 93 per cent. These readings can be interpreted that you have a 90 per cent confidence that the reliability is greater than 86 per cent and a 50 per cent confidence that it is greater than 93 per cent.

15.3.5 Warning. By definition, the nonparametric reliability is not a figure of merit which can be extrapolated in the time domain. It applies to the test time interval only. In order to be able to extrapolate this reliability, parametric considerations must be added. The most common of these is the assumption that the exponential failure law applies. The author emphasizes that the results of this assumption can be serious, particularly for the producer if there is not engineering proof that the assumption is valid. A major function of reliability engineering is thus to make sure that the exponential does apply before reliability measurements are extrapolated into predictions.

15.4 PART RELIABILITY ACCEPTANCE TESTING

15.4.1 General. The reliability of parts depends on the values for two types of failure, random catastrophic (chance) and degradation failures. The same acceptance tests may provide information for both types, but the data must be analyzed differently. The following sections discuss the two types separately and then show how combined tests can be planned. The following sections also deal with the problems associated with the sample size available for test and whether or not the test is destructive.

15.4.2 The Ideal Reliability Test for Parts. The ideal reliability test for parts is based on the following conditions:

1. The test lot is large.
2. The test lot is expendable and can be tested to failure.
3. The test environmental conditions exactly simulate the final end-use conditions.
4. No time limits are imposed on the test.
5. Complete automatic measurement facilities are available for continuous monitoring of the major part parameters for each part.
6. The complete life history of each item is obtained and analyzed.

From a test of this nature the following reliability information can be established with a very high degree of confidence:

1. The existence of any early life debugging or burn-in period, its duration and severity being accurately measured.
2. The presence of any "flat" portion of a life characteristic curve during which the failure rate for the lot is nearly constant and the failure events are random in time. The value of this random failure rate would be accurately measured.
3. The individual degradation curves for each of the items would be accurately determined so that final wear-out time is measured to specified acceptance limits.
4. The mean time to wear-out for the lot and the distribution about this mean would be measured.

This type of test was used to evaluate the parts used in the American undersea telephone repeater amplifiers. A large percentage of each lot of parts used for this service was expended in comprehensive parts tests. Most modern projects do not allow the luxury of this ideal test. Frequently, the test lot is small, the test time is limited, the parts are not expendable, and the actual conditions of use cannot be simulated completely for the test. In spite of these obstacles, the same types of information are needed. The problem is to establish simplified tests to provide approximate answers with a known confidence.

15.4.3 Determining Debugging Period. To determine the length of the debugging period, run as large a sample as possible for several hundred hours under conditions as close to maximum rating as possible. Record the failure rate per short interval of time over the entire test period. If debugging or burn-in exists, the failure rate will decrease and level off at some low rate. Subsequent reliability testing should begin after the failure rate becomes constant. This may be from zero to 500 hours, but it is usually less than 150 hours for modern, good parts. Many suppliers age their parts so that all outgoing lots are stabilized at time of shipment. No infant mortality can be detected on these parts, and reliability testing can begin immediately.

15.4.4 Measuring Random Chance Failures. The chance failure rate is accelerated by electrical and environmental stress. Therefore, the test conditions to achieve the fastest and most accurate results are the maximum rating for each part. Short tests to maximum ratings are not generally considered to be destructive tests.

Test as large a sample as possible and keep a running record of the failure rate to ascertain that the wear-out period is not being reached during the test. The wear-out period is defined in the AGREE report as beginning when the failure rate reaches twice the constant value. Large short-time fluctuations in the failure rate are logical and to be expected because of the random nature of the failures. These are smoothed to obtain the average failure rate.

Determine the sample size required based on the length of time provided for the test and the confidence required in the failure rate measured. The greatest-likelihood point estimate can be determined as follows:

$$\text{Failure rate}_{60} = (f/Nt) \times 10^{-5} \qquad \%/1{,}000 \text{ hours} \tag{15.19}$$

where N is the number of parts in the test sample, t is the length of the test, in hours, and f is the number of failures observed. (*Note.* For most practical tests the value of f is so small compared to the sample size N that the negligible error resulting from a reducing sample size as parts fail can be ignored. Do not attempt to keep the sample size constant by replacing failed parts. To do so may introduce a more serious error resulting from nonuniform aging of different parts in the test sample and can introduce failure characteristics from other populations of parts.)

Equation (15.19) provides a point estimate at about a 60 per cent confidence level. If this level is adequate, the test plan can be established as follows: Suppose the requirement is to plan a chance-failure-rate test to measure at approximately 60 per cent confidence the λ_p for a lot of 1,000 parts. How long will the test take if the failure rate is no better than 0.1%/1,000 hours? From Eq. (15.19):

$$\bar{\lambda}_{60} = (f/Nt) \times 10^{-5}$$

If $f = 1$ $\quad t = f/(N\lambda \times 10^5) = 1/(1{,}000)(0.1 \times 10^5) = 1{,}000$ hours
If $f = 2$ $\quad t = 2/0.01 = 2{,}000$ hours
If $f = 3$ $\quad t = 3{,}000$ hours
Etc.

The product Nt is an important factor in this equation, and for a quick look at possible test plans for various sample sizes and failure numbers the author has developed Tables 15.6*a* and 15.6*b*. These data can be illustrated as shown on the family of curves in Fig. 15.4. From these curves any intermediate value of λ_p at approximately 60 per cent confidence for various values of f and Nt can be found. From this graph or from the values shown in the Table 15.6*a* it can be seen that, to demonstrate a failure rate of 0.001%/1,000 hours, an Nt value of 100 million is required. In other words, if an acceptance test to demonstrate reliability to this level in a 5,000-hour test period were desired, a total sample size N of 20,000 pieces would be required and no more than one failure could occur during the test. Some relief from this problem can be achieved by demonstrating the 0.001%/1,000 hour failure rate to a lower confidence level.

Table 15.6a Failure Rate vs. Test-item-hours *Nt* Data

Test-item-hours *Nt*	Failure rate λ, %/1,000 hours (greatest-likelihood estimate), at approximately 60% confidence for *f* failures							
	1	2	3	4	5	6	7	8
10K	10	20	30	40	50	60	70	80
20K	5	10	15	20	25	30	35	40
30K	3.33	6.66	10	13.3	16.6	20	23.3	26.6
50K	2	4	6	8	10	12	14	16
100K	1	2	3	4	5	6	7	8
200K	0.5	1	1.5	2.0	2.5	3.0	3.5	4.0
300K	0.33	0.66	1.0	1.33	1.66	2.0	2.33	2.66
500K	0.2	0.4	0.6	0.8	1.0	1.2	1.4	1.6
1,000K	0.1	0.2	0.3	0.4	0.5	0.6	0.7	0.8
2,000K	0.05	0.1	0.15	0.20	0.25	0.30	0.35	0.40
3,000K	0.033	0.06	0.10	0.13	0.16	0.20	0.23	0.26
5,000K	0.02	0.04	0.06	0.08	0.10	0.12	0.14	0.16
10,000K	0.01	0.02	0.03	0.04	0.05	0.06	0.07	0.08
20,000K	0.005	0.01	0.015	0.020	0.025	0.030	0.035	0.040
30,000K	0.003	0.006	0.010	0.013	0.016	0.020	0.023	0.026
50,000K	0.002	0.004	0.006	0.008	0.010	0.012	0.014	0.016
100,000K	0.001	0.002	0.003	0.004	0.005	0.006	0.007	0.008

Table 15.6b Failure Rate vs. Test-item-hours *Nt* Data

Test-item-hours *Nt*	Failure rate λ, %/1,000 hours (greatest-likelihood estimate), at approximately 60% confidence for *f* failures						
	9	10	11	12	13	14	15
10K	90	100	110	120	130	140	150
20K	45	50	55	60	65	70	75
30K	30	33.3	36.6	40	43.3	46.6	50
50K	18	20	22	24	26	28	30
100K	9	10	11	12	13	14	15
200K	4.5	5.0	5.5	6.0	6.5	7.0	7.5
300K	3.0	3.33	3.66	4.0	4.33	4.66	5.0
500K	1.8	2.0	2.2	2.4	2.6	2.8	3.0
1,000K	0.9	1.0	1.1	1.2	1.3	1.4	1.5
2,000K	0.45	0.50	0.55	0.60	0.65	0.70	0.75
3,000K	0.30	0.33	0.36	0.40	0.43	0.46	0.5
5,000K	0.18	0.20	0.22	0.24	0.26	0.28	0.30
10,000K	0.09	0.1	0.11	0.12	0.13	0.14	0.15
20,000K	0.045	0.050	0.055	0.060	0.065	0.070	0.075
30,000K	0.030	0.033	0.036	0.040	0.043	0.046	0.050
50,000K	0.018	0.020	0.022	0.024	0.026	0.028	0.030
100,000K	0.009	0.01	0.011	0.012	0.013	0.014	0.015

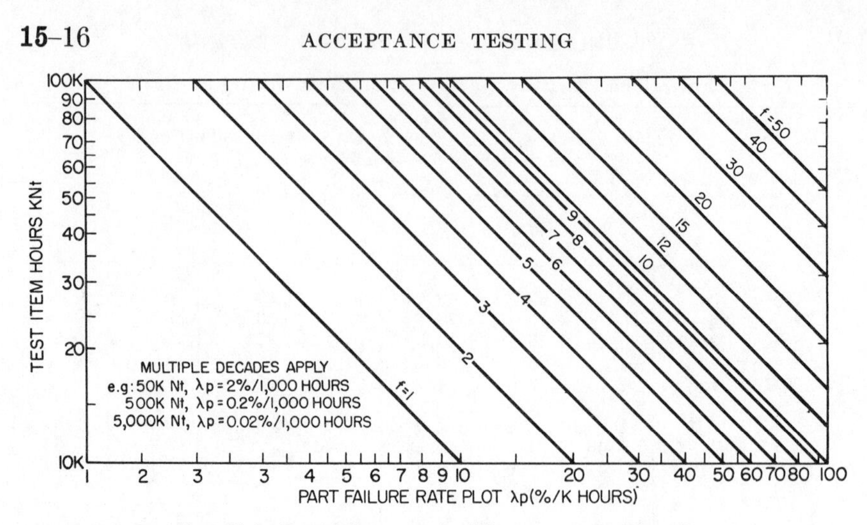

Fig. 15.4 Part failure rate λ_p (60 per cent confidence).

By using the reciprocal of the mean life found by using the multipliers of Table 15.3 or by direct derivation from the χ^2 table using the one-sided confidence equation (15.10), it is possible to show the effect of confidence level on sample size as in Fig. 15.5. This shows that for a test performed to the first failure with an item-test-hours Nt of 100,000 the results can be interpreted as follows:

λ_p of 3%/1,000 hours with confidence of 95%
λ_p of 1%/1,000 hours with confidence of 60%
λ_p of 0.1%/1,000 hours with confidence of 10%

A similar chart, Fig. 15.6, shows the closeness of the curves when 10 failures are observed. Figure 15.7 shows the part failure rates for various numbers of failures from 1 to 15 at a 90 per cent confidence. Figure 15.8 gives a similar display for various numbers of failures from 1 to 15 at a 10 per cent confidence.

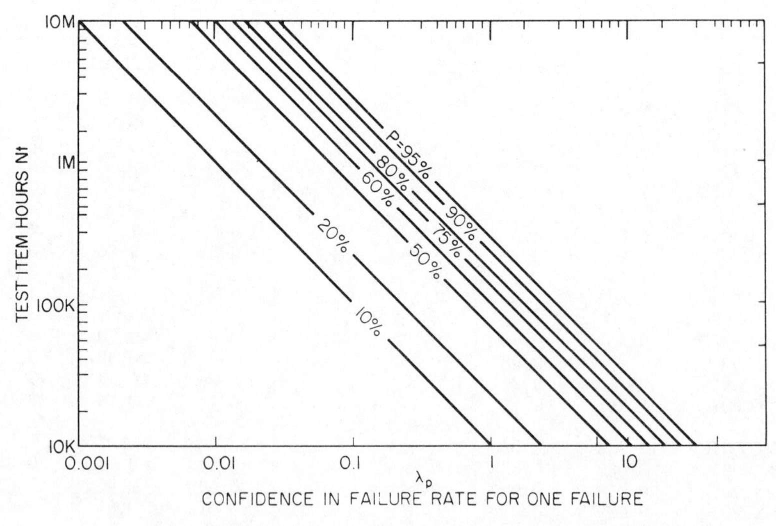

Fig. 15.5 Failure rate of parts, %/1,000 hours, for one failure.

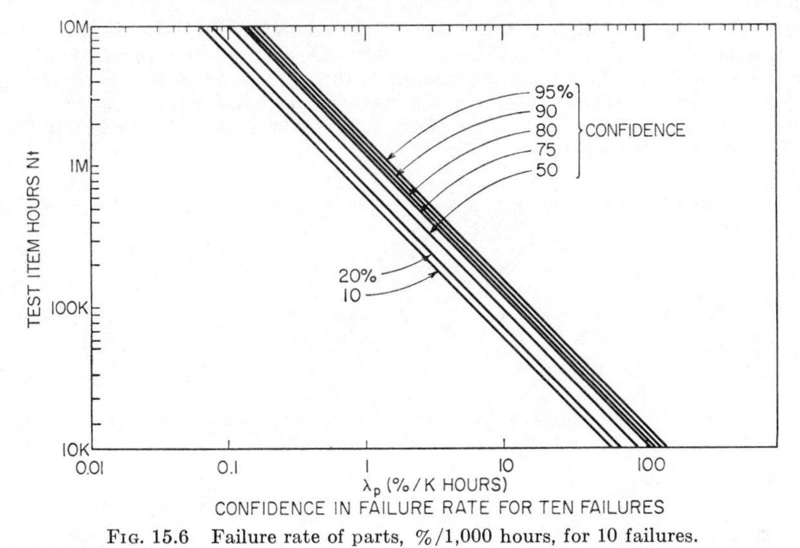

FIG. 15.6 Failure rate of parts, %/1,000 hours, for 10 failures.

From these curves it is obvious that the very low failure rates required of parts in modern complex systems cannot be demonstrated in short acceptance tests. The alternative usually used is to substitute an accelerated test on a moderate-sized sample to confirm uniformity of high-quality product which has previously proved reliable or to achieve an accelerated failure rate. Another alternative now being used as the basis for such specifications as the new 38,000 series (USAF) is to base the total product failure rate on cumulative data obtained over a period of time from tests on many lots. Still another trick being used in these new specifications is to base the confidence determination on a 10 per cent producer's risk. This is the same as saying that the part manufacturer has a specified confidence that only 10 per cent of his product will fail the acceptance test. This type of control is favorable to the parts manufacturer but gives little protection to the equipment manufacturer or to the user.

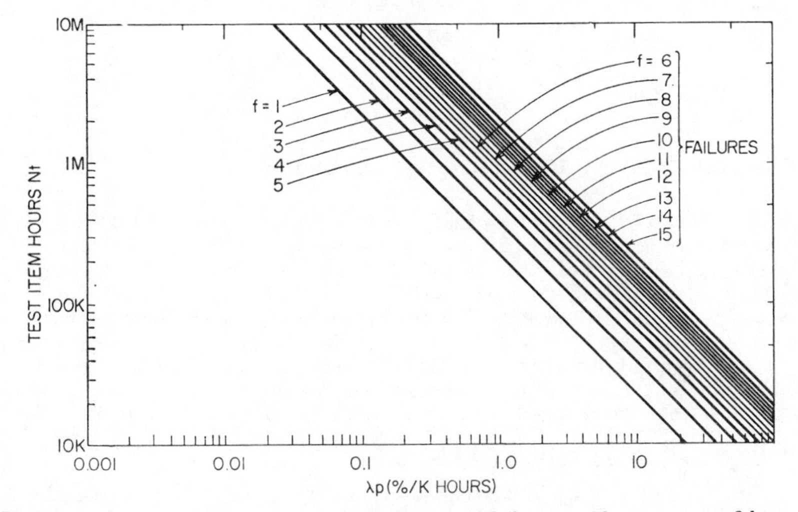

FIG. 15.7 Part failure rates for various numbers of failures at 90 per cent confidence.

A table for failure-rate tests for lot acceptance taken from MIL-R-38100A (USAF) is reproduced as Table 15.7. This table shows the failure-rate-level symbol (at 60 per cent confidence) for the failure-rate grades from 1.0%/1,000 hours to 0.0001%/1,000 hours and the sample size required to be tested for 1,000 hours. The acceptance number for all these tests is 1. Therefore, if two failures occur, the lot is reject at a 60 per cent confidence level.

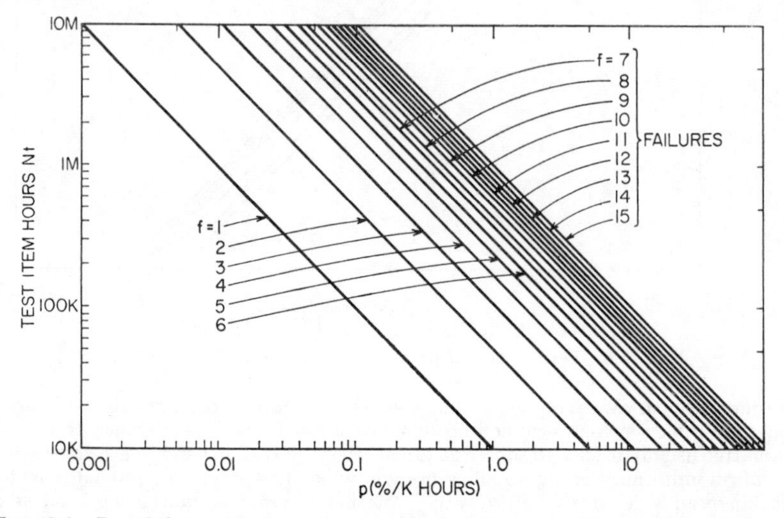

Fig. 15.8 Part failure rates for various numbers of failures at 10 per cent confidence.

Table 15.7 Sampling for Lot-by-Lot Acceptance Test*

Failure-rate-level symbol	Failure rate, %/1,000 hours	Sample size
M	1.0	110
P	0.1	150
R	0.01	300
S	0.001	750
T	0.0001	1,500

* From MIL-R-38100A (USAF), acceptance number $C = 1$.

From Table 15.7 it can be seen that if 1,500 parts are tested for 1,000 hours and only one failure occurs, the part is acceptable as a level-*T* part having an assigned failure rate per 1,000 hours of 0.0001%. The weakness in this approach is clear when the consumer's most likely estimate of reliability is calculated from these same figures. The total $Nt = 1.5 \times 10^6$ hours. For one failure the approximate 60 per cent confidence failure rate is 0.066%/1,000 hours. If it is assumed that a second failure might have occurred just after the 1,000-hour test was completed, the demonstrated reliability (approximately 60 per cent confidence) would be 0.13%/1,000 hours and not the 0.0001%/1,000 hours allowed by the MIL-R-38100A specification.

15.4.5 Accelerated Tests for Acceptance

A. Derating Factor Tests. One common type of accelerated test stresses the test sample to the maximum ratings for the part. Acceleration factors are then applied to achieve a probable failure rate which would have been applicable at considerably derated conditions. For example, paper capacitors commonly exhibit a fifth-power acceleration factor with voltage. Most other parts exhibit close to a third-power acceleration factor. A standard third power is frequently used for acceptance tests. For example, suppose a test is performed to demonstrate a failure rate of 1.0%/1,000 hours while operated at full rated voltage. This could be interpreted as the equivalent of 0.008%/1,000 hours at 20 per cent of the full voltage rating. This is calculated as follows

$$\text{Derated failure rate } d = \frac{\text{full rating}}{(\text{rated voltage/derated voltage})^3} \tag{15.20}$$

$$d = \frac{1.0\%/K \text{ hours}}{(V_R/0.2V_R)^3} = \frac{1.0}{5^3} = 0.008\%/K \text{ hours}$$

B. Step-Stress Tests. A second important form of accelerated test is known as the step-stress test. This type reveals the uniformity and strength of a product but does not normally yield failure-rate data. The step-stress test repeatedly employs

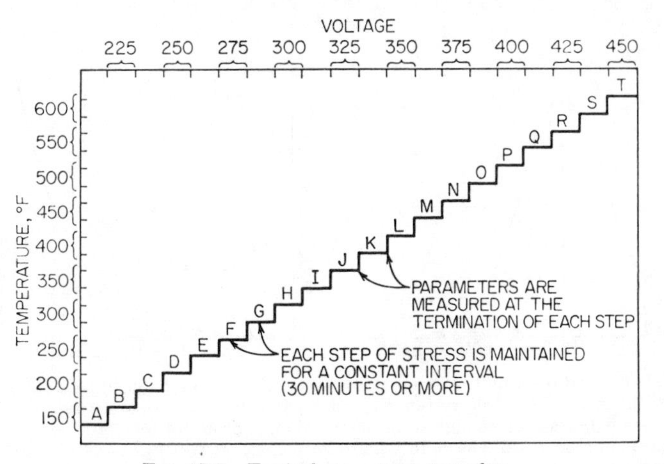

FIG. 15.9 Typical step-stress test plan.

increased stresses according to a prearranged test plan. One or more types of stress such as temperature and voltage can be combined in this test with increments of time. After testing at each step or level of stress for the prescribed interval of time, the parameters are measured and the number of rejects or failures is determined. The increased stressing is continued according to the plan until the entire sample has failed.

A typical test plan is illustrated in Fig. 15.9, and typical resulting data are plotted in Fig. 15.10. The resultant data plot forms a "fingerprint" of the strength characteristics of the parts. Any changes or differences in the materials, processes, or design are quickly revealed by the changes they promote in the step-stress data plot or fingerprint.

The SS test is relatively quick and does not require a large sample to be expended. A sample size of from 20 to 40 is adequate. Many important decisions have been made on the basis of 20 items tested. Above 40 items little additional information is to be expected.

The conditions of environment and electrical stress to be imposed at each step are planned to start at near the maximum rating for the item being tested and be increased regularly according to the plan until 100 per cent failure of the sample results. The failure data are then smoothed and plotted as a density distribution to reveal the step-stress fingerprint.

The two plots for lots I and II in Fig. 15.10 reveal the strength and uniformity characteristics for the two lots. This illustrates using the SS fingerprint to compare parts of the same type from different manufacturers or parts from different lots of the same type from the same manufacturer. Another valuable use of the SS fingerprint is to measure the extent to which supposedly nondestructive tests cause part degradation. In this service a sample from a large lot is SS-fingerprinted at the start of 100 per cent screening tests. A second sample from the same lot is SS-fingerprinted at the end of the screening tests. Any degradation caused by the screening tests will cause a shift in the before and after fingerprints. This type of testing is sometimes called an SS-fingerprint delta test or, simply, SS delta test.

A second form of analysis of step-stress test data provides information about the number of failure mechanisms involved and the stress levels at which these mechanisms come into action. This information is helpful for interpreting the SS fingerprint and also for designing long accelerated life tests. Accelerated tests for producing valid failure-rate data must be performed at stress levels which accelerate only the failure mode normally encountered at conventional operating conditions. In other words,

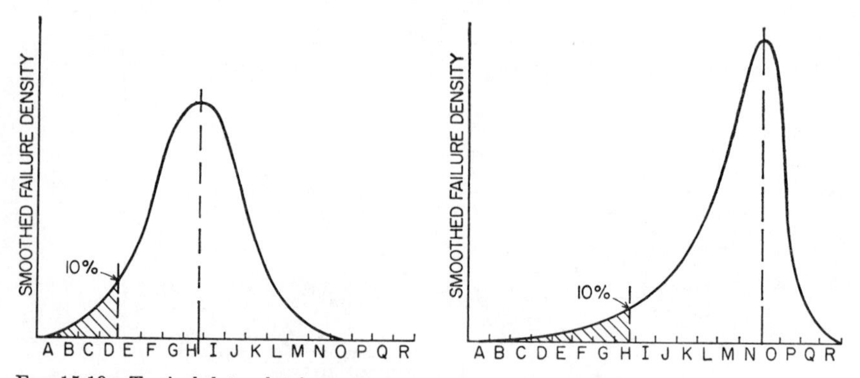

FIG. 15.10 Typical data plot for step-stress test. (*Left*) Lot I (30 units): (1) 10 per cent units failed at low stress (below step *E*); conclusion, low stress safety factor. (2) Wide dispersion; some units good to high steps; conclusion, poor uniformity of product. (3) Peak of distribution at step *H* (lot characteristic fingerprint). (*Right*) Lot II (30 units): (1) Step *H* was reached before 10 per cent failed; conclusion, superior stress safety factor. (2) Narrow dispersion; conclusion, more uniform product. (3) Peak of distribution at step *O* (lot characteristic fingerprint).

if an accelerated test employs too high stress so that new failure modes are introduced, then the accelerated failure data may have no correlation with failure rates to be expected under normal stress conditions.

It should be pointed out that the shape of the SS fingerprint is a powerful analytical tool, because the stress conditions employed are deliberately chosen severe enough to excite more than one level of failure mechanism. This is why it was stated earlier that the step-stress test should not be used to obtain failure-rate data. Rather, it analyzes the strength and character of the materials used in the product and the uniformity of the production process. Any change in the materials or process can result in a different SS fingerprint which is immediately detectable. Although these changes may not affect the long-life reliability, they definitely void any prior qualification obtained for the part by conventional tests. A change in the fingerprint indicates the need for a requalification of the part to more conventional stress conditions.

C. Failure-mechanism Detection. Whenever stress conditions rise to a level which causes chemical decomposition or physical changes in the materials involved, different failure mechanisms come into play. For example, if the test temperature is raised to a level causing chemical decomposition of a dielectric material, then the failure rate will suddenly increase because of the new material characteristics involved. This sudden change of slope of the failure rate can be detected visually if the step-stress failure data are plotted on a sheet of arithmetic-probability graph paper. The cumulative per cent failures are generally plotted on the probability scale against stress on the linear scale. If the curve joining the points is a straight line, it is assumed that only one failure mechanism is present and that the failure rate observed is a function of the stress applied. If the line joining the points is discontinuous, a new failure mechanism has been introduced at the stress level where the curve changes slope.

Frequently, failure-mechanism tests are performed for each critical type of stress independently and the steps are made small so that the plotted data yield a smooth curve. A typical example is given in Fig. 15.11. The straight line from A to B indicates one failure mechanism. The break at B indicates the start of a new failure mechanism such as charring of insulation or breakdown from material ionization. Tests at high voltage frequently cause a mechanism of failure resulting from corona discharge. No internal corona occurs until a certain critical voltage is reached. At this point internal degradation occurs and changes the slope of the data plot. A valuable conclusion which can be drawn from the data plotted in Fig. 15.11 is that

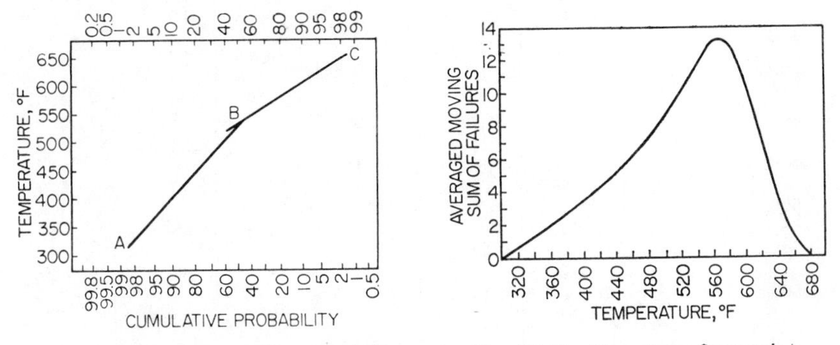

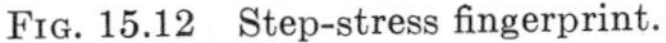

Fig. 15.11 Failure-mechanism probability plot.

Fig. 15.12 Step-stress fingerprint.

temperature-accelerated tests for these parts will yield valid failure-rate information only if the test temperature is not allowed to rise above 530°C.

The same data can be plotted on a linear scale to reveal the fingerprint as shown in Fig. 15.12. Table 15.8 illustrates a simple treatment of the data in preparation for plotting. The number of failures f observed at each test interval is listed in column 2 opposite the temperature at each step listed in column 1. The data can be smoothed in any desired fashion. In this particular case a 65° moving sum was obtained and listed in column 3. Since the plot is to reveal shape only and not absolute values, it is not necessary to obtain a moving average. However, the user is cautioned to use the same smoothing technique for all SS fingerprints which are to be compared. Further simple smoothing is provided in Table 15.8 by listing in column 4 the averaged intervals from column 3. The data in columns 1 and 4 are plotted on Fig. 15.12.

The failure-mechanism-plot data for Fig. 15.11 are treated as in the last four columns of Table 15.8. Column 5 sums the cumulative failures from column 2. The intervals are averaged and listed in column 6. The percentage of the total sample of 30 items represented by each entry in column 6 is listed in column 7. The intermediate temperature points corresponding to the smoothed data points are obtained from column 1 and entered in column 8. The data in columns 5 and 8 are plotted in Fig. 15.11.

In a majority of cases the discontinuity on the failure-mechanism plot (Fig. 15.11) will indicate an increased failure rate at the introduction of the new mechanism.

Table 15.8 Data Chart

Temp, °F (1)	Observed failures f (2)	65° moving sum $\Sigma 65°$ (3)	Averaged intervals (4)*	Cumul. failures Σf (5)	Averaged intervals (6)	Per cent of sample (30) (7)†	Temp, °C (8)
320	0	...	0.5	0			
		1	....	...	0.5	1.66	330
340	1	...	1.5	1			
		2	....	...	1.0	3.3	350
360	0	...	2.5	1			
		3	....	...	1.5	5	370
380	1	...	3	2			
		3	....	...	2.5	8.3	390
400	1	...	3.5	3			
		4	....	...	3.5	11.6	410
420	1	...	4.5	4			
		5	....	...	4.5	15	430
440	1	...	5.5	5			
		6	....	...	6	20	450
460	2	...	6.5	7			
		7	....	...	8	26.6	470
480	2	...	7.5	9			
		8	....	...	10	33.3	490
500	2	...	8.5	11			
		9	....	...	12	40	510
520	2	...	9.5	13			
		11	....	...	14.5	48	530
540	3	...	12	16			
		13	....	...	18	60	550
560	4	...	13.5	20			
		14	....	...	22	73.3	570
580	4	...	13	24			
		12	....	...	25.5	85	590
600	3	...	10.5	27			
		9	....	...	27.5	91.5	610
620	1	...	7	28			
		5	....	...	28.5	95	630
640	1	...	3.5	29			
		2	....	...	29.5	98	650
660	0	...	1.0				

* Fingerprint linear plot (Fig. 15.8)
† Mechanism of failure probability plot (Fig. 15.11)

However, this is not always so. Some conditions will indicate a discontinuity caused by a decreased failure rate. This can result from a situation in which the high temperature applied as an environmental stress performs a baking or annealing service. For example, the epoxy binder in a dielectric or a composition resistor may be more thoroughly cured and thus improved. Actual data which illustrate this are treated in Table 15.9 and Figs. 15.13 and 15.14.

In Fig. 15.13 the lines *AB* and *CD* represent the same failure mechanism as offset by a curing of the epoxy binder. At point *D* a new mechanism such as carbonization sets in. When this combination of failure mechanisms was brought to the attention of the part manufacturer, his first step toward product improvement was to provide a higher curing temperature. This eliminated the offset condition shown between *B* and *C* on Fig. 15.13. He later altered his chemical mixture, and this raised the temperature point of the second mechanism (*D*) to above 250°C. This resulted in a more stable and more reliable product.

Table 15.9 Data Chart

Temp, °F (1)	Observed failures f (2)	45° moving sum $\Sigma 45°$ (3)	Averaged intervals (4)*	Cumul. failures Σf (5)	Averaged (smoothed) intervals (6)	Per cent of sample (40) (7)†	Temp, °F (8)
320	0	0	....	0			
			0	...	0	0	333
340	0	0	....	0			
			0.5	...	0	0	350
360	0	1	....	0			
			1.5	...	0.5	1.25	370
380	1	2	....	1			
			3	...	1.5	3.75	390
400	1	4	....	2			
			3.5	...	3	7.5	410
420	2	3	....	4			
			5	...	4	10	430
440	0	7	....	4			
			9.5	...	6.5	16.2	450
460	5	12	....	9			
			22.5	...	12.5	31.2	470
480	7	33	....	16			
			29.5	...	31.5	53.8	490
500	11	26	....	27			
			24.5	...	31	77.4	510
520	8	23	....	35			
			18	...	37	92.5	530
540	4	13	....	39			
			9	...	39.5	98.8	550
560	1	5	....	40			
			3	...	40	100	570
580	0	1	....	40			
			0.5	...	40	100	
600	0	0	....	40			
			0				
620	0	0	....				
			0				
640	0	0	....				
			0				
660	0	0	....				

* Fingerprint linear plot (Fig. 15.14)
† Multiple mechanism of failure (Fig. 15.13)

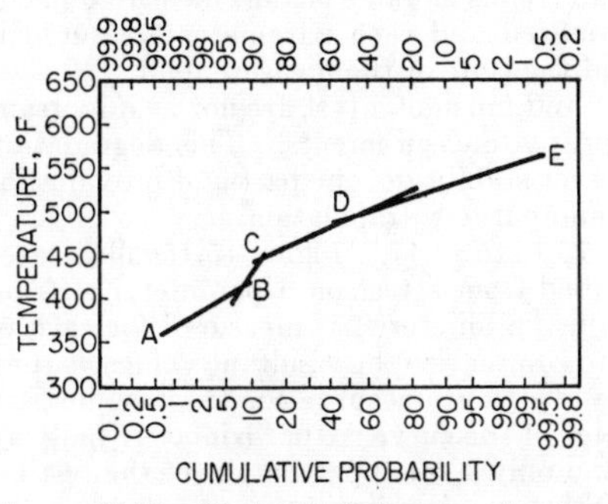

Fig. 15.13 Multiple-failure-mechanism plot.

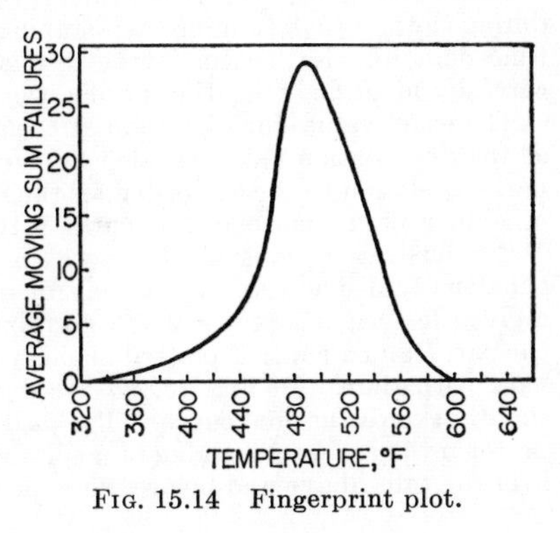

Fig. 15.14 Fingerprint plot.

15.4.6 Degradation Tests

A. General. Most of the reliability tests previously described have related to chance or random catastrophic failures. These are valuable for reliability prediction when the design safety factors are large enough to eliminate possible circuit malfunction caused by part-parameter drift. Degradation tests can be used to determine how serious is the change in value of key parameters during power-on periods.

Most parts have from one to three parameters which must not drift beyond certain limits during the reliable-life period. These limits are usually established during the circuit-design worst-case analysis. These drift limits can then be used to establish plans for degradation testing. The part degradation tests are used to evaluate the probability of degradation or drift beyond the acceptable limits during a specified mission life. These tests can be relatively short time tests to predict the ultimate failure point nondestructively.

B. Degradation Test Plan. The degradation test is usually from 1,000 to 1,500 hours long and is performed on 100 per cent of each lot. The parts are cycled power-on-off according to a normal-use schedule. The load and temperature conditions are

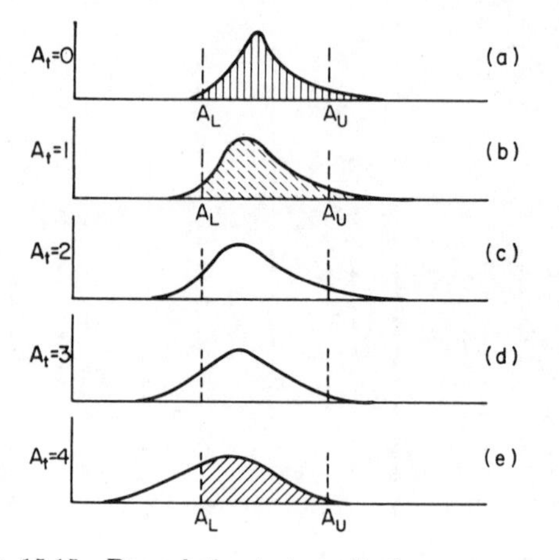

FIG. 15.15 Degradation test results for parameter A.

preferably set at the rated maximum for the parts. Less information is gained when typical average conditions only are used; however, some people prefer this milder test for fear of damaging the parts. The degradation test itself can be used to determine if the maximum conditions are too severe.

Degradation tests consist of the repeated measurement of the critical parameters during the test to determine their stability and trends of value change measured in the time domain. Each item in the test is serialized and each parameter measured is carefully identified with the specific item and the time of the measurement.

The exact values for each item at the start and finish of a test are not as important as the degradation path revealed by the repeated measurements. This degradation path in relation to the limits defines the inherent stability of the part and provides for reliability prediction based on curve fitting and curve extrapolation.

For illustration, consider the curves of Fig. 15.15 to 15.17. Figure 15.15*a* illustrates the density distribution which may be obtained from a test on a parameter A from a given lot of parts at $t = 0$. A certain major parameter A is measured for each of the parts, and a curve is plotted in this way to summarize the resultant values as they vary with time. In Fig. 15.15*a* the density distribution curve for the test lot is a slightly skewed normal curve. The peak value of the curve is approximately halfway between the acceptable-value range limits, and only a small percentage of the test lot is in the tails above and below the upper and lower reject limits A_U and A_L.

Now suppose this same test is repeated for the same parts following a burn-in reliability test. At $t = 1$ the density distribution curve may be as illustrated in Fig. 15.15*b*. The peak has shifted slightly toward the lower limit and the curve is now broader at the mode.

The broadening of the curve is a typical consequence of aging and results in a higher percentage of the test lot falling outside the acceptance limits. The same effect of

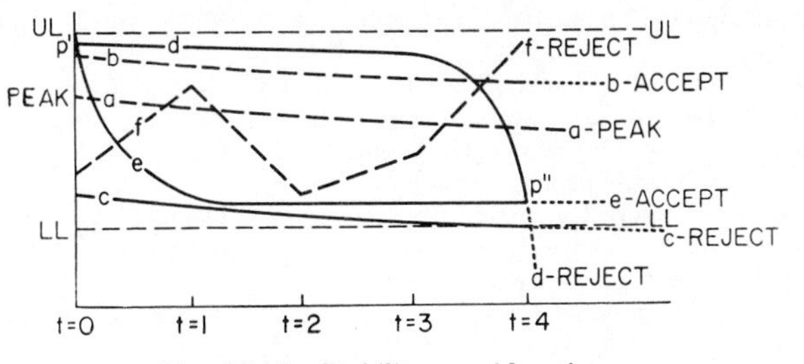

FIG. 15.16 Stability considerations.

shifting downward and broadening of the peak for subsequent measurement intervals through $t = 4$ is shown in Fig. 15.15*c*. By this time the peak of the curve is still within the acceptance band but a large percentage of the parts have shifted in value to be lower than the lower reject limit. From this illustration it can be seen that the value of the peak and the dispersion of the density distribution curve tell very little about the individual reliability of the parts in the test lot.

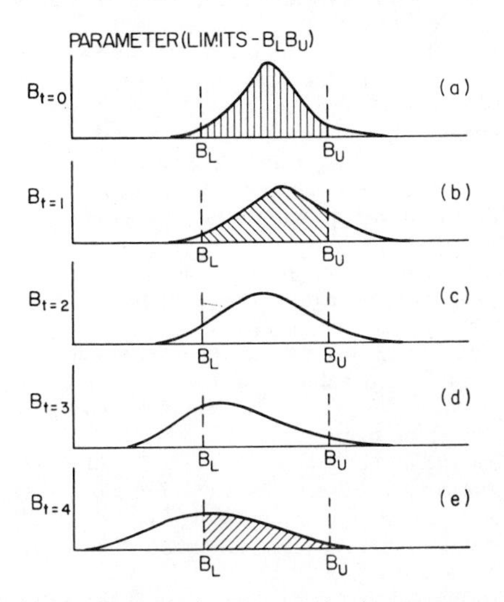

FIG. 15.17 Degradation test results for parameter B.

The test data must be recorded for each part at each test interval, and degradation curves (or their equivalent) must be plotted for each part. The significance of this can be seen by the illustrative curves in Fig. 15.16. The central dotted curve a in Fig. 15.16 is a trace of the peak value shown in Fig. 15.15 as it shifts downward during successive test intervals. This curve may or may not have reliability significance for parts depending on their value at $t = 0$. For example, both curves b and c show the same degradation trend as the peak value. However, curve b starts near the upper limit and curve *c* starts near the lower limit at $t = 0$. Thus at $t = 4$, although both curves are still within the acceptance limits, the part with curve *c* degradation should be rejected as probably unreliable.

The extrapolation of the curve fitted to the successive test points indicates that end of life for successful performance by the part with curve *c* will be reached shortly after time $t = 4$. On the other hand, the part with curve *b* can suffer many times the amount of degradation experienced by time $t = 4$ before it approaches the lower acceptance limit. This shows that the trend of the peak value does not tell the degradation story.

Consider the reliability characteristics revealed by the plots of curves *d* and *e*. Both these parts start out at time $t = 0$ at the same value P^1 and degrade by time $t = 4$ to the same value P^{11}. They thus exhibit the same total degradation. However, simple extrapolations of the degradation curves show that part *d* suffered very little degradation until near the end of the test, when it appears to be completely worn out. A short time later the extrapolated curve would pass well into the rejection region of the plot. On the other hand, curve *e* showed considerable initial degradation until a stable operating value P^{11} was reached. After that time, very little additional degradation occurred. This is a common burn-in characteristic of good parts. The indication is that an extrapolated curve *e* would show considerable time lapse before failure occurred.

The curve *f* shows a typical unstable part which cannot be depended upon for reliable performance. Even though all the measured values are within the acceptance limits throughout the test to time $t = 4$, this unit should be rejected as potentially unreliable. An extension of this same principle requires the rejection of any part whose test value goes outside the acceptable limits at any time during the reliability test.

C. Critical Parameters. All the various parameters for a part may exhibit different degradation and stability characteristics. For example, the series of density distribution curves shown in Fig. 15.17 for parameter *B* may be from the same test on the same parts as that which produced the degradation illustrated in Fig. 15.15 for parameter *A*. Here the peak value of the curve is shown to shift upward during early portions of the test before a decided downward swing sets in. For any particular parameter the reject criteria relating to fitting and extrapolating degradation curves apply as shown in Fig. 15.16. If only one parameter is critical to circuit performance, then the acceptance test can be based on the revealed degradation curve for that critical parameter.

If the exact part application is not known, an effective reliability acceptance test can be based on the principle that the inherent reliability of most present population parts can be predicated on the two most critical parameters. In other words, for most parts there are two critical tell-tale parameters which can be selected. If both these parameters are stable and exhibit no dangerous degradation trends, then the part can be accepted. If either of the two tell-tale parameters is reject, then the reliability of the part must be suspect.

D. Interpreting Degradation Test Results. Several methods are being used for interpreting degradation test results. They involve various systems, both machine and manual, for linear and nonlinear extrapolation and related assessment of the meaning of the degradation curves. The following discussion is intended to clarify several interrelated factors.

Assume all the *N* supposedly identical parts constituting a lot degrade at a rate such that one-third of the total fail each 1,000 hours after the test. This unrealistic situation is illustrated in Fig. 15.18. The slope of the trend revealed during the 1,000-hour degradation test can be extrapolated to reveal that the degradation failure rate for these parts is expected to be 33⅓%/1,000 hours.

The distribution of failures from degradation in the time domain can never be expected to be as uniform as shown in Fig. 15.8. However, the important point is that the interpretation of degradation data must consider not only the part degradation in time but the life requirements of the end-use application. For example, if the operation is to last for 3,000 hours, the chance of any of the parts illustrated in

Fig. 15.18 surviving to the end is very small. If the operation is very short, a major percentage of the parts may survive.

A more realistic situation is shown in Fig. 15.19. The degradation trends during the 1,000-hour test might be extrapolated in time to indicate that the life expectancy of 98 per cent of the lot is greater than 10,000 hours, 2 per cent of the lot can be expected to fail beyond the 5,000-hour point, and only one part is likely to fail at a time between 3,000 and 4,000 hours. If the total life use requirement for these parts is 1,000 hours, then the probability of a degradation failure occurring during the operation is very small. One way to interpret this situation is that the design limits on the acceptable

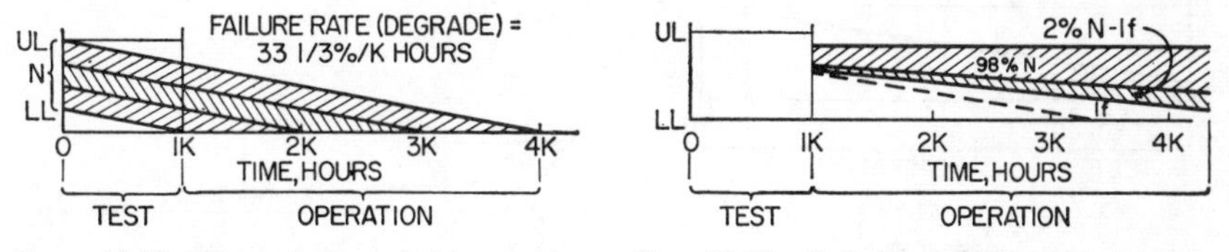

Fig. 15.18 Degradation interpretation (not practical).

Fig. 15.19 Reliable 1,000-hour operation.

lower operating value of this key parameter provide such a large safety factor for the part characteristics that degradation-type failures can be ignored. In this case reliability prediction is valid when based solely on the catastrophic or random-type failure.

An interesting and valuable use of degradation tests is to achieve a nondestructive screening of all parts plus a characteristic "fingerprint" which can be used to compare subsequent lots of supposedly identical parts. The few parts which indicate short life expectancy can be weeded out, and data extrapolated from the rest can be plotted to reveal the density distribution at end of life in the time domain. The probable life expectancies can be grouped in convenient-size cells and a curve can be plotted. Any

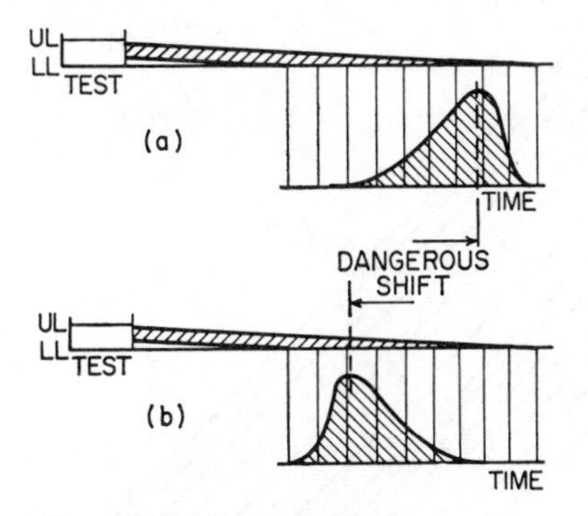

Fig. 15.20 Degradation fingerprint.

variation in test conditions or part strength can result in a different distribution. This is illustrated in Fig. 15.20. The two plots 15.20*a* and 15.20*b* are of different lots from the same supplier. The difference in distribution density should be cause for immediate investigation. The supplier may have changed his design or materials or process, and serious differences in random-failure characteristics might result. Even if the fingerprint indicates only minor change, subsequent plots should be analyzed for potentially dangerous degradation trends.

The preceding discussion has dealt only with the lower limit. Obviously, some part parameters, such as diode leakage, may generally drift upward with age. In these cases only the upper limit need be considered. In other cases both upward and

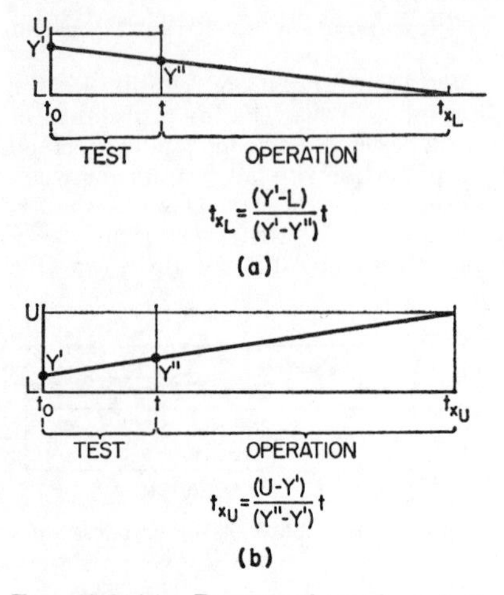

FIG. 15.21(*a*) Downward-trend analysis. (*b*) Upward-trend analysis.

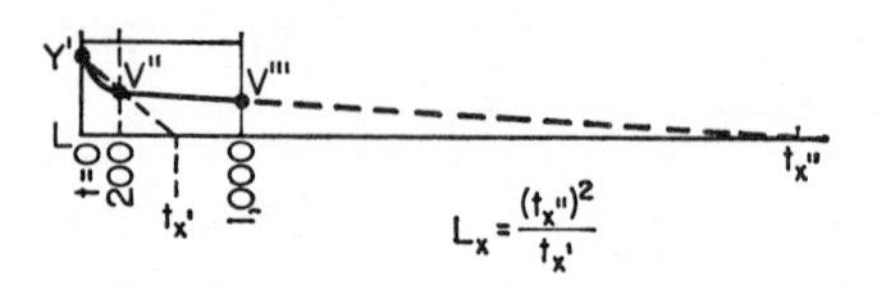

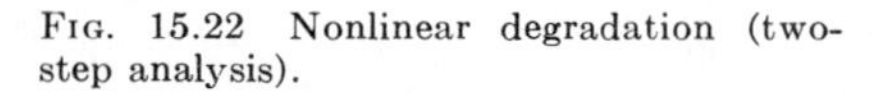

FIG. 15.22 Nonlinear degradation (two-step analysis).

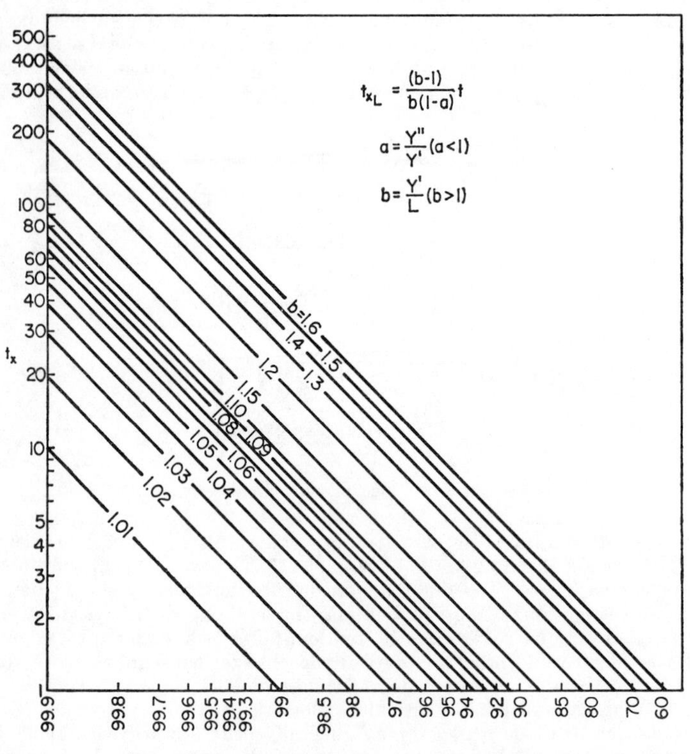

FIG. 15.23 End-of-life downward-trend plot.

downward variation are common and both upper and lower limits must be considered in the analysis.

E. Analyzing the Data. The simplifying assumption common with most degradation tests is that the trend revealed during the test period is considered to progress linearly during the subsequent operating period until an acceptable limit value is

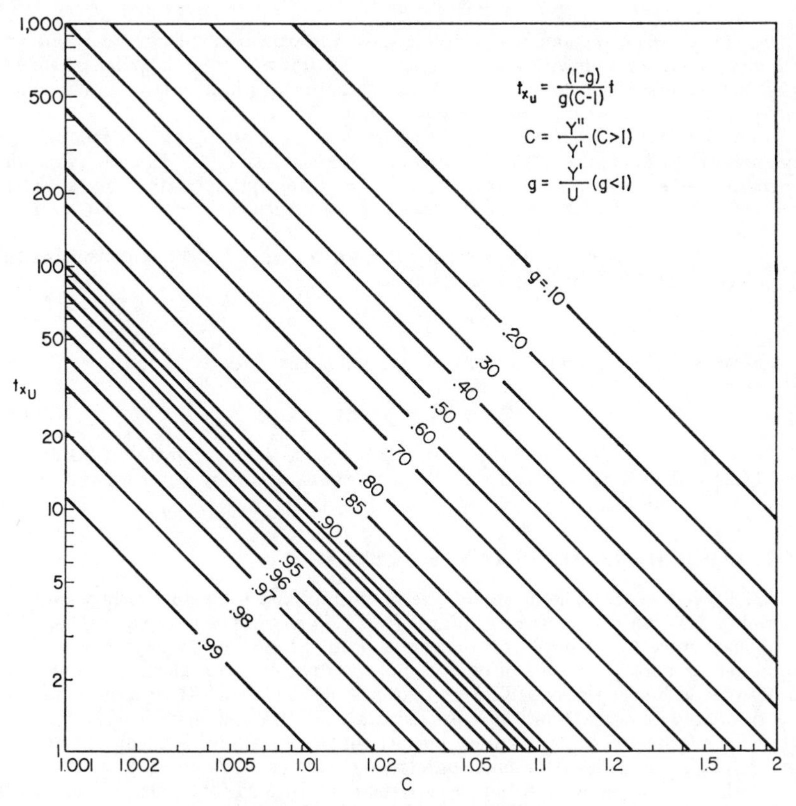

FIG. 15.24 End-of-life upward-trend plot.

exceeded. If a linear trend is observed during the test, the probable end of life for the part is found by using the relationships

$$t_{xL} = (Y' - L)t/(Y' - Y'') \qquad (15.21)$$

and

$$t_{xU} = (U - Y')t/(Y'' - Y') \qquad (15.22)$$

where t_{xL} = length of life to point that critical parameter value exceeds lower limit
t_{xU} = length of life to point that critical parameter value exceeds upper limit
t = time length, in hours, of degradation test
L = value of the lower limit
U = value of the upper limit
Y' = initial value of parameter at time zero
Y'' = value of parameter at end of test time t

This solution is illustrated in Fig. 15.21 for downward and upward trends.

If the degradation is not linear during the test, two segments are analyzed separately. One segment of the trend is analyzed during the first 200 hours of the test, and the second is analyzed during the last half or two-thirds of the test. For example, suppose the degradation occurred as shown in Fig. 15.22. Measurements are made at

t_0, t_{200}, t_{500}, and $t_{1,000}$ hours. In this case an infant degradation leveled off. If similar parts exhibit the same burn-in, the degradation test should not be started until after 250 hours of power-on. The value of the trend for this illustrated case is simply the ratio of t_x'' to t_x', and a figure of merit for the longevity measured by the degradation test is

$$\text{Longevity } L_x = (t_x'')^2/t_x' \tag{15.23}$$

This equation can be rationalized as follows: An upward curvature in a downward trend is beneficial, as is a downward curvature in an upward trend. When there is no curvature (linear degradation), the longevity figure of merit is the same as though only one interval of degradation were measured.

Equations for t_{xL} (15.21) and t_{xU} (15.22) may be programmed into a computer and the longevity L_x figure of merit computed automatically. This is particularly desirable for use with large-lot testing. However, for manual data analysis or where a computer is not available it may be easier to work with ratios of Y'' to Y' and ratios of Y' to L or U.

From Eq. (15.21) for t_{xL}, if a is substituted for the ratio Y''/Y' and b is substituted for the ratio Y'/L, we have the equation

$$t_{xL} = [(b - 1)/b(1 - a)]t \tag{15.24}$$

Also, if c is substituted for Y''/Y' and g is substituted for Y'/U in the equation for t_{xU}, we have

$$t_{xU} = [(1 - g)/g(c - 1)]t \tag{15.25}$$

The author has worked out the two plots for these values shown in the Figs. 15.23 and 15.24. The t_x can be read directly from these charts by entering at the proper a, b, c, and g values.

15.5 EQUIPMENT ACCEPTANCE TESTING

15.5.1 General. The figure of merit usually used for measuring equipment reliability is mean time between failures (MTBF), or simply m in equations. Other symbols such as θ, ϕ, and ψ are sometimes used. Reliability acceptance testing for equipment generally consists of operational tests performed under simulated end-use conditions with acceptable MTBF and confidence specified. If we assume here that the simulated end-use conditions are accurate to the test specifications, then full attention can be given to interpreting short-time test data in terms of probable long-time reliability with its attendant confidence measures. (See Secs. 8 and 10.)

If early or infant-mortality failures are removed by a suitable burn-in period and if wear-out failures are removed by suitable preventive maintenance and replacement, then only chance failures remain. The objective is to measure the chance-failure rate over a short period of time and then affix to this a probability or measure of confidence that the mean of the short-time sample is typical of a normal life period. In other words, from a test, what can be said about the value of MTBF and with what confidence can it be assumed that this same value would be obtained if averaged over the total usable life of the equipment?

First, let us clarify that the reliability figure of merit for the equipment is the MTBF and not the confidence. The confidence applies only to the interpretation of the test result. One cannot specify an equipment to have a certain MTBF and a certain confidence. One can only specify that an equipment will have a certain inherent MTBF which will be proved to a certain confidence by a specified test.

A question commonly asked is, "How much testing is required to be 90 per cent confident that the MTBF is greater than x hours?" This question cannot be answered without additional information. For example, if the actual MTBF is less than x hours, then no amount of testing can prove that it is greater than it is. If the actual MTBF is just barely over the value x, then a great deal more testing is required than if the actual MTBF is much greater than the x value.

The test measures the most likely value of MTBF, and the amount of statistical data obtained during the test must be evaluated to determine the confidence which can be placed on the measurement. When this has been done, the following statements can be made: "The best estimate of the MTBF is *B* hours; but, based on the amount of data, we can be 90 per cent sure, for example, that it is not more than an upper limit of *A* hours and 90 per cent sure that it is not less than a lower limit of *C* hours." This defines an 80 per cent double-sided confidence interval between the limits *A* and *C*. In other words, there is an 80 per cent confidence that the true value lies between the values of *A* and *C*.

Usually, for acceptance testing, the single-sided description stating the cumulative probability that a measured MTBF is greater than a certain specified minimum value has the greatest usefulness. Both these approaches are illustrated in the following specific examples.

15.5.2 Conventional Testing. Suppose we run a reliability test on an equipment for 600 hours and observe three failures. The best estimate of the MTBF is 600 divided by 3, or 200 hours. Intuition should indicate that if we ran another test of 600 hours we would not likely observe exactly 3 failures. In other words, even if the equipment actual MTBF is 200 hours, each of a whole series of 600-hour tests might, by chance, yield a different result. When the total operating time is divided by the total failures, the average figure would be 200 hours if the actual MTBF were 200 hours. However, to obtain by chance a figure of 200 hours in a short, 600-hour test is not proof of this fact.

To find the upper and lower limits of the two-sided confidence interval for this example, refer to Table 15.2. Enter this table at row $f = 3$ and find the 90 per cent confidence upper limit K_U to be 3.68 and the lower limit K_L to be 0.476. When these K values are multiplied by the best estimate of 200 hours, the upper limit (90 per cent confidence band) is found to be 736 hours and the lower limit is 95.2 hours. In other words, we can say that we have a 90 per cent confidence that the true value lies between 95 and 736 hours.

If we determine the upper and lower limits for the 60 per cent confidence band from Table 15.2 at row $f = 3$, K_U is 1.95 and K_L is 0.698. When these values are multiplied by the best estimate of 200 hours, we find that there is a 60 per cent confidence that the true value lies between 140 and 390 hours.

From these examples it is apparent that the confidence band is wide when only enough test time is devoted to allow just a few failures. For the most effective testing, at least ten failure incidents should be allowed. For example, suppose we test three equipments for a period of 1,000 hours and observe a total of 10 failures. The total number of equipment hours is thus 3,000, and the most likely estimate of reliability is 3,000 divided by 10, or 300 hours MTBF. From Table 15.2 the 90 per cent K_U for $f = 10$ is 1.84 and the K_L is 0.637. Thus there is a 90 per cent confidence that the actual MTBF lies between 552 and 191 hours. This result can also be interpreted that there is a 95 per cent confidence that the lower limit of the true MTBF is 191 hours. In other words, there is a 95 per cent confidence that the true MTBF is greater than 191 hours.

This brings to mind that it is most frequently desired to plan equipment acceptance tests to prove with a known confidence that the MTBF is greater than a certain specified figure. The lower-limit half of Table 15.2 can be used for this purpose as just described. However, a simpler method using Table 15.3 will give the same answer directly. For example, suppose we want to prove with a 95 per cent confidence that the MTBF is greater than 191 hours. If we set 10 failures as our acceptance number, how many equipment test hours are required? From Table 15.3 the w value at 95 per cent confidence for $f = 10$ is 15.70. When this constant is multiplied by the specified 191 hours, we again arrive at the figure of 3,000 equipment test hours required. This may be supplied as one equipment tested for 3,000 hours or any combination of equipment times hours totaling 3,000 (for example, 6 equipments tested for 500 hours with $f_A = 10$).

In these examples the assumption involved is that the exponential failure law applies. Some people have objected to using this type of test analysis because they are not sure that the exponential assumption is valid. This fear is generally not pertinent to the decision whether or not to use the test and Table 15.3 if the equipment passes the test. Fortunately, the test is more likely to reject the equipment if the exponential law does not apply and if the failure rates are increasing with time. In other words, from the standpoint of the user's risk the exponential law assumption is a fail-safe one in most cases. (See Sec. 3.)

From the standpoint of the producer, the opposite is true. The exponential assumption can lead to erroneously rejected equipment if the failure rate is not constant. The protective recourse for the producer is twofold. First, he can analyze and test his equipment before acceptance testing to make sure that the exponential failure law does apply and redesign or rebuild until it does, or, second, he can take the chance that it does and negotiate for longer test times of more equipments so that the confidence interval will be narrower. Under these conditions the chances are less that a few random failures will cause rejection. Under no circumstances should test plans that can cause rejection on one or two failures be used. From six to ten failures is the preferred acceptance number level.

15.5.3 Consumer's Risk Tests. The preceding discussion describes how to use the tables to analyze the results from conventional reliability tests. The author has devised some simple charts for the consumer to use in planning for reliability tests. The plot shown in Fig. 15.25 gives the equipment reliability at 90 per cent confidence

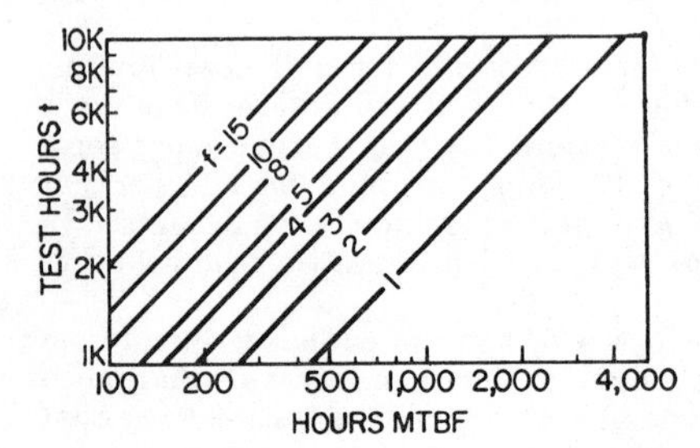

Fig. 15.25 Equipment reliability MTBF at 90 per cent confidence.

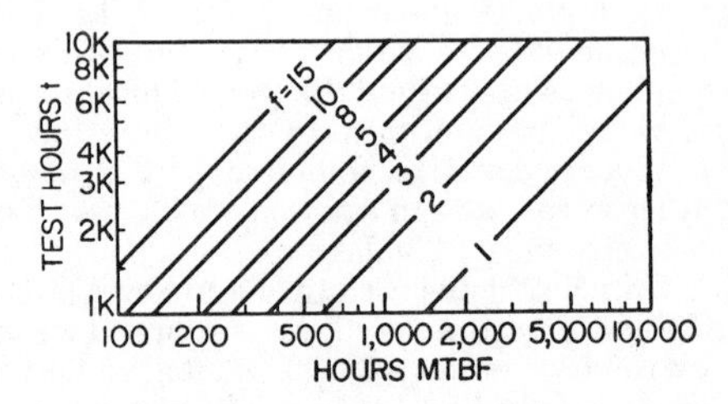

Fig. 15.26 Equipment reliability MTBF at 50 per cent confidence.

for various test hours and acceptable numbers of failure. The plot in Fig. 15.26 is a similar display but for test results at 50 per cent confidence. These two confidence levels can cover the majority of acceptance testing cases. For other levels perform the calculations as indicated previously, using Tables 15.2 and 15.3 and Fig. 15.3.

15.5.4 Sequential Acceptance Tests

A. General. Sequential testing differs from other test procedures in that the length of test is not established before the test begins but depends upon what happens during the test. The test sample is tested while subjected to a prescribed environment and duty cycle until the preassigned limitations on the risks of making wrong decisions based on the cumulative test evidence have been satisfied. The ratio of quantity of failures to the length of test at any test interval is interpreted according to a sequential-analysis test plan. Conspicuously good items are accepted quickly; conspicuously bad items are rejected quickly; and items of intermediate quality require more extensive testing.

The major advantage in using sequential test procedure is that it results in less testing on the average than other testing procedures when the preassigned limitations on the risks of making both kinds of wrong decisions are the same for both tests. The chief disadvantage is that the test time required to reach a decision cannot be determined prior to testing. Also, it might be argued that higher-level personnel are required to run the test and to interpret the test result.

The sequential test is designed to determine whether or not the MTBF (m) of a population with a constant failure rate is at least the specified number of hours M_R. The normal use of this procedure is to determine the failure rate of a population before its wear-out period, which begins at time T_w. Thus the units under test should be replaced by new units before time T_w. The sample is tested until the test results decide whether to accept or reject the lot. In this procedure two types of error are possible:

Error type 1. That the population is accepted when the m is less than M_R hours. The probability of making this error depends upon how low the true m is. The probability of accepting very bad populations is very small. The chance of accepting those with an m just under M_R hours is larger. The maximum probability of making this error is the buyer's risk β.

Error type 2. That the population is rejected when the m is greater than or equal to M_R hours. The maximum probability of this error is the vendor's or producer's risk α.

Both risks cannot be limited if a decision is to be made within a reasonable amount of time. Since M_R hours is the desired minimum MTBF, the first type of error is considered to be the most serious and is usually limited. An error similar to that of the second type, namely, rejection of a population which has an m greater than some upper acceptance limit M_A, can be limited by choosing M_A greater than M_R. As M_A is made larger, the average time required to come to a decision becomes greater. If the ratio of M_A/M_R is set at too large a figure, the advantages of sequential testing can be lost. The choice recommended by the DOD-AGREE report[1] is that M_A be 50 per cent more than M_R. This means that the ratio $M_A/M_R = 1.5$. The sequential test plan proposed by AGREE is the one most commonly used in industry. It is also based on the recommended values for both α and β of 0.10, that is, 10 per cent risk for both the producer and the consumer.

B. The AGREE Sequential Test. The AGREE sequential testing plan is based on the acceptance and rejection numbers of failures as shown in Table 15.10. The first column lists normalized test time. This is found by multiplying the number of equipments on test N by the time of test t at a specific test interval and dividing this product by the specified MTBF M_R, in hours, i.e.,

$$t_m = Nt/M_R \tag{15.26}$$

The second and third columns list the rejection and acceptance numbers of failures, respectively. The fourth column brackets the number of failures at each test interval (column 1) for which the testing must be continued. The data in this table can be plotted in the form of upper and lower acceptance limits as shown in Fig. 15.27. This plot is based on the limit equations developed by the AGREE for the specified conditions as follows:

$$t_{nA} = 0.81F_A + 4.4 \qquad \text{accept limit} \tag{15.27}$$

$$t_{nR} = 0.81F_R - 4.4 \qquad \text{reject limit} \tag{15.28}$$

It is emphasized that the sequential test, like any other test, is meaningful only if the test conditions simulate actual end-use conditions. These conditions include environmental factors such as temperature and vibration, electrical and mechanical load conditions, and operational simulation. The latter includes such factors as duty cycle and maintenance procedures. Four levels of environmental conditions were described by AGREE. Whenever the AGREE sequential test is specified, it is essential that the applicable environmental stress level also be specified. These levels will be described following this example of a typical sequential test based on Table 15.10 and Fig. 15.27.

Suppose it is desired to apply the AGREE sequential analysis to accept or reject an equipment with a specified MTBF of 200 hours.

Given. Required $M_R = 200$ hours; number of equipments to be tested $N = 2$.

Start of Test. The two equipments are installed in a test station, and the prescribed environments and operational conditions are applied. The operating effectiveness is continually monitored, and each part failure is quickly detected and repaired. The operating test time is recorded at each failure event, and the data are recorded. A typical set of data might be that given in Table 15.11.

[1] Advisory Group Reliability of Electronic Equipment OASD-(R&E), June 4, 1957.

Table 15.10 Sequential Test Accept-Reject Criteria

Normalized test time, hours	Total relevant failures for			Normalized test time, hours	Total relevant failures for		
	Reject decision	Accept decision	Continue-test decision		Reject decision	Accept decision	Continue-test decision
(1)	(2)	(3)	(4)	(1)	(2)	(3)	(4)
0.5	6	...	0–5	16.7	26	...	16–25
1.3	7	...	0–6	17.4	...	16	17–26
2.1	8	...	0–7	17.5	27	...	17–26
2.9	9	...	0–8	18.2	...	17	18–27
3.7	10	...	0–9	18.3	28	...	18–27
4.4	...	0	1–10	19.0	...	18	19–28
4.5	11	...	1–10	19.1	29	...	19–28
5.2	...	1	2–11	19.8	...	19	20–29
5.3	12	...	2–11	19.9	30	...	20–29
6.0	...	2	3–12	20.6	...	20	21–30
6.1	13	...	3–12	20.7	31	...	21–30
6.8	...	3	4–13	21.4	...	21	22–31
6.9	14	...	4–13	21.5	32	...	22–31
7.6	...	4	5–14	22.2	...	22	22–32
7.8	15	...	5–14	22.3	33	...	23–32
8.5	...	5	6–15	23.0	...	23	24–33
8.6	16	...	6–15	23.1	34	...	24–33
9.3	...	6	7–16	23.8	...	24	25–34
9.4	17	...	7–16	24.0	35	...	25–34
10.1	...	7	8–17	24.7	...	25	26–35
10.2	18	...	8–17	24.8	36	...	26–35
10.9	...	8	9–18	25.5	...	26	27–36
11.0	19	...	9–18	25.6	37	...	27–36
11.7	...	9	10–19	26.3	...	27	28–37
11.8	20	...	10–19	26.4	38	...	28–37
12.5	...	10	11–20	27.1	...	28	29–38
12.6	21	...	11–20	27.2	39	...	29–38
13.3	...	11	12–21	27.9	...	29	30–39
13.4	22	...	12–21	28.0	40	...	30–39
14.1	...	12	13–22	28.7	...	30	31–40
14.2	23	...	13–22	29.5	...	31	32–40
14.9	...	13	14–23	30.3	...	32	33–40
15.0	24	...	14–23	31.1	...	33	34–40
15.7	...	14	15–24	31.9	...	34	35–40
15.9	25	...	15–24	32.8	...	35	36–40
16.6	...	15	16–25	33.0	41	40	

Note 1. Column 1 is the total test time (operating time), in hours, accumulated by all equipments in a test group under test divided by the contract specified MTBF, in hours.

Note 2. When the total number of relevant failures accumulated by all equipments in a test group during the corresponding normalized test-time period shown in column 1 (*a*) equals or exceeds the number in column 2, a *reject* decision is made, (*b*) equals or is less than the number in column 3, an *accept* decision is made, (*c*) falls in the range shown in column 4, a *continue-test* decision is made.

Note 3. A test group may be one or more equipments. A reject or accept decision on a test group is applicable to all equipments of the group from which the test group was selected.

These data, t_n and f, are plotted in solid lines, marked case A, on the graph of Fig. 15.27. Another analysis of these same data can be made to show the comparative results which might have been obtained from conventional analysis at each failure event. The upper and lower limits of an 80 per cent confidence band are shown as derived by using K_U and K_L factors from Table 15.2. These data are given in Table 15.12.

It can be seen from these data that had any acceptance number below seven failures been assigned for use in a conventional test, the equipment would very likely have

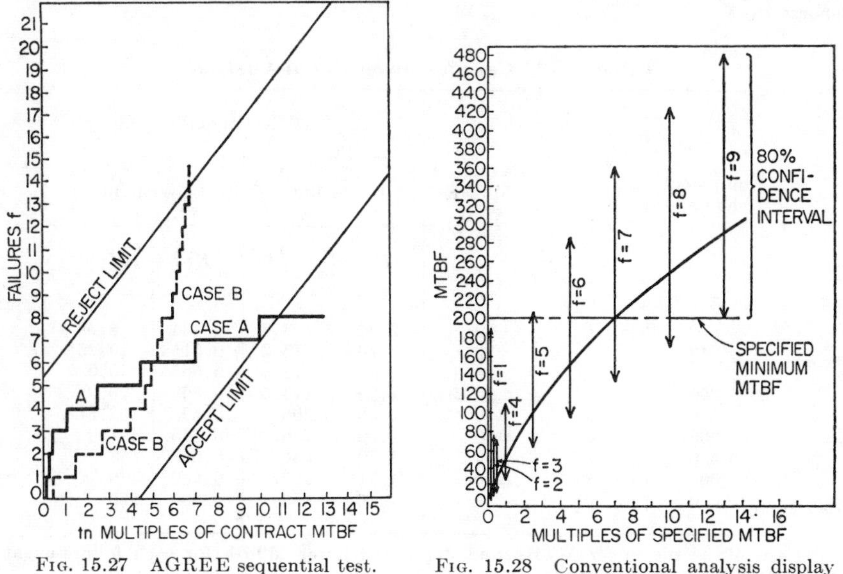

Fig. 15.27 AGREE sequential test.

Fig. 15.28 Conventional analysis display of sequential test data, case A.

been rejected; whereas, when the sequential test was employed, no decision was reached until the ninth failure, when the data give clear evidence that the equipment should be accepted. This evidence can be illustrated in another way as shown in Fig. 15.28. Here the vertical bars delineate the 80 per cent confidence interval.

Table 15.11 Case A Sequential Data

Failures f	Test time t, hours	Equipment hours Nt	Normalized test time $tn = Nt/200$	Sequential test conclusion (from Figs. 15.19 and 15.20)
1	10	20	0.1	Continue
2	20	40	0.2	Continue
3	40	80	0.4	Continue
4	100	200	1.0	Continue
5	250	500	2.5	Continue
6	450	900	4.5	Continue
7	700	1,400	7.0	Continue
8	1,000	2,000	10.0	Continue
9	1,300	2,600	13.0	Accept

Another example wherein the same conditions existed but led to a different result is given as case *B*. The data are tabulated in Table 15.13 and are plotted as case *B* in Fig. 15.27. This plot shows a different trend in the failure rate from that exhibited by case *A*. This can be analyzed as shown in Table 15.14.

In example of case *B* it can be seen that if four or five failures had been established as an acceptance number for a conventional test, the equipment would very likely have been accepted, whereas the sequential test delayed making a conclusion until the evidence for rejection was overwhelmingly valid. This test analysis reveals a case of equipment degradation with time which would not have been revealed by a shorter test.

Table 15.12 Case A Conventional Analysis

Equipment hours	Failures f	Estimated MTBF Nt/f	80% confidence band			
			Upper limit		Lower limit	
			K_U	M_U*	K_L	M_L*
20	1	20	9.44	188.8	0.434	8.68
40	2	20	3.76	75.2	0.515	10.30
80	3	26.6	2.72	72.4	0.565	15.0
200	4	50	2.29	114.5	0.598	29.9
500	5	100	2.06	206	0.625	62.5
900	6	150	1.90	285	0.645	97.0
1,400	7	200	1.80	360	0.667	133
2,000	8	250	1.71	427	0.680	170
2,600	9	290	1.66	482	0.690	200

* Here M_U = the upper MTBF and M_L = the lower MTBF for each failure-event interval (80% confidence band).

Table 15.13 Case B Sequential Data

Failures f	Test time t, hours	Equipment hours Nt	Normalized $tn = Nt/200$	Sequential test conclusion (from Figs. 15.19 and 15.20)
1	50	100	0.5	Continue
2	150	300	1.5	Continue
3	270	540	2.7	Continue
4	400	800	4.0	Continue
5	470	940	4.7	Continue
6	500	1,000	5.0	Continue
7	530	1,060	5.3	Continue
8	550	1,100	5.5	Continue
9	600	1,200	6.0	Continue
10	620	1,240	6.2	Continue
11	640	1,280	6.4	Continue
12	650	1,300	6.5	Continue
13	660	1,320	6.6	Continue
14	680	1,360	6.8	Reject

Table 15.14 Case B Conventional Analysis

Equipment hours Nt	Failures f	Estimated MTBF Nt/f	80% confidence band			
			Upper limit		Lower limit	
			K_U	M_U	K_L	M_L
100	1	100	9.44	944	0.434	43.4
300	2	150	3.76	566	0.515	77.0
540	3	180	2.72	490	0.565	102
800	4	200	2.29	458	0.598	119.6
940	5	188	2.06	386	0.625	116
1,000	6	166	1.90	315	0.645	107
1,060	7	151	1.80	272	0.667	101
1,100	8	138	1.71	236	0.680	94
1,200	9	132	1.66	218	0.690	91
1,240	10	124	1.61	200	0.704	87
1,280	11	116	1.56	181	0.714	83
1,300	12	108	1.53	165	0.720	78
1,320	13	102	1.50	153	0.730	74
1,360	14	97	1.48	144	0.736	71

15.5.5 Environmental Simulation. It was mentioned that the AGREE sequential analysis test plan was to be used with the test sample subjected to one of four sets of standardized environmental conditions. These conditions have been modified slightly and frequently specified as in MIL-R-22973. Here five levels are established as shown in the following table.

Environmental Conditions

	Test Level I
Temperature.............	25 ± 5°C (68 to 86°F)
Vibration................	None
Heating and cooling cycles	None
Input voltage............	Nominal (within range specified for equipment)
	Test Level II
Temperature.............	40 ± 5°C (95 to 113°F)
Vibration................	2 g at any nonresonant frequency between 20 and 60 cps; measured at the mounting points on the equipment
Heating and cooling cycles	None
Input voltage............	Maximum specified voltage, 0 to 2% at maximum temperature; minimum specified voltage, 2 to 0% at minimum temperature

The vibration requirement may be met by mounting the equipment solidly, i.e., without the equipment shock vibration mounts, to a strong flat plate which is supported by vibration mounts and to which is attached a suitable motor with an asymmetric weight which will meet the requirements of amplitude and frequency. The direction of vibration is not critical. The duration of vibration shall be at least ten minutes out of every hour of heating-cycle time.

	Test Level III
Chamber temperature....	−54 to 55°C (−65 to 131°F)
Vibration................	Same as test level II
Heating cycle............	Time to stabilize at the high temperature, by actual measurement, plus 3 hours unless specified otherwise
Cooling cycle............	Time to stabilize at the low temperature
Input voltage............	Same as test level II

This test requires the use of an environmental test chamber which can change in 2 hours from −54°C to +55°C or two chambers and a means for rapidly moving the equipment. The cooling-cycle time will be determined by the size and complexity of the equipment. The heating-cycle time will be determined by the time to stabilize, the duty cycle, and the length of the work day. If the equipment is hermetically sealed or is intended to operate when dripping wet, no precautions need be taken during the chamber transfer. It is not intended to penalize the equipment with high humidity or condensation. Regardless of the environmental cycling system employed, at least four complete cycles per 24 hours are desired.

Test Level IV

Temperature.............	−65 to 71°C (−85 to 160°F)
Vibration................	Same as test level II
Heating and cooling cycles	Same as test level III
Input voltage............	Same as test level II

This test is the same as test level III, but it is intended for use with equipment that will be subjected to a more extreme temperature range. It is important in both test level III and test level IV that the equipment be turned on as soon as it is placed cold in the hot chamber or as soon as the heat is turned on.

Test Level V

Temperature.............	50° ± 5°C (113 to 131°F)
Altitude..................	Normal (0 to 5,000 ft)
Humidity................	Room ambient (up to 90%)
Vibration................	Same as test level II
Input voltage............	Nominal (within range specified for equipment)
Heating and cooling cycles	None

In addition to these environmental values, a chart describing a specified on-off and hot-cold duty cycle is frequently employed. This is reproduced as Fig. 15.29.

15.6 SYSTEM ACCEPTANCE TESTING

15.6.1 General. Because of the many possible variations of and limitations on system testing, this discussion will be limited to the rather ideal case wherein a system is available and dedicated to extended reliability tests under simulated- or actual-use conditions in an accessible location.

It is likely that more is expected of the system test than to determine only the MTBF. If this were not the case, the general considerations described under equipment testing would apply. It will be assumed here that the system acceptance test is to determine the availability of the system for effective and successful use. As pointed out in Sec. 1, the availability index A is given by

$$A = \bar{t}_o/(\bar{t}_o + \bar{t}_r) \qquad (15.29)$$

where A = system availability
$\bar{t}_o$ = average operating or up time
$\bar{t}_r$ = average repair or down time

A simple numerical value for A, the system availability, can be obtained by carefully recording the operating- and down-time intervals and then solving Eq. (15.29) for the average values $\bar{t}_o$ and $\bar{t}_r$. However, considerably more information can be derived from the tests by an analysis of the data for the purpose of establishing time dependencies and the relation of malfunction and repair difficulties to design deficiencies.

15.6.2 Time Dependencies. Simple graphical analysis using linear time plots provides a gross amount of information about time dependencies. The data should be smoothed by using a moving average and then plotted. The optimum base length to be used for the moving average is about 3 times the MTBF. Such a linear plot may reveal the familiar bathtub curve previously described; see Fig. 15.30 for an example. The early debugging period may or may not be apparent depending on the quality and stability of the parts used in the system. The "flat" portion of the operating period may or may not exist depending on the safety factors incorporated in the design. Systems which can malfunction after only slight degradation of components and parts will exhibit a very short stable operating period before the wear-out curve commences. By definition, the wear-out point at which the longevity period is terminated is at a failure rate value λ_L of twice λ_S:

$$L = t_{(\lambda_L = 2\lambda_S)} \tag{15.30}$$

The system test should provide adequate data so that these descriptive values can be established.

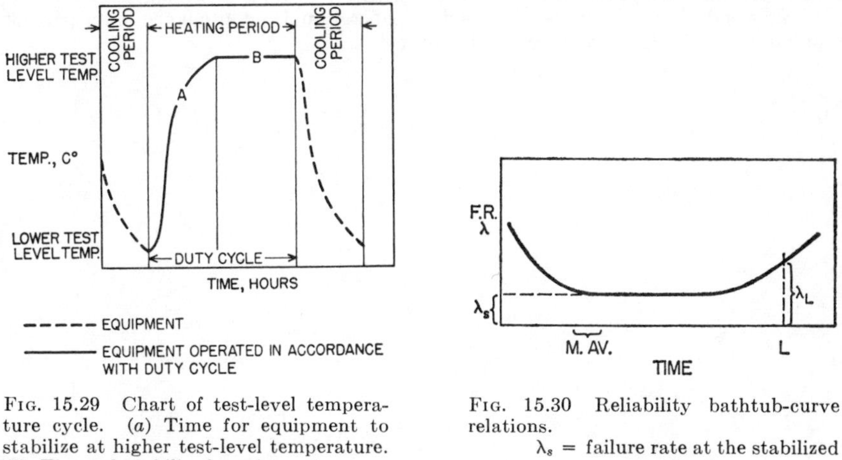

FIG. 15.29 Chart of test-level temperature cycle. (*a*) Time for equipment to stabilize at higher test-level temperature. (*b*) Time of stabilized equipment operation at higher test-level temperature.

FIG. 15.30 Reliability bathtub-curve relations.
λ_s = failure rate at the stabilized constant level.
M.AV. = moving-average base = 3 MTBF $= 3/\lambda$
$L = t(\lambda L = 2\lambda_s)$

15.6.3 Dependent Failures. A linear time plot of the failure data can aid in the interpretation of failure dependency. Consistent patterns in the time domain can frequently be spotted by the eye from a linear plot much quicker than with any other form of analysis. Figure 15.31 illustrates a typical example. The consistent pattern of failure intervals circled were dependent failures. Each time event *A* occurred, this overloaded and damaged the parts which subsequently failed as events *B* and *C*. For purposes of estimating the potential system reliability after the basic design or part deficiencies which caused event *A* have been eliminated, the data can be censored, leaving out these known interrelated dependent events, and the data can

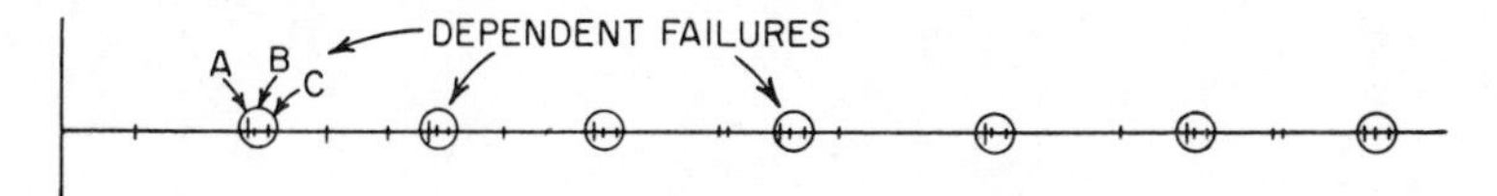

FIG. 15.31 Linear time plot of system failure data.

be replotted. For acceptance purposes the inherent reliability can be established by counting each clump of dependent failures as one failure event.

It is no doubt obvious that the preceding discussion for system test and data analysis is equally applicable to the test and evaluation of complete equipments. In these cases the equipments are treated as simple independent systems.

15.6.4 System Interaction Effects. No discussion of system acceptance testing would be complete without mention of system interaction effects. The validity of results from any system test can depend almost entirely on the extent to which these interaction factors are taken into account. Without going into detail here, the following list summarizes some of the important factors to be considered in the acceptance test.

1. Common power sources: voltage surges, poor regulation, conducted r-f interference, loss of power
2. Radiated interference: radio frequency, low frequency, electromagnetic, acoustic noise
3. Environmental interactions: heat, vibration, shock, physical, other
4. Function interaction: beam collision, frequency beating, jamming, safety
5. Unspecified conditions: environmental, operational, maintenance, transporting and handling, storage, make-ready, checkout

Section 16

ORGANIZING FOR RELIABILITY AND QUALITY CONTROL

J. Y. McCLURE

DIRECTOR, RELIABILITY, QUALITY CONTROL, VALUE CONTROL
GENERAL DYNAMICS CORPORATION, NEW YORK

Mr. McClure is well known in quality and reliability circles, having served as Executive Director, Treasurer, Vice President, and President of the American Society for Quality Control. He has served on many other national committees, including the position as Chairman of the Western Region AIA Quality Control Committee, Chairman of General Dynamics' Panel on Reliability, and Chairman of AIA Reliability Committee.

Mr. McClure attended Southern Illinois University and Curtiss-Wright Technical Institute. He served as an instructor in the University of California War Training Classes. He is a Fellow of ASQC and a member of the American Society for Metals, American Society for Non-Destructive Testing, and the National Management Association.

After employment with Lockheed Aircraft Corporation, Consolidated Aircraft Corporation, W. M. Bill Schoenfeldt Racing Planes, and Northrop Aircraft Company, Mr. McClure became associated with the General Dynamics Corporation in 1940. Since then he has held many important positions at various locations with that organization leading to his present position as Director of Reliability, Quality Control and Value Control at the Corporate level.

ORGANIZING FOR RELIABILITY AND QUALITY CONTROL

J. Y. McClure

CONTENTS

16.1 SCOPE OF THIS SECTION

This section is devoted to the careful consideration of the many factors which are important in planning and implementing, as well as making continuing refinements in, reliability and quality control organizations. Refinements in organization, systems, and procedures are frequently needed in an active organization to assure maximum effectiveness. Changing products lines and management objectives, new manufacturing techniques, and the challenge of competition require a continuing search for improved management of reliability and quality control operations. Remember that the development of optimum organization structure and effectiveness is a means to an end, not an end in itself.

16.2 BASIC CONSIDERATIONS

16.2a Impact of Organized Competition. The managing of reliability and quality control areas under the impact of today's organized world competition is a highly complex and challenging task. Management's reliability and quality control ingenuity in surmounting the technological developments required for plant equipment, process controls, and manufactured hardware requires a close working relationship between all producer- and user-organization elements concerned.

The techniques and applications of reliability and quality control are rapidly advancing and changing on an international basis. Industry views the use of higher performance and reliability standards as scientific management tools for securing major advantage over their competition. The application of these modern sciences to military equipment, space systems, and commercial products offers both challenge and opportunity to those responsible for organization effectiveness. The use of intensified reliability and quality programs as a means to improving product designs, proving hardware capability, and reducing costs offers far-reaching opportunity for innovations in organization and methods.

Certainly, American domestic competition, as well as foreign, has witnessed the emphasis being placed upon reliability, performance, and price by industry, the military, and National Aeronautic and Space Administration (NASA) since World War II. Even our automobile and other manufacturers have seen the economic necessity to assure their customers reliable and quality product by providing guarantees which they have continuously increased until certain automobile manufacturers now give service policies or warranties of 5 years or 50,000 miles. Manufacturers of many other lines of products are also providing increasing levels of confidence, assurance, and warranties for their products.

The effects of the increasing complexity, reliability, schedule, and cost competition on the reliability and quality control organization have required that all top management be aware of the most logical cost-saving areas and be assured that the product is as dependable as possible under the allowable conditions of contract or competition.

To manufacture an excellent-quality product with a very high numerical reliability sometimes requires much more money than a customer is willing to pay. Therefore, since high reliability and acceptable product costs are often initially difficult to achieve, it becomes necessary that timely management decisions be made regarding reliability, schedule, and cost trade-offs. These decisions require the use of very exacting and cautiously selected information and careful organization of implementing actions in order to obtain the most value for the money expended. Many facets of organization are involved. They will be analyzed in detail in subsequent portions of this section.

16.2b Management Objectives. The management objectives in organizing the reliability and quality control department should be to design and develop an organizational plan that will provide the controls necessary to assure that the services and products of the parent organization meet contractual requirements. These management objectives may be stated in many different ways, but in essence they probably all have the same meaning. For example, the end objective in organizing the quality control and reliability department is to assure that competitively priced services and hardware that meet or exceed the customer's requirements are provided.

Of course, there must be an optimum balance between the quality and reliability aspects of a product and its cost; otherwise, the industry may price itself out of the range that the customer is willing or has the ability to pay. Also, in some instances the customer may deliberately elect to sacrifice some reliability assurance for schedule reasons. Deliberate actions are required of management in order to accomplish its planned objectives for a program effectively and to assure that any trade-offs affecting product reliability and maintenance are clearly understood by the producer and customer.

Management is responsible for the business enterprise showing a profit. It is in this area that quality control and reliability have the responsibility to assist top management by assuring that planned actions are met in the design, manufacture, and use phases of the hardware. The company that develops a reputation for the manufacture of reliable products within budget will usually grow and prosper. Certainly a manufacturing or service enterprise of high integrity and enthusiasm will increase the prosperity and security of the organization and employees, as well as contribute to the social well-being of the community and nation.

Management of each organization element must be flexible and able to react quickly to meet the demands of any possible competition or new customer requirement. The ability to react quickly, objectively, and effectively to quality and reliability challenges and to anticipate these needs before difficulties arise is an organization characteristic most desired for the reasons noted in the following subsection. Quality control and reliability departments have a responsibility to minimize warranty and customer service complaints by planned preventive actions as well as timely corrective-action coordinations. A satisfied customer is a most important contributing factor to the continuance of the manufacturing enterprise and the achievement of management objectives.

16.3 TOP MANAGEMENT'S ROLE IN RELIABILITY AND QUALITY CONTROL PROGRAMS

Management must provide the controls needed to assure that all quality attributes affecting reliability, maintainability, safety, and cost comply with commitments and satisfy the customer's requirements. Tersely stated, management must have well-planned policies, effective program planning, timely scheduling, and technical training. Management must clearly *state and support its objectives and policies* for accomplishing the product quality and reliability and assign responsibility for accomplishment to appropriate functions throughout the organization.

Top management's basic objective is to provide and maintain quality and reliability organizations capable of efficiently accomplishing the necessary inspection, test, and

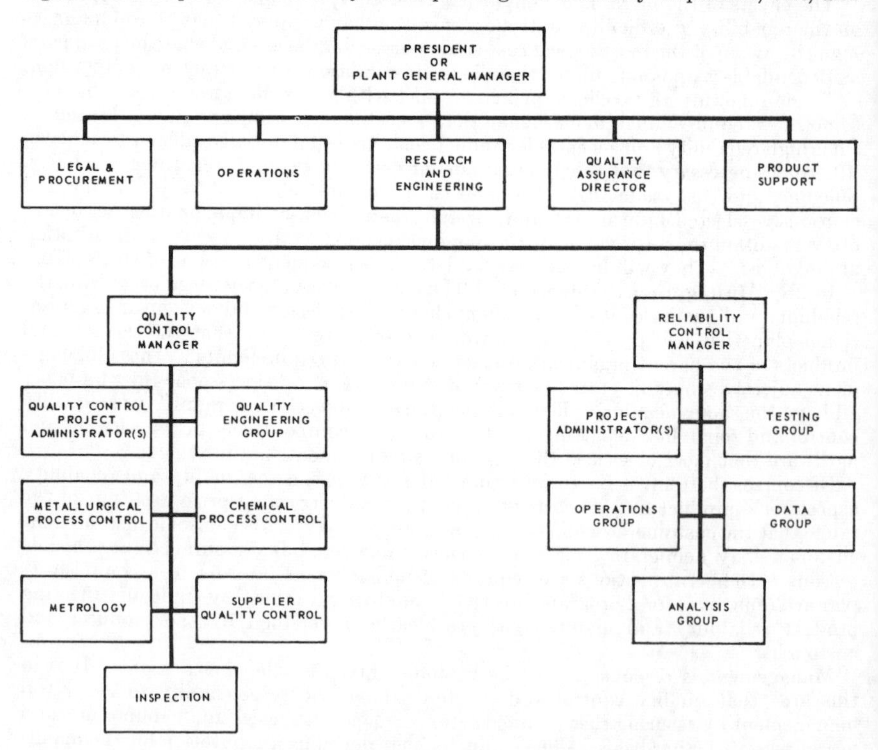

FIG. 16.1 Top-management organization.

analytical laboratory services to assure that all products satisfy the specified requirements of quality and reliability. The quality control organization must support these objectives in a timely, objective, and helpful manner. *Improved product performance and lower costs must be continually emphasized,* and the results must be made visible to management.

Figure 16.1 depicts a top-management organization which shows the responsible management of the *combined* quality control and reliability control departments. This arrangement provides for the entire function to be headed by a director, with the quality control and reliability control functions headed by managers. In this manner the necessary coordination, services, and assurances at the equally important policy-setting operating levels of the various programs are kept on the policy course and not

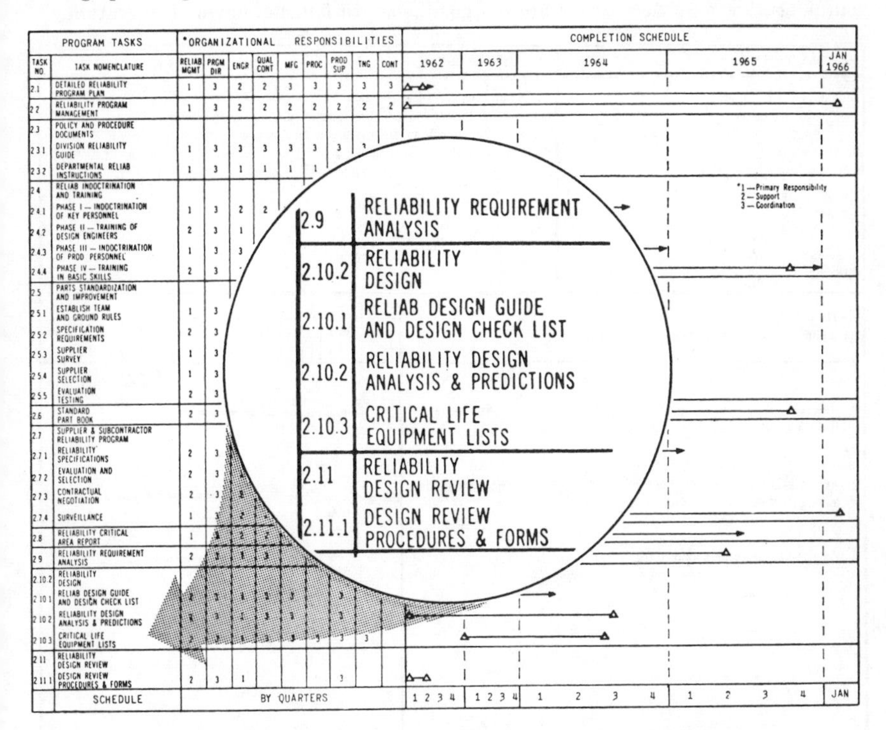

FIG. 16.2 Typical program planning and time-phase-scheduling chart.

allowed to "drift off" to the detriment of any one aspect. Advantages of this combined quality control and reliability organization are that top management has one point of communication and the overhead costs of combined R&QC organization may be lower than for separate organizations.

16.3a Time-phase Planning, Scheduling, and Implementation. The importance of reliability and quality control management control through detailed scheduling of each item of the reliability and quality task must be emphasized. Care must be exercised to sequence reliability and quality program elements to coincide with related total program plans. For example, it would not be practical to request a major change in existing procedures when the contract is nearing completion and the return will not justify the effort expended. Nor would it be practical to expect the accomplishment of tests in nonessential areas of operation when the cost of the test equipment would not be justified by the service the equipment would provide. However, the purchase and installation of equipment for assurance may more than justify itself when compared with the potential impact of equipment failure in customer operations.

Time-phase programming provides task administration and completion control which cannot be otherwise obtained. Figure 16.2 depicts a typical program-planning and time-phase-scheduling chart. Diligent adherence to the time schedule is essential to successful implementation of the reliability and quality program. Figure 16.3 shows a chart for implementation of the individual tasks related to a program time-phase plan.

Management follow-up and evaluation of reliability and quality control program progress should be accomplished by use of audits and simple reports that are specifically designed for the purpose. These management reports serve as decision-making tools and forewarn management in the event progress becomes static. Timely management action must be readily available and applied as needed to many areas of the manufacturing sequences to maintain a good, smooth-flowing, low-cost operation.

RESPONSIBILITY	IMPLEMENTATION
Chief of Quality Systems Reporting Frequency: Semi-Monthly	Quality procedures will be initiated in accordance with program policy and specification required to authorize and establish the method of achieving optimum product quality. These procedures will assure that quality functions are performed uniformly in obtaining the required level of product quality and that the reliability and maintainability are not downgraded during the design, development, procurement, and manufacturing.

ACTION REQUIRED (Explain each milestone required to accomplish task)	DEC	JAN 1963	FEB	MAR	APR	MAY	JUN	JUL	AUG
1. Review existing procedures: Process Standards, Quality Control Instructions, and Serial Memos for applicability to program.	START	PROGRESS ALONG SPAN TIME					ESTIMATED COMPLETION		
2. Revise existing procedures Process Standards, Quality Control Instructions, and Serial Memos for applicability to program.									
3. Prepare new procedures as required.							CONTINUOUS ACTION		

Fig. 16.3 Individual task implementation time-phased plan.

16.3b Management Selection of Key Personnel. Management must recognize and choose the type of persons that are needed to fill the key positions in the reliability and quality control organization. Management must know that these selected people will be able to work closely with and motivate others to accomplish their respective tasks. Top-management philosophy establishes the climate for employee motivation throughout the enterprise.

Top management must be organizationally situated to apprise, counsel, and instruct the middle management that reports to them. All levels of management must maintain clear two-way communications and motivate others without destroying initiative and creativity. Raymond B. Fosdick has said of the challenges in management:[1]

[1] *The Old Savage in the New Civilization*, Doubleday, Doran & Company, Inc., New York, 1928, p. 50.

This is the challenge that we face in our generation. It is a challenge the answer to which cannot be postponed. That answer calls for boldness, for a spirit of daring, for a certain scorn of the past, for a fearless facing of the present facts. It involves the analysis and reconsideration of the worth and utility of human institutions and practices. It means a fundamental appraisal of things that have hitherto been regarded as more or less sacrosanct.

When top management can report improvements in progress, whether it be in implementing a new program or during the actual manufacturing process, the chances are good that the operations of the particular departments are contributing effectively to assuring a fair profit for the business enterprise.

16.4 COST EFFECTIVENESS CONSIDERATIONS

16.4a Organization Responsibility. Responsibility for costs within the reliability and quality control organizations can be most effectively accomplished when specific, capable individuals are charged with coordinating all matters relating to cost analysis and budget control. However, the assignment of coordination responsibility to these individuals must not be allowed to detract from the duty of each member of the reliability and quality control organization to maintain a high level of cost effectiveness.

The cost-control function within the reliability and quality control organization is most frequently located within the quality control Administrative Group, the Quality Control Systems Group, or the Quality Control Engineering Group. Regardless of which group is given the responsibility, the director of reliability and quality control and his department managers must maintain very close and continuing communications with the responsible individuals. Timely analysis of trends and decisions and guidance should be provided frequently.

16.4b Timely Cost Planning. The reliability and quality control management team has value to the total organization that is related directly to its favorable impact on product reliability, performance, and costs. Its contribution to the organized task is of greatest value when performance, reliability, and maintainability of the product are optimized with total program costs.

Although many individuals cooperatively contribute to the overall performance-schedule-cost-profit objective, it is necessary that the executive authority of R&QC management enter into the cycle whenever the desired voluntary cooperation in other branches of the organization falters or the need for new ground rules and policy decisions becomes evident.

1. *Product quality assurance is most economically secured when the conditions which might lead to loss of sale, customer rejection, or excessive warranty costs are predicted, prevented, or corrected at the earliest possible time.* This factor of timeliness applies throughout the design and production processes—from conception to delivery. The nature of the product and organization will determine the earliest points in the sequences of events at which reliability and quality control management may most effectively allocate its total efforts and investment in the product.

2. Some organizations may profitably allocate a significant part of their quality control effort upstream of their own material-receiving inspection. Prevention of costly in-plant delays due to defective materials is the prime objective. Vendor selection and the specification of reliability and quality control requirements in the procurement documents are very important. These activities are best carried out within the terms of specific contractual provisions, the compromise of which will frequently include warranty and incentive allowances. Proper application of incentive contract provisions—usually keyed to reliability and value-engineering provisions in government prime contracts and warranty provisions in industrial contracts—may lead to rewarding savings in the overall reliability–quality control organization effort. (Procurement quality control is covered elsewhere in this volume.)

3. A direct contribution to total organization cost effectiveness by reliability and quality control involves participation in customer service and training. The prevention of field failures by instructing the user in proper methods of calibration, test, and preventive maintenance is a profitable supplement to the conventional

quality control task. This is normally accomplished most economically through the use of proven "customer-motivating" service publications distributed with the product at the time of delivery. For high-cost custom installations, periodic service inspection by technically oriented quality assurance personnel may be most effective. This function may be located in the quality control engineering group. Added advantages accrue to the organization through the feedback of customer complaints to the quality control and reliability functions for fast corrective action.

16.4c Incentive Contracts—Reliability—Maintainability. The abrupt deemphasis of cost-plus-fixed-fee military contracting has focused attention upon the incentive contract as a means for assuring effective management interest in achieving product reliability and maintenance commitments. With this medium, a specified scale of incentive—and sometimes penalty—is applied as a factor in the total contract price. Penalty scales are usually applied at lower rates than incentive scales and may be omitted in competitive fixed-price contracts.

1. Incentive clauses and payments based on design performance will usually be determined from engineering certification of test data and the quality control historical records of actual performance tests.

2. The functional organization structure of a company will determine the role of the quality and reliability records systems in verifying the extent of achieved reliability and maintainability. The achieved reliability and maintainability verification will be accomplished under prescribed conditions of product use, and price adjustments are made, based on calculated incentives.

3. Value engineering/value control profit-sharing incentives must be planned cautiously, and they should be made contingent upon a mutually understood baseline which embraces all of the conventional parameters of improved performance, reliability, availability, and maintainability. These incentives are easier to apply in procurement where little or no research and development is involved and in which it is clear to all contracting parties exactly what latitude of redesign is to be allowed.

A new measurement of contractual fulfillment is introduced when reliability incentive provisions condition the acceptability of products. Utmost care must be employed when negotiating the terms of such contracts; for the conditions frequently resolve sizable profit-margin differentials—sometimes in excess of +10 per cent. Obviously, issues adverse to personal or corporate integrity cannot be injected into the resolution of the incentive-penalty application, nor, on the other hand, can an ultimate decision rest upon the recorded judgment of inspectors with inadequate training and experience. Hence, reliability and quality control personnel involved in contract negotiation and final determination of compliance must possess both the technical judgment and the tactful personality required for the factual determination of incentive realization. If these persons are not already endowed with the respective reliability or quality control administrative authority, then they must certainly be vested with the authority for evaluating contract performance for the company. It is preferable that the same persons who accomplish incentive negotiations monitor the methods and results of incentive performance appraisals.

16.4d Cost Analysis and Budgeting. Every product merits an analysis of the total task to be performed within the allowed costs. The estimation of costs for every function must be quite close to the final actual costs of the specific function if effective results are to be achieved. It is apparent that the general readjustment (usually arbitrary cuts) of budgetary estimates by top management will be in those areas where the departmental estimates and accounting reports of past performance on similar programs are in obvious disagreement.

When it is considered that the combined direct and overhead cost of the total reliability and quality control program may equal a typical plant's net profit of 5 to 8 per cent, the importance of accurate cost estimates is obvious. The cost of rework and scrap in a typical manufacturing plant must be given careful attention. The costs frequently fall in the range of 3 to 15 per cent of plant direct-labor costs. Too often, these costs are not known by top management. Proper use of accurate data by the R&QC organization can be extremely beneficial to all concerned. The need to establish clear methods and guidelines for the use of all reliability and quality control personnel concerned with cost estimating is patent. Figure 16.4 shows one company's cost analysis and budget sheet used in documenting its reports to management.

16.4e Equipment and Facility Costs. Cost estimation of the equipment and facilities required for standards and calibration, process control, inspection, and test is another essential task for reliability and quality control engineers. Applicable staff and line personnel should be given the opportunity to take part in the planning of all equipment and facilities expansion, retirement, or replacement.

Great care must be exercised to determine that adequate justification exists for the addition or replacement of facilities. Improved product reliability and lower costs must be tangible and measurable. Savings predicted should offset the cost of new

PROJECT BUDGET SUMMARY

PAGE ____ OF ____

PER FORECAST DATED ____________

____________ RESPONSIBILITY

____________ ANALYSIS

CONTRACT NUMBER	DESCRIPTION	* TYPE CONT.	% COMP.	INDICATED AT COMPLETION			DIRECT LABOR HOURS /UNDER BUDGET		LABOR AND OVERHEAD RATES /UNDER BUDGET			OTHER CHARGES	BETTER/	NOTE
				TOTAL COST	/UNDER BUDGET		HOURS	EQUIV, DOLLARS	PER HOUR		AMOUNT	UNDER BUDGET	THIS MONTH	
					AMOUNT	%			LABOR	O.H.				

* CPFF - COST PLUS FIXED FEE; CPIF - COST PLUS INCENTIVE FEE; FPI - FIXED PRICE INCENTIVE; FP - FIXED PRICE

Fig. 16.4 Cost analysis and budget sheet.

equipment and facilities within a period prescribed by top management and the board of directors.

16.4f Cost Records. Reliability and quality control organizations have the responsibility for generating and maintaining the important segments of product records of rework and scrap costs, testing costs, warranty costs, etc., upon which pricing structures, company procedures, redesign, and even critical litigation have been founded. The cost of these record-keeping and data-processing activities must certainly be compared with their worth to the company. The responsibility for this falls upon those who implement and make the system work.

Cost estimation for this requirement must include the consideration of savings through the use of automated data-processing equipment, the ever-increasing cost of records storage and data retrieval, the nature of any contractual requirement for data reproduction and translation, participation in data centers such as the Battelle Memorial Institute Electronic Component Reliability Center or the Interservice Data Exchange Program (IDEP), and the extent to which the in-plant data system must be made compatible with the systems established by the customer or subcontractors.

16.4g Quality and Reliability Cost Control. To control cost in the quality and reliability programs, careful long-range planning must be exercised by management. This planning must be accomplished by those to whom top management has delegated the responsibility and who will be held accountable for the implementation of the plans. The controlling of these long-range plans at the time of implementation is one of the basic principles of cost control.

Study programs, research and development programs, production programs, prevention, assessment, rework, and scrap cost estimates should all be made in the long-range plans whereby proper budgeting may be forecast and arrangements made.

16.5 EFFECTS OF COMPANY SIZE ON ORGANIZATION DECISIONS

16.5a Design Surveillance. The problems of designing and building a reliable product are related to the type of product, the size of the company, and the degree to which company management is motivated by the business personality and policies of its highest ranking official. The product of a one-man business has exactly those properties deemed necessary as he conceives, designs, makes, inspects, and sells his products. His integrity and growth are most usually based on a foundation of product reliability and his personal guarantee of customer satisfaction. Within the bounds that one man can motivate and control a company as it expands from the one man to a hundred, a thousand, ten thousand or more, the effectiveness of management is directly correlated to the successful incorporation of these same traits in all elements of the organization. The extent to which this is so is a measure of management effectiveness.

Many problems of managing a business result from lack of communication and differences of interpretation—and excluding any deliberate efforts by personnel to obstruct action desired by top management. It follows that any type of management organization is representative of the method considered most effective in assuring excellent operations with a minimum established level of communications. One requirement is immediately apparent: top management presumes at every instant that the *capability* for correct interpretation and action is possessed by *every employee* and that communication is the means by which responsibility will be delegated.

16.5b Program Planning. Product uniformity requires communication of instructions through policy memoranda, standard practices, process control standards, engineering releases, planning cards, and inspection acceptance records. These latter must be understood by craftsmen and technicians and must be coordinated and enforced at the lowest management level. Management effectiveness in establishing (or restoring) constancy through discrete delegation of authority and in providing smoothness of business transactions and continuity of profit margins is progressively more difficult as the organization increases in size. This adverse effect of organization size is of prime importance to reliability and quality control engineers and supervision.

The larger companies with a versatile R&QC staff frequently have more flexibility in handling contracts for specific programs which involve heavy planning and monitoring functions during early program phases. Small organizations may offset this advantage, if their product line is less complex, by the use of consultants who are temporarily assigned to the task of tailoring a program which will meet both customer and management demands. The consultants should bring knowledge and experience in standardized reliability-program building blocks, math models, and integrated test programs to the small organization's program planning.

16.5c Process-control Monitoring. Process-control monitoring is more difficult in larger companies because of state-of-the-art difficulties and sequential communication delays related to the number of manufacturing operations required and the geographical location of the work stations. The cost of materials and schedule loss from an error in process application will often far exceed the cost of very few capable personnel assigned to perform the monitoring of critical processes.

16.5d Training. The cost of training operators and quality control personnel must be offset through a higher quality of end product, increased productivity, and improved schedules. The large company with a training staff and multiple production programs will have a distinct advantage if it will arrange initial instruction for a few key employees, who are then used as instructors for many more employees as they are moved to the new program. Subcontracted training will cost the smaller company more per capita, but it will impose less full-time overhead-cost burden. This becomes a strong advantage in bidding for new contracts.

16.5e Design Aid. A product improvement (value-engineering) program for the small company will usually be most helpful in optimizing the product design for production over a considerable period. In the small company the product-improvement or design-review team may well be composed of capable individuals selected from supervision, engineering, and quality control.

16.5f Product Improvement. Improving the product of a large company is sometimes more difficult because of a higher order of complexity and the need to examine many related product characteristics. Produceability, reliability, and maintainability may be subjected to varying trade-offs in order to achieve an optimum balance of total product cost effectiveness. In many examples, particularly in Defense Department procurement, the extended life of the product usually affords opportunity to make improvements of real significance in reducing maintenance and spares costs. Generally, the difficulties of product improvement by large companies relate to the static and dynamic inertia of their operations. As new model designs become scheduled, competitive innovations must be undertaken, using scheduled design reviews as a normal course of business. Representatives from key departments, including quality control, may be required on a full-time basis for this purpose.

16.5g Delegation of Authority—Single-plant Operations. One objective of a manager is to operate his plant facility in such manner that his subordinates will quickly learn to make the *same* decisions in problem dispositions as the manager would make under the same circumstances. In a single-plant organization the manager may delegate more authority and responsibility to individuals on his staff than would be the case if a staff of corporate executives or consultants were available for guidance. This often requires that reliability and quality control engineers in small companies cover a wider range in their analysis and decisions than individuals in large or multiplant operations.

The continual problem of management is how best to convey to all employees the degree of flexibility with which they may make decisions and under what circumstances they should carry urgent problems up the ladder of authority. In general, management may expect little initiative in employee decision making until it has established guiding policies and precedents, such as quality control standards and procedures. The employee may then anticipate problems and respond quickly to prevent or solve difficulties. The single-plant, centralized organization involves the least risk in testing the degree of such delegation that may safely be employed.

16.5h Line and Staff Organization. Another organizational precept divides the structure into line and staff functions. The *line functions* are those which are essential to the immediate workings of the company, and they are carried out routinely in the chain of executive command. The *staff functions* do not take part in this routine, but examine it minutely and advise the plant or department manager when, why, and how elements of the plant operation should be altered. The highest executive (plant or departmental) remains the focal point of all decisions, with the *line* functions implementing those he has elected to implement and the *staff* functions directing attention to the best choices and supporting data for his decisions. The combined line and staff type of organization structure is best suited to a plant of medium or large size whose products are complex or diversified and whose field of operations is highly competitive.

The line and staff type of organization demands a strong manager who can make decisions quickly and effectively. His staff must be highly capable and diplomatic in order to be the most effective. This is particularly true with regard to reliability and quality control staff men.

16.5i The Functional Organization. The *functional* type of organization channels administrative orders and instructions for accomplishment through employees especially skilled in specified functions of the organization and hence in terms of what is to be done in order to turn out the product. A functional-organization chart may be developed from any simple list of all things which are done by the company. Figure 16.5 shows one such functional organization chart.

The functional organization must include a one-to-one correspondence between delegated responsibilities and the necessary tasks, but otherwise may be compressed or expanded at will. A compression of the structure has the advantages of making effective use of time and manpower by combining categories of specialization and shortening the communication link from manager to employee. It is generally accepted that the normal *maximum* expansion of organization structure should be no more than six to ten (dependent on management skills) subdivisions of authority or responsibility at any level of the organization and that no employee should be more than six levels of authority removed from top management. Reliability and quality control organizations usually function best when the height of organization pyramid is low and the base is broad.

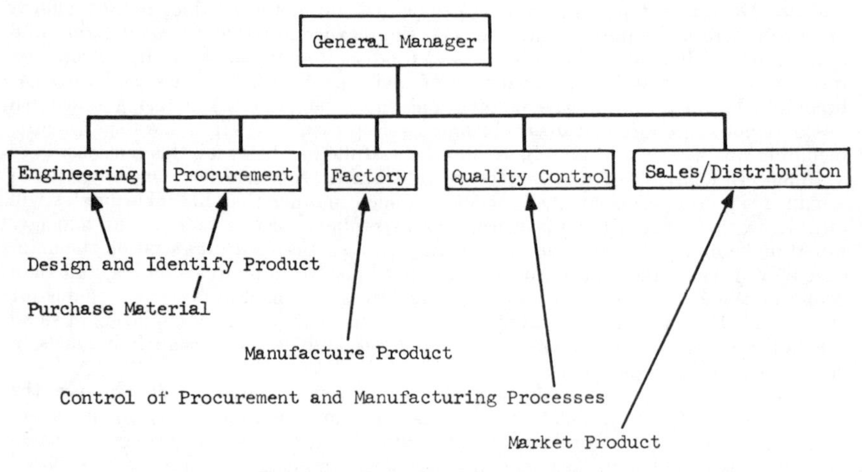

FIG. 16.5 Functional organization.

16.6 MULTIPLANT ORGANIZATION CONSIDERATIONS

16.6a Clear Lines of Authority. The multiplant organization in which the geographically separated divisions are neither autonomized nor projectized will face a compounded communication problem which will make clear lines of authority essential. Efficient reliability and quality control operations may be more difficult. Each executive must know exactly which functional responsibilities will be resolved by an extra-plant manager of a function, which are to be resolved by the in-plant manager, and how executive decisions will be disposed of or implemented when communication is not established. This problem should be minimized through corporate policy standards which clearly delegate responsibility and authority relative to division operations.

16.6b Policy Coordination. Reliability and quality control policy directive coordination is more easily performed when corporate policy documents are used as a "constitution" or as guides. The amending or revising of division standards for accord with corporate standards should involve a closed-loop communication in which each division policy change is sent back to the issuing secretary as a form of check sheet to acknowledge compliance.

16.6c Staff Coordination. Multiplant organizations in which staff members are also line executives in separate plants have greater communications problems. In such circumstances, the coordination of *staff* matters must be complementary with the division *line* responsibility. If the staff function is completely extra-division, the acquisition of staff recommendations will often involve expensive executive liaison between divisions. This must be optimized through the excellent performance of the division reliability and quality control personnel with whom staff personnel must consult.

16.6d Project Coordination—Multiplant Operations. When project organization cannot be carried out entirely within a given division, it becomes necessary for an extra-plant management to tie the project elements together. Project management will be headquartered at the facility whose task is more critical or at which most of the project is to be performed, with assistant project management delegated to the fewest remote locations practical. Reliability and quality control operations should closely parallel the project organization in retaining a clear line of authority and communication to the project and basic organization headquarters.

16.6e Evaluation Techniques. Project management evaluation of reliability and quality trends is sometimes difficult in multiplant operation, making necessary the complete, concise reporting from the appropriate division line management supplemented by audits by the extra-division project or staff man. The quality assurance executive responsible for multiplant projects is further burdened by the reduced level of his personal contacts normally possible in evaluating changes in policy or procedure which affect a broad base of corporate effort. For these reasons the evaluation of policies or administrative decisions at the multiplant level should be keyed to extremely reliable reports and analysis which directly reveal the potential effect upon profit (or loss).

16.6f Autonomous Divisions, Either Single or Multiplant. Very large corporations, in which the number of levels of supervision and the number of subordinates reporting to each level of authority could become excessive, find it desirable to relieve some of the stress of communications and authority delegation by making each major subdivision autonomous. Multiplant corporate executives affect division operations in a staff capacity only, and the division manager may set up his own organization structure as permitted by the total corporate policy document. The corporate management will usually evaluate autonomous division management on the basis of gross sales performance, product quality trends, and the profit or loss experience. The autonomous-division concept is convenient for more effective reliability and quality control organization, policy, and standards because they may be more readily tailored to fit the specific products of the division.

16.6g Need for Periodic and Accurate Reporting. The acceptance of autonomy, however, also requires the acceptance of the responsibility to maintain a clear, periodic reporting of reliability and quality trends and costs which accurately shows the status of the division performance. The management disciplines needed for the improvement of internal operations are normally applied from within the division, so that factors which may affect changes in reliability, quality, future profits, or sales will be acted upon in a timely and effective manner. A fast-reacting, timely reporting system is essential.

16.6h Analysis of Reports. The analysis of the data contained in reports will often result in reliability and quality control recommendations for departmental or management corrective action. These may be projected as parameters which influence raw-material procurement, rework, scrap, material conservation, idle time of men and machines, off-station rescheduling, and dollar costs. Effective reports will bring reliability and quality control functions into a close liaison with industrial engineering, planning, scheduling, pricing, production supervision, traffic, and procurement. Conversely, inaccurate or too voluminous reports will counteract the effectiveness of an otherwise good reliability and quality control organization.

16.6i Corporate Surveillance of Autonomous Divisions. The top management of an autonomous division will constantly be under the surveillance of the corporate staff function, even though an operational hands-off policy is maintained. The corporate level of management normally remains silent when targets are equaled or exceeded, but should quickly recognize potential or real difficulties and provide guidance within the division when needed.

16.6j Follow-up. It is important that follow-up be maintained with regard to improvement recommendations made within a plant by its own, corporate, or customer personnel. However, the corporate policy might be sustained within the divisions when it is apparent that perpetuation of management remains dependent upon the maintenance of profitable performance.

The rapid organizational advancement often experienced by reliability and quality control personnel is a result of their knowledge, enthusiasm, accuracy, timeliness, and integrity.

16.7 TYPICAL RELIABILITY AND QUALITY CONTROL ORGANIZATIONS

16.7a Combined Functions. The advantages of combined reliability and quality control functions must be carefully studied. Points to be considered include matters such as the capabilities of available executives and supervision and the possible organization policies limiting the number of executives reporting to the president or general manager. In a typical organization the span of control limitations frequently results in combined functions of the type shown in Fig. 16.6. Note that, in this typical organization, provision is made for program (project) planning and control in complete coordination with the normal "line" organization functions.

One very capable man in charge of the complementary functions of reliability and quality control can be very effective, provided that complete management support and teamwork are attained. The active cooperation of design, procurement, and manufacturing supervision is essential for this type of organization. Advantages of combined functions include:

1. Improved timeliness and effectiveness through integrated planning and implementation of important functions and tasks.
2. Minimum duplication (or omission) of tasks due to departmental interface difficulties.
3. Improved budgeting; reduced administrative costs.
4. Improved communications effectiveness.

Disadvantages which must be considered are:

1. A scarcity of the very high caliber men required to organize and administer the combined functions.
2. Difficulties in securing the full support of other departments in functions such as design review, failure analysis, and corrective action.
3. Problems experienced in assuring that the *planning* and *monitoring* elements of the combined-centralized function do not detract from the "doing" responsibilities of other departments.

16.7b Major Considerations. The location of reliability and quality control in the total organization is the result of several major considerations. Some of these consist of management checks and balances designed to assure that organization functions are kept in their true perspective and to assure the delivery of product or services of adequate quality on schedule and within budget.

More uniformity in reliability and quality control cost effectiveness usually results when the responsible executives report to top management at the same level with design, procurement, manufacturing, etc.

Organization considerations by management must include careful consideration of the levels of essentiality or criticality of the products, including design difficulties, ease of manufacture, performance, maintainability, and financial liability. In the manufacture of complex industrial equipment, home appliances and vehicles much organization variation is found to exist, with various arrangements in the different plants of some multiplant companies.

Customer complaints and warranty claim liabilities can be more effectively reduced if the reliability–quality control organization is integrated, oriented toward cost

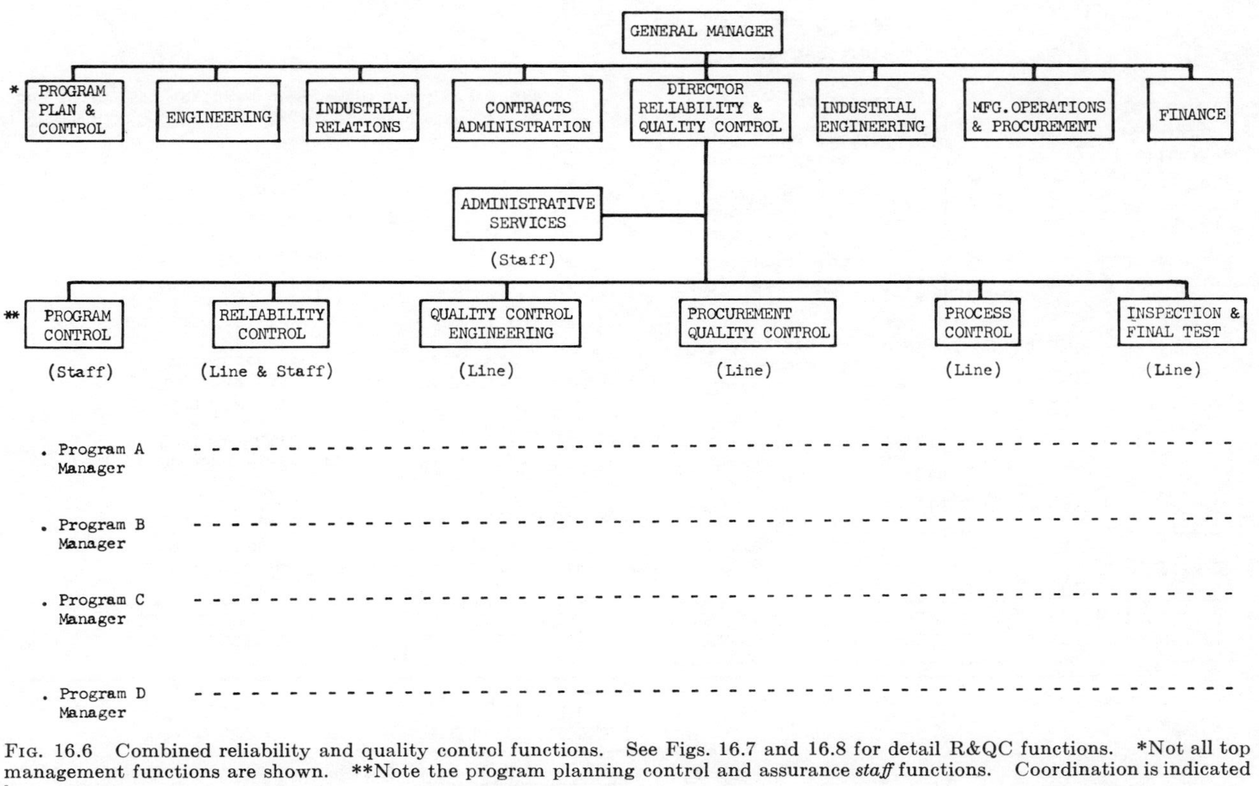

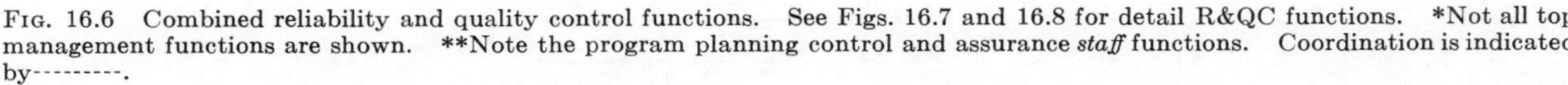

FIG. 16.6 Combined reliability and quality control functions. See Figs. 16.7 and 16.8 for detail R&QC functions. *Not all top management functions are shown. **Note the program planning control and assurance *staff* functions. Coordination is indicated by---------.

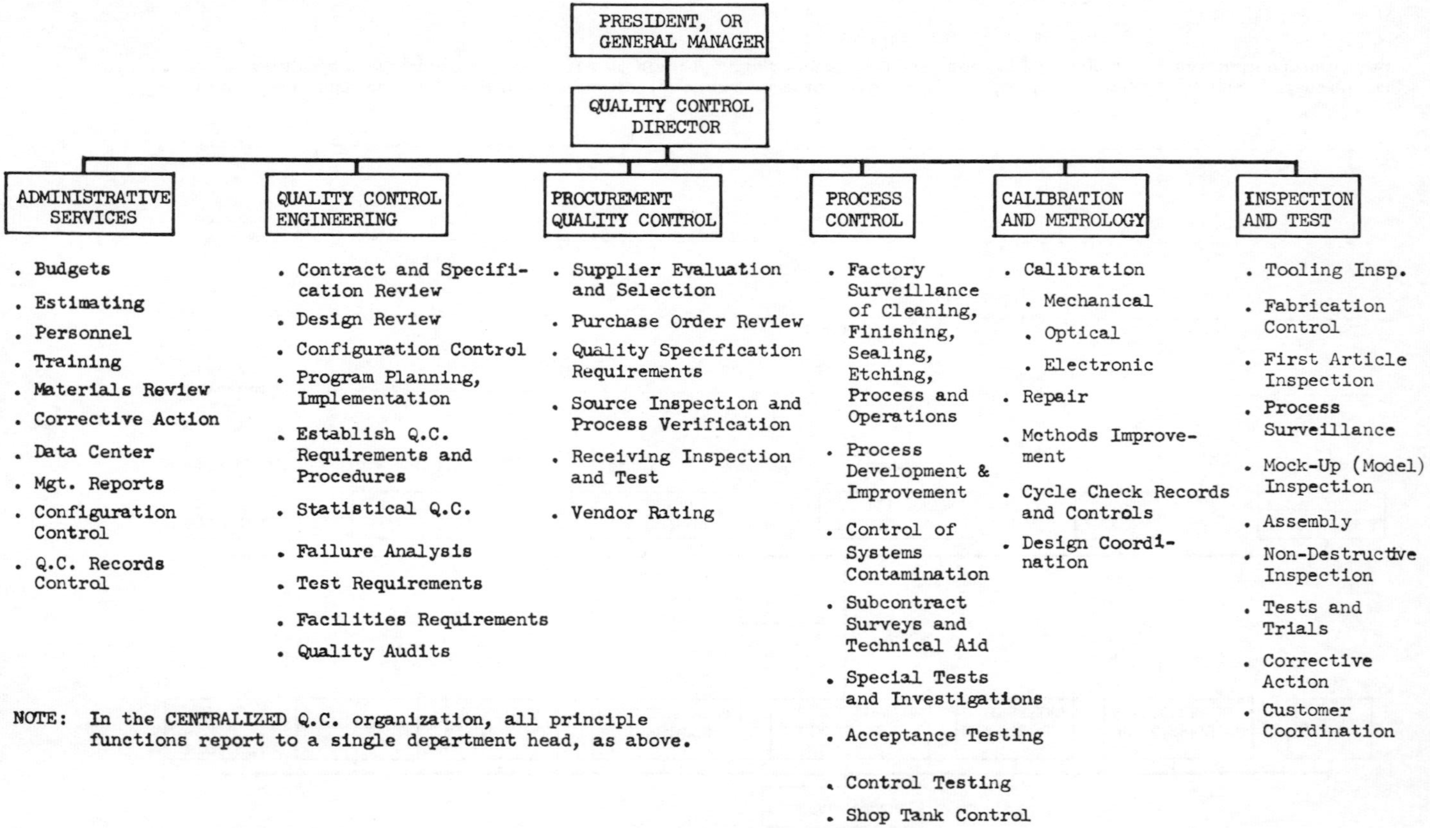

PRESIDENT, OR GENERAL MANAGER

QUALITY CONTROL DIRECTOR

ADMINISTRATIVE SERVICES
- Budgets
- Estimating
- Personnel
- Training
- Materials Review
- Corrective Action
- Data Center
- Mgt. Reports
- Configuration Control
- Q.C. Records Control

QUALITY CONTROL ENGINEERING
- Contract and Specification Review
- Design Review
- Configuration Control
- Program Planning, Implementation
- Establish Q.C. Requirements and Procedures
- Statistical Q.C.
- Failure Analysis
- Test Requirements
- Facilities Requirements
- Quality Audits

PROCUREMENT QUALITY CONTROL
- Supplier Evaluation and Selection
- Purchase Order Review
- Quality Specification Requirements
- Source Inspection and Process Verification
- Receiving Inspection and Test
- Vendor Rating

PROCESS CONTROL
- Factory Surveillance of Cleaning, Finishing, Sealing, Etching, Process and Operations
- Process Development & Improvement
- Control of Systems Contamination
- Subcontract Surveys and Technical Aid
- Special Tests and Investigations
- Acceptance Testing
- Control Testing
- Shop Tank Control

CALIBRATION AND METROLOGY
- Calibration
  - Mechanical
  - Optical
  - Electronic
- Repair
- Methods Improvement
- Cycle Check Records and Controls
- Design Coordination

INSPECTION AND TEST
- Tooling Insp.
- Fabrication Control
- First Article Inspection
- Process Surveillance
- Mock-Up (Model) Inspection
- Assembly
- Non-Destructive Inspection
- Tests and Trials
- Corrective Action
- Customer Coordination

NOTE: In the CENTRALIZED Q.C. organization, all principle functions report to a single department head, as above.

FIG. 16.7 Typical quality control functional organization (centralized).

effectiveness, and of sufficient stature and authority to get results. The returns from reduced in-plant scrap and rework and lower warranty costs may equal 2 to 5 per cent, or more, of total sales. These savings, plus the increased business levels brought on by customer confidence, have tremendously benefited certain manufacturers of products such as autos and appliances.

The manufacturer of a single line (brand) of well-known American autos finds it economic to employ over 800 reliability and quality control personnel. The manufacturer of a low-cost German compact auto indicates an even heavier level of effective quality control effort. In each case sales and profit, with minimum user complaints, indicate a successful blend of industrial functions and management checks and balances.

16.7c Typical Quality Control Organization—Centralized. The *centralized* quality control organization is so designated because all related functions are grouped under one department head, usually reporting to the plant president or general manager. Figure 16.7 is typical of the centralized quality control organization structure and allocation of functional responsibilities. Figure 16.8 portrays the separate (but centralized) reliability functional organization and responsibilities.

The typical quality control organization for large and small manufacturing plants must be staffed to accomplish a broad range of administrative and technical tasks and functions such as design review, manufacturing process control, and the complete spectrum of planned (and unplanned) inspections and tests required to assure the production and delivery of products which will provide long and faithful service. The centralized organization offers many advantages and some disadvantages.

16.7d Advantages of Centralized Organization. Advantages include increased versatility, flexibility, and economy through the use of qualified personnel on multiple assignments, more uniform application of policy and procedural controls, and less "reinvention of the wheel." In medium or small companies with centralized quality control administration, certain of the quality control functions may be accomplished by one or more individuals on the chief engineer's staff. For example, it is frequently economical and practical to:

1. Develop quality control engineers who, by previous qualification as chemists, metallurgists, electronic engineers, etc., are capable of a broad spectrum of in-house analysis, consultation, and problem solutions.
2. Train inspection personnel to perform multiple quality control functions.
3. Hire the calibration of standards by a qualified laboratory.
4. Secure vendor surveys by a reputable firm with strategically located and highly qualified specialists.

16.7e Disadvantages of Centralized Organization. Some disadvantages are experienced with the centralized type of organization. It is very important that these be considered at the outset and that offsetting controls, savings, or other advantageous factors be developed. Typical disadvantages are:

1. A tendency to develop higher overhead costs owing to added levels of supervision and clerical functions
2. Higher inertia in developing specific program cost effectiveness refinements
3. Reduced sensitivity to customer reactions, owing to "filtering" of complaints
4. Vulnerability to excessive management budget slashing, owing to the increased visibility and apparent size of the centralized organization

16.7f Project Orientation. Customers for products of major importance frequently require the producer to utilize the project or program manager concept. Provision for strict program planning and control is implicit in this concept. Reliability and quality control functions can benefit greatly when an adequate balance of organizational and technical checks and balances is achieved. However, great care must be exercised to prevent high costs and potential confusion through the duplication of functions. Figure 16.9 illustrates a typical organization with provisions for *project control* in conjunction with line organization.

16.7g Projectization Case History: Minneapolis-Honeywell. A case history of the successful development of an organization for effective multiple program

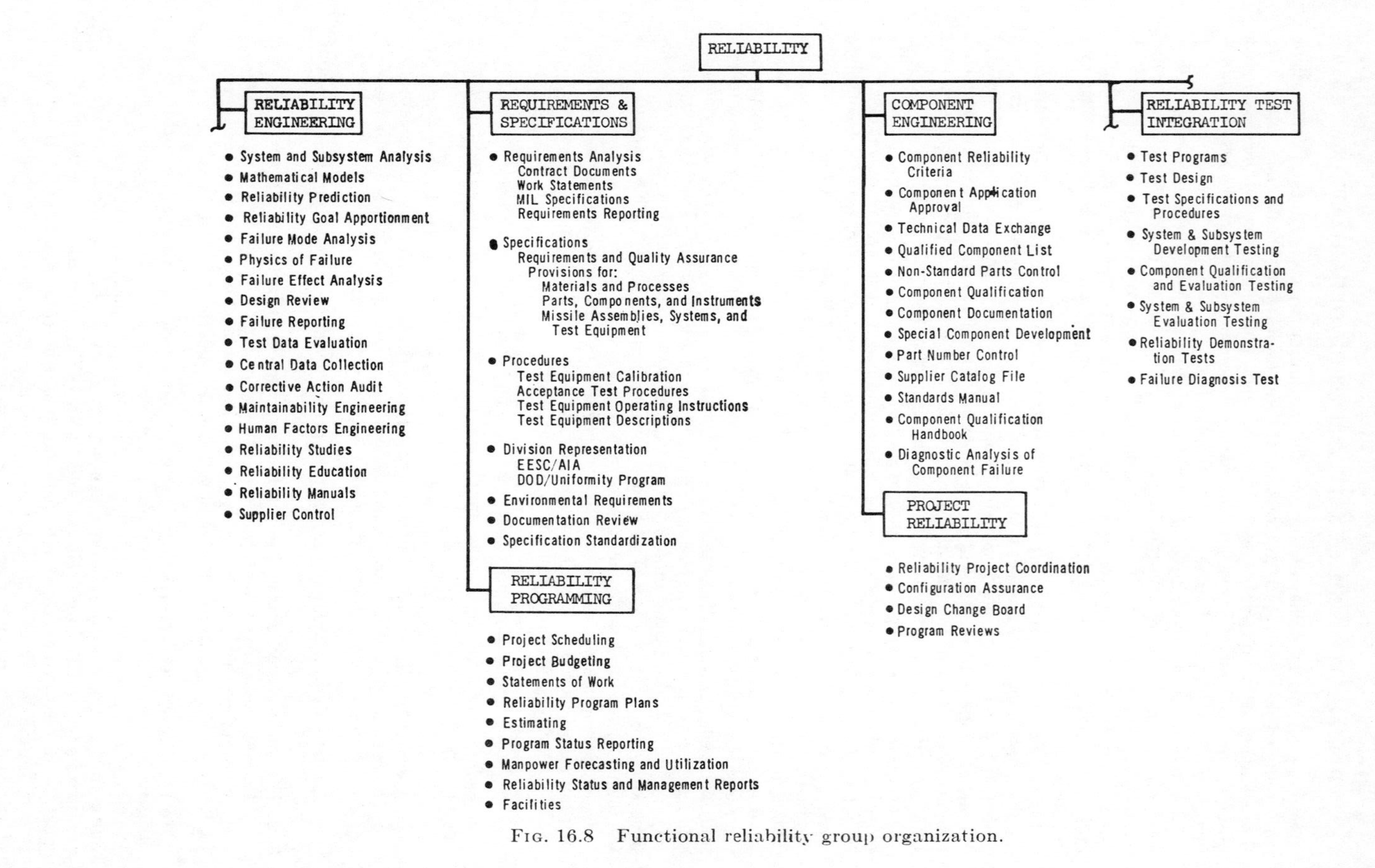

Fig. 16.8 Functional reliability group organization.

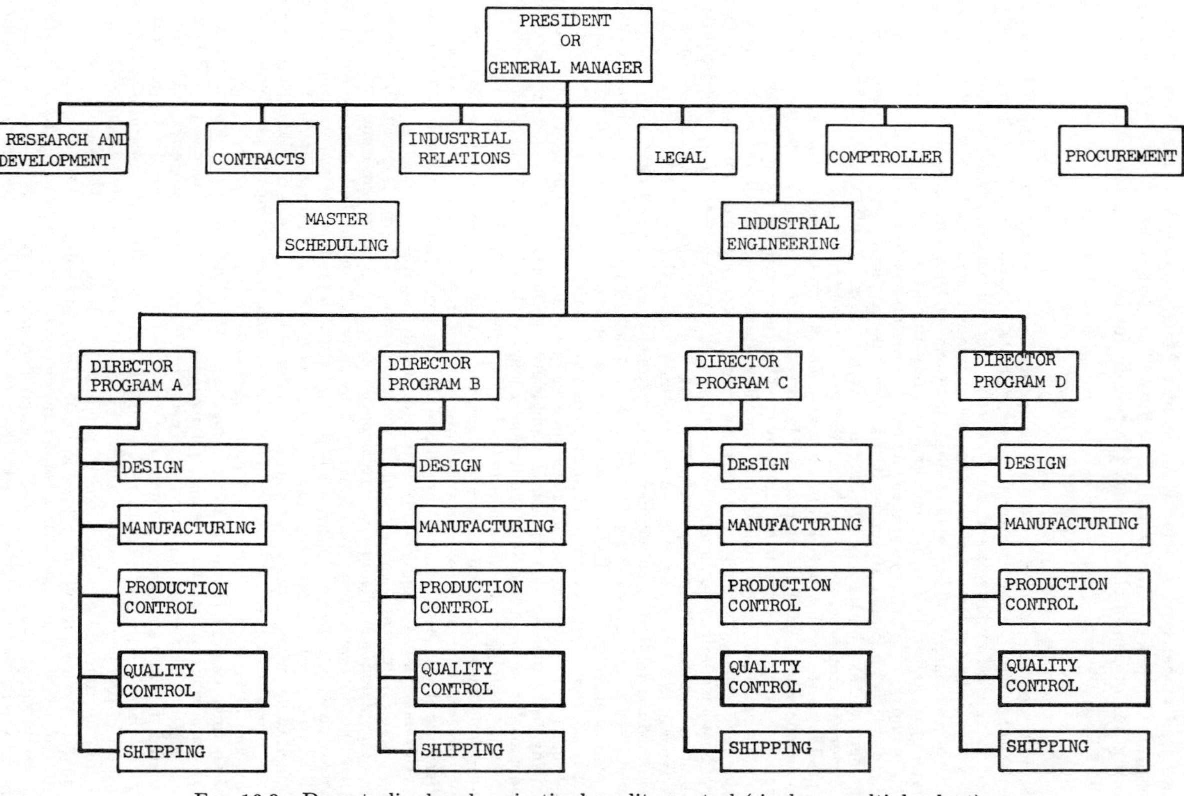

FIG. 16.9 Decentralized and projectized quality control (single or multiple plant).

control is demonstrated by the Management Matrix concept of the Minneapolis-Honeywell Company. Mr. Orestes B. Johnson[1] of that company has provided the following description of this concept, the reasons for its development, and the many factors considered:

The Management Matrix

The adroitness of a company to remain competitive and maintain its profit level requires more than the ability to engineer and produce products in quantity. The matrix technique applied to decision making provides an objective means for solving various management problems. Quality assurance of a product or system is a significant factor in the growth pattern of a company. The departmental functions, policies and responsibilities dictate the type of organizational structure which can best fulfill the objectives of the consumer and the company. At the top management level, the matrix technique is useful in determining the organizational structure based upon the responsibilities delegated to each department and as a basis for penetrating new market areas. In all cases, the effectiveness of the management process is directly related to profitability through consumer assurance that product performance and quality are maximized within the negotiated cost structure.

Management of a department responsible for administration of the quality assurance program in a division of a company primarily oriented to research, development and production of diversified aerospace products and systems requires special planning, techniques and philosophy. The management must have the capability to continually maintain the proper level of customer satisfaction and evaluate product performance even though the products and systems are usually required to perform at limits bounded by the state-of-the-art. In extreme cases, development and production may be nearly parallel since the products have immediate application in manned and unmanned space flight. In general, each product or system has performance requirements in scope and magnitude such that the product assurance requirements specified are as diverse as the product line, depending upon the customer documents or procurement agency involved in the contract.

The solution, to the stated conditions, must be one of dynamic planning of the steps in organizing to accomplish the department objectives. Elements of the matrix can then be sequentially incorporated into the organizational structure in logically phased steps. The matrix planning is always an evolutionary process to eliminate the administrative stresses associated with revolutionary changes due to new business and profound requirements. A continual audit of the structure, contract requirements and personnel skill requirements should be conducted to validate the effectiveness of the organization in cost and performance and its applicability with program demands.

The organization described has applied this philosophy to its quality assurance operations. The planning has shown its effectiveness through the transition from aircraft into aerospace products and systems business. This organization has had the planned capability to cope with the increased requirements for extended product performance during the transition and has the added capability to deal with further requirements which will be demanded in upcoming programs. Future business analyses, reviews of advanced product designs, corresponding test requirements and applications of the company's technology are heavily considered in the management planning matrix.

The planning was initiated by the Department Director some years ago. Comprehensive studies were made of market trends and industry-government relationships. Product quality responsibilities were delineated into primary and secondary categories. The study culminated in a management document defining the organizational matrix in these categorical terms. It was then coordinated with all concerned departments responsible for any aspect of product quality and concurred by top level management. The plan incorporated the fundamental considerations of product complexity, product assurance requirements, and the cost controls which would be necessary to manage future programs.

With the study as a basis, several organizations were proposed as functional matrices which would satisfy the program hypotheses detailed in the study. At one extreme, total decentralization into projects was considered. Under this plan, complete responsibility for all quality assurance would be entirely vested in each project. At the other extreme, a single quality organization, composed of all programs, was proposed. Although each of the proposed approaches possessed the capability of assuring the quality of the product, a weighting of factors and trade-offs between objectives to be attained defined the appropriate structure which had the greatest flexibility and capability to control the diversified products and systems. A composite of a strong central quality assurance organization, which had functional responsibilities for control of individual products, those special products for the projects which are produced within the product sections, and several specialized project

[1] Special Assistant to the Director of Quality.

sections with complete responsibility and authority for project requirements, was established. A skeleton organization is shown in Fig. 16.10.

This skeleton structure indicates the major functional responsibilities as they relate to achievement of the department's objectives. The quality engineering section has responsibility for planning. As such, it has authority to develop the formal quality plans which are applied to the product sections. The inspection function is responsible for acceptance of all products and implementation of those functions detailed in the quality plan and inspection instructions. The procurement section is responsible for acceptance of supplier products and coordination of requirements with the suppliers. The central staff is responsible for inter and intradepartmental coordination and liaison. The special projects sections are responsible for achievement of project quality requirements.

Figure 16.10 indicates that the project quality managers are retained under functional control of the director, although they are responsible to the program directors or managers for satisfaction of program requirements. The nature of product diversification is such that some requirements must be implemented by sections outside the projects. The projects specify the requirement and maintain final authority and responsibility for compliance. As an example, the procurement quality assurance section, a procurement department–quality department oriented function, controls the quality of procured items, both at receiving inspection and in the supplier's facility. Program demands are specified to procurement quality assurance and performed therein. Responsibility is retained by the project manager with subsequent management authority vested in the department director.

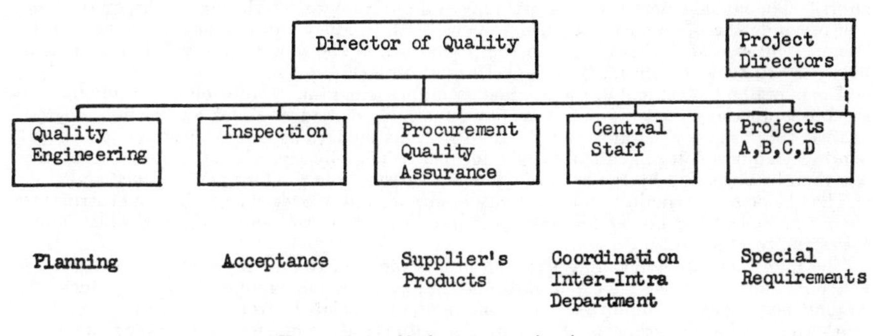

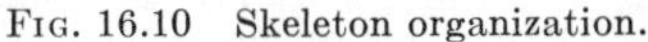
FIG. 16.10 Skeleton organization.

Staff assistance is provided by the central staff to the entire department. The various requirements are reviewed by the staff for uniformity of procedures and compliance with contractual requirements. A review of all major procedures provides a check and balance upon the department.

A study of programs determined the need for an operational analysis since the interface relations between the sections for each contract would have to be established during the proposal stage. Each new program is placed in the organization after a decision has been made as to the need for establishing it as a project. Several factors are considered and the methodology of decision theory is applied. The following factors are considered as the most heavily weighted.

Customer Requirement. Certain programs are of such magnitude that management and communications must extend in an unbroken line through all levels of procurement. The need for a specific organizational structure is a customer requirement. This does not assure that all activities will be performed by the project but that authority and responsibility for compliance with requirements is maintained by the project.

Special Requirements. The product or system and/or contractual requirements are so specific and different that existing procedures cannot suffice.

Schedule. This objective requires special attention. A large schedule requires appropriate manpower to evaluate acceptability of the production flow. In some cases, the personnel performing acceptance must be certified in special ways or have specific talents.

Product Complexity and Skill Levels. Product complexity (processes, test techniques, production fabrication) and skill levels are such that the product is significantly different from related products.

Dollar Volume as a Function of Time. The ratio of program cost/time is high. This implies a concentrated program effort is required.

Manpower Availability. The program requirements for specialized manpower are such that this factor is considered. This objective is not heavily weighted since it is related to attainment of other objectives.

These objectives are weighted in terms of the various courses of action using the matrix approach to establish a decision. This approach has a basic purpose of analyzing the array of actions and depicting the decision in mathematical terms.

The management function then utilizes this tool for planning and action in performance of its activities. The organization matrix provides the mechanism for management in an expeditious manner and efficient departmental control commensurate with this company's products and philosophies.

Customer procurement documents specify and direct that quality and reliability assurance programs are maintained. The proper management of these programs is the master control which assures the customer that all requirements will be met. Customer interpretations of the government quality and reliability specifications are explicitly stated as contractual requirements. The methods used to implement the requirements and evaluate the level of compliance with the contract are particularly significant in the planned management of these programs.

The quality and reliability programs are functionally so related that to achieve the objectives of both programs, it has been mandatory to define the responsibilities of each department into areas of control. These functions are all inclusive as to control of the product. The study of departmental and program functions and responsibilities provides a basis for defining those areas where overlapping between departments could lead to duplication of effort. The management is one of planning and implementation through techniques which will preclude the possibility of noncompliance with product requirements. Delegation of responsibilities and authority within the framework of the quality and reliability plans is necessary to achieve continuity of objectives at all levels.

The overall management of the product assurance program, in addition to planning, must establish objective goals as milestones and methods of auditing and evaluating progress against this base. Programs of redress to prevent deviations from objectives, goals and contract requirements have special significance. The effectiveness of product assurance can then be measured by the level of customer satisfaction, product performance and cost.

The placement of quality and reliability assurance in the overall organizational structure should be considered on the basis of optimum product control and assurance which minimizes the total program costs.

The management review and planning process applied to new products (or revisions to existing products) must rely upon information and direction supplied by the product line committees. These committees are composed of specialists from engineering, quality, marketing and production as shown in Fig. 16.11. Recommendations emanating from their actions are usually technical, but are relevant to the management decision to pursue diverse business opportunities when the technical information substantiates the capability of an organization to realistically comply with all specifications and other requirements. The product line matrix is one step in the systematic decision process which is commanded by a continuous sequence of management controls.

The effective use of the matrix technique in decision making can be a useful tool which focuses attention upon all program requirements and allows the decision maker to efficiently trade-off or heavily weight those sections which contribute the greatest to program overall success. The technique is adaptable through all levels of management and provides a documented analysis for the decision maker to use in re-evaluating his original decision in the light of new information.

16.7h Decentralized Organization—Separate Functions. Decentralization is somewhat of a hybrid between the centralized and project-type organizations in which very large organizations retain the policy and planning functions at the highest management level and delegate increased executive authority to managers of a small number of product lines. This concept provides a wider latitude for policy making by top management, but it is completely dependent upon the effective choice of executives at the operational level for accurate policy evaluation. The size of organization which may practically decentralize is that wherein—with consistent operating profits—the top executive finds it expedient to delegate the performance of many executive functions, especially those peculiar to the administration of multiple large programs (Fig. 16.12).

The clue to the success of decentralization is whether projected profit levels can be maintained as more executive authority is delegated. It has the advantage of requiring less technical accomplishment in the subdivided specialty lines by top manage-

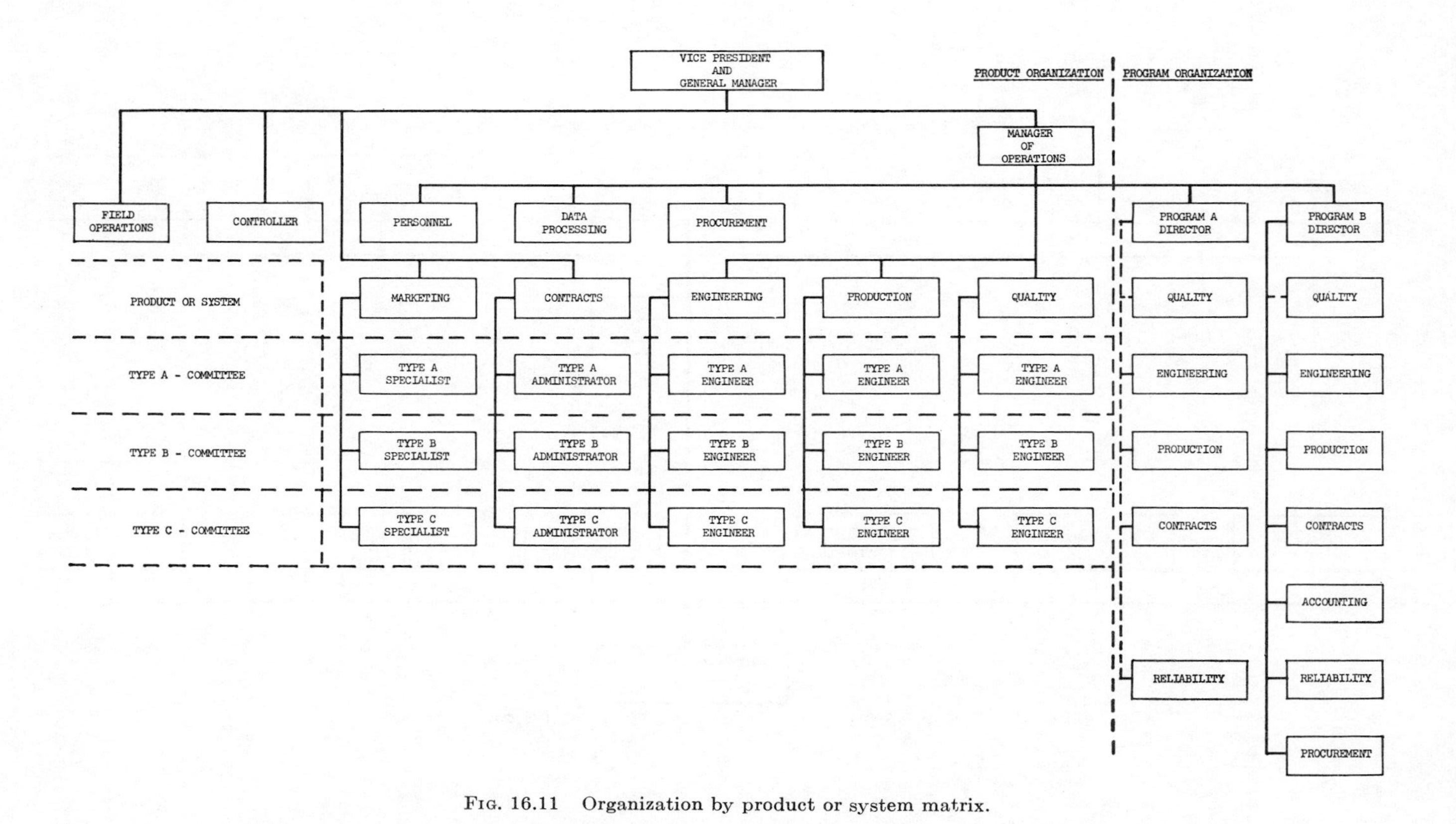

Fig. 16.11 Organization by product or system matrix.

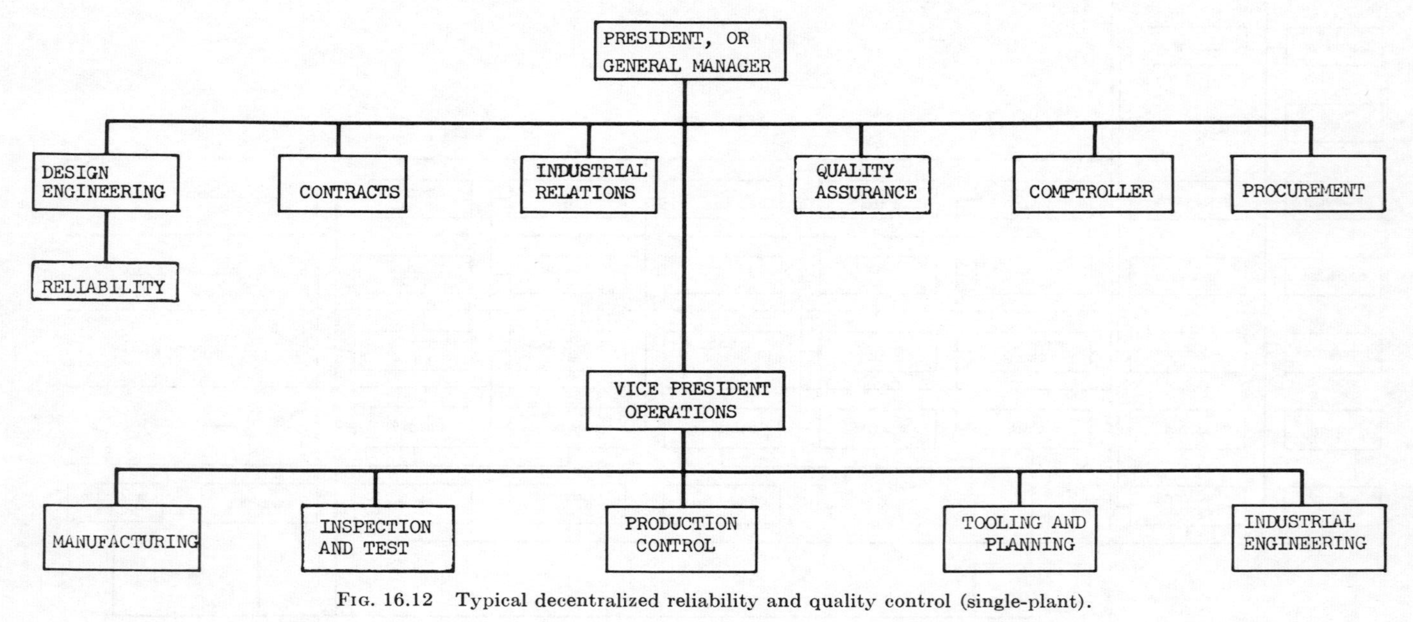

FIG. 16.12 Typical decentralized reliability and quality control (single-plant).

ment, and it thereby facilitates product diversification. It has the disadvantage that there is built-in lag between cause and effect of mismanagement at the project level, so that sizable product losses or warranty liabilities may accrue before audit or customer reaction indicates the need for management corrective action.

16.8 FUNCTIONAL ORGANIZATION RESPONSIBILITY ANALYSIS

The specific requirements should be made clear for the various assignments of responsibilities for reliability and quality control among engineering design, procurement, manufacturing planning, and the reliability and quality control departments.

The quality control department should be organized to utilize to the maximum all of the abilities and experience of its personnel in the achievement of the reliability and quality objectives. These objectives are to assure that competitively priced products and their services meet or exceed contractual requirements.

The quality objectives can best be attained by first establishing an organization capable of defining and implementing the appropriate concepts, procedures, and actions. These concepts can vary slightly in wording from company to company but basically, industry-wide, the concepts probably remain very nearly the same.

16.8a Organization—Quality Control Director. The quality control director reporting to the president or general manager (Fig. 16.7) is responsible for establishing and administering an effective, time-phased quality program which is planned, developed, and implemented in coordination with other program functions. The quality department should assure by a persistent continuing audit that the quality contractual requirements are being maintained.

16.8b Process Control—Responsibilities. The established functional responsibilities for the control of processes are shown in Fig. 16.7. The organization of process control is established to maintain surveillance over manufacturing processes and materials directly related to production, repair, and the development of parts.

Process control should establish material procurement specifications applicable to materials of the manufactured end product within engineering specifications when further refinement is necessary for manufacturing economy and maintenance of quality. The process control department should assist the general procurement department or affected project material department, upon request, to evaluate a prospective vendor's metallurgical, chemical, bonding, or plastics capability. The vendor's processing equipment, facilities, controls, and methods should also be examined to assure conformance to requirements.

Tests should be conducted on samples of supply items, as required or upon request, for the purpose of determining satisfactory materials and multiple procurement sources. Chemical, metallurgical, bonding, and plastics analyses pertaining to manufacturing operations should also be performed as required.

It may be necessary to enforce observance of process standards of manufacturing instructions by shutting down the process to maintain quality. Certification tests to certify processing equipment and personnel are also a functional responsibility which should be assumed by the process control department.

16.8c Procurement Quality Control. Procurement quality control should start prior to the time a component or subsystem requirement on vendor service first becomes apparent. Procurement quality control should provide inspection planning for all vendor-supplied material. This would include the quality specification requirements usually prepared in advance for the vendor's information and action.

A vendor should be surveyed to assure his quality capabilities. Procurement quality control should also administer the vendor-rating list, maintain a list of qualified vendors, and have a system for the review of purchase orders. The vendor's processes should be verified, and source inspection of all vendor-supplied material should be provided.

A planned system of coordinated effort in the prevention and investigation of and corrective action on vendor quality nonconformities must also be designed and maintained by those responsible for vendor activities. (See Sec. 16.19.)

16.8d Calibration and Metrology. The basic functional responsibility of the calibration and metrology area is to maintain all commercial and special equipment used within the company and provide reference and working standards for certification and calibration. Provisions should be made to transfer and maintain measurement standards traceable to the National Bureau of Standards.

Quality control should assist engineering and operations in determining that the most efficient equipment, consistent with costs, is used to perform the required tests. Assistance in establishing precision and state-of-the-art measurement techniques and equipment should also be provided. A system of assuring that instruments in use conform to primary standards of recognized measurement values should be maintained, along with a procedure for periodic recalibration. (See Sec. 16.14.)

16.8e Inspection and Test. The functional responsibilities of inspection and test should be concerned with the physical and functional inspection of incoming production and test materials and parts, as well as those parts and assemblies produced by the factory's fabrication and assembly departments, for quality and conformance to specification. Procedures should also be maintained for the inspection, and tryout if deemed necessary, of purchased and in-plant-manufactured production and tooling tools.

A system for the acceptance or rejection of inspected parts, assemblies, and operations on the basis of conformance or nonconformance to applicable requirements must be maintained. This system should be coordinated with other departments in order that timely materials review action may be obtained on rejected parts and problem areas and that appropriate corrective action will be taken.

16.8f Quality Control Engineering. Quality engineering functional responsibilities are very important to the operations of an economical quality program. Quality engineering should conduct studies on various quality control projects as directed by their management. They should be prepared to develop, plan, and implement new systems or adapt existing systems for controlling and assuring product quality. Analyses to solve quality problems and improve quality control systems should be conducted.

Statistical analysis services by quality engineering should be provided for quality control sections and other departments when approved by management. This would include statistical sampling plans, as well as the designing of statistical experiments to identify problem conditions, evaluate quality performance, improve processes and quality standards, raise equipment reliability levels, and reduce testing for lower operating cost.

Quality engineering should continually perform audits to assure the effectiveness of the quality control systems. The auditors document the deficiencies noted and initiate requests for corrective/preventive action and follow-up, as required, to assure timely corrective/preventive action. (See Sec. 16.17.)

16.8g Administrative Services. Quality control administrative services should develop, estimate, and coordinate budget requirements and forecasts for direct and indirect manpower levels and material dollar allocations. They should estimate necessary space and facility requirements as well as control and report on expenditures and provide data on status and overall budget position.

Administrative services should also advise and assist line management in matters of personnel procurement and utilization; preparation and processing of personnel documents, classifications, and utilization; and interpretation of company personnel and security policies.

A data center should be maintained for the purpose of processing quality control records and data generated both in and out of plant. These data can then be utilized to prepare management reports and trend charts and maintain control of product configuration.

16.8h Organization—Reliability Function.[1] The initial effort in any reliability program is the development of a reliability-program plan outlining the tasks

[1] Information was adapted in part from General Dynamics/Pomona publication "Reliability Functional Organization and Mauler Reliability Program Plan Summary," B-18641-R1, September, 1963.

required to attain reliability goals satisfactory to the customer. A functional organization (Fig. 16.8) with specific functions or responsibilities must be defined.

The next few paragraphs describe the functional responsibilities related to reliability engineering, requirements and specifications, reliability programming, component engineering, project reliability, and reliability test integration as indicated in Fig. 16.8.

16.8i Reliability Engineering. Reliability engineering should conduct *system and subsystem studies* to determine the functional relationships of components and subassemblies. Inherent reliability weaknesses and potential problem areas should be analyzed and defined.

Based on the system and subsystem analyses, reliability *mathematical models* which define the reliability relationship of components and subassemblies should be prepared. These models should describe how the reliability of lower-level assemblies combines to form subsystem and system reliability. The model should also include considerations such as partial and total redundancy as well as operational modes. (See Secs. 2, 5, and 10.)

Reliability goals based on the specific contract requirements and reliability-prediction studies should be established. The established *goals* are *apportioned* to the major subsystems according to complexity and predicted failure rates.

Failure-mode analyses on critical components and subassemblies should be performed. These studies attempt to define the type and physical manner of failures which will occur in components and subassemblies and their relative frequency. The results of reliability tests are correlated to define the primary failure modes under environments.

Physics of failure or component failure criteria on functional and physical parameters should be provided. Chemical and physical failure mechanisms as well as failure stresses and component failure rates should be defined.

The effects of component and subassembly tolerance limits and failure modes on system performance and reliability should be analyzed and evaluated. Component and subassembly interaction problems prior to system assembly should also be defined.

Reliability engineering should maintain a design-review board which will periodically evaluate critical designs to provide additional assurance that the equipment is capable of achieving the performance and reliability requirements. The reliability design-review function should utilize circuit-analysis information, schematics, failure modes, parts list, cost data, reliability predictions and goals, available test data, and specifications. A design-review checklist for use by design personnel and the review board in evaluating and improving designs should be provided.

A reliability monitoring function is necessary to assure that *failures are reported* and that the failure reports provide sufficient information and failure analysis to define the failure adequately and to facilitate corrective-action audit procedures. (See Sec. 9.)

Statistical evaluation of test data resulting from reliability tests should be performed. This function should also provide statistical data analysis for engineering and test groups upon request. If data analysis is requested prior to test, additional assistance should be provided in the design of the test to assure the proper type and quantity of data are collected. (See Secs. 4 and 6.) This function should provide for a central data agency to collect, correlate, and analyze failure data. This can be accomplished with the assistance of computer facilities for automated control of large quantities of data. Periodic reports of significant failure trends should be made available to management.

A corrective-action audit function which initiates and monitors corrective action and reliability improvement should be implemented. This activity should monitor and report on current significant problem areas as well as the status of the corrective-action procedure relating to the problem areas.

A maintainability engineering function is needed to define spares allocation and predictions of mean time to repair/replace. System-availability figures can be developed from these studies. (See Sec. 11.)

Methods for defining the *human factors* and their effect on system parameters such as accuracy, availability, environmental extremes, and overall weapon system effectiveness should be designed. Statistical evaluations of system operation with special emphasis on the human element need to be conducted. (See Sec. 12.)

Reliability engineering performs special studies to develop and modify reliability methods and procedures of analysis. Reliability studies of the product based upon the results of reliability assessment and test should be performed. Formal presentations, manuals, and handbooks on the basic reliability philosophy and techniques should be prepared. These reliability philosophies should be disseminated to develop reliability-conscious design and fabrication personnel. Reliability manuals and handbooks which define the basic reliability concepts, techniques, and procedures should be prepared.

Supplier reliability programs and reliability demonstration procedures with respect to completeness and conformance to system reliability requirements must be evaluated. There is also a need to provide reliability guidelines and handbooks for supplier use in establishing and conducting adequate reliability programs and procedures.

16.8j Requirements and Specifications Group. Contract documents, work statements, and specifications for technical and reliability requirements must be reviewed and interpreted. Recommended changes for the resolution of conflicts in specifications and requirements should be reported.

Specifications to provide a clear, accurate description of the technical requirements and quality assurance provisions for materials and processes, parts, components and instruments, assemblies, systems, and test equipment must be developed. Whenever qualification approval is required, the specifications should include appropriate qualification tests and prescribed methods of inspection.

Procedures and instructions for conducting acceptance inspection and test of systems, subsystems, and components by verifying compliance with design and general quality provisions of the associated hardware specifications must be provided. Many large organizations find it desirable to have a representative on such committees as the AIA Electronic Equipment Specifications Committee or the DOD-Specifications Uniformity Program for the purpose of participating in specification and contract-requirement standardization projects and the national surveys related thereto. Reports should be prepared and presented before the committee or the DOD uniformity group regarding projects and their status.

The analysis of contractual and system environmental requirements and the preparation of environmental requirements charts and documents for subsystems and components should be accomplished. The review and approval of interim procurement and acceptance documents prepared by other groups (for expedience) during developmental phases of various programs should also be accomplished. The general design documentation in support of reliability and design should be reviewed. General standardization and simplification of specification requirements and formats should be accomplished.

16.8k Reliability Programming. Reliability programming should develop and maintain a master reliability schedule, by project and task, indicating major milestones and progress. Incoming work should be evaluated for schedule requirements and effect on overall reliability commitments.

Project budgeting should maintain and evaluate budget records with regard to funding, expenditures, and progress. It should also assure adequate funding for projects assigned, forecast budget status at project completion, and issue periodic budget status reports.

Statements of work by task should be evaluated to assure that every phase of an overall reliability requirement is adequately detailed. Changes in statements of work and/or tasks should be evaluated to properly reflect the impact on established reliability programs, schedules, budgets, manpower, and facilities. Schedule and budget estimate requests for reliability work or changes to existing work should be provided.

A broad basic reliability-program plan which reflects the latest state-of-the-art advances in reliability, together with specific *project* reliability-program plans developed to meet specific project reliability requirements, should be maintained. Periodic status reports assessing reliability progress in terms of specific program plans should be prepared.

Manpower requirements by function and specialty in terms of known and forecasted workload should be forecasted. Periodic status and management reports assessing overall reliability activities should be prepared and issued. A specification and procedure release system should be maintained and controlled. Reliability requirements for space, equipment, and coordinate requirements should be accomplished with the cooperation of appropriate organizations.

16.8l Component Engineering. The technique for ensuring *component reliability* through evaluation and testing should be established. The integration of these techniques into the procurement specifications for the components should be provided for.

A technical consultation service should be provided for the designers who assist in circuit and design analysis to ensure proper application of components. Information received from agencies such as GMDEP, IDEP, and BATTELLE ECRC should be furnished, and a method for the distribution of this information should be provided. A qualified-component list which describes the component type, rating, qualification status, and procurement specification and designates the approved suppliers should be prepared and maintained. (See Sec. 6.)

A controlling function to ensure that the use of *nonstandard parts* in a design is both minimized and justified should be provided. The design should be evaluated, and wherever necessary, a suitable standard part should be recommended. A component-application handbook should be provided design engineers. The specific and extensive application data will facilitate the design process and aid the designer in proper component application.

A diagnostic analysis function to establish component failure modes, as well as to define component reliability weaknesses, should be instigated. In some instances the analysis may detect improper part application or a design weakness.

Continuous assessment of the designer's needs for special components and assistance in developing methods for providing these components should be provided. Liaison with suppliers for the development of special components should be conducted.

A supplier catalog file which defines the components offered by each supplier as well as component ratings and application data should be maintained. A standards manual that defines those components which have been standardized should also be maintained. The manual should provide additional information such as physical dimensions, component ratings, and application data. (See Sec. 6.)

Component qualification requirements for inclusion in the procurement specifications should be established. When adequate qualification data are not available, qualification tests should be conducted and a report on the results of the tests should be submitted. The documentation of component technical requirements and quality assurance provisions should be established. Assurance that the necessary procurement specifications and drawings are documented for all components should be provided.

16.8m Reliability Testing. Reliability test programs that are consistent with the system reliability requirements and the contractual demonstration procedures should be developed. Test plans, test conduction, and analysis should be established in accordance with the overall reliability-program plan.

Proposed test items should be evaluated and test designs utilizing statistical inference techniques such as sequential analysis, life testing, regression analysis, and test to failure should be prepared. Test specifications and procedures should be prepared to assure that items are tested in accordance with system reliability and performance requirements.

Reliability developmental testing on system and subsystems should be performed to define existing and potential reliability weaknesses and problem areas. Comparison tests to assist design personnel in the selection of proposed subassemblies should be also accomplished. (See Sec. 8 for complete details of testing programs.)

16.9 ADMINISTRATIVE ORGANIZATION FOR RELIABILITY AND QUALITY CONTROL

The magnitude and scope of many quality control and reliability organizations are such that it becomes necessary for the director to assign the basic responsibilities for a number of department-wide administrative functions to a small group of specialists. This makes it possible to concentrate the responsibility for a number of functions which have much influence on the effectiveness and cost of the overall quality and reliability functions within a small group of people. These individuals, in the routine accomplishment of their duties, must coordinate with the director and all applicable supervision throughout the quality control organization and with the applicable key people in other functions of the parent organization. Much of the work consists of the formulation of short- and long-range plans, including the required facilities and personnel requirements, policy directives, procedures, and standards. These items are covered in more detail below.

16.9a Policy Directives and Procedures. It is important that quality control and reliability policy directives be clear, concise, and up to date. Furthermore, it is necessary that these directives be in consonance with other directives and policy documents of the plant or parent organization and with the contractual requirements of the customers. It is frequently desirable to provide departmental instructions and procedures which provide supplemental simplified interpretations of major policy directives and contractual requirements. These detail instructions too must be in consonance with other directives. It is often desirable for the individuals who write procedures to audit the application and use of the procedures periodically to assure that they are adequate and up to date and provide optimum cost effectiveness.

16.9b Time-phased Planning. The time-phased planning of reliability and quality control functions is of paramount importance. This planning should encompass both short- and long-range product design and manufacturing plans and should be periodically reviewed and updated. The following basic factors should be given careful consideration:

A. Basic Organization. Some realignment of the organization may occasionally be necessary in order to improve administrative or supervisory effectiveness commensurate with changing types of products or services and to secure maximum economy. This need for change is most likely to occur when there is a sizable expansion or reduction in production activities. Timely adjustment of the organization is frequently the key to the successful and economic accomplishment of the quality control function.

B. Project Organizations. A number of government procurement agencies require that major procurement contracts be placed under "project or program direction." The purpose of this requirement is to assure that such contracts receive a very high level and adequate amount of administrative attention.

It is usually desirable for the quality control department to parallel the rest of the organization in the project concept. Great care must be exercised, however, to be certain that multiple organization levels and redundant responsibilities in cost are not created through the projectization concept. In this regard, the quality control administrative group is often utilized to assure maximum commonality of internal services, and functions, plus clear communications and minimum expenses.

C. Facilities. The advance planning for the acquisition (and phase-out) of quality control equipment and facilities is an important administrative function. This is emphasized because of the potential beneficial or adverse impact of facilities and test-equipment cost on the competitive position of the organization. Careful administration in the planning and procurement of facilities can well result in tremendous long-range savings through the use of improved nondestructive testing equipment and automated checkout equipment. Under certain conditions the elimination of requirements for facilities, through redesign, or the use of other quality assurance techniques can be achieved by knowledgeable quality control administrative personnel.

16.9c Quality Control Personnel Planning. Personnel planning is a vital portion of the quality control administrative functions. The objective is to have the right quantity of adequately trained personnel available exactly when needed. As with other quality control administrative functions, this obviously requires a considerable amount of planning and detail follow-up in order to achieve effective results. A number of the important factors to be considered are discussed below.

A. Prediction. The prediction of the requirements for quality control personnel requires careful advance coordination by the quality control administrative personnel with design and manufacturing supervision, this in order to learn the requirements of new designs as related to inspection methods, techniques, and required skill levels and to determine the manufacturing phase and timing wherein the new types of personnel skills will be required.

B. Position Descriptions. The establishment of position descriptions for all quality control jobs can be very helpful in assuring the proper selection of suitable personnel. Further, the establishment of position descriptions makes it possible to eliminate duplication of functions and to provide quality control coverage when functions may not have been adequately described in the past. Position descriptions should be written or tailored to fit the type of organization, products, and manufacturing processes involved.

C. Training. Training program planning and coordination is another essential element in the quality control administrative operations. In many plants, training normally is a function of the industrial relations department. This does not, however, relieve the quality control director of his responsibility to assure that adequately trained personnel are utilized at all times. Therefore, methods of periodic evaluation of quality control employee job knowledge for current and future products must be utilized.

Frequently, quality control administrative personnel must take the lead in establishing minimum requirements for training and certification of manufacturing employees as well as quality control personnel. This does not mean, however, that the actual training will be accomplished by quality control (unless they possess capable instructors meeting the requirements of the industrial relations training program). It is important to reemphasize that effective and economical quality control can best be accomplished through the use of very adequately trained and motivated employees, and the quality control administrative group is often in a position to recognize and initiate the necessary action for timely job knowledge evaluation and training.

16.9d Communication and Coordination. Effective communication and coordination techniques are extremely important if the quality control operation is to provide timely preventive and corrective action. A portion of these communications are accomplished through the dissemination of basic policy, procedures, and standards as described earlier. It is equally important, however, that very clear modes of communication and coordination be established throughout the internal organization, with the suppliers of raw materials, and with the end customers or users of the products (or services). A portion of the quality control administrative function is to assure that this flow of information, both internal and external, is accomplished in a very timely and effective fashion with adequate feedback to assure corrective-action measures.

The adequacy of reliability and quality control trend reports, and their corrective-action effectiveness, must be assured by the administrative group. This does not necessarily mean that this group will actually provide the reports, but the group will, at least, monitor their timeliness, accuracy, and effectiveness.

The quality assurance audit function, in some organizations, is placed in the quality control administrative group. The advantages of this are that the group is normally constituted of very highly skilled quality control engineers and technicians of excellent versatility. These men are very frequently capable of performing the audit function to assure that the policies, procedures, etc., with which they are so well acquainted, are actually being adhered to (or revised as necessary for greatest effectiveness). The director of quality control and reliability must be kept advised at all times concerning the results of audits. He will normally summarize certain audit results for higher management personnel to keep those personnel aware of important changing conditions. (See Secs. 16.17 and 16.18.)

16.9e Motivation Programs. Motivation programs for accelerating the achievement of highly dependable, low-cost products are an essential element of a well-balanced quality control and reliability program. Such motivation programs must be very broad in scope, covering all facets of supplier, in-plant, and customer areas of activity. Much ingenuity is required in order to assure that these programs are effective and timely.

The quality control engineer responsible for this activity must be unusually ingenious and capable of working closely with the industrial relations, public relations, and top management personnel. Carefully planned programs must be coordinated with the plant and presented for top-management sponsorship in such a manner that top management may implement the programs with minimum of effort and secure complete support and enthusiasm of all personnel.

Quality control supervision should expect to stay well in the background in publicity relating to this type of motivation program. Every effort should be made to provide recognition in company and city newspapers, trade magazines, etc., concerning outstanding accomplishments by manufacturing and other personnel. Quality posters and periodic awards for outstanding individual or departmental quality performance are very helpful.

Effective accomplishment of the motivation functions and services by the quality control administrative group can provide added effectiveness to the quality control program. In all of this, it is important that cost comparisons, including savings achieved, be continually reviewed to ensure that an adequate payoff in the form of savings to justify the program expense is being achieved (see Sec. 16.12).

16.10 RELIABILITY AND QUALITY CONTROL ENGINEERING—METHODS/SYSTEMS

It is important in the management of reliability and quality control programs that a single philosophy of optimum product quality and cost be preserved. Usually this can be done most easily if the responsibility for both reliability and quality assurance is placed in one man. A significant number of industrial and government organizations have done this, with the principal authority being delegated to a director of product assurance (or director of reliability and quality assurance). In some organizations this man has also been given additional responsibilities for such functions as value control, maintainability, and product safety.

The effectiveness of the reliability and quality control engineering group in providing advanced planning for methods and systems is an important consideration of the total planning effort. This planning must include the development of adequate attention to preventive effort, starting with design review of the products and continuing through procurement, production, test, and delivery. This planning must include adequate provisions for trend analysis, corrective action, and reporting.

The capability of the quality control engineering staff in developing new methods of nondestructive inspection and innovations in the control of processes and in systems of closed-loop failure reporting must be continually examined and improved in order to provide utmost value to the total organization.

The facilities required for process-control measurement and test must be completely compatible with product design and the manufacturing plan. If the company is to continually upgrade its product reputation and business growth, it is almost inevitable that high standards of quality and reliability will be a strong competitive factor. In order to meet this competition for highly reliable products, the average company must:

1. Commit itself to more extensive capital outlays for the increased automation of production and quality control facilities in order to reduce product unit cost and increase the yield of acceptable items.

2. Use much innovation to meet precise quality requirements with unique and effective process and machine controls.

16.10a Quality Control Methods Planning. The skill levels required of personnel who carry out the reliability and quality control efforts vary extremely in terms of the complexity of the equipments under observation. Little difficulty is anticipated when these requirements remain on the same level, but the expansion to product lines with which the company is less familiar may cause a part of the skill level to diverge, thus building in labor-management problems when the character of the task is again changed. Obviously, the company cannot afford to retain college graduates for routine mechanical part inspection, nor can it permit non-high-school graduates to perform reliability demonstration tests upon infrared scan detectors. Product specialization will usually minimize the degree of the skill-level problem in terms of unknown future business, but it is nonetheless a condition which management must resolve.

16.10b Training Programs. Reliability and quality control training programs will ordinarily be seen as a major challenge only for the larger companies, and they will generally impose cost and schedule problems on contracts for which research and development programs are necessary. A large company may retain its own educational and training staff, in which case the cost of instructor training and investment in educational aids will be compared to the cost of contracting for the training task with established institutions, or occasionally with a special-equipment supplier. A smaller company will ordinarily contract its training program to other organizations and in many cases will send its employees to the plant of an equipment supplier or a large customer.

16.10c Process-control Requirements and Test Planning. The absolute necessity for complete documentation and reporting appears in setting up process-control requirements and test plans. Many process-control standards involve destructive testing. The loss of positive test conclusions through inadequate process standards, insufficient training of personnel, or failure to record results could bring unbearable costs and schedule delays upon an organization. A continuous monitoring and quality control audit procedure is helpful in maintaining awareness of new technical developments and provides an extra contact for gathering suggestions for improvement from employees.

16.10d Inspection and Test Work-flow Diagrams. Inspection and test work-flow diagrams are useful for a rapid management appraisal of a development or production operation. These may be made even more effective by combining task diagrams with the program schedule. One major military contractor incorporates schedule and man-load requirements into his production flow diagram as shown in Fig. 16.13.

Man-load requirements incorporated into work-flow diagrams provide management with a useful history upon which to base its work-load forecasting, whether with the view of matching existing personnel with the demands of firm future business or of increasing the employment to meet the demands of expanded research, development, or production programs. Since payroll allocation is usually the largest individual budgetary item for an industrial organization, it is apparent that an unbalance in tasks vs. available manpower could seriously tax the resources of a company more quickly than any other single factor. While this is not primarily a task for reliability and quality control, it is apparent that histories gathered by these functions would provide much needed information for the broader aspects of management.

16.10e Automated Test and Measurement Equipment. Companies which employ typical mass-production methods and techniques will compare the values of inspection and test performed by people and by machines. The decision involves a prediction of the future business of the company as well as the relative merits of the choices and whether possible immediate rewards in the use of automated equipments will be offset in the future costs of labor.

The decision whether to abandon human versatility for the speed and exclusion of human error offered by the machine will usually hinge upon whether the computed savings available from the automated process can be offset in the then-known future business in which the process will be used. The character of business is a prime determining factor. A food-processing company could reasonably prorate automation costs over decades of anticipated business, whereas producers of novelty products, with considerable inventory capital risk, would expect to pay for the automated process within the initial venture. Almost all automated inspection-process equipments are custom-made, and there is practically no recoverable equity from such equipments other than that resulting from product sales.

Of course, automated test equipment must promise such distinct improvement over the equivalent human effort that its performance factor is practically measurable in the overall process reliability number. If this cannot be clearly determined, the decision is easily resolved against the automation feature. Certain processes are resolved in favor of the automated inspection or test equipment when safety considerations dictate, e.g., explosives manufacture, handling of radioactive materials, and certain critical chemical and biological interactions.

The design criteria of test and measurement equipment are set by the requirements of the product to be tested. As dimensional and performance tolerances of the product are tightened, the design problems of the measurement equipment are increased exponentially. Management must know that its own engineering has determined that

PAGE ___ OF ___

CREW LOAD CHART

APPROVED *[signature]* DATE *5-15-62*
APPROVED *[signature]* DATE *5-15-62*
PREPARED *[signature]* DATE *5-15-62*
INDUSTRIAL ENGINEERING DEPT

PROGRAM ___ ITEM 602 (PRIMARY)
DEPARTMENT 57 STATION 2 (COST CENTER 04)

SHIP NUMBER *65* 7 DAY CYCLE
ACTUAL START DATE *5-16*
ACTUAL COMPLETION DATE ___
TOTAL TASK (MAN HOURS) 1120

SCHEDULE DATE: *5-16*, *5-17*, *5-20*, *5-21*, *5-22*, *5-23*, *5-24*
HOURS: 8, 16, 24, 32, 40, 48, 56, 64, 72, 80, 88, 96, 104, 112

	1ST SHIFT	2ND SHIFT	Tasks (in schedule order)
P & H	*MOORE* 1.	*ROE* 1.	54815 INST TUBE 5.0 AFT; 54564 INST HYD LINE; 54855 INST TUBE; 54246 INST TUBING WING; 54113 DRAIN VALVE; 55178 HYD LINE; 54984 ADJ ACTUATOR L/H; 55040 INST PIPE; 55061 INST PIPE; 54983 ADJ ACT R/H; 54390 PIPING 5.0 FWD; 55121 TUBE INSTL
	STEVENS 2.	*MILLS* 2.	54395 HYDRAULIC PNEUMATIC L.E. R/H; 54393 HYDRAULIC PNEUMATIC L.E. L/H; 55097 TUBING R/H; 54009 BRAKE M.L.G. EQUIP R/H; 30713 PIPING M.L.G. R/H; 30727 HYD RESERVOIR R/H; 54396 STABLE TABLE AREA HYD; 54394 STABLE TABLE AREA HYD; 30757 CHAFF DISPENSER PIPING R/H
	BURNS 3.	*LEWIS* 3.	55098 TUBING L/H; 54010 BRAKE M.L.G. EQUIP L/H; 30714 PIPING M.L.G. L/H; 30728 HYD RESERVOIR L/H; 30756 CHAFF DISPENSER PIPING L/H
	ADAMS 4.	*CLARK* 4.	36177 INSTALL ELEVON TUBING R/H; 30721 INSTALL ELEVON TUBING L/H; 30723 INST TUBING WING L/H; 30724 INST TUBING WING R/H; 54380 PIPING 5.0 FWD
STRUCTURE	*ROY* 5.	*WEBB* 5.	30743 INSTALL CENTER & AFT W.W. FAIRING L/H; 30744 INSTALL CENTER & AFT W.W. FAIRING R/H; 54991 BRKT L/H; 54987 INST RADOME L/H; 54440 - 45020 MANIFOLD REFUEL LINE; 30541 POD CK; 54841 INST VALVE; 54796 FUEL TRAP; 54442 DUCT INSTL STABLE TABLE AREA
	STRONG 6.	*BENTON* 6.	54992 BRKT R/H; 54988 INST RADOME R/H; 54813 SAFETY VALVE; 54819 COLD AIR DUCT; 54903 STABLE COOL; 54859 INST SERVO LINE; 54443 DUCT INSTL STABLE TABLE AREA
	RUBIN 7.	*SIMS* 7.	55091 BRKT; 55010 DUCT INSTL ASTRO TRACK; 54018 TURBINE COMPRESS; 54019 TURBINE COMPRESS; 54862 INST DRAIN LINE; 56081 INST DUCT 5.0 FWD; 30668 PRESS SENSORS; 30652 DRAG STRUT; 30651 DRAG STRUT
	WILLIAMS 8.	*ROSS* 8.	RMV COWL; 30790 #2 FUEL SYS; 30796 HEAT OUTLET; 30799 EXCH DUCT; 30794 HEAT EXCH; 30790 COMPLETE FUEL SYS; INST COWL; RMV COWL; 30790 #3 FUEL SYS; 30796 HEAT OUTLET; 30799 EXCH DUCT; 30794 HEAT EXCH; 30790 COMPLETE #3 SYS; INST COWL
ELEC.	*SLACK* 9.	*MINOR* 9.	55078 CHAFF DISPENSER HARNESS L/H; 55079 CHAFF DISPENSER HARNESS R/H; 30280 4E1885 HARN; 30281 4E1880 HARN; 30330 4D30073 HARN; 30332 4E 1319; 30543 ROUTE W.W. HARN L/H; 30333 4E1316 HARN; 30441 4E1327 HARN; 30442 4E 1328 HARN; HARNESS CLEAN UP
	LILES 10.	*HUEY* 10.	30102 4E1318 HARN; 30103 4E1321 HARN; 30104 4E1331 HARN; 30602 4D203 HARN; 30650 BRKT; 30555 4D30075 HARN; 30651 BRKT'S; 30880 4E1012 HARN; 30992 4E1880 HARN; 30993 4E1710 HARN; 30652 4E111 HARN; 30653 4E112 HARN; 30482 4D30074 HARN; 30542 ROUTE W.W. HARN R/H; 30899 4D11034 HARN; 30900 4D30075 HARN

PERCENT ACTUAL
PERCENT SCHED: 0, 14.3, 28.6, 42.9, 57.2, 71.5, 85.8, 100

FIG. 16.13 Daily accomplishments crew load chart.

such equipments are compatible with the products to be tested and also that they are adaptable to future changes in the product line.

16.11 RELIABILITY AND QUALITY CONTROL FACILITIES AND EQUIPMENT

The nature of the reliability and quality control activity imposes an added burden upon the planning which must precede the provision of facilities and equipment. The managers of plant engineering and facilities functions are under constant pressure to hold down the costs of space, equipment, and material, as well as the cost of personnel. In the natural optimism for and self-confidence in the organization and its product, quality and reliability methods and equipment requirements are sometimes taken for granted.

To anticipate the necessary provisions for product assurance in advance of the final (production) design and manufacturing places reliability and quality in superposition with profits. Many companies whose contracts are primarily with the Department of Defense are so organized that these requirements are accounted as overhead; hence, the procurement of facilities and equipment whose cost was not fully provided for in final contract negotiations is exceedingly difficult, except where justified by cost savings or accepted as nonreimbursed overhead. Advance planning of all such costs is necessary if management is contractually responsible for reliability and quality performance, and certainly to whatever modicum the company feels ethically bound in the absence of a specification.

It is desirable that the provisions for reliability and quality control facilities and equipment be made in close cooperation with the company's engineering design group; if feasible, the planning should be made during the concept and preliminary design phase of the product, and certainly in conjunction with plans for new plant locations or structural additions to the existing plant. It is important that any particular requirements for test equipment be given to management so that they can be provided in the planning layout of new facilities.

Coordination of reliability and quality control with design engineering results in a knowledge of what the product is intended to do. This information and the contract specifications will allow the setting up of economical quality control sampling plans and appropriate reliability demonstration test levels, thereby determining the appropriate facilities and test equipment.

This liaison enhances the compatibility of test tolerances at all stages of product inspection and permits an orderly expansion of generic tolerances from part supplier to assembly producer to consumer.

16.11a Funding and Schedules. The critical demands of advance planning for reliability and quality control equipment appear in the funding and scheduling of the production master plan. Equipments which require long lead procurement—and sometimes purchase-guarantee escrows prior to delivery—must be included within the master schedule to minimize the terms of loan capital provided for this purpose. Similarly, the funding requirements for facilities must be evaluated, for these will include such considerations as inspection-area lighting, temperature, humidity, air conditioning, clean-room air control and flow distribution, special disposal and sanitation installations, personnel safety provisions, and mobile access into all such areas.

16.11b Equipment Specifications. To unify the management of reliability and quality control organizations, it is to the advantage of each that test-equipment procurement specifications be generated within the organization. In this way no other operating group can establish the boundaries of test and inspection by indirection and reliability and quality tests can be established over the full design spectrum of the product. Also, for companies with multiple product lines, reliability and quality control management can see the entire test picture and can advise the purchase of equipments compatible with any tests which may be required. In this manner fewer equipments of greater capability may involve less capital expenditure than more equipments of limited and singular capability. If the company procurement policy does not allow the generation of reliability and quality control specifications, the management of those functions will certainly elect to advise the procuring group of its judgment through appropriate intracompany communication.

16.11c Reliability and Quality Control Design of Test Equipment. In some organizations the reliability and quality control groups have been given the responsibility for test-equipment design. This requires that very capable engineers be made responsible for this effort. When adequately staffed, certain advantages may accrue through this organization policy. These advantages include improved timeliness and effectiveness of test equipment, greater emphasis on automation, improved supplier coordination, improved integration of all test functions, and optimum emphasis on nondestructive inspection and test methods.

Disadvantages may develop if capable equipment-design personnel are not available to staff the equipment-design function. This frequently leads to the use of alternate or makeshift types of equipment, which do not provide optimum cost effectiveness. It must be recognized that an additional level of coordination with product engineering is required in order to assure maximum compatibility of the test equipment with a product. This is very important, particularly for products which have frequent model changes.

16.12 TRAINING

The performace of personnel who define, design, procure, manufacture, construct, test, repair, and operate equipment is inherently variable because of inequalities in skills, knowledge, personality, education, and training. This performance variability affects the quality of equipment and manufactured products. Advance planning for training is an activity that should coincide with the advance product goals set by the management of the company.

The plan of action by management for the advance planning of the goals rests and is dependent on the company's resources such as facilities, tools, raw materials, personnel, productive capacity, sales outlets, and public relations nature of previous business and the expectations for possible future business. Because business is subject to change, it is rather difficult to predetermine definite training courses during the early product-planning stage. But when a product becomes firm business and specifications are known, training plans must be activated on a time-phased basis.

16.12a Reliability and Quality Control Training. The purpose of reliability and quality control training is to communicate skills, methods, ideas, objectives, and attitudes to all personnel levels in an organization. Effective training incorporates the

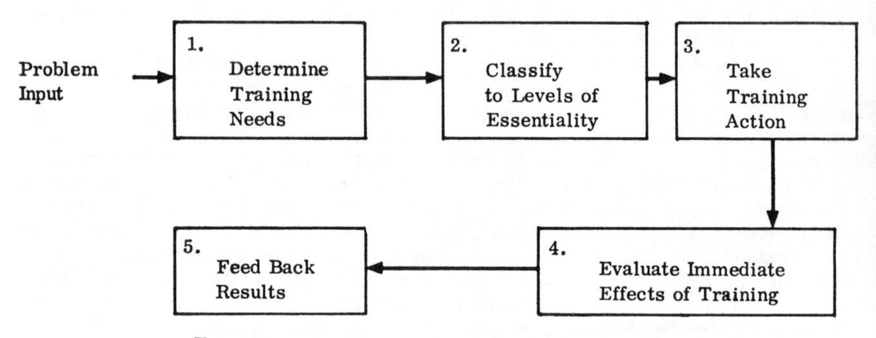

Fig. 16.14 Systematic planning and training cycle.

identifying, measuring, and supplying of the training needs that develop day-by-day in various activities. Reliability and quality control management should assure the accomplishment of education programs to indoctrinate all personnel whose work relates to the product's reliability. The assigned personnel must understand the value of their individual contributions to the product and be motivated to provide excellent results.

The need for additional specialized training can be evaluated by consulting the sources of information concerning any new task. The program plan certainly should indicate the various operations which require accomplishment. With the various operations and responsibilities known, the job performance and qualification requirements relating to the task should be explored.

One of the duties of the quality assurance engineer should be to ensure that supervisory personnel become aware of the training needs of their workers and to make certain that means are devised and used to determine exactly what, when, and how training is to be implemented and made effective (Fig. 16.14).

16.12b How to Measure the Employee's Knowledge and Specific Training Needed. In order to measure the employee's knowledge and determine what specific training is needed, we can use what might be called job knowledge quotient or JKQ. Job knowledge quotient is a series of test questions designed to be answered by employees. Different sets of questions can be made applicable to specific areas required of job knowledge.

An employee's experience and background provide management with an indication of the needs of training that can be expected. Once the information is gathered and analyzed, there should be an understanding of how much and what kind of training the employee needs. Training needs comprise the skills, knowledge, information, and attitudes which individuals require to meet reliability and quality specifications. Changing demands often alter requirements of skill and knowledge and increase training needs.

Subjective measurement devices and techniques are available to identify and gauge these needs. These techniques and devices are:

1. Job or activity analysis
2. Tests or examinations
3. Questionnaire or improvement-checklist surveys
4. Purposeful observations and consultations based on history of errors
5. Reliability and quality control reports and audits
6. State-of-the-art surveillance and review for changes
7. Merit and performance ratings.

The difference between what is required and what the employee is capable of performing represents specific training required:

$$R - P = T$$

where R = requirements of task
P = what employee is capable of performing
T = specific training required

The primary objectives of reliability and quality control training and indoctrination are to:

1. Promote reliability and quality control consciousness in all personnel engaged on the project.

2. Emphasize to personnel in engineering, manufacturing, reliability quality control, purchasing, etc., the specific effects of their particular jobs in contributing to or detracting from system reliability.

3. Afford all personnel with sufficient knowledge and understanding of the specific and general factors affecting system reliability to assure the incorporation of good reliability techniques into the design and manufacture of equipment.

4. Assure that all reliability and quality control personnel are capable of performing their tasks effectively and efficiently.

5. Concentrate attention on those areas of activity considered to be particularly amenable to a reliability improvement effort.

To implement these objectives, a formal reliability training course should be instituted for engineering and quality control personnel involved in the analysis, design, manufacture, and test of complex products. The main topics of the course may include the following: introduction to reliability probability theory, reliability parameters and distributions, techniques, design applications, contract interpretation and compliance proofs, failure reporting and corrective actions, new-product assurance, and manufacturing procedures.

Formal training should be given to those who "need to know" new methods and techniques. Naturally, there may be some personnel who are not concerned with certain phases of training and have no specific need to attend all particular training sessions. Economy must be observed in training budgets, as in other elements of the organization. Detail training programs should also be established to train and certify specific types of personnel employed on technical and production jobs, which might be in the area of reliability, monitoring, nondestructive testing, acceptance testing, qualification testing, welding operators, soldering personnel, and inspection.

16.12c Informal Training.[1] Informal training (on the job) occurs throughout industry when any member of management gives instructions to his subordinates. Skill in such communication is important in achieving desired actions. Motivation for quality and reliability is a daily task and is the result of organized effort. It

[1] Training information adapted for this section was derived from the *Quality/Reliability Assurance Training Program Notebook*, vols. 1 and 2, and used with the permission of General Dynamics/Electric Boat, Groton, Conn.

Schedule # 105

SYSTEM: STRUCTURES

COURSE:		# EMP.	DEPT.
Number		10	180-2
Outline #		1	182
Type	SPEC	2	12
Length	20 HRS	8	29
Start	18 MAY 64		
End	22 MAY 64		
Hrs/Day	4/5		
Class Time			
Originator			
Instructor			
Coordinator			
REMARKS:			

Schedule # 106

SYSTEM: LANDING GEAR W&B

COURSE:		# EMP.	DEPT.
Number		10	180-2
Outline #		1	182
Type	SPEC	3	50/51
Length	10 HRS		
Start	18 MAY 64		
End	22 MAY 64		
Hrs/Day	2/5		
Class Time			
Originator			
Instructor			
Coordinator			
REMARKS:			

Schedule # 107

SYSTEM: LANDING GEAR W&B

COURSE:		# EMP.	DEPT.
Number		3	12
Outline #		4	27
Type	SPEC	8	29
Length	10 HRS		
Start	25 MAY 64		
End	28 MAY 64		
Hrs/Day	3/4		
Class Time			
Originator			
Instructor			
Coordinator			
REMARKS:			

Schedule # 108

SYSTEM: LG, WHEELS & BRAKES

COURSE:		# EMP.	DEPT.
		5	24-1
Outline #		8	94
Type	FAM	1	107
Length	5 HRS		
Start	5 MAY 64		
End	6 MAY 64		
Hrs/Day	3-2/5		
Class Time			
Originator			
Instructor			
Coordinator			
REMARKS:			

Schedule # 425

SYSTEM: HYDRAULICS

COURSE:		# EMP	DEPT.
Number		2	178
Outline #		9	94
Type	FAM		
Length	5 HRS		
Start	8 SEPT 64		
End	9 SEPT 64		
Hrs/Day	3-2/2		
Class Time			
Originator			
Instructor			
Coordinator			
REMARKS:			

Schedule # 426

SYSTEM: PNEUMATICS

COURSE:		# EMP.	DEPT.
Number		2	178
Outline #			
Type	FAM		
Length	5 HRS		
Start	11 SEPT 64		
End	11 SEPT 64		
Hrs/Day	2/1		
Class Time			
Originator			
Instructor			
Coordinator			
REMARKS:			

FIG. 16.15 Training schedule.

requires the measurement of progress and gives frequent feedback to employees of the quality of job they are doing. Control charts provide a scoreboard of personnel performance. This feedback of information, when coupled with plans for corrective-action patterns, will promote desired motivation.

Motivation, explanation, performance, review, and follow-up are four basic steps that may serve supervisors in making on-the-job training effective.

Motivation. Trainees must know why they are being trained, what they are expected to learn, and how they can learn quickly and efficiently. Interest and concern for the training intended must be aroused.

Explanation. Supervisors must explain and discuss the complete task in relation to other tasks and the information required to complete the task. They must show what task is to be completed and how and why.

Performance. Trainee performs the task or job required. Supervisor points out errors committed and explains methods of avoiding errors.

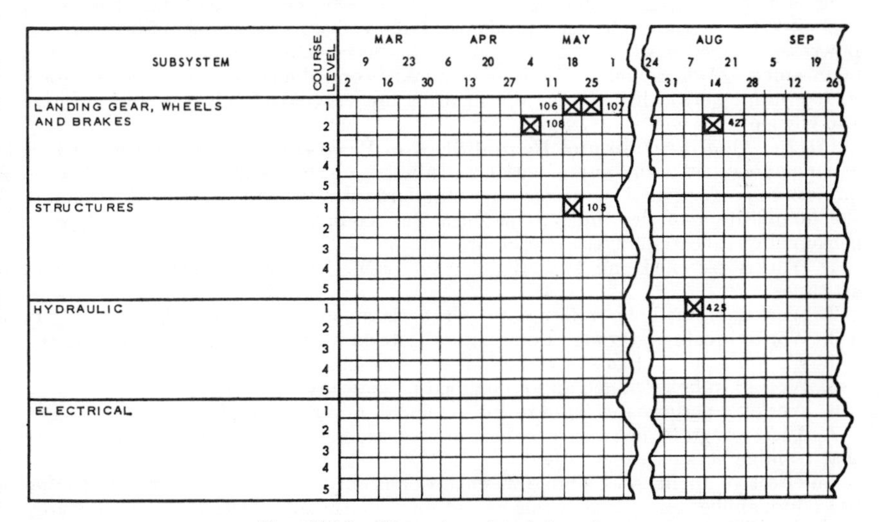

FIG. 16.16 Time-phased training plan.

Review and follow-up. Follow-up in which errors are corrected, questions and further explanations are made, and further training is given where necessary.

16.12d Formal Training. Formal training occurs when skills, experience, ideas, and information are organized into a classroom curriculum to achieve desired levels of skills and understanding. The objectives in training programs must be stated, and they must be realistic. The applicable subject matter must be organized and accurate, and methods must be suited to subject matter. Instructors must be qualified and experienced, and proper evaluation and feedback for curriculum improvement must be provided. Schedules must be realistic and planned to have personnel trained as the task is implemented. Figures 16.15 and 16.16 show a time-phased training program plan.

16.12e Training Schedule. Figure 16.16 indicates a portion of a simulated time-phased training program. In this example class schedules are presented; they indicate the calender week in which the various classes are scheduled. Each schedule block is identified by an adjacent schedule number. Figure 16.15 gives details of starting and ending dates, duration, and hours.

In most instances courses on subsystems are offered on several different levels. Level 1 is a thorough course of detailed theory intended for those requiring complete specialized knowledge of the system. Level 2 is a familiarization course of approximately one-half the duration of level 1, suitable for mechanics, technicians, and inspectors, while level 3 is an orientation-type course involving little system theory and is about 15 per cent of the duration of level 1. Classes below level 3 will be packaged (owing to their short duration) into electronic and mechanical subsystem indoctrinations. Starting dates for the various classes should be realistically based on current-schedule task plans and departmental needs.

16.12f Certification Programs. Reliability and quality control individuals and management should make certain that specific procedures provide for the control of special processes. Such special processes are frequently required to be certified by customer or government representatives in accordance with their operating policies. It is characteristic of special processes that the result of the inspection and test of the product after completion of the process is often insufficient or too late. For this reason, the effort to produce the required control should be made during the in-process procedure.

The responsible quality department may certify the conformance of process procedures as a means of establishing the quality level when sufficient data have been evaluated to determine the acceptability of the process. Products employing critical processes require more attention in order to be certain that sufficient control is maintained. Added effort by process or quality control engineers may be necessary in determining criticality and in implementing process controls which will be effective and efficient.

16.12g Qualification and Requalification Programs. Since the performance of personnel varies with comprehension and ability, job standards should be established for desired levels of quality work. Such standards should be the basis for determining the qualification of personnel to perform the required work. The basic steps in personnel qualification determination are:

1. Worker screened and evaluated for applicable background, prerequisites, and attitudes.
2. Worker trained, both formally in the classroom and informally on the job.
3. Worker qualified through examination or tests based on validated skills criteria.
4. Worker's performance to be periodically evaluated.
5. Worker to be retrained and requalified periodically.
6. Worker's performance is recorded, reviewed, and analyzed for retraining.
7. Audit program instituted to ensure that each step is complete and make sure that training actions are effective.

An effective qualifications program ensures not only that an adequate qualification procedure applies but also that this procedure is continually upgraded to meet changing specifications. Additionally, an effective qualifications program ensures that the variability of each factor of the qualification procedure is controlled to a desired degree.

16.12h Evaluation of Training. Evaluation of training is necessary to determine whether trainees have or have not reached predetermined goals. The basis of effective evaluation is the observation and measurement of some performance before planned training and measurement of the same performance after training. A comparison of the results evaluates training. Evaluation is based upon a record of all available evidence which shows the degree to which training objectives were or were not realized, the improvements effected, and the ultimate effects on production activities. Training which involves measurement of errors, defects, failures, waste, or speed and productivity can be evaluated and measured objectively with the before and after approach. Training which involves mental skills and long-term development will involve subjective measurement.

The following factors can be used to evaluate training for both mental and physical skills:

1. Statistical measurement of before and after performance recorded on control charts
2. Checklist enumeration of improvements before and after performance
3. Recorded changes in job-performance ratings by supervisor
4. Written tests and examinations

5. Tabulation and analysis of quality control and reliability reports
6. Comparison with simulated control groups
7. Comparison with personnel case histories
8. Number of hours spent in training

16.12i Guidelines for Effective Evaluation

1. Evaluation must seek out successes as well as failures.
2. Evaluation must start with specific skill objectives to be achieved.
3. Evaluation must be built around a systematic long-term, continuous plan as required.
4. Evaluation must determine the degree to which training resulted in sufficient learning.
5. Evaluation should be made immediately before there are significant losses from other sources.
6. Evaluations tailored to one's own activities are better than the use of ready-made ones by outsiders.

Training records should be maintained in a manner similar to production- or inventory-record maintenance. These records, when accumulated over a period of time, should represent an inventory of skills and a distribution of variability in both professional and trade skills existing in the facility.

16.13 PRODUCT CONFIGURATION AND CHANGE CONTROL

Management must be assured of the incorporation of proper design changes for special-purpose products, such as the customer options in automobiles and changes effecting improved safety and performance in aircraft and marine vessels. The need for the compatibility of service equipment and manuals with the manufactured product by a certain model designation is apparent.

Control must be emphasized from early design conception to end use in order to prevent failures and improve the value of the product and assure management that performance warranties will be achieved. One can readily visualize that, with fewer failures extending over greater periods of time, the maintenance, spares, and technical manual programs will be more effective and less expensive.

16.13a Product Configuration Organization. Configuration management may be referred to as those procedural concepts by which a uniform system of configuration identification, control, and accounting is established and maintained for all systems, equipment, and components.

The organization of configuration management must be arranged to serve to the best advantage the assignment of responsibilities and duties through the agencies available to them. These agencies must be able to integrate or coordinate most effectively in a homogeneous manner the many interfaces that are involved in configuration management.

Management of the various programs administer the assigned configuration task. Engineering designs the various products to meet the requirements specified by the customer. Procurement and manufacturing are activated by engineering drawings and specifications. Quality assurance verifies and audits engineering, procurement, and manufacturing compliance with customer requirements and performance specifications. When inspection accepts the product to engineering drawings, quality assurance is assured that interfaces have been established and that no degradation of quality or reliability of product has taken place during the manufacture of product.

16.13b Method of Integrated Electronic Data-processing Systems for Configuration Control. The first contract for an aerial weapons system was issued to Wilbur and Orville Wright by the U.S. Government for one heavier-than-air flying machine in accordance with Signal Corps Specification No. 486 dated December 23, 1907. The general requirements for this weapons system were rather simple and straightforward, but even then the contractor could be penalized or rewarded depending on his performance to contract. In the total weapon system of the Wright brothers were what we now refer to as end items of equipment. Determining the configuration was simple, and documentation was not required and probably not necessary. If the Wright brothers had been asked what the configuration of their airplane was, they could have probably said, "That's easy, linen, pianowire, wood, fish glue, nuts, and bolts."

Not too long ago it was necessary on occasion to refer back to the research-and-development configuration status of certain operational missiles in order to assure accurate configuration for tactical operation. It has also been found that similar problems existed in certain manned aircraft in the past. This past experience in missiles and manned aircraft along with a sincere desire to provide the customer with a weapon system which can be operated efficiently has prompted certain industries[1] to provide the ultimate in configuration information and control.

The following complex elements represent the basic points of good logistic support, including configuration-identification documents, which must form a part of overall logistics:

Configuration-identification index
Configuration status accounting report
Time compliance technical orders
Technical manuals
Spares
Training
Maintenance of aircraft
Armament

It must be remembered that an integrated data-processing system can perform the configuration-control functions which will most economically support engineering design, procurement, quality control, reliability, and ground-support equipment departments. Through this centralized function, various levels of management and the customer receive important status reports on product configuration in a timely and accurate manner.

As previously stated, configuration management is a formal set of procedural *concepts* by which a uniform system of *configuration identification, configuration control,* and *configuration accounting* is established and maintained for all customer systems, equipment, and components of systems and equipment. These concepts are basic. However, it has been stated that[2]

> It is the right arm and nervous system for the program director, giving him the official detailed information necessary to adequately define the weapon system being supplied to the customer and it gives him the management tools necessary to bring engineering change information to the individual having decision-making authority. It also provides the Manager with a system to disseminate the decisions to all who need them. It gives the Manager the Management System that supports a continuing follow-up to make certain that the decisions have been carried out and allows him to know the actual status of accomplishment on a continuing basis.

The new document referred to is the Air Force System Command Manual 375-1, dated June 1, 1962, called Configuration Management During the Acquisition Phase, which means until delivery of the last end item by the contractor. This document is clearly written, is one of the best on the subject of configuration management, and lends itself readily to further expansion.

It should be noted that many of the requirements called for in AFS CM 375-1 existed previously in some form. The basic objective was to gather all such requirements into one document and then clearly define previous desires of the customer.

[1] General Dynamics/Astronautics, San Diego, Calif., has been an outstanding leader in the development of configuration control as we know it today. A refined electronic data processing for configuration status accounting is now in use at General Dynamics/Fort Worth. (See Sec. 7.)

[2] This was stated by Lt. Col. Ballis, USAF Office, Deputy Chief of Staff/System, who, under General Shreiver's orders, established the customer's basic configuration management policies and requirements from which a new document was developed.

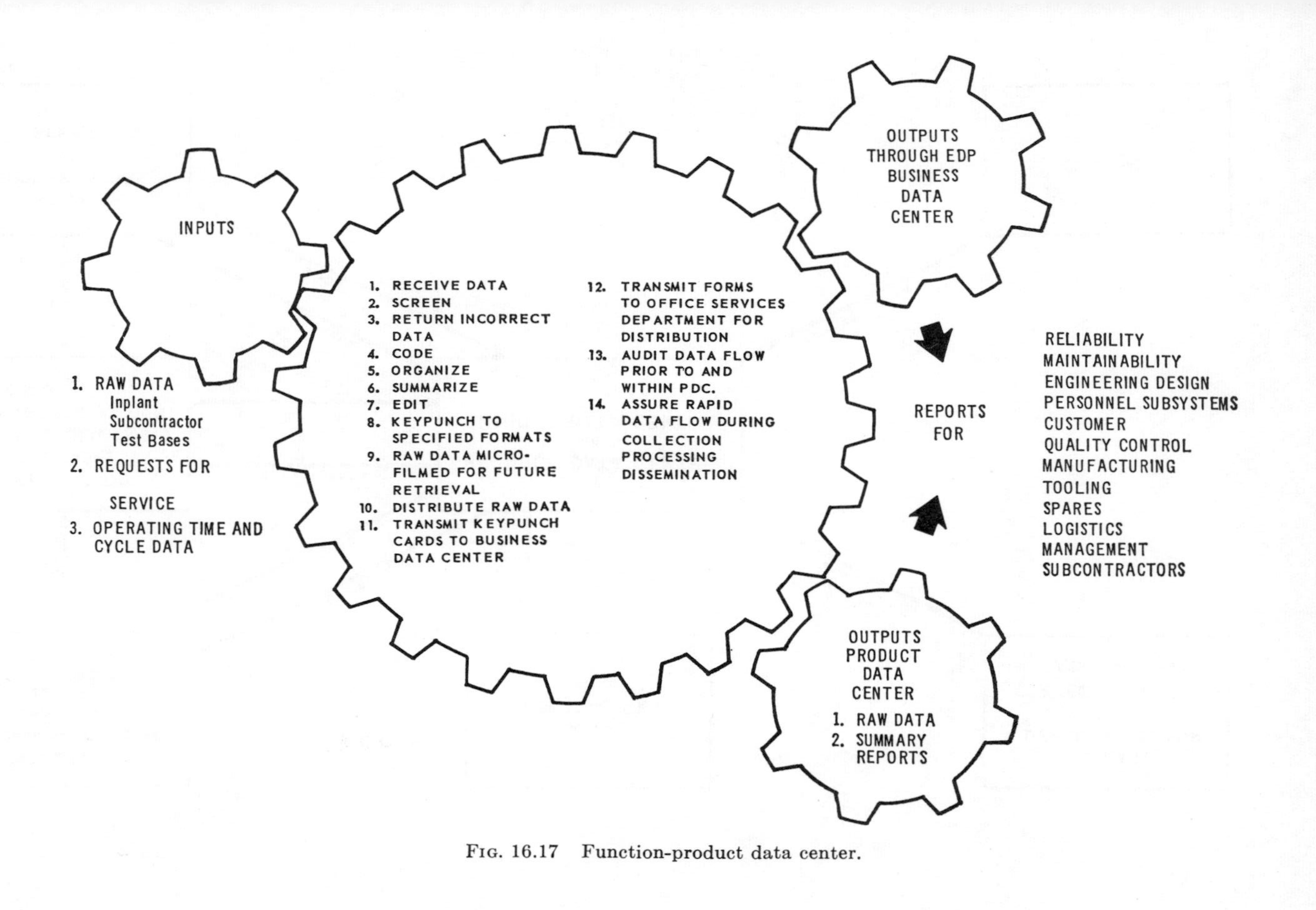

FIG. 16.17 Function-product data center.

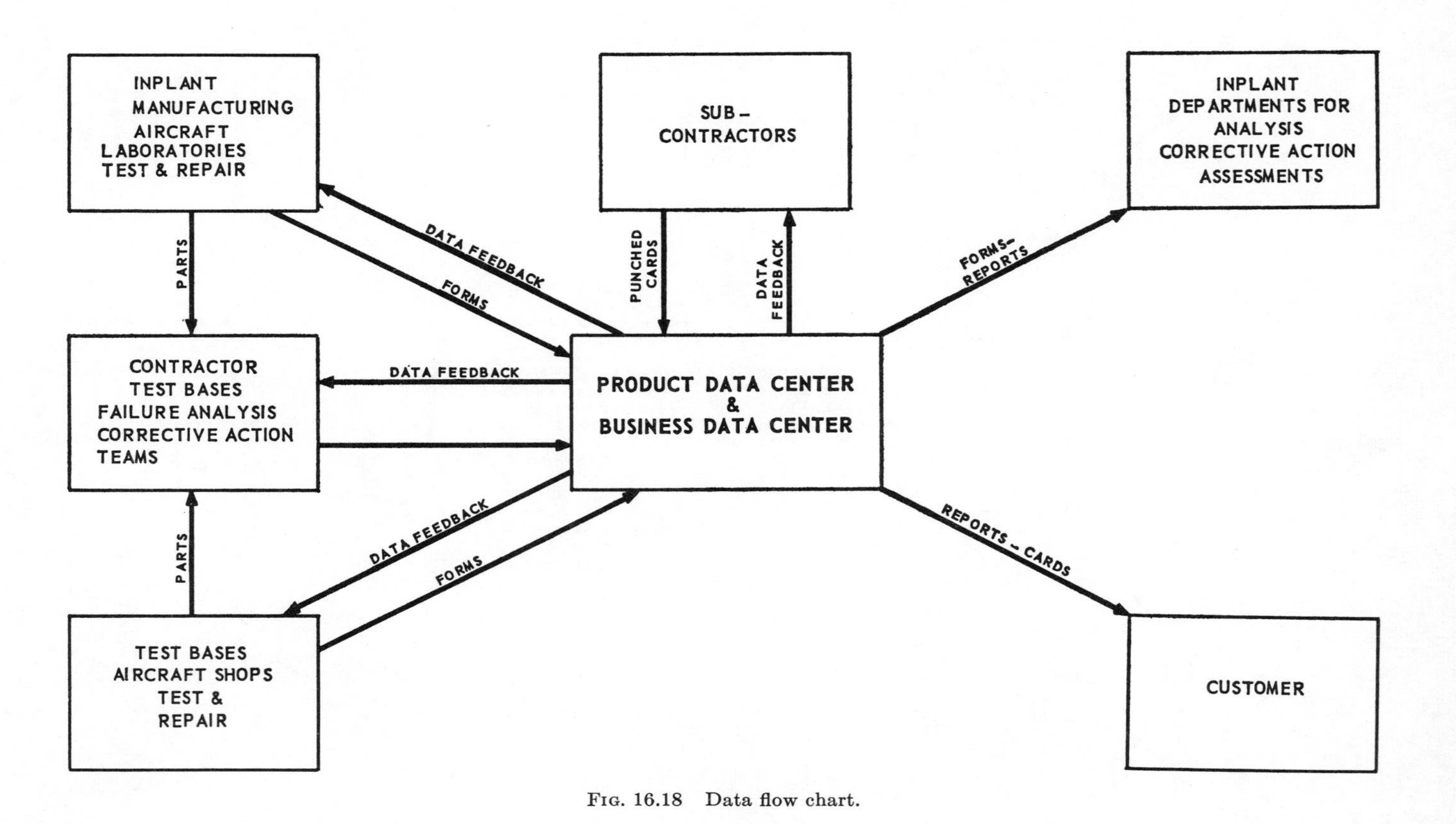

FIG. 16.18 Data flow chart.

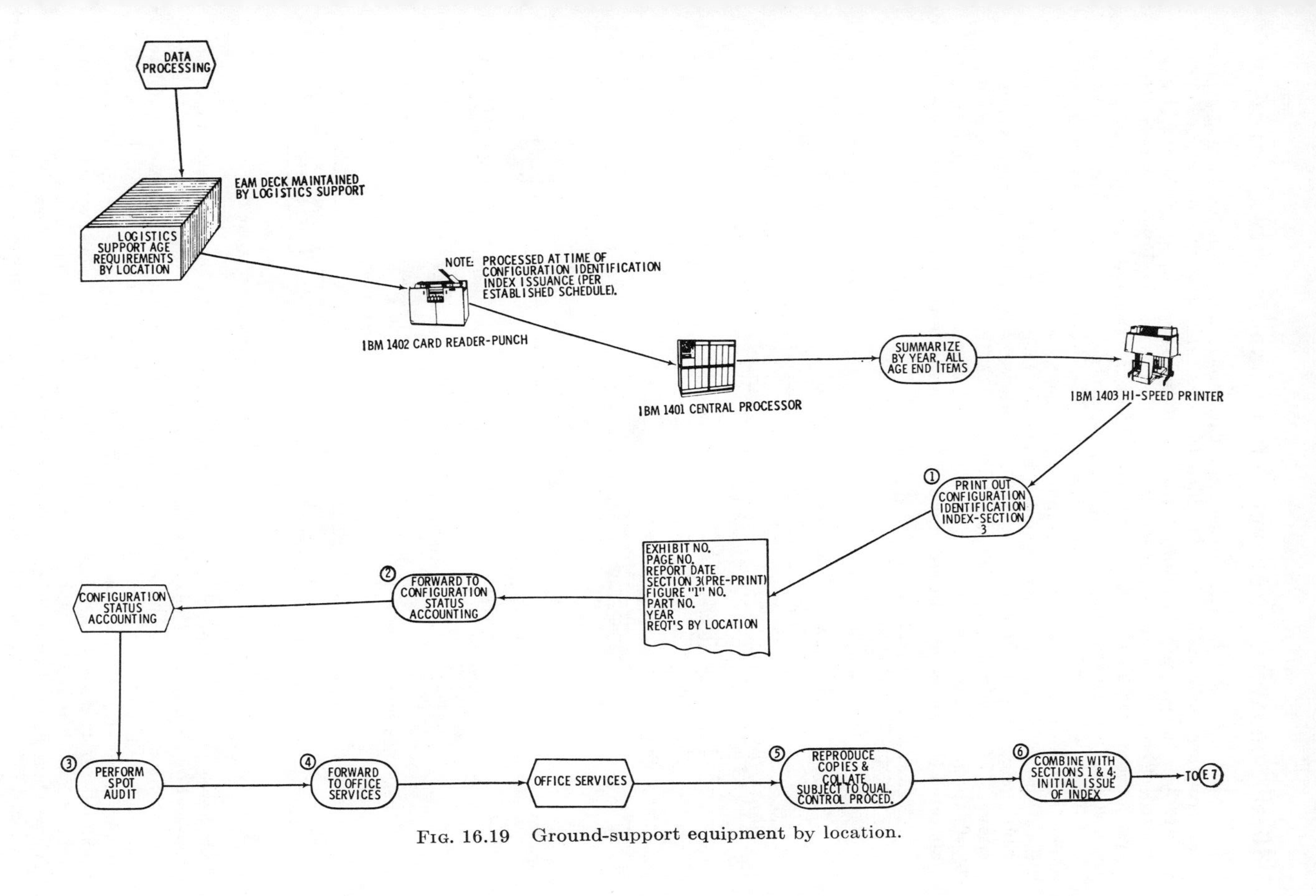

Fig. 16.19 Ground-support equipment by location.

It also set up a comprehensive electronic data-processing status accounting system for recording configuration from the inception of a program to its conclusion.

16.13c Investigate Possible Data-input Sources. It is mandatory that all related electronic accounting machine (EAM) or electronic data-processing (EDP) systems and programs—existing or proposed—be investigated as possible data-input sources. This is necessary to avoid duplication of effort in source-document search and input encoding. The example shown in Fig. 16.19 is an actual case in point. The logistic support—aerospace ground equipment section maintains an EAM card deck containing all AGE requirements, including their location. By a simple IBM 1401 pass, selected data are extracted and processed, resulting in the issuance of section 3 of the configuration identification index–AGE requirements by location.

In the November, 1962, issue of *Systems and Logistics News for Aerospace Industry* appeared this comment: "Contractor's systems for collecting maintenance data from

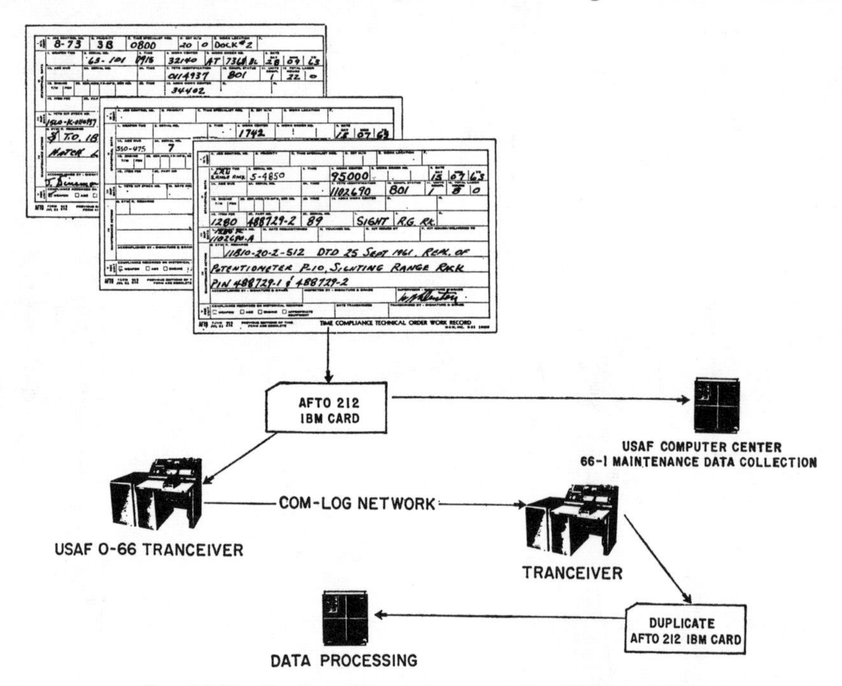

FIG. 16.20 Configuration-status accounting EDP system.

Air Force field organization duplicated the Air Force data collection system. The duplication resulted from lack of availability of data from Air Force in the form desired by some contractors." This indicates that a number of contractors lacked complete knowledge of sophisticated EDP techniques or the operation of the Air Force 66-1 Maintenance Data Collection System. When contractors have to transcribe data manually from various Air Force reports generated by the 66-1 system, the data reflected in their reports will be from 30 to 90 days old at the time of publication. The system should be tied in effectively with the existing Air Force system. Shown in Fig. 16.20 is a schematic picturing the addition of TCTO accomplishment data into the configuration-status accounting system. This approach eliminates manual encoding and duplicate keypunching, and as a result the data reflected in these index and status reports will be current. This same technique can be applied to end-item removals and replacements. It will be noted that this company proposes to receive duplicate AFTO-212/211 cards over the com-log network which utilizes the IBM 65/66 data transceiver.

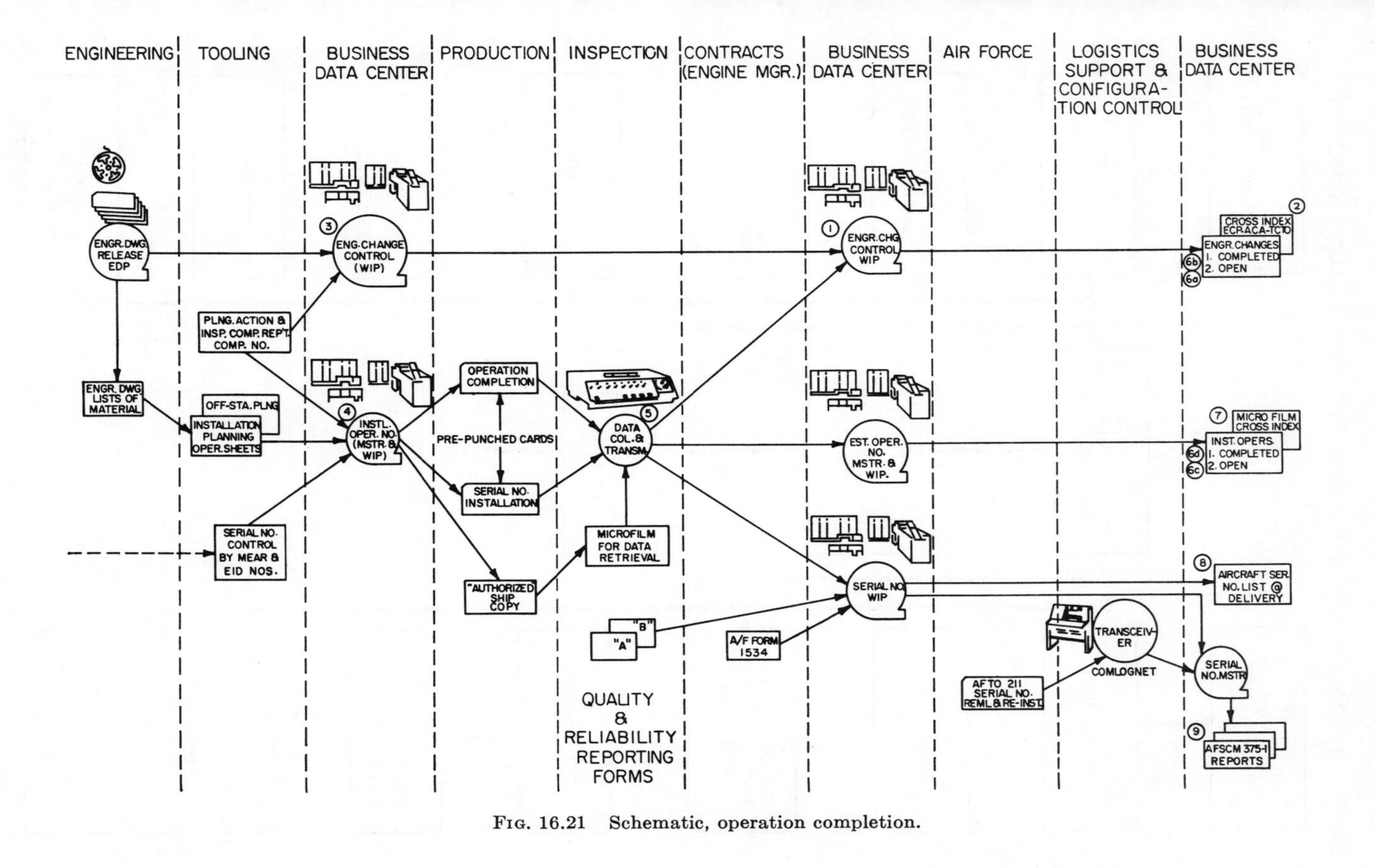

FIG. 16.21 Schematic, operation completion.

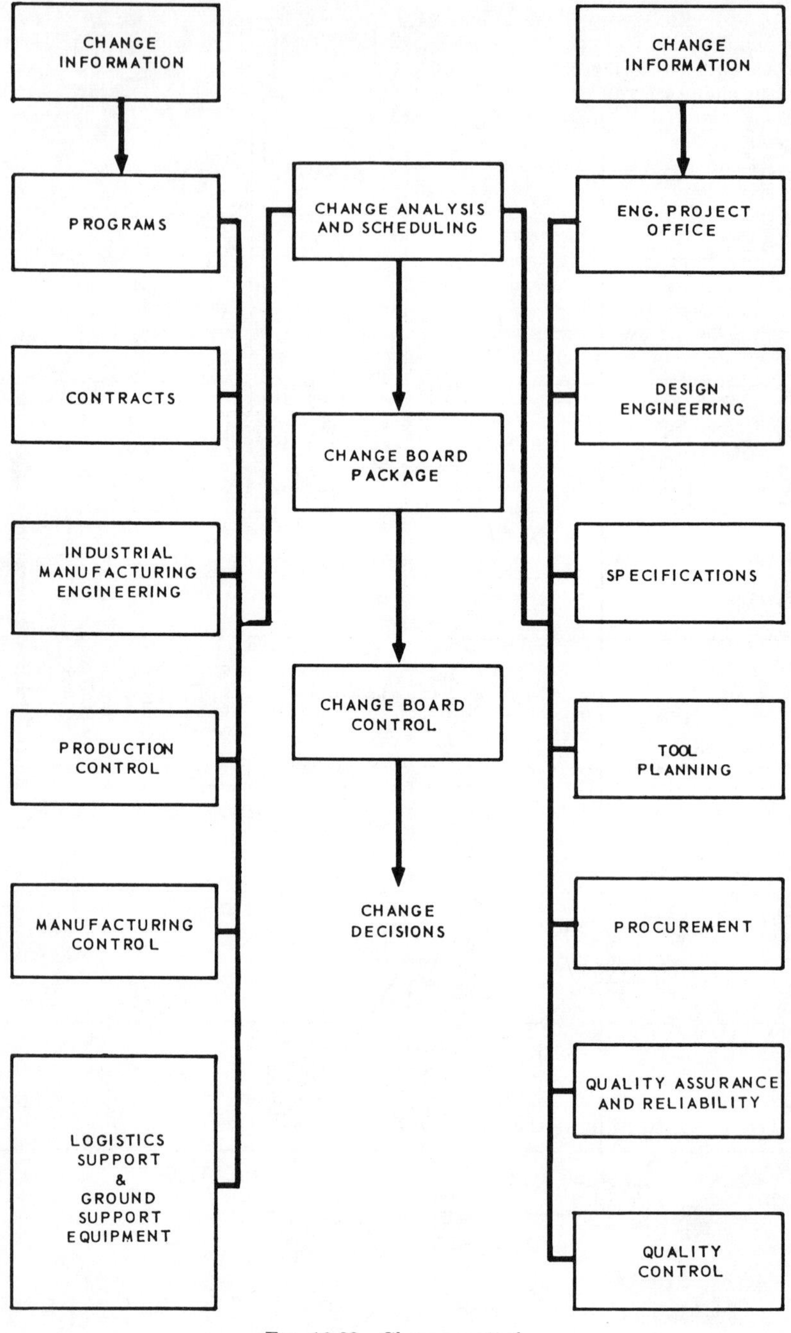

FIG. 16.22 Change control.

16.13d Electronic-data-processing Control System. Figure 16.21 shows a schematic indicating operation completion and the serial number EDP control system. It is explained as follows:

The engineering-drawing-release EDP system has as its main output new or revised lists of material. An engineering-change-control (ACA) file is the first phase in this system. The airplane change analysis (ACA) file applies effectivity dates to every usage and parts list record entering engineering release and also sets up a work-in-process file for engineering-change verification.

Tool planning prepares regular and off-station planning (installation operations) and planning action and inspection completion reports, as well as notes the operation number, item, and station where a serially controlled end item is to be installed.

The business data center performs file maintenance activities on the engineering-change control and installation-operation files. The business data center also prepares prepunched cards for operation-completion reporting and serial-number installation, as well as prepares an "authorized" ship copy listing.

As production completes operations or installs serially controlled parts, these are recorded by means of the prepunched cards in data-collection and -transmission units (readers).

Inspection certifies operation completion and serial-number installation. Serial-number removal, rework, and/or reinstallation are recorded on quality assurance data forms (A&B).

Contractor prepares Air Force Form 1534 to report location and status of engines assigned to contractor.

The business data center also performs file maintenance activities on engineering-change control, installation operation, and serial-number files.

16.13e Change Control. Change control may be defined as a process by which management recognizes the need for a change in the product. Since the possibility of designing and manufacturing a perfect product on the initial effort is unlikely and most companies constantly strive for this perfection, there is a need for methods to process and control the desired changes.

Change recommendations may derive from many sources, so management must develop the method for recommending the needed change, proposing, analyzing, evaluating, and approving the change, and verifying change incorporation as scheduled. A flow chart regarding change control from the various sources is depicted in Fig. 16.22.

16.14 METROLOGY AND STANDARDS LABORATORIES[1]

16.14a Importance of Calibration. Calibrations and measurements provide a foundation for quality and reliability work in all of its phases. They help to assure uniformity and accuracy of electrical, physical, and dimensional measurements and tests among research, design, engineering, and manufacturing operation within a company facility as well as between the company and customer organizations, contractors, and vendors.

Test-equipment items, coupled with the increased accuracies and ranges for these instruments, cause problems for quality and reliability people in all related industries. An example of what happens when the proper attention is not given to calibrations is given in Fig. 16.23, which details the out-of-specification conditions found for a group of instruments used to perform special valve certification for pressure vessels. In this particular group, 58.5 per cent, or 66 out of 113 instruments, were found to be outside the specifications.

Some of the specific reasons why it is important to have good calibrations and measurements are as follows:

1. Compatibility of parts, assemblies, and systems
2. To check out parts, assemblies, and systems which have been repaired

[1] Much of the information and many of the illustrations in this metrology and standards section were taken from *Quality/Reliability Assurance Training Program Notebook*, vol. 2, and used with permission of General Dynamics/Electric Boat.

3. To obtain peak performance of all systems in use
4. To avoid damage during performance checkout runs because of improperly calibrated or functioning instruments

Very important, from the organization and administration viewpoint, are the technicians who are the ultimate users of the calibrated instruments, because they are another link in the chain of measurements and calibrations which prevails in the facility. Any weak link in this chain is bad; and if these people make errors in their measurements, the good work done in the standards laboratories is nullified. The naive assumption that anybody can make measurements is unwarranted in view of the measurement discrepancies which arise from time to time and the damage and abuse to which delicate instruments are subjected. Nowhere are these results more obvious than in the instruments which standards laboratories receive for calibration.

Calibration and repair work can often be greatly simplified and the accuracy level of measurements work greatly improved through careful training in measurements and

OUT OF SPECIFICATION CONDITIONS FOUND FOR A GROUP OF INSTRUMENTS USED TO PERFORM SPECIAL VALVE CERTIFICATION TESTS FOR SHIPS.

Instrument Type	Total	Within Specs	Outside Specs
Pressure Gages	32	16	16
Pressure Transducers	52	20	32
Gage Savers	16		16
Calibration Resistor	1	1	
Power Supply	1	1	
Frequency Standard	1	1	
Oscillograph & Recorder	1		1
Oscillograph Galvanometer	8	8	
DC Digital Voltmeter	1		1
Totals	113	47	66

FIG. 16.23 Outside specifications (58 per cent).

instrumentation. Training cuts down on damage and abuse of instruments, is a major factor which helps assure correct test data, and reduces testing and measurement time. Management and administrative personnel should be qualified in:

1. The importance of calibrations and measurements
2. How to administer a good program of calibrations and measurements
3. How to use measurement agreement and other evaluation techniques to find out how good or bad the calibration and measurement program is

Accuracy cannot be determined by blind faith in instrument readings, by repeatability of these instrument readings, or by sensitivity of the instruments used. Instead, accuracy must be determined by calibration, coupled with a proper analysis and evaluation of the various sources of error which may have arisen in the course of making the calibration. A good calibration program provides a solid basis for confidence in measurements made. It is the job of the standards laboratories to supply this confidence through the calibrations which are made.

16.14b Traceability. One of the essential features of calibration is that it is a process whereby an instrument or a standard is compared with a standard which is noted at a higher level of accuracy. This higher-level standard is an instrument or standard which has been calibrated at the National Bureau of Standards or at least calibrated or referenced to instruments or standards whose history relates their accuracy to the NBS instruments.

Traceability alone will not assure a compatible system of measurements and standards, because it is only one of several types of efforts. However, traceability is a very important and fundamental factor, because a rigorous consideration of traceability and all of its ramifications leads one to understand better the factors which go to make up a compatible system of measurements and standards.

There are various types of traceability: direct traceability, indirect traceability, frequency-conversion traceability, range-extension traceability, and derived traceability, as shown in Fig. 16.24. These various types of traceability are necessary to describe the ways by which we get standards for the wide variety of measurement categories, accuracies, magnitude ranges, and frequency ranges in use today.

However, in addition to the various types of traceable standards just mentioned, there are some independent types of standards which should also be considered and used where necessary. Reference to Fig. 16.25 will clarify what is meant. Although

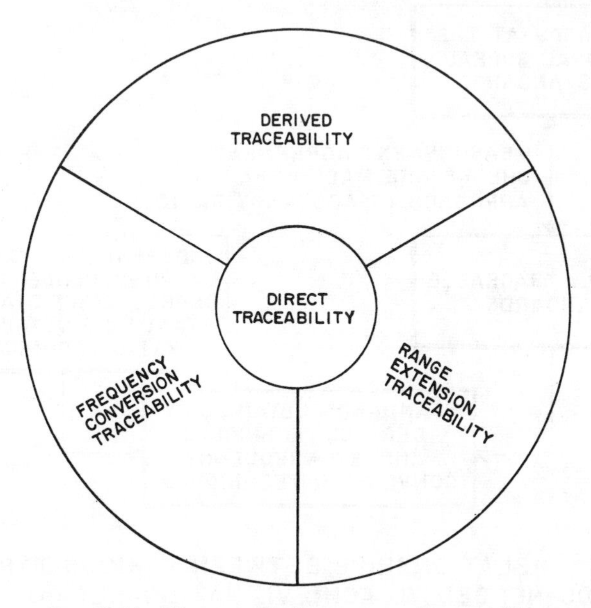

FIG. 16.24 Direct and indirect types of traceability.

there may be other general classes of independent standards, only two are shown here. One of these is the class of independently reproducible standards: standards which are dependent upon accepted values of natural physical constants. Typical examples are the krypton-86 orange light source for dimensional calibrations and the cesium-beam type of microwave frequency standard. The other general class of independent standards covers standards obtained by self-calibration, using ratio-calibration techniques, as described in reports of the National Bureau of Standards and others.

The technique known as "measurement agreement" provides a mechanism for improving the confidence level for standards established by indirectly traceable methods by ratio techniques (described later), or even standards which are in the independently reproducible category. By measurement agreement we mean the establishment or enhancement of the confidence level for the accuracy of a calibration by showing that similar results are obtained with two or more calibration facilities. This is commonly done by calibrating, with two or more calibration facilities, a stable sample representing the quantity in question.

For example, NBS Boulder and NBS Washington have achieved measurement agreement of approximately 1.2 parts per million for 1-ohm resistance calibrations by having three Thomas 1-ohm resistors calibrated at both facilities. Measurement agreement also provides a means of improving the accuracy of standards by making it possible to approach 1:1 equality with the next higher standards or calibration laboratory. This may be particularly important in those cases when NBS does offer calibration service in a particular category but when accuracy requirements are essentially the same as the national standards at NBS. In this case one can often establish his own standards by some indirectly traceable or ratio-type technique, then cross-check with NBS by measurement agreement and make necessary adjustments to approach 1:1 equality with NBS.

This 1:1 ratio is commonly referred to as an accuracy ratio, which, in turn, can be defined as the ratio of the accuracy of the calibration system with which a calibration is performed to the accuracy quoted for the standard or instrument calibrated with

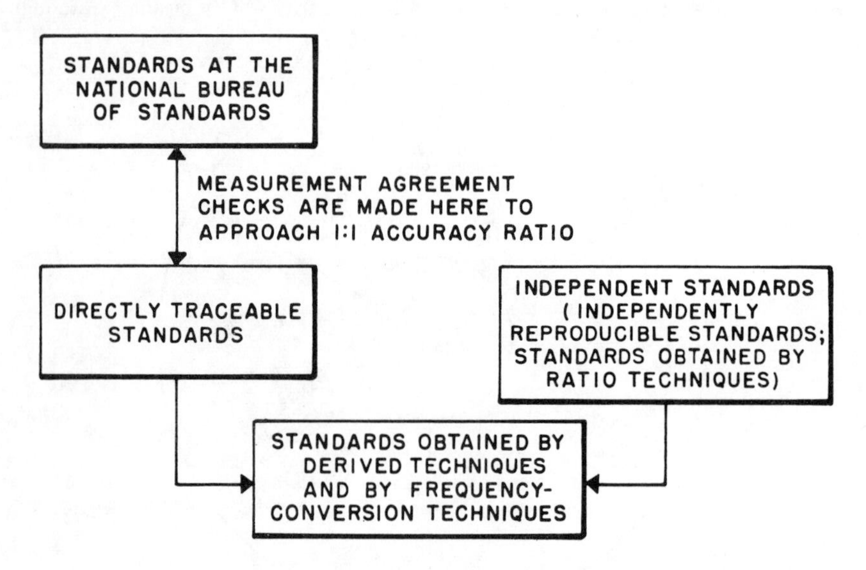

Fig. 16.25 Relationship between types of standards.

that system. Generally, the accuracy ratio is of the order of 2:1 or, in some cases, even as high as 10:1. The reason for this is that there may be some degradation of accuracy in the comparison equipment which is used with the standards against which comparisons are being made.

Evaluations of instruments and standards should, whenever possible, be made in the purchase of the measurement standard or instrument. The next best thing is to make evaluations as part of incoming acceptance testing and initial calibrations. However, even after an instrument is purchased, and regardless of whether the organization buys it or whether it is furnished, it is generally desirable to make evaluations on a continuing basis during the life of the instrument. Such evaluations need not necessarily be a special effort, because the periodic recalibrations which are necessary for most instruments and standards supply much of the information required. Other information can be obtained from repair records which tell the frequency of repair and whether the repairs were major or superficial.

In addition to in-house experience with use, repairs, calibration stabilities, etc., a laboratory evaluation is particularly desirable. It not only provides a common basis for comparison of several similar types of instruments but also permits a comparison in terms of specific measurement or calibration needs. Another advantage of such a laboratory evaluation is that it provides a very good way to tell how well human-engineered and easy to use the item is. Figure 16.26 indicates the Navy and prime-contractor calibration program structure as related to shipyard calibration work.

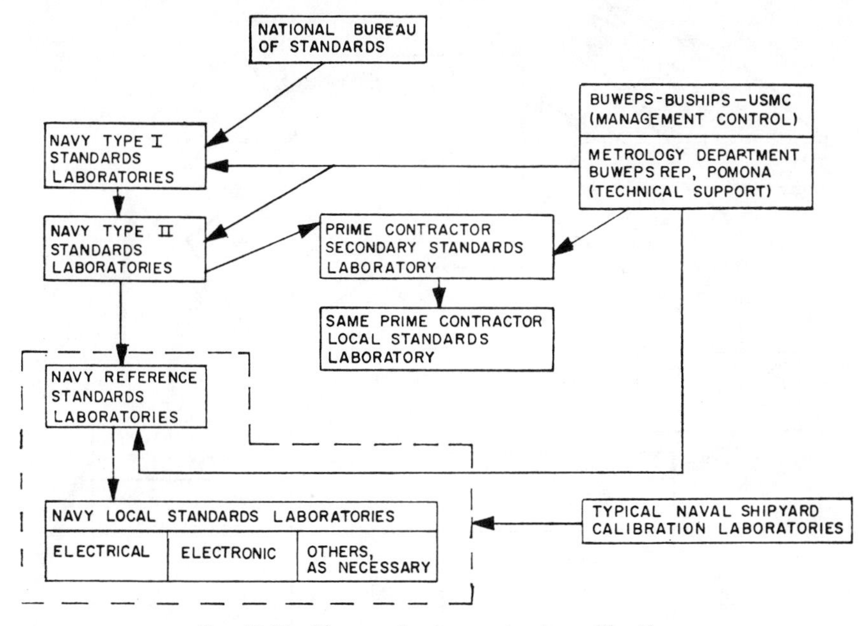

FIG. 16.26 Navy and prime-contractor calibration.

16.15 PREVENTIVE- AND CORRECTIVE-ACTION PROGRAM

The prevention of unreliability must be given serious consideration in organizing for the reliability and quality control programs. When effective prevention procedures are established for the purpose of preventing errors from occurring, the task of corrective action will be greatly reduced. Savings will be realized not only in the corrective-action area but also in the failure and appraisal cost areas. Even though preventive costs may possibly be slightly higher in the early program, a reduction in defects will easily justify the slight additional expense. It has a chain reaction on reliability and quality cost and creates a reduced need for routine inspection and test activities.

The reliability and quality of the manufactured product represent the ability to meet the design engineers' requirements, and the engineers' requirements are the result of the customer's desires that have been translated into reliability requirements and written into specifications. The design engineer is very interested in obtaining information during the testing, manufacturing, and use phases of the hardware that will allow him to improve the reliability in the early product design. To do this means that it is essential that we develop quantitative methods to aid the engineer in predicting reliability, as well as techniques for handling engineering problems which constitute possible sources of trouble. Figure 16.27 depicts a closed-loop program through which possible improvement in process can be detected and related to engineering design terms.

16.15a Review of Engineering Design, Contracts, and Specifications. One of the more important functional processes in the prevention of costly errors and rework is the review of contracts, government specifications, and engineering designs to determine the quality and reliability requirements and define their tasks. After the tasks have been defined, it is important that the quality and reliability control requirements be incorporated into the applicable quality and reliability control specifications and procedures control.

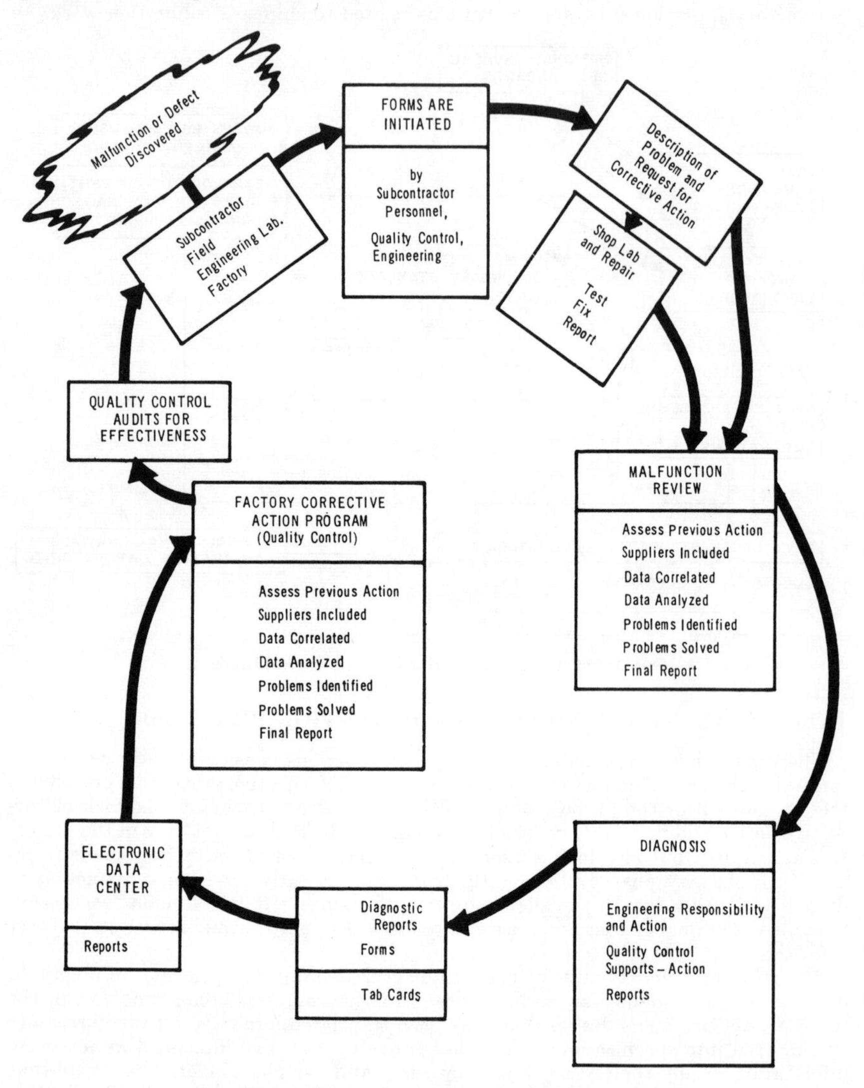

Fig. 16.27 Corrective-action closed-loop program.

One company has accomplished the prevention task of reviewing the contracts, specifications, and engineering drawings by requiring that reliability and quality control receive copies of these documents along with engineering-design-released drawings. The engineering drawings are received by the quality control department from the manufacturing production control department. Drawings are then screened by quality control for items affecting quality of product, and the selected ones are forwarded to the applicable section of quality control or reliability for review. The drawings are reviewed by these quality control and reliability specialists to assure that applicable requirements have been incorporated. A document review card, which has spaces for accepting the drawings or listing deficiencies, is attached to the drawings. In the event that deficiencies exist, engineering design is notified through proper procedure and the document review card is held in a suspense file until deficiencies have been corrected. Similar procedures are utilized for the effective control of specifications and contracts.

16.15b Corrective-action Program. The quality control plan should establish the steps for the identification of quality and reliability problems and the appropriate corrective action. Basic causes for poor quality or reliability may include:

1. The basic quality problems, both quality level and conformance to standards
2. Process or employee at fault
3. Corrective measures for other processes; i.e., people, machine, etc.

The correction of the quality-level type of problem should be the responsibility of departments other than those responsible for meeting requirements. Correction of these deviations from product conformance to standards should be the responsibility of line personnel who are responsible for meeting product requirements. Quality problems should be identified by type to determine the best method of analysis.

Follow-up action of recommended corrective actions should be performed to check results and to measure progress. The quality control engineer should concentrate on the areas in which more assurance of quality will give the greatest return. Figure 16.28 illustrates a corrective- or preventive-action report form used by one company to document desired action on discrepancies.

The reliability and quality control function can contribute substantially to the reduction of cost by systematic audits of the delay experienced in production owing to rework caused by workmanship errors, lack of adequate plans, and unrealistic specification requirements. An evaluation of the causes of delays traceable to poor quality and reliability decisions can lead to improved corrective-action procedures.

Quality control and reliability should be concerned with the analysis of failure reports and defects reports, and every failure report should result in effective corrective action. (See Secs. 7 and 9.)

16.15c Warranty-program Organization Responsibilities. The warranty-program responsibilities to the customer must be recognized by the management of design, manufacturing, quality control, and product support, by the comptroller, and by top management. A product that meets all the expectations of the customer is probably the best sales-promotion device a manufacturer can possess.

The design engineer will always be responsible for the design and its inherent reliability and hardware characteristics. But it must be remembered that quality control and reliability must bring to the attention of the design engineer the actual modes of failure and corrective action that will prove of value to him.

Manufacturing has a warranty responsibility in that it must actually build the product according to the designer's blueprints. When products meet all expectations the first time, a great deal of money is saved in possible rework and manpower as the product moves into the customer's possession.

Product support (field service) must maintain close liaison with the customer throughout the life of the product. Reports from product support personnel at customer bases are valuable sources of information because the hardware is then being put to its ultimate use under the conditions for which it has been designed and built. Prompt improvements and corrective actions resulting from feedback information can do much in the cementing of contractor-consumer relationships.

Top management and comptroller warranty responsibilities lie in the areas of directing the quality and reliability organization and of properly budgeting for adequate staffing of qualified personnel and necessary facilities and equipments.

CORRECTIVE OR PREVENTIVE ACTION REPORT

PROBLEM NO.

SERIAL NO.

TO ________ DEPT. ________ DATE ________

FROM ________ DEPT. ________ MAIL ZONE ________

Please state cause and furnish corrective and/or preventive action for the condition(s) causing the following discrepancy(s).

DISCREPANCY

REFERENCES

SIGNED ________

REPLY DUE ________

CAUSE

ACTION TAKEN

EFFECTIVITY OF ACTION (Date, A/C Number, etc.) ________

SIGNATURE (Supervision) ________ DATE ________

FIG. 16.28 Corrective- or preventive-action report.

16.16 TECHNICAL DATA AND PUBLICATIONS QUALITY

The management objectives in establishing a control for technical data and publications should be to provide a system that is satisfactory both to its own operations and to the customer. Management should make certain that the system is implemented on a scheduled timely basis. It is also a responsibility of management to review by a scheduled management reporting system the results of the system in order to assure that the implemented system is performing the expected task and that it remains effective.

A clear-cut understanding between the supplier and the customer regarding requirements is necessary in order to be assured of correct and accurate technical data and their timely publication. This is particularly important when the prime-contractor–subcontractor–final customer relationship exists. Conferences with the supplier, final customer, and prime contractor in attendance are necessary to assure that the requirements are discussed and clarified. These requirements should be written in a

Contract No. ____________ Purchase Order No. ____________
Reference Technical Manual(s) ____________

PRELIMINARY ☐ FINAL ☐

	ELEMENTS	COMPLETE	INCOMPLETE	INITIALS	DATE
1.	Applicable Government Specifications				
2.	Applicable Government Standards				
3.	Applicable Government Regulations				
4.	Applicable Contractor Specifications				
5.	Applicable Contractor Standards				
6.	Applicable Contractor Test Procedures				
7.	Authorized Deviations Incorporated				
8.	Contract Change Notices Incorporated				
9.	Correct Security Classification				
10.	Format to Specification Requirements				
11.	Text to Specification Requirements				
12.	I.P.B.'s to Specification Requirements				
13.	Required Technical Data Incorporated				
14.	Required Engineering Data Incorporated				
15.	Government Nomenclature Assigned				
16.	Commercial Nomenclature Assigned				
17.	Parts Provisioning Incorporated				
18.	Mechanical Illustrations Incorporated				
19.	Graphic Art Presentation				
20.	Schematic and Wiring Diagram				
21.	Packaging Provisions				
22.	Engineering Approval of Technical Content				
23.	Review of Preliminary Copy				
24.	Review of Type Copy				
25.	Review of Final Negatives and				
26.	Technical Writer Approval				
27.	Technical Editor Acceptance				
28.	Approved for Submittal for Government Inspection				

FIG. 16.29 Technical publications final-acceptance checklist.

statement of work or in a specification for the technical manuals and data relating to a specific product or system.

Technical data and publications quality control personnel should be people who are experienced in the field of technical data and publications. College graduates or the equivalent who have the technical and editorial background to work intelligently with design engineers are best qualified to meet the requirements for assuring the quality of technical data and related publications. They should be people who are thorough in application and possess perseverance and seriousness of purpose.

Technical publications are usually jointly prepared by the technical publications and engineering departments. In some cases the technical publications department is a division of the engineering department; in other organizations, management may find it desirable to place the technical publications operation under the manager of logistics support or customer service department.

Engineering furnishes the technical information and support for the publications. The publications are jointly reviewed for technical content and adequacy by the

T.O. ________ Program ________ Handbook Title ________
Configuration to ITTFL Part No. ________ Issue Dated ________
Military Specifications ________ Tech. Writer ________

Internal Inspection Supervisor ________
Editor ________
Eng. Configuration ________
Technical Accuracy ________

	WRITER				EDITOR				QUALITY CONTROL			
	INS	REJ	INS	ACCEPT	INS	REJ	INS	ACCEPT	INS	REJ	INS	ACCEPT
T.O. Number	☐	☐	☐	☐	☐	☐	☐	☐	☐	☐	☐	☐
Security Classification	☐	☐	☐	☐	☐	☐	☐	☐	☐	☐	☐	☐
Format												
Section Index	☐	☐	☐	☐	☐	☐	☐	☐	☐	☐	☐	☐
Paragraph Index	☐	☐	☐	☐	☐	☐	☐	☐	☐	☐	☐	☐
Marginal Copy	☐	☐	☐	☐	☐	☐	☐	☐	☐	☐	☐	☐
Page Limits	☐	☐	☐	☐	☐	☐	☐	☐	☐	☐	☐	☐
Type & Type Size	☐	☐	☐	☐	☐	☐	☐	☐	☐	☐	☐	☐
Illustrations & Tables	☐	☐	☐	☐	☐	☐	☐	☐	☐	☐	☐	☐
Quality of Printing												
Text	☐	☐	☐	☐	☐	☐	☐	☐	☐	☐	☐	☐
Line Art	☐	☐	☐	☐	☐	☐	☐	☐	☐	☐	☐	☐
Photos	☐	☐	☐	☐	☐	☐	☐	☐	☐	☐	☐	☐
Figure Numbers and Layout	☐	☐	☐	☐	☐	☐	☐	☐	☐	☐	☐	☐
Page and Paragraph Numbers	☐	☐	☐	☐	☐	☐	☐	☐	☐	☐	☐	☐
Title Page	☐	☐	☐	☐	☐	☐	☐	☐	☐	☐	☐	☐
"A" Page	☐	☐	☐	☐	☐	☐	☐	☐	☐	☐	☐	☐
Table of Contents	☐	☐	☐	☐	☐	☐	☐	☐	☐	☐	☐	☐
Alphabetical Index (LPB)	☐	☐	☐	☐	☐	☐	☐	☐	☐	☐	☐	☐
Numerical Index (IPB)	☐	☐	☐	☐	☐	☐	☐	☐	☐	☐	☐	☐
Reference Designation Index (IPB)	☐	☐	☐	☐	☐	☐	☐	☐	☐	☐	☐	☐
Grammar	☐	☐	☐	☐	☐	☐	☐	☐	☐	☐	☐	☐
Clarity and Expression	☐	☐	☐	☐	☐	☐	☐	☐	☐	☐	☐	☐
Consistency in Terminology	☐	☐	☐	☐	☐	☐	☐	☐	☐	☐	☐	☐
Correct Photos & Overlays	☐	☐	☐	☐	☐	☐	☐	☐	☐	☐	☐	☐
Abbreviations Symbols	☐	☐	☐	☐	☐	☐	☐	☐	☐	☐	☐	☐
Reference Designations	☐	☐	☐	☐	☐	☐	☐	☐	☐	☐	☐	☐
Uniformity of Numbering of Paragraphs, Figures, Tables	☐	☐	☐	☐	☐	☐	☐	☐	☐	☐	☐	☐
Warning & Caution Notes	☐	☐	☐	☐	☐	☐	☐	☐	☐	☐	☐	☐
Negative Quality	☐	☐	☐	☐	☐	☐	☐	☐	☐	☐	☐	☐
Group Assembly Parts List	☐	☐	☐	☐	☐	☐	☐	☐	☐	☐	☐	☐
Reproduction Assembly Sheet	☐	☐	☐	☐	☐	☐	☐	☐	☐	☐	☐	☐
Line Drawings	☐	☐	☐	☐	☐	☐	☐	☐	☐	☐	☐	☐

Comments:

FIG. 16.30 Quality control checklist for technical publications.

preparing technical publications and engineering departments. Specific types of checklists may be developed to assure uniformity, correct content, and adequacy of the technical publications. These may be specified for by the subcontractor's publications editor and publications quality control. Figures 16.29 and 16.30 depict examples of each of the forms mentioned. Some customers require retention of such checklists for a period of time.

Prior to release, the applicable section of quality control should review and participate in proofing the technical publications for acceptability in meeting configuration control and quality standards requirements. Quality control should also coordinate the necessary changes and corrective action with responsible departments to assure quality of publications and compatibility with customer maintenance and use capabilities.

Technical publications are best validated and verified by the supplier when the instructions are used to operate and maintain systems and equipments by actual demonstration. Information contained in the instructions must reflect the contract configuration of the product and be easily understood by using personnel. Adequacy and clarity of the instructions for equipment operation must be compatible with human factors and safety criteria.

Data records of the publication discrepancies must be maintained in order that corrective action, follow-up, and reports may be accomplished. Technical-publication discrepancies may be recorded on a designed discrepancy form and this information translated for processing by data-processing machines. From the resulting printout, quality control may make statistical analyses and formulate their recommendations for corrective action and follow-up by management.

16.17 ORGANIZED AUDITS OF RELIABILITY AND QUALITY CONTROL PROGRAM PERFORMANCE

The advantages of a dual examination of a product by management have been recognized for many years with the continuing use of quality control programs; similarly the value of reexamination of performance with a reliability program has recently been acknowledged. Management's assurance of quality becomes a dual examination only when management has reexamined the quality control and reliability functions in the same way that these functions reexamined the product. The purpose of an organized audit is, then, the delivery of a second and more comprehensive report to management that all is well. The audit is an organized examination and official verification of methods and procedures by which the quality engineer may identify problem areas, determine the effectiveness of planning, and report his observations to management.

If the product line is closely knit and products are relatively complex, a company may find it advantageous to form a small staff of auditors—on call from appropriate departments—whose experience and training particularly qualify them for examining methods from a remote perspective. In all circumstances the personnel selected for performing quality audits must be capable, experienced, friendly, and diplomatic, but must have an aggressive and inquisitive nature. They must be familiar with the products in the areas they are assigned to survey and must have a thorough knowledge of processes and procedures applied.

Although the choice of subjects to be audited is always a problem, the following lend themselves to the auditing process:

1. Design-engineering review
2. Inspection and test effectiveness
3. Configuration control
4. Process controls
5. Employee qualification and training effectiveness
6. Adequacy and timeliness of instructions and procedures
7. Effectiveness of the quality control plan
8. Material identification and control
9. Technical manuals and data
10. Tool and test-equipment calibration

The auditing of inspection, test, and process-control functions is a vital part of the quality program and offers evaluation of the adherence to engineering and customer standards. Quality control and reliability audits of the indirect quality functions, i.e., testing, sampling, supervisory checks, and recorded observations of workers, are valuable tools for product assurance. Quality control engineers will not have answers for all problems, because their audit is but a document performing as a sensory aid. Many problems confirmed by the audit must be referred to other branches of the organization for solution.

QUALITY ASSURANCE AUDIT

ASSEMBLY PROCEDURES ______________________ SECTION

AUDITOR ______________________ DATE: ______________________

		YES	NO	REAUDIT		REMARKS
				YES	NO	
1.	Is the acceptance of rework/repairs in accordance with applicable procedures? QCDI G-1 *					
2.	Are inspection criteria used to prevent repetitive discrepancies? Are they adequate? QCSM 1-0-1					
3	Are bailed property procedures complied with? QCDI G-27 & I-40 *					
4.	Is the control of sealed areas adequate and being followed? QCDI I-27					
5.	Are inspection checks of stock rooms made in accordance with QCDI I-3? *					
6.	Is torque putty and lacquer being used per existing requirements? QCDI K-18 *					
7.	Is procedure adequate and being followed for final assembly completion acceptance? QCDI I-31 *					
8.	Is control adequate to assure that all areas are free from loose hardware prior to acceptance?					
9.	Is Break of Inspection utilized for purposes intended? QCDI I-27 *					
10.	Are Unsatisfactory Reports submitted in accordance with QCDI G-33? *					
11.	Are procedures governing Reliability-Maintainability data in effect? QCDI G-42 *					

*(Quality Control Departmental Instructions)

A. B. Smith
Certification of Acceptability on ReAudit

Fig. 16.31 Quality assurance audit.

It is implicit that the specific intent of an audit and the time of its execution always remain variable and individual. Advance knowledge would result in advance preparation within the area to be surveyed, and the audit would reveal only the deficiencies in the preparation. Recommendations made after an examination of a biased environment might easily add more problems than were originally subject to the audit.

Many audits will be initiated at the direction of top management or on recommendation of an advisory staff. Industrial-engineering sections or value-analysis teams may solicit the support of the quality-engineering function which has established its capability in increasing the effectiveness of the total quality control effort. Events which may signal a management directive for an audit include customer complaints, adverse profit-sales ratios, unfavorable trends in production or inspection management charts, employee suggestions, and quality-safety interfaces exposed by industrial accidents. Of course, an audit ordered by top management will clearly vest the quality control engineer with all the authority required. An aggressive management may occasionally call for an audit without any signs of effectiveness simply to substantiate that operations are proceeding as intended, or perhaps to confirm that it has properly delegated the task of quality assurance.

Audits performed as a normal quality assurance responsibility and at the direction of its own management may have some routine characteristics, but they may still require nonperiodic performance to eliminate the factor of advance preparation. The selection of what to audit—within the general areas already listed—will depend upon:

1. The variations which cause shop scrap, inspection rejections, and functional failures
2. Departures from a specified planning operation
3. Misinterpretation of blueprints and engineering orders
4. Differences in records maintained by production and quality control personnel
5. Complexity or excessiveness of processes
6. Repeatability and accuracy of inspection and test operations

A typical quality audit is shown in Fig. 16.31. The quality control audit is both a reporting and an action device. Its utility lies in correcting the deficiencies which it reveals. Since the role of quality control is to document experiences, the role of the auditor becomes twofold: to document his original examination and to document his follow-up examination. The audit is complete only when it has *detected,* and then *corrected,* any deficiencies. A well-documented audit will be countersigned by the quality control engineer's supervisor.

Audits and certified reaudits by quality control or reliability generate reports of official findings to the appropriate level of management. Unless otherwise directed, the report will also be delivered to supervision of all areas examined. Recommendations for improvement are an essential element of the report and should be made in brief, pointed terms from which a clear management decision may be reached.

16.18 REPORTS FOR MANAGEMENT USE

Managers must control the work for which they are responsible and are held accountable. This requires accurate information for evaluating routine and special situations. A manager can make the best decisions if he is provided with timely reports which are simple and easily understood. Timely reports will support preventive action and prevent excessive losses. The ratio of cost will frequently be 1:10, or more, in favor of prevention through timely decisions and action.

Since the scheduled task must be carried out according to the requirements of the performance standards set by the manager, it is the responsibility of those so delegated to keep him aware of the progress being made. The responsibility of the department manager in controlling the task involves four areas of action. He must make certain there are understandable and acceptable milestones or standards for measurement of the task as it progresses. The *milestones* should be *established* as an integral part of the initial reliability and quality control plans required to implement the task. The *progress* of a planned task has to be *measured* and documented accurately. The *results* of these measurements then have to be *interpreted* and evaluated, whereby performance may be compared with the milestones or estimates originally set and planned for. These measurements may be in terms of the dollar value of rejects, repairs, failures, activities completed, warranty claims, reliability demonstration, and reliability achievement. If there is a variance from the original plan, the manager must see that the necessary corrective action is accomplished to realign the task on the desired course. In this manner management is in complete control and aware by documented evidence that the achieved reliability and quality of hardware are measuring up to the requirements set forth in the contractual commitments to the customer.

Reports to management are normally the result of a selected compilation of evidence gathered by various people. Functional organization responsibilities should be established in order that these evidences can be documented accurately on the correct format. In order to have an effective data-gathering and -reporting system, this function should be independent of the operating (product-oriented) reliability and quality control groups. The overall coordination of much information affecting optimum reliability and quality of product is required. The group, by necessity, must maintain close contact with the production personnel and the products and services which they produce.

The real test for a reliability and quality control organization comes when management needs to know whether the reliability and quality standard of the hardware is being maintained during the manufacturing process.[1] Managers on each level are obligated to provide succinct reports of progress, problems, and planned activities to their superior. It is of necessity that they be aware of what their superior desires in a report. A reliability and quality manager has important coordination and reporting to do, since the reliability and quality control activity involves many different phases of the manufacturing task and customer reaction. Reporting may differ with the type of organization. Since each manager has a different organization environment, the approach he desires to take in reporting to his superior must be carefully tailored to fit. Some of the reports that are found in a typical industrial facility are:

1. Periodic reliability and quality control program planning and progress reports
2. Reliability milestones and achievement reports
3. Rework and scrap (preferably referencing dollar savings or losses, because they are tangible)
4. Reliability status reports (indicates predicted and assessed reliability regarding systems)
5. Warranty-claim trends

Annual report. Should concern itself with dollar savings and losses and operations information, e.g., the departmental operations cost. This can be broken down into yearly comparisons of added tools and equipments, number of lots inspected, percentages of lots found defective, and increases or decreases of line and field complaints.

Monthly reports. This report or chart can use the total productive labor cost (all hourly rated production help as a base). Total defective and rework cost can then be estimated as a percentage of the total labor dollar. The cost rises as defectives increase, thereby increasing the percentage. If more productive labor hours are added without an increase in the defective rate, the percentage rate will naturally be lowered. The monthly report should contain charts relating to defective lots and, where necessary to pinpoint areas of trouble, should indicate total defects by types. Analysis of field complaints and returned lots should also be included. Reports concerning vendor difficulties and their ratings can be of value to the purchasing department.

Weekly and *daily* reports concerning the defects occurring in the shop are valuable to departmental foremen.[2]

[1] For a description of quality control data uses see the tabulation, under The Purposes of Inspection, in J. M. Juran (ed.), *Quality Control Handbook*, 2d ed., McGraw-Hill Book Company, New York, 1962, p. 195.

[2] For a more complete listing of quality control reports refer to Henry J. Jacobson, "Quality Control Management of Small Business," *Ind. Quality Control*, p. 9, March, 1963.

Figure 16.32 shows a monthly report of the quality status of one company's total facility, and Fig. 16.33 provides the report of quality status relating to indirect departments and code charges. Figure 16.34 indicates a reliability status report relating to the assessed and predicted mean time between failure (MTBF) of an aircraft flight control. The proper reporting of particular situations at the right time to the right people can correct a difficult problem as well as possibly prevent further trouble.

Management must take advantage of the new scientific methods, e.g., electronic data processing and program evaluation and review techniques (PERT). The PERT (program evaluation and review technique) system has proved to be a very useful tool for managers, technical administrators, and engineers of reliability and quality control. PERT has been used effectively in planning, scheduling, and

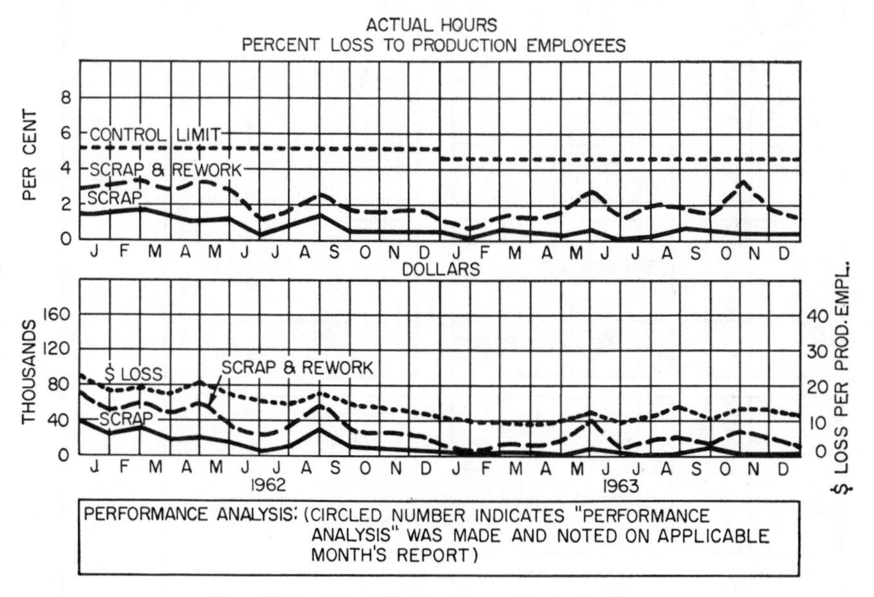

Fig. 16.32 Total facility (actual hours) report.

controlling the time elements involved in utilization of resources, facilities, and manpower on a scheduled plan. PERT has been defined as "a system that quantifies estimates of time for the completion of tasks in the approved plan for meeting significant end objectives."[1] Time is used as a common basis for scheduling and implementing planned sequences of operations and facility and resource applications, as well as the required technical performance.

In order to apply PERT to the reliability and quality control program plan, a reliability monitoring index should be developed for the various levels of management to monitor the approved tasks. By use of the monitoring index the reliability and quality control program tasks are specified by designing a PERT network relating to the events and activities of the planned program. The PERT network system can then be used to measure schedule commitments against time as well as audit the quality of results. A related computer program may be altered slightly to print out indicated problem areas whereby management may be made aware at an early

[1] "PERT/Cost System," Willard Fazar, Vice President, Management and Systems Engineering, Herner and Company, Washington, D.C. For a more detailed discussion of PERT see "The Role of PERT in Missile and Space Management," *Trans. Am. Soc. Quality Control*, pp. 79–84, 1963.

	ENGINEERING		MFG. CONTROL		TOOLING		QUALITY CONTROL		CODE CONTRACTOR		TEST & MAINTENANCE		SET-UP	
	%	$ COST	%	$ COST	%	$ COST	%	$ COST	%	$ COST	%	$ COST	%	$ COST
1962														
J	.10	1,011	.02	257	.16	2,703	.003	49	.87	14,440	.13	2,200	.43	7,040
F	.06	1,134	.01	116	.29	5,178	.00	00	.80	14,455	.15	2,633	.41	7,523
M	.06	969	.02	363	.10	1,702	.01	111	.90	15,023	.09	1,531	.20	3,398
A	.06	1,026	.03	481	.11	1,936	.002	37	.69	12,465	.40	7,347	.50	9,031
M	.03	391	.02	238	.08	1,066	.00	00	.90	11,572	.26	3,380	.21	2,718
J	.02	327	.003	70	.04	840	.01	301	.19	4,028	.14	2,976	.20	4,100
J	.01	74	.02	327	.13	2,133	.004	66	.41	6,662	.06	1,008	.18	2,879
A	.16	3,560 *	.01	141	.06	1,420	.01	104	1.19	26,529 **	.15	3,403	.23	5,237
S	.06	945	.00	00	.04	576	.01	130	.52	8,614	.08	1,381	.12	1,964
O	.08	1,328	.04	695	.11	1,770	.00	000	.54	8,533	.07	1,044	.06	956
N	.05	797	.003	49	.03	565	.01	166	.49	8,304	.26	4,348	.13	2,287
D	.00	00	.002	24	.13	1,700	.00	0C	.37	4,895	.20	2,658	.08	1,053
TOTAL	.06	11,562	.01	2,761	.11	21,589	.005	964	.65	135,520	.17	33,909	.23	48,186
1963														
J	.02	299	.00	00	.06	970	.00	00	.18	3,033	.20	3,463	.00	00
F	.002	31	.001	6	.26	3,299	.00	00	.52	6,592	.01	125	.02	210
M	.008	101	.048	592	.03	353	.00	00	.42	5,144	.17	2,090	.00	0C
A	.076	899	.002	26	.07	843	.00	00	.65	7,689	.03	323	.01	148
M	.035	525	.006	84	.053	784	.008	124	1.89	28,172	.00	00	.025	366
J	.035	366	.002	20	.152	1603	.008	81	0.29	3,102	.00	00	.00	00
J	.094	950	.017	173	.184	1,852	.012	122	0.71	7,128	.00	00	.00	00
A	.044	536	.001	6	.058	713	.002	26	0.60	7,377	.02	204	.00	00
S	.044	407	.004	38	.034	312	.093	858	0.46	4,273	.00	00	.00	00
O	.105	957	.002	16	.446	4,066	.014	125	0.67	6,144	.01	103	.00	00
N	.027	305	.00	00	.137	1,566	.000	00	0.00	0,00	.04	458	.00	00
D	.000	000	.00	00	.213	1,579	.011	85	0.11	85	.00	00	.001	9
TOTAL														

* NOTE: QUALITY PERFORMANCE EXCEEDED ESTABLISHED LIMITS, AND ANALYSIS WAS MADE AND NOTED ON APPLICABLE MONTH'S REPORT.

PERFORMANCE ANALYSIS:

Fig. 16.33 Indirect department and code charges.

moment of the dangers to the quality of hardware and the possible inadequate support of the reliability objectives previously planned.

As previously mentioned, the first step in the PERT/Cost System is the construction of a network consisting of the activities (project tasks) to be performed and the events or milestones to be attained. The network reflects the carefully developed plan for accomplishing the interrelationships and interdependencies in the work to be performed. By its use, the effects of any schedule slippages on the entire project can be readily determined.

Figure 16.35 shows a typical PERT network with three time estimates (optimistic, most likely, and pessimistic) for each activity. The critical path (the longest path to completion) is shown by the gray line. All other paths on the network are

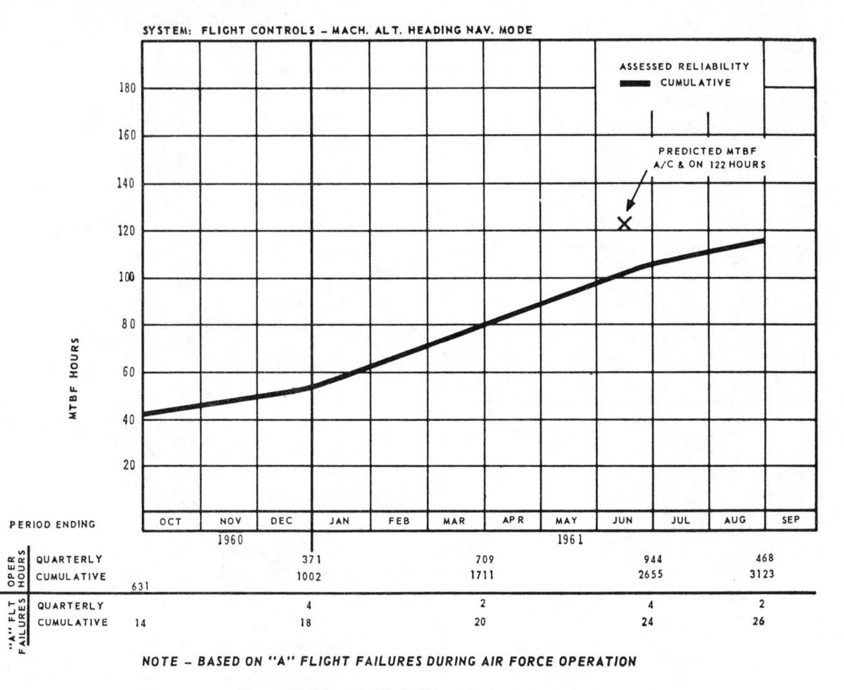

FIG. 16.34 Reliability-status report.

called slack paths. The value t_e is the expected time for each activity, statistically derived from the three time estimates for each activity.

After the network has been prepared and time estimates have been developed for the network activities, the manager will establish a schedule. This schedule will be based on the critical-path calculations, the directed dates, and the manager's judgment concerning the goals he established for accomplishing the activities.[1]

16.19 SURVEILLANCE BY CUSTOMER

The increased complexity of today's products has made it necessary for each customer to place more responsibility on each supplier and to devise means of assuring that the supplier's organization, procedures, equipment, and personnel are adequate to produce the desired product. For defense contractors, this responsibility often

[1] A more complete and detailed description of "The PERT/Cost System" may be obtained from Government Printing Office, Washington 25, D.C.

includes the purchase of systems and subsystems which were formerly supplied as government-furnished equipment and which must meet all government requirements. With this increased procurement responsibility also goes the responsibility for assuring maximum achievement of inherent design reliability.

A key task of reliability and quality control engineers working in the procurement area must now include additional emphasis on vendor evaluation, source-inspection criteria and records, control of processes, failure analysis, and vendor corrective action. The value of this added effort is equally effective for industrial and military products, with savings often far exceeding the added reliability and quality control costs.

The details of administering procurement quality and reliability programs exhibit similarities and differences. Quality control and reliability men are obligated to extend the requirements of the prime contract into the subcontracts and assure

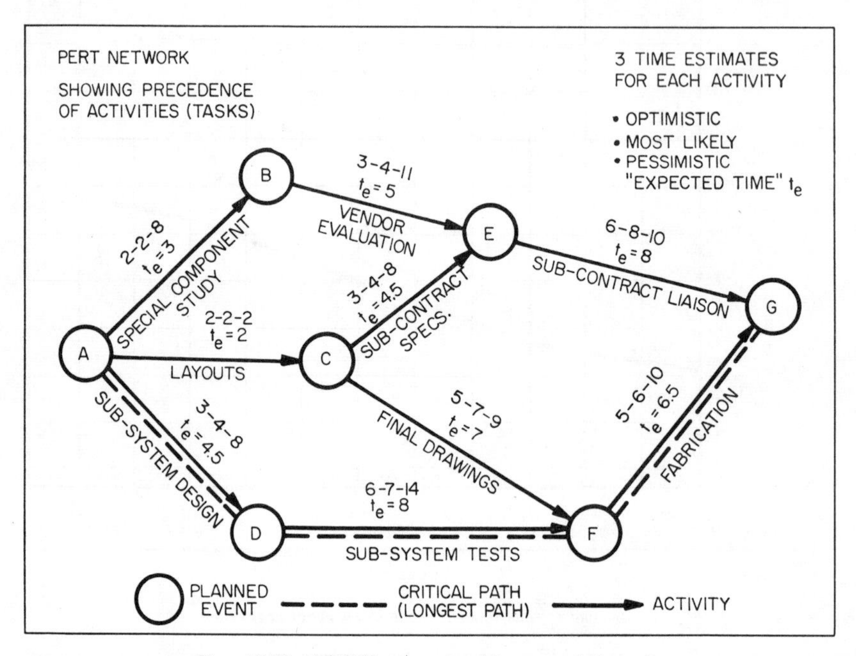

FIG. 16.35 PERT network of "expected times."

adequate control for all applicable items. Usually, quality control and reliability personnel participate in separate (but coordinated) efforts in the preparation of subcontract requirements and work statements, because each has many specialized requirements to be included in the work statements and contracts. Both groups assist in negotiating with the subcontractors by providing consultation and advice to the buyers and others.

Quality control must review all purchase orders for production materials to be certain that they contain all the necessary quality requirements and references and that the purchase orders specify the proper place for inspection: source, destination, or none. The reliability functional review of purchase orders assures the inclusion of "policy" items and those items with assigned reliability values and reliability demonstration requirements.

Both reliability and quality control engineers may be called on to perform on-site evaluation surveys and participate in the final selection of vendors. Figure 16.36 indicates a small part of one company's procurement quality control vendor appraisal

PROCUREMENT QUALITY CONTROL
VENDOR APPRAISAL SUMMARY

VENDOR
COMMODITY

Please Check Applicable Box Which Best Describes this Vendors Performance.

Date

This Column is for Tabulator's use only

8 Quality Control Procedure Evaluation
9 Quality Control Performance Evaluation

					SCORE	
A. QUALITY CONTROL ORGANIZATION	Evaluators Comments:					
Not Applicable	Formal Quality Control Organization and Program/ Performance	Quality Control Organization and program/performance Requires Improvement	Extensive Revisions required in quality organization and program/performance	No formal quality control organization and program/ performance	8	9
Procedure 0 / Performance 0	Procedure 1 / Performance 1	Procedure 2 / Performance 2	Procedure 3 / Performance 3	Procedure 4 / Performance 4		
B. DRAWING AND CONTRACT CHANGE CONTROL	Evaluators Comments					
Not Applicable	Effective Drawing and change control procedure/ performance	Drawing and Change Control procedure/performance requires improvement	Extensive Revisions required in drawing and change control procedure/performance	No drawing and change control procedure/performance	8	9
Procedure 0 / Performance 0	Procedure 1 / Performance 1	Procedure 2 / Performance 2	Procedure 3 / Performance 3	Procedure 4 / Performance 4		
C. PURCHASE ORDER, DRAWING AND SPEC'N REVIEWED BY Q.C.	Evaluators Comments:					
Not Applicable	Effective procedure/ performance for review of Purchase Orders, Drawings and Specifications	Procedure / performance for review of Purchase Orders Drawings and specification requires improvement	Extensive revisions required in procedure/performance for review of Purchase Orders, Drawings and Specifications	No procedure/performance for review of Purchase Orders, Drawings and specifications	8	9
Procedure 0 / Performance 0	Procedure 1 / Performance 1	Procedure 2 / Performance 2	Procedure 3 / Performance 3	Procedure 4 / Performance 4		
D. QUALITY CONTROL PLANNING	Evaluators Comments					
Not Applicable	Effective Quality Control Planning Procedure/ Performance	Procedure/ Performance for Quality Control Planning requires improvement	Extensive revisions required in the Quality Control Procedure/Performance	No Procedure/Performance for Quality Control Planning	8	9
Procedure 0 / Performance 0	Procedure 1 / Performance 1	Procedure 2 / Performance 2	Procedure 3 / Performance 3	Procedure 4 / Performance 4		
E. DETERMINATION & IMPLEMENTATION of Recorded Characteristics	Evaluators Comments:					
Not Applicable	Effective Procedure, Performance for Determination and implementation of recorded characteristics	Procedure/Performance for Determination and Implementation of recorded characteristics. Requies improvement	Extensive Revisions required in the determination and implementation of recorded characteristics Procedure/Performance	No Procedure/Performance for determination and implementation of recorded characteristics	8	9
Procedure 0 / Performance 0	Procedure 1 / Performance 1	Procedure 2 / Performance 2	Procedure 3 / Performance 3	Procedure 4 / Performance 4		
F. Control of in-process inspection including Quality Control Release:	Evaluators Comments:					
Not Applicable	Effective Procedure/Performance for Control of inspection, including Quality Control release	Procedure/Performance for control of in-process inspection, including Quality Control Release requires improvement	Extensive Revisions required in the Procedure/ Performance for control of in-process inspection including Quality Control Release	No Procedure/Performance for control of in-process inspection including Quality Control Release	8	9
Procedure 0 / Performance 0	Procedure 1 / Performance 1	Procedure 2 / Performance 2	Procedure 3 / Performance 3	Procedure 4 / Performance 4		
				Page Total		
				Actual Score		
				Applicable Sections		

FIG. 16.36 Procurement quality control vendor appraisal summary.

summary. This form, along with supporting-detail checklist, is used to evaluate the capability of a vendor's organization.

It should be noted that there are many areas of common reliability and quality control interest. Major differences in the reliability and quality control monitoring of vendor performance result from specific areas of interest. The quality interests lie in the surveillance of the vendor's entire manufacturing and quality control system, and the reliability interests lie in monitoring the vendor's design, testing, and performance demonstration activities. Reliability requirements may include warranty provisions in the contracts based on numerically specified trouble-free service and may also include contractual incentive-penalty measures based on mean time between failures and maintenance requirements. Quality control will want to provide for preventive measures and the initial acceptance (or rejection) of hardware, whereas the reliability usually depends on product tests and use experience, in the specified environments, for finite periods of time.

Many managerial organization arrangements have been developed by industry to carry out the responsibilities relating to procurement quality control and reliability. Quality control and reliability have been separated but placed on the same level as other management functions reporting to top management. This permits recognition of these two organizations and allows them to express their viewpoints directly to top-level management. This separate organization method has caused some difficulty

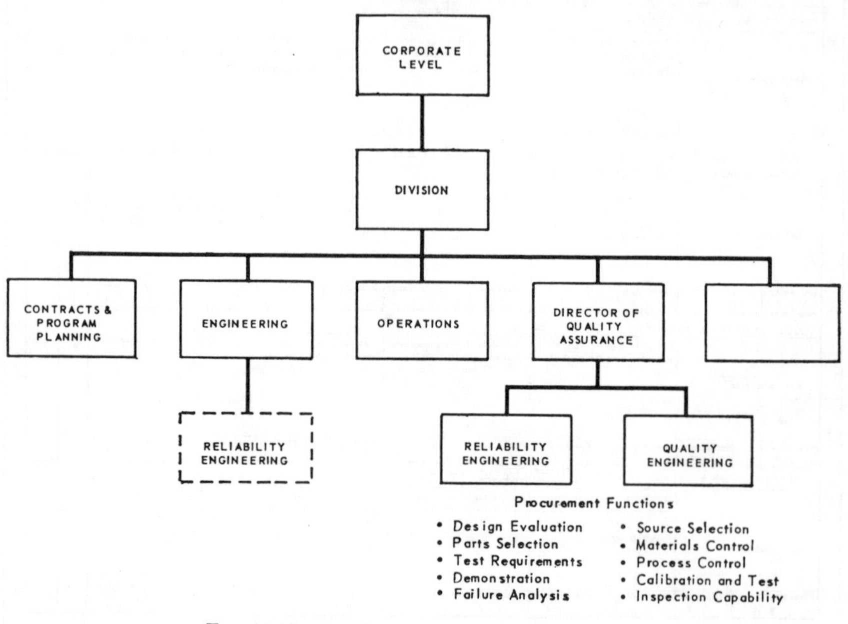

FIG. 16.37 Combined procurement functions.

for managers, however, in securing coordinated communications in-plant and with the vendors, especially when one group is not informed as to what the other group is trying to accomplish.

Figure 16.37 depicts quality assurance as the top-level reporting agency to top management. Quality engineering and reliability engineering are arranged in order that they may report on equal footing to the director of quality assurance. These people will probably be working in the same area and rather close together, which provides for easy solutions of the problems arising through communications. The director of quality assurance also has an opportunity to be aware of what is going on at the floor level other than through the reports that he receives.

16.20 MAKING USE OF PROFESSIONAL AND TECHNICAL SOCIETIES

Industrial organizations have become aware that the exchange of ideas within the industry is necessary to keep pace with advancement of the state of the art in reliability and quality control. One way that companies and employees contribute to the exchange is their participation in the activities of the many professional and technical societies whose conventions, symposia, and specialized publications supply the communication link.

Most companies encourage employees to participate in professional-society programs, because they afford the opportunity to meet others with similar interests and to broaden their knowledge by exchanging information in the specialty field.

Some companies maintain sustaining memberships in the societies, supply brochures for their meetings, and rent exhibit or display space at the symposia. Employee incentive is offered by the paying of dues and special fees for those who become officers and directors and by the contributions of time, per diem, and transportation for attendance at distant events.

While a company receives some measure of recognition and prestige within the industry, its best gauge of the worth of its participation is the employee development and the broadening of employee ability to organize, administrate, and expand technical ingenuity.

Section 17

COST ASPECTS OF A RELIABILITY AND QUALITY ASSURANCE PROGRAM

L. J. KNIGHT

CHIEF, QUALITY CONTROL GROUP, NORTHROP SPACE LABORATORIES
NORTHROP CORPORATION, HAWTHORNE, CALIFORNIA

Mr. Knight has had twenty-two years experience in the field of quality control, primarily in the aircraft, missile, and aerospace ground equipment fields. He has had primary responsibility for both assembly and final-test operations in these fields.

Currently with the Northrop Space Laboratories, Mr. Knight is Chief of the Quality Control Group. Most recently his work has concentrated on the problems of vendor-vendee programs, quality-cost analysis, and communication methods relative to quality and reliability specifications and demonstration.

Mr. Knight received the B.A. degree from the University of Southern California in 1955 and is currently pursuing graduate study there. He is a senior member of the American Society for Quality Control and served as chairman of that organization's Vendor-Vendee Technical Committee. He has previously served on other Society committees and has presented a number of reports and papers before the Society's meetings.

COST ASPECTS OF A RELIABILITY AND QUALITY ASSURANCE PROGRAM

L. J. Knight

CONTENTS

The relative position of quality and reliability as opposed to schedule and cost is presently being expressed quantitatively, in contractual terms, through incentive provisions which apportion the emphasis on cost, delivery, and operational performance in accordance with the requirements of the particular contract. The measurement and management analysis of reliability and quality costs have not kept pace with the advancing technology of industrial cost accounting. Detailed quality-cost information in many defense industries is almost nonexistent. Without adequate cost data it is extremely difficult to make valid cost effectiveness comparisons of alternative program requirements. All too often, management has considered quality costs primarily in terms of total program funding. Predicted quality assurance and reliability expenditures for new contracts are often determined through the application of standard ratios such as inspection to manufacturing personnel or a predetermined percentage of the total estimated manufacturing hours.

During the past few years, military specifications and contract requirements have made it mandatory that the contractor demonstrate his attention to the requirements for quality assurance and reliability. In all too many instances this has had the effect of requiring the contractor to provide the basic framework of a quality and reliability assurance program, but without the essential proof of program effectiveness. Demands for high quality and reliability have been accompanied by either stated or implied understanding that cost considerations are of secondary importance. This inevitably has led the contractor to do everything possible to improve the quality of his product with the additional costs accruing to the contracting agency. Such conditions do not, however, relieve management of its concern and responsibility for managing the resources at its disposal in the most efficient manner possible.

The purpose of this section is to discuss applications of the *value vs. cost* concept and to provide some guidelines for the development of an effective cost-analysis program. How does management determine the amount of contract funding that should be allocated for quality and reliability assurance activities? Once it has determined that, management must decide how that budget is to be distributed among the various elements of the total program. The contribution of each element or major activity toward the attainment of established goals must be determined. The goals themselves must be measured and evaluated in terms of how much their attainment is worth to the customer. The cost of the various activities must be measured in such a way that the contribution can be weighed against the expenditure.

The mandate to establish and maintain a program for the control and assurance of quality and reliability, a program that is both effective and economical, presents a real challenge for the program manager. Current trends within the Department of Defense emphasize the importance of reducing total quality and reliability costs by establishing the proper balance of preventive, appraisal, and failure costs.

17.1 BASIC QUALITY COSTS

To obtain the most effective quality and reliability program at a reasonable cost, one must identify all costs and establish their cause-and-effect relationship. The basic principle underlying this approach to product assurance is that there are two basic categories of costs.

17.1a Quality and Reliability Input Costs. (Prevention, evaluation, administration, and quality program performance audits) Included in this category are those costs deliberately incurred in an effort to achieve and control an assured level of quality and reliability. Those activities which make a direct contribution by assuring that manpower, materials, and factory capacity are not wasted in the form of defective products are *preventive* in nature. Included in the preventive function are program elements such as the planning, coordination, and establishment of all program activities into the total quality assurance and reliability effort; the establishment of system reliability requirements based on the reliability objectives established by contract; the review of design drawings, specifications, and processes; the selection of component parts in terms of failure rates, performance, and endurance; and the establishment of quality assurance and reliability data systems.

The effectiveness of the prevention activities can be evaluated in terms of higher rate of acceptable product, fewer failures, scrap reduction, smaller maintenance costs, fewer customer complaints, and lower field-service costs

Activities which tend only to measure results are commonly called *evaluation* costs. This type of activity includes program elements such as the review and analysis of reliability demonstration test results; inspection and test of incoming materials; evaluation of in-process controls; supplier product quality and reliability rating; evaluation testing of production articles; calibration of gauges, measuring devices, and inspection-test equipment; and final-inspection and off-site test programs.

Administration costs are all indirect costs associated with the management of a quality assurance and reliability program.

Quality audit costs include all costs of performing independent, important audits on the adequacy of quality control plans and procedures, test methods, process controls, and reinspection of work previously accepted.

17.1b Deficiency Costs. (Scrap, rework, additional inspection, parts-failure analysis, handling of customer complaints) Included in this category are those activities which involve the handling of nonconforming purchased materials (includes deficiency feedback to suppliers), in-plant failures, and failures encountered after customer acceptance.

The basic principle underlying selection of those two basic categories is that quality and reliability input costs must bear a direct relationship to deficiency costs. There must also exist an effective prevention and evaluation policy which will optimize the total distribution of prevention, evaluation, and deficiency costs.

17.2 BASIC STRUCTURE OF ACCOUNTS

The following system of accounts covers those activities which are considered reliability, quality control engineering, quality control (includes quality control management), and inspection.

100 Preventive Costs

110 Procurement. Cost associated with the planning and maintenance of a procurement control program.

111 Supplier Surveys. Costs incurred by personnel conducting surveys of a supplier's process facilities and quality control system to ensure that he is capable of supplying articles which meet all quality requirements. Also included in this subaccount are the costs involving periodic resurveys.

112 Procurement Document Review. Costs incurred by personnel reviewing subcontracts and purchase orders to ensure that applicable drawings, specifications, and quality requirements are clearly and precisely transmitted.

113 Coordination of Contractor-Supplier Acceptance Plans. Costs incurred by personnel involved in preproduction coordination activities with supplier's quality control organization. This coordination activity includes the review and analysis of the supplier's process controls and inspection test plans and the establishment of data requirements.

114 Receiving Checklists. Costs incurred in the development of receiving-inspection checklists to specify characteristics to be inspected, variables data requirements, inspection-test equipment, and the amount of inspection effort.

115 Appraisal of Supplier Reliability Practices. Costs incurred by personnel monitoring supplier's practices to ensure fully qualified parts. Also included in this account are the costs involving the evaluation and approval of the supplier's reliability program and participation in technical review meetings with the supplier.

116 Reliability Data Review. Costs incurred by personnel reviewing supplier's data for compliance with reliability requirements and the evaluation of the supplier's achieved reliability vs. predicted growth.

117 Review of Supplier Design. Costs incurred by personnel conducting design reviews of supplier's drawings and the monitoring of supplier's parts and materials procurement specifications and procedures.

120 Product Assurance Preproduction and Technical Support Activities. Costs associated with quality control and reliability engineering activities during the design and development phase and technical support of a preventive nature during the production phase.

121 Drawing and Specification Review. Costs incurred in the review of drawings and specifications to ensure that characteristics which influence the quality and reliability of the product are clearly delineated. This activity includes the classification-of-characteristics program.

122 Preparation of Specifications. Costs incurred in determining and establishing specifications that define the environment to which the product should be subjected.

123 Establishment of Test Environments. Cost of personnel defining the environmental limits for each test and developing the test method to simulate the environment, Also included in this account are the costs involved in defining criteria for conducting the test, including the length of the test, definition of failure, establishment of monitoring points, and power requirements.

124 Reliability Demonstration Plan. Cost of personnel formulating a reliability-demonstration plan.

125 Process Control. Costs associated with the analysis of process specifications, drawings, and manufacturing processes. This activity includes process-capability studies and the preparation of process-control plans and procedures.

126 Quality Planning. Costs incurred in the preparation of inspection plans which provide adequate inspection points throughout the manufacturing cycle and specify the sequence of inspections and tests. This activity includes test-plan analysis.

127 Personnel Certification. Cost incurred by personnel involved in the certification of manufacturing personnel responsible for controlling special processes or for performing fabrication, inspection, and test operations of a specialized nature having a significant effect upon quality and reliability. This also includes periodic recertification costs.

128 **Equipment Qualification.** Cost incurred by personnel involved in the qualification of equipment to ensure performance in a satisfactory and consistent manner. This also includes periodic requalification costs.

200 Evaluation Costs

210 **Inspection and Test of Purchased Materials.** Cost incurred by personnel involved in receiving inspection and test of incoming purchased materials and units of hardware. This also includes the cost of itinerant and resident source inspectors.

211 **Evaluation Testing of Incoming Product.** Cost incurred by personnel performing tests such as verification of qualification, life tests, use tests, and marginal testing on selected purchased components. This also includes the cost of all material consumed in destructive-type tests performed on incoming product.

212 **In-process Inspection.** Cost incurred by personnel performing inspections in the manufacturing cycle (tooling, fabrication, and assembly inspection).

213 **Testing.** Cost incurred by personnel witnessing or performing functional tests throughout the manufacturing cycle. This does not include testing of incoming purchased articles.

214 **Calibration of Gauges and Inspection Test Equipment.** Cost incurred by personnel comparing and adjusting dimensional, physical, optical, and input-output measuring devices to standards. Also included in this subaccount are the costs associated with the establishment of calibration procedures and the maintenance of records on the recalibration status, condition, and repairs for each unit of inspection, measuring, and test equipment.

215 **Preservation, Packaging, and Shipping Inspection.** Cost incurred by personnel verifying that completed articles have been preserved, packaged, identified, and marked in accordance with applicable procedures and specifications. Also included in this account are the costs associated with the verification of shipping documents and packing sheets.

216 **Reliability Demonstration Tests.** Cost incurred by personnel involved in the evaluation of reliability demonstration tests and the analysis of test results.

217 **Field Testing.** Cost incurred by personnel witnessing or performing field testing at remote sites prior to final acceptance by the customer. Included in this account are field evaluation tests and prelaunch and launch activities conducted by the contractor in conjunction with the customer.

300 Administration Costs

This includes the costs incurred in the management and support of a reliability and quality assurance program other than those included under prevention, evaluation, or deficiency costs.

310 **Personnel.** Costs associated with secretarial, clerical, and managerial reliability and quality assurance personnel. Supervisor costs for various reliability and quality assurance activities should be included in the appropriate activity accounts.

311 **Reliability and Quality Data Handling.** Costs associated with the analysis and compilation of reliability and quality data from all sources on materials, parts, and products. Such costs as the preparation of quality performance charts and reliability progress reports to management and the customer are included in this account.

312 **Miscellaneous Support Activities.** Costs incurred by personnel engaged in evaluating customer requirements and preparing quality- and reliability-program plans for requests for proposals. Such costs as the preparation

and revision of quality assurance and reliability policies and procedures are included in this account.

313 **Plans and Budget.** Cost incurred by personnel involved in the establishment and control of budget, preparation of task statements, establishment of schedules, and manpower requirements. This also includes the cost of monitoring the implementation of quality-assurance- and reliability-program plans.

314 **Technical Data.** Cost incurred by personnel involved in the development of procedures, preferred parts listing, preferred parts handbook, and material specifications to aid designers.

400 Quality Audit

This includes the cost incurred by personnel performing impartial, objective evaluations of quality-program procedures, process controls, previously inspected articles, and compliance with established requirements. Also included in this account are the costs associated with the preparation of audit reports and follow-up to ensure correction of deficiencies.

500 Deficiency Costs

This includes the cost associated with rework, scrap, additional inspection, in-plant failures, and customer complaints.

510 **Procurement Deficiencies.** Costs incurred by personnel involved in handling nonconforming purchased materials. This account includes the cost of liaison activity with suppliers in eliminating nonconformance to drawing and specification requirements and quality characteristics.

520 **In-plant Deficiencies.** Cost associated with defective materials and components that fail to meet drawing, specification, and quality requirements.

524 **Scrap.** Cost of defective materials and components scrapped for nonconformance to drawing or specification. It does not include materials scrapped for other reasons such as engineering design changes or surplus stock.

525 **Rework.** Cost associated with rework due to nonconformance or materials and components to drawing or specification.

526 **Materials Review Board.** Cost associated with the review, control, and disposition of nonconforming material and components.

527 **Failure Analysis.** Cost of analysis and examination of deficient or failed articles to determine the nature and basic cause of the failure. Such costs as inspection, test, analysis, disassembly, and failure-analysis reports are included in this account.

528 **Customer Complaints.** Cost of analyzing the modes of failure of finished products which have been delivered to the customer. Such costs as the investigation and follow-up action on customer complaints are included in this account.

The foregoing system of accounts is easily adaptable to present computer-programming methods. Reorganization of certain accounts can be accomplished to meet the particular requirements of individual contractors.

A variation of the foregoing system of accounts is presented to illustrate the flexibility inherent in the system. The following adaptation of this accounting method for manual timekeeping purposes preserves the general intent of the system.

The need for flexibility in the interpretation and application of the foregoing system of accounts to specific contractual situations is both recognized and recommended. *It is often difficult to categorize precisely those borderline cases of preventive costs which also exhibit certain evaluation characteristics.* Questionable areas of this nature will require close scrutiny to ensure that consistency in categorization is maintained throughout the quality-cost accounting system.

17.3 DESCRIPTION OF MANUAL TIMEKEEPING SYSTEM

The system described in this section illustrates the accumulation of quality assurance and reliability *input costs* and *deficiency costs* on a manual basis. The metal-bonding

DAILY TIMEKEEPING REPORT

ORGAN - SHIFT - MAN NO. | J. Smith NAME | DATE

A/C	SALES ORDER	DASH	SECTION	JOB NO.				DRAWING NUMBER	HRS	10ths
61	10000	10	4120	7	6	1	1		2	0
38	2000	10	4611	7	6	1	2		2	0
47	3000	12	3912	7	6	1	3		2	0
15	4000	14	4214	7	6	1	4		2	0

Fig. 17.1 Daily timekeeping report.

department was selected for this illustration owing to the diversification and proximity of manufacturing and processing techniques within the work area. The subaccount structure in this particular instance has been consolidated and designed for ultimate

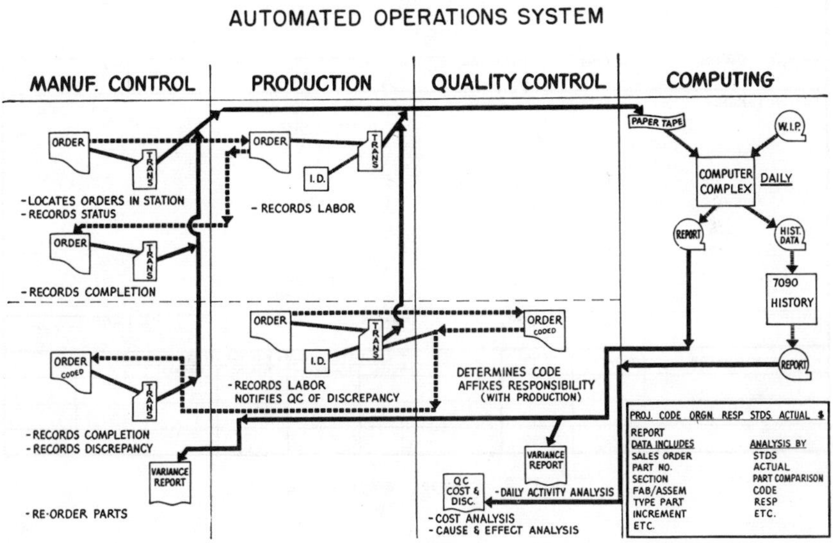

Fig. 17.2 Automated operations system.

use as a part of the work organization's number. This was done to ensure compatibility with existing computer programs related to costing, payroll, and other business-data systems.

Activities involving quality costs were reported daily through the use of a manual timekeeping system utilizing the job number and organization number as illustrated in Fig. 17.1. The daily timekeeping record (Fig. 17.1) contains job numbers of

CORRECTIVE ACTION REQUEST
FORM 27-651 (R.12-62)

NONFUNCTIONAL C 03547

DIRECT TO:	ORGN. NO.	DATE	PART OR UNIT SERIAL NO.

DOCUMENT/PART NO.	DOCUMENT/PART NAME

AIRCRAFT/UNIT	SALES ORDER RELEASE	SECTION	PROJECT	VENDOR	PURCHASE ORDER NO.

ORIGINATOR	ORGN. NO.	EXT.	DOCUMENTARY REFERENCE

DISCREPANCY/REQUESTED ACTION:

APPROVAL SIGNATURE	DATE

INVESTIGATION FINDINGS:

ORGN. RESP.	TASK CODE	CAUSE CODE	DEFECT CODE	SIGNATURE	DATE

ACTION TAKEN:

EFFECTIVITY	AIRCRAFT/UNIT	SIGNATURE	DATE

INVESTIGATOR	ORGN. NO.	HOURS EXPENDED	ESTIMATED HOURS FOR CORRECTIVE ACTION						
			ENGR.	PROCESS	TOOL DESIGN	PLANNING	TOOL FAB.		

FIG. 17.3 Corrective-action request.

special significance, e.g., job number 7611. The first digit, 7, indicates that J. Smith is assigned to the quality control department. The second digit, 6, indicates that he is also in the quality engineering group. The third digit, 1, shows that the major account is prevention. The fourth digit indicates the function performed or subaccount: 1, specification preparation; 2, equipment qualification and requalification; 3, personnel certification; and 4, requalification testing, etc.

In this instance the existing business-data computer programs are keyed to the first two digits of the organization number. Therefore, whenever the manual system is replaced with a mechanized system (Fig. 17.2), special reports will be generated without additional passes and with a minimum of additional programming.

A corrective-action request form (Fig. 17.3) was designed to record deficiency data and the time expended to investigate a deficiency and to record the cause and corrective action taken. The corrective-action request was utilized to record deficiencies requiring investigative action. Included in this category were those conditions involving scrap and processing errors and other discrepant conditions that could not be readily resolved. Whenever the investigation involved the services of more than one organization, the time expended by each of the participating organizations was noted in the hours-expended block on the corrective action request. The cost of rework was obtained from inspection call sheets and rejection-tag forms. Cost computations for deficiency data were made by utilizing appropriate average rates and factors from financial records, standard hours from the master operation file, and material cost provided by the cost-estimating organization.

The general accounts and job-number definitions that follow were adapted for use in this quality-assurance cost system.

Account 1000—Preventive Costs

Job No. 7611 Specification Preparation. Time spent by personnel in the coordination and preparation of specifications and process-control bulletins.

Job No. 7612 Equipment Qualification and Recertification. Process-control qualification and recertification of processing equipment and development of control devices.

Job No. 7613 Personnel Training and Certification Program. Includes the cost of certifying that personnel performing critical or special processes are properly trained and certified.

Job No. 7614 Requalification. Activities performed by technical or quality assurance personnel to determine that reproducibility factors of assembly meet original qualification requirements of tools and process. *Note.* In the event a requalification assembly fails, the requalification activity will be assigned as part of this job number. The assembly will be handled on a corrective action request and all subsequent requalification activities will be considered as a deficiency cost.

Job No. 7615 Process-control Activities. Consists only of planned activities related to the control of processes, in which a record is maintained of the elements checked and the status of control found, e.g., solution control, temperature control, adhesive disbursement control, test-specimen preparation not related to the production assemblies, preventive maintenance on bonding tools, equipment, etc.

Job No. 7317 Surveillance. Activities wherein the correction of discrepancies found would prevent potential product deficiencies, e.g., faulty handling equipment and poor stocking practices.

Account 2000—Evaluation Costs

Job No. 7621 Laboratory Testing. Time test personnel spend in testing specimens representative of the production assemblies and materials. Includes cost of test preparation and recording of test data on laboratory reports or shop travelers.

Job No. 7622 Quality Planning. Performed by personnel in the quality planning function who translate product design and customer quality requirements into planned location of inspection checkpoints, the kind and amount of inspection to be performed, characteristics to be measured, equipment to be used, and records to be kept. Preparation of visual aids or visual physical standards.

Job No. 7623 Technical Assistance. Shop liaison expenditures for technical support of in-process control activities, e.g., interpretation of specifications and test data.

Job No. 7324 Acceptance Testing. Time spent in normal evaluation of production assemblies per production orders, drawings, and process specifications.

Account 3000—Administration Costs

Job No. 7631 Management. For the purpose of this application, quality-engineering administration costs associated with the metal-bond process were estimated.

Job No. 7632 Quality Assurance Procedures. Includes the time spent by personnel in the development, implementation, and revision of quality assurance procedures.

Job No. 7633 Data Analysis. Includes the time spent by personnel in the analysis of quality and reliability data for purposes of control and the preparation of quality assurance status reports.

Job No. 7334 Management. Includes quality control administrative costs incurred in support of the quality assurance program. Included in this job number are those administrative cost elements not logically a part of preventive, evaluation, or audit functions.

Account 4000—Quality Audit

Job No. 7641 Systems and Procedures. Costs incurred by personnel auditing the adequacy of quality assurance program procedures and systems.

Job No. 7642 Product. Costs incurred by personnel performing reinspection of work previously accepted by quality control in the area.

Account 5000—Deficiency Costs

Job No. 7651 Failure-analysis Cost. Includes the time spent by personnel performing cause or failure analysis of in-plant deficiencies; includes diagnostic, disassembly, and test functions.

Job No. 7652 Review of Discrepant Material. Includes the time spent by quality-engineering personnel serving in a technical advisory capacity on the materials review board.

Job No. 7351 Failure Analysis. All time spent determining the cause of in-plant failures; includes investigative, test, and disassembly functions.

Job No. 7354 Rework Cost. Includes the cost of additional inspection necessitated by product deficiencies.

Job No. 7050 Materials Review. Includes all costs incurred in handling nonconforming material and in rendering dispositions on defective assemblies.

All quality assurance input deficiency cost data were summarized, computed for element of cost, and reported on a weekly basis by the cost accounting organization.

17.4 FORMS AND REPORTS

Periodic quality-cost reports and charts are issued on a weekly and monthly basis to various levels of management for the purpose of indicating trends and providing composite quality-cost data. The quality-cost detail summary report, shown in Fig. 17.4, provides a summary of quality costs by project and major account numbers. The job numbers or subaccounts indicate the work organization and particular function performed.

The relative distribution of quality costs on a quarterly basis is shown in Fig. 17.5. This method of presentation clearly illustrates the actual proportionment of the two basic categories of cost (quality input and deficiency) itemized in the detail summary report. The quality-cost chart (Fig. 17.6) reflects the costs associated with evaluation, deficiency, prevention, and administration plotted on a weekly basis to illustrate graphically the relative distribution of these major accounts.

The scrap-cost analysis chart (Fig. 17.7) provides a breakdown of total scrap dollars by automated operations control (AOC) stations. The two principal elements of this basic information are then correlated to illustrate the defect-cause relationship.

QUALITY COST DETAIL SUMMARY REPORT				Month of December	
Account 1000 - Prevention	Project A	Project B	Job No. Sub-Total	Orgn. Sub-Total	Account Total
Job 7611 Specification Preparation	27.16		27.16		
Job 7612 Equip. Qual. & Re-Cert.	18.10	18.10	36.20		
Job 7613 Personnel Training & Cert.	466.05	502.41	968.46		
Job 7614 Requalification	1,268.99	366.88	1,635.87		
Job 7615 Process Control	948.50	963.88	1,912.38	4,580.07	
Job 7317 Surveillance	777.35	693.38	1,470.73	1,470.73	
TOTAL PREVENTION EXPENDITURE	3,506.15	2,544.65			44,001.80
Account 2000 Evaluation					
Job 7621 Laboratory Testing	267.01	678.79	945.80		
Job 7622 Quality Planning					
Job 7623 Technical Assistance	915.91	1,320.46	2,236.37	3,182.17	
Job 7324 Acceptance Testing	6,077.96	6,729.05	12,807.01		
Job 7325 Nondestructive Test	2,065.67		2,065.67	14,872.68	
Specimen Tests	11,400.33	14,545.44	25,946.77	25,946.77	
TOTAL EVALUATION EXPENDITURE	20,726.88	23,274.74			44,001.62
Account 3000 Administration					
Job 7631 Management	520.02	470.55	990.57		
Job 7632 Quality Control Procedure	79.33	29.75	109.08		
Job 7633 Data Analysis	105.95	164.76	270.71	1,370.36	
Job 7334 Management	249.66	199.77	449.43		
TOTAL ADMINISTRATION EXPENDITURE	954.96	864.83			1,819.79
Account 4000 Quality Audit					
Job 7641 Systems and Procedures					
Job 7642 Product Audit					
TOTAL QUALITY AUDIT EXPENDITURE					
Account 5000 Deficiency					
Job 7651 Failure Analysis	401.10	773.14	1,184.24		
Job 7652 Materials Review Board				1,184.24	
Job 7351 Failure Analysis					
Job 7354 Reinspection (Rework)	320.63	297.52	618.16	618.16	
Job 7050 Materials Review Insp.	978.24	734.71	1,712.95	1,712.95	
Investigations: Mfg. Orgn.	64.44	102.53	166.97	166.97	
Investigations: Tooling Orgn.	142.93	66.36	209.29	209.29	
Investigations: Other Orgns.		334.82	344.82	334.82	
Scrap - Prod. Raw Material	706.26	1,342.14	2,048.40	2,048.40	
Scrap - Allocated Material	879.00	2,058.36	2,937.36	2,937.36	
Scrap - Labor (Lost Hours) Dollars	2,409.25	7,259.15	9,668.40	9,663.40	
Rework - Production Orders	633.47	393.21	1,026.68	1,026.68	
TOTAL DEFICIENCY EXPENDITURE	6,544.17	13,363.10			19,907.27
TOTAL QUALITY COST	31,359.44	39,970.70			71,779.48

FIG. 17.4 Quality-cost detail summary report.

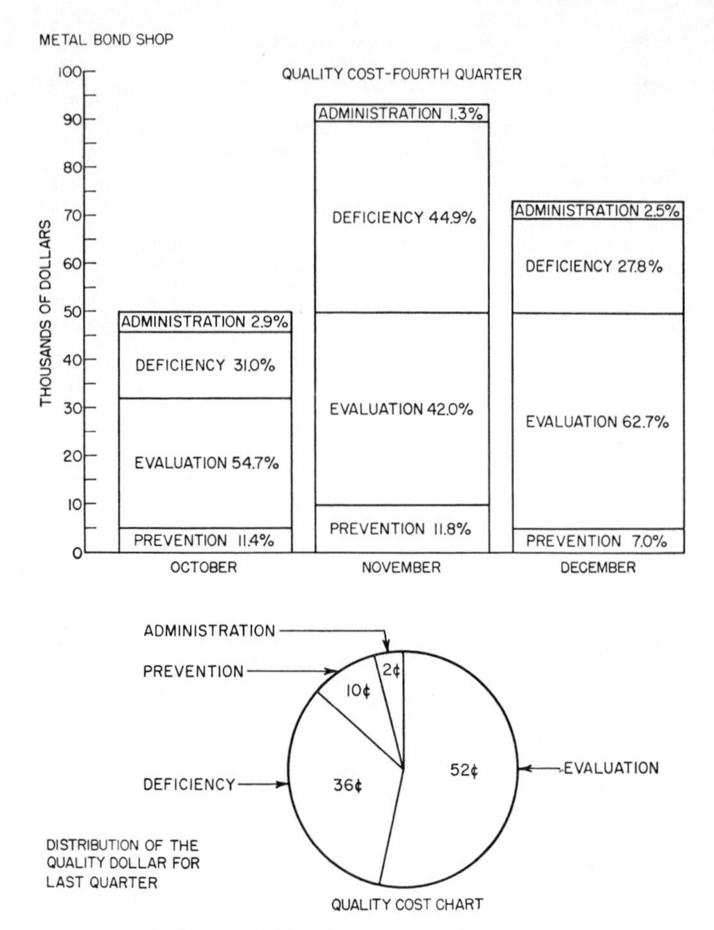

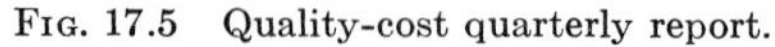

FIG. 17.5 Quality-cost quarterly report.

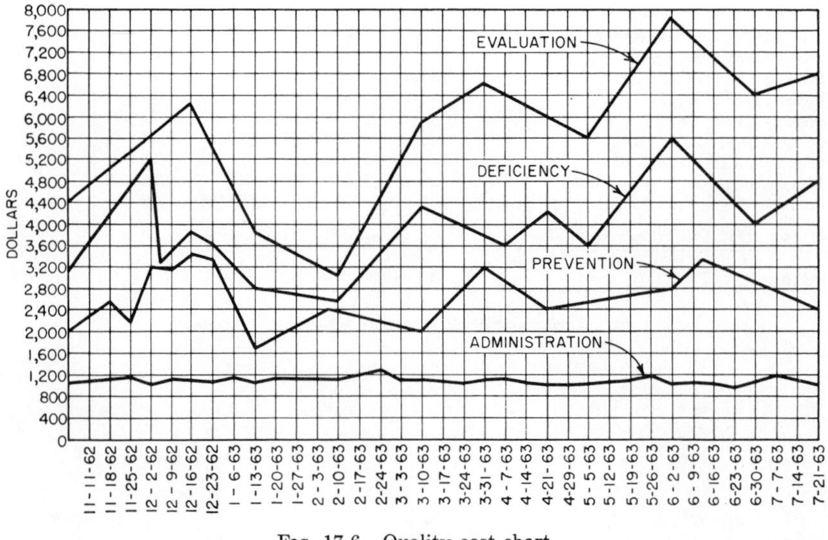

FIG. 17.6 Quality-cost chart.

SCRAP COST DATA — DEFECT-CAUSE ANALYSIS

METAL BOND SHOP

AOC STA	MANUFACTURING AREA	SCRAP COSTS PROJ A	SCRAP COSTS PROJ B	SCRAP COSTS PROJ C	CAUSE SUMMARY: Tools & Mainten-ance	Inadequate Procedure	Handling and Storage	Equip Failure	Defective Parts	Tempera-ture Control	Unknown	Operator Error	TOTAL SCRAP COSTS	% of Total
064	Machine & Router Polyglycol	686.04	530.28		228.49			407.44	227.26			353.13	1216.32	15.0
070	Tanks, Adhesives, Bake & Pre-Cure	22.28	144.73							57.10		209.91	267.01	3.0
071	Autoclaves and Saws	368.39	525.72			316.73						577.38	894.11	11.0
072	Hydro Crush	765.24	595.34	446.99	446.99	463.65	361.59			57.10	538.32		1807.57	22.0
073	Sub-Ass'y and Core Splice	107.60	35.84						20.11			123.33	193.44	2.0
074	Pre-Fit	964.00	1247.42	40.03	626.09		135.36		1490.00				2251.45	28.0
075	Assembly	177.09	996.11		692.16		177.09		17.59		286.36		1173.20	15.0
076	Final and Repair		357.29		195.94							161.35	357.29	4.0
	TOTAL	3190.64	4432.73	487.02	2189.67	720.38	674.04	407.44	1754.96	114.12	824.68	1425.10	$8110.39	
	PER CENT OF TOTAL	39.3	54.6	6.1	27.0	8.9	8.3	5.0	21.6	1.4	10.1	17.7		

A.O.C STATION SUMMARY

064 50% of dollar loss due to machine

Damage caused by worn equipment 30% of dollar loss due to core damage caused by incorrect polyglycol setup

072 Tool proofing and procedural development account for 90% of total dollar loss

074 Major portion of dollar loss due to defective parts received from stock-stains, corrosion and scratches

075 50% of dollar loss due to voids and thick gluelines

DEFECT SUMMARY

Defect	Tools & Mainten-ance	Inadequate Procedure	Handling and Storage	Equip Failure	Defective Parts	Tempera-ture Control	Unknown	Operator Error	TOTAL SCRAP COSTS	% of Total
Contour Surface Variations								208.99	208.99	2.6
Exterior damage			185.44		664.67			368.39	1218.50	15.1
Gaps, steps interference								87.49	87.49	1.1
Delaminations and Voids	517.88	169.84				114.12	286.36	145.79	1233.99	15.3
Glueline Thickness	370.22							81.93	452.15	5.6
Machine Damage	228.49	403.65		115.65				115.65	863.44	10.6
Damaged Core	446.99		471.73	291.79	20.11		384.09	294.58	1909.29	23.5
Dimensional	629.09				991.82		154.23	122.28	1894.42	23.3
Pre-Cure & Cure		146.89							146.89	1.8
Contamination Stains			16.87		60.77				77.64	.9
Mislocated					17.59				17.59	.2
								TOTAL	$8110.39	

Related Quality Cost: Failure Analysis and Investigations for the month - $1,249.64

FIG. 17.7 Scrap-cost data.

17.5 INHERENT VALUE OF RELIABILITY

Industry has been criticized by government agents for neglecting reliability and quality assurance requirements in the design, development, and production of equipment. It is the customer's contention that reliability is not to be treated as a supplementary package or a commodity to be procured at additional cost if and when deemed necessary. In the preparation of bid proposals, the contractor is now required to include the quantitative reliability levels he expects to achieve at the component, subsystem, and system levels. Specific, minimum acceptable reliability requirements are now found in government contracts for systems and associated material. Greater emphasis is being placed on the application of incentive-type contracts as a means of

QUALITY PROGRAM ELEMENT CHECK SHEET

Page 6 of 12

TEST PLANNING AND EVALUATION	Start Date	Completion Date	Man-hours Allocated	Responsible Department	Method of Accomplishment
1. Establish environmental criteria	Nov. 62	March 63		Reliability Engineering	Program Plan
2. Establish type and quantity of test items	Nov. 62	June 64			Contract
3. Establish equipment need dates for test items	Nov. 62	June 64		"	Coordinate with Test Office
4. Write reliability test specifications	Jan. 63	Nov. 65		"	Specification
5. Participate in determining requirements for test equipments.	Nov. 62	Aug. 63		"	Test Specification
6. Draft procurement orders for test items	Nov. 62	June 66			Procurement Orders
7. Monitor test item delivery schedule	Nov. 62	June 66		"	Correspondence Meetings
8. Approve reliability test procedures	Jan. 63	End of Contract		"	"
9. Monitor testing activity	Jan. 63	"		"	Liaison with Test Activity
10. Review test reports					
11. Evaluate test results	Mar. 63	"		"	Publish Reports
12. Evaluate results of all test program for experience retention	Mar. 63	"		"	Reports
13. Investigate test failures	Mar. 63	"		"	Failure Analysis Reports
14. Prepare diagnostic test plans	Jan. 63	"		"	Test Plan
15. Apply test results to obtain design changes	June 63	"		"	Program Directive
16. Revise reliability specifications	June 63	"		"	Test Specifications
17. Formulate and document reliability demonstration plan	Oct. 62	Dec. 62		"	Demonstration Plan

FIG. 17.8 Quality-program-element check sheet.

improving reliability. This means that contractors will be rewarded for producing highly reliable equipment and systems and will be penalized if they fall short of their goals.

To achieve these reliability goals, reliability procedures and disciplines must be applied in each phase of program development from the requirements identification phase until delivery of the final product to the customer. The optimum allocation of resources to the reliability effort in the conceptual and design-development stages will increase product effectiveness, minimize spares requirements for systems support, and lower maintenance costs. It has been estimated that maintenance expenditures in the Department of Defense exceed $8 billion per year—a figure approaching 25 per cent of the total defense budget.

17.6 SIGNIFICANT ADVANTAGES OF RELIABILITY- AND QUALITY-COST ACCOUNTABILITY

The identification and control of reliability and quality assurance costs provide the means for better management decisions. Interrelated cost elements and cause-

and-effect relationships are brought into proper perspective. Accumulated cost data assist management in budgeting the funds necessary for a reliability and quality assurance program commensurate with cost, schedule, and customer requirements. With the cost figures broken down into significant program elements the program manager is in a position to measure with reasonable accuracy the dollar value of each reliability and quality assurance program element. Expenditures, for example, of reliability and quality control engineering activities in drawing and specification review, test-plan analysis, parts selection, supplier coordination, etc., can be evaluated in terms of lower deficiency costs, the application of more effective and efficient quality assurance techniques, and an earlier attainment of equipment reliability.

The accumulation and analysis of deficiency costs provide a measure of the effectiveness of the contractor's corrective-action system. For example, the identification and prompt correction of deficiencies due to human error may highlight basic causes such as inadequate operator skill levels; deficient in-process controls; lack of preciseness, clarity, and completeness of drawings and specifications; and inadequate training.

17.7 RELIABILITY AND QUALITY ASSURANCE PROGRAMMING

The analysis of reliability and quality assurance cost data derived from previous contracts through a system of accounts provides the program manager with reliable guidelines for establishing optimum reliability and quality assurance policies when requests for proposals or new contracts are being considered. An important aspect of the preplanning effort is the prudent assignment of manpower and equipment for establishment and implementation of reliability and quality assurance programs. One method of allocating manpower to each significant element of the reliability program for a new contract is depicted in a representative section from the reliability task list (Fig. 17.8). In a highly competitive situation, it is obviously in a contractor's best interest to minimize overall costs and maintain the highest reliability and quality assurance consistent with available resources.

APPENDIXES

APPENDIX A: TABLES

Table A.1 Summation of Terms of Poisson's Exponential Binomial Limit*
1,000 × probability of c or fewer occurrences of event that has average number of occurrences equal to c' or np'

c' or np' \ c	0	1	2	3	4	5	6	7	8	9
0.02	980	1,000								
0.04	961	999	1,000							
0.06	942	998	1,000							
0.08	923	997	1,000							
0.10	905	995	1,000							
0.15	861	990	999	1,000						
0.20	819	982	999	1,000						
0.25	779	974	998	1,000						
0.30	741	963	996	1,000						
0.35	705	951	994	1,000						
0.40	670	938	992	999	1,000					
0.45	638	925	989	999	1,000					
0.50	607	910	986	998	1,000					
0.55	577	894	982	998	1,000					
0.60	549	878	977	997	1,000					
0.65	522	861	972	996	999	1,000				
0.70	497	844	966	994	999	1,000				
0.75	472	827	959	993	999	1,000				
0.80	449	809	953	991	999	1,000				
0.85	427	791	945	989	998	1,000				
0.90	407	772	937	987	998	1,000				
0.95	387	754	929	984	997	1,000				
1.00	368	736	920	981	996	999	1,000			
1.1	333	699	900	974	995	999	1,000			
1.2	301	663	879	966	992	998	1,000			
1.3	273	627	857	957	989	998	1,000			
1.4	247	592	833	946	986	997	999	1,000		
1.5	223	558	809	934	981	996	999	1,000		
1.6	202	525	783	921	976	994	999	1,000		
1.7	183	493	757	907	970	992	998	1,000		
1.8	165	463	731	891	964	990	997	999	1,000	
1.9	150	434	704	875	956	987	997	999	1,000	
2.0	135	406	677	857	947	983	995	999	1,000	

* From *Statistical Quality Control*, 3d ed., by Eugene L. Grant. Copyright, 1964. McGraw-Hill Book Company. Used by permission.

Table A.1 Summation of Terms of Poisson's Exponential Binomial Limit
(Continued)

c' or np' \ c	0	1	2	3	4	5	6	7	8	9
2.2	111	355	623	819	928	975	993	998	1,000	
2.4	091	308	570	779	904	964	988	997	999	1,000
2.6	074	267	518	736	877	951	983	995	999	1,000
2.8	061	231	469	692	848	935	976	992	998	999
3.0	050	199	423	647	815	916	966	988	996	999
3.2	041	171	380	603	781	895	955	983	994	998
3.4	033	147	340	558	744	871	942	977	992	997
3.6	027	126	303	515	706	844	927	969	988	996
3.8	022	107	269	473	668	816	909	960	984	994
4.0	018	092	238	433	629	785	889	949	979	992
4.2	015	078	210	395	590	753	867	936	972	989
4.4	012	066	185	359	551	720	844	921	964	985
4.6	010	056	163	326	513	686	818	905	955	980
4.8	008	048	143	294	476	651	791	887	944	975
5.0	007	040	125	265	440	616	762	867	932	968
5.2	006	034	109	238	406	581	732	845	918	960
5.4	005	029	095	213	373	546	702	822	903	951
5.6	004	024	082	191	342	512	670	797	886	941
5.8	003	021	072	170	313	478	638	771	867	929
6.0	002	017	062	151	285	446	606	744	847	916

c' or np' \ c	10	11	12	13	14	15	16
2.8	1,000						
3.0	1,000						
3.2	1,000						
3.4	999	1,000					
3.6	999	1,000					
3.8	998	999	1,000				
4.0	997	999	1,000				
4.2	996	999	1,000				
4.4	994	998	999	1,000			
4.6	992	997	999	1,000			
4.8	990	996	999	1,000			
5.0	986	995	998	999	1,000		
5.2	982	993	997	999	1,000		
5.4	977	990	996	999	1,000		
5.6	972	988	995	998	999	1,000	
5.8	965	984	993	997	999	1,000	
6.0	957	980	991	996	999	999	1,000

Table A.1 Summation of Terms of Poisson's Exponential Binomial Limit *(Continued)*

c' or np' \ c	0	1	2	3	4	5	6	7	8	9
6.2	002	015	054	134	259	414	574	716	826	902
6.4	002	012	046	119	235	384	542	687	803	886
6.6	001	010	040	105	213	355	511	658	780	869
6.8	001	009	034	093	192	327	480	628	755	850
7.0	001	007	030	082	173	301	450	599	729	830
7.2	001	006	025	072	156	276	420	569	703	810
7.4	001	005	022	063	140	253	392	539	676	788
7.6	001	004	019	055	125	231	365	510	648	765
7.8	000	004	016	048	112	210	338	481	620	741
8.0	000	003	014	042	100	191	313	453	593	717
8.5	000	002	009	030	074	150	256	386	523	653
9.0	000	001	006	021	055	116	207	324	456	587
9.5	000	001	004	015	040	089	165	269	392	522
10.0	000	000	003	010	029	067	130	220	333	458

c' or np' \ c	10	11	12	13	14	15	16	17	18	19
6.2	949	975	989	995	998	999	1,000			
6.4	939	969	986	994	997	999	1,000			
6.6	927	963	982	992	997	999	999	1,000		
6.8	915	955	978	990	996	998	999	1,000		
7.0	901	947	973	987	994	998	999	1,000		
7.2	887	937	967	984	993	997	999	999	1,000	
7.4	871	926	961	980	991	996	998	999	1,000	
7.6	854	915	954	976	989	995	998	999	1,000	
7.8	835	902	945	971	986	993	997	999	1,000	
8.0	816	888	936	966	983	992	996	998	999	1,000
8.5	763	849	909	949	973	986	993	997	999	999
9.0	706	803	876	926	959	978	989	995	998	999
9.5	645	752	836	898	940	967	982	991	996	998
10.0	583	697	792	864	917	951	973	986	993	997

c' or np' \ c	20	21	22
8.5	1,000		
9.0	1,000		
9.5	999	1,000	
10.0	998	999	1,000

Table A.1 Summation of Terms of Poisson's Exponential Binomial Limit
(Continued)

c' or np' \ c	0	1	2	3	4	5	6	7	8	9
10.5	000	000	002	007	021	050	102	179	279	397
11.0	000	000	001	005	015	038	079	143	232	341
11.5	000	000	001	003	011	028	060	114	191	289
12.0	000	000	001	002	008	020	046	090	155	242
12.5	000	000	000	002	005	015	035	070	125	201
13.0	000	000	000	001	004	011	026	054	100	166
13.5	000	000	000	001	003	008	019	041	079	135
14.0	000	000	000	000	002	006	014	032	062	109
14.5	000	000	000	000	001	004	010	024	048	088
15.0	000	000	000	000	001	003	008	018	037	070
	10	11	12	13	14	15	16	17	18	19
10.5	521	639	742	825	888	932	960	978	988	994
11.0	460	579	689	781	854	907	944	968	982	991
11.5	402	520	633	733	815	878	924	954	974	986
12.0	347	462	576	682	772	844	899	937	963	979
12.5	297	406	519	628	725	806	869	916	948	969
13.0	252	353	463	573	675	764	835	890	930	957
13.5	211	304	409	518	623	718	798	861	908	942
14.0	176	260	358	464	570	669	756	827	883	923
14.5	145	220	311	413	518	619	711	790	853	901
15.0	118	185	268	363	466	568	664	749	819	875
	20	21	22	23	24	25	26	27	28	29
10.5	997	999	999	1,000						
11.0	995	998	999	1,000						
11.5	992	996	998	999	1,000					
12.0	988	994	997	999	999	1,000				
12.5	983	991	995	998	999	999	1,000			
13.0	975	986	992	996	998	999	1,000			
13.5	965	980	989	994	997	998	999	1,000		
14.0	952	971	983	991	995	997	999	999	1,000	
14.5	936	960	976	986	992	996	998	999	999	1,000
15.0	917	947	967	981	989	994	997	998	999	1,000

Table A.1 Summation of Terms of Poisson's Exponential Binomial Limit
(Continued)

c' or np' \ c	4	5	6	7	8	9	10	11	12	13
16	000	001	004	010	022	043	077	127	193	275
17	000	001	002	005	013	026	049	085	135	201
18	000	000	001	003	007	015	030	055	092	143
19	000	000	001	002	004	009	018	035	061	098
20	000	000	000	001	002	005	011	021	039	066
21	000	000	000	000	001	003	006	013	025	043
22	000	000	000	000	001	002	004	008	015	028
23	000	000	000	000	000	001	002	004	009	017
24	000	000	000	000	000	000	001	003	005	011
25	000	000	000	000	000	000	001	001	003	006
	14	15	16	17	18	19	20	21	22	23
16	368	467	566	659	742	812	868	911	942	963
17	281	371	468	564	655	736	805	861	905	937
18	208	287	375	469	562	651	731	799	855	899
19	150	215	292	378	469	561	647	725	793	849
20	105	157	221	297	381	470	559	644	721	787
21	072	111	163	227	302	384	471	558	640	716
22	048	077	117	169	232	306	387	472	556	637
23	031	052	082	123	175	238	310	389	472	555
24	020	034	056	087	128	180	243	314	392	473
25	012	022	038	060	092	134	185	247	318	394
	24	25	26	27	28	29	30	31	32	33
16	978	987	993	996	998	999	999	1,000		
17	959	975	985	991	995	997	999	999	1,000	
18	932	955	972	983	990	994	997	998	999	1,000
19	893	927	951	969	980	988	993	996	998	999
20	843	888	922	948	966	978	987	992	995	997
21	782	838	883	917	944	963	976	985	991	994
22	712	777	832	877	913	940	959	973	983	989
23	635	708	772	827	873	908	936	956	971	981
24	554	632	704	768	823	868	904	932	953	969
25	473	553	629	700	763	818	863	900	929	950
	34	35	36	37	38	39	40	41	42	43
19	999	1,000								
20	999	999	1,000							
21	997	998	999	999	1,000					
22	994	996	998	999	999	1,000				
23	988	993	996	997	999	999	1,000			
24	979	987	992	995	997	998	999	999	1,000	
25	966	978	985	991	994	997	998	999	999	1,000

Table A.2 Confidence Limits for a Proportion (One-sided), $n \leq 30$

(For confidence limits for $n > 30$, see Charts B.1 to B.4)

If the observed proportion is r/n, enter the table with n and r for an upper one-sided limit. For a lower one-sided limit, enter the table with n and $n - r$ and subtract the table entry from 1.

r	90%	95%	99%	r	90%	95%	99%	r	90%	95%	99%
	$n = 2$				$n = 3$				$n = 4$		
0	0.684	0.776	0.900	0	0.536	0.632	0.785−	0	0.438	0.527	0.684
1	0.949	0.975−	0.995−	1	0.804	0.865−	0.941	1	0.680	0.751	0.859
				2	0.965+	0.983	0.997	2	0.857	0.902	0.958
								3	0.974	0.987	0.997
	$n = 5$				$n = 6$				$n = 7$		
0	0.369	0.451	0.602	0	0.319	0.393	0.536	0	0.280	0.348	0.482
1	0.584	0.657	0.778	1	0.510	0.582	0.706	1	0.453	0.521	0.643
2	0.753	0.811	0.894	2	0.667	0.729	0.827	2	0.596	0.659	0.764
3	0.888	0.924	0.967	3	0.799	0.847	0.915+	3	0.721	0.775−	0.858
4	0.979	0.990	0.998	4	0.907	0.937	0.973	4	0.830	0.871	0.929
				5	0.983	0.991	0.998	5	0.921	0.947	0.977
								6	0.985+	0.993	0.999
	$n = 8$				$n = 9$				$n = 10$		
0	0.250	0.312	0.438	0	0.226	0.283	0.401	0	0.206	0.259	0.369
1	0.406	0.471	0.590	1	0.368	0.429	0.544	1	0.337	0.394	0.504
2	0.538	0.600	0.707	2	0.490	0.550	0.656	2	0.450	0.507	0.612
3	0.655+	0.711	0.802	3	0.599	0.655+	0.750	3	0.552	0.607	0.703
4	0.760	0.807	0.879	4	0.699	0.749	0.829	4	0.646	0.696	0.782
5	0.853	0.889	0.939	5	0.790	0.831	0.895−	5	0.733	0.778	0.850
6	0.931	0.954	0.980	6	0.871	0.902	0.947	6	0.812	0.850	0.907
7	0.987	0.994	0.999	7	0.939	0.959	0.983	7	0.884	0.913	0.952
				8	0.988	0.994	0.999	8	0.945+	0.963	0.984
								9	0.990	0.995−	0.999
	$n = 11$				$n = 12$				$n = 13$		
0	0.189	0.238	0.342	0	0.175−	0.221	0.319	0	0.162	0.206	0.298
1	0.310	0.364	0.470	1	0.287	0.339	0.440	1	0.268	0.316	0.413
2	0.415+	0.470	0.572	2	0.386	0.438	0.537	2	0.360	0.410	0.506
3	0.511	0.564	0.660	3	0.475+	0.527	0.622	3	0.444	0.495−	0.588
4	0.599	0.650	0.738	4	0.559	0.609	0.698	4	0.523	0.573	0.661
5	0.682	0.729	0.806	5	0.638	0.685−	0.765+	5	0.598	0.645+	0.727
6	0.759	0.800	0.866	6	0.712	0.755−	0.825+	6	0.669	0.713	0.787
7	0.831	0.865−	0.916	7	0.781	0.819	0.879	7	0.736	0.776	0.841
8	0.895+	0.921	0.957	8	0.846	0.877	0.924	8	0.799	0.834	0.889
9	0.951	0.967	0.986	9	0.904	0.928	0.961	9	0.858	0.887	0.931
10	0.990	0.995+	0.999	10	0.955−	0.970	0.987	10	0.912	0.934	0.964
				11	0.991	0.996	0.999	11	0.958	0.972	0.988
								12	0.992	0.996	0.999
	$n = 14$				$n = 15$				$n = 16$		
0	0.152	0.193	0.280	0	0.142	0.181	0.264	0	0.134	0.171	0.250
1	0.251	0.297	0.389	1	0.236	0.279	0.368	1	0.222	0.264	0.349
2	0.337	0.385+	0.478	2	0.317	0.363	0.453	2	0.300	0.344	0.430
3	0.417	0.466	0.557	3	0.393	0.440	0.529	3	0.371	0.417	0.503
4	0.492	0.540	0.627	4	0.464	0.511	0.597	4	0.439	0.484	0.569
5	0.563	0.610	0.692	5	0.532	0.577	0.660	5	0.504	0.548	0.630

This table was computed by Robert S. Gardner and is reproduced, by permission of the author and the publisher, from E. L. Crow, F. A. Davis, and M. W. Maxfield, *Statistics Manual*, China Lake, Calif., U.S. Naval Ordnance Test Station, NAVORD Report 3369 (NOTS 948), pp. 262–265.

Table A.2 Confidence Limits for a Proportion (One-sided), $n \leq 30$ *(Continued)*

r	90%	95%	99%	r	90%	95%	99%	r	90%	95%	99%
	n = 14 (Continued)				n = 15 (Continued)				n = 16 (Continued)		
6	0.631	0.675−	0.751	6	0.596	0.640	0.718	6	0.565+	0.609	0.687
7	0.695+	0.736	0.805+	7	0.658	0.700	0.771	7	0.625−	0.667	0.739
8	0.757	0.794	0.854	8	0.718	0.756	0.821	8	0.682	0.721	0.788
9	0.815−	0.847	0.898	9	0.774	0.809	0.865+	9	0.737	0.773	0.834
10	0.869	0.896	0.936	10	0.828	0.858	0.906	10	0.790	0.822	0.875−
11	0.919	0.939	0.967	11	0.878	0.903	0.941	11	0.839	0.868	0.912
12	0.961	0.974	0.989	12	0.924	0.943	0.969	12	0.886	0.910	0.945−
13	0.993	0.996	0.999	13	0.964	0.976	0.990	13	0.929	0.947	0.971
				14	0.993	0.997	0.999	14	0.966	0.977	0.990
								15	0.993	0.997	0.999
	n = 17				n = 18				n = 19		
0	0.127	0.162	0.237	0	0.120	0.153	0.226	0	0.114	0.146	0.215+
1	0.210	0.250	0.332	1	0.199	0.238	0.316	1	0.190	0.226	0.302
2	0.284	0.326	0.410	2	0.269	0.310	0.391	2	0.257	0.296	0.374
3	0.352	0.396	0.480	3	0.334	0.377	0.458	3	0.319	0.359	0.439
4	0.416	0.461	0.543	4	0.396	0.439	0.520	4	0.378	0.419	0.498
5	0.478	0.522	0.603	5	0.455+	0.498	0.577	5	0.434	0.476	0.554
6	0.537	0.580	0.658	6	0.512	0.554	0.631	6	0.489	0.530	0.606
7	0.594	0.636	0.709	7	0.567	0.608	0.681	7	0.541	0.582	0.655+
8	0.650	0.689	0.758	8	0.620	0.659	0.729	8	0.592	0.632	0.702
9	0.703	0.740	0.803	9	0.671	0.709	0.774	9	0.642	0.680	0.746
10	0.754	0.788	0.845−	10	0.721	0.756	0.816	10	0.690	0.726	0.788
11	0.803	0.834	0.883	11	0.769	0.801	0.855−	11	0.737	0.770	0.827
12	0.849	0.876	0.918	12	0.815−	0.844	0.890	12	0.782	0.812	0.863
13	0.893	0.915+	0.948	13	0.858	0.884	0.923	13	0.825−	0.853	0.897
14	0.933	0.950	0.973	14	0.899	0.920	0.951	14	0.866	0.890	0.927
15	0.968	0.979	0.991	15	0.937	0.953	0.975−	15	0.905−	0.925−	0.954
16	0.994	0.997	0.999	16	0.970	0.980	0.992	16	0.941	0.956	0.976
				17	0.994	0.997	0.999	17	0.972	0.981	0.992
								18	0.994	0.997	0.999
	n = 20				n = 21				n = 22		
0	0.109	0.139	0.206	0	0.104	0.133	0.197	0	0.099	0.127	0.189
1	0.181	0.216	0.289	1	0.173	0.207	0.277	1	0.166	0.198	0.266
2	0.245−	0.283	0.358	2	0.234	0.271	0.344	2	0.224	0.259	0.330
3	0.304	0.344	0.421	3	0.291	0.329	0.404	3	0.279	0.316	0.389
4	0.361	0.401	0.478	4	0.345+	0.384	0.460	4	0.331	0.369	0.443
5	0.415−	0.456	0.532	5	0.397	0.437	0.512	5	0.381	0.420	0.493
6	0.467	0.508	0.583	6	0.448	0.487	0.561	6	0.430	0.468	0.541
7	0.518	0.558	0.631	7	0.497	0.536	0.608	7	0.477	0.515+	0.587
8	0.567	0.606	0.677	8	0.544	0.583	0.653	8	0.523	0.561	0.630
9	0.615+	0.653	0.720	9	0.590	0.628	0.695+	9	0.568	0.605−	0.672
10	0.662	0.698	0.761	10	0.636	0.672	0.736	10	0.611	0.647	0.712
11	0.707	0.741	0.800	11	0.679	0.714	0.774	11	0.654	0.689	0.750
12	0.751	0.783	0.837	12	0.722	0.755+	0.811	12	0.695+	0.729	0.786
13	0.793	0.823	0.871	13	0.764	0.794	0.845+	13	0.736	0.767	0.821
14	0.834	0.860	0.902	14	0.804	0.832	0.878	14	0.775+	0.804	0.853
15	0.873	0.896	0.931	15	0.842	0.868	0.908	15	0.813	0.840	0.884
16	0.910	0.929	0.956	16	0.879	0.901	0.935−	16	0.850	0.874	0.912
17	0.944	0.958	0.977	17	0.914	0.932	0.959	17	0.885+	0.906	0.938
18	0.973	0.982	0.992	18	0.946	0.960	0.978	18	0.918	0.935+	0.961
19	0.995−	0.997	0.999	19	0.974	0.983	0.993	19	0.949	0.962	0.979
				20	0.995−	0.988	1.000	20	0.976	0.984	0.993
								21	0.995+	0.998	1.000

Table A.2 Confidence Limits for a Proportion (One-sided), $n \leq 30$
(Continued)

r	90%	95%	99%
	$n = 23$		
0	0.095+	0.122	0.181
1	0.159	0.190	0.256
2	0.215+	0.249	0.318
3	0.268	0.304	0.374
4	0.318	0.355−	0.427
5	0.366	0.404	0.476
6	0.413	0.451	0.522
7	0.459	0.496	0.567
8	0.503	0.540	0.609
9	0.546	0.583	0.650
10	0.589	0.625−	0.689
11	0.630	0.665−	0.727
12	0.670	0.704	0.763
13	0.710	0.742	0.797
14	0.748	0.778	0.829
15	0.786	0.814	0.860
16	0.822	0.848	0.889
17	0.857	0.880	0.916
18	0.890	0.910	0.941
19	0.922	0.938	0.962
20	0.951	0.963	0.980
21	0.977	0.984	0.993
22	0.995+	0.998	1.000

r	90%	95%	99%
	$n = 24$		
0	0.091	0.117	0.175−
1	0.153	0.183	0.246
2	0.207	0.240	0.307
3	0.258	0.292	0.361
4	0.306	0.342	0.412
5	0.352	0.389	0.460
6	0.398	0.435−	0.505−
7	0.442	0.479	0.548
8	0.484	0.521	0.590
9	0.526	0.563	0.630
10	0.567	0.603	0.668
11	0.608	0.642	0.705−
12	0.647	0.681	0.740
13	0.685+	0.718	0.774
14	0.723	0.754	0.806
15	0.759	0.788	0.837
16	0.795+	0.822	0.867
17	0.830	0.854	0.894
18	0.863	0.885+	0.920
19	0.895+	0.914	0.943
20	0.925+	0.941	0.964
21	0.953	0.965+	0.981
22	0.978	0.985−	0.994
23	0.996	0.998	1.000

r	90%	95%	99%
	$n = 25$		
0	0.088	0.113	0.168
1	0.147	0.176	0.237
2	0.199	0.231	0.296
3	0.248	0.282	0.349
4	0.295−	0.330	0.398
5	0.340	0.375+	0.444
6	0.383	0.420	0.488
7	0.426	0.462	0.531
8	0.467	0.504	0.571
9	0.508	0.544	0.610
10	0.548	0.583	0.648
11	0.587	0.621	0.684
12	0.625−	0.659	0.719
13	0.662	0.695−	0.752
14	0.699	0.730	0.784
15	0.735−	0.764	0.815+
16	0.770	0.798	0.845+
17	0.804	0.830	0.873
18	0.837	0.861	0.899
19	0.869	0.890	0.923
20	0.899	0.918	0.946
21	0.928	0.943	0.966
22	0.955+	0.966	0.982
23	0.979	0.986	0.994
24	0.996	0.998	1.000

r	90%	95%	99%
	$n = 26$		
0	0.085−	0.109	0.162
1	0.142	0.170	0.229
2	0.192	0.223	0.286
3	0.239	0.272	0.337
4	0.284	0.318	0.385−
5	0.328	0.363	0.430
6	0.370	0.405+	0.473
7	0.411	0.447	0.514
8	0.451	0.487	0.554
9	0.491	0.526	0.592
10	0.529	0.564	0.628
11	0.567	0.602	0.664
12	0.604	0.638	0.698
13	0.641	0.673	0.731
14	0.676	0.708	0.763
15	0.711	0.742	0.794
16	0.746	0.774	0.823
17	0.779	0.806	0.851
18	0.812	0.837	0.878
19	0.843	0.866	0.903
20	0.874	0.894	0.927
21	0.903	0.921	0.948
22	0.931	0.946	0.967
23	0.957	0.968	0.983
24	0.979	0.986	0.994
25	0.996	0.998	1.000

r	90%	95%	99%
	$n = 27$		
0	0.082	0.105+	0.157
1	0.137	0.164	0.222
2	0.185+	0.215+	0.277
3	0.231	0.263	0.326
4	0.275−	0.308	0.373
5	0.317	0.351	0.417
6	0.358	0.392	0.458
7	0.397	0.432	0.498
8	0.436	0.471	0.537
9	0.475−	0.509	0.574
10	0.512	0.547	0.610
11	0.549	0.583	0.645+
12	0.585−	0.618	0.679
13	0.620	0.653	0.711
14	0.655+	0.687	0.743
15	0.689	0.720	0.773
16	0.723	0.752	0.802
17	0.756	0.783	0.831
18	0.788	0.814	0.857
19	0.819	0.843	0.883
20	0.849	0.871	0.907
21	0.879	0.899	0.930
22	0.907	0.924	0.950
23	0.934	0.948	0.968
24	0.958	0.969	0.983
25	0.980	0.987	0.994
26	0.996	0.998	1.000

r	90%	95%	99%
	$n = 28$		
0	0.079	0.101	0.152
1	0.132	0.159	0.215−
2	0.179	0.208	0.268
3	0.223	0.254	0.316
4	0.265+	0.298	0.361
5	0.306	0.339	0.404
6	0.346	0.380	0.445−
7	0.385−	0.419	0.484
8	0.422	0.457	0.521
9	0.459	0.494	0.558
10	0.496	0.530	0.593
11	0.532	0.565−	0.627
12	0.567	0.600	0.660
13	0.601	0.634	0.692
14	0.635+	0.667	0.723
15	0.669	0.699	0.753
16	0.701	0.731	0.782
17	0.733	0.762	0.810
18	0.765−	0.792	0.837
19	0.796	0.821	0.863
20	0.826	0.849	0.888
21	0.855+	0.876	0.911
22	0.883	0.902	0.932
23	0.911	0.927	0.952
24	0.936	0.950	0.969
25	0.960	0.970	0.984
26	0.981	0.987	0.995−
27	0.996	0.998	1.000

Table A.2 Confidence Limits for a Proportion (One-sided), $n \leq 30$ *(Continued)*

r	90%	95%	99%	r	90%	95%	99%
	$n = 29$				$n = 30$		
0	0.076	0.098	0.147	0	0.074	0.095+	0.142
1	0.128	0.153	0.208	1	0.124	0.149	0.202
2	0.173	0.202	0.260	2	0.168	0.195+	0.252
3	0.216	0.246	0.307	3	0.209	0.239	0.298
4	0.257	0.288	0.350	4	0.249	0.280	0.340
5	0.297	0.329	0.392	5	0.287	0.319	0.381
6	0.335−	0.368	0.432	6	0.325−	0.357	0.420
7	0.372	0.406	0.470	7	0.361	0.394	0.457
8	0.409	0.443	0.507	8	0.397	0.430	0.493
9	0.445+	0.479	0.542	9	0.432	0.465+	0.527
10	0.481	0.514	0.577	10	0.466	0.499	0.561
11	0.515+	0.549	0.610	11	0.500	0.533	0.594
12	0.550	0.583	0.643	12	0.533	0.566	0.626
13	0.583	0.616	0.674	13	0.566	0.598	0.657
14	0.616	0.648	0.705−	14	0.599	0.630	0.687
15	0.649	0.680	0.734	15	0.630	0.661	0.716
16	0.681	0.711	0.763	16	0.662	0.692	0.744
17	0.712	0.741	0.791	17	0.692	0.721	0.772
18	0.743	0.771	0.818	18	0.723	0.750	0.799
19	0.774	0.800	0.843	19	0.752	0.779	0.824
20	0.803	0.828	0.868	20	0.782	0.807	0.849
21	0.832	0.855−	0.892	21	0.810	0.834	0.873
22	0.860	0.881	0.914	22	0.838	0.860	0.896
23	0.888	0.906	0.935−	23	0.865+	0.885+	0.917
24	0.914	0.930	0.954	24	0.891	0.909	0.937
25	0.938	0.951	0.970	25	0.917	0.932	0.955+
26	0.961	0.971	0.985	26	0.941	0.953	0.972
27	0.982	0.988	0.995−	27	0.963	0.972	0.985+
28	0.996	0.998	1.000	28	0.982	0.988	0.995+
				29	0.996	0.998	1.000

Table A.3 Confidence Limits for a Proportion (Two-sided), $n \leq 30$*

(For confidence limits for $n > 30$, see Charts B.1 to B.4)

Upper limits are in boldface. The observed proportion in a random sample is r/n.

r	90%		95%		99%	
			$n = 1$			
0	0	**0.900**	0	**0.950**	0	**0.990**
1	0.100	**1**	0.050	**1**	0.010	**1**
			$n = 3$			
0	0	**0.536**	0	**0.632**	0	**0.785−**
1	0.035−	**0.804**	0.017	**0.865−**	0.003	**0.941**
2	0.196	**0.965+**	0.135−	**0.983**	0.059	**0.997**
3	0.464	**1**	0.368	**1**	0.215+	**1**
			$n = 5$			
0	0	**0.379**	0	**0.500**	0	**0.602**
1	0.021	**0.621**	0.010	**0.657**	0.002	**0.778**
2	0.112	**0.753**	0.076	**0.811**	0.033	**0.894**
3	0.247	**0.888**	0.189	**0.924**	0.106	**0.967**
4	0.379	**0.979**	0.343	**0.990**	0.222	**0.998**
5	0.621	**1**	0.500	**1**	0.398	**1**
			$n = 7$			
0	0	**0.316**	0	**0.377**	0	**0.500**
1	0.015	**0.500**	0.007	**0.554**	0.001	**0.643**
2	0.079	**0.684**	0.053	**0.659**	0.023	**0.764**
3	0.170	**0.721**	0.129	**0.775**	0.071	**0.858**
4	0.279	**0.830**	0.225+	**0.871**	0.142	**0.929**
5	0.316	**0.921**	0.341	**0.947**	0.236	**0.977**
6	0.500	**0.985+**	0.446	**0.993**	0.357	**0.999**
7	0.684	**1**	0.623	**1**	0.500	**1**
			$n = 9$			
0	0	**0.232**	0	**0.289**	0	**0.402**
1	0.012	**0.391**	0.006	**0.443**	0.001	**0.598**
2	0.061	**0.515+**	0.041	**0.558**	0.017	**0.656**
3	0.129	**0.610**	0.098	**0.711**	0.053	**0.750**
4	0.210	**0.768**	0.169	**0.749**	0.105+	**0.829**
5	0.232	**0.790**	0.251	**0.831**	0.171	**0.895−**
6	0.390	**0.871**	0.289	**0.902**	0.250	**0.947**
7	0.485−	**0.939**	0.442	**0.959**	0.344	**0.983**
8	0.609	**0.988**	0.557	**0.994**	0.402	**0.999**
9	0.768	**1**	0.711	**1**	0.598	**1**
			$n = 11$			
0	0	**0.197**	0	**0.250**	0	**0.359**
1	0.010	**0.315+**	0.005−	**0.369**	0.001	**0.509**
2	0.049	**0.423**	0.033	**0.500**	0.014	**0.593**
3	0.105−	**0.577**	0.079	**0.631**	0.043	**0.660**
4	0.169	**0.685+**	0.135+	**0.667**	0.084	**0.738**
5	0.197	**0.698**	0.200	**0.750**	0.134	**0.806**
6	0.302	**0.803**	0.250	**0.800**	0.194	**0.866**
7	0.315+	**0.831**	0.333	**0.865−**	0.262	**0.916**
8	0.423	**0.895+**	0.369	**0.921**	0.340	**0.957**
9	0.577	**0.951**	0.500	**0.967**	0.407	**0.986**
10	0.685−	**0.990**	0.631	**0.995+**	0.500	**0.999**
11	0.803	**1**	0.750	**1**	0.641	**1**

r	90%		95%		99%	
			$n = 2$			
0	0	**0.684**	0	**0.776**	0	**0.900**
1	0.051	**0.949**	0.025+	**0.975−**	0.005+	**0.995−**
2	0.316	**1**	0.224	**1**	0.100	**1**
			$n = 4$			
0	0	**0.500**	0	**0.527**	0	**0.684**
1	0.026	**0.680**	0.013	**0.751**	0.003	**0.859**
2	0.143	**0.857**	0.098	**0.902**	0.042	**0.958**
3	0.320	**0.974**	0.249	**0.987**	0.141	**0.997**
4	0.500	**1**	0.473	**1**	0.316	**1**
			$n = 6$			
0	0	**0.345−**	0	**0.402**	0	**0.536**
1	0.017	**0.542**	0.009	**0.598**	0.002	**0.706**
2	0.093	**0.667**	0.063	**0.729**	0.027	**0.827**
3	0.201	**0.799**	0.153	**0.847**	0.085−	**0.915+**
4	0.333	**0.907**	0.271	**0.937**	0.173	**0.973**
5	0.458	**0.983**	0.402	**0.991**	0.294	**0.998**
6	0.655+	**1**	0.598	**1**	0.464	**1**
			$n = 8$			
0	0	**0.255−**	0	**0.315+**	0	**0.451**
1	0.013	**0.418**	0.006	**0.500**	0.001	**0.590**
2	0.069	**0.582**	0.046	**0.685−**	0.020	**0.707**
3	0.147	**0.745+**	0.111	**0.711**	0.061	**0.802**
4	0.240	**0.760**	0.193	**0.807**	0.121	**0.879**
5	0.255−	**0.853**	0.289	**0.889**	0.198	**0.939**
6	0.418	**0.931**	0.315+	**0.954**	0.293	**0.980**
7	0.582	**0.987**	0.500	**0.994**	0.410	**0.999**
8	0.745+	**1**	0.685−	**1**	0.549	**1**
			$n = 10$			
0	0	**0.222**	0	**0.267**	0	**0.376**
1	0.010	**0.352**	0.005+	**0.397**	0.001	**0.512**
2	0.055−	**0.500**	0.037	**0.603**	0.016	**0.624**
3	0.116	**0.648**	0.087	**0.619**	0.048	**0.703**
4	0.188	**0.659**	0.150	**0.733**	0.093	**0.782**
5	0.222	**0.778**	0.222	**0.778**	0.150	**0.850**
6	0.341	**0.812**	0.267	**0.850**	0.218	**0.907**
7	0.352	**0.884**	0.381	**0.913**	0.297	**0.952**
8	0.500	**0.945+**	0.397	**0.963**	0.376	**0.984**
9	0.648	**0.990**	0.603	**0.995−**	0.488	**0.999**
10	0.778	**1**	0.733	**1**	0.624	**1**
			$n = 12$			
0	0	**0.184**	0	**0.236**	0	**0.321**
1	0.009	**0.294**	0.004	**0.346**	0.001	**0.445+**
2	0.045+	**0.398**	0.030	**0.450**	0.013	**0.555−**
3	0.096	**0.500**	0.072	**0.550**	0.039	**0.679**
4	0.154	**0.602**	0.123	**0.654**	0.076	**0.698**
5	0.184	**0.706**	0.181	**0.706**	0.121	**0.765+**
6	0.271	**0.729**	0.236	**0.764**	0.175−	**0.825+**
7	0.294	**0.816**	0.294	**0.819**	0.235−	**0.879**
8	0.398	**0.846**	0.346	**0.877**	0.302	**0.924**
9	0.500	**0.904**	0.450	**0.928**	0.321	**0.961**
10	0.602	**0.955−**	0.550	**0.970**	0.445+	**0.987**
11	0.706	**0.991**	0.654	**0.996**	0.555−	**0.999**
12	0.816	**1**	0.764	**1**	0.679	**1**

* This table was calculated by Edwin L. Crow, Eleanor G. Crow, and Robert S. Gardner and is reproduced, by permission of the authors and the publisher, from E. L. Crow, F. A. Davis, and M. W. Maxfield, *Statistics Manual*, China Lake, Calif., U.S. Naval Ordnance Test Station, NAVORD Report 3369 (NOTS 948), 1955, pp. 257–261.

Table A.3 Confidence Limits for a Proportion (Two-sided), $n \leq 30$
(Continued)

$n = 13$

r	90%		95%		99%	
0	0	**0.173**	0	**0.225+**	0	**0.302**
1	0.008	**0.276**	0.004	**0.327**	0.001	**0.429**
2	0.042	**0.379**	0.028	**0.434**	0.012	**0.523**
3	0.088	**0.470**	0.066	**0.520**	0.036	**0.594**
4	0.142	**0.545−**	0.113	**0.587**	0.069	**0.698**
5	0.173	**0.621**	0.166	**0.673**	0.111	**0.727**
6	0.246	**0.724**	0.224	**0.740**	0.159	**0.787**
7	0.276	**0.754**	0.260	**0.776**	0.213	**0.841**
8	0.379	**0.827**	0.327	**0.834**	0.273	**0.889**
9	0.455+	**0.858**	0.413	**0.887**	0.302	**0.931**
10	0.530	**0.912**	0.480	**0.934**	0.406	**0.964**
11	0.621	**0.958**	0.566	**0.972**	0.477	**0.988**
12	0.724	**0.992**	0.673	**0.996**	0.571	**0.999**
13	0.827	**1**	0.775−	**1**	0.698	**1**

$n = 14$

r	90%		95%		99%	
0	0	**0.163**	0	**0.207**	0	**0.286**
1	0.007	**0.261**	0.004	**0.312**	0.001	**0.392**
2	0.039	**0.365+**	0.026	**0.389**	0.011	**0.500**
3	0.081	**0.422**	0.061	**0.500**	0.033	**0.608**
4	0.131	**0.578**	0.104	**0.611**	0.064	**0.636**
5	0.163	**0.594**	0.153	**0.629**	0.102	**0.714**
6	0.224	**0.645+**	0.206	**0.688**	0.146	**0.751**
7	0.261	**0.739**	0.207	**0.793**	0.195−	**0.805+**
8	0.355−	**0.776**	0.312	**0.794**	0.249	**0.854**
9	0.406	**0.837**	0.371	**0.847**	0.286	**0.898**
10	0.422	**0.869**	0.389	**0.896**	0.364	**0.936**
11	0.578	**0.919**	0.500	**0.939**	0.392	**0.967**
12	0.635−	**0.961**	0.611	**0.974**	0.500	**0.989**
13	0.739	**0.993**	0.688	**0.996**	0.608	**0.999**
14	0.837	**1**	0.793	**1**	0.714	**1**

$n = 15$

r	90%		95%		99%	
0	0	**0.154**	0	**0.191**	0	**0.273**
1	0.007	**0.247**	0.003	**0.302**	0.001	**0.373**
2	0.036	**0.326**	0.024	**0.369**	0.010	**0.461**
3	0.076	**0.400**	0.057	**0.448**	0.031	**0.539**
4	0.122	**0.500**	0.097	**0.552**	0.059	**0.627**
5	0.154	**0.600**	0.142	**0.631**	0.094	**0.672**
6	0.205+	**0.674**	0.191	**0.668**	0.135−	**0.727**
7	0.247	**0.675−**	0.192	**0.706**	0.179	**0.771**
8	0.325+	**0.753**	0.294	**0.808**	0.229	**0.821**
9	0.326	**0.795−**	0.332	**0.809**	0.273	**0.865+**
10	0.400	**0.846**	0.369	**0.838**	0.328	**0.906**
11	0.500	**0.878**	0.448	**0.903**	0.373	**0.941**
12	0.600	**0.924**	0.552	**0.943**	0.461	**0.969**
13	0.674	**0.964**	0.631	**0.976**	0.539	**0.990**
14	0.753	**0.993**	0.698	**0.997**	0.627	**0.999**
15	0.846	**1**	0.809	**1**	0.727	**1**

$n = 16$

r	90%		95%		99%	
0	0	**0.147**	0	**0.179**	0	**0.264**
1	0.007	**0.235+**	0.003	**0.273**	0.001	**0.357**
2	0.034	**0.305+**	0.023	**0.352**	0.010	**0.451**
3	0.071	**0.381**	0.053	**0.429**	0.029	**0.525−**
4	0.114	**0.450**	0.090	**0.500**	0.055+	**0.579**
5	0.147	**0.550**	0.132	**0.571**	0.088	**0.643**
6	0.189	**0.619**	0.178	**0.648**	0.125+	**0.705−**
7	0.235+	**0.695−**	0.179	**0.727**	0.166	**0.739**
8	0.299	**0.701**	0.272	**0.728**	0.212	**0.788**
9	0.305+	**0.765−**	0.273	**0.821**	0.261	**0.834**
10	0.381	**0.811**	0.352	**0.822**	0.295+	**0.875−**
11	0.450	**0.853**	0.429	**0.868**	0.357	**0.912**
12	0.550	**0.886**	0.500	**0.910**	0.421	**0.945−**
13	0.619	**0.929**	0.571	**0.947**	0.475+	**0.971**
14	0.695−	**0.966**	0.648	**0.977**	0.549	**0.990**
15	0.765−	**0.993**	0.727	**0.997**	0.643	**0.999**
16	0.853	**1**	0.821	**1**	0.736	**1**

$n = 17$

r	90%		95%		99%	
0	0	**0.140**	0	**0.167**	0	**0.243**
1	0.006	**0.225+**	0.003	**0.254**	0.001	**0.346**
2	0.032	**0.290**	0.021	**0.337**	0.009	**0.413**
3	0.067	**0.364**	0.050	**0.417**	0.027	**0.500**
4	0.107	**0.432**	0.085−	**0.489**	0.052	**0.587**
5	0.140	**0.500**	0.124	**0.544**	0.082	**0.620**
6	0.175+	**0.568**	0.166	**0.594**	0.117	**0.662**
7	0.225+	**0.636**	0.167	**0.663**	0.155+	**0.757**
8	0.277	**0.710**	0.253	**0.746**	0.197	**0.758**
9	0.290	**0.723**	0.254	**0.747**	0.242	**0.803**
10	0.364	**0.775−**	0.338	**0.833**	0.243	**0.845**
11	0.432	**0.825−**	0.406	**0.834**	0.338	**0.883**
12	0.500	**0.860**	0.456	**0.876**	0.380	**0.918**
13	0.568	**0.893**	0.511	**0.915+**	0.413	**0.948**
14	0.636	**0.933**	0.583	**0.950**	0.500	**0.973**
15	0.710	**0.968**	0.663	**0.979**	0.587	**0.991**
16	0.775−	**0.994**	0.746	**0.997**	0.654	**0.999**
17	0.860	**1**	0.833	**1**	0.757	**1**

$n = 18$

r	90%		95%		99%	
0	0	**0.135−**	0	**0.157**	0	**0.228**
1	0.006	**0.216**	0.003	**0.242**	0.001	**0.318**
2	0.030	**0.277**	0.020	**0.325−**	0.008	**0.397**
3	0.063	**0.349**	0.047	**0.381**	0.025+	**0.466**
4	0.101	**0.419**	0.080	**0.444**	0.049	**0.534**
5	0.135−	**0.482**	0.116	**0.556**	0.077	**0.603**
6	0.163	**0.536**	0.156	**0.619**	0.110	**0.682**
7	0.216	**0.584**	0.157	**0.625+**	0.145+	**0.686**
8	0.257	**0.651**	0.236	**0.675+**	0.184	**0.772**
9	0.277	**0.723**	0.242	**0.758**	0.226	**0.774**
10	0.349	**0.743**	0.325−	**0.764**	0.228	**0.816**
11	0.416	**0.784**	0.375−	**0.843**	0.314	**0.855−**
12	0.464	**0.837**	0.381	**0.844**	0.318	**0.890**
13	0.518	**0.865+**	0.444	**0.884**	0.397	**0.923**
14	0.581	**0.899**	0.556	**0.920**	0.466	**0.951**
15	0.651	**0.937**	0.619	**0.953**	0.534	**0.975−**
16	0.723	**0.970**	0.675+	**0.980**	0.603	**0.992**
17	0.784	**0.994**	0.758	**0.997**	0.682	**0.999**
18	0.865+	**1**	0.843	**1**	0.772	**1**

Table A.3 Confidence Limits for a Proportion (Two-sided), $n \leq 30$ *(Continued)*

$n = 19$

r	90%		95%		99%	
0	0	**0.130**	0	**0.150**	0	**0.218**
1	0.006	**0.209**	0.003	**0.232**	0.001	**0.305+**
2	0.028	**0.265+**	0.019	**0.316**	0.008	**0.383**
3	0.059	**0.337**	0.044	**0.365−**	0.024	**0.455+**
4	0.095+	**0.387**	0.075+	**0.426**	0.046	**0.515+**
5	0.130	**0.440**	0.110	**0.500**	0.073	**0.564**
6	0.151	**0.560**	0.147	**0.574**	0.103	**0.617**
7	0.209	**0.613**	0.150	**0.635+**	0.137	**0.695−**
8	0.238	**0.614**	0.222	**0.655+**	0.173	**0.707**
9	0.265+	**0.663**	0.232	**0.688**	0.212	**0.782**
10	0.337	**0.735−**	0.312	**0.768**	0.218	**0.788**
11	0.386	**0.762**	0.345−	**0.778**	0.293	**0.827**
12	0.387	**0.791**	0.365−	**0.850**	0.305+	**0.863**
13	0.440	**0.849**	0.426	**0.853**	0.383	**0.897**
14	0.560	**0.870**	0.500	**0.890**	0.436	**0.927**
15	0.613	**0.905−**	0.574	**0.925−**	0.485−	**0.954**
16	0.668	**0.941**	0.635+	**0.956**	0.545−	**0.976**
17	0.735−	**0.972**	0.684	**0.981**	0.617	**0.992**
18	0.791	**0.994**	0.768	**0.997**	0.695−	**0.999**
19	0.870	**1**	0.850	**1**	0.782	**1**

$n = 20$

r	90%		95%		99%	
0	0	**0.126**	0	**0.143**	0	**0.209**
1	0.005+	**0.203**	0.003	**0.222**	0.001	**0.293**
2	0.027	**0.255−**	0.018	**0.294**	0.008	**0.375−**
3	0.056	**0.328**	0.042	**0.351**	0.023	**0.424**
4	0.090	**0.367**	0.071	**0.411**	0.044	**0.500**
5	0.126	**0.422**	0.104	**0.467**	0.069	**0.576**
6	0.141	**0.500**	0.140	**0.533**	0.098	**0.601**
7	0.201	**0.578**	0.143	**0.589**	0.129	**0.637**
8	0.221	**0.633**	0.209	**0.649**	0.163	**0.707**
9	0.255−	**0.642**	0.222	**0.706**	0.200	**0.726**
10	0.325	**0.675+**	0.293	**0.707**	0.209	**0.791**
11	0.358	**0.745+**	0.294	**0.778**	0.274	**0.800**
12	0.367	**0.779**	0.351	**0.791**	0.293	**0.837**
13	0.422	**0.799**	0.411	**0.857**	0.363	**0.871**
14	0.500	**0.859**	0.467	**0.860**	0.399	**0.902**
15	0.578	**0.874**	0.533	**0.896**	0.424	**0.931**
16	0.633	**0.910**	0.589	**0.929**	0.500	**0.956**
17	0.672	**0.944**	0.649	**0.958**	0.576	**0.977**
18	0.745+	**0.973**	0.706	**0.982**	0.625+	**0.992**
19	0.797	**0.995−**	0.778	**0.997**	0.707	**0.999**
20	0.874	**1**	0.857	**1**	0.791	**1**

$n = 21$

r	90%		95%		99%	
0	0	**0.123**	0	**0.137**	0	**0.201**
1	0.005+	**0.197**	0.002	**0.213**	0.000	**0.283**
2	0.026	**0.245−**	0.017	**0.277**	0.007	**0.347**
3	0.054	**0.307**	0.040	**0.338**	0.022	**0.409**
4	0.086	**0.353**	0.068	**0.398**	0.041	**0.466**
5	0.121	**0.407**	0.099	**0.455+**	0.065+	**0.534**
6	0.130	**0.458**	0.132	**0.506**	0.092	**0.591**
7	0.191	**0.542**	0.137	**0.551**	0.122	**0.653**
8	0.192	**0.593**	0.197	**0.602**	0.155−	**0.661**
9	0.245−	**0.647**	0.213	**0.662**	0.189	**0.717**
10	0.306	**0.693**	0.276	**0.723**	0.201	**0.743**
11	0.307	**0.694**	0.277	**0.724**	0.257	**0.799**
12	0.353	**0.755+**	0.338	**0.787**	0.283	**0.811**
13	0.407	**0.808**	0.398	**0.803**	0.339	**0.845+**
14	0.458	**0.809**	0.449	**0.863**	0.347	**0.878**
15	0.542	**0.870**	0.494	**0.868**	0.409	**0.908**
16	0.593	**0.879**	0.545−	**0.901**	0.466	**0.935−**
17	0.647	**0.914**	0.602	**0.932**	0.534	**0.959**
18	0.693	**0.946**	0.662	**0.960**	0.591	**0.978**
19	0.755+	**0.974**	0.723	**0.983**	0.653	**0.993**
20	0.808	**0.995−**	0.787	**0.998**	0.717	**1.000**
21	0.877	**1**	0.863	**1**	0.799	**1**

$n = 22$

r	90%		95%		99%	
0	0	**0.116**	0	**0.132**	0	**0.194**
1	0.005−	**0.182**	0.002	**0.205+**	0.000	**0.273**
2	0.024	**0.236**	0.016	**0.264**	0.007	**0.334**
3	0.051	**0.289**	0.038	**0.326**	0.021	**0.396**
4	0.082	**0.340**	0.065−	**0.389**	0.039	**0.454**
5	0.115−	**0.393**	0.094	**0.424**	0.062	**0.505−**
6	0.116	**0.444**	0.126	**0.500**	0.088	**0.550**
7	0.181	**0.500**	0.132	**0.576**	0.116	**0.604**
8	0.182	**0.556**	0.187	**0.582**	0.147	**0.666**
9	0.236	**0.607**	0.205+	**0.617**	0.179	**0.682**
10	0.289	**0.660**	0.260	**0.674**	0.194	**0.727**
11	0.290	**0.710**	0.264	**0.736**	0.242	**0.758**
12	0.340	**0.711**	0.326	**0.740**	0.273	**0.806**
13	0.393	**0.764**	0.383	**0.795−**	0.318	**0.821**
14	0.444	**0.818**	0.418	**0.813**	0.334	**0.853**
15	0.500	**0.819**	0.424	**0.868**	0.396	**0.884**
16	0.556	**0.884**	0.500	**0.874**	0.450	**0.912**
17	0.607	**0.885+**	0.576	**0.906**	0.495+	**0.938**
18	0.660	**0.918**	0.611	**0.935+**	0.546	**0.961**
19	0.711	**0.949**	0.674	**0.962**	0.604	**0.979**
20	0.764	**0.976**	0.736	**0.984**	0.666	**0.993**
21	0.818	**0.995+**	0.795−	**0.998**	0.727	**1.000**
22	0.884	**1**	0.868	**1**	0.806	**1**

Table A.3 Confidence Limits for a Proportion (Two-sided), $n \leq 30$ *(Continued)*

$n = 23$

r	90%		95%		99%	
0	0	**0.111**	0	**0.127**	0	**0.187**
1	0.005−	**0.174**	0.002	**0.198**	0.000	**0.365+**
2	0.023	**0.228**	0.016	**0.255**	0.007	**0.323**
3	0.049	**0.274**	0.037	**0.317**	0.020	**0.386**
4	0.078	**0.328**	0.062	**0.361**	0.038	**0.429**
5	0.110	**0.381**	0.090	**0.409**	0.059	**0.500**
6	0.111	**0.431**	0.120	**0.457**	0.084	**0.571**
7	0.173	**0.479**	0.127	**0.543**	0.111	**0.580**
8	0.174	**0.522**	0.178	**0.591**	0.140	**0.616**
9	0.228	**0.569**	0.198	**0.639**	0.171	**0.677**
10	0.273	**0.619**	0.247	**0.640**	0.187	**0.702**
11	0.274	**0.672**	0.255−	**0.683**	0.229	**0.735−**
12	0.328	**0.726**	0.317	**0.745+**	0.265+	**0.771**
13	0.381	**0.727**	0.360	**0.753**	0.298	**0.813**
14	0.431	**0.772**	0.361	**0.802**	0.323	**0.829**
15	0.478	**0.826**	0.409	**0.822**	0.384	**0.860**
16	0.521	**0.827**	0.457	**0.873**	0.420	**0.889**
17	0.569	**0.889**	0.543	**0.880**	0.429	**0.916**
18	0.619	**0.890**	0.591	**0.910**	0.500	**0.941**
19	0.672	**0.922**	0.639	**0.938**	0.571	**0.962**
20	0.726	**0.951**	0.683	**0.963**	0.614	**0.980**
21	0.772	**0.977**	0.745+	**0.984**	0.677	**0.993**
22	0.826	**0.995+**	0.802	**0.998**	0.735−	**1.000**
23	0.889	**1**	0.873	**1**	0.813	**1**

$n = 24$

r	90%		95%		99%	
0	0	**0.105+**	0	**0.122**	0	**0.181**
1	0.004	**0.165+**	0.002	**0.191**	0.000	**0.259**
2	0.022	**0.221**	0.015+	**0.246**	0.006	**0.313**
3	0.047	**0.264**	0.035−	**0.308**	0.019	**0.364**
4	0.075−	**0.317**	0.059	**0.347**	0.036	**0.416**
5	0.105−	**0.370**	0.086	**0.396**	0.057	**0.464**
6	0.105+	**0.423**	0.115−	**0.443**	0.080	**0.536**
7	0.165−	**0.448**	0.122	**0.500**	0.106	**0.584**
8	0.165+	**0.552**	0.169	**0.557**	0.133	**0.636**
9	0.221	**0.553**	0.191	**0.604**	0.163	**0.638**
10	0.259	**0.587**	0.234	**0.653**	0.181	**0.687**
11	0.264	**0.630**	0.246	**0.661**	0.216	**0.720**
12	0.317	**0.683**	0.308	**0.692**	0.257	**0.743**
13	0.370	**0.736**	0.339	**0.754**	0.280	**0.784**
14	0.413	**0.741**	0.347	**0.766**	0.313	**0.819**
15	0.447	**0.779**	0.396	**0.809**	0.362	**0.837**
16	0.448	**0.835−**	0.443	**0.831**	0.364	**0.867**
17	0.552	**0.835+**	0.500	**0.878**	0.416	**0.894**
18	0.577	**0.895−**	0.557	**0.885+**	0.464	**0.920**
19	0.630	**0.895+**	0.604	**0.914**	0.536	**0.943**
20	0.683	**0.925+**	0.653	**0.941**	0.584	**0.964**
21	0.736	**0.953**	0.692	**0.965+**	0.636	**0.981**
22	0.779	**0.978**	0.754	**0.985−**	0.687	**0.994**
23	0.835−	**0.996**	0.809	**0.998**	0.741	**1.000**
24	0.895−	**1**	0.878	**1**	0.819	**1**

$n = 25$

r	90%		95%		99%	
0	0	**0.102**	0	**0.118**	0	**0.175+**
1	0.004	**0.159**	0.002	**0.185+**	0.000	**0.246**
2	0.021	**0.214**	0.014	**0.238**	0.006	**0.305−**
3	0.045−	**0.255−**	0.034	**0.303**	0.018	**0.352**
4	0.072	**0.307**	0.057	**0.336**	0.034	**0.403**
5	0.101	**0.362**	0.082	**0.384**	0.054	**0.451**
6	0.102	**0.390**	0.110	**0.431**	0.077	**0.500**
7	0.158	**0.432**	0.118	**0.475−**	0.101	**0.549**
8	0.159	**0.500**	0.161	**0.525+**	0.127	**0.597**
9	0.214	**0.568**	0.185+	**0.569**	0.155+	**0.648**
10	0.246	**0.610**	0.222	**0.616**	0.175+	**0.658**
11	0.255	**0.611**	0.238	**0.664**	0.205+	**0.695+**
12	0.307	**0.640**	0.296	**0.683**	0.245+	**0.754**
13	0.360	**0.693**	0.317	**0.704**	0.246	**0.755−**
14	0.389	**0.745+**	0.336	**0.762**	0.305−	**0.795−**
15	0.390	**0.754**	0.384	**0.778**	0.342	**0.825−**
16	0.432	**0.786**	0.431	**0.815−**	0.352	**0.845−**
17	0.500	**0.841**	0.475−	**0.839**	0.403	**0.873**
18	0.568	**0.842**	0.525+	**0.882**	0.451	**0.899**
19	0.610	**0.898**	0.569	**0.890**	0.500	**0.923**
20	0.638	**0.899**	0.616	**0.918**	0.549	**0.946**
21	0.693	**0.928**	0.664	**0.943**	0.597	**0.966**
22	0.745+	**0.955+**	0.697	**0.966**	0.648	**0.982**
23	0.786	**0.979**	0.762	**0.986**	0.695+	**0.994**
24	0.841	**0.996**	0.815−	**0.998**	0.751	**1.000**
25	0.898	**1**	0.882	**1**	0.825−	**1**

$n = 26$

r	90%		95%		99%	
0	0	**0.098**	0	**0.114**	0	**0.170**
1	0.004	**0.152**	0.002	**0.180**	0.000	**0.235−**
2	0.021	**0.209**	0.014	**0.230**	0.006	**0.298**
3	0.043	**0.247**	0.032	**0.283**	0.017	**0.342**
4	0.069	**0.299**	0.054	**0.325+**	0.033	**0.393**
5	0.097	**0.343**	0.079	**0.374**	0.052	**0.442**
6	0.098	**0.377**	0.106	**0.421**	0.073	**0.487**
7	0.151	**0.419**	0.114	**0.465−**	0.097	**0.526**
8	0.152	**0.460**	0.154	**0.506**	0.122	**0.562**
9	0.209	**0.540**	0.180	**0.542**	0.149	**0.607**
10	0.233	**0.581**	0.212	**0.579**	0.170	**0.658**
11	0.247	**0.623**	0.230	**0.626**	0.195−	**0.678**
12	0.299	**0.657**	0.282	**0.675−**	0.234	**0.702**
13	0.342	**0.658**	0.283	**0.717**	0.235−	**0.765+**
14	0.343	**0.701**	0.325+	**0.718**	0.298	**0.766**
15	0.377	**0.753**	0.374	**0.770**	0.322	**0.805+**
16	0.419	**0.767**	0.421	**0.788**	0.342	**0.830**
17	0.460	**0.791**	0.458	**0.820**	0.393	**0.851**
18	0.540	**0.848**	0.494	**0.846**	0.438	**0.878**
19	0.581	**0.849**	0.535−	**0.886**	0.474	**0.903**
20	0.623	**0.902**	0.579	**0.894**	0.513	**0.927**
21	0.657	**0.903**	0.626	**0.921**	0.558	**0.948**
22	0.701	**0.931**	0.675−	**0.946**	0.607	**0.967**
23	0.753	**0.957**	0.717	**0.968**	0.658	**0.983**
24	0.791	**0.979**	0.770	**0.986**	0.702	**0.994**
25	0.848	**0.996**	0.820	**0.998**	0.765+	**1.000**
26	0.902	**1**	0.886	**1**	0.830	**1**

Table A.3 Confidence Limits for a Proportion (Two-sided), $n \leq 30$
(Continued)

$n = 27$

r	90%		95%		99%	
0	0	**0.093**	0	**0.110**	0	**0.166**
1	0.004	**0.146**	0.002	**0.175−**	0.000	**0.225−**
2	0.020	**0.204**	0.013	**0.223**	0.006	**0.297**
3	0.042	**0.239**	0.031	**0.270**	0.017	**0.332**
4	0.066	**0.291**	0.052	**0.316**	0.032	**0.384**
5	0.093	**0.327**	0.076	**0.364**	0.050	**0.419**
6	0.094	**0.365+**	0.101	**0.415−**	0.070	**0.461**
7	0.145+	**0.407**	0.110	**0.437**	0.093	**0.539**
8	0.146	**0.447**	0.148	**0.500**	0.117	**0.581**
9	0.204	**0.500**	0.175−	**0.563**	0.143	**0.587**
10	0.221	**0.553**	0.202	**0.570**	0.166	**0.617**
11	0.239	**0.593**	0.223	**0.598**	0.185−	**0.668**
12	0.291	**0.635−**	0.269	**0.636**	0.224	**0.702**
13	0.326	**0.673**	0.270	**0.684**	0.225−	**0.716**
14	0.327	**0.674**	0.316	**0.730**	0.284	**0.775+**
15	0.365+	**0.709**	0.364	**0.731**	0.298	**0.776**
16	0.407	**0.761**	0.402	**0.777**	0.332	**0.815+**
17	0.447	**0.779**	0.430	**0.798**	0.383	**0.834**
18	0.500	**0.796**	0.437	**0.825+**	0.413	**0.857**
19	0.553	**0.854**	0.500	**0.852**	0.419	**0.883**
20	0.593	**0.855−**	0.563	**0.890**	0.461	**0.907**
21	0.635−	**0.906**	0.585+	**0.899**	0.539	**0.930**
22	0.673	**0.907**	0.636	**0.924**	0.581	**0.950**
23	0.709	**0.934**	0.684	**0.948**	0.616	**0.968**
24	0.761	**0.958**	0.730	**0.969**	0.668	**0.983**
25	0.796	**0.980**	0.777	**0.987**	0.703	**0.994**
26	0.854	**0.996**	0.825+	**0.998**	0.775+	**1.000**
27	0.907	**1**	0.890	**1**	0.834	**1**

$n = 28$

r	90%		95%		99%	
0	0	**0.090**	0	**0.106**	0	**0.162**
1	0.004	**0.140**	0.002	**0.170**	0.000	**0.218**
2	0.019	**0.201**	0.013	**0.217**	0.005+	**0.273**
3	0.040	**0.232**	0.030	**0.259**	0.016	**0.323**
4	0.064	**0.284**	0.050	**0.307**	0.031	**0.365−**
5	0.089	**0.312**	0.073	**0.357**	0.048	**0.408**
6	0.090	**0.355−**	0.098	**0.384**	0.068	**0.449**
7	0.139	**0.396**	0.106	**0.424**	0.089	**0.500**
8	0.140	**0.435+**	0.142	**0.463**	0.112	**0.551**
9	0.197	**0.473**	0.170	**0.537**	0.137	**0.592**
10	0.208	**0.527**	0.192	**0.576**	0.162	**0.635+**
11	0.232	**0.565−**	0.217	**0.616**	0.175+	**0.636**
12	0.284	**0.604**	0.258	**0.619**	0.214	**0.677**
13	0.310	**0.645+**	0.259	**0.645+**	0.218	**0.727**
14	0.312	**0.688**	0.307	**0.693**	0.272	**0.728**
15	0.335−	**0.690**	0.355−	**0.741**	0.273	**0.782**
16	0.396	**0.716**	0.381	**0.742**	0.323	**0.786**
17	0.435+	**0.768**	0.384	**0.783**	0.364	**0.825−**
18	0.473	**0.792**	0.424	**0.808**	0.365−	**0.838**
19	0.527	**0.803**	0.463	**0.830**	0.408	**0.863**
20	0.565−	**0.860**	0.537	**0.858**	0.449	**0.888**
21	0.604	**0.861**	0.576	**0.894**	0.500	**0.911**
22	0.645+	**0.910**	0.616	**0.902**	0.551	**0.932**
23	0.688	**0.911**	0.643	**0.927**	0.592	**0.952**
24	0.716	**0.936**	0.693	**0.950**	0.635+	**0.969**
25	0.768	**0.960**	0.741	**0.970**	0.677	**0.984**
26	0.799	**0.981**	0.783	**0.987**	0.727	**0.995−**
27	0.860	**0.996**	0.830	**0.998**	0.782	**1.000**
28	0.910	**1**	0.894	**1**	0.838	**1**

$n = 29$

r	90%		95%		99%	
0	0	**0.087**	0	**0.103**	0	**0.160**
1	0.004	**0.135−**	0.002	**0.166**	0.000	**0.211**
2	0.018	**0.190**	0.012	**0.211**	0.005+	**0.263**
3	0.039	**0.225−**	0.029	**0.251**	0.015+	**0.316**
4	0.062	**0.279**	0.049	**0.299**	0.030	**0.354**
5	0.086	**0.303**	0.070	**0.340**	0.046	**0.397**
6	0.087	**0.345−**	0.094	**0.374**	0.065+	**0.438**
7	0.134	**0.385+**	0.103	**0.413**	0.086	**0.477**
8	0.135−	**0.425−**	0.136	**0.451**	0.108	**0.523**
9	0.189	**0.463**	0.166	**0.500**	0.132	**0.562**
10	0.190	**0.500**	0.184	**0.549**	0.157	**0.603**
11	0.225−	**0.537**	0.211	**0.587**	0.165+	**0.646**
12	0.276	**0.575+**	0.247	**0.626**	0.206	**0.654**
13	0.294	**0.615−**	0.251	**0.660**	0.211	**0.684**
14	0.308	**0.655+**	0.299	**0.661**	0.260	**0.737**
15	0.345−	**0.697**	0.339	**0.701**	0.263	**0.740**
16	0.385+	**0.706**	0.340	**0.749**	0.316	**0.789**
17	0.425−	**0.724**	0.374	**0.753**	0.346	**0.794**
18	0.463	**0.775+**	0.413	**0.789**	0.354	**0.835−**
19	0.500	**0.810**	0.451	**0.816**	0.397	**0.843**
20	0.537	**0.811**	0.500	**0.834**	0.438	**0.868**
21	0.575+	**0.865+**	0.549	**0.864**	0.477	**0.892**
22	0.615−	**0.866**	0.587	**0.897**	0.523	**0.914**
23	0.655+	**0.913**	0.626	**0.906**	0.562	**0.935−**
24	0.697	**0.914**	0.660	**0.930**	0.603	**0.954**
25	0.721	**0.938**	0.701	**0.951**	0.646	**0.970**
26	0.775+	**0.961**	0.749	**0.971**	0.684	**0.985−**
27	0.810	**0.982**	0.789	**0.988**	0.737	**0.995−**
28	0.865+	**0.996**	0.834	**0.998**	0.789	**1.000**
29	0.913	**1**	0.897	**1**	0.840	**1**

$n = 30$

r	90%		95%		99%	
0	0	**0.084**	0	**0.100**	0	**0.152**
1	0.004	**0.130**	0.002	**0.163**	0.000	**0.206**
2	0.018	**0.183**	0.012	**0.205+**	0.005+	**0.256**
3	0.037	**0.219**	0.028	**0.244**	0.015−	**0.310**
4	0.059	**0.266**	0.047	**0.292**	0.028	**0.345−**
5	0.083	**0.295−**	0.068	**0.325−**	0.045−	**0.388**
6	0.084	**0.336**	0.091	**0.364**	0.063	**0.430**
7	0.129	**0.376**	0.100	**0.403**	0.083	**0.469**
8	0.130	**0.416**	0.131	**0.440**	0.104	**0.505+**
9	0.182	**0.455+**	0.163	**0.476**	0.127	**0.538**
10	0.183	**0.492**	0.175+	**0.524**	0.151	**0.570**
11	0.219	**0.524**	0.205+	**0.560**	0.152	**0.612**
12	0.265−	**0.554**	0.236	**0.597**	0.198	**0.655+**
13	0.266	**0.584**	0.244	**0.636**	0.206	**0.671**
14	0.295−	**0.624**	0.292	**0.675+**	0.294	**0.692**
15	0.336	**0.664**	0.324	**0.676**	0.256	**0.744**
16	0.376	**0.705+**	0.325−	**0.708**	0.308	**0.751**
17	0.416	**0.734**	0.364	**0.756**	0.329	**0.794**
18	0.446	**0.735+**	0.403	**0.764**	0.345−	**0.802**
19	0.476	**0.781**	0.440	**0.795−**	0.388	**0.848**
20	0.508	**0.817**	0.476	**0.825−**	0.430	**0.849**
21	0.545−	**0.818**	0.524	**0.837**	0.462	**0.873**
22	0.584	**0.870**	0.560	**0.869**	0.495−	**0.896**
23	0.624	**0.871**	0.597	**0.900**	0.531	**0.917**
24	0.664	**0.916**	0.636	**0.909**	0.570	**0.937**
25	0.705+	**0.917**	0.675+	**0.932**	0.612	**0.955+**
26	0.734	**0.941**	0.708	**0.953**	0.655+	**0.972**
27	0.781	**0.963**	0.756	**0.972**	0.690	**0.985+**
28	0.817	**0.982**	0.795−	**0.988**	0.744	**0.995−**
29	0.870	**0.996**	0.837	**0.998**	0.794	**1.000**
30	0.916	**1**	0.900	**1**	0.848	**1**

Table A.4 Areas under the Normal Curve*

Proportion of total area under the curve that is under the portion of the curve from $-\infty$ to $(X_i - \bar{X}')/\sigma'$. (X_i represents any desired value of the variable X.)

$\frac{X_i - \bar{X}'}{\sigma'}$	0.09	0.08	0.07	0.06	0.05	0.04	0.03	0.02	0.01	0.00
−3.5	0.00017	0.00017	0.00018	0.00019	0.00019	0.00020	0.00021	0.00022	0.00022	0.00023
−3.4	0.00024	0.00025	0.00026	0.00027	0.00028	0.00029	0.00030	0.00031	0.00033	0.00034
−3.3	0.00035	0.00036	0.00038	0.00039	0.00040	0.00042	0.00043	0.00045	0.00047	0.00048
−3.2	0.00050	0.00052	0.00054	0.00056	0.00058	0.00060	0.00062	0.00064	0.00066	0.00069
−3.1	0.00071	0.00074	0.00076	0.00079	0.00082	0.00085	0.00087	0.00090	0.00094	0.00097
−3.0	0.00100	0.00104	0.00107	0.00111	0.00114	0.00118	0.00122	0.00126	0.00131	0.00135
−2.9	0.0014	0.0014	0.0015	0.0015	0.0016	0.0016	0.0017	0.0017	0.0018	0.0019
−2.8	0.0019	0.0020	0.0021	0.0021	0.0022	0.0023	0.0023	0.0024	0.0025	0.0026
−2.7	0.0026	0.0027	0.0028	0.0029	0.0030	0.0031	0.0032	0.0033	0.0034	0.0035
−2.6	0.0036	0.0037	0.0038	0.0039	0.0040	0.0041	0.0043	0.0044	0.0045	0.0047
−2.5	0.0048	0.0049	0.0051	0.0052	0.0054	0.0055	0.0057	0.0059	0.0060	0.0062
−2.4	0.0064	0.0066	0.0068	0.0069	0.0071	0.0073	0.0075	0.0078	0.0080	0.0082
−2.3	0.0084	0.0087	0.0089	0.0091	0.0094	0.0096	0.0099	0.0102	0.0104	0.0107
−2.2	0.0110	0.0113	0.0116	0.0119	0.0122	0.0125	0.0129	0.0132	0.0136	0.0139
−2.1	0.0143	0.0146	0.0150	0.0154	0.0158	0.0162	0.0166	0.0170	0.0174	0.0179
−2.0	0.0183	0.0188	0.0192	0.0197	0.0202	0.0207	0.0212	0.0217	0.0222	0.0228
−1.9	0.0233	0.0239	0.0244	0.0250	0.0256	0.0262	0.0268	0.0274	0.0281	0.0287
−1.8	0.0294	0.0301	0.0307	0.0314	0.0322	0.0329	0.0336	0.0344	0.0351	0.0359
−1.7	0.0367	0.0375	0.0384	0.0392	0.0401	0.0409	0.0418	0.0427	0.0436	0.0446
−1.6	0.0455	0.0465	0.0475	0.0485	0.0495	0.0505	0.0516	0.0526	0.0537	0.0548
−1.5	0.0559	0.0571	0.0582	0.0594	0.0606	0.0618	0.0630	0.0643	0.0655	0.0668
−1.4	0.0681	0.0694	0.0708	0.0721	0.0735	0.0749	0.0764	0.0778	0.0793	0.0808
−1.3	0.0823	0.0838	0.0853	0.0869	0.0885	0.0901	0.0918	0.0934	0.0951	0.0968
−1.2	0.0985	0.1003	0.1020	0.1038	0.1057	0.1075	0.1093	0.1112	0.1131	0.1151
−1.1	0.1170	0.1190	0.1210	0.1230	0.1251	0.1271	0.1292	0.1314	0.1335	0.1357
−1.0	0.1379	0.1401	0.1423	0.1446	0.1469	0.1492	0.1515	0.1539	0.1562	0.1587
−0.9	0.1611	0.1635	0.1660	0.1685	0.1711	0.1736	0.1762	0.1788	0.1814	0.1841
−0.8	0.1867	0.1894	0.1922	0.1949	0.1977	0.2005	0.2033	0.2061	0.2090	0.2119
−0.7	0.2148	0.2177	0.2207	0.2236	0.2266	0.2297	0.2327	0.2358	0.2389	0.2420
−0.6	0.2451	0.2483	0.2514	0.2546	0.2578	0.2611	0.2643	0.2676	0.2709	0.2743
−0.5	0.2776	0.2810	0.2843	0.2877	0.2912	0.2946	0.2981	0.3015	0.3050	0.3085
−0.4	0.3121	0.3156	0.3192	0.3228	0.3264	0.3300	0.3336	0.3372	0.3409	0.3446
−0.3	0.3483	0.3520	0.3557	0.3594	0.3632	0.3669	0.3707	0.3745	0.3783	0.3821
−0.2	0.3859	0.3897	0.3936	0.3974	0.4013	0.4052	0.4090	0.4129	0.4168	0.4207
−0.1	0.4247	0.4286	0.4325	0.4364	0.4404	0.4443	0.4483	0.4522	0.4562	0.4602
−0.0	0.4641	0.4681	0.4721	0.4761	0.4801	0.4840	0.4880	0.4920	0.4960	0.5000

* From *Statistical Quality Control*, 3d ed., by Eugene L. Grant. Copyright 1964. McGraw-Hill Book Company. Used by permission.

Table A.4 Areas under the Normal Curve (*Continued*)

$\frac{X_i - \bar{X}'}{\sigma'}$	0.00	0.01	0.02	0.03	0.04	0.05	0.06	0.07	0.08	0.09
+0.0	0.5000	0.5040	0.5080	0.5120	0.5160	0.5199	0.5239	0.5279	0.5319	0.5359
+0.1	0.5398	0.5438	0.5478	0.5517	0.5557	0.5596	0.5636	0.5675	0.5714	0.5753
+0.2	0.5793	0.5832	0.5871	0.5910	0.5948	0.5987	0.6026	0.6064	0.6103	0.6141
+0.3	0.6179	0.6217	0.6255	0.6293	0.6331	0.6368	0.6406	0.6443	0.6480	0.6517
+0.4	0.6554	0.6591	0.6628	0.6664	0.6700	0.6736	0.6772	0.6808	0.6844	0.6879
+0.5	0.6915	0.6950	0.6985	0.7019	0.7054	0.7088	0.7123	0.7157	0.7190	0.7224
+0.6	0.7257	0.7291	0.7324	0.7357	0.7389	0.7422	0.7454	0.7486	0.7517	0.7549
+0.7	0.7580	0.7611	0.7642	0.7673	0.7704	0.7734	0.7764	0.7794	0.7823	0.7852
+0.8	0.7881	0.7910	0.7939	0.7967	0.7995	0.8023	0.8051	0.8079	0.8106	0.8133
+0.9	0.8159	0.8186	0.8212	0.8238	0.8264	0.8289	0.8315	0.8340	0.8365	0.8389
+1.0	0.8413	0.8438	0.8461	0.8485	0.8508	0.8531	0.8554	0.8577	0.8599	0.8621
+1.1	0.8643	0.8665	0.8686	0.8708	0.8729	0.8749	0.8770	0.8790	0.8810	0.8830
+1.2	0.8849	0.8869	0.8888	0.8907	0.8925	0.8944	0.8962	0.8980	0.8997	0.9015
+1.3	0.9032	0.9049	0.9066	0.9082	0.9099	0.9115	0.9131	0.9147	0.9162	0.9177
+1.4	0.9192	0.9207	0.9222	0.9236	0.9251	0.9265	0.9279	0.9292	0.9306	0.9319
+1.5	0.9332	0.9345	0.9357	0.9370	0.9382	0.9394	0.9406	0.9418	0.9429	0.9441
+1.6	0.9452	0.9463	0.9474	0.9484	0.9495	0.9505	0.9515	0.9525	0.9535	0.9545
+1.7	0.9554	0.9564	0.9573	0.9582	0.9591	0.9599	0.9608	0.9616	0.9625	0.9633
+1.8	0.9641	0.9649	0.9656	0.9664	0.9671	0.9678	0.9686	0.9693	0.9699	0.9706
+1.9	0.9713	0.9719	0.9726	0.9732	0.9738	0.9744	0.9750	0.9756	0.9761	0.9767
+2.0	0.9773	0.9778	0.9783	0.9788	0.9793	0.9798	0.9803	0.9808	0.9812	0.9817
+2.1	0.9821	0.9826	0.9830	0.9834	0.9838	0.9842	0.9846	0.9850	0.9854	0.9857
+2.2	0.9861	0.9864	0.9868	0.9871	0.9875	0.9878	0.9881	0.9884	0.9887	0.9890
+2.3	0.9893	0.9896	0.9898	0.9901	0.9904	0.9906	0.9909	0.9911	0.9913	0.9916
+2.4	0.9918	0.9920	0.9922	0.9925	0.9927	0.9929	0.9931	0.9932	0.9934	0.9936
+2.5	0.9938	0.9940	0.9941	0.9943	0.9945	0.9946	0.9948	0.9949	0.9951	0.9952
+2.6	0.9953	0.9955	0.9956	0.9957	0.9959	0.9960	0.9961	0.9962	0.9963	0.9964
+2.7	0.9965	0.9966	0.9967	0.9968	0.9969	0.9970	0.9971	0.9972	0.9973	0.9974
+2.8	0.9974	0.9975	0.9976	0.9977	0.9977	0.9978	0.9979	0.9979	0.9980	0.9981
+2.9	0.9981	0.9982	0.9983	0.9983	0.9984	0.9984	0.9985	0.9985	0.9986	0.9986
+3.0	0.99865	0.99869	0.99874	0.99878	0.99882	0.99886	0.99889	0.99893	0.99896	0.99900
+3.1	0.99903	0.99906	0.99910	0.99913	0.99915	0.99918	0.99921	0.99924	0.99926	0.99929
+3.2	0.99931	0.99934	0.99936	0.99938	0.99940	0.99942	0.99944	0.99946	0.99948	0.99950
+3.3	0.99952	0.99953	0.99955	0.99957	0.99958	0.99960	0.99961	0.99962	0.99964	0.99965
+3.4	0.99966	0.99967	0.99969	0.99970	0.99971	0.99972	0.99973	0.99974	0.99975	0.99976
+3.5	0.99977	0.99978	0.99978	0.99979	0.99980	0.99981	0.99981	0.99982	0.99983	0.99983

Table A.5 Percentiles of the t Distribution*

df	$t_{.60}$	$t_{.70}$	$t_{.80}$	$t_{.90}$	$t_{.95}$	$t_{.975}$	$t_{.99}$	$t_{.995}$
1	.325	.727	1.376	3.078	6.314	12.706	31.821	63.657
2	.289	.617	1.061	1.886	2.920	4.303	6.965	9.925
3	.277	.584	.978	1.638	2.353	3.182	4.541	5.841
4	.271	.569	.941	1.533	2.132	2.776	3.747	4.604
5	.267	.559	.920	1.476	2.015	2.571	3.365	4.032
6	.265	.553	.906	1.440	1.943	2.447	3.143	3.707
7	.263	.549	.896	1.415	1.895	2.365	2.998	3.499
8	.262	.546	.889	1.397	1.860	2.306	2.896	3.355
9	.261	.543	.883	1.383	1.833	2.262	2.821	3.250
10	.260	.542	.879	1.372	1.812	2.228	2.764	3.169
11	.260	.540	.876	1.363	1.796	2.201	2.718	3.106
12	.259	.539	.873	1.356	1.782	2.179	2.681	3.055
13	.259	.538	.870	1.350	1.771	2.160	2.650	3.012
14	.258	.537	.868	1.345	1.761	2.145	2.624	2.977
15	.258	.536	.866	1.341	1.753	2.131	2.602	2.947
16	.258	.535	.865	1.337	1.746	2.120	2.583	2.921
17	.257	.534	.863	1.333	1.740	2.110	2.567	2.898
18	.257	.534	.862	1.330	1.734	2.101	2.552	2.878
19	.257	.533	.861	1.328	1.729	2.093	2.539	2.861
20	.257	.533	.860	1.325	1.725	2.086	2.528	2.845
21	.257	.532	.859	1.323	1.721	2.080	2.518	2.831
22	.256	.532	.858	1.321	1.717	2.074	2.508	2.819
23	.256	.532	.858	1.319	1.714	2.069	2.500	2.807
24	.256	.531	.857	1.318	1.711	2.064	2.492	2.797
25	.256	.531	.856	1.316	1.708	2.060	2.485	2.787
26	.256	.531	.856	1.315	1.706	2.056	2.479	2.779
27	.256	.531	.855	1.314	1.703	2.052	2.473	2.771
28	.256	.530	.855	1.313	1.701	2.048	2.467	2.763
29	.256	.530	.854	1.311	1.699	2.045	2.462	2.756
30	.256	.530	.854	1.310	1.697	2.042	2.457	2.750
40	.255	.529	.851	1.303	1.684	2.021	2.423	2.704
60	.254	.527	.848	1.296	1.671	2.000	2.390	2.660
120	.254	.526	.845	1.289	1.658	1.980	2.358	2.617
∞	.253	.524	.842	1.282	1.645	1.960	2.326	2.576
df	$-t_{.40}$	$-t_{.30}$	$-t_{.20}$	$-t_{.10}$	$-t_{.05}$	$-t_{.025}$	$-t_{.01}$	$-t_{.005}$

When the table is read from the foot, the tabled values are to be prefixed with a negative sign. Interpolation should be performed using the reciprocals of the degrees of freedom.

* The data of this table extracted from Table III of Fisher and Yates, *Statistical Tables*, with the permission of the authors and publishers, Oliver & Boyd, Ltd., Edinburgh and London. From *Introduction to Statistical Analysis*, 2d ed., by W. J. Dixon and F. J. Massey, Jr. Copyright, 1957. McGraw-Hill Book Company. Used by permission.

Table A.6 Percentiles of the χ^2 Distribution*

df	Per Cent									
	.5	1	2.5	5	10	90	95	97.5	99	99.5
1	.000039	.00016	.00098	.0039	.0158	2.71	3.84	5.02	6.63	7.88
2	.0100	.0201	.0506	.1026	.2107	4.61	5.99	7.38	9.21	10.60
3	.0717	.115	.216	.352	.584	6.25	7.81	9.35	11.34	12.84
4	.207	.297	.484	.711	1.064	7.78	9.49	11.14	13.28	14.86
5	.412	.554	.831	1.15	1.61	9.24	11.07	12.83	15.09	16.75
6	.676	.872	1.24	1.64	2.20	10.64	12.59	14.45	16.81	18.55
7	.989	1.24	1.69	2.17	2.83	12.02	14.07	16.01	18.48	20.28
8	1.34	1.65	2.18	2.73	3.49	13.36	15.51	17.53	20.09	21.96
9	1.73	2.09	2.70	3.33	4.17	14.68	16.92	19.02	21.67	23.59
10	2.16	2.56	3.25	3.94	4.87	15.99	18.31	20.48	23.21	25.19
11	2.60	3.05	3.82	4.57	5.58	17.28	19.68	21.92	24.73	26.76
12	3.07	3.57	4.40	5.23	6.30	18.55	21.03	23.34	26.22	28.30
13	3.57	4.11	5.01	5.89	7.04	19.81	22.36	24.74	27.69	29.82
14	4.07	4.66	5.63	6.57	7.79	21.06	23.68	26.12	29.14	31.32
15	4.60	5.23	6.26	7.26	8.55	22.31	25.00	27.49	30.58	32.80
16	5.14	5.81	6.91	7.96	9.31	23.54	26.30	28.85	32.00	34.27
18	6.26	7.01	8.23	9.39	10.86	25.99	28.87	31.53	34.81	37.16
20	7.43	8.26	9.59	10.85	12.44	28.41	31.41	34.17	37.57	40.00
24	9.89	10.86	12.40	13.85	15.66	33.20	36.42	39.36	42.98	45.56
30	13.79	14.95	16.79	18.49	20.60	40.26	43.77	46.98	50.89	53.67
40	20.71	22.16	24.43	26.51	29.05	51.81	55.76	59.34	63.69	66.77
60	35.53	37.48	40.48	43.19	46.46	74.40	79.08	83.30	88.38	91.95
120	83.85	86.92	91.58	95.70	100.62	140.23	146.57	152.21	158.95	163.64

For large values of degrees of freedom the approximate formula

$$\chi_\alpha^2 = n\left(1 - \frac{2}{9n} + z_\alpha\sqrt{\frac{2}{9n}}\right)^3$$

where z_α is the normal deviate and n is the number of degrees of freedom, may be used. For example $\chi_{.99}^2 = 60[1 - .00370 + 2.326(.06086)]^3 = 60(1.1379)^3 = 88.4$ for the 99th percentile for 60 degrees of freedom.

* From *Introduction to Statistical Analysis*, 2d ed., by W. J. Dixon and F. J. Massey, Jr. Copyright, 1957. McGraw-Hill Book Company. Used by permission.

Table A.7 Percentiles of the $F(\nu_1, \nu_2)$ Distribution with Degrees of Freedom ν_1 for the Numerator and ν_2 for the Denominator*

ν_2	Cum. Prop. \ ν_1	1	2	3	4	5	6	7	8	9	10	11	12	Cum. Prop.
1	.0005	$.0^{6}62$	$.0^{3}50$	$.0^{2}38$	$.0^{2}94$	.016	.022	.027	.032	.036	.039	.042	.045	.0005
	.001	$.0^{5}25$	$.0^{2}10$	$.0^{2}60$	.013	.021	.028	.034	.039	.044	.048	.051	.054	.001
	.005	$.0^{4}62$	$.0^{2}51$	.018	.032	.044	.054	.062	.068	.073	.078	.082	.085	.005
	.010	$.0^{3}25$	.010	.029	.047	.062	.073	.082	.089	.095	.100	.104	.107	.010
	.025	$.0^{2}15$	.026	.057	.082	.100	.113	.124	.132	.139	.144	.149	.153	.025
	.05	$.0^{2}62$	.054	.099	.130	.151	.167	.179	.188	.195	.201	.207	.211	.05
	.10	.025	.117	.181	.220	.246	.265	.279	.289	.298	.304	.310	.315	.10
	.25	.172	.389	.494	.553	.591	.617	.637	.650	.661	.670	.680	.684	.25
	.50	1.00	1.50	1.71	1.82	1.89	1.94	1.98	2.00	2.03	2.04	2.05	2.07	.50
	.75	5.83	7.50	8.20	8.58	8.82	8.98	9.10	9.19	9.26	9.32	9.36	9.41	.75
	.90	39.9	49.5	53.6	55.8	57.2	58.2	58.9	59.4	59.9	60.2	60.5	60.7	.90
	.95	161	200	216	225	230	234	237	239	241	242	243	244	.95
	.975	648	800	864	900	922	937	948	957	963	969	973	977	.975
	.99	405^{1}	500^{1}	540^{1}	562^{1}	576^{1}	586^{1}	593^{1}	598^{1}	602^{1}	606^{1}	608^{1}	611^{1}	.99
	.995	162^{2}	200^{2}	216^{2}	225^{2}	231^{2}	234^{2}	237^{2}	239^{2}	241^{2}	242^{2}	243^{2}	244^{2}	.995
	.999	406^{3}	500^{3}	540^{3}	562^{3}	576^{3}	586^{3}	593^{3}	598^{3}	602^{3}	606^{3}	609^{3}	611^{3}	.999
	.9995	162^{4}	200^{4}	216^{4}	225^{4}	231^{4}	234^{4}	237^{4}	239^{4}	241^{4}	242^{4}	243^{4}	244^{4}	.9995
2	.0005	$.0^{6}50$	$.0^{3}50$	$.0^{2}42$	.011	.020	.029	.037	.044	.050	.056	.061	.065	.0005
	.001	$.0^{5}20$	$.0^{2}10$	$.0^{2}68$	.016	.027	.037	.046	.054	.061	.067	.072	.077	.001
	.005	$.0^{4}50$	$.0^{2}50$	.020	.038	.055	.069	.081	.091	.099	.106	.112	.118	.005
	.01	$.0^{3}20$	.010	.032	.056	.075	.092	.105	.116	.125	.132	.139	.144	.01
	.025	$.0^{2}13$	.026	.062	.094	.119	.138	.153	.165	.175	.183	.190	.196	.025
	.05	$.0^{2}50$	.053	.105	.144	.173	.194	.211	.224	.235	.244	.251	.257	.05
	.10	.020	.111	.183	.231	.265	.289	.307	.321	.333	.342	.350	.356	.10
	.25	.133	.333	.439	.500	.540	.568	.588	.604	.616	.626	.633	.641	.25
	.50	.667	1.00	1.13	1.21	1.25	1.28	1.30	1.32	1.33	1.34	1.35	1.36	.50
	.75	2.57	3.00	3.15	3.23	3.28	3.31	3.34	3.35	3.37	3.38	3.39	3.39	.75
	.90	8.53	9.00	9.16	9.24	9.29	9.33	9.35	9.37	9.38	9.39	9.40	9.41	.90
	.95	18.5	19.0	19.2	19.2	19.3	19.3	19.4	19.4	19.4	19.4	19.4	19.4	.95
	.975	38.5	39.0	39.2	39.2	39.3	39.3	39.4	39.4	39.4	39.4	39.4	39.4	.975
	.99	98.5	99.0	99.2	99.2	99.3	99.3	99.4	99.4	99.4	99.4	99.4	99.4	.99
	.995	198	199	199	199	199	199	199	199	199	199	199	199	.995
	.999	998	999	999	999	999	999	999	999	999	999	999	999	.999
	.9995	200^{1}	200^{1}	200^{1}	200^{1}	200^{1}	200^{1}	200^{1}	200^{1}	200^{1}	200^{1}	200^{1}	200^{1}	.9995
3	.0005	$.0^{6}46$	$.0^{3}50$	$.0^{2}44$	.012	.023	.033	.043	.052	.060	.067	.074	.079	.0005
	.001	$.0^{5}19$	$.0^{2}10$	$.0^{2}71$	.018	.030	.042	.053	.063	.072	.079	.086	.093	.001
	.005	$.0^{4}46$	$.0^{2}50$	.021	.041	.060	.077	.092	.104	.115	.124	.132	.138	.005
	.01	$.0^{3}19$	.010	.034	.060	.083	.102	.118	.132	.143	.153	.161	.168	.01
	.025	$.0^{2}12$	.026	.065	.100	.129	.152	.170	.185	.197	.207	.216	.224	.025
	.05	$.0^{2}46$	.052	.108	.152	.185	.210	.230	.246	.259	.270	.279	.287	.05
	.10	.019	.109	.185	.239	.276	.304	.325	.342	.356	.367	.376	.384	.10
	.25	.122	.317	.424	.489	.531	.561	.582	.600	.613	.624	.633	.641	.25
	.50	.585	.881	1.00	1.06	1.10	1.13	1.15	1.16	1.17	1.18	1.19	1.20	.50
	.75	2.02	2.28	2.36	2.39	2.41	2.42	2.43	2.44	2.44	2.44	2.45	2.45	.75
	.90	5.54	5.46	5.39	5.34	5.31	5.28	5.27	5.25	5.24	5.23	5.22	5.22	.90
	.95	10.1	9.55	9.28	9.12	9.01	8.94	8.89	8.85	8.81	8.79	8.76	8.74	.95
	.975	17.4	16.0	15.4	15.1	14.9	14.7	14.6	14.5	14.5	14.4	14.4	14.3	.975
	.99	34.1	30.8	29.5	28.7	28.2	27.9	27.7	27.5	27.3	27.2	27.1	27.1	.99
	.995	55.6	49.8	47.5	46.2	45.4	44.8	44.4	44.1	43.9	43.7	43.5	43.4	.995
	.999	167	149	141	137	135	133	132	131	130	129	129	128	.999
	.9995	266	237	225	218	214	211	209	208	207	206	204	204	.9995

Read $.0^{3}56$ as .00056, 200^{1} as 2000, 162^{4} as 1620000, etc.

* From *Introduction to Statistical Analysis*, 2d ed., by W. J. Dixon and F. J. Massey, Jr.

Table A.7 Percentiles of the $F(\nu_1, \nu_2)$ Distribution with Degrees of Freedom ν_1 for the Numerator and ν_2 for the Denominator *(Continued)*

ν_1 / Cum. Prop.	15	20	24	30	40	50	60	100	120	200	500	∞	Cum. Prop.	ν_2
.0005	.051	.058	062	.066	.069	.072	.074	.077	.078	.080	.081	.083	.0005	1
.001	.060	.067	.071	.075	.079	.082	.084	.087	.088	.089	.091	.092	.001	
.005	.093	.101	.105	.109	.113	.116	.118	.121	.122	.124	.126	.127	.005	
.01	.115	.124	.128	.132	.137	.139	.141	.145	.146	.148	.150	.151	.01	
.025	.161	.170	.175	.180	.184	.187	.189	.193	.194	.196	.198	.199	.025	
.05	.220	.230	.235	.240	.245	.248	.250	.254	.255	.257	.259	.261	.05	
.10	.325	.336	.342	.347	.353	.356	.358	.362	.364	.366	.368	.370	.10	
.25	.698	.712	.719	.727	.734	.738	.741	.747	.749	.752	.754	.756	.25	
.50	2.09	2.12	2.13	2.15	2.16	2.17	2.17	2.18	2.18	2.19	2.19	2.20	.50	
.75	9.49	9.58	9.63	9.67	9.71	9.74	9.76	9.78	9.80	9.82	9.84	9.85	.75	
.90	61.2	61.7	62.0	62.3	62.5	62.7	62.8	63.0	63.1	63.2	63.3	63.3	.90	
.95	246	248	249	250	251	252	252	253	253	254	254	254	.95	
.975	985	993	997	100^{1}	101^{1}	101^{1}	101^{1}	101^{1}	101^{1}	102^{1}	102^{1}	102^{1}	.975	
.99	616^{1}	621^{1}	623^{1}	626^{1}	629^{1}	630^{1}	631^{1}	633^{1}	634^{1}	635^{1}	636^{1}	637^{1}	.99	
.995	246^{2}	248^{2}	249^{2}	250^{2}	251^{2}	252^{2}	253^{2}	253^{2}	254^{2}	254^{2}	254^{2}	255^{2}	.995	
.999	616^{3}	621^{3}	623^{3}	626^{3}	629^{3}	630^{3}	631^{3}	633^{3}	634^{3}	635^{3}	636^{3}	637^{3}	.999	
.9995	246^{4}	248^{4}	249^{4}	250^{4}	251^{4}	252^{4}	252^{4}	253^{4}	253^{4}	253^{4}	254^{4}	254^{4}	.9995	
.0005	.076	.088	.094	.101	.108	.113	.116	.122	.124	.127	.130	.132	.0005	2
.001	.088	.100	.107	.114	.121	.126	.129	.135	.137	.140	.143	.145	.001	
.005	.130	.143	.150	.157	.165	.169	.173	.179	.181	.184	.187	.189	.005	
.01	.157	.171	.178	.186	.193	.198	.201	.207	.209	.212	.215	.217	.01	
.025	.210	.224	.232	.239	.247	.251	.255	.261	.263	.266	.269	.271	.025	
.05	.272	.286	.294	.302	.309	.314	.317	.324	.326	.329	.332	.334	.05	
.10	.371	.386	.394	.402	.410	.415	.418	.424	.426	.429	.433	.434	.10	
.25	.657	.672	.680	.689	.697	.702	.705	.711	.713	.716	.719	.721	.25	
.50	1.38	1.39	1.40	1.41	1.42	1.42	1.43	1.43	1.43	1.44	1.44	1.44	.50	
.75	3.41	3.43	3.43	3.44	3.45	3.45	3.46	3.47	3.47	3.48	3.48	3.48	.75	
.90	9.42	9.44	9.45	9.46	9.47	9.47	9.47	9.48	9.48	9.49	9.49	9.49	.90	
.95	19.4	19.4	19.5	19.5	19.5	19.5	19.5	19.5	19.5	19.5	19.5	19.5	.95	
.975	39.4	39.4	39.5	39.5	39.5	39.5	39.5	39.5	39.5	39.5	39.5	39.5	.975	
.99	99.4	99.4	99.5	99.5	99.5	99.5	99.5	99.5	99.5	99.5	99.5	99.5	.99	
.995	199	199	199	199	199	199	199	199	199	199	199	200	.995	
.999	999	999	999	999	999	999	999	999	999	999	999	999	.999	
.9995	200^{1}	200^{1}	200^{1}	200^{1}	200^{1}	200^{1}	200^{1}	200^{1}	200^{1}	200^{1}	200^{1}	200^{1}	.9995	
.0005	.093	.109	.117	.127	.136	.143	.147	.156	.158	.162	.166	.169	.0005	3
.001	.107	.123	.132	.142	.152	.158	.162	.171	.173	.177	.181	.184	.001	
.005	.154	.172	.181	.191	.201	.207	.211	.220	.222	.227	.231	.234	.005	
.01	.185	.203	.212	.222	.232	.238	.242	.251	.253	.258	.262	.264	.01	
.025	.241	.259	.269	.279	.289	.295	.299	.308	.310	.314	.318	.321	.025	
.05	.304	.323	.332	.342	.352	.358	.363	.370	.373	.377	.382	.384	.05	
.10	.402	.420	.430	.439	.449	.455	.459	.467	.469	.474	.476	.480	.10	
.25	.658	.675	.684	.693	.702	.708	.711	.719	.721	.724	.728	.730	.25	
.50	1.21	1.23	1.23	1.24	1.25	1.25	1.25	1.26	1.26	1.26	1.27	1.27	.50	
.75	2.46	2.46	2.46	2.47	2.47	2.47	2.47	2.47	2.47	2.47	2.47	2.47	.75	
.90	5.20	5.18	5.18	5.17	5.16	5.15	5.15	5.14	5.14	5.14	5.14	5.13	.90	
.95	8.70	8.66	8.63	8.62	8.59	8.58	8.57	8.55	8.55	8.54	8.53	8.53	.95	
.975	14.3	14.2	14.1	14.1	14.0	14.0	14.0	14.0	13.9	13.9	13.9	13.9	.975	
.99	26.9	26.7	26.6	26.5	26.4	26.4	26.3	26.2	26.2	26.2	26.1	26.1	.99	
.995	43.1	42.8	42.6	42.5	42.3	42.2	42.1	42.0	42.0	41.9	41.9	41.8	.995	
.999	127	126	126	125	125	125	124	124	124	124	124	123	.999	
.9995	203	201	200	199	199	198	198	197	197	197	196	196	.9995	

Table A.7 Percentiles of the $F(\nu_1, \nu_2)$ Distribution with Degrees of Freedom ν_1 for the Numerator and ν_2 for the Denominator (*Continued*)

ν_2	Cum. Prop. \ ν_1	1	2	3	4	5	6	7	8	9	10	11	12	Cum. Prop.
4	.0005	$.0^{6}44$	$.0^{3}50$	$.0^{2}46$	.013	.024	.036	.047	.057	.066	.075	.082	.089	.0005
	.001	$.0^{5}18$	$.0^{2}10$	$.0^{2}73$	.019	.032	.046	.058	.069	.079	.089	.097	.104	.001
	.005	$.0^{4}44$	$.0^{2}50$	.022	.043	.064	.083	.100	.114	.126	.137	.145	.153	.005
	.01	$.0^{3}18$	.010	.035	.063	.088	.109	.127	.143	.156	.167	.176	.185	.01
	.025	$.0^{2}11$	.026	.066	.104	.135	.161	.181	.198	.212	.224	.234	.243	.025
	.05	$.0^{2}44$	.052	.110	.157	.193	.221	.243	.261	.275	.288	.298	.307	.05
	.10	.018	.108	.187	.243	.284	.314	.338	.356	.371	.384	.394	.403	.10
	.25	.117	.309	.418	.484	.528	.560	.583	.601	.615	.627	.637	.645	.25
	.50	.549	.828	.941	1.00	1.04	1.06	1.08	1.09	1.10	1.11	1.12	1.13	.50
	.75	1.81	2.00	2.05	2.06	2.07	2.08	2.08	2.08	2.08	2.08	2.08	2.08	.75
	.90	4.54	4.32	4.19	4.11	4.05	4.01	3.98	3.95	3.94	3.92	3.91	3.90	.90
	.95	7.71	6.94	6.59	6.39	6.26	6.16	6.09	6.04	6.00	5.96	5.94	5.91	.95
	.975	12.2	10.6	9.98	9.60	9.36	9.20	9.07	8.98	8.90	8.84	8.79	8.75	.975
	.99	21.2	18.0	16.7	16.0	15.5	15.2	15.0	14.8	14.7	14.5	14.4	14.4	.99
	.995	31.3	26.3	24.3	23.2	22.5	22.0	21.6	21.4	21.1	21.0	20.8	20.7	.995
	.999	74.1	61.2	56.2	53.4	51.7	50.5	49.7	49.0	48.5	48.0	47.7	47.4	.999
	.9995	106	87.4	80.1	76.1	73.6	71.9	70.6	69.7	68.9	68.3	67.8	67.4	.9995
5	.0005	$.0^{6}43$	$.0^{3}50$	$.0^{2}47$	.014	.025	.038	.050	.061	.070	.081	.089	.096	.0005
	.001	$.0^{5}17$	$.0^{2}10$	$.0^{2}75$	.019	.034	.048	.062	.074	.085	.095	.104	.112	.001
	.005	$.0^{4}43$	$.0^{2}50$	.022	.045	.067	.087	.105	.120	.134	.146	.156	.165	.005
	.01	$.0^{3}17$	.010	.035	.064	.091	.114	.134	.151	.165	.177	.188	.197	.01
	.025	$.0^{2}11$	.025	.067	.107	.140	.167	.189	.208	.223	.236	.248	.257	.025
	.05	$.0^{2}43$	.052	.111	.160	.198	.228	.252	.271	.287	.301	.313	.322	.05
	.10	.017	.108	.188	.247	.290	.322	.347	.367	.383	.397	.408	.418	.10
	.25	.113	.305	.415	.483	.528	.560	.584	.604	.618	.631	.641	.650	.25
	.50	.528	.799	.907	.965	1.00	1.02	1.04	1.05	1.06	1.07	1.08	1.09	.50
	.75	1.69	1.85	1.88	1.89	1.89	1.89	1.89	1.89	1.89	1.89	1.89	1.89	.75
	.90	4.06	3.78	3.62	3.52	3.45	3.40	3.37	3.34	3.32	3.30	3.28	3.27	.90
	.95	6.61	5.79	5.41	5.19	5.05	4.95	4.88	4.82	4.77	4.74	4.71	4.68	.95
	.975	10.0	8.43	7.76	7.39	7.15	6.98	6.85	6.76	6.68	6.62	6.57	6.52	.975
	.99	16.3	13.3	12.1	11.4	11.0	10.7	10.5	10.3	10.2	10.1	9.96	9.89	.99
	.995	22.8	18.3	16.5	15.6	14.9	14.5	14.2	14.0	13.8	13.6	13.5	13.4	.995
	.999	47.2	37.1	33.2	31.1	29.7	28.8	28.2	27.6	27.2	26.9	26.6	26.4	.999
	.9995	63.6	49.8	44.4	41.5	39.7	38.5	37.6	36.9	36.4	35.9	35.6	35.2	.9995
6	.0005	$.0^{6}43$	$.0^{3}50$	$.0^{2}47$	.014	.026	.039	.052	.064	.075	.085	.094	.103	.0005
	.001	$.0^{5}17$	$.0^{2}10$	$.0^{2}75$	.020	.035	.050	.064	.078	.090	.101	.111	.119	.001
	.005	$.0^{4}43$	$.0^{2}50$	.022	.045	.069	.090	.109	.126	.140	.153	.164	.174	.005
	.01	$.0^{3}17$	.010	.036	.066	.094	.118	.139	.157	.172	.186	.197	.207	.01
	.025	$.0^{2}11$	.025	.068	.109	.143	.172	.195	.215	.231	.246	.258	.268	.025
	.05	$.0^{2}43$	.052	.112	.162	.202	.233	.259	.279	.296	.311	.324	.334	.05
	.10	.017	.107	.189	.249	.294	.327	.354	.375	.392	.406	.418	.429	.10
	.25	.111	.302	.413	.481	.524	.561	.586	.606	.622	.635	.645	.654	.25
	.50	.515	.780	.886	.942	.977	1.00	1.02	1.03	1.04	1.05	1.05	1.06	.50
	.75	1.62	1.76	1.78	1.79	1.79	1.78	1.78	1.78	1.77	1.77	1.77	1.77	.75
	.90	3.78	3.46	3.29	3.18	3.11	3.05	3.01	2.98	2.96	2.94	2.92	2.90	.90
	.95	5.99	5.14	4.76	4.53	4.39	4.28	4.21	4.15	4.10	4.06	4.03	4.00	.95
	.975	8.81	7.26	6.60	6.23	5.99	5.82	5.70	5.60	5.52	5.46	5.41	5.37	.975
	.99	13.7	10.9	9.78	9.15	8.75	8.47	8.26	8.10	7.98	7.87	7.79	7.72	.99
	.995	18.6	14.5	12.9	12.0	11.5	11.1	10.8	10.6	10.4	10.2	10.1	10.0	.995
	.999	35.5	27.0	23.7	21.9	20.8	20.0	19.5	19.0	18.7	18.4	18.2	18.0	.999
	.9995	46.1	34.8	30.4	28.1	26.6	25.6	24.9	24.3	23.9	23.5	23.2	23.0	.9995

Table A.7 Percentiles of the $F(\nu_1, \nu_2)$ Distribution with Degrees of Freedom ν_1 for the Numerator and ν_2 for the Denominator *(Continued)*

Cum. Prop. \ ν_1	15	20	24	30	40	50	60	100	120	200	500	∞	Cum. Prop.	ν_2
.0005	.105	.125	.135	.147	.159	.166	.172	.183	.186	.191	.196	.200	.0005	**4**
.001	.121	.141	.152	.163	.176	.183	.188	.200	.202	.208	.213	.217	.001	
.005	.172	.193	.204	.216	.229	.237	.242	.253	.255	.260	.266	.269	.005	
.01	.204	.226	.237	.249	.261	.269	.274	.285	.287	.293	.298	.301	.01	
.025	.263	.284	.296	.308	.320	.327	.332	.342	.346	.351	.356	.359	.025	
.05	.327	.349	.360	.372	.384	.391	.396	.407	.409	.413	.418	.422	.05	
.10	.424	.445	.456	.467	.478	.485	.490	.500	.502	.508	.510	.514	.10	
.25	.664	.683	.692	.702	.712	.718	.722	.731	.733	.737	.740	.743	.25	
.50	1.14	1.15	1.16	1.16	1.17	1.18	1.18	1.18	1.18	1.19	1.19	1.19	.50	
.75	2.08	2.08	2.08	2.08	2.08	2.08	2.08	2.08	2.08	2.08	2.08	2.08	.75	
.90	3.87	3.84	3.83	3.82	3.80	3.80	3.79	3.78	3.78	3.77	3.76	3.76	.90	
.95	5.86	5.80	5.77	5.75	5.72	5.70	5.69	5.66	5.66	5.65	5.64	5.63	.95	
.975	8.66	8.56	8.51	8.46	8.41	8.38	8.36	8.32	8.31	8.29	8.27	8.26	.975	
.99	14.2	14.0	13.9	13.8	13.7	13.7	13.7	13.6	13.6	13.5	13.5	13.5	.99	
.995	20.4	20.2	20.0	19.9	19.8	19.7	19.6	19.5	19.5	19.4	19.4	19.3	.995	
.999	46.8	46.1	45.8	45.4	45.1	44.9	44.7	44.5	44.4	44.3	44.1	44.0	.999	
.9995	66.5	65.5	65.1	64.6	64.1	63.8	63.6	63.2	63.1	62.9	62.7	62.6	.9995	
.0005	.115	.137	.150	.163	.177	.186	.192	.205	.209	.216	.222	.226	.0005	**5**
.001	.132	.155	.167	.181	.195	.204	.210	.223	.227	.233	.239	.244	.001	
.005	.186	.210	.223	.237	.251	.260	.266	.279	.282	.288	.294	.299	.005	
.01	.219	.244	.257	.270	.285	.293	.299	.312	.315	.322	.328	.331	.01	
.025	.280	.304	.317	.330	.344	.353	.359	.370	.374	.380	.386	.390	.025	
.05	.345	.369	.382	.395	.408	.417	.422	.432	.437	.442	.448	.452	.05	
.10	.440	.463	.476	.488	.501	.508	.514	.524	.527	.532	.538	.541	.10	
.25	.669	.690	.700	.711	.722	.728	.732	.741	.743	.748	.752	.755	.25	
.50	1.10	1.11	1.12	1.12	1.13	1.13	1.14	1.14	1.14	1.15	1.15	1.15	.50	
.75	1.89	1.88	1.88	1.88	1.88	1.88	1.87	1.87	1.87	1.87	1.87	1.87	.75	
.90	3.24	3.21	3.19	3.17	3.16	3.15	3.14	3.13	3.12	3.12	3.11	3.10	.90	
.95	4.62	4.56	4.53	4.50	4.46	4.44	4.43	4.41	4.40	4.39	4.37	4.36	.95	
.975	6.43	6.33	6.28	6.23	6.18	6.14	6.12	6.08	6.07	6.05	6.03	6.02	.975	
.99	9.72	9.55	9.47	9.38	9.29	9.24	9.20	9.13	9.11	9.08	9.04	9.02	.99	
.995	13.1	12.9	12.8	12.7	12.5	12.5	12.4	12.3	12.3	12.2	12.2	12.1	.995	
.999	25.9	25.4	25.1	24.9	24.6	24.4	24.3	24.1	24.1	23.9	23.8	23.8	.999	
.9995	34.6	33.9	33.5	33.1	32.7	32.5	32.3	32.1	32.0	31.8	31.7	31.6	.9995	
.0005	.123	.148	.162	.177	.193	.203	.210	.225	.229	.236	.244	.249	.0005	**6**
.001	.141	.166	.180	.195	.211	.222	.229	.243	.247	.255	.262	.267	.001	
.005	.197	.224	.238	.253	.269	.279	.286	.301	.304	.312	.318	.324	.005	
.01	.232	.258	.273	.288	.304	.313	.321	.334	.338	.346	.352	.357	.01	
.025	.293	.320	.334	.349	.364	.375	.381	.394	.398	.405	.412	.415	.025	
.05	.358	.385	.399	.413	.428	.437	.444	.457	.460	.467	.472	.476	.05	
.10	.453	.478	.491	.505	.519	.526	.533	.546	.548	.556	.559	.564	.10	
.25	.675	.696	.707	.718	.729	.736	.741	.751	.753	.758	.762	.765	.25	
.50	1.07	1.08	1.09	1.10	1.10	1.11	1.11	1.11	1.12	1.12	1.12	1.12	.50	
.75	1.76	1.76	1.75	1.75	1.75	1.75	1.74	1.74	1.74	1.74	1.74	1.74	.75	
.90	2.87	2.84	2.82	2.80	2.78	2.77	2.76	2.75	2.74	2.73	2.73	2.72	.90	
.95	3.94	3.87	3.84	3.81	3.77	3.75	3.74	3.71	3.70	3.69	3.68	3.67	.95	
.975	5.27	5.17	5.12	5.07	5.01	4.98	4.96	4.92	4.90	4.88	4.86	4.85	.975	
.99	7.56	7.40	7.31	7.23	7.14	7.09	7.06	6.99	6.97	6.93	6.90	6.88	.99	
.995	9.81	9.59	9.47	9.36	9.24	9.17	9.12	9.03	9.00	8.95	8.91	8.88	.995	
.999	17.6	17.1	16.9	16.7	16.4	16.3	16.2	16.0	16.0	15.9	15.8	15.7	.999	
.9995	22.4	21.9	21.7	21.4	21.1	20.9	20.7	20.5	20.4	20.3	20.2	20.1	.9995	

Table A.7 Percentiles of the $F(\nu_1, \nu_2)$ Distribution with Degrees of Freedom ν_1 for the Numerator and ν_2 for the Denominator *(Continued)*

ν_2	Cum. Prop. \ ν_1	1	2	3	4	5	6	7	8	9	10	11	12	Cum. Prop.
7	.0005	$.0^{6}42$	$.0^{3}50$	$.0^{2}48$	.014	.027	.040	.053	.066	.078	.088	.099	.108	.0005
	.001	$.0^{5}17$	$.0^{2}10$	$.0^{2}76$	.020	.035	.051	.067	.081	.093	.105	.115	.125	.001
	.005	$.0^{4}42$	$.0^{2}50$	.023	.046	.070	.093	.113	.130	.145	.159	.171	.181	.005
	.01	$.0^{3}17$	.010	.036	.067	.096	.121	.143	.162	.178	.192	.205	.216	.01
	.025	$.0^{2}10$	.025	.068	.110	.146	.176	.200	.221	.238	.253	.266	.277	.025
	.05	$.0^{2}42$	.052	.113	.164	.205	.238	.264	.286	.304	.319	.332	.343	.05
	.10	.017	.107	.190	.251	.297	.332	.359	.381	.399	.414	.427	.438	.10
	.25	.110	.300	.412	.481	.528	.562	.588	.608	.624	.637	.649	.658	.25
	.50	.506	.767	.871	.926	.960	.983	1.00	1.01	1.02	1.03	1.04	1.04	.50
	.75	1.57	1.70	1.72	1.72	1.71	1.71	1.70	1.70	1.69	1.69	1.69	1.68	.75
	.90	3.59	3.26	3.07	2.96	2.88	2.83	2.78	2.75	2.72	2.70	2.68	2.67	.90
	.95	5.59	4.74	4.35	4.12	3.97	3.87	3.79	3.73	3.68	3.64	3.60	3.57	.95
	.975	8.07	6.54	5.89	5.52	5.29	5.12	4.99	4.90	4.82	4.76	4.71	4.67	.975
	.99	12.2	9.55	8.45	7.85	7.46	7.19	6.99	6.84	6.72	6.62	6.54	6.47	.99
	.995	16.2	12.4	10.9	10.0	9.52	9.16	8.89	8.68	8.51	8.38	8.27	8.18	.995
	.999	29.2	21.7	18.8	17.2	16.2	15.5	15.0	14.6	14.3	14.1	13.9	13.7	.999
	.9995	37.0	27.2	23.5	21.4	20.2	19.3	18.7	18.2	17.8	17.5	17.2	17.0	.9995
8	.0005	$.0^{6}42$	$.0^{3}50$	$.0^{2}48$	.014	.027	.041	.055	.068	.081	.092	.102	.112	.0005
	.001	$.0^{5}17$	$.0^{2}10$	$.0^{2}76$	.020	.036	.053	.068	.083	.096	.109	.120	.130	.001
	.005	$.0^{4}42$	$.0^{2}50$	.027	.047	.072	.095	.115	.133	.149	.164	.176	.187	.005
	.01	$.0^{3}17$	.010	.036	.068	.097	.123	.146	.166	.183	.198	.211	.222	.01
	.025	$.0^{2}10$	.025	.069	.111	.148	.179	.204	.226	.244	.259	.273	.285	.025
	.05	$.0^{2}42$	.052	.113	.166	.208	.241	.268	.291	.310	.326	.339	.351	.05
	.10	.017	.107	.190	.253	.299	.335	.363	.386	.405	.421	.435	.445	.10
	.25	.109	.298	.411	.481	.529	.563	.589	.610	.627	.640	.654	.661	.25
	.50	.499	.757	.860	.915	.948	.971	.988	1.00	1.01	1.02	1.02	1.03	.50
	.75	1.54	1.66	1.67	1.66	1.66	1.65	1.64	1.64	1.64	1.63	1.63	1.62	.75
	.90	3.46	3.11	2.92	2.81	2.73	2.67	2.62	2.59	2.56	2.54	2.52	2.50	.90
	.95	5.32	4.46	4.07	3.84	3.69	3.58	3.50	3.44	3.39	3.35	3.31	3.28	.95
	.975	7.57	6.06	5.42	5.05	4.82	4.65	4.53	4.43	4.36	4.30	4.24	4.20	.975
	.99	11.3	8.65	7.59	7.01	6.63	6.37	6.18	6.03	5.91	5.81	5.73	5.67	.99
	.995	14.7	11.0	9.60	8.81	8.30	7.95	7.69	7.50	7.34	7.21	7.10	7.01	.995
	.999	25.4	18.5	15.8	14.4	13.5	12.9	12.4	12.0	11.8	11.5	11.4	11.2	.999
	.9995	31.6	22.8	19.4	17.6	16.4	15.7	15.1	14.6	14.3	14.0	13.8	13.6	.9995
9	.0005	$.0^{6}41$	$.0^{3}50$	$.0^{2}48$	.015	.027	.042	.056	.070	.083	.094	.105	.115	.0005
	.001	$.0^{5}17$	$.0^{2}10$	$.0^{2}77$	.021	.037	.054	.070	.085	.099	.112	.123	.134	.001
	.005	$.0^{4}42$	$.0^{2}50$	.023	.047	.073	.096	.117	.136	.153	.168	.181	.192	.005
	.01	$.0^{3}17$	.010	.037	.068	.098	.125	.149	.169	.187	.202	.216	.228	.01
	.025	$.0^{2}10$	.025	.069	.112	.150	.181	.207	.230	.248	.265	.279	.291	.025
	.05	$.0^{2}40$	.052	.113	.167	.210	.244	.272	.296	.315	.331	.345	.358	.05
	.10	.017	.107	.191	.254	.302	.338	.367	.390	.410	.426	.441	.452	.10
	.25	.108	.297	.410	.480	.529	.564	.591	.612	.629	.643	.654	.664	.25
	.50	.494	.749	.852	.906	.939	.962	.978	.990	1.00	1.01	1.01	1.02	.50
	.75	1.51	1.62	1.63	1.63	1.62	1.61	1.60	1.60	1.59	1.59	1.58	1.58	.75
	.90	3.36	3.01	2.81	2.69	2.61	2.55	2.51	2.47	2.44	2.42	2.40	2.38	.90
	.95	5.12	4.26	3.86	3.63	3.48	3.37	3.29	3.23	3.18	3.14	3.10	3.07	.95
	.975	7.21	5.71	5.08	4.72	4.48	4.32	4.20	4.10	4.03	3.96	3.91	3.87	.975
	.99	10.6	8.02	6.99	6.42	6.06	5.80	5.61	5.47	5.35	5.26	5.18	5.11	.99
	.995	13.6	10.1	8.72	7.96	7.47	7.13	6.88	6.69	6.54	6.42	6.31	6.23	.995
	.999	22.9	16.4	13.9	12.6	11.7	11.1	10.7	10.4	10.1	9.89	9.71	9.57	.999
	.9995	28.0	19.9	16.8	15.1	14.1	13.3	12.8	12.4	12.1	11.8	11.6	11.4	.9995

Table A.7 Percentiles of the $F(\nu_1, \nu_2)$ Distribution with Degrees of Freedom ν_1 for the Numerator and ν_2 for the Denominator *(Continued)*

Cum. Prop. \ ν_1	15	20	24	30	40	50	60	100	120	200	500	∞	Cum. Prop.	ν_2
.0005	.130	.157	.172	.188	.206	.217	.225	.242	.246	.255	.263	.268	.0005	7
.001	.148	.176	.191	.208	.225	.237	.245	.261	.266	.274	.282	.288	.001	
.005	.206	.235	.251	.267	.285	.296	.304	.319	.324	.332	.340	.345	.005	
.01	.241	.270	.286	.303	.320	.331	.339	.355	.358	.366	.373	.379	.01	
.025	.304	.333	.348	.364	.381	.392	.399	.413	.418	.426	.433	.437	.025	
.05	.369	.398	.413	.428	.445	.455	.461	.476	.479	.485	.493	.498	.05	
.10	.463	.491	.504	.519	.534	.543	.550	.562	.566	.571	.578	.582	.10	
.25	.679	.702	.713	.725	.737	.745	.749	.760	.762	.767	.772	.775	.25	
.50	1.05	1.07	1.07	1.08	1.08	1.09	1.09	1.10	1.10	1.10	1.10	1.10	.50	
.75	1.68	1.67	1.67	1.66	1.66	1.66	1.65	1.65	1.65	1.65	1.65	1.65	.75	
.90	2.63	2.59	2.58	2.56	2.54	2.52	2.51	2.50	2.49	2.48	2.48	2.47	.90	
.95	3.51	3.44	3.41	3.38	3.34	3.32	3.30	3.27	3.27	3.25	3.24	3.23	.95	
.975	4.57	4.47	4.42	4.36	4.31	4.28	4.25	4.21	4.20	4.18	4.16	4.14	.975	
.99	6.31	6.16	6.07	5.99	5.91	5.86	5.82	5.75	5.74	5.70	5.67	5.65	.99	
.995	7.97	7.75	7.65	7.53	7.42	7.35	7.31	7.22	7.19	7.15	7.10	7.08	.995	
.999	13.3	12.9	12.7	12.5	12.3	12.2	12.1	11.9	11.9	11.8	11.7	11.7	.999	
.9995	16.5	16.0	15.7	15.5	15.2	15.1	15.0	14.7	14.7	14.6	14.5	14.4	.9995	
.0005	.136	.164	.181	.198	.218	.230	.239	.257	.262	.271	.281	.287	.0005	8
.001	.155	.184	.200	.218	.238	.250	.259	.277	.282	.292	.300	.306	.001	
.005	.214	.244	.261	.279	.299	.311	.319	.337	.341	.351	.358	.364	.005	
.01	.250	.281	.297	.315	.334	.346	.354	.372	.376	.385	.392	.398	.01	
.025	.313	.343	.360	.377	.395	.407	.415	.431	.435	.442	.450	.456	.025	
.05	.379	.409	.425	.441	.459	.469	.477	.493	.496	.505	.510	.516	.05	
.10	.472	.500	.515	.531	.547	.556	.563	.578	.581	.588	.595	.599	.10	
.25	.684	.707	.718	.730	.743	.751	.756	.767	.769	.775	.780	.783	.25	
.50	1.04	1.05	1.06	1.07	1.07	1.07	1.08	1.08	1.08	1.09	1.09	1.09	.50	
.75	1.62	1.61	1.60	1.60	1.59	1.59	1.59	1.58	1.58	1.58	1.58	1.58	.75	
.90	2.46	2.42	2.40	2.38	2.36	2.35	2.34	2.32	2.32	2.31	2.30	2.29	.90	
.95	3.22	3.15	3.12	3.08	3.04	3.02	3.01	2.97	2.97	2.95	2.94	2.93	.95	
.975	4.10	4.00	3.95	3.89	3.84	3.81	3.78	3.74	3.73	3.70	3.68	3.67	.975	
.99	5.52	5.36	5.28	5.20	5.12	5.07	5.03	4.96	4.95	4.91	4.88	4.86	.99	
.995	6.81	6.61	6.50	6.40	6.29	6.22	6.18	6.09	6.06	6.02	5.98	5.95	.995	
.999	10.8	10.5	10.3	10.1	9.92	9.80	9.73	9.57	9.54	9.46	9.39	9.34	.999	
.9995	13.1	12.7	12.5	12.2	12.0	11.8	11.8	11.6	11.5	11.4	11.4	11.3	.9995	
.0005	.141	.171	.188	.207	.228	.242	.251	.270	.276	.287	.297	.303	.0005	9
.001	.160	.191	.208	.228	.249	.262	.271	.291	.296	.307	.316	.323	.001	
.005	.220	.253	.271	.290	.310	.324	.332	.351	.356	.366	.376	.382	.005	
.01	.257	.289	.307	.326	.346	.358	.368	.386	.391	.400	.410	.415	.01	
.025	.320	.352	.370	.388	.408	.420	.428	.446	.450	.459	.467	.473	.025	
.05	.386	.418	.435	.452	.471	.483	.490	.508	.510	.518	.526	.532	.05	
.10	.479	.509	.525	.541	.558	.568	.575	.588	.594	.602	.610	.613	.10	
.25	.687	.711	.723	.736	.749	.757	.762	.773	.776	.782	.787	.791	.25	
.50	1.03	1.04	1.05	1.05	1.06	1.06	1.07	1.07	1.07	1.08	1.08	1.08	.50	
.75	1.57	1.56	1.56	1.55	1.55	1.54	1.54	1.53	1.53	1.53	1.53	1.53	.75	
.90	2.34	2.30	2.28	2.25	2.23	2.22	2.21	2.19	2.18	2.17	2.17	2.16	.90	
.95	3.01	2.94	2.90	2.86	2.83	2.80	2.79	2.76	2.75	2.73	2.72	2.71	.95	
.975	3.77	3.67	3.61	3.56	3.51	3.47	3.45	3.40	3.39	3.37	3.35	3.33	.975	
.99	4.96	4.81	4.73	4.65	4.57	4.52	4.48	4.42	4.40	4.36	4.33	4.31	.99	
.995	6.03	5.83	5.73	5.62	5.52	5.45	5.41	5.32	5.30	5.26	5.21	5.19	.995	
.999	9.24	8.90	8.72	8.55	8.37	8.26	8.19	8.04	8.00	7.93	7.86	7.81	.999	
.9995	11.0	10.6	10.4	10.2	9.94	9.80	9.71	9.53	9.49	9.40	9.32	9.26	.9995	

Table A.7 Percentiles of the $F(\nu_1,\nu_2)$ Distribution with Degrees of Freedom ν_1 for the Numerator and ν_2 for the Denominator *(Continued)*

ν_2	Cum. Prop. \ ν_1	1	2	3	4	5	6	7	8	9	10	11	12	Cum. Prop.
10	.0005	$.0^{6}41$	$.0^{3}50$	$.0^{2}49$	.015	.028	.043	.057	.071	.085	.097	.108	.119	.0005
	.001	$.0^{5}17$	$.0^{2}10$	$.0^{2}77$	.021	.037	.054	.071	.087	.101	.114	.126	.137	.001
	.005	$.0^{4}41$	$.0^{2}50$	.023	.048	.073	.098	.119	.139	.156	.171	.185	.197	.005
	.01	$.0^{3}17$	.010	.037	.069	.100	.127	.151	.172	.190	.206	.220	.233	.01
	.025	$.0^{2}10$	.025	.069	.113	.151	.183	.210	.233	.252	.269	.283	.296	.025
	.05	$.0^{2}41$	.052	.114	.168	.211	.246	.275	.299	.319	.336	.351	.363	.05
	.10	.017	.106	.191	.255	.303	.340	.370	.394	.414	.430	.444	.457	.10
	.25	.107	.296	.409	.480	.529	.565	.592	.613	.631	.645	.657	.667	.25
	.50	.490	.743	.845	.899	.932	.954	.971	.983	.992	1.00	1.01	1.01	.50
	.75	1.49	1.60	1.60	1.59	1.59	1.58	1.57	1.56	1.56	1.55	1.55	1.54	.75
	.90	3.28	2.92	2.73	2.61	2.52	2.46	2.41	2.38	2.35	2.32	2.30	2.28	.90
	.95	4.96	4.10	3.71	3.48	3.33	3.22	3.14	3.07	3.02	2.98	2.94	2.91	.95
	.975	6.94	5.46	4.83	4.47	4.24	4.07	3.95	3.85	3.78	3.72	3.66	3.62	.975
	.99	10.0	7.56	6.55	5.99	5.64	5.39	5.20	5.06	4.94	4.85	4.77	4.71	.99
	.995	12.8	9.43	8.08	7.34	6.87	6.54	6.30	6.12	5.97	5.85	5.75	5.66	.995
	.999	21.0	14.9	12.6	11.3	10.5	9.92	9.52	9.20	8.96	8.75	8.58	8.44	.999
	.9995	25.5	17.9	15.0	13.4	12.4	11.8	11.3	10.9	10.6	10.3	10.1	9.93	.9995
11	.0005	$.0^{6}41$	$.0^{3}50$	$.0^{2}49$	.015	.028	.043	.058	.072	.086	.099	.111	.121	.0005
	.001	$.0^{5}16$	$.0^{2}10$	$.0^{2}78$	.021	.038	.055	.072	.088	.103	.116	.129	.140	.001
	.005	$.0^{4}40$	$.0^{2}50$	.023	.048	.074	.099	.121	.141	.158	.174	.188	.200	.005
	.01	$.0^{3}16$	.010	.037	.069	.100	.128	.153	.175	.193	.210	.224	.237	.01
	.025	$.0^{2}10$	.025	.069	.114	.152	.185	.212	.236	.256	.273	.288	.301	.025
	.05	$.0^{2}41$	.052	.114	.168	.212	.248	.278	.302	.323	.340	.355	.368	.05
	.10	.017	.106	.192	.256	.305	.342	.373	.397	.417	.435	.448	.461	.10
	.25	.107	.295	.408	.481	.529	.565	.592	.614	.633	.645	.658	.667	.25
	.50	.486	.739	.840	.893	.926	.948	.964	.977	.986	.994	1.00	1.01	.50
	.75	1.47	1.58	1.58	1.57	1.56	1.55	1.54	1.53	1.53	1.52	1.52	1.51	.75
	.90	3.23	2.86	2.66	2.54	2.45	2.39	2.34	2.30	2.27	2.25	2.23	2.21	.90
	.95	4.84	3.98	3.59	3.36	3.20	3.09	3.01	2.95	2.90	2.85	2.82	2.79	.95
	.975	6.72	5.26	4.63	4.28	4.04	3.88	3.76	3.66	3.59	3.53	3.47	3.43	.975
	.99	9.65	7.21	6.22	5.67	5.32	5.07	4.89	4.74	4.63	4.54	4.46	4.40	.99
	.995	12.2	8.91	7.60	6.88	6.42	6.10	5.86	5.68	5.54	5.42	5.32	5.24	.995
	.999	19.7	13.8	11.6	10.3	9.58	9.05	8.66	8.35	8.12	7.92	7.76	7.62	.999
	.9995	23.6	16.4	13.6	12.2	11.2	10.6	10.1	9.76	9.48	9.24	9.04	8.88	.9995
12	.0005	$.0^{6}41$	$.0^{3}50$	$.0^{2}49$	.015	.028	.044	.058	.073	.087	.101	.113	.124	.0005
	.001	$.0^{5}16$	$.0^{2}10$	$.0^{2}78$	.021	.038	.056	.073	.089	.104	.118	.131	.143	.001
	.005	$.0^{4}39$	$.0^{2}50$	.023	.048	.075	.100	.122	.143	.161	.177	.191	.204	.005
	.01	$.0^{3}16$	.010	.037	.070	.101	.130	.155	.176	.196	.212	.227	.241	.01
	.025	$.0^{2}10$	.025	.070	.114	.153	.186	.214	.238	.259	.276	.292	.305	.025
	.05	$.0^{2}41$	.052	.114	.169	.214	.250	.280	.305	.325	.343	.358	.372	.05
	.10	.016	.106	.192	.257	.306	.344	.375	.400	.420	.438	.452	.466	.10
	.25	.106	.295	.408	.480	.530	.566	.594	.616	.633	.649	.662	.671	.25
	.50	.484	.735	.835	.888	.921	.943	.959	.972	.981	.989	.995	1.00	.50
	.75	1.46	1.56	1.56	1.55	1.54	1.53	1.52	1.51	1.51	1.50	1.50	1.49	.75
	.90	3.18	2.81	2.61	2.48	2.39	2.33	2.28	2.24	2.21	2.19	2.17	2.15	.90
	.95	4.75	3.89	3.49	3.26	3.11	3.00	2.91	2.85	2.80	2.75	2.72	2.69	.95
	.975	6.55	5.10	4.47	4.12	3.89	3.73	3.61	3.51	3.44	3.37	3.32	3.28	.975
	.99	9.33	6.93	5.95	5.41	5.06	4.82	4.64	4.50	4.39	4.30	4.22	4.16	.99
	.995	11.8	8.51	7.23	6.52	6.07	5.76	5.52	5.35	5.20	5.09	4.99	4.91	.995
	.999	18.6	13.0	10.8	9.63	8.89	8.38	8.00	7.71	7.48	7.29	7.14	7.01	.999
	.9995	22.2	15.3	12.7	11.2	10.4	9.74	9.28	8.94	8.66	8.43	8.24	8.08	.9995

Table A.7 Percentiles of the $F(\nu_1, \nu_2)$ Distribution with Degrees of Freedom ν_1 for the Numerator and ν_2 for the Denominator *(Continued)*

Cum. Prop. \ ν_1	15	20	24	30	40	50	60	100	120	200	500	∞	Cum. Prop.	ν_2
.0005	.145	.177	.195	.215	.238	.251	.262	.282	.288	.299	.311	.319	.0005	10
.001	.164	.197	.216	.236	.258	.272	.282	.303	.309	.321	.331	.338	.001	
.005	.226	.260	.279	.299	.321	.334	.344	.365	.370	.380	.391	.397	.005	
.01	.263	.297	.316	.336	.357	.370	.380	.400	.405	.415	.424	.431	.01	
.025	.327	.360	.379	.398	.419	.431	.441	.459	.464	.474	.483	.488	.025	
.05	.393	.426	.444	.462	.481	.493	.502	.518	.523	.532	.541	.546	.05	
.10	.486	.516	.532	.549	.567	.578	.586	.602	.605	.614	.621	.625	.10	
.25	.691	.714	.727	.740	.754	.762	.767	.779	.782	.788	.793	.797	.25	
.50	1.02	1.03	1.04	1.05	1.05	1.06	1.06	1.06	1.06	1.07	1.07	1.07	.50	
.75	1.53	1.52	1.52	1.51	1.51	1.50	1.50	1.49	1.49	1.49	1.48	1.48	.75	
.90	2.24	2.20	2.18	2.16	2.13	2.12	2.11	2.09	2.08	2.07	2.06	2.06	.90	
.95	2.85	2.77	2.74	2.70	2.66	2.64	2.62	2.59	2.58	2.56	2.55	2.54	.95	
.975	3.52	3.42	3.37	3.31	3.26	3.22	3.20	3.15	3.14	3.12	3.09	3.08	.975	
.99	4.56	4.41	4.33	4.25	4.17	4.12	4.08	4.01	4.00	3.96	3.93	3.91	.99	
.995	5.47	5.27	5.17	5.07	4.97	4.90	4.86	4.77	4.75	4.71	4.67	4.64	.995	
.999	8.13	7.80	7.64	7.47	7.30	7.19	7.12	6.98	6.94	6.87	6.81	6.76	.999	
.9995	9.56	9.16	8.96	8.75	8.54	8.42	8.33	8.16	8.12	8.04	7.96	7.90	.9995	
.0005	.148	.182	.201	.222	.246	.261	.271	.293	.299	.312	.324	.331	.0005	11
.001	.168	.202	.222	.243	.266	.282	.292	.313	.320	.332	.343	.353	.001	
.005	.231	.266	.286	.308	.330	.345	.355	.376	.382	.394	.403	.412	.005	
.01	.268	.304	.324	.344	.366	.380	.391	.412	.417	.427	.439	.444	.01	
.025	.332	.368	.386	.407	.429	.442	.450	.472	.476	.485	.495	.503	.025	
.05	.398	.433	.452	.469	.490	.503	.513	.529	.535	.543	.552	.559	.05	
.10	.490	.524	.541	.559	.578	.588	.595	.614	.617	.625	.633	.637	.10	
.25	.694	.719	.730	.744	.758	.767	.773	.780	.788	.794	.799	.803	.25	
.50	1.02	1.03	1.03	1.04	1.05	1.05	1.05	1.06	1.06	1.06	1.06	1.06	.50	
.75	1.50	1.49	1.49	1.48	1.47	1.47	1.47	1.46	1.46	1.46	1.45	1.45	.75	
.90	2.17	2.12	2.10	2.08	2.05	2.04	2.03	2.00	2.00	1.99	1.98	1.97	.90	
.95	2.72	2.65	2.61	2.57	2.53	2.51	2.49	2.46	2.45	2.43	2.42	2.40	.95	
.975	3.33	3.23	3.17	3.12	3.06	3.03	3.00	2.96	2.94	2.92	2.90	2.88	.975	
.99	4.25	4.10	4.02	3.94	3.86	3.81	3.78	3.71	3.69	3.66	3.62	3.60	.99	
.995	5.05	4.86	4.76	4.65	4.55	4.49	4.45	4.36	4.34	4.29	4.25	4.23	.995	
.999	7.32	7.01	6.85	6.68	6.52	6.41	6.35	6.21	6.17	6.10	6.04	6.00	.999	
.9995	8.52	8.14	7.94	7.75	7.55	7.43	7.35	7.18	7.14	7.06	6.98	6.93	.9995	
.0005	.152	.186	.206	.228	.253	.269	.280	.305	.311	.323	.337	.345	.0005	12
.001	.172	.207	.228	.250	.275	.291	.302	.326	.332	.344	.357	.365	.001	
.005	.235	.272	.292	.315	.339	.355	.365	.388	.393	.405	.417	.424	.005	
.01	.273	.310	.330	.352	.375	.391	.401	.422	.428	.441	.450	.458	.01	
.025	.337	.374	.394	.416	.437	.450	.461	.481	.487	.498	.508	.514	.025	
.05	.404	.439	.458	.478	.499	.513	.522	.541	.545	.556	.565	.571	.05	
.10	.496	.528	.546	.564	.583	.595	.604	.621	.625	.633	.641	.647	.10	
.25	.695	.721	.734	.748	.762	.771	.777	.789	.792	.799	.804	.808	.25	
.50	1.01	1.02	1.03	1.03	1.04	1.04	1.05	1.05	1.05	1.05	1.06	1.06	.50	
.75	1.48	1.47	1.46	1.45	1.45	1.44	1.44	1.43	1.43	1.43	1.42	1.42	.75	
.90	2.11	2.06	2.04	2.01	1.99	1.97	1.96	1.94	1.93	1.92	1.91	1.90	.90	
.95	2.62	2.54	2.51	2.47	2.43	2.40	2.38	2.35	2.34	2.32	2.31	2.30	.95	
.975	3.18	3.07	3.02	2.96	2.91	2.87	2.85	2.80	2.79	2.76	2.74	2.72	.975	
.99	4.01	3.86	3.78	3.70	3.62	3.57	3.54	3.47	3.45	3.41	3.38	3.36	.99	
.995	4.72	4.53	4.43	4.33	4.23	4.17	4.12	4.04	4.01	3.97	3.93	3.90	.995	
.999	6.71	6.40	6.25	6.09	5.93	5.83	5.76	5.63	5.59	5.52	5.46	5.42	.999	
.9995	7.74	7.37	7.18	7.00	6.80	6.68	6.61	6.45	6.41	6.33	6.25	6.20	.9995	

Table A.7 Percentiles of the $F(\nu_1,\nu_2)$ Distribution with Degrees of Freedom ν_1 for the Numerator and ν_2 for the Denominator *(Continued)*

ν_2	ν_1 / Cum. Prop.	1	2	3	4	5	6	7	8	9	10	11	12	Cum. Prop.
15	.0005	$.0^{6}41$	$.0^{3}50$	$.0^{2}49$	.015	.029	.045	.061	.076	.091	.105	.117	.129	.0005
	.001	$.0^{5}16$	$.0^{2}10$	$.0^{2}79$	.021	.039	.057	.075	.092	.108	.123	.137	.149	.001
	.005	$.0^{4}39$	$.0^{2}50$	.023	.049	.076	.102	.125	.147	.166	.183	.198	.212	.005
	.01	$.0^{3}16$	.010	.037	.070	.103	.132	.158	.181	.202	.219	.235	.249	.01
	.025	$.0^{2}10$	.025	.070	.116	.156	.190	.219	.244	.265	.284	.300	.315	.025
	.05	$.0^{2}41$	.051	.115	.170	.216	.254	.285	.311	.333	.351	.368	.382	.05
	.10	.016	.106	.192	.258	.309	.348	.380	.406	.427	.446	.461	.475	.10
	.25	.105	.293	.407	.480	.531	.568	.596	.618	.637	.652	.667	.676	.25
	.50	.478	.726	.826	.878	.911	.933	.948	.960	.970	.977	.984	.989	.50
	.75	1.43	1.52	1.52	1.51	1.49	1.48	1.47	1.46	1.46	1.45	1.44	1.44	.75
	.90	3.07	2.70	2.49	2.36	2.27	2.21	2.16	2.12	2.09	2.06	2.04	2.02	.90
	.95	4.54	3.68	3.29	3.06	2.90	2.79	2.71	2.64	2.59	2.54	2.51	2.48	.95
	.975	6.20	4.76	4.15	3.80	3.58	3.41	3.29	3.20	3.12	3.06	3.01	2.96	.975
	.99	8.68	6.36	5.42	4.89	4.56	4.32	4.14	4.00	3.89	3.80	3.73	3.67	.99
	.995	10.8	7.70	6.48	5.80	5.37	5.07	4.85	4.67	4.54	4.42	4.33	4.25	.995
	.999	16.6	11.3	9.34	8.25	7.57	7.09	6.74	6.47	6.26	6.08	5.93	5.81	.999
	.9995	19.5	13.2	10.8	9.48	8.66	8.10	7.68	7.36	7.11	6.91	6.75	6.60	.9995
20	.0005	$.0^{6}40$	$.0^{3}50$	$.0^{2}50$	.015	.029	.046	.063	.079	.094	.109	.123	.136	.0005
	.001	$.0^{5}16$	$.0^{2}10$	$.0^{2}79$	.022	.039	.058	.077	.095	.112	.128	.143	.156	.001
	.005	$.0^{4}39$	$.0^{2}50$	.023	.050	.077	.104	.129	.151	.171	.190	.206	.221	.005
	.01	$.0^{3}16$	.010	.037	.071	.105	.135	.162	.187	.208	.227	.244	.259	.01
	.025	$.0^{2}10$	.025	.071	.117	.158	.193	.224	.250	.273	.292	.310	.325	.025
	.05	$.0^{2}40$	.051	.115	.172	.219	.258	.290	.318	.340	.360	.377	.393	.05
	.10	.016	.106	.193	.260	.312	.353	.385	.412	.435	.454	.472	.485	.10
	.25	.104	.292	.407	.480	.531	.569	.598	.622	.641	.656	.671	.681	.25
	.50	.472	.718	.816	.868	.900	.922	.938	.950	.959	.966	.972	.977	.50
	.75	1.40	1.49	1.48	1.47	1.45	1.44	1.43	1.42	1.41	1.40	1.39	1.39	.75
	.90	2.97	2.59	2.38	2.25	2.16	2.09	2.04	2.00	1.96	1.94	1.91	1.89	.90
	.95	4.35	3.49	3.10	2.87	2.71	2.60	2.51	2.45	2.39	2.35	2.31	2.28	.95
	.975	5.87	4.46	3.86	3.51	3.29	3.13	3.01	2.91	2.84	2.77	2.72	2.68	.975
	.99	8.10	5.85	4.94	4.43	4.10	3.87	3.70	3.56	3.46	3.37	3.29	3.23	.99
	.995	9.94	6.99	5.82	5.17	4.76	4.47	4.26	4.09	3.96	3.85	3.76	3.68	.995
	.999	14.8	9.95	8.10	7.10	6.46	6.02	5.69	5.44	5.24	5.08	4.94	4.82	.999
	.9995	17.2	11.4	9.20	8.02	7.28	6.76	6.38	6.08	5.85	5.66	5.51	5.38	.9995
24	.0005	$.0^{6}40$	$.0^{3}50$	$.0^{2}50$	.015	.030	.046	.064	.080	.096	.112	.126	.139	.0005
	.001	$.0^{5}16$	$.0^{2}10$	$.0^{2}79$	.022	.040	.059	.079	.097	.115	.131	.146	.160	.001
	.005	$.0^{4}40$	$.0^{2}50$	.023	.050	.078	.106	.131	.154	.175	.193	.210	.226	.005
	.01	$.0^{3}16$	.010	.038	.072	.106	.137	.165	.189	.211	.231	.249	.264	.01
	.025	$.0^{2}10$	.025	.071	.117	.159	.195	.227	.253	.277	.297	.315	.331	.025
	.05	$.0^{2}40$	.051	.116	.173	.221	.260	.293	.321	.345	.365	.383	.399	.05
	.10	.016	.106	.193	.261	.313	.355	.388	.416	.439	.459	.476	.491	.10
	.25	.104	.291	.406	.480	.532	.570	.600	.623	.643	.659	.671	.684	.25
	.50	.469	.714	.812	.863	.895	.917	.932	.944	.953	.961	.967	.972	.50
	.75	1.39	1.47	1.46	1.44	1.43	1.41	1.40	1.39	1.38	1.38	1.37	1.36	.75
	.90	2.93	2.54	2.33	2.19	2.10	2.04	1.98	1.94	1.91	1.88	1.85	1.83	.90
	.95	4.26	3.40	3.01	2.78	2.62	2.51	2.42	2.36	2.30	2.25	2.21	2.18	.95
	.975	5.72	4.32	3.72	3.38	3.15	2.99	2.87	2.78	2.70	2.64	2.59	2.54	.975
	.99	7.82	5.61	4.72	4.22	3.90	3.67	3.50	3.36	3.26	3.17	3.09	3.03	.99
	.995	9.55	6.66	5.52	4.89	4.49	4.20	3.99	3.83	3.69	3.59	3.50	3.42	.995
	.999	14.0	9.34	7.55	6.59	5.98	5.55	5.23	4.99	4.80	4.64	4.50	4.39	.999
	.9995	16.2	10.6	8.52	7.39	6.68	6.18	5.82	5.54	5.31	5.13	4.98	4.85	.9995

Table A.7 Percentiles of the $F(\nu_1, \nu_2)$ Distribution with Degrees of Freedom ν_1 for the Numerator and ν_2 for the Denominator *(Continued)*

Cum. Prop. \ ν_1	15	20	24	30	40	50	60	100	120	200	500	∞	Cum. Prop.	ν_2
.0005	.159	.197	.220	.244	.272	.290	.303	.330	.339	.353	.368	.377	.0005	**15**
.001	.181	.219	.242	.266	.294	.313	.325	.352	.360	.375	.388	.398	.001	
.005	.246	.286	.308	.333	.360	.377	.389	.415	.422	.435	.448	.457	.005	
.01	.284	.324	.346	.370	.397	.413	.425	.450	.456	.469	.483	.490	.01	
.025	.349	.389	.410	.433	.458	.474	.485	.508	.514	.526	.538	.546	.025	
.05	.416	.454	.474	.496	.519	.535	.545	.565	.571	.581	.592	.600	.05	
.10	.507	.542	.561	.581	.602	.614	.624	.641	.647	.658	.667	.672	.10	
.25	.701	.728	.742	.757	.772	.782	.788	.802	.805	.812	.818	.822	.25	
.50	1.00	1.01	1.02	1.02	1.03	1.03	1.03	1.04	1.04	1.04	1.04	1.05	.50	
.75	1.43	1.41	1.41	1.40	1.39	1.39	1.38	1.38	1.37	1.37	1.36	1.36	.75	
.90	1.97	1.92	1.90	1.87	1.85	1.83	1.82	1.79	1.79	1.77	1.76	1.76	.90	
.95	2.40	2.33	2.39	2.25	2.20	2.18	2.16	2.12	2.11	2.10	2.08	2.07	.95	
.975	2.86	2.76	2.70	2.64	2.59	2.55	2.52	2.47	2.46	2.44	2.41	2.40	.975	
.99	3.52	3.37	3.29	3.21	3.13	3.08	3.05	2.98	2.96	2.92	2.89	2.87	.99	
.995	4.07	3.88	3.79	3.69	3.59	3.52	3.48	3.39	3.37	3.33	3.29	3.26	.995	
.999	5.54	5.25	5.10	4.95	4.80	4.70	4.64	4.51	4.47	4.41	4.35	4.31	.999	
.9995	6.27	5.93	5.75	5.58	5.40	5.29	5.21	5.06	5.02	4.94	4.87	4.83	.9995	
.0005	.169	.211	.235	.263	.295	.316	.331	.364	.375	.391	.408	.422	.0005	**20**
.001	.191	.233	.258	.286	.318	.339	.354	.386	.395	.413	.429	.441	.001	
.005	.258	.301	.327	.354	.385	.405	.419	.448	.457	.474	.490	.500	.005	
.01	.297	.340	.365	.392	.422	.441	.455	.483	.491	.508	.521	.532	.01	
.025	.363	.406	.430	.456	.484	.503	.514	.541	.548	.562	.575	.585	.025	
.05	.430	.471	.493	.518	.544	.562	.572	.595	.603	.617	.629	.637	.05	
.10	.520	.557	.578	.600	.623	.637	.648	.671	.675	.685	.694	.704	.10	
.25	.708	.736	.751	.767	.784	.794	.801	.816	.820	.827	.835	.840	.25	
.50	.989	1.00	1.01	1.01	1.02	1.02	1.02	1.03	1.03	1.03	1.03	1.03	.50	
.75	1.37	1.36	1.35	1.34	1.33	1.33	1.32	1.31	1.31	1.30	1.30	1.29	.75	
.90	1.84	1.79	1.77	1.74	1.71	1.69	1.68	1.65	1.64	1.63	1.62	1.61	.90	
.95	2.20	2.12	2.08	2.04	1.99	1.97	1.95	1.91	1.90	1.88	1.86	1.84	.95	
.975	2.57	2.46	2.41	2.35	2.29	2.25	2.22	2.17	2.16	2.13	2.10	2.09	.975	
.99	3.09	2.94	2.86	2.78	2.69	2.64	2.61	2.54	2.52	2.48	2.44	2.42	.99	
.995	3.50	3.32	3.22	3.12	3.02	2.96	2.92	2.83	2.81	2.76	2.72	2.69	.995	
.999	4.56	4.29	4.15	4.01	3.86	3.77	3.70	3.58	3.54	3.48	3.42	3.38	.999	
.9995	5.07	4.75	4.58	4.42	4.24	4.15	4.07	3.93	3.90	3.82	3.75	3.70	.9995	
.0005	.174	.218	.244	.274	.309	.331	.349	.384	.395	.416	.434	.449	.0005	**24**
.001	.196	.241	.268	.298	.332	.354	.371	.405	.417	.437	.455	.469	.001	
.005	.264	.310	.337	.367	.400	.422	.437	.469	.479	.498	.515	.527	.005	
.01	.304	.350	.376	.405	.437	.459	.473	.505	.513	.529	.546	.558	.01	
.025	.370	.415	.441	.468	.498	.518	.531	.562	.568	.585	.599	.610	.025	
.05	.437	.480	.504	.530	.558	.575	.588	.613	.622	.637	.649	.659	.05	
.10	.527	.566	.588	.611	.635	.651	.662	.685	.691	.704	.715	.723	.10	
.25	.712	.741	.757	.773	.791	.802	.809	.825	.829	.837	.844	.850	.25	
.50	.983	.994	1.00	1.01	1.01	1.02	1.02	1.02	1.02	1.02	1.03	1.03	.50	
.75	1.35	1.33	1.32	1.31	1.30	1.29	1.29	1.28	1.28	1.27	1.27	1.26	.75	
.90	1.78	1.73	1.70	1.67	1.64	1.62	1.61	1.58	1.57	1.56	1.54	1.53	.90	
.95	2.11	2.03	1.98	1.94	1.89	1.86	1.84	1.80	1.79	1.77	1.75	1.73	.95	
.975	2.44	2.33	2.27	2.21	2.15	2.11	2.08	2.02	2.01	1.98	1.95	1.94	.975	
.99	2.89	2.74	2.66	2.58	2.49	2.44	2.40	2.33	2.31	2.27	2.24	2.21	.99	
.995	3.25	3.06	2.97	2.87	2.77	2.70	2.66	2.57	2.55	2.50	2.46	2.43	.995	
.999	4.14	3.87	3.74	3.59	3.45	3.35	3.29	3.16	3.14	3.07	3.01	2.97	.999	
.9995	4.55	4.25	4.09	3.93	3.76	3.66	3.59	3.44	3.41	3.33	3.27	3.22	.9995	

Table A.7 Percentiles of the $F(\nu_1, \nu_2)$ Distribution with Degrees of Freedom ν_1 for the Numerator and ν_2 for the Denominator *(Continued)*

ν_2	Cum. Prop \ ν_1	1	2	3	4	5	6	7	8	9	10	11	12	Cum. Prop.
30	.0005	$.0^{6}40$	$.0^{3}50$	$.0^{2}50$	.015	.030	.047	.065	.082	.098	.114	.129	.143	.0005
	.001	$.0^{5}16$	$.0^{2}10$	$.0^{2}80$	.022	.040	.060	.080	.099	.117	.134	.150	.164	.001
	.005	$.0^{4}40$	$.0^{2}50$	.024	.050	.079	.107	.133	.156	.178	.197	.215	.231	.005
	.01	$.0^{3}16$	.010	.038	.072	.107	.138	.167	.192	.215	.235	.254	.270	.01
	.025	$.0^{2}10$	.025	.071	.118	.161	.197	.229	.257	.281	.302	.321	.337	.025
	.05	$.0^{2}40$	.051	.116	.174	.222	.263	.296	.325	.349	.370	.389	.406	.05
	.10	.016	.106	.193	.262	.315	.357	.391	.420	.443	.464	.481	.497	.10
	.25	.103	.290	.406	.480	.532	.571	.601	.625	.645	.661	.676	.688	.25
	.50	.466	.709	.807	.858	.890	.912	.927	.939	.948	.955	.961	.966	.50
	.75	1.38	1.45	1.44	1.42	1.41	1.39	1.38	1.37	1.36	1.35	1.35	1.34	.75
	.90	2.88	2.49	2.28	2.14	2.05	1.98	1.93	1.88	1.85	1.82	1.79	1.77	.90
	.95	4.17	3.32	2.92	2.69	2.53	2.42	2.33	2.27	2.21	2.16	2.13	2.09	.95
	.975	5.57	4.18	3.59	3.25	3.03	2.87	2.75	2.65	2.57	2.51	2.46	2.41	.975
	.99	7.56	5.39	4.51	4.02	3.70	3.47	3.30	3.17	3.07	2.98	2.91	2.84	.99
	.995	9.18	6.35	5.24	4.62	4.23	3.95	3.74	3.58	3.45	3.34	3.25	3.18	.995
	.999	13.3	8.77	7.05	6.12	5.53	5.12	4.82	4.58	4.39	4.24	4.11	4.00	.999
	.9995	15.2	9.90	7.90	6.82	6.14	5.66	5.31	5.04	4.82	4.65	4.51	4.38	.9995
40	.0005	$.0^{6}40$	$.0^{3}50$	$.0^{2}50$	.016	.030	.048	.066	.084	.100	.117	.132	.147	.0005
	.001	$.0^{5}16$	$.0^{2}10$	$.0^{2}80$	.022	.042	.061	.081	.101	.119	.137	.153	.169	.001
	.005	$.0^{4}40$	$.0^{2}50$	.024	.051	.080	.108	.135	.159	.181	.201	.220	.237	.005
	.01	$.0^{3}16$	.010	.038	.073	.108	.140	.169	.195	.219	.240	.259	.276	.01
	.025	$.0^{3}99$	.025	.071	.119	.162	.199	.232	.260	.285	.307	.327	.344	.025
	.05	$.0^{2}40$	.051	.116	.175	.224	.265	.299	.329	.354	.376	.395	.412	.05
	.10	.016	.106	.194	.263	.317	.360	.394	.424	.448	.469	.488	.504	.10
	.25	.103	.290	.405	.480	.533	.572	.603	.627	.647	.664	.680	.691	.25
	.50	.463	.705	.802	.854	.885	.907	.922	.934	.943	.950	.956	.961	.50
	.75	1.36	1.44	1.42	1.40	1.39	1.37	1.36	1.35	1.34	1.33	1.32	1.31	.75
	.90	2.84	2.44	2.23	2.09	2.00	1.93	1.87	1.83	1.79	1.76	1.73	1.71	.90
	.95	4.08	3.23	2.84	2.61	2.45	2.34	2.25	2.18	2.12	2.08	2.04	2.00	.95
	.975	5.42	4.05	3.46	3.13	2.90	2.74	2.62	2.53	2.45	2.39	2.33	2.29	.975
	.99	7.31	5.18	4.31	3.83	3.51	3.29	3.12	2.99	2.89	2.80	2.73	2.66	.99
	.995	8.83	6.07	4.98	4.37	3.99	3.71	3.51	3.35	3.22	3.12	3.03	2.95	.995
	.999	12.6	8.25	6.60	5.70	5.13	4.73	4.44	4.21	4.02	3.87	3.75	3.64	.999
	.9995	14.4	9.25	7.33	6.30	5.64	5.19	4.85	4.59	4.38	4.21	4.07	3.95	.9995
60	.0005	$.0^{6}40$	$.0^{3}50$	$.0^{2}51$	.016	.031	.048	.067	.085	.103	.120	.136	.152	.0005
	.001	$.0^{5}16$	$.0^{2}10$	$.0^{2}80$	.022	.041	.062	.083	.103	.122	.140	.157	.174	.001
	.005	$.0^{4}40$	$.0^{2}50$	.024	.051	.081	.110	.137	.162	.185	.206	.225	.243	.005
	.01	$.0^{3}16$	.010	.038	.073	.109	.142	.172	.199	.223	.245	.265	.283	.01
	.025	$.0^{3}99$	.025	.071	.120	.163	.202	.235	.264	.290	.313	.333	.351	.025
	.05	$.0^{2}40$	.051	.116	.176	.226	.267	.303	.333	.359	.382	.402	.419	.05
	.10	.016	.106	.194	.264	.318	.362	.398	.428	.453	.475	.493	.510	.10
	.25	.102	.289	.405	.480	.534	.573	.604	.629	.650	.667	.680	.695	.25
	.50	.461	.701	.798	.849	.880	.901	.917	.928	.937	.945	.951	.956	.50
	.75	1.35	1.42	1.41	1.38	1.37	1.35	1.33	1.32	1.31	1.30	1.29	1.29	.75
	.90	2.79	2.39	2.18	2.04	1.95	1.87	1.82	1.77	1.74	1.71	1.68	1.66	.90
	.95	4.00	3.15	2.76	2.53	2.37	2.25	2.17	2.10	2.04	1.99	1.95	1.92	.95
	.975	5.29	3.93	3.34	3.01	2.79	2.63	2.51	2.41	2.33	2.27	2.22	2.17	.975
	.99	7.08	4.98	4.13	3.65	3.34	3.12	2.95	2.82	2.72	2.63	2.56	2.50	.99
	.995	8.49	5.80	4.73	4.14	3.76	3.49	3.29	3.13	3.01	2.90	2.82	2.74	.995
	.999	12.0	7.76	6.17	5.31	4.76	4.37	4.09	3.87	3.69	3.54	3.43	3.31	.999
	.9995	13.6	8.65	6.81	5.82	5.20	4.76	4.44	4.18	3.98	3.82	3.69	3.57	.9995

Table A.7 Percentiles of the $F(\nu_1, \nu_2)$ Distribution with Degrees of Freedom ν_1 for the Numerator and ν_2 for the Denominator *(Continued)*

Cum. Prop. \ ν_1	15	20	24	30	40	50	60	100	120	200	500	∞	Cum. Prop.	ν_2
.0005	.179	.226	.254	.287	.325	.350	.369	.410	.420	.444	.467	.483	.0005	**30**
.001	.202	.250	.278	.311	.348	.373	.391	.431	.442	.465	.488	.503	.001	
.005	.271	.320	.349	.381	.416	.441	.457	.495	.504	.524	.543	.559	.005	
.01	.311	.360	.388	.419	.454	.476	.493	.529	.538	.559	.575	.590	.01	
.025	.378	.426	.453	.482	.515	.535	.551	.585	.592	.610	.625	.639	.025	
.05	.445	.490	.516	.543	.573	.592	.606	.637	.644	.658	.676	.685	.05	
.10	.534	.575	.598	.623	.649	.667	.678	.704	.710	.725	.735	.746	.10	
.25	.716	.746	.763	.780	.798	.810	.818	.835	.839	.848	.856	.862	.25	
.50	.978	.989	.994	1.00	1.01	1.01	1.01	1.02	1.02	1.02	1.02	1.02	.50	
.75	1.32	1.30	1.29	1.28	1.27	1.26	1.26	1.25	1.24	1.24	1.23	1.23	.75	
.90	1.72	1.67	1.64	1.61	1.57	1.55	1.54	1.51	1.50	1.48	1.47	1.46	.90	
.95	2.01	1.93	1.89	1.84	1.79	1.76	1.74	1.70	1.68	1.66	1.64	1.62	.95	
.975	2.31	2.20	2.14	2.07	2.01	1.97	1.94	1.88	1.87	1.84	1.81	1.79	.975	
.99	2.70	2.55	2.47	2.39	2.30	2.25	2.21	2.13	2.11	2.07	2.03	2.01	.99	
.995	3.01	2.82	2.73	2.63	2.52	2.46	2.42	2.32	2.30	2.25	2.21	2.18	.995	
.999	3.75	3.49	3.36	3.22	3.07	2.98	2.92	2.79	2.76	2.69	2.63	2.59	.999	
.9995	4.10	3.80	3.65	3.48	3.32	3.22	3.15	3.00	2.97	2.89	2.82	2.78	.9995	
.0005	.185	.236	.266	.301	.343	.373	.393	.441	.453	.480	.504	.525	.0005	**40**
.001	.209	.259	.290	.326	.367	.396	.415	.461	.473	.500	.524	.545	.001	
.005	.279	.331	.362	.396	.436	.463	.481	.524	.534	.559	.581	.599	.005	
.01	.319	.371	.401	.435	.473	.498	.516	.556	.567	.592	.613	.628	.01	
.025	.387	.437	.466	.498	.533	.556	.573	.610	620	.641	.662	.674	.025	
.05	.454	.502	.529	.558	.591	.613	.627	.658	.669	.685	.704	.717	.05	
.10	.542	.585	.609	.636	.664	.683	.696	.724	.731	.747	.762	.772	.10	
.25	.720	.752	.769	.787	.806	.819	.828	.846	.851	.861	.870	.877	.25	
.50	.972	.983	.989	.994	1.00	1.00	1.01	1.01	1.01	1.01	1.02	1.02	.50	
.75	1.30	1.28	1.26	1.25	1.24	1.23	1.22	1.21	1.21	1.20	1.19	1.19	.75	
.90	1.66	1.61	1.57	1.54	1.51	1.48	1.47	1.43	1.42	1.41	1.39	1.38	.90	
.95	1.92	1.84	1.79	1.74	1.69	1.66	1.64	1.59	1.58	1.55	1.53	1.51	.95	
.975	2.18	2.07	2.01	1.94	1.88	1.83	1.80	1.74	1.72	1.69	1.66	1.64	.975	
.99	2.52	2.37	2.29	2.20	2.11	2.06	2.02	1.94	1.92	1.87	1.83	1.80	.99	
.995	2.78	2.60	2.50	2.40	2.30	2.23	2.18	2.09	2.06	2.01	1.96	1.93	.995	
.999	3.40	3.15	3.01	2.87	2.73	2.64	2.57	2.44	2.41	2.34	2.28	2.23	.999	
.9995	3.68	3.39	3.24	3.08	2.92	2.82	2.74	2.60	2.57	2.49	2.41	2.37	.9995	
.0005	.192	.246	.278	.318	.365	.398	.421	.478	.493	.527	.561	.585	.0005	**60**
.001	.216	.270	.304	.343	.389	.421	.444	.497	.512	.545	.579	.602	.001	
.005	.287	.343	.376	.414	.458	.488	.510	.559	.572	.602	.633	.652	.005	
.01	.328	.383	.416	.453	.495	.524	.545	.592	.604	.633	.658	.679	.01	
.025	.396	.450	.481	.515	.555	.581	.600	.641	.654	.680	.704	.720	.025	
.05	.463	.514	.543	.575	.611	.633	.652	.690	.700	.719	.746	.759	.05	
.10	.550	.596	.622	.650	.682	.703	.717	.750	.758	.776	.793	.806	.10	
.25	.725	.758	.776	.796	.816	.830	.840	.860	.865	.877	.888	.896	.25	
.50	.967	.978	.983	.989	.994	.998	1.00	1.00	1.01	1.01	1.01	1.01	.50	
.75	1.27	1.25	1.24	1.22	1.21	1.20	1.19	1.17	1.17	1.16	1.15	1.15	.75	
.90	1.60	1.54	1.51	1.48	1.44	1.41	1.40	1.36	1.35	1.33	1.31	1.29	.90	
.95	1.84	1.75	1.70	1.65	1.59	1.56	1.53	1.48	1.47	1.44	1.41	1.39	.95	
.975	2.06	1.94	1.88	1.82	1.74	1.70	1.67	1.60	1.58	1.54	1.51	1.48	.975	
.99	2.35	2.20	2.12	2.03	1.94	1.88	1.84	1.75	1.73	1.68	1.63	1.60	.99	
.995	2.57	2.39	2.29	2.19	2.08	2.01	1.96	1.86	1.83	1.78	1.73	1.69	.995	
.999	3.08	2.83	2.69	2.56	2.41	2.31	2.25	2.11	2.09	2.01	1.93	1.89	.999	
.9995	3.30	3.02	2.87	2.71	2.55	2.45	2.38	2.23	2.19	2.11	2.03	1.98	.9995	

Table A.7 Percentiles of the $F(\nu_1, \nu_2)$ Distribution with Degrees of Freedom ν_1 for the Numerator and ν_2 for the Denominator *(Continued)*

ν_2	ν_1 / Cum. Prop.	1	2	3	4	5	6	7	8	9	10	11	12	Cum. Prop.
120	.0005	$.0^{6}40$	$.0^{3}50$	$.0^{2}51$	.016	.031	.049	.067	.087	.105	.123	.140	.156	.0005
	.001	$.0^{5}16$	$.0^{2}10$	$.0^{2}81$	.023	.042	.063	.084	.105	.125	.144	.162	.179	.001
	.005	$.0^{4}39$	$.0^{2}50$	.024	.051	.081	.111	.139	.165	.189	.211	.230	.249	.005
	.01	$.0^{3}16$	.010	.038	.074	.110	.143	.174	.202	.227	.250	.271	.290	.01
	.025	$.0^{3}99$	.025	.072	.120	.165	.204	.238	.268	.295	.318	.340	.359	.025
	.05	$.0^{2}39$	.051	.117	.177	.227	.270	.306	.337	.364	.388	.408	.427	.05
	.10	.016	.105	.194	.265	.320	.365	.401	.432	.458	.480	.500	.518	.10
	.25	.102	.288	.405	.481	.534	.574	.606	.631	.652	.670	.685	.699	.25
	.50	.458	.697	.793	.844	.875	.896	.912	.923	.932	.939	.945	.950	.50
	.75	1.34	1.40	1.39	1.37	1.35	1.33	1.31	1.30	1.29	1.28	1.27	1.26	.75
	.90	2.75	2.35	2.13	1.99	1.90	1.82	1.77	1.72	1.68	1.65	1.62	1.60	.90
	.95	3.92	3.07	2.68	2.45	2.29	2.18	2.09	2.02	1.96	1.91	1.87	1.83	.95
	.975	5.15	3.80	3.23	2.89	2.67	2.52	2.39	2.30	2.22	2.16	2.10	2.05	.975
	.99	6.85	4.79	3.95	3.48	3.17	2.96	2.79	2.66	2.56	2.47	2.40	2.34	.99
	.995	8.18	5.54	4.50	3.92	3.55	3.28	3.09	2.93	2.81	2.71	2.62	2.54	.995
	.999	11.4	7.32	5.79	4.95	4.42	4.04	3.77	3.55	3.38	3.24	3.12	3.02	.999
	.9995	12.8	8.10	6.34	5.39	4.79	4.37	4.07	3.82	3.63	3.47	3.34	3.22	.9995
∞	.0005	$.0^{6}39$	$.0^{3}50$	$.0^{2}51$	.016	.032	.050	.069	.088	.108	.127	.144	.161	.0005
	.001	$.0^{5}16$	$.0^{2}10$	$.0^{2}81$	.023	.042	.063	.085	.107	.128	.148	.167	.185	.001
	.005	$.0^{4}39$	$.0^{2}50$	.024	.052	.082	.113	.141	.168	.193	.216	.236	.256	.005
	.01	$.0^{3}16$	.010	.038	.074	.111	.145	.177	.206	.232	.256	.278	.298	.01
	.025	$.0^{3}98$	.025	.072	.121	.166	.206	.241	.272	.300	.325	.347	.367	.025
	.05	$.0^{2}39$	.051	.117	.178	.229	.273	.310	.342	.369	.394	.417	.436	.05
	.10	.016	.105	.195	.266	.322	.367	.405	.436	.463	.487	.508	.525	.10
	.25	.102	.288	.404	.481	.535	.576	.608	.634	.655	.674	.690	.703	.25
	.50	.455	.693	.789	.839	.870	.891	.907	.918	.927	.934	.939	.945	.50
	.75	1.32	1.39	1.37	1.35	1.33	1.31	1.29	1.28	1.27	1.25	1.24	1.24	.75
	.90	2.71	2.30	2.08	1.94	1.85	1.77	1.72	1.67	1.63	1.60	1.57	1.55	.90
	.95	3.84	3.00	2.60	2.37	2.21	2.10	2.01	1.94	1.88	1.83	1.79	1.75	.95
	.975	5.02	3.69	3.12	2.79	2.57	2.41	2.29	2.19	2.11	2.05	1.99	1.94	.975
	.99	6.63	4.61	3.78	3.32	3.02	2.80	2.64	2.51	2.41	2.32	2.25	2.18	.99
	.995	7.88	5.30	4.28	3.72	3.35	3.09	2.90	2.74	2.62	2.52	2.43	2.36	.995
	.999	10.8	6.91	5.42	4.62	4.10	3.74	3.47	3.27	3.10	2.96	2.84	2.74	.999
	.9995	12.1	7.60	5.91	5.00	4.42	4.02	3.72	3.48	3.30	3.14	3.02	2.90	.9995

For sample sizes larger than, say, 30, a fairly good approximation to the F distribution percentiles can be obtained from

$$\log_{10} F_\alpha(\nu_1, \nu_2) \approx \left(\frac{a}{\sqrt{h - b}}\right) - cg$$

where $h = 2\nu_1\nu_2/(\nu_1 + \nu_2)$, $g = (\nu_2 - \nu_1)/\nu_1\nu_2$, and a, b, c are functions of α given below:

α	.50	.75	.90	.95	.975	.99	.995	.999	.9995
a	0	0.5859	1.1131	1.4287	1.7023	2.0206	2.2373	2.6841	2.8580
b	—	0.58	0.77	0.95	1.14	1.40	1.61	2.09	2.30
c	0.290	0.355	0.527	0.681	0.846	1.073	1.250	1.672	1.857

Table A.7 Percentiles of the $F(\nu_1, \nu_2)$ Distribution with Degrees of Freedom ν_1 for the Numerator and ν_2 for the Denominator (*Continued*)

Cum. Prop. \ ν_1	15	20	24	30	40	50	60	100	120	200	500	∞	Cum. Prop.	ν_2
.0005	.199	.256	.293	.338	.390	.429	.458	.524	.543	.578	.614	.676	.0005	**120**
.001	.223	.282	.319	.363	.415	.453	.480	.542	.568	.595	.631	.691	.001	
.005	.297	.356	.393	.434	.484	.520	.545	.605	.623	.661	.702	.733	.005	
.01	.338	.397	.433	.474	.522	.556	.579	.636	.652	.688	.725	.755	.01	
.025	.406	.464	.498	.536	.580	.611	.633	.684	.698	.729	.762	.789	.025	
.05	.473	.527	.559	.594	.634	.661	.682	.727	.740	.767	.785	.819	.05	
.10	.560	.609	.636	.667	.702	.726	.742	.781	.791	.815	.838	.855	.10	
.25	.730	.765	.784	.805	.828	.843	.853	.877	.884	.897	.911	.923	.25	
.50	.961	.972	.978	.983	.989	.992	.994	1.00	1.00	1.00	1.01	1.01	.50	
.75	1.24	1.22	1.21	1.19	1.18	1.17	1.16	1.14	1.13	1.12	1.11	1.10	.75	
.90	1.55	1.48	1.45	1.41	1.37	1.34	1.32	1.27	1.26	1.24	1.21	1.19	.90	
.95	1.75	1.66	1.61	1.55	1.50	1.46	1.43	1.37	1.35	1.32	1.28	1.25	.95	
.975	1.95	1.82	1.76	1.69	1.61	1.56	1.53	1.45	1.43	1.39	1.34	1.31	.975	
.99	2.19	2.03	1.95	1.86	1.76	1.70	1.66	1.56	1.53	1.48	1.42	1.38	.99	
.995	2.37	2.19	2.09	1.98	1.87	1.80	1.75	1.64	1.61	1.54	1.48	1.43	.995	
.999	2.78	2.53	2.40	2.26	2.11	2.02	1.95	1.82	1.76	1.70	1.62	1.54	.999	
.9995	2.96	2.67	2.53	2.38	2.21	2.11	2.01	1.88	1.84	1.75	1.67	1.60	.9995	
.0005	.207	.270	.311	.360	.422	.469	.505	.599	.624	.704	.804	1.00	.0005	∞
.001	.232	.296	.338	.386	.448	.493	.527	.617	.649	.719	.819	1.00	.001	
.005	.307	.372	.412	.460	.518	.559	.592	.671	.699	.762	.843	1.00	.005	
.01	.349	.413	.452	.499	.554	.595	.625	.699	.724	.782	.858	1.00	.01	
.025	.418	.480	.517	.560	.611	.645	.675	.741	.763	.813	.878	1.00	.025	
.05	.484	.543	.577	.617	.663	.694	.720	.781	.797	.840	.896	1.00	.05	
.10	.570	.622	.652	.687	.726	.752	.774	.826	.838	.877	.919	1.00	.10	
.25	.736	.773	.793	.816	.842	.860	.872	.901	.910	.932	.957	1.00	.25	
.50	.956	.967	.972	.978	.983	.987	.989	.993	.994	.997	.999	1.00	.50	
.75	1.22	1.19	1.18	1.16	1.14	1.13	1.12	1.09	1.08	1.07	1.04	1.00	.75	
.90	1.49	1.42	1.38	1.34	1.30	1.26	1.24	1.18	1.17	1.13	1.08	1.00	.90	
.95	1.67	1.57	1.52	1.46	1.39	1.35	1.32	1.24	1.22	1.17	1.11	1.00	.95	
.975	1.83	1.71	1.64	1.57	1.48	1.43	1.39	1.30	1.27	1.21	1.13	1.00	.975	
.99	2.04	1.88	1.79	1.70	1.59	1.52	1.47	1.36	1.32	1.25	1.15	1.00	.99	
.995	2.19	2.00	1.90	1.79	1.67	1.59	1.53	1.40	1.36	1.28	1.17	1.00	.995	
.999	2.51	2.27	2.13	1.99	1.84	1.73	1.66	1.49	1.45	1.34	1.21	1.00	.999	
.9995	2.65	2.37	2.22	2.07	1.91	1.79	1.71	1.53	1.48	1.36	1.22	1.00	.9995	

The values given in this table are abstracted with permission from the following sources:

1. All values for ν_1, ν_2 equal to 50, 100, 200, 500 are from A. Hald, *Statistical Tables and Formulas*, John Wiley & Sons, Inc., New York, 1952.
2. For cumulative proportions .5, .75, .9, .95, .975, .99, .995 most of the values are from M. Merrington and C. M. Thompson, *Biometrika*, vol. 33 (1943), p. 73.
3. For cumulative proportions .999 the values are from C. Colcord and L. S. Deming, *Sankhyā*, vol. 2 (1936), p. 423.
4. For cum. prop. $= \alpha < .5$ the values are the reciprocals of values for $1 - \alpha$ (with ν_1 and ν_2 interchanged). The values in Merrington and Thompson and in Colcord and Deming are to five significant figures, and it is hoped (but not expected) that the reciprocals are correct as given. The values in Hald are to three significant figures, and the reciprocals are probably accurate within one to two digits in the third significant figure except for those values very close to unity, where they may be off four to five digits in the third significant figure.
5. Gaps remaining in the table after using the above sources were filled in by interpolation.

$$\alpha = \frac{(\nu_1/\nu_2)^{\frac{1}{2}\nu_1}}{\beta(\frac{1}{2}\nu_1, \frac{1}{2}\nu_2)} \int_{-\infty}^{F_\alpha} F^{\frac{1}{2}\nu_1 - 1} \left(1 + \frac{\nu_1 F}{\nu_2}\right)^{-(\nu_1+\nu_2)/2} dF$$

Table A.8 Tolerance Factors for Normal Distributions (One-sided)*

Factors K such that the probability is γ that at least a proportion $1 - \alpha$ of the distribution will be less than $\overline{X} + Ks$ (or greater than $\overline{X} - Ks$), where $\overline{X}$ and s are estimates of the mean and the standard deviation computed from a sample of size n.

n \ α	$\gamma = 0.75$					$\gamma = 0.90$				
	0.25	0.10	0.05	0.01	0.001	0.25	0.10	0.05	0.01	0.001
3	1.464	2.501	3.152	4.396	5.805	2.602	4.258	5.310	7.340	9.651
4	1.256	2.134	2.680	3.726	4.910	1.972	3.187	3.957	5.437	7.128
5	1.152	1.961	2.463	3.421	4.507	1.698	2.742	3.400	4.666	6.112
6	1.087	1.860	2.336	3.243	4.273	1.540	2.494	3.091	4.242	5.556
7	1.043	1.791	2.250	3.126	4.118	1.435	2.333	2.894	3.972	5.201
8	1.010	1.740	2.190	3.042	4.008	1.360	2.219	2.755	3.783	4.955
9	0.984	1.702	2.141	2.977	3.924	1.302	2.133	2.649	3.641	4.772
10	0.964	1.671	2.103	2.927	3.858	1.257	2.065	2.568	3.532	4.629
11	0.947	1.646	2.073	2.885	3.804	1.219	2.012	2.503	3.444	4.515
12	0.933	1.624	2.048	2.851	3.760	1.188	1.966	2.448	3.371	4.420
13	0.919	1.606	2.026	2.822	3.722	1.162	1.928	2.403	3.310	4.341
14	0.909	1.591	2.007	2.796	3.690	1.139	1.895	2.363	3.257	4.274
15	0.899	1.577	1.991	2.776	3.661	1.119	1.866	2.329	3.212	4.215
16	0.891	1.566	1.977	2.756	3.637	1.101	1.842	2.299	3.172	4.164
17	0.883	1.554	1.964	2.739	3.615	1.085	1.820	2.272	3.136	4.118
18	0.876	1.544	1.951	2.723	3.595	1.071	1.800	2.249	3.106	4.078
19	0.870	1.536	1.942	2.710	3.577	1.058	1.781	2.228	3.078	4.041
20	0.865	1.528	1.933	2.697	3.561	1.046	1.765	2.208	3.052	4.009
21	0.859	1.520	1.923	2.686	3.545	1.035	1.750	2.190	3.028	3.979
22	0.854	1.514	1.916	2.675	3.532	1.025	1.736	2.174	3.007	3.952
23	0.849	1.508	1.907	2.665	3.520	1.016	1.724	2.159	2.987	3.927
24	0.845	1.502	1.901	2.656	3.509	1.007	1.712	2.145	2.969	3.904
25	0.842	1.496	1.895	2.647	3.497	0.999	1.702	2.132	2.952	3.882
30	0.825	1.475	1.869	2.613	3.454	0.966	1.657	2.080	2.884	3.794
35	0.812	1.458	1.849	2.588	3.421	0.942	1.623	2.041	2.833	3.730
40	0.803	1.445	1.834	2.568	3.395	0.923	1.598	2.010	2.793	3.679
45	0.795	1.435	1.821	2.552	3.375	0.908	1.577	1.986	2.762	3.638
50	0.788	1.426	1.811	2.538	3.358	0.894	1.560	1.965	2.735	3.604

n \ α	$\gamma = 0.95$					$\gamma = 0.99$				
	0.25	0.10	0.05	0.01	0.001	0.25	0.10	0.05	0.01	0.001
3	3.804	6.158	7.655	10.552	13.857					
4	2.619	4.163	5.145	7.042	9.215					
5	2.149	3.407	4.202	5.741	7.501					
6	1.895	3.006	3.707	5.062	6.612	2.849	4.408	5.409	7.334	9.540
7	1.732	2.755	3.399	4.641	6.061	2.490	3.856	4.730	6.411	8.348
8	1.617	2.582	3.188	4.353	5.686	2.252	3.496	4.287	5.811	7.566
9	1.532	2.454	3.031	4.143	5.414	2.085	3.242	3.971	5.389	7.014
10	1.465	2.355	2.911	3.981	5.203	1.954	3.048	3.739	5.075	6.603
11	1.411	2.275	2.815	3.852	5.036	1.854	2.897	3.557	4.828	6.284
12	1.366	2.210	2.736	3.747	4.900	1.771	2.773	3.410	4.633	6.032
13	1.329	2.155	2.670	3.659	4.787	1.702	2.677	3.290	4.472	5.826
14	1.296	2.108	2.614	3.585	4.690	1.645	2.592	3.189	4.336	5.651
15	1.268	2.068	2.566	3.520	4.607	1.596	2.521	3.102	4.224	5.507
16	1.242	2.032	2.523	3.463	4.534	1.553	2.458	3.028	4.124	5.374
17	1.220	2.001	2.486	3.415	4.471	1.514	2.405	2.962	4.038	5.268
18	1.200	1.974	2.453	3.370	4.415	1.481	2.357	2.906	3.961	5.167
19	1.183	1.949	2.423	3.331	4.364	1.450	2.315	2.855	3.893	5.078
20	1.167	1.926	2.396	3.295	4.319	1.424	2.275	2.807	3.832	5.003
21	1.152	1.905	2.371	3.262	4.276	1.397	2.241	2.768	3.776	4.932
22	1.138	1.887	2.350	3.233	4.238	1.376	2.208	2.729	3.727	4.866
23	1.126	1.869	2.329	3.206	4.204	1.355	2.179	2.693	3.680	4.806
24	1.114	1.853	2.309	3.181	4.171	1.336	2.154	2.663	3.638	4.755
25	1.103	1.838	2.292	3.158	4.143	1.319	2.129	2.632	3.601	4.706
30	1.059	1.778	2.220	3.064	4.022	1.249	2.029	2.516	3.446	4.508
35	1.025	1.732	2.166	2.994	3.934	1.195	1.957	2.431	3.334	4.364
40	0.999	1.697	2.126	2.941	3.866	1.154	1.902	2.365	3.250	4.255
45	0.978	1.669	2.092	2.897	3.811	1.122	1.857	2.313	3.181	4.168
50	0.961	1.646	2.065	2.863	3.766	1.096	1.821	2.296	3.124	4.096

* Reproduced from "Tables for One-sided Statistical Tolerance Limits," by Gerald J. Lieberman, *Industrial Quality Control*, vol. XIV, no. 10, p. 8, April, 1958, with the permission of the author and journal.

Table A.9 Tolerance Factors for Normal Distributions (Two-sided)*

N \ P	$\gamma = 0.75$					$\gamma = 0.90$					$\gamma = 0.95$					$\gamma = 0.99$				
	0.75	0.90	0.95	0.99	0.999	0.75	0.90	0.95	0.99	0.999	0.75	0.90	0.95	0.99	0.999	0.75	0.90	0.95	0.99	0.999
2	4.498	6.301	7.414	9.531	11.920	11.407	15.978	18.800	24.167	30.227	22.858	32.019	37.674	48.430	60.573	114.363	160.193	188.491	242.300	303.054
3	2.501	3.538	4.187	5.431	6.844	4.132	5.847	6.919	8.974	11.309	5.922	8.380	9.916	12.861	16.208	13.378	18.930	22.401	29.055	36.616
4	2.035	2.892	3.431	4.471	5.657	2.932	4.166	4.943	6.440	8.149	3.779	5.369	6.370	8.299	10.502	6.614	9.398	11.150	14.527	18.383
5	1.825	2.599	3.088	4.033	5.117	2.454	3.494	4.152	5.423	6.879	3.002	4.275	5.079	6.634	8.415	4.643	6.612	7.855	10.260	13.015
6	1.704	2.429	2.889	3.779	4.802	2.196	3.131	3.723	4.870	6.188	2.604	3.712	4.414	5.775	7.337	3.743	5.337	6.345	8.301	10.548
7	1.624	2.318	2.757	3.611	4.593	2.034	2.902	3.452	4.521	5.750	2.361	3.369	4.007	5.248	6.676	3.233	4.613	5.488	7.187	9.142
8	1.568	2.238	2.663	3.491	4.444	1.921	2.743	3.264	4.278	5.446	2.197	3.136	3.732	4.891	6.226	2.905	4.147	4.936	6.468	8.234
9	1.525	2.178	2.593	3.400	4.330	1.839	2.626	3.125	4.098	5.220	2.078	2.967	3.532	4.631	5.899	2.677	3.822	4.550	5.966	7.600
10	1.492	2.131	2.537	3.328	4.241	1.775	2.535	3.018	3.959	5.046	1.987	2.839	3.379	4.433	5.649	2.508	3.582	4.265	5.594	7.129
11	1.465	2.093	2.493	3.271	4.169	1.724	2.463	2.933	3.849	4.906	1.916	2.737	3.259	4.277	5.452	2.378	3.397	4.045	5.308	6.766
12	1.443	2.062	2.456	3.223	4.110	1.683	2.404	2.863	3.758	4.792	1.858	2.655	3.162	4.150	5.291	2.274	3.250	3.870	5.079	6.477
13	1.425	2.036	2.424	3.183	4.059	1.648	2.355	2.805	3.682	4.697	1.810	2.587	3.081	4.044	5.158	2.190	3.130	3.727	4.893	6.240
14	1.409	2.013	2.398	3.148	4.016	1.619	2.314	2.756	3.618	4.615	1.770	2.529	3.012	3.955	5.045	2.120	3.029	3.608	4.737	6.043
15	1.395	1.994	2.375	3.118	3.979	1.594	2.278	2.713	3.562	4.545	1.735	2.480	2.954	3.878	4.949	2.060	2.945	3.507	4.605	5.876
16	1.383	1.977	2.355	3.092	3.946	1.572	2.246	2.676	3.514	4.484	1.705	2.437	2.903	3.812	4.865	2.009	2.872	3.421	4.492	5.732
17	1.372	1.962	2.337	3.069	3.917	1.552	2.219	2.643	3.471	4.430	1.679	2.400	2.858	3.754	4.791	1.965	2.808	3.345	4.393	5.607
18	1.363	1.948	2.321	3.048	3.891	1.535	2.194	2.614	3.433	4.382	1.655	2.366	2.819	3.702	4.725	1.926	2.753	3.279	4.307	5.497
19	1.355	1.936	2.307	3.030	3.867	1.520	2.172	2.588	3.399	4.339	1.635	2.337	2.784	3.656	4.667	1.891	2.703	3.221	4.230	5.399
20	1.347	1.925	2.294	3.013	3.846	1.506	2.152	2.564	3.368	4.300	1.616	2.310	2.752	3.615	4.614	1.860	2.659	3.168	4.161	5.312
21	1.340	1.915	2.282	2.998	3.827	1.493	2.135	2.543	3.340	4.264	1.599	2.286	2.723	3.577	4.567	1.833	2.620	3.121	4.100	5.234
22	1.334	1.906	2.271	2.984	3.809	1.482	2.118	2.524	3.315	4.232	1.584	2.264	2.697	3.543	4.523	1.808	2.584	3.078	4.044	5.163
23	1.328	1.898	2.261	2.971	3.793	1.471	2.103	2.506	3.292	4.203	1.570	2.244	2.673	3.512	4.484	1.785	2.551	3.040	3.993	5.098
24	1.322	1.891	2.252	2.950	3.778	1.462	2.089	2.480	3.270	4.176	1.557	2.225	2.651	3.483	4.447	1.764	2.522	3.004	3.947	5.039
25	1.317	1.883	2.244	2.948	3.764	1.453	2.077	2.474	3.251	4.151	1.545	2.208	2.631	3.457	4.413	1.745	2.494	2.972	3.904	4.985
26	1.313	1.877	2.236	2.938	3.751	1.444	2.065	2.460	3.232	4.127	1.534	2.193	2.612	3.432	4.382	1.727	2.460	2.941	3.865	4.935
27	1.309	1.871	2.229	2.929	3.740	1.437	2.054	2.447	3.215	4.106	1.523	2.178	2.595	3.409	4.353	1.711	2.446	2.914	3.828	4.888
30	1.297	1.855	2.210	2.904	3.708	1.417	2.025	2.413	3.170	4.049	1.497	2.140	2.549	3.350	4.278	1.668	2.385	2.841	3.733	4.768
35	1.283	1.834	2.185	2.871	3.667	1.390	1.988	2.368	3.112	3.974	1.462	2.090	2.490	3.272	4.179	1.613	2.306	2.748	3.611	4.611
40	1.271	1.818	2.166	2.846	3.635	1.370	1.959	2.334	3.066	3.917	1.435	2.052	2.445	3.213	4.104	1.571	2.247	2.677	3.518	4.493
45	1.262	1.805	2.150	2.826	3.609	1.354	1.935	2.306	3.030	3.871	1.414	2.021	2.408	3.165	4.042	1.539	2.200	2.621	3.444	4.399
50	1.255	1.794	2.138	2.809	3.588	1.340	1.916	2.284	3.001	3.833	1.396	1.996	2.379	3.126	3.993	1.512	2.162	2.576	3.385	4.323
55	1.249	1.785	2.127	2.795	3.571	1.329	1.901	2.265	2.976	3.801	1.382	1.976	2.354	3.094	3.951	1.490	2.130	2.538	3.335	4.260
60	1.243	1.778	2.118	2.784	3.556	1.320	1.887	2.248	2.955	3.774	1.369	1.958	2.333	3.066	3.916	1.471	2.103	2.506	3.293	4.206

Table A.9 Tolerance Factors for Normal Distributions (Two-sided)* (*Continued*)

N \ P	$\gamma = 0.75$					$\gamma = 0.90$					$\gamma = 0.95$					$\gamma = 0.99$				
	0.75	0.90	0.95	0.99	0.999	0.75	0.90	0.95	0.99	0.999	0.75	0.90	0.95	0.99	0.999	0.75	0.90	0.95	0.99	0.999
65	1.239	1.771	2.110	2.773	3.543	1.312	1.875	2.235	2.937	3.751	1.359	1.943	2.315	3.042	3.886	1.455	2.080	2.478	3.257	4.160
70	1.235	1.765	2.104	2.764	3.531	1.304	1.865	2.222	2.920	3.730	1.349	1.929	2.299	3.021	3.859	1.440	2.060	2.454	3.225	4.120
75	1.231	1.760	2.098	2.757	3.521	1.298	1.856	2.211	2.906	3.712	1.341	1.917	2.285	3.002	3.835	1.428	2.042	2.433	3.197	4.084
80	1.228	1.756	2.092	2.749	3.512	1.292	1.848	2.202	2.894	3.696	1.334	1.907	2.272	2.986	3.814	1.417	2.026	2.414	3.173	4.053
85	1.225	1.752	2.087	2.743	3.504	1.287	1.841	2.193	2.882	3.682	1.327	1.897	2.261	2.971	3.795	1.407	2.012	2.397	3.150	4.024
90	1.223	1.748	2.083	2.737	3.497	1.283	1.834	2.185	2.872	3.669	1.321	1.889	2.251	2.958	3.778	1.398	1.999	2.382	3.130	3.999
95	1.220	1.745	2.079	2.732	3.490	1.278	1.828	2.178	2.863	3.657	1.315	1.881	2.241	2.945	3.763	1.390	1.987	2.368	3.112	3.976
100	1.218	1.742	2.075	2.727	3.484	1.275	1.822	2.172	2.854	3.646	1.311	1.874	2.233	2.934	3.748	1.383	1.977	2.355	3.096	3.954
110	1.214	1.736	2.069	2.719	3.473	1.268	1.813	2.160	2.839	3.626	1.302	1.861	2.218	2.915	3.723	1.369	1.958	2.333	3.066	3.917
120	1.211	1.732	2.063	2.712	3.464	1.262	1.804	2.150	2.826	3.610	1.294	1.850	2.205	2.898	3.702	1.358	1.942	2.314	3.041	3.885
130	1.208	1.728	2.059	2.705	3.456	1.257	1.797	2.141	2.814	3.595	1.288	1.841	2.194	2.883	3.683	1.349	1.928	2.298	3.019	3.857
140	1.206	1.724	2.054	2.700	3.449	1.252	1.791	2.134	2.804	3.582	1.282	1.833	2.184	2.870	3.666	1.340	1.916	2.283	3.000	3.833
150	1.204	1.721	2.051	2.695	3.443	1.248	1.785	2.127	2.795	3.571	1.277	1.825	2.175	2.859	3.652	1.332	1.905	2.270	2.983	3.811
160	1.202	1.718	2.047	2.691	3.437	1.245	1.780	2.121	2.787	3.561	1.272	1.819	2.167	2.848	3.638	1.326	1.896	2.259	2.968	3.792
170	1.200	1.716	2.044	2.687	3.432	1.242	1.775	2.116	2.780	3.552	1.268	1.813	2.160	2.839	3.627	1.320	1.887	2.248	2.955	3.774
180	1.198	1.713	2.042	2.683	3.427	1.239	1.771	2.111	2.774	3.543	1.264	1.808	2.154	2.831	3.616	1.314	1.879	2.239	2.942	3.759
190	1.197	1.711	2.039	2.680	3.423	1.236	1.767	2.106	2.768	3.536	1.261	1.803	2.148	2.823	3.606	1.309	1.872	2.230	2.931	3.744
200	1.195	1.709	2.037	2.677	3.419	1.234	1.764	2.102	2.762	3.529	1.258	1.798	2.143	2.816	3.597	1.304	1.865	2.222	2.921	3.731
250	1.190	1.702	2.028	2.665	3.404	1.224	1.750	2.085	2.740	3.501	1.245	1.780	2.121	2.788	3.561	1.286	1.839	2.191	2.880	3.678
300	1.186	1.696	2.021	2.656	3.393	1.217	1.740	2.073	2.725	3.481	1.236	1.767	2.106	2.767	3.535	1.273	1.820	2.169	2.850	3.641
400	1.181	1.688	2.012	2.644	3.378	1.207	1.726	2.057	2.703	3.452	1.223	1.749	2.084	2.739	3.499	1.255	1.794	2.138	2.809	3.589
500	1.177	1.683	2.006	2.636	3.368	1.201	1.717	2.046	2.689	3.434	1.215	1.737	2.070	2.721	3.475	1.243	1.777	2.117	2.783	3.555
600	1.175	1.680	2.002	2.631	3.360	1.196	1.710	2.038	2.678	3.421	1.209	1.729	2.060	2.707	3.458	1.234	1.764	2.102	2.763	3.530
700	1.173	1.677	1.998	2.626	3.355	1.192	1.705	2.032	2.670	3.411	1.204	1.722	2.052	2.697	3.445	1.227	1.755	2.091	2.748	3.511
800	1.171	1.675	1.996	2.623	3.350	1.189	1.701	2.027	2.663	3.402	1.201	1.717	2.046	2.688	3.434	1.222	1.747	2.082	2.736	3.495
900	1.170	1.673	1.993	2.620	3.347	1.187	1.697	2.023	2.658	3.396	1.198	1.712	2.040	2.682	3.426	1.218	1.741	2.075	2.726	3.483
1000	1.169	1.671	1.992	2.617	3.344	1.185	1.695	2.019	2.654	3.390	1.195	1.709	2.036	2.676	3.418	1.214	1.736	2.068	2.718	3.472
∞	1.150	1.645	1.960	2.576	3.291	1.150	1.645	1.960	2.576	3.291	1.150	1.645	1.960	2.576	3.291	1.150	1.645	1.960	2.576	3.291

Table A.10 Some Useful Numerical Constants

$\pi = 3.14159\ 26536$	$\log_{10}2 = 0.30102\ 99957$
$\log_{10}\pi = 0.49714\ 98727$	$\ln_e 2 = 0.69314\ 71806$
$\ln_e \pi = 1.14472\ 98858$	$e = 2.71828\ 18285$
$1/\pi = 0.31830\ 98862$	$\sqrt{e} = 1.64872\ 12707$
$\pi^2 = 9.86960\ 44010$	$\log_{10}e = 0.43429\ 44819$
$\sqrt{\pi} = 1.77245\ 38509$	$\ln_e 10 = 2.30258\ 50930$

Table A.11 Some Powers of the First Twenty-five Integers

n	n^2	n^3	$\sqrt{n}$	$1/\sqrt{n}$	$1/\sqrt{n(n-1)}$
2	4	8	1.414214	0.707107	0.707107
3	9	27	1.732051	0.577350	0.408248
4	16	64	2.000000	0.500000	0.288675
5	25	125	2.236068	0.447214	0.223607
6	36	216	2.449490	0.408248	0.182574
7	49	343	2.645751	0.377964	0.154303
8	64	512	2.828427	0.353553	0.133631
9	81	729	3.000000	0.333333	0.117851
10	100	1,000	3.162278	0.316228	0.105409
11	121	1,331	3.316625	0.301511	0.095346
12	144	1,728	3.464102	0.288678	0.087039
13	169	2,197	3.605551	0.277350	0.080064
14	196	2,744	3.741657	0.267261	0.074125
15	225	3,375	3.872983	0.258199	0.069007
16	256	4,096	4.000000	0.250000	0.064550
17	289	4,913	4.123106	0.242536	0.060634
18	324	5,832	4.242641	0.235702	0.057166
19	361	6,859	4.358899	0.229416	0.054074
20	400	8,000	4.472136	0.223607	0.051299
21	441	9,261	4.582576	0.218218	0.048795
22	484	10,648	4.690416	0.213201	0.046524
23	529	12,167	4.795832	0.208514	0.044455
24	576	13,824	4.898979	0.204124	0.042563
25	625	15,625	5.000000	0.200000	0.040825

Table A.12 n **Factorial for** $n = 1(1)15$

n	$n!$
1	1
2	2
3	6
4	24
5	120
6	720
7	5,040
8	40,320
9	362,880
10	3,628,800
11	39,916,800
12	479,001,600
13	6,227,020,800
14	87,178,291,200
15	130,767,436,800

Table A.13 Useful Identities and Series

$$(a+b)^n = \sum_{i=0}^{n} C(n,i)a^{n-i}b^i$$

$$(1+X)^n = \sum_{i=0}^{n} C(n,i)X^i$$

$$\frac{(1-X)^n}{1-X} = \sum_{i=0}^{n-1} X^i$$

$$\frac{1}{(1-X)^n} = \sum_{i=0}^{\infty} C(n+i-1,\, i)X^i$$

$$\frac{X}{(1-X)^2} = \sum_{i=1}^{n} iX^i \qquad |X| < 1$$

$$e^X = \sum_{i=0}^{\infty} \frac{X^i}{i!}$$

$$\sum_{i=1}^{n} i = n(n+1)/2$$

$$\sum_{i=1}^{n} (2i-1) = n$$

$$\sum_{i=1}^{n} i^2 = n(n+1)(2n+1)/6$$

$$\sum_{i=1}^{n} i^3 = n^2(n+1)^2/4 = \left(\sum_{i=1}^{n} i\right)^2$$

$$\sum_{i=1}^{n} i^4 = \sum_{i=1}^{n} i^2 \left[6 \sum_{i=1}^{n} i - 1\right]/5$$

Table A.14 Values of e^{-X}

X	0	1	2	3	4	5	6	7	8	9
0.00	1.00000	0.99900	0.99800	0.99700	0.99600	0.99501	0.99401	0.99302	0.99203	0.99104
0.01	0.99004	0.98906	0.98807	0.98708	0.98609	0.98511	0.98412	0.98314	0.98216	0.98117
0.02	0.98019	0.97921	0.97824	0.97726	0.97628	0.97350	0.97433	0.97336	0.97238	0.97141
0.03	0.97044	0.96947	0.96850	0.96753	0.96657	0.96560	0.96464	0.96367	0.96271	0.96175
0.04	0.96078	0.95982	0.95886	0.95791	0.95695	0.95599	0.95504	0.95408	0.95313	0.95218
0.05	0.95122	0.95027	0.94932	0.94838	0.94743	0.94648	0.94553	0.94459	0.94364	0.94270
0.06	0.94176	0.94082	0.93988	0.93894	0.93800	0.93706	0.93613	0.93519	0.93426	0.93332
0.07	0.93239	0.93146	0.93053	0.92960	0.92867	0.92774	0.92681	0.92588	0.92496	0.92404
0.08	0.92312	0.92219	0.92127	0.92035	0.91943	0.91851	0.91759	0.91668	0.91576	0.91485
0.09	0.91393	0.91302	0.91211	0.91119	0.91028	0.90937	0.90846	0.90755	0.90665	0.90574
0.10	0.90484	0.90393	0.90302	0.90212	0.90122	0.90032	0.89942	0.89853	0.89763	0.89673

X	1	2	3	4	5	6	7	8	9	10
e^{-X}	0.36788	0.13534	0.04979	0.01832	0.00674	0.00248	0.00091	0.00034	0.00012	0.00005

Note. For other values of e^{-x} use the relationship $e^{-(x_1+x_2)} = e^{-x_1}e^{-x_2}$.
Example. $e^{-0.24} = e^{-0.10}e^{-0.10}e^{-0.04} = (0.90484)^2(0.96078) = 0.7866$

APPENDIX B: CHARTS

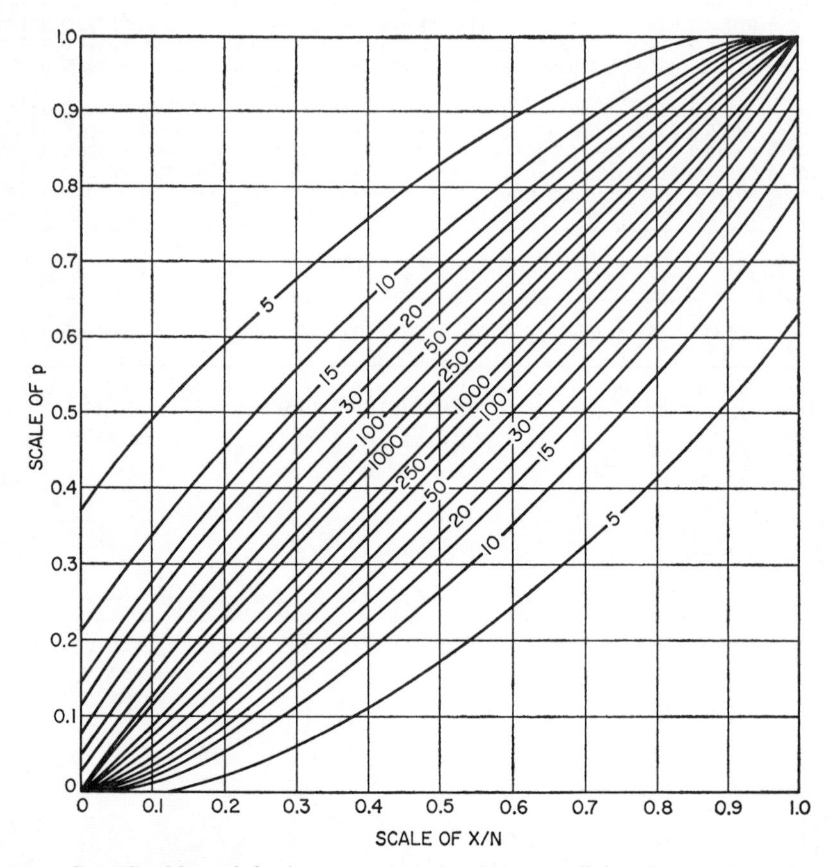

CHART B.1 Confidence belts for proportions (confidence coefficient 0.80). (*From Introduction to Statistical Analysis, 2d ed., by W. J. Dixon and F. J. Massey, Jr. Copyright, 1957. McGraw-Hill Book Company. Used by permission.*)

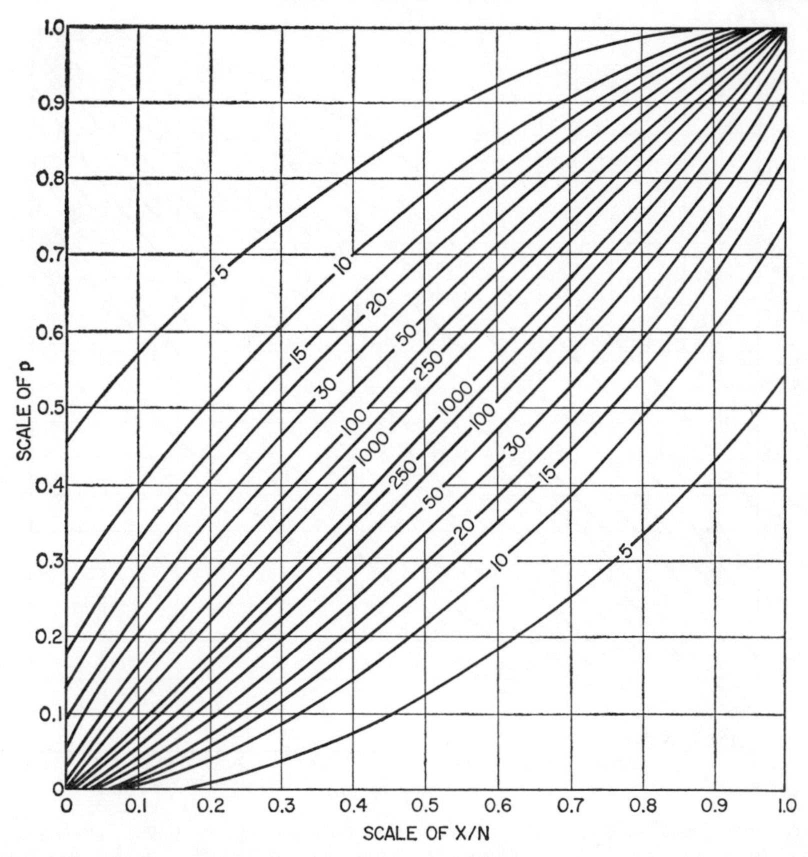

CHART B.2 Confidence belts for proportions (confidence coefficient 0.90). (*From Introduction to Statistical Analysis, 2d ed., by W. J. Dixon and F. J. Massey, Jr. Copyright, 1957. McGraw-Hill Book Company. Used by permission.*)

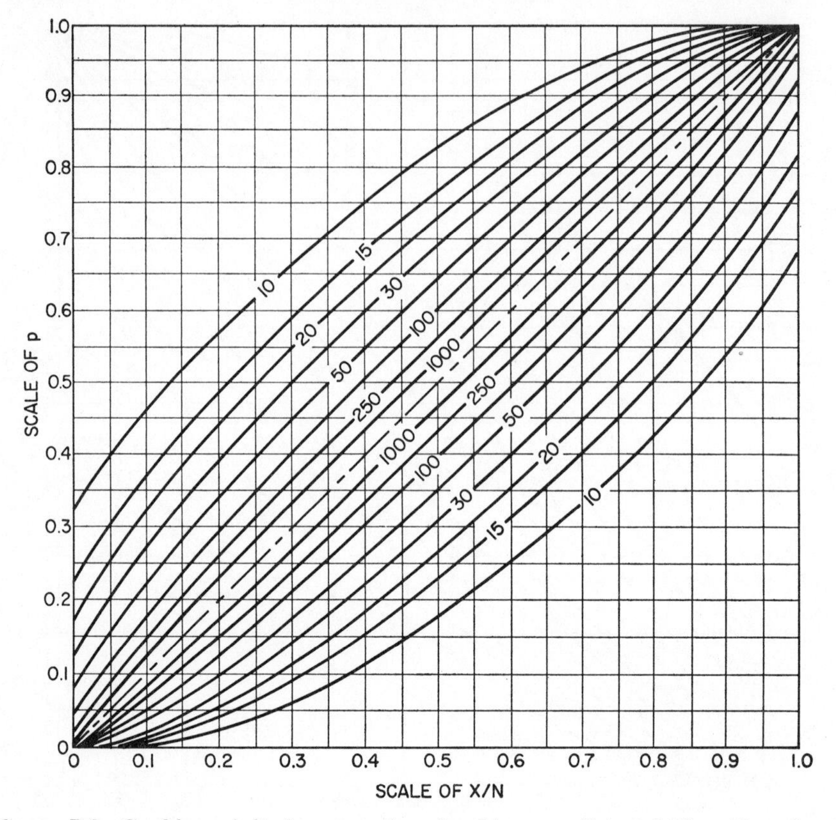

CHART B.3 Confidence belts for proportions (confidence coefficient 0.95). (*Reproduced from "The Use of Confidence or Fiducial Limits Illustrated in the Case of the Binomial," by Prof. E. S. Pearson and C. J. Clopper, with permission from Biometrika, vol.* 26, *p.* 404, 1934. *Also reproduced from Introduction to Statistical Analysis,* 2d *ed., by W. J. Dixon and F. J. Massey, Jr., copyright,* 1957, *by permission of McGraw-Hill Book Company.*)

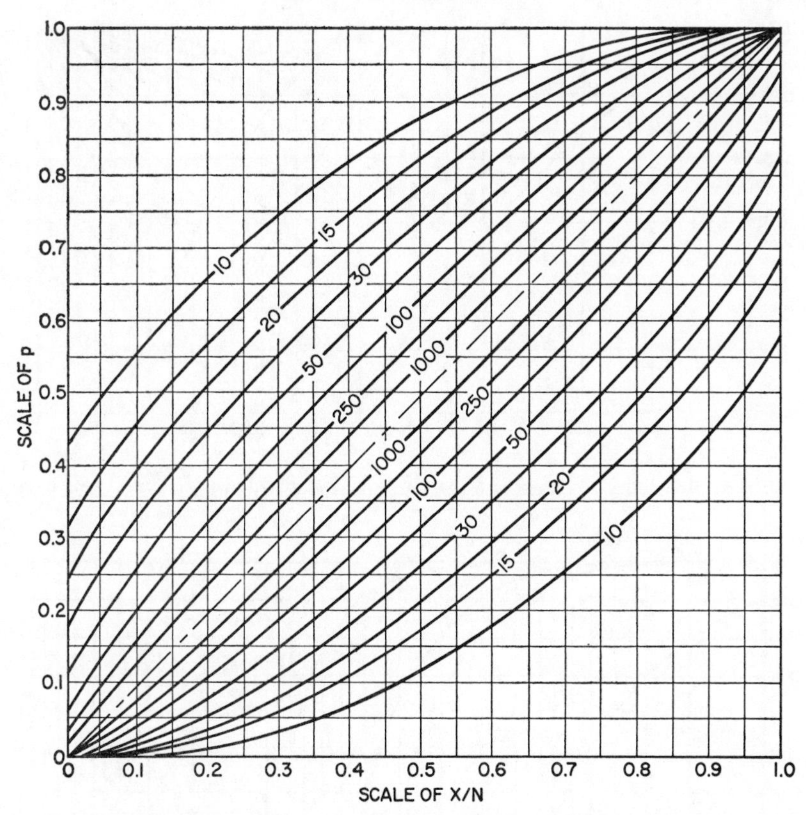

CHART B.4 Confidence belts for proportions (confidence coefficient 0.99). (*Reproduced from "The Use of Confidence or Fiducial Limits Illustrated in the Case of the Binomial," by Prof. E. S. Pearson and C. J. Clopper, with permission from Biometrika, vol.* 26, *p.* 404, 1934. *Also reproduced from Introduction to Statistical Analysis, 2d ed., by W. J. Dixon and F. J. Massey, Jr., copyright,* 1957, *by permission of McGraw-Hill Book Company.*)

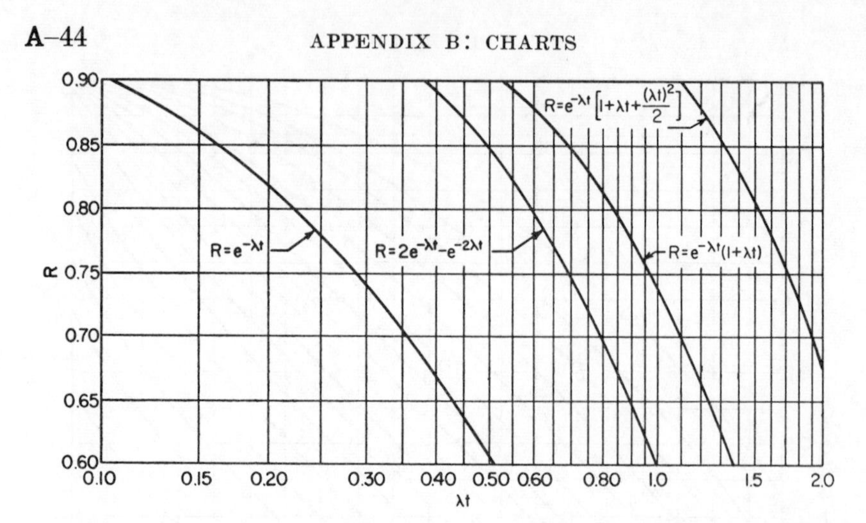

CHART B.5 Reliability R versus λt for the exponential distribution ($0.60 \leq R \leq 0.90$).

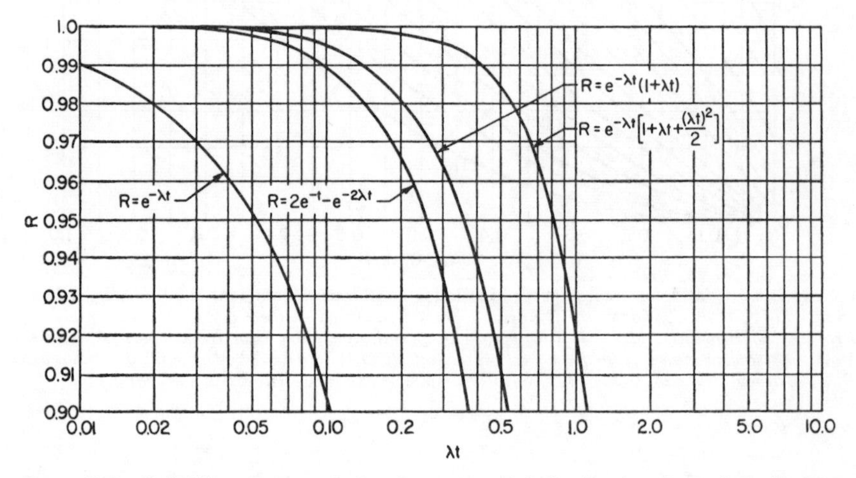

CHART B.6 Reliability R versus λt for the exponential distribution ($0.90 \leq R \leq 1.00$).

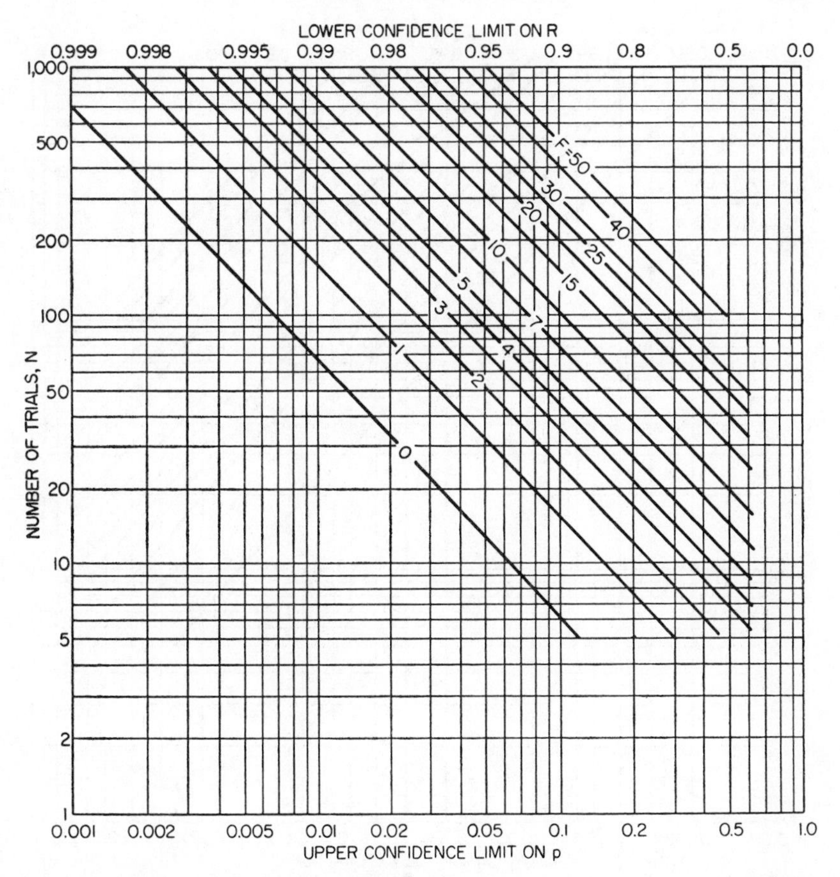

CHART B.7 Confidence limits on reliability. Upper confidence limit on unreliability (1 minus lower confidence limit on reliability), number of trials N, observed failures F, confidence coefficient $\gamma = 0.50$. (*David K. Lloyd and Myron Lipow, Reliability: Management, Methods and Mathematics.* © 1962, *by permission of Prentice-Hall, Inc., Englewood Cliffs, New Jersey.*)

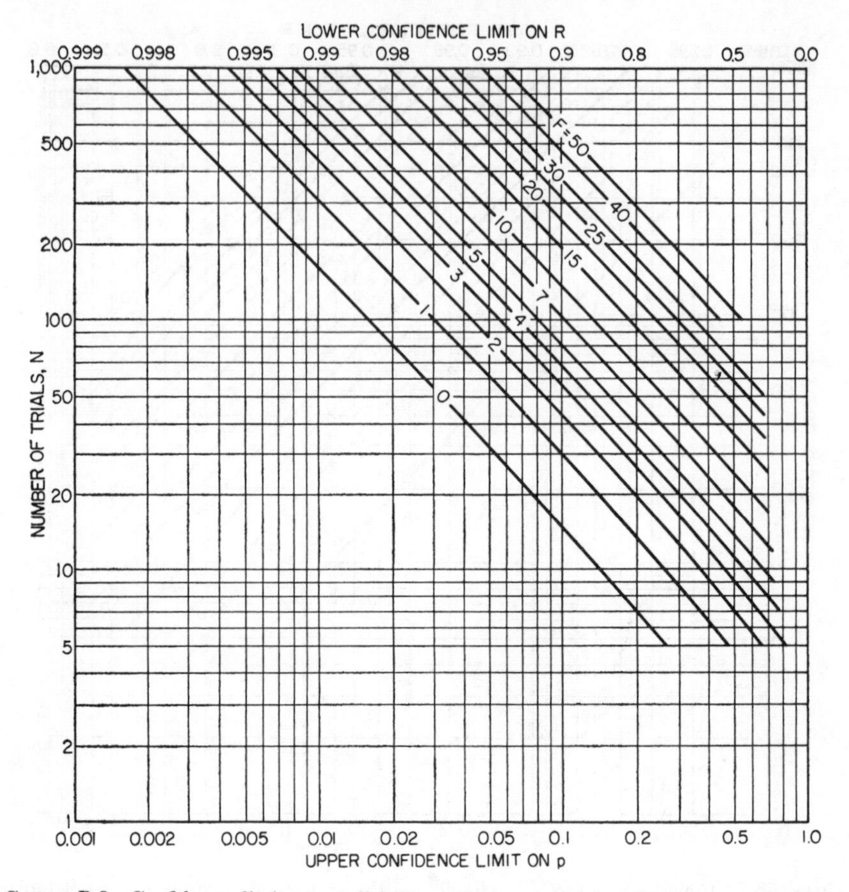

CHART B.8 Confidence limits on reliability. Upper confidence limit on unreliability (1 minus lower confidence limit on reliability), number of trials N, observed failures F, confidence coefficient $\gamma = 0.80$. (*David K. Lloyd and Myron Lipow, Reliability: Management, Methods and Mathematics.* © *1962, by permission of Prentice-Hall, Inc., Englewood Cliffs, New Jersey.*)

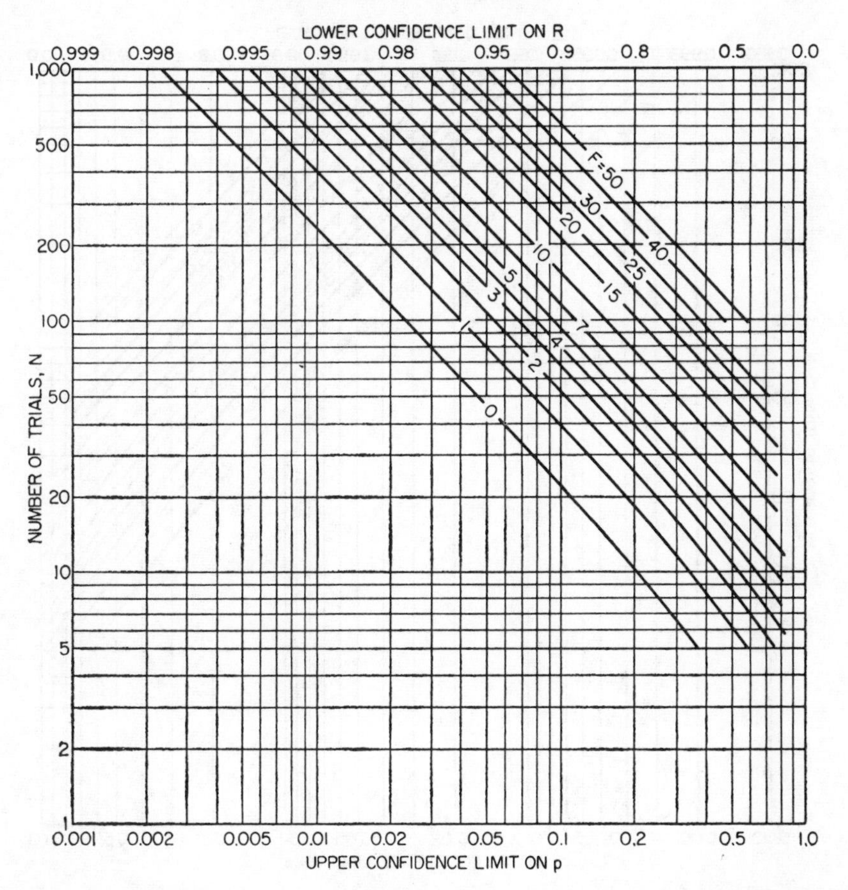

CHART B.9 Confidence limits on reliability. Upper confidence limit on unreliability (1 minus lower confidence limit on reliability), number of trials N, observed failures F, confidence coefficient $\gamma = 0.90$. (*David K. Lloyd and Myron Lipow, Reliability: Management, Methods and Mathematics.* © 1962, *by permission of Prentice-Hall, Inc., Englewood Cliffs, New Jersey.*)

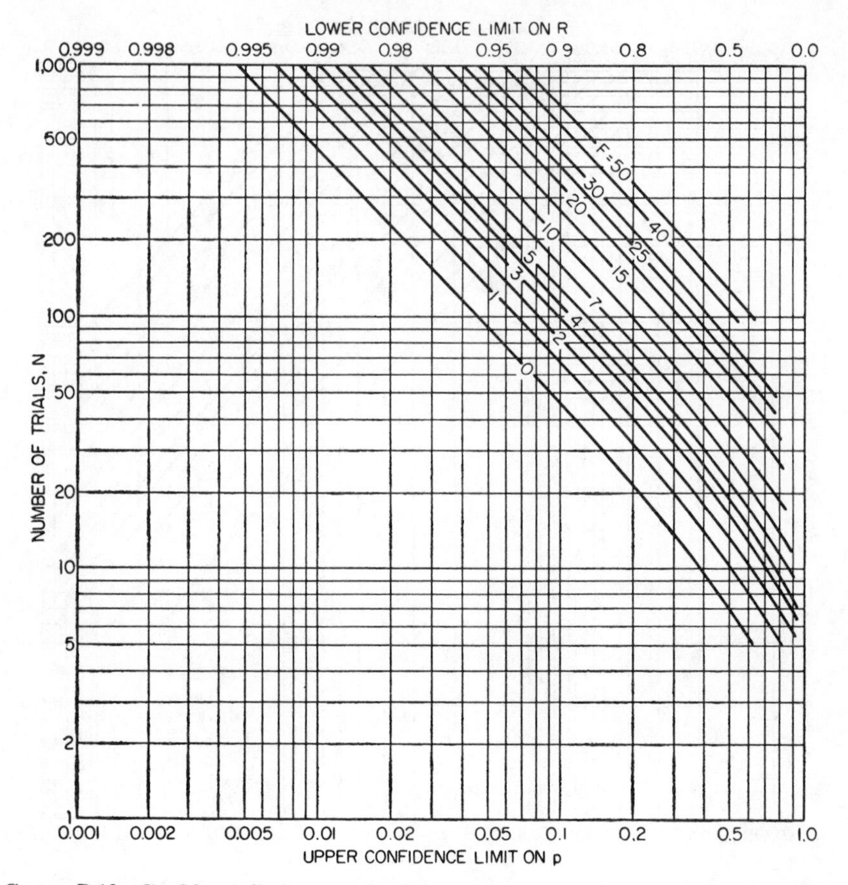

Chart B.10 Confidence limits on reliability. Upper confidence limit on unreliability (1 minus lower confidence limit on reliability), number of trials N, observed failures F, confidence coefficient $\gamma = 0.95$. (*David K. Lloyd and Myron Lipow, Reliability: Management, Methods and Mathematics.* © 1962, *by permission of Prentice-Hall, Inc., Englewood Cliffs, New Jersey.*)

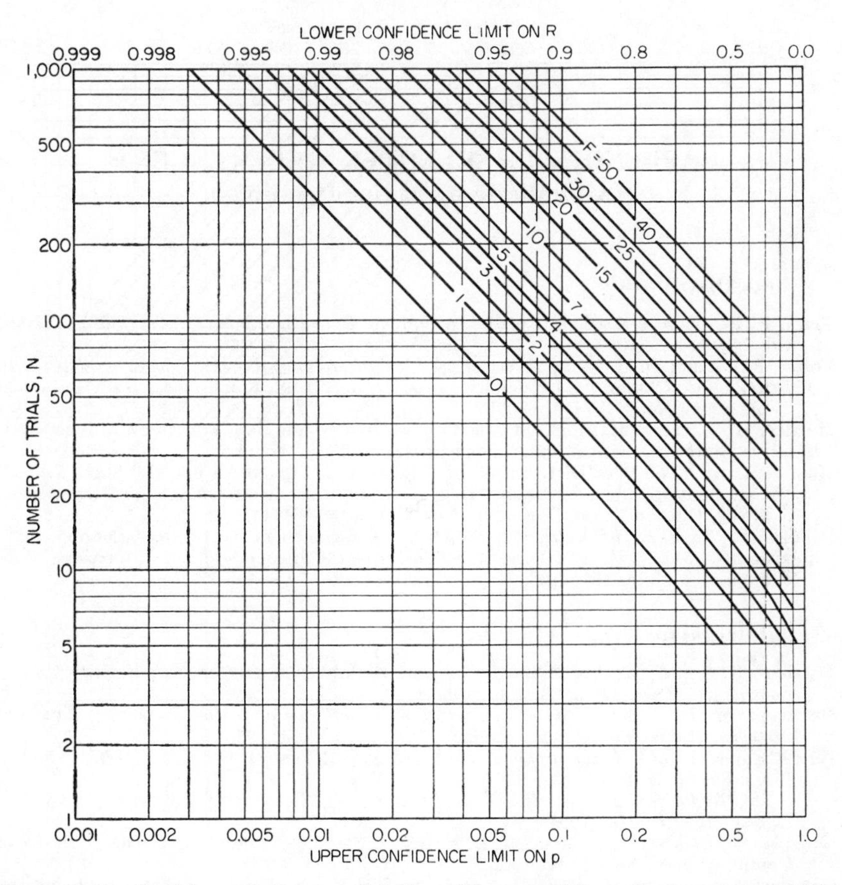

CHART B.11 Confidence limits on reliability. Upper confidence limit on unreliability (1 minus lower confidence limit on reliability), number of trials N, observed failures F, confidence coefficient $\gamma = 0.99$. (*David K. Lloyd and Myron Lipow, Reliability: Management, Methods and Mathematics.* © *1962, by permission of Prentice-Hall, Inc., Englewood Cliffs, New Jersey.*)

APPENDIX C: SELECTED REFERENCES
(Alphabetic by Major Headings)

Beta Distribution

Clark, R. E., "Percentage Points of the Incomplete Beta Function," *J. Am. Statist. Assoc.*, vol. 148, p. 831, 1953.

Foster, F. G., and D. H. Rees, *Tables of the Upper Percentage Points of the Generalized Beta Distribution*, Cambridge University Press, London (from *Biometrika*, "New Statistical Table Series No. XXVI"), 1960.

Hartley, H. O., and E. R. Fitch, "A Chart for the Incomplete Beta Function and the Cumulative Binomial Distribution," *Biometrika*, vol. 38, p. 423, 1951.

Kao, J. H. K., "The Beta Distribution in Reliability and Quality Control," *Tech. Rept.* 2, Department of Industrial Engineering, Cornell University, Ithaca, N.Y., Department of Navy, Office of Naval Research, Contract Nonr-401(43).

Pearson, K., *Tables of Incomplete Beta Function*, Cambridge University Press, London, 1932.

Thompson, C. M., "Tables of Percentage Points of the Incomplete Beta Function," *Biometrika*, vol. 32, p. 151, 1941.

Bibliography

Balaban, H. S., "A Selected Bibliography on Reliability," *IRE Trans. Reliability Quality Control*, vol. RQC-11, p. 86, July, 1962.

Buckland, W. R., and R. Fox, *Bibliography of Basic Texts and Monographs on Statistical Methods*, Hafner Publishing Company, Inc., New York, 1963.

Butterbaugh, G. I., *A Bibliography of Statistical Quality Control*, University of Washington Press, Seattle, 1946.

———, *A Bibliography of Statistical Quality Control—Supplement*, University of Washington Press, Seattle, 1951.

Doig, A. G., and M. G. Kendall, *Bibliography on Statistical Literature*, Hafner Publishing Company, Inc., New York, 1962.

Greenwood, J. A., and H. O. Hartley, *Guide to Tables in Mathematical Statistics*, Princeton University Press, Princeton, N.J., 1962.

Mendelhall, W., "A Bibliography on Life Testing and Related Topics," *Biometrika*, vol. 45, p. 521, 1958.

National Aeronautics and Space Administration, *Reliability Abstracts and Technical Reviews*, 1961–1962, National Aeronautics and Space Administration, Washington, D.C., 1962.

Binomial Distribution

Bahadur, R. R., "Some Approximations to the Binomial Distribution Function," *Ann. Math. Statist.*, vol. 31, p. 43, 1960.

Blischke, W. R., "Moment Estimators for the Parameters of a Mixture of Two Binomial Distributions," *Ann. Math. Statist.*, vol. 33, p. 444, 1962.

Blischke, W. R., "Estimating Parameters of Mixtures of Binomial Distributions," *J. Am. Statist. Assoc.*, vol. 59, p. 510, 1964.

Buehler, R. J., "Confidence Intervals for the Product of Two Binomial Parameters," *J. Am. Statist. Assoc.*, vol. 52, p. 482, 1953.

Clopper, C. J., and E. S. Pearson, "The Use of Confidence or Fiducial Limits Illustrated in the Case of the Binomial," *Biometrika*, vol. 26, p. 404, 1934.

Crow, E. L., "Confidence Intervals for Proportions," *Biometrika*, vol. 43, p. 423, 1956.

Eagle, E. L., and R. L. Hatfield, *Binomial Confidence Tables for Reliability and Quality Control Applications* (50% confidence PB 181479, 80% confidence PB 181480, 90%

confidence PB 181481, 95% confidence PB 181482, 99% confidence PB 181483), U.S. Department of Commerce, Office of Technical Services, Washington, D.C., 1962.

Harvard University, *Cumulative Binomial Probability Distribution*, Harvard University Press, Cambridge, Mass., 1955.

National Bureau of Standards, *Tables of the Binomial Probability Distribution*, Applied Mathematics Series 6, Government Printing Office, Washington, D.C., 1950.

Pachares, J., "Tables of Confidence Limits for the Binomial Distribution," *J. Am. Statist. Assoc.*, vol. 55, p. 521, 1960.

Romig, H. G., 50–100 *Binomial Tables*, John Wiley & Sons, Inc., New York, 1953.

Steck, G. P., *Upper Confidence Limits for the Failure Probability of Complex Networks*, *Sandia Res. Rept. SC*-4133, Sandia Corp., Albuquerque, N.M., 1957. (Office of Technical Services, Washington, D.C.)

Weintraub, Sol, *Tables of the Cumulative Binomial Probability Distribution for Small Values of p*, The Free Press of Glencoe, New York, 1963.

Bivariate Normal Distribution

National Bureau of Standards, *Tables of the Bivariate Normal Distribution Function and Related Functions*, Applied Mathematics Series 50, Government Printing Office, Washington, D.C., 1959.

Owen, D. B., "Tables for Computing Bivariate Normal Probabilities," *Ann. Math. Statist.*, vol. 27, p. 1075, 1956.

Definitions

Kendall, M. G., and W. R. Buckland, *A Dictionary of Statistical Terms*, Oliver & Boyd Ltd., Edinburgh and London, 1957.

Ryerson, C. M., "Glossary and Dictionary of Terms Relating Specifically to Reliability," *Proc. Third Natl. Symp. Reliability Quality Control*, Washington, D.C., 1957, p. 59.

Engineering Statistics and Quality Control (Books)

Bowker, A. H., and G. J. Lieberman, *Engineering Statistics*, Prentice-Hall, Inc., Englewood Cliffs, N.J., 1959.

Burr, I. W., *Engineering Statistics and Quality Control*, McGraw-Hill Book Company, New York, 1953.

Duncan, A. J., *Quality Control and Industrial Statistics*, Richard D. Irwin, Inc., Homewood, Ill., 1952.

Grant, E. L., *Statistical Quality Control*, 3d ed., McGraw-Hill Book Company, New York, 1964.

Hald, A., *Statistical Theory with Engineering Applications*, John Wiley & Sons, Inc., New York, 1952.

Juran, J. M. (ed.), *Quality Control Handbook*, 2d ed., McGraw-Hill Book Company, New York, 1962.

Shewhart, W. A., *Economic Control of Quality of Manufactured Product*, D. Van Nostrand Company, Inc., Princeton, N.J., 1931.

Simon, L. E., *An Engineers Manual of Statistical Method*, John Wiley & Sons, Inc., New York, 1941.

Wine, R. L., *Statistics for Scientists and Engineers*, Prentice-Hall, Inc., Englewood Cliffs, N.J., 1964.

Experimental Design

Anderson, R. L., and T. A. Bancroft, *Statistical Theory in Research*, McGraw-Hill Book Company, New York, 1952.

Barnett, E. H., "Introduction to Evolutionary Operations," *Ind. Eng. Chem.*, vol. 52, p. 500, 1960.

Box, G. E. P., "Evolutionary Operation: A Method for Increasing Productivity," *Appl. Statist.*, vol. 6, no. 2, 1957.

——— and J. S. Hunter, "Multi-factor Experimental Designs for Exploring Response Surfaces," *Ann. Math. Statist.*, vol. 28, p. 195, 1957.

Bradley, R. A., "Determination of Optimum Operating Conditions by Experimental Methods, Part I, Mathematics and Statistics Fundamental to the Fitting of Response Surfaces," *Ind. Qual. Control*, vol. 15, no. 1, p. 16, 1958.

Brownlee, K. A., *Industrial Experimentation*, 2d ed., Chemical Publishing Company, Inc., New York, 1948.
Chew, V. (ed.), *Experimental Designs in Industry*, John Wiley & Sons, Inc., New York, 1957.
Cochran, W., "Some Consequences When the Assumptions for the Analysis of Variance Are Not Satisfied," *Biometrics*, vol. 3, p. 22, 1947.
Cochran, W. G., and G. M. Cox, *Experimental Designs*, 2d ed., John Wiley & Sons, Inc., New York, 1957.
Cox, D. R., *Planning of Experiments*, John Wiley & Sons, Inc., New York, 1958.
Crump, S. L., "The Estimation of Variance Components in the Analysis of Variance," *Biometrics*, vol. 2, p. 7, 1946.
Davies, O. L. (ed.), *The Design and Analysis of Industrial Experiments*, Oliver & Boyd Ltd., Edinburgh and London, 1954.
Eisenhart, C., "The Assumptions Underlying the Analysis of Variance," *Biometrics*, vol. 3, p. 1, 1947.
Federer, W. T., *Experimental Designs*, The Macmillan Company, New York, 1955.
Finney, D. J., *Experimental Design and Its Statistical Basis*, The University of Chicago Press, Chicago, 1955.
Fisher, R. A., *The Design of Experiments*, 7th ed., Hafner Publishing Company, Inc., New York, 1960.
Hicks, C. R., "The Fundamental Analysis of Variance, Part I, The Analysis of Variance Model," *Ind. Qual. Control*, vol. 13, no. 2, 1956; "Part II, The Components of Variance Model and the Mixed Model," *ibid.*, vol. 13, no. 3, 1956; "Part III, Nested Designs in Analysis of Variance," *ibid.*, vol. 13, no. 4, 1956. Available in reprint form, American Society for Quality Control, Milwaukee, Wis.
———, *Introduction to Experimental Design*, Holt, Rinehart and Winston, Inc., New York, 1963.
Hunter, J. S., "Determination of Optimum Operating Conditions by Experimental Methods," *Ind. Qual. Control*, vol. 15, 1958.
Kempthorne, O., *The Design and Analysis of Experiments*, John Wiley & Sons, Inc., New York, 1952.
Mann, H. B., *Analysis and Design of Experiments*, Dover Publications, Inc., New York, 1949.
McCall, C. H., Jr., "Linear Contrasts, Parts I, II, and III," *Ind. Qual. Control*, vol. 17, July–September, 1960.
Ostle, B., *Statistics in Research*, 2d ed., The Iowa State University Press, Ames, Iowa, 1963.
Quenouille, M. H., *The Design and Analysis of Experiments*, Hafner Publishing Company, Inc., New York, 1953.
Satterthwaite, F. E., "Random Balance Experimentation," *Technometrics*, vol. 1, no. 2, p. 111, 1959.
Scheffe, H., *The Analysis of Variance*, John Wiley & Sons, Inc., New York, 1959.
Snedecor, G. W., *Statistical Methods*, 5th ed., The Iowa State University Press, Ames, Iowa, 1956.
Thompson, C. M., and Maxine Merrington, "Tables for Testing the Homogeneity of a Set of Expected Variances," *Biometrika*, vol. 33, p. 296, 1943.

Exponential Distribution

Department of Defense, *Sampling Procedures and Tables for Life and Reliability Testing (Based on Exponential Distribution)*, H-108, Government Printing Office, Washington, D.C., 1961.
Epstein, B., "Truncated Life Tests in the Exponential Case," *Ann. Math. Statist.*, vol. 25, p. 555, 1954.
———, "Exponential Distribution and Its Role in Life Testing," *Ind. Qual. Control*, vol. 15, no. 6, p. 4, 1958.
———, "Tests for the Validity of the Assumption That the Underlying Distribution of Life is Exponential," *Technometrics*, vol. 2, p. 83, February, 1960; vol. 2, p. 167, May, 1960.
——— and M. Sobel, "Some Theorems Relevant to Life Testing from an Exponential Distribution," *Ann. Math. Statist.*, vol. 25, p. 373, 1954.
——— and ———, "Sequential Life Tests in the Exponential Case," *Ann. Math. Statist.*, vol. 26, p. 82, 1955.
Pugh, E. L., "The Best Estimate of Reliability in the Exponential Case," *Operations Res.*, vol. 11, p. 57, 1963.

Rider, P. R., "The Method of Moments Applied to a Mixture of Exponential Distributions," *Ann. Math. Statist.*, vol. 32, p. 143, 1961.

Gamma Distribution

Chapman, D. C., "Estimating the Parameters of a Truncated Gamma Distribution," *Ann. Math. Statist.*, vol. 27, p. 498, 1956.
Greenwood, J. A., and D. Durand, "Aids for Fitting the Gamma Distribution by Maximum Likelihood," *Technometrics*, vol. 2, p. 55, February, 1960.
Gupta, S. S., "Order Statistics from the Gamma Distribution," *Technometrics*, vol. 2, p. 243, 1960.
——— and Phyllis Groll, "Gamma Distribution in Acceptance Sampling Based on Life Tests," *J. Am. Statist. Assoc.*, vol. 56, p. 942, 1961.
Lentner, M. M., and R. J. Buehler, "Some Inferences about Gamma Parameters with an Application to a Reliability Problem," *J. Am. Statist. Assoc.*, vol. 58, p. 670, 1963.
Pearson, K., *Tables of Incomplete Gamma Function*, Cambridge University Press, London, 1922.

Hypergeometric Distribution

Lieberman, G. J., and D. B. Owen, *Tables of the Hypergeometric Distribution*, Stanford University Press, Stanford, Calif., 1961.

Life Testing

Bartholomew, D. J., "A Problem in Life Testing," *J. Am. Statist. Assoc.*, vol. 52, p. 350, 1957.
Epstein, B., "Statistical Developments in Life Testing," *Proc. Third Natl. Symp. Reliability Quality Control*, Washington, D.C., 1957, p. 106.
———, "Statistical Problems in Life Testing," *Quality Control Conf. Papers, Seventh Ann. Conv., Am. Soc. Quality Control.*
——— and M. Sobel, "Life Testing," *J. Am. Statist. Assoc.*, vol. 48, p. 485, 1953.
Kao, J. H. K., "The Design and Analysis of Life-testing Experiments," 1958, *Middle Atlantic Conf., Am. Soc. Quality Control*, p. 217.
Zelen, M., "Problems in Life Testing: Factorial Experiments," *Trans. Thirteenth Quality Control Conf., Am. Soc. Quality Control*, 1958, p. 21.
———, "Factorial Experiments in Life Testing," *Technometrics*, vol. 1, p. 269, 1959.

Log-normal Distribution

Aitchison, J., and J. Brown, *The Log-normal Distribution*, Cambridge University Press, London, 1951.
Goldthwaite, L. R., "Failure Rate Study for the Lognormal Lifetime Model," *Proc. Seventh Natl. Symp. Reliability Quality Control*, Philadelphia, 1961, p. 208.
Guild, R. L., "Correlation of Conventional and Accelerated Test Conditions for Heater Burnouts by the Log-normal Distribution," *Ind. Qual. Control*, vol. 19, p. 27, November, 1952.
Kamins, M., "Two Notes on the Log-normal Distribution," *Memo RM-3781-PR*, The Rand Corporation, Santa Monica, Calif., August, 1963.

Mathematical Handbook

Korn, G. A., and Theresa M. Korn, *Mathematical Handbook for Scientists and Engineers*, McGraw-Hill Book Company, New York, 1961.

Multinomial Distribution

Johnson, N. L., "An Approximation to the Multinomial Distribution, Some Properties and Applications," *Biometrika*, vol. 47, p. 93, 1960.
Johnson, N. L., and D. H. Young, "Some Applications of Two Approximations to the Multinomial Distribution," *Biometrika*, vol. 47, p. 463, 1960.
Quesenberry, C. P., and D. C. Hurst, "Large Sample Simultaneous Confidence Intervals for Multinomial Proportions," *Technometrics*, vol. 6, p. 191, 1964.

Nonparametric Statistics

Fraser, D. A. S., *Nonparametric Methods in Statistics*, John Wiley & Sons, Inc., New York, 1957.

Siegel, S., *Nonparametric Statistics for the Behavioral Sciences*, McGraw-Hill Book Company, New York, 1956.

Normal Distribution

Gupta, S. S., "Life Test Sampling Plans for Normal and Lognormal Distributions," *Technometrics*, vol. 4, p. 151, 1962.

National Bureau of Standards, *Table of the Normal Probability Functions*, Government Printing Office, Washington, D.C.

Order Statistics

Epstein, B., "Application of the Theory of Extreme Values in Fracture Problems," *J. Am. Statist. Assoc.*, vol. 43, p. 403, 1948.

———, "Elements of the Theory of Extreme Values," *Technometrics*, vol. 2, p. 27, February, 1960.

——— and H. Brooks, "The Theory of Extreme Values and Its Implications in the Study of the Dielectric Strength," *J. Appl. Physics*, vol. 19, p. 544, 1948.

Fisher, R. A., and L. H. C. Tippett, "Limiting Forms of the Frequency Distribution of the Largest or Smallest Member of a Sample," *Proc. Cambridge Phil. Soc.*, vol. 24, p. 280, 1928.

Gumbel, E. J., *Statistical Theory of Extreme Values and Some Practical Applications*, National Bureau of Standards, Applied Mathematics Series 33, Government Printing Office, Washington 25, D.C., 1954.

———, *Statistics of Extremes*, John Wiley & Sons, Inc., New York, 1958.

Lieblein, J., "A New Method of Analyzing Extreme-value Data," *Natl. Advisory Comm. Aeron. Tech. Note* 3053, 1954.

National Bureau of Standards, *Probability Tables for Analysis of Extreme Value Data*, Applied Mathematics Series 22, Government Printing Office, Washington, D.C., 1953.

Sarhan, A. E., "Estimation of the Mean and Standard Deviation by Order Statistics," *Ann. Math. Statist.*, vol. 25, p. 317, 1954; vol. 26, p. 576, 1955.

——— and B. G. Greenberg (eds.), *Contributions to Order Statistics*, John Wiley & Sons, Inc., New York, 1962.

Wilks, S. S., "Order Statistics," *Bull. Am. Math. Soc.*, vol. 54, p. 6, 1948.

Poisson Distribution

Cohen, A. C., "Estimating Parameters of a Modified Poisson Distribution," *J. Am. Statist Assoc.*, vol. 55, p. 139, 1960.

Crow, E. L., and R. S. Gardner, *Table of Confidence Limits for the Expectation of Poisson Variable*, Cambridge University Press, London, 1960.

General Electric Company, *Tables of the Individual and Cumulative Terms of the Poisson Distribution*, D. Van Nostrand Company, Inc., Princeton, N.J., 1962.

Molina, E. C., *Tables of Poisson's Exponential Limit*, D. Van Nostrand Company, Inc., Princeton, N.J., 1945.

Rider, P. R., "Estimating the Parameters of Mixed Poisson, Binomial, and Weibull Distributions by the Method of Moments," *Bull. Intern. Statist. Inst.*, vol. 39, p. 225, 1962.

Weiss, H., "Estimation of Reliability in a Complex System with a Poisson-type Failure," *Operations Res.*, vol. 4, p. 532, 1956.

Probability (Books)

Feller, W., *An Introduction to Probability Theory and Its Application*, vol. 1, 2d ed., John Wiley & Sons, Inc., New York, 1957.

Fry, T. C., *Probability and Its Engineering Uses*, D. Van Nostrand Company, Inc., New York, 1928.

Munroe, M. E., *The Theory of Probability*, McGraw-Hill Book Company, New York, 1951.

Parzen, E., *Modern Probability Theory and Its Applications*, John Wiley & Sons, Inc., New York, 1960.

Uspensky, J. V., *Introduction to Mathematical Probability*, McGraw-Hill Book Company, New York, 1937.
Von Mises, R., *Probability, Statistics and Truth*, The Macmillan Company, New York, 1939.

Redundancy

Aroian, L. A., and R. H. Meyers, "Redundancy Considerations in Space and Satellite Systems," *Proc. Seventh Natl. Symp. Reliability Quality Control*, Philadelphia, 1961, p. 292.
Black, G., and F. Proschan, "An Optimal Redundancy," *J. Operations Res. Soc. Am.*, vol. 7, p. 581, 1959.
Flehinger, B. J., "Reliability Improvement through Redundancy at Various Systems Levels," *IRE Natl. Conv. Record*, part 6, 1958.
Greveling, C. J., "Increasing the Reliability of Electronic Equipment by Use of Redundant Circuits," *Proc. IRE*, vol. 44, p. 509, 1956.
Kneale, S. G., "Reliability of Parallel System with Repair and Switching," *Proc. Seventh Nat. Symp. Reliability Quality Control*, Philadelphia, 1961, p. 129.
Madansky, A., "Approximate Confidence Limits for the Reliability of Series and Parallel Systems," *Rept. RM*-2552, The Rand Corporation, Santa Monica, Calif., April, 1960.
Proschan, F., and T. A. Bray, "Optimum Redundancy under Multiple Constraints," *AD*-408 393, Office of Technical Services, Washington, D.C., May, 1963.
Wilcox, R. H., and W. C. Mann (eds.), *Redundancy Techniques for Computing Systems*, Spartan Books, Washington, D.C., 1962.

Regression and Correlation

Acton, F. S., *Analysis of Straight-line Data*, John Wiley & Sons, Inc., New York, 1959.
Bartlett, M. S., "Fitting a Straight Line When Both Variables Are Subject to Error," *Biometrics*, vol. 5, no. 3, p. 207, 1949.
Eisenhart, C., "The Interpretation of Certain Regression Methods and Their Use in Biological and Industrial Research," *Ann. Math. Statist.*, vol. 10, no. 2, p. 162, 1939.
Ezekiel, M., *Methods of Correlation Analysis*, John Wiley & Sons, Inc., New York, 1941.
Fisher, R. A., "On the Probable Error of a Coefficient of Correlation Deduced from Small Samples," *Metron*, vol. 1, no. 4, p. 3, 1921.
Graybill, F. A., *An Introduction to Linear Statistical Models*, vol. 1, McGraw-Hill Book Company, New York, 1961.
Hader, R. J., and A. H. Grandage, "Simple and Multiple Regression Analyses," in V. Chew (ed.), *Experimental Designs in Industry*, John Wiley & Sons, Inc., New York, 1958.
Hartley, H. O., "The Modified Gauss-Newton Method for the Fitting of Nonlinear Regression Functions by Least Squares," *Technometrics*, vol. 13, p. 269, 1961.
Kendall, M. G., *Rank Correlation Methods*, Charles Griffin & Company, Ltd., London, 1948.
Kramer, C. Y., "Simplified Computations for Multiple Regression," *Ind. Qual. Control*, vol. 13, no. 8, p. 8, 1957.
Madansky, A., "The Fitting of Straight Lines When Both Variables Are Subject to Error," *J. Am. Statist. Assoc.*, vol. 54, p. 173, 1959.
Mandel, J., "Fitting a Straight Line to Certain Types of Cumulative Data," *J. Am. Statist. Assoc.*, vol. 52, p. 552, 1958.
Treloar, A. E., *Correlation Analysis*, Burgess Publishing Company, Minneapolis, 1942.

Reliability (Articles and Pamphlets)

Barlow, R. E., and L. C. Hunter, "Mathematical Models for System Reliability," *Sylvania Technologist*, vol. 13, nos. 1 and 2, 1960.
Basu, A. P., "Estimates of Reliability for Some Distributions Useful in Life Testing," *Technometrics*, vol. 6, p. 215, 1964.
Connor, W. S., "Interpreting Reliability by Fitting Theoretical Distributions to Failure Data," *Ind. Eng. Chem.*, vol. 52, p. 75A, February, 1960; p. 71A, April, 1960.
Davis, D. J., "An Analysis of Some Failure Data," *J. Am. Statist. Assoc.*, vol. 47, p. 113, 1952.
Drenick, R. F., "Mathematical Aspects of the Reliability Problem," *J. Soc. Ind. Appl. Math.*, vol. 8, p. 125, 1960.
Garver, D. P., Jr., "Random Hazard in Reliability Problems," *Technometrics*, vol. 5, p. 211, 1963.

Harter, H. L., "Some Aspects of Reliability and Life Testing," *J. Electron. Div., Am. Soc. Quality Control*, vol. 3, no. 1, p. 5, 1964.
Herd, G. R., "Estimation of Reliability Functions," *ARINC Monograph* 3, ARINC Research Corp., Washington, D.C., May, 1956.
Herd, G. R., "Some Statistical Concepts and Techniques for Reliability Analysis and Prediction," *Proc. Fifth Natl. Symp. Reliability Quality Control*, Philadelphia, 1959, p. 126.
Luebbert, W. F., "Principles and Concepts of Reliability for Equipment and Systems, Part II, Simple Models for Failure of Complex Equipment," *TR*-91, Stanford University, Stanford, Calif., Aug. 18, 1955.
Weiss, L., "On Estimating Scale and Location Parameters," *J. Am. Statist. Assoc.*, vol. 58, p. 658, 1963.

Reliability (Books)

AGREE, *Reliability of Military Electronic Equipment*, Government Printing Office, Washington, D.C., 1957.
ASQC, *Production and Field Reliability*, American Society for Quality Control, Milwaukee, Wis., 1959.
ASQC, *Research and Development Reliability*, American Society for Quality Control, Milwaukee, Wis., 1961.
Bazovsky, I., *Reliability Theory and Practice*, Prentice-Hall, Inc., Englewood Cliffs, N.J., 1962.
Calabro, S. R., *Reliability Principles and Practices*, McGraw-Hill Book Company, New York, 1962.
Chorafas, D. N., *Statistical Processes and Reliability Engineering*, D. Van Nostrand Company, Inc., Princeton, N.J., 1960.
Dummer, G. W. A., and N. Griffin, *Electronic Equipment Reliability*, John Wiley & Sons, Inc., New York, 1960.
Gyrna, F. M., Jr. et al. (eds.), *Reliability Training Text*, American Society for Quality Control, Milwaukee, Wis., and Institute of Radio Engineers, New York, 1960.
Haviland, R. P., *Engineering Reliability and Long Life Design*, D. Van Nostrand Company, Inc., Princeton, N.J., 1964.
Henney, V. (ed.), *Reliability Factors for Ground Electronic Equipment*, McGraw-Hill Book Company, New York, 1956.
Landers, R. R., *Reliability and Product Assurance*, Prentice-Hall, Inc., Englewood Cliffs, N.J., 1963.
Leake, C. E., *Understanding Reliability*, Pasadena Lithographers, Pasadena, Calif., 1960.
Lloyd, D. K., and M. Lipow, *Reliability: Management, Methods, and Mathematics*, Prentice-Hall, Inc., Englewood Cliffs, N.J., 1962.
Myers, R. H., K. L. Wong, and H. M. Gordy (eds.), *Reliability Engineering for Electronic Systems*, John Wiley & Sons, Inc., New York, 1964.
Pieruschka, E., *Principles of Reliability*, Prentice-Hall, Inc., Englewood Cliffs, N.J., 1963.
RADC, *RADC Reliability Handbook*, ASTIA Document No. AD-148868, RADC, Rome, N.Y., RADC-TR-58-111, 1959. (Available from U.S. Department of Commerce, Washington, D.C.)
Sandler, G. H., *System Reliability Engineering*, Prentice-Hall, Inc., Englewood Cliffs, N.J., 1963.
Shwap, J. E., and H. J. Sullivan (eds.), *Semiconductor Reliability*, Engineering Publishers, Elizabeth, N.J., 1961.
Society of Automotive Engineers, *Reliability Control in Aerospace Equipment Development*, The Macmillan Company, New York, 1963.
Zelen, M. (ed.), *Statistical Theory in Reliability* (Proceedings of Advanced Seminar, University of Wisconsin, 1962), The University of Wisconsin Press, Madison, Wis., 1963.

Sequential Analysis

Wald, A., *Sequential Analysis*, John Wiley & Sons, Inc., New York, 1947.

Statistics (Books)

Cramer, H., *Mathematical Methods of Statistics*, Princeton University Press, Princeton, N.J., 1946.
Crow, E. L., F. A. Davis, and Margaret Maxfield, *Statistics Manual*, Dover Publications, Inc., New York, 1960.

Croxton, F. E., and D. J. Cowden, *Applied General Statistics*, 2d ed., Prentice-Hall, Inc., Englewood Cliffs, N.J., 1955.

Dixon, W. J., and F. J. Massey, Jr., *Introduction to Statistical Analysis*, 2d ed., McGraw-Hill Book Company, New York, 1957.

Ehrenfeld, S., and S. B. Littauer, *Introduction to Statistical Method*, McGraw-Hill Book Company, New York, 1964.

Freeman, H. A., *Introduction to Statistical Inference*, Addison-Wesley Publishing Company, Inc., Reading, Mass., 1963.

Freund, J. M., *Modern Elementary Statistics*, 2d ed., Prentice-Hall, Inc., Englewood Cliffs, N.J., 1960.

Hoel, P. G., *Elementary Statistics*, John Wiley & Sons, Inc., New York, 1960.

Hogg, R. V., and A. T. Craig, *Introduction to Mathematical Statistics*, The Macmillan Company, New York, 1959.

Kendall, M. G., *The Advanced Theory of Statistics*, vols. 1 and 2, Charles Griffin & Company, Ltd., London, 1948.

Lehmann, E. L., *Testing Statistical Hypotheses*, John Wiley & Sons, Inc., New York, 1959.

Lindgren, B. W., *Statistical Theory*, The Macmillan Company, New York, 1962.

——— and G. McElrath, *Introduction to Probability and Statistics*, The Macmillan Company, New York, 1959.

Mood, A. M., *Introduction to the Theory of Statistics*, McGraw-Hill Book Company, New York, 1950.

——— and F. A. Graybill, *Introduction to the Theory of Statistics*, 2d ed., McGraw-Hill Book Company, New York, 1963.

Natrella, Mary Gibbons, "Experimental Statistics," *Natl. Bur. Std. (U.S.)*, *Handbook* 91, 1963.

Statistical Research Group, Columbia University, *Techniques of Statistical Analysis*, McGraw-Hill Book Company, New York, 1947.

Statistical Tables (General)

Fisher, R. A., and F. Yates, *Statistical Tables for Biological, Agriculture and Medical Research*, 5th ed., Oliver & Boyd Ltd., Edinburgh and London, 1957.

Hald, A., *Statistical Tables and Formulas*, John Wiley & Sons, Inc., New York, 1952.

Owen, D. B., *Handbook of Statistical Tables*, Addison-Wesley Publishing Company, Inc., Reading, Mass., 1962.

National Bureau of Standards, *Tables of Probability Functions*, Applied Mathematics Series 23, Government Printing Office, Washington, D.C., 1953.

Pearson, E. S., and H. O. Hartley, *Biometrika Tables for Statisticians*, vol. 1, Cambridge University Press, London, 1958.

Rand Corporation, *A Million Random Digits with 100,000 Normal Deviates*, The Free Press of Glencoe, New York, 1955.

Systems Engineering

Goode, H. H., and R. E. Machol, *System Engineering*, McGraw-Hill Book Company, New York, 1957.

Tolerance Limits

Owen, D. B., "Tables of Factors for One-sided Tolerance Limits for a Normal Distribution," *Sandia Monograph SCR*-13, Sandia Corp., Albuquerque, N.M., April, 1958. (Office of Technical Services, Department of Commerce, Washington, D.C.)

Weibull Distribution

Department of Defense, "Sampling Procedures and Tables for Life and Reliability Testing Based on the Weibull Distribution (Mean Life Criterion)," *TR*-3, Office of the Assistant Secretary of Defense (Supply and Logistics), Washington, D.C., Sept. 30, 1961.

———, "Sampling Procedures and Tables for Life and Reliability Testing Based on the Weibull Distribution (Hazard Rate Criterion)," *TR*-4, Office of the Assistant Secretary of Defense (Supply and Logistics), Washington, D.C., Feb. 28, 1962.

Goode, H., and J. H. K. Kao, "Sampling Procedures and Tables for Life and Reliability Testing Based on the Weibull Distribution (Hazard Rate Criterion)," *Eighth Natl. Symp. Reliability Quality Control*, Washington, D.C., 1962, p. 37.

Govindarajulu, Z., and M. Joshi, "Best Linear Unbiased Estimation of Location and Scale Parameters of Weibull Distribution Using Ordered Observations," *AD*-409 685, Office of Technical Services, Washington, D.C., November, 1962.

Kao, J. H. K., "The Weibull Distribution in Reliability Studies," *Tech. Rept.* 33, *Res. Rept. EE* 343, Cornell University, Ithaca, N.Y., 1957.

———, "Computer Methods for Estimating Weibull Parameters in Reliability Studies," *IRE Trans. Reliability Quality Control*, vol. PGRQC-13, p. 15, July, 1958.

———, "A Graphical Estimation of Mixed Weibull Parameters in Life Testing of Electron Tubes," *Technometrics*, vol. 1, p. 389, 1959.

Mendenhall, W., and E. H. Lehman, Jr., "An Approximation to the Negative Moments of the Positive Binomial Useful in Life Testing," *Technometrics*, vol. 2, p. 227, 1960.

Menon, M. V., "Estimation of the Shape and Scale Parameters of the Weibull Distribution," *Technometrics*, vol. 5, no. 2, p. 175, 1963.

Rider, P. R., "Estimating the Parameters of Mixed Poisson, Binomial and Weibull Distributions by the Method of Moments," *Bull. Intern. Statist. Inst.*, vol. 39, part 2, p. 225, 1962.

Weibull, W., "A Statistical Distribution Function of Wide Applicability," *J. Appl. Mech.*, vol. 18, p. 293, 1951.

———, "A Statistical Representation of Fatigue Failure in Solids," *Roy. Inst. Tech* (*Stockholm*), November, 1954.

INDEX